Developments in Petroleum Science
Volume 64

Core Analysis: A Best Practice Guide

Series Editor
John Cubitt
Holt, Wales

Developments in Petroleum Science
Volume 64

Core Analysis: A Best Practice Guide

Colin McPhee
LR Senergy Ltd., UK

Jules Reed
LR Senergy Ltd., UK

Izaskun Zubizarreta
LR Senergy Ltd., UK

ELSEVIER
AMSTERDAM • BOSTON • HEIDELBERG • LONDON • NEW YORK • OXFORD
PARIS • SAN DIEGO • SAN FRANCISCO • SINGAPORE • SYDNEY • TOKYO

Elsevier
Radarweg 29, PO Box 211, 1000 AE Amsterdam, Netherlands
The Boulevard, Langford Lane, Kidlington, Oxford OX5 1GB, UK
225 Wyman Street, Waltham, MA 02451, USA

First edition 2015

ISBN: 978-0-444-63533-4
ISSN: 0376-7361

For information on all Elsevier publications
visit our web site at store.elsevier.com

Contents

Series Editors Preface xvii
Preface xix

1. Best Practice in Coring and Core Analysis 1

1.1 Core Analysis Data: The Foundation of Formation Evaluation 1
1.2 Core Analysis Data Uncertainty 4
1.2.1 Reasons and Consequences 4
1.2.2 Reducing Uncertainty 8
1.3 Core Analysis Management Framework 8
1.3.1 Core Analysis Planning and Design 8
1.3.2 Programme Design Considerations 9
1.3.3 Core Analysis Focal Points 9
1.3.4 Real-Time Quality Control 11
1.4 Best Practice in Core Analysis: An Overview 11
1.4.1 Coring, Core Handling and Core Processing 12
1.4.2 Sample Preparation 12
1.4.3 Routine Core Analysis 12
1.4.4 Special Core Analysis 12
1.4.5 Geomechanics Tests 13
1.4.6 Quality Control Procedures and Diagnostics 13
1.4.7 Example Core Analysis Programmes 14
1.4.8 Benefits 14
References 15

2. Wellsite Core Acquisition, Handling and Transportation 17

2.1 Coring Systems 17
2.1.1 Conventional Full-Diameter Coring Systems 18
2.1.2 Wireline-Retrievable Cores 27
2.1.3 Gel Coring Systems 28
2.1.4 Liquid and Gas Retention Coring Systems 28
2.1.5 Oriented Coring 35
2.1.6 Sidewall Cores 37
2.2 Conventional Coring Operations 40
2.2.1 Health, Safety and Environmental Considerations 41
2.2.2 Coring Team 42
2.2.3 Managing Coring Risks 42

2.3 Coring Fluids 45
2.3.1 Mud Types 45
2.3.2 Mud Tracers 45
2.4 Core Damage and Core Fluid/Petrophysical Property Alteration 48
2.4.1 Fluid Saturation Alteration 48
2.4.2 Stress Release 52
2.4.3 Wettability Alteration 57
2.5 Best Practice in Wellsite Handling 66
2.5.1 Core Recovery on Rig Floor 66
2.5.2 Linear Layout and Marking 68
2.5.3 Wellsite Gamma-Ray Scanning 70
2.5.4 Wellsite Liner Cutting 70
2.5.5 Wellsite Core Stabilisation and Preservation 72
2.5.6 Wellsite Sampling 77
2.5.7 Core Transportation 79
2.5.8 Coring Report 81
2.6 Special Handling Considerations for Difficult Rock Types 81
2.6.1 Unconsolidated Core 81
2.6.2 Carbonates 82
2.6.3 Shaly Sands 83
2.6.4 Shales 83
2.6.5 Coal 84
References 84
Recommended Reading 87

3. Core Laboratory Processing and Screening 89

3.1 Introduction 89
3.2 Core Receipt and Cutting 91
3.3 CT Scanning 91
3.4 Gamma Ray Logging 95
3.5 Removal from Liners 100
3.6 Core Viewing and Sample Selection 102
3.7 Sample Preservation 104
3.7.1 Dry Preservation 104
3.7.2 Wet Preservation 106
3.8 Core Plugging 106
3.8.1 Plug Samples 106
3.8.2 Drill Press and Plugging Fluids 108
3.8.3 Plug Orientation 110
3.8.4 Plug Allocation 112
3.8.5 Plug Trimming 116
3.9 Core Slabbing 117
3.10 Core Resination 118
3.11 Core Photography and Imaging 120
3.11.1 Conventional Imaging 121
3.11.2 360° Core Imaging 124
3.12 Weak or Unconsolidated Core Processing 125
3.12.1 Core Receipt and Cutting 125

3.12.2 CT Scanning 125
3.12.3 Core Gamma Ray Scanning 129
3.12.4 Core Slabbing 129
3.12.5 Core Viewing and Sample Selection 129
3.12.6 Core Plugging and Plug Protection 130
3.12.7 Core Photography 133
3.12.8 Core Preservation 133
References 134
Recommended Reading 134

4. Core Sample Preparation 135

4.1 Introduction 135
4.2 Cleaning 136
4.2.1 Solvents 136
4.2.2 Common Cleaning Methods 138
4.2.3 Core Cleaning Methods: Advantages and Drawbacks 146
4.3 Core Drying 146
4.3.1 Conventional (and Vacuum) Oven Drying 146
4.3.2 Humidity Oven Drying 146
4.3.3 Critical Point Drying 150
4.3.4 Flow-Through Drying 152
4.3.5 Core Drying Methods: Advantages and Drawbacks 153
4.4 Quality Control Issues, Checks and Diagnostics 153
4.5 Clays and Clay Damage Mechanisms 156
4.5.1 Clay Mineral Structures 158
4.5.2 Clay Types 160
4.5.3 Cation Exchange Capacity 161
4.5.4 Clay Morphology and Rock Property Controls 163
4.5.5 Clay Damage Mechanisms 165
4.5.6 Testing Without Drying 168
4.6 Core Conditioning for Porosity Measurements 170
4.6.1 Porosity Evaluation 170
4.6.2 Total and Effective Porosity Concepts 170
4.6.3 Log and Core Porosity 172
4.7 Special Considerations in Core Preparation 175
4.7.1 Wettability 175
4.7.2 Carbonates and Chalks 176
4.7.3 Halite 176
References 178
Recommended Reading/Viewing 179

5. Routine Core Analysis 181

5.1 Introduction 181
5.2 Fluid Saturation Measurements 182
5.2.1 Retort Method 183
5.2.2 Dean–Stark Analysis 187
5.3 Porosity Measurements 195
5.3.1 Helium Grain Volume and Grain Density 199

5.3.2 Helium Pore Volume 204
5.3.3 Bulk Volume 210
5.3.4 Liquid Saturation Porosity 216
5.3.5 Accuracy and Repeatability of Porosity Measurements 219
5.4 Permeability Measurements 221
5.4.1 Definitions 221
5.4.2 Darcy's Law 221
5.4.3 Non-Darcy Flow: Klinkenberg Effects 224
5.4.4 Non-Darcy Flow: Forchheimer Effect 228
5.4.5 Steady-State Permeability Measurements 229
5.4.6 Unsteady-State Permeability Measurements 239
5.4.7 Steady-State Liquid (Absolute) Permeability Measurements 247
5.4.8 Probe or Profile Permeability Measurements 252
5.5 Whole Core Analysis Measurements 262
5.5.1 Sample Preparation 262
5.5.2 Fluid Saturations 263
5.5.3 Porosity 263
5.5.4 Gas Permeability 263
References 266
Recommended Reading/Viewing 267

6. Preparation for Special Core Analysis 269

6.1 Fluid Preparation and Characterisation 269
6.1.1 Formation Water Samples and Brine Preparation 269
6.1.2 Oil 276
6.1.3 Gas 279
6.1.4 Data Requirements 280
6.2 Interfacial Tension 281
6.2.1 Methods to Determine IFT 282
6.3 Sample Selection for SCAL 286
6.3.1 Sample Provenance 286
6.3.2 CT Scanning and Photography 286
6.3.3 Petrophysical Properties 292
6.4 Reservoir Stress Estimation 300
6.4.1 Test Cell Design 300
6.4.2 Effective Stress Concepts 300
6.4.3 Net Confining Stress Loading Conditions 301
6.4.4 Reservoir Stress Data Sources 304
References 311
Recommended Reading 311

7. Wettability and Wettability Tests 313

7.1 Introduction 313
7.2 Contact Angle Method 315
7.2.1 Sample Preparation 316
7.2.2 Test Equipment 317

7.2.3 Test Procedures 319
7.2.4 Results 320
7.2.5 Data Reporting Requirements 320
7.2.6 Contact Angle Summary 320
7.3 Amott (Amott–Harvey) Method 322
7.3.1 Sample Preparation 323
7.3.2 Test Conditions 325
7.3.3 Test Equipment 327
7.3.4 Test Procedures 327
7.3.5 Amott–Harvey Wettability Index Calculation 329
7.3.6 Data Reporting Requirements 330
7.3.7 Amott (Amott–Harvey) Summary 331
7.4 USBM Method 332
7.4.1 Sample Preparation 334
7.4.2 Test Equipment 334
7.4.3 Key Processes 334
7.4.4 Test Procedures 334
7.4.5 USBM Index Calculation 336
7.4.6 Data Reporting Requirements 337
7.4.7 USBM Summary 338
7.5 Combined Amott–USBM (Combination) Method 339
7.5.1 Sample Preparation 340
7.5.2 Test Equipment 340
7.5.3 Test Procedures 340
7.5.4 Data Reporting Requirements 341
7.5.5 USBM and Amott–Harvey Index Calculation 341
7.5.6 Combined Amott–USBM Summary 342
References 344
Recommended Reading 345

8. Electrical Property Tests 347

8.1 Introduction 348
8.1.1 Resistivity Logs 348
8.1.2 Water Saturation Interpretation—Archie's Law 349
8.1.3 Water Saturation Interpretation—Shaly Sands 352
8.1.4 Water Saturation Interpretation Parameters from Core 353
8.2 FRF Tests 354
8.2.1 Sample Selection and Preparation 354
8.2.2 Test Equipment 356
8.2.3 Test Procedures—Ambient (Base) Conditions 359
8.2.4 Test Procedures—Stress (Overburden) Conditions 360
8.2.5 Data Utilisation 366
8.2.6 Data Reporting Requirements 369
8.2.7 Advantages and Drawbacks/Issues 369
8.2.8 FRF Test Quality Control Issues, Checks and Diagnostics 371
8.3 Resistivity Index Tests 378
8.3.1 Test Equipment 379

8.3.2 Test Procedures 382
8.3.3 Data Assignment and Utilisation 387
8.3.4 Saturation Exponent: Laboratory Artefacts 390
8.3.5 Saturation Exponent: Rock Type and Pore/Fluid System Controls 398
8.3.6 Data Reporting Requirements 406
8.3.7 Advantages and Drawbacks/Issues 409
8.3.8 RI Test Quality Control Issues, Checks and Diagnostics 409
8.4 Waxman–Smits Parameters 416
8.4.1 Shaly Sand Models 416
8.4.2 Waxman–Smits Equation 420
8.4.3 CEC and Qv Tests 424
8.4.4 Data Utilisation and Assignment 430
8.4.5 Data Reporting Requirements 433
8.4.6 Advantages and Drawbacks/Issues 433
8.4.7 Qv, m^* and n^* Test Quality Control Issues, Checks and Diagnostics 433
8.5 Alternative Method for Saturation Exponent Determination 439
References 444
Recommended Reading 447

9. Capillary Pressure 449

9.1 Introduction 449
9.2 Primary Drainage Capillary Pressure 452
9.3 High Speed Centrifuge 453
9.3.1 Sample Preparation 454
9.3.2 Key Processes 455
9.3.3 Test Procedures (Primary Drainage) 458
9.3.4 Data Reporting Requirements 462
9.4 Porous Plate (Semi-Permeable Membrane): Primary Drainage 471
9.4.1 Sample Preparation 472
9.4.2 Key Processes 472
9.4.3 Test Procedures 474
9.4.4 Alternatives to Porous Plate Membranes 478
9.4.5 Data Reporting Requirements 479
9.5 Mercury–Air (MICP) 484
9.5.1 Sample Preparation 486
9.5.2 Test Equipment 486
9.5.3 Test Procedures 488
9.5.4 Pore Throat Size Distribution 490
9.5.5 Typical Output 493
9.5.6 Data Reporting Requirements 493
9.6 Capillary Pressure Methods: Equilibration Times 501
9.7 Summary of Drainage Capillary Pressure Methods 501
9.8 Data Corrections 502
9.8.1 Closure Corrections 502
9.8.2 Stress Corrections 508

9.8.3 Clay-Bound Water Corrections 508
9.8.4 Fluid Corrections 510
9.8.5 J Function 513
References 516
Recommended Reading 517

10. Relative Permeability 519

10.1 Introduction 520
10.1.1 Section Outline 520
10.1.2 Definitions 521
10.1.3 Irreducible Water Saturation 523
10.1.4 Residual Oil Saturations: Water–Oil System 525
10.1.5 Residual Oil Saturations: Gas–Oil System 527
10.1.6 Residual Gas Saturation: Water–Gas System 528
10.1.7 Critical Gas Saturation: Gas–Oil System 530
10.1.8 Critical Gas Saturation: Water–Gas System 530
10.1.9 Critical Water Saturation 532
10.1.10 Critical Oil Saturation: Gas Condensate Systems 533
10.1.11 Wettability 535
10.1.12 Corey Exponents 541
10.2 Measurement Techniques 543
10.2.1 USS Tests 543
10.2.2 SS Tests 545
10.2.3 Centrifuge Tests 547
10.2.4 Curve Definition as a Function of Test Method 547
10.2.5 Test Equipment 548
10.2.6 *In Situ* Saturation Monitoring (ISSM) 551
10.3 Test Data Interpretation Methodology 554
10.3.1 Unsteady State 554
10.3.2 Steady State 557
10.3.3 Centrifuge 557
10.3.4 Flow Rate and Capillary End Effects 558
10.3.5 Coreflood Simulation 562
10.4 Sample Selection, Test State and Test Conditions 565
10.4.1 Sample Selection 565
10.4.2 Test State 568
10.4.3 Overburden Stress 568
10.4.4 Initial Water Saturation 569
10.5 Measurement Descriptions 572
10.5.1 Unsteady State Water-Oil Relative Permeability (USSWO) 574
10.5.2 Steady State Water–Oil Relative Permeability (SSWO) 584
10.5.3 Centrifuge Water–Oil Relative Permeability (CentKrWO) 589
10.5.4 Unsteady State Gas–Oil Relative Permeability (USSGO) 592
10.5.5 Steady State Gas–Oil Relative Permeability (SSGO) 598
10.5.6 Centrifuge Gas–Oil Relative Permeability (CentKrGO) 603
10.5.7 Gas–Oil Relative Permeability Curve Generation 606

10.5.8 Unsteady State Gas–Water Relative Permeability (USSGW) 607
10.5.9 Steady State Gas–Water Relative Permeability (SSGW) 610
10.5.10 Unsteady State Water–Gas Relative Permeability (USSWG) 613
10.5.11 Steady State Water–Gas Relative Permeability (SSWG) 618
10.5.12 Centrifuge Water–Gas Relative Permeability (CentKrWG) 622
10.5.13 Residual Gas Saturation: Imbibition Centrifuge Capillary Pressure (Sgr-CentPc) 626
10.5.14 Counter-current Imbibition Residual (Trapped) Gas Saturation (CCI-Sgr) 629
10.6 Critical Gas Saturation (Depressurisation) 634
10.6.1 Sample Preparation 635
10.6.2 Test Equipment 635
10.6.3 Test Procedures 636
10.7 Experimental Data Provision 639
10.7.1 Samples and Fluids (for All Tests) 639
10.7.2 Drainage Capillary Pressure (Desaturation to Swir) 640
10.7.3 Centrifuge Relative Permeability (Water–Oil, Gas–Oil, Decane–Water) 641
10.7.4 Counter-Current Imbibition 642
10.7.5 USS and SS Flood Tests (Water–Oil, Gas–Oil and Water–Gas) 642
10.8 Summary and Recommendations 643
10.8.1 Oil Reservoirs 644
10.8.2 Gas and Gas Condensate Reservoirs 647
References 650
Recommended Reading 651

11. Nuclear Magnetic Resonance (NMR) 655

11.1 Introduction 655
11.2 Nuclear Spin Relaxation in Rocks 657
11.3 NMR Relaxation and Pore Size 659
11.4 Porosity from NMR 661
11.5 Clay-Bound Water 662
11.6 Permeability Estimation 662
11.7 NMR Tests on Core 663
11.7.1 Sample Preparation 663
11.7.2 Key Processes 664
11.7.3 Test Equipment 664
11.7.4 Recommended NMR Measurement Parameters 665
11.7.5 Test Procedures 666
11.7.6 Data Reporting Requirements 667
11.8 NMR Core Measurement Summary 668
References 668
Recommended Reading 669

12. Geomechanics Tests 671

12.1 Introduction 672
12.2 Sample Selection and Preparation 673
12.2.1 Test Sites 674
12.2.2 Sample Orientation 678
12.2.3 Plugging Fluids 679
12.2.4 Sample Dimensions 679
12.2.5 Sample Saturation 680
12.3 Unconfined Compressive Strength Tests 681
12.3.1 Purpose and Sample Requirements 681
12.3.2 Test Equipment 681
12.3.3 Test Procedures 682
12.3.4 Data Utilisation 682
12.3.5 Data Reporting Requirements 683
12.3.6 Advantages and Drawbacks/Issues 683
12.3.7 UCS Quality Control Issues, Checks and Diagnostics 684
12.4 Triaxial Compression Strength Tests 687
12.4.1 Purpose and Sample Requirements 687
12.4.2 Test Equipment 688
12.4.3 Test Procedures 688
12.4.4 Data Utilisation 695
12.4.5 Data Reporting Requirements 700
12.4.6 Advantages and Drawbacks/Issues 700
12.4.7 Triaxial Test Quality Control Issues, Checks and Diagnostics 700
12.5 Triaxial Testing of Shales 706
12.5.1 Purpose and Sample Requirements 706
12.5.2 Sample Preparation 706
12.5.3 Test Equipment 709
12.5.4 Test Procedures 709
12.5.5 Data Utilisation 712
12.5.6 Data Reporting Requirements 713
12.5.7 Advantages and Drawbacks/Issues 714
12.5.8 Shale Triaxial Test Quality Control Issues, Checks and Diagnostics 714
12.6 Thick-Wall Cylinder Tests 717
12.6.1 Purpose and Sample Requirements 717
12.6.2 Test Equipment 719
12.6.3 Test Procedures (Standard TWC) 722
12.6.4 Data Utilisation 722
12.6.5 Data Reporting Requirements 724
12.6.6 Advantages and Drawbacks/Issues 724
12.6.7 TWC Test Quality Control Issues, Checks and Diagnostics 725
12.7 Tensile Strength Tests 728
12.7.1 Purpose and Sample Requirements 728
12.7.2 Test Equipment and Procedures 729

12.7.3 Data Utilisation 730
12.7.4 Data Reporting Requirements 731
12.7.5 Advantages and Drawbacks/Issues 731
12.7.6 Tensile Strength Test Quality Control Issues, Checks and Diagnostics 732
12.8 Acoustic Velocity (Travel Time) Tests 732
12.8.1 Purpose and Sample Requirements 732
12.8.2 Sample Preparation 733
12.8.3 Test Equipment 733
12.8.4 Test Procedures 735
12.8.5 Data Utilisation 735
12.8.6 Data Reporting Requirements 736
12.8.7 Advantages and Drawbacks/Issues 736
12.8.8 ATT Test Quality Control Issues, Checks and Diagnostics 736
12.9 DSCA Tests 739
12.9.1 Purpose and Sample Requirements 739
12.9.2 Test Equipment 739
12.9.3 Test Procedures 740
12.9.4 Data Utilisation 742
12.9.5 Data Reporting Requirements 744
12.9.6 Advantages and Drawbacks/Issues 744
12.9.7 DSCA Test Quality Control Issues, Checks and Diagnostics 744
12.10 Pore Volume Compressibility Tests 746
12.10.1 Purpose and Compressibility Definitions 746
12.10.2 Compressibility Test Loading Conditions 747
12.10.3 Sample Preparation 751
12.10.4 Uniaxial K_0 Test Equipment 751
12.10.5 Uniaxial K_0 Test Procedures 751
12.10.6 Data Utilisation 754
12.10.7 Data Reporting Requirements 757
12.10.8 Advantages and Drawbacks/Issues 757
12.10.9 Uniaxial K_0 Compressibility Test Quality Control Issues, Checks and Diagnostics 757
12.11 Particle Size Analysis Tests 761
12.11.1 Purpose 761
12.11.2 Mechanical Particle Size Analysis 762
12.11.3 Laser Particle Size Analysis 769
References 776
Recommended Reading 778

13. Example of a Core Analysis Programme 781

13.1 Introduction 781
13.2 Core Analysis Focal Point 782
13.3 Design and Management 783

13.4 Lithological Considerations for RCA and SCAL 785
13.4.1 Clean Consolidated Sandstones 785
13.4.2 Unconsolidated Sandstones 785
13.4.3 Shaly Sands 786
13.4.4 Carbonates 786
13.4.5 Vuggy Carbonates 786
13.4.6 Low-Permeability Reservoirs 787
13.4.7 Fractured Reservoirs 787
13.5 Routine Core Analysis 787
13.5.1 Recommended RCA Tests 789
13.6 SCAL Programme 793
13.6.1 Gas Field 793
13.6.2 Oil Field 801

Index 811

Series Editor's Preface

This is the fifth book in the *Developments in Petroleum Science* series since it incorporated the *Handbook of Petroleum Exploration and Production* in 2013. After books on geophysics, stratigraphic reservoir characterisation, petrophysics and integrated sand management, we now switch our attention to core analysis.

In the ongoing search for oil and gas around the world, much of our understanding of the nature and distribution of hydrocarbon-bearing reservoirs is obtained through remotely sensing the subsurface. We are all familiar with the use of modern seismic, gravity, magnetic and electromagnetic techniques in the exploration for these reservoirs, but these techniques have also been enhanced to provide ongoing or time-lapse reservoir monitoring capabilities during production of reservoir fluids. These techniques for remote sensing are also used in drilling and petrophysics to understand the nature and distribution of various types of rocks and fluids.

However, in the evaluation of oil and gas reservoir rock properties, there are no techniques that allow us to directly see or measure the properties of these subsurface reservoirs, with a single exception. Core analysis is the one and only method of quantitatively measuring reservoir properties at surface or simulated subsurface conditions by the rigorous laboratory testing of cores obtained through drilling, be it for conventional or unconventional oil and gas reservoirs. It enables both exploration geoscientists and reservoir development teams to provide quantitative control of their exploration or production models and to determine the quantity of hydrocarbons in place and the speed that which it could be produced. Ultimately this can lead to a better understanding of commercial viability of the in-place hydrocarbons.

This book reviews the vitally important area of core analysis and the best laboratory and field practices needed to obtain the highest quality data with minimal errors, thereby maximizing the value of the data to exploration geoscientists and reservoir engineers. It should therefore be essential reading for those professionals active in hydrocarbon field exploration or production worldwide. In particular as the authors state, this book 'should provide the foundation or ground truth upon which formation evaluation rests'.

John Cubitt
Holt, Wales

Preface

In the evaluation of oil and gas reservoirs, core analysis provides the only direct and quantitative measurement of the reservoir properties and should provide the foundation or ground truth upon which formation evaluation rests. Geologists, reservoir engineers, and petrophysicists rely on accurate and representative core analysis data to provide key data input for building static and dynamic reservoir models. However, variable laboratory test and data reporting standards have resulted in variable data quality. Coupled with an inconsistent and inexperienced approach to the design and management of the laboratory test programmes, this has led to a situation where, in the authors' extensive global experience, around 70% of legacy special core analysis (SCAL) data are not fit for purpose.

There are few books on core analysis and its applications, and only one industry-recognised recommended practice, which although largely still relevant and appropriate, does not fully cover SCAL and has not been updated for almost over 25 years. The objective of this book is to provide the end users of core analysis data with the essential information that they need to enable a more pro-active, more coherent, and more consistent approach to programme design, data acquisition, and quality control of core analysis data. The authors are actively engaged in the design and management of laboratory core analysis programmes and in interpreting and using the data. Frustrated by the lack of standards and the often inadequate utility of core analysis data, we have written practical and pragmatic guidelines for core analysis best practice that should maximise data quality, encourage more effective engagement between the end users and the data acquisition laboratories, and thereby add value to core analysis investments.

Obtaining high-quality, undamaged core is the essential prerequisite for representative and reliable core analysis data. We therefore start by describing the principal coring systems and fluids, and how to identify and minimise damage to the core on its journey from the reservoir to the test laboratory. Analysis is usually performed on plug samples cut from core, so we discuss the methods and equipment used to cut plugs and prepare them for analysis and their impact on the measured properties. Specialist preparation methods for wettability conditioning and preservation of delicate mineralogies are described.

Routine core analysis typically involves "as received" fluid saturation measurements, and porosity and permeability measurements on dry plugs, which are used to characterise the reservoir properties and for log-core integration. As the test equipment and methods vary between laboratories, we list the advantages and drawbacks/issues associated with each test and summarise quality control checks and diagnostics.

SCAL measures fundamental static and dynamic reservoir rock properties, often at reservoir-appropriate fluid, stress, and temperature conditions. Correct planning for SCAL tests is fundamental to the acquisition of high-quality, formation-representative data, so we focus on: the preparation and characterisation of test fluids; how to select and screen representative test plugs; and how to determine the equivalent reservoir effective stress. Formation wettability is a fundamental rock property. We describe laboratory wettability tests but, more importantly, the controls on native wettability and how wettability can be altered during coring, core recovery, and core handling, and the methods used to try and restore native wettability conditions.

We then focus on the principal SCAL tests: porosity and permeability at overburden stress; electrical properties for clean formations and shaly sands; capillary pressure; nuclear magnetic resonance (NMR); and relative permeability. We explain the various core test conditions, test methods, equipment, and procedures, and data reporting requirements. The advantages and drawbacks/issues of each suite of tests are described, and the quality control checks and diagnostics are summarised. We clarify the corrections and refinement protocols used to process laboratory data for static and dynamic model input.

As core also supplies a source for geomechanical measurements for faster and safer drilling and better completions, we include a chapter on rock mechanics tests and particle size analysis tests. Finally, we provide generic examples of core analysis test programmes that the reader can use as guide or template to design and specify the workflow for core test programmes on oil and gas reservoirs.

Following the best practice guidelines in this book will benefit petrophysicists, geoscientists, reservoir engineers, and production engineers who use the data but are often unaware of the uncertainty range. These are merely guidelines, gained from years of experience, and are necessarily generic: each reservoir is different and the data required to characterise and aid field development planning should be considered and discussed with recognised specialists before finalising the detailed experimental programme. Nevertheless, we hope the reader will gain a better understanding of core analysis procedures and methods; be better placed to judge the quality of core analysis data; and ultimately be able to reduce uncertainty in static and dynamic reservoir models.

This book is based on material from many sources including hundreds of laboratory reports; publications from the Society of Petroleum Engineers, the Society of Professional Well Log Analysts, the Society of Core Analysts; academic institutions; as well as oil company and service company websites. All reasonable effort has been made to trace and refer to the sources of these materials.

We would like to recognise the cooperation of the oil companies and core analysis laboratories who have helped us develop these best practice guidelines. Labs aim to deliver excellent work often under trying circumstances. We are also indebted to the various companies and organisations who have given us permission to include their material in this work, and the management of LR Senergy for their support. This book would not have been possible without the contributions, encouragement, and assistance of the following people: Mel Boulby, Mike Byrne, Tim Conroy, Gill Daniels, Bob Harrison, Lynne Harrower, Michelle Hubbard, Elise Johnston, Rick Lemanczyk, Michele Loseto, David Milton-Tayler, Iain Morrison, Phil McCurdy, Georgia McKendrick, Vivien MacKinlay, John Owens, Max Podolyak, Chris Reed, Beth Reid, Yelitza Sorrentino, Maria Velazco, Graham Webber, and Phanthip Wongtui. Finally, we are thankful for the encouragement and patience of Elsevier editorial staff, particularly Derek Coleman and the series editor, John Cubitt.

Colin McPhee
Edinburgh, UK

Jules Reed
Aberdeen, UK

Izaskun Zubizarreta
Aberdeen, UK
June 2015

Chapter 1

Best Practice in Coring and Core Analysis

Chapter Outline

1.1 Core Analysis Data: The Foundation of Formation Evaluation 1
1.2 Core Analysis Data Uncertainty 4
1.2.1 Reasons and Consequences 4
1.2.2 Reducing Uncertainty 8
1.3 Core Analysis Management Framework 8
1.3.1 Core Analysis Planning and Design 8
1.3.2 Programme Design Considerations 9
1.3.3 Core Analysis Focal Points 9
1.3.4 Real-Time Quality Control 11
1.4 Best Practice in Core Analysis: An Overview 11
1.4.1 Coring, Core Handling and Core Processing 12
1.4.2 Sample Preparation 12
1.4.3 Routine Core Analysis 12
1.4.4 Special Core Analysis 12
1.4.5 Geomechanics Tests 13
1.4.6 Quality Control Procedures and Diagnostics 13
1.4.7 Example Core Analysis Programmes 14
1.4.8 Benefits 14
References 15

1.1 CORE ANALYSIS DATA: THE FOUNDATION OF FORMATION EVALUATION

The primary goal of geologists and petrophysicists is to estimate the volume of hydrocarbons initially in place in a reservoir. The primary goal of the reservoir engineer is to understand the physics of the reservoir-fluid system so that the ultimate recovery of hydrocarbons is maximised in the most economic matter. Both require a detailed knowledge of the reservoir geometry, structure and the interaction between the reservoir and the fluids, either in place, or which may be introduced into the reservoir. In reservoir modelling, a computer model of the reservoir is constructed by geologists, petrophysicists and geophysicists to provide a description of the reservoir which is principally used to determine the volumes of hydrocarbon in place. This is normally referred to as a static model. Reservoir simulation models are constructed

Developments in Petroleum Science, Vol. 64. http://dx.doi.org/10.1016/B978-0-444-63533-4.00001-9

by reservoir engineers to describe and map the hydrocarbon recovery processes under different production mechanisms. These dynamic models are principally used to determine reserves and recovery factors and to predict hydrocarbon production profiles for economic analysis.

Both static and dynamic reservoir models draw on a variety of disparate data sources including regional geology, seismic, sedimentological modelling, drilling data, wireline and logging/measurement while drilling data, fluid pressures and rock and fluid property data. The nature and quality of the model input data change throughout the lifetime of a field, so it is important to constantly review data quality to minimise uncertainties and to include data quality assessment in reservoir modelling. The quantity and quality of data used for both static and dynamic reservoir modelling must always be fit for purpose and match the field development objectives.

Core is normally the only part of the (relatively) undisturbed reservoir formation we can actually see, touch and feel at the surface. Consequently, core analysis data should be the "ground truth", or the foundation upon which integrated formation evaluation and reservoir characterisation rest. All other data sources are essentially remote, so reliable and representative core analysis data are essential to calibrate and validate other data.

For example, the volume of stock tank oil initially in place (OIIP) in a reservoir can be determined from

$$\mathrm{OIIP} = \mathrm{GRV}\left[\frac{N}{G}\right]\phi(1-\mathrm{Sw})\frac{1}{B_\mathrm{o}} \tag{1.1}$$

Determination of the gross rock volume (GRV) and gross factor (G) in the net/gross ratio (N/G) is the primary responsibility of geophysicists and geologists. The reservoir engineer is responsible for oil formation volume factor (B_o) from pressure, volume temperature (PVT) experiments. The petrophysicist is responsible for net (N), porosity (ϕ) and water saturation (Sw) where data input relies principally on logs. Reservoir net thickness is normally defined by a permeability cut-off, and high-resolution permeability data are only possible from core. Porosity interpretation (e.g. from density logs) should be verified by, or calibrated against, stressed core porosities. Resistivity log interpretation requires core electrical property measurements to quantitatively determine water saturation in clean formations, and normalised cation exchange capacity is required to correct formation resistivity response for the presence of conductive clays. Water saturation can be determined directly by extracting water from core using Dean Stark or retort methods or indirectly, from primary drainage capillary pressure measurements.

The typical core analysis tests which are offered by commercial core analysis vendors and used as data input in petrophysical static models are summarised in Table 1.1. Historically these tests were carried out only at ambient conditions (low or no confining stress; ambient laboratory temperature), but most commercial laboratories can now provide these tests at more representative reservoir-appropriate stress, fluid and temperature conditions.

TABLE 1.1 Typical Core Analysis Data Input to Volumetric Calculations (Static Models)

Parameter	Data Source	Test Methods
Net	Permeability	Air permeability (ambient or reservoir stress) Klinkenberg permeability (ambient or reservoir stress) Brine (water) permeability (ambient or reservoir stress) Probe permeability (ambient conditions)
Porosity	Density porosity	Helium porosity (ambient or reservoir stress) Resaturation porosity (ambient or reservoir stress)
Water saturation	Electrical parameters	Formation resistivity factor (ambient or reservoir stress) Resistivity index (ambient or reservoir-appropriate conditions) Wet chemistry cation exchange capacity (CEC) for shaly sands (ambient conditions) Multiple-salinity tests (normalised CEC) at ambient or reservoir stress
	Primary drainage capillary pressure	Low-pressure mercury injection (ambient or reservoir stress) High-pressure mercury injection (ambient conditions) Gas–water or oil–water porous plate (ambient or reservoir-appropriate conditions) Gas–water or oil–water centrifuge (ambient conditions or limited reservoir stress)
	Direct measurement	Retort extraction (ambient conditions) Dean-Stark extraction (ambient conditions)

As Dake (1991) points out, "determination of the recovery factor is the most important single task of the reservoir engineer". Recovery factors may be determined on purely technical criteria, but, more probably, on economic or environmental terms. For example, hydrocarbon recovery efficiency in a waterflood in an oil reservoir is largely governed by the mobility ratio:

$$M = \frac{M_{\text{rw}}}{M_{\text{ro}}} = \frac{\text{krw}}{\text{kro}} \frac{\mu_{\text{w}}}{\mu_{\text{o}}} \tag{1.2}$$

where M_{rw} and M_{ro} are the relative mobilities of water and oil, respectively. The parameters kro and kro are the relative permeabilities to oil and water, and μ_w and μ_o are water and oil viscosities, respectively. If kro and krw are relative permeabilities at residual phase saturations (kro at irreducible water saturation, Swir, and krw at residual oil saturation, Sro), then M is defined as the endpoint mobility ratio. If this value is ≤ 1, then in a waterflood, oil can travel through the reservoir at a speed greater than or equal to water. This "piston-like" displacement results in a sharp interface between the fluids which minimises oil bypass and results in a very efficient displacement. However, if M is >1, water can travel faster than oil leading to earlier water breakthrough at producers, and a poor or unstable displacement resulting in oil bypass. It may take injection of several barrels of water to recover one barrel of oil. Relative permeabilities of oil, gas and water therefore provide key input data in reservoir dynamic modelling.

The typical core analysis tests which are offered by commercial core analysis vendors and used as data input in dynamic reservoir models are summarised in Table 1.2. This is not meant to be exhaustive—rather it lists the principal tests to determine the key modelling parameters. Combinations of two or more tests are normally required to fully define the parameters required to predict recovery factors under different drive mechanisms and to describe fluid movement for saturated and under-saturated oil reservoirs, lean and rich gas reservoirs and gas condensate reservoirs. Historically these tests were carried out only at ambient (low or no confining stress; ambient laboratory temperature) conditions using synthetic fluids, but many commercial laboratories now provide these tests at more representative reservoir-appropriate stress, fluid and temperature conditions.

1.2 CORE ANALYSIS DATA UNCERTAINTY

Potentially, core analysis is the only direct and quantitative measurement of the "intact" reservoir properties and should underpin static and dynamic modelling. Harrison (2009) argues that taking and analysing core in increasingly more complex reservoirs have never been more important. Core and core analysis data confirm lithology and mineralogy; calibrate estimates of fundamental rock properties; show how fluids occupy and flow in pore space; and supply a source for geomechanical measurements for faster and safer drilling and better completions. Harrison points out that "logs alone cannot characterise the reservoir if knowledge of the rock is absent so subsequent modelling must rely on un-calibrated and unverified log-derived correlations". The predictable consequence of not having core analysis data is greater uncertainty.

1.2.1 Reasons and Consequences

Based on the authors' extensive experience in the acquisition, verification and interpretation of core analysis data of different vintages from across the world,

TABLE 1.2 Typical Core Analysis Data Input to Reservoir Dynamic Modelling

Reservoir Fluid	Drive Mechanism	Parameters	Test Methods
Oil	Waterflood	kro and krw endpoints	Unsteady-state waterflood to Sro
		Intermediate kro and krw	Unsteady-state water→oil relative permeability Steady-state water→oil relative permeability Centrifuge water→oil (kro only)
	Solution gas Gas cap expansion	krg and kro endpoints	Unsteady-state gasflood to Sro
		Intermediate krg and kro	Unsteady-state gas→oil relative permeability Steady-state gas→oil relative permeability (at Swir) Centrifuge gas→oil (kro only)
	Imbibition capillary pressure	Spontaneous	Water→oil porous plate
		Forced	Water→oil porous plate (limited curve) Water→oil centrifuge (full curve)
	All	Wettability	Contact angle measurement Amott method USBM method Combined Amott/USBM
Gas	Gas storage/ injection	Drainage krg endpoint	Unsteady-state gasflood to Swir Centrifuge gas→water to Swir
	Aquifer drive Water imbibition	Imbibition krg and krw endpoints	Unsteady-state water flood to Sgr

Continued

TABLE 1.2 Typical Core Analysis Data Input to Reservoir Dynamic Modelling—Cont'd

Reservoir Fluid	Drive Mechanism	Parameters	Test Methods
			Centrifuge gas→water to Swir
		Intermediate krg and krw	Unsteady-state water→gas relative permeability Steady-state water→gas relative permeability Centrifuge water→gas or decane (krg only)
	Water imbibition capillary pressure	Spontaneous	Water→gas porous plate Water→gas or decane centrifuge Countercurrent imbibition (Sgr only)
		Forced	Water→gas/decane centrifuge
Oil and gas	Natural depletion	Pore volume compressibility	Helium porosity versus stress Saturated porosity versus stress Uniaxial (K_0) pore volume compressibility

we conservatively estimate that *70% of core analysis data are unfit for purpose* due to their unreliability, inapplicability or inappropriateness. This has resulted from a combination of the following factors:

1. Poor inter-laboratory data comparability due to the lack of standardisation of test procedures and the sensitivity of core data to different test methods. For example, there are four principal methods used to determine porosity, yet these can give completely different results depending on the core plug size and shape.
2. The lack of thought given to the programme test design by the commissioning end users including appropriateness of specified core tests; the

reliability of the data and their applicability; and the lack of understanding of the practical difficulties faced by core analysis laboratories and the technical and commercial constraints they must work under.

3. Historically inadequate reporting standards which give little real information on the provenance of test data and their interpretation, and the failure to provide experimental data to check and verify the results.
4. Strong market competition which has required the commercial core analysis vendors to produce data more reliably, for less money and with faster turnaround times. Competitive pressures have also limited investment in equipment and R&D, contributed to low staff retention and constrained management succession planning in some commercial laboratories.

In a multiple laboratory comparative study, Amabeoku and BinNasser (2012) found that some core service contractors provided very poor quality data on some of the tests and concluded that many laboratories do not have quality control protocols in-house and just report data acquired. They recommended very strict supervision of, and engagement with, the vendor throughout the data acquisition programme. McPhee and Arthur (1988) found inexplicable relative permeability data discrepancies between commercial laboratories testing exactly the same core material and fluids. One laboratory produced residual oil saturation data that were impossible to reproduce and appeared to have manipulated relative permeability endpoints to fit their own expectations. There is a recommended practice for routine core analysis (RCA): API RP 40 (1988). However, while providing an exhaustive description of core acquisition, handling and screening and core analysis test methods, it arguably does not address what is now considered to be best practice and is limited to RCA tests. It does not include special core analysis (SCAL) tests. The apparent reluctance in the industry to harmonise or standardise core analysis measurements is often puzzling to the outsider—compared to soil mechanics and rock mechanics testing for civil engineering purposes, for example. Maintaining a commercially competitive edge and the belief that each reservoir core is fundamentally distinct and needs to be treated differently may explain the lack of apparent enthusiasm to standardise.

Nevertheless, it is the end user (the client who contracts the core programme) rather than the lack of test standardisation, or the poor performance of the test laboratory, who is often more culpable. Too often core analysis programmes are ill considered, badly designed and poorly supervised, and the results are only crudely integrated with other well and reservoir data. The results, in terms of data acquired, are often unrepresentative or contradictory.

It may be no surprise therefore that it remains an uphill struggle to convince management in some companies that the project benefits from the knowledge gained from core analysis. Yet, with proper planning and management of the coring and analysis processes, core data should be and can be the ground truth for formation evaluation.

1.2.2 Reducing Uncertainty

The examples of laboratory artefacts presented here have a significant impact on petrophysical interpretation and reservoir characterisation. Yet with experience, learnings and the appropriate diagnostic tools, these uncertainties are recognisable and manageable.

The purpose of this book is to help minimise data uncertainties by detailing best practice in core analysis data acquisition, quality control and data interpretation. Incorporating the practices and procedures described herein within an integrated core analysis management framework will bring significant benefits to both the end users and the test laboratories. In particular, amalgamating best practice procedures with pro-active project management will

- enable improved communication and learnings between the test laboratories and the end users;
- generate a better understanding of core analysis methods and data sensitivities;
- provide a more coherent and consistent approach to data acquisition;
- reduce uncertainties and data redundancy;
- facilitate a full data audit trail.

Demonstrably this will bring better equity and unitisation positions, enable easier and more efficient presentation of core analysis plans and results to partners and most significantly provide added value from core analysis investments. There should be a marked improvement in both data quality and technical communication with core analysis vendors. In the authors' experience, data redundancy rates can be reduced from 70% to <10% (McPhee, 2012).

1.3 CORE ANALYSIS MANAGEMENT FRAMEWORK

1.3.1 Core Analysis Planning and Design

Coring and core analysis are often poorly planned. On too many occasions, the engineer or geoscientist simply copies and pastes from previous programmes without any real understanding of the data that the tests will acquire, or how applicable or representative the data will be. However, they still want control of the core analysis programme and are often reluctant to take advice from the core analysis vendor as they feel they may be oversold tests they do not think they need. This is compounded by the traditional master (client)–servant (laboratory) relationship where the "customer is always right". A deficient or ill-considered test programme results in under-utilisation, poor appreciation and mis-application of the resultant core data. Proper design, planning and supervision of a core analysis programme can do much to reduce the data redundancy rate. Petrophysicists, geologists, reservoir engineers and drilling and completions engineers all have a role to play in the planning team, which must include the laboratory that will carry out the work.

Engaging the test laboratories through regular meetings and lab visits during the planning, design and data acquisition phases of ongoing projects means they become more aligned with the client stakeholder objectives and better understand how important their data really are in field development planning and decisions.

1.3.2 Programme Design Considerations

The key questions and factors that should be addressed prior to embarking on acquiring core and designing a test programme are as follows:

1. Are there areas of concern or anomalies or suspicious data in the existing core database that need to be resolved? How closely do the core, log and test data agree for the well in question and the reservoir in general?
2. Are special precautions or measures required to cut, recover, handle and transport core? For example, what core-tripping rates are required to prevent damage to core? Do we need to take samples at wellsite? Should the core be stabilised for transport? What is the risk of wettability contamination from the coring fluid?
3. How do we prepare the core for analysis? Will the preparation methods cause any alteration in the petrophysical properties and how do we quantify this?
4. What core analysis tests do we actually need? It is important not to select tests from a "menu" or to "do what we have always done". Each core is different and each core analysis programme is unique. For example, reservoir engineers should ask themselves why they require full relative permeability curves. In some circumstances, just endpoints may suffice.
5. Is the contractor interpretation correct and can operators improve it? Laboratory interpretation of experimental data can be subjective and less than rigorous. This misplaced belief in the veracity of laboratory reported data is analogous to a petrophysicist or reservoir engineer relying entirely on the logging company's or the testing company's interpretation of wireline log or time/pressure/rate data. Best practice dictates that they check the vendor data and provide their own interpretation—yet this is rarely the case for core analysis data. As Harrison (2010) states when discussing log interpretation: "end users must not abdicate their responsibility for the interpretation". It is also true, if not more so, for core data.

1.3.3 Core Analysis Focal Points

Recurring themes in many core laboratory audits are the need to improve communication between the laboratory and the client, and client education in core analysis acquisition and interpretation. In particular, many vendors have stated that they are too often faced with conflicting and contradictory

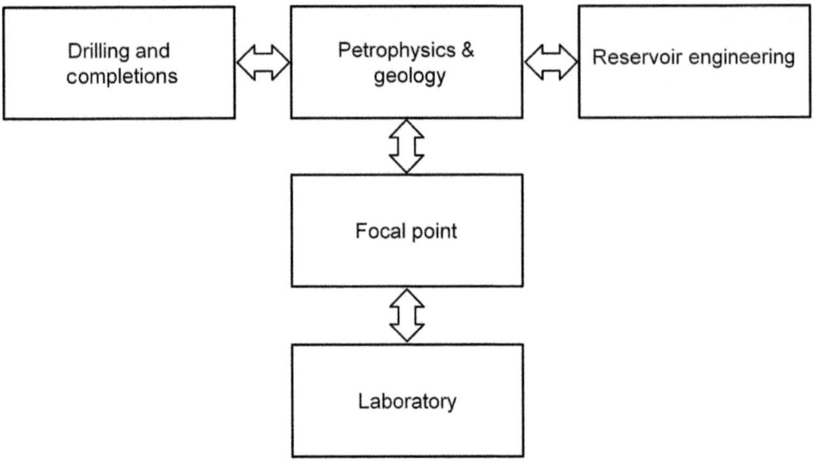

FIGURE 1.1 Core analysis focal point in core analysis management.

instructions from within the operator's different discipline functions, and would prefer to deal with a single, knowledgeable, core analysis focal point who understands the applications and limitations of core analysis tests. This should ensure that the data quality requirements are maintained and, more importantly, that the data are fit for purpose. The client focal point is the liaison between the client's different subsurface disciplines and the laboratory, as indicated in Fig. 1.1.

Amongst the key focal point responsibilities are as follows:

- **i.** Design and costing of the test programmes, with the assistance of the laboratory;
- **ii.** Preparing cost justifications to management;
- **iii.** Coordination with drilling and wellsite engineers and laboratory staff to review core drilling, core recovery and wellsite handling, storage and transportation procedures;
- **iv.** Design and specification of the test and reporting procedures to be adopted in the scope of work including deliverables, milestones and project reporting;
- **v.** Reviewing contractor performance against initially set goals, objectives and deliverables;
- **vi.** Analysis and checking of the contractors' data as soon as possible after they are received;
- **vii.** Preparing a final report on the core analysis study, which reconciles core results with other well and reservoir data and provides appropriately interpreted and reliable core analysis data that can be used for static and dynamic models.

The laboratory should also appoint a project manager with the key responsibilities of client expectation management; organising and controlling the work; and documentation of the project requirements, test specifications, lab worksheets and analysed data. Peer review of both intermediate and final data is essential before delivery to the client.

1.3.4 Real-Time Quality Control

Regular monitoring of vendors' performance and the provision of, and checking, experimental data can ensure that any problems or unusual, anomalous or inconsistent results can be identified as soon as possible, so that they can be rectified before the test programme is completed. Thereafter, it is too late, and costs are often incurred in retesting which can lead to largely undeserved but lingering resentment over laboratory performance. Contractor supervision and quality control will ensure that complete records of the test methods and procedures together with laboratory experimental data are available, so that there is a complete audit trail. The increasing use of digital data transmission and internet client "dropbox" sites has significantly improved quality assurance and timely decision making.

1.4 BEST PRACTICE IN CORE ANALYSIS: AN OVERVIEW

The first challenge in the design of a core analysis programme is to obtain representative core material for the laboratory to test. Hence, the coring programme should be planned in detail taking into consideration core retrieval from reservoir to surface, surface handling, preservation and transportation to the laboratory. Once the core is in the laboratory, different methods are available to obtain a variety of measurements, each test having defined strengths and weaknesses. The advantages and drawbacks/issues associated with each principal core analysis test will be explained in detail, as a precise understanding of these tests and methods is important for the evaluation and integration of the core data with the log-reservoir simulation data.

This book sets out to provide what the authors believe to represent current best practice in core analysis. It provides a practical overview of core analysis, from coring at the wellsite through laboratory testing to results and data presentation. It is dedicated to the description and understanding of the various techniques and procedures required for the analysis of both siliciclastic and carbonate reservoirs in commercial core analysis and rock mechanics laboratories.

Certain commercial entities, equipment, or materials may be identified in this book in order to describe an experimental procedure or concept. This is not to imply recommendation or endorsement by the authors, nor is it intended to imply that the entities, equipment, or materials are the best available for the purpose.

1.4.1 Coring, Core Handling and Core Processing

Obtaining high-quality, undamaged core is an essential prerequisite for representative and reliable core analysis data. Even the best petrophysical measurements will prove futile if the core has been damaged or altered before the tests are actually performed. Full details and explanations of the principal coring systems, and recommendations on avoiding core damage at wellsite and during transportation to the test laboratory, are provided.

1.4.2 Sample Preparation

For most core analysis tests, the core samples must be initially cleaned and dried to remove oil and water, as well as evaporated salts, mud filtrate and wettability contaminants. Selection of the most appropriate sample preparation method depends on the lithology, petrography, potential reservoir wettability and possible wettability contamination from drilling mud, as well as the objectives of the analysis. Details of the principal core/plug cleaning and drying methods and equipment, and typical cleaning solvents are explained. Specialist sample preparation for porosity measurements, wettability conditioning and preservation of authigenic clays and halite cements are described.

1.4.3 Routine Core Analysis

Routine—or basic or conventional core analysis—principally involves measurements of fluid saturations on core, and porosity and absolute permeability measurements using single-phase fluids normally at ambient conditions on dry cores. The data are principally used to characterise the reservoir properties and for log–core integration. Recommended test equipment, test procedures and data reporting requirements are provided for the principal RCA tests.

1.4.4 Special Core Analysis

The methods and procedures to prepare and characterise test fluids used in SCAL (oil, water and gas), and how to select representative test samples and representative stress conditions, are described.

The principal core analysis tests which are commonly grouped under the generic description of SCAL are explained in detail. These include:

- porosity at stress;
- formation resistivity factor;
- resistivity index;
- cation exchange capacity tests (wet chemistry cation exchange capacity and multiple salinity Qv);
- drainage and imbibition capillary pressure tests using mercury injection (low and high pressure), porous plate and centrifuge methods;

- contact angle, Amott and United States Bureau of Mines (USBM) wettability tests;
- water–oil, gas–oil and gas–water and water–gas relative permeability tests using unsteady-state, steady-state and centrifuge techniques;
- nuclear magnetic resonance tests to calibrate log responses.

1.4.5 Geomechanics Tests

Geomechanics has an important role to play in assessing formation integrity during well construction and completion and in the response of the reservoir to oil production, water injection and depletion. The tests most commonly used to determine fundamental rock mechanics parameters are described including:

- unconfined compressive strength,
- thick wall cylinder,
- tensile strength,
- triaxial tests,
- pore volume compressibility,
- elastic moduli,
- particle size analyses.

1.4.6 Quality Control Procedures and Diagnostics

As even small laboratory artefacts and measurement uncertainties have a significant impact on key core data inputs, laboratory SCAL reports must include a detailed description of the work performed, the equipment and procedures used and details of the methods used by the lab in analysing the data. Details of appropriate instrument calibration data, and the equations used to generate the analysed data from "raw" measurements, should also be provided after each test or suite of tests so that data can be checked in "real time".

Laboratory data reporting requirements for both experimental and interpreted data that can be used to check or verify the laboratory results, and to provide an alternative interpretation of the data, are listed. Laboratories often charge extra for compiling this information, and although the data might never be required, they can prove invaluable in audit trailing and unitisation.

Understanding the history of test samples can be crucial in evaluating data quality—especially in formations of sensitive to stress cycling and rock/fluid incompatibilities—yet deciphering sample history records from standard report presentations can be challenging. Figure 1.2 presents an example of a customisable single-page SCAL sample history sheet that charts the history of each sample throughout the sample preparation and testing sequence and provides sample photographs before and after testing. Key results can also be captured by this format. More effective knowledge sharing—highlighting unusual or anomalous data in the report text, tables and figures—will allow

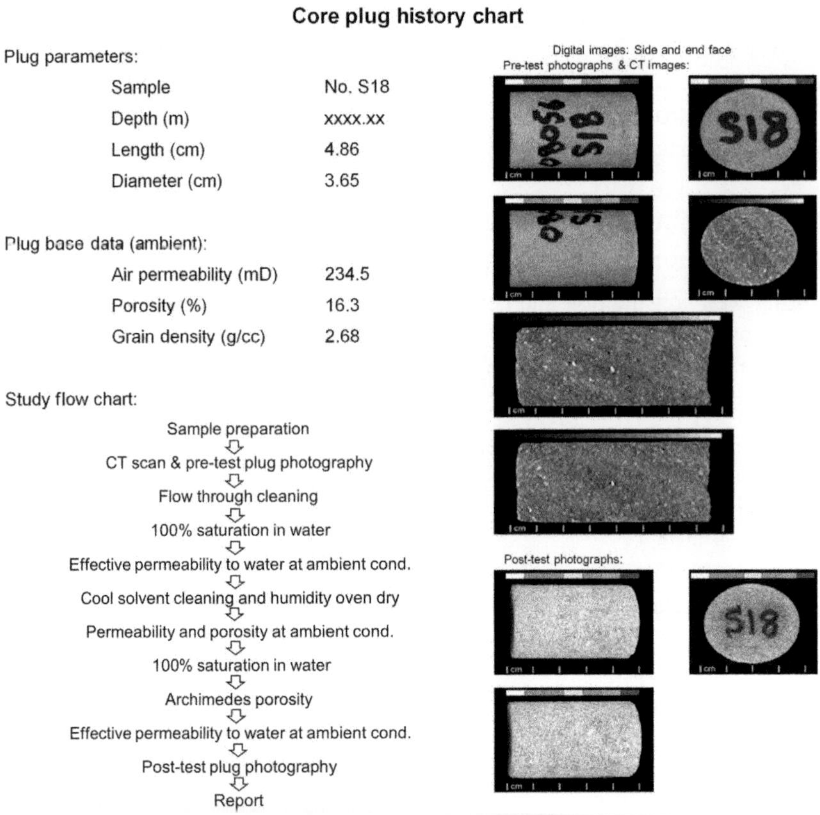

FIGURE 1.2 Example core plug history sheet. *Courtesy of Woodside Energy Limited.*

the end user to gain and benefit from the lab's undoubted expertise and experience in similar lithologies.

Quality control diagnostic procedures and data presentations which help identify potentially suspect or anomalous data are presented for most RCA and SCAL tests.

1.4.7 Example Core Analysis Programmes

Templates for typical RCA and SCAL test programmes are provided. These can be used as blueprints that can be adapted and modified to help plan, design and specify core test programmes for oil and gas reservoirs.

1.4.8 Benefits

The primary objective of this guide to best practice in core analysis is to provide reservoir engineers, geoscientists and petrophysicists with a solid grounding in the acquisition, evaluation and implementation of reliable and representative RCA and SCAL data, and the essential knowledge to be able to judge the quality and reliability of core analysis data that are used in reservoir evaluation and characterisation.

REFERENCES

Amabeoku, M., BinNasser, R.H., 2012. Quality control/quality assurance of core analysis data from multiple commercial laboratories. In: SPE Saudi Arabian Section Technical Symposium and Exhibition, April, 2012, SPE 160868.

American Petroleum Institute (API), 1988. Recommended Practice RP 40. Recommended Practices for Core Analysis, second ed. API Publishing Services, Washington, DC.

Dake, L.P., 1991. Fundamentals of Reservoir Engineering. Elsevier Science Publishers, Amsterdam.

Harrison, B., 2009. Formation evaluation. SPE J. Petrol. Technol. 61, 34.

Harrison, B., 2010. Formation evaluation. SPE J. Petrol. Technol. 62, 42.

McPhee, C.A., 2012. The core analysis elephant in the formation evaluation room. In: SPE Annual Technical Conference and Exhibition, San Antonio, Texas, USA, 8–10, October, SPE 158087.

McPhee, C.A., Arthur, K.G., 1988. Relative permeability measurements: an inter-laboratory comparison. In: European Petroleum Conference, London, October, 1994, SPE 28826.

Chapter 2

Wellsite Core Acquisition, Handling and Transportation

Chapter Outline

2.1 Coring Systems **17**
- 2.1.1 Conventional Full-Diameter Coring Systems 18
- 2.1.2 Wireline-Retrievable Cores 27
- 2.1.3 Gel Coring Systems 28
- 2.1.4 Liquid and Gas Retention Coring Systems 28
- 2.1.5 Oriented Coring 35
- 2.1.6 Sidewall Cores 37

2.2 Conventional Coring Operations **40**
- 2.2.1 Health, Safety and Environmental Considerations 41
- 2.2.2 Coring Team 42
- 2.2.3 Managing Coring Risks 42

2.3 Coring Fluids **45**
- 2.3.1 Mud Types 45
- 2.3.2 Mud Tracers 45

2.4 Core Damage and Core Fluid/Petrophysical Property Alteration **48**
- 2.4.1 Fluid Saturation Alteration 48
- 2.4.2 Stress Release 52
- 2.4.3 Wettability Alteration 57

2.5 Best Practice in Wellsite Handling **66**
- 2.5.1 Core Recovery on Rig Floor 66
- 2.5.2 Linear Layout and Marking 68
- 2.5.3 Wellsite Gamma-Ray Scanning 70
- 2.5.4 Wellsite Liner Cutting 70
- 2.5.5 Wellsite Core Stabilisation and Preservation 72
- 2.5.6 Wellsite Sampling 77
- 2.5.7 Core Transportation 79
- 2.5.8 Coring Report 81

2.6 Special Handling Considerations for Difficult Rock Types **81**
- 2.6.1 Unconsolidated Core 81
- 2.6.2 Carbonates 82
- 2.6.3 Shaly Sands 83
- 2.6.4 Shales 83
- 2.6.5 Coal 84

References **84**

Recommended Reading **87**

2.1 CORING SYSTEMS

There are several types of cores that can be recovered from the well. Conventional coring systems include full-diameter cores and wireline-retrievable cores. In addition, specialist systems are available that retain fluids or

Developments in Petroleum Science, Vol. 64. http://dx.doi.org/10.1016/B978-0-444-63533-4.00002-0

pressure, and that can be used to orient cores. Sidewall coring systems are used to sample the formation where fully- or semi-continuous coring operations are logistically or economically difficult.

2.1.1 Conventional Full-Diameter Coring Systems

Full-diameter coring systems are the most common source of core for analysis. Drilling is halted at the designated coring point and the drillstring is pulled out of hole (POOH). The drillbit is removed and a rotary coring bit is attached in its place. The rotary coring bit consists of solid metal with diamonds or tungsten for cutting, but unlike a drillbit, a rotary coring bit has a hollow centre. Core bits are selected based on the formation types, the requirement to minimise drilling mud-filtrate invasion, optimised core recovery, penetration rates and efficiency. Some examples of coring bits are shown in Figs. 2.1 (rotary coring bit) and 2.2 (polycrystalline diamond compact or PDC bit).

The length of conventional core barrels can range from around 30 to over 400 ft. The core diameter depends on the hole size, as indicated in Table 2.1. For example, if the reservoir section is to be drilled in 8½″ hole size, then core diameters are around 3½″ – 4″. In most cases, larger core diameters mean that fluid invasion from the drilling mud filtrate is unlikely to penetrate deep into the core. Smaller diameter cores are likely to be exposed to more filtrate invasion and potential damage than larger diameter cores, and only small-diameter plug samples might be able to be taken.

FIGURE 2.1 Example rotary coring bit for full-diameter cores. *Courtesy of Borehole Research Group, Lamont-Doherty Earth Observatory of Columbia University.*

FIGURE 2.2 Example PDC core bit. *Image courtesy of Baker Hughes.*

TABLE 2.1 Typical Full-Diameter Core Sizes and Hole Diameter

Hole Size (in.)	Core Size (in.)
5¾–6½	$2^9/_{16}$
6½–8	3½
8–9¼	4
9¼″ – 11	5¼
11–12½	5¼

The hardness (compressive strength) and variability of the rocks to be cored have the greatest influence on core bit selection. These include natural diamond bits and thermally stable diamond bits for hard or very hard formations, and polycrystalline diamond compact (PDC) bits for very soft to moderately hard formations.

When drilling with conventional core bits (e.g. Fig. 2.3), filtrate invasion occurs ahead of the bit (referred to as the filtrate bank), in the throat of the bit and in the core barrel during coring and on exposure to a static column of mud in the hole on core recovery. The bit cutters cut into the external and internal filter cakes thereby exposing the formation to rapid filtrate intrusion which can result in a complete flushing of the core with filtrate. Oil-based mud (OBM) filtrates can cause significant alteration in native wettability and filtrate invasion can alter the native formation fluid saturations in the core. Low-invasion core bits are often used to minimise filtrate invasion (Fig. 2.4). These bits are designed to cut as fast as possible without breaking

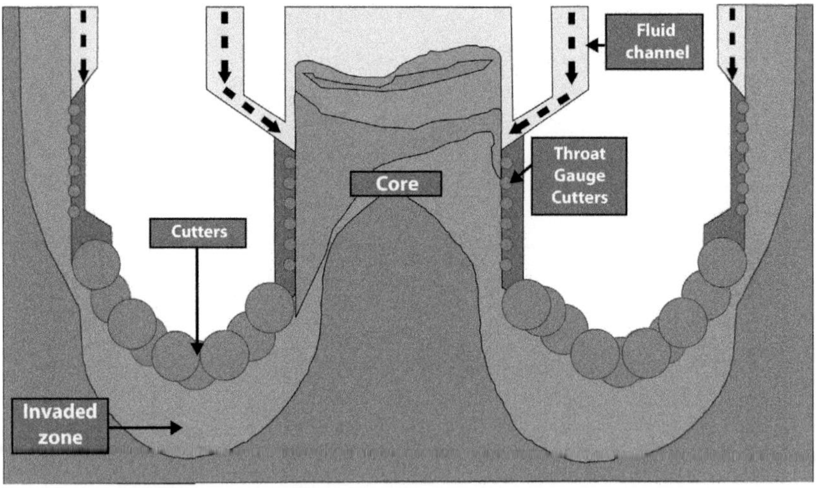

FIGURE 2.3 Schematic of a conventional core bit.

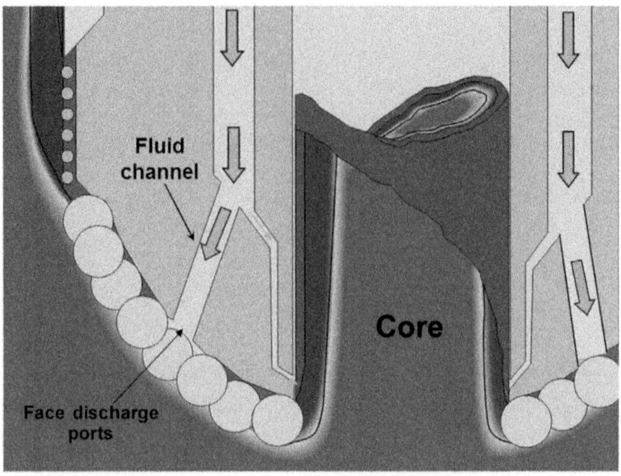

FIGURE 2.4 Schematic of a low-invasion core bit. *Image courtesy of Baker Hughes.*

the formation apart as a faster penetration rate reduces the mud exposure time. The cutters are designed to produce a deep cut that removes the initial filtrate spurt, and the core head is designed to sweep mud away from the core.

The core barrel is made up of an inner and outer barrel separated by ball bearings, which allow the inner barrel to remain stationary and retain the core sample while the outer barrel is rotated by the drillstring and cuts the core. Figures 2.5 and 2.6 provide examples of a typical assembly.

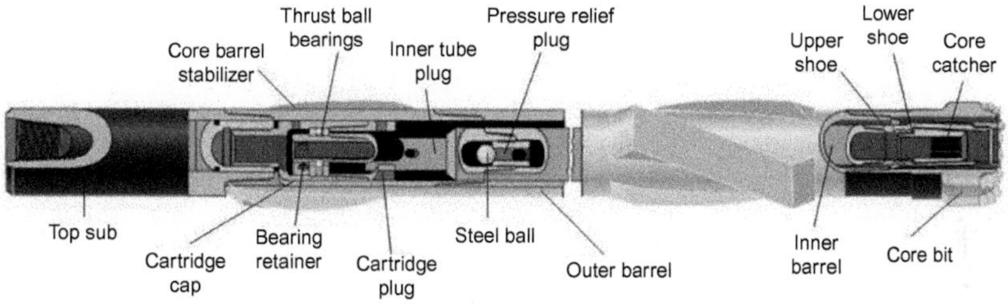

FIGURE 2.5 Example core barrel assembly. *Image courtesy of Baker Hughes.*

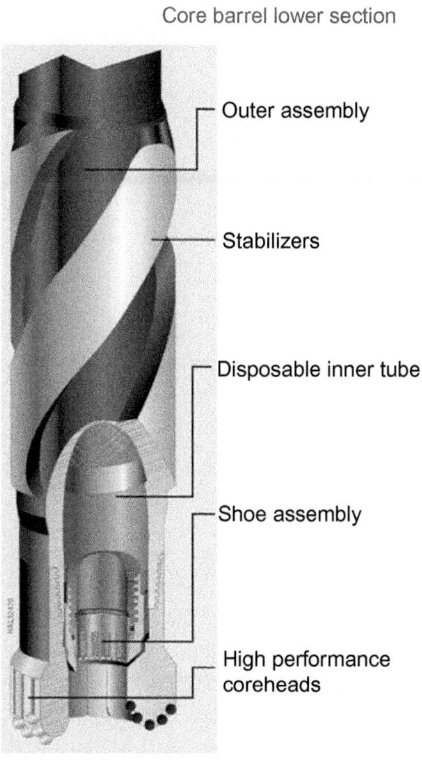

FIGURE 2.6 Lower barrel assembly. *Courtesy of Halliburton.*

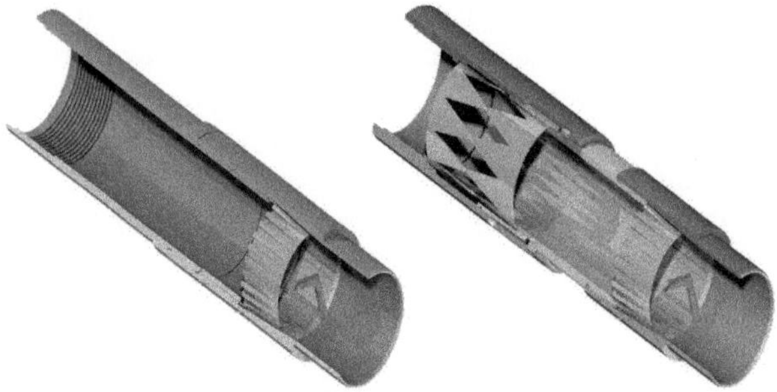

FIGURE 2.7 Spring catcher (left) and full-closure catcher (right). *Courtesy of ALS Oil and Gas.*

The core catcher is located within the core barrel and serves the primary purpose of stopping the core falling out of the core barrel during tripping out of hole or rotating off bottom. It has finger-like springs which are designed to prevent the core from sliding out of the core barrel during the trip to surface. Conventional spring core catchers merely provide a diameter restriction but unconsolidated formations present a challenge as weak formations may not be retained: a reliable full-closure mechanism is required to retain the core. Examples of core catchers are shown in Fig. 2.7.

Before disposable core liners were introduced, core was extruded from the barrel directly into pre-marked wooden or cardboard boxes. This enabled the core lithology and features to be described at wellsite but exposed the core to the atmosphere and to the drilling operations in a largely uncontrolled environment. Core liners were developed primarily to support the core to help prevent damage, and to protect the core at wellsite by preventing fluid loss/ingress, atmospheric contamination and tampering. The liner is held in place by the core catcher assembly and friction and is typically 30 ft. (9.1 m) long. Liners give support to the core and also serve as packaging to the laboratory, so there is no need to transfer the core from the inner tubes to sample boxes at wellsite.

The three principal types of disposable core liner tubes that are principally used nowadays are:

1. fibreglass/plastic
2. aluminium solid liners
3. aluminium clam shell liners

Fibreglass liners are principally used for coring unconsolidated formations. An example is shown in Fig. 2.8. The glazed bore provides low-friction core entry, which is especially suitable for soft or fractured formations. Fibreglass liners have temperature (typically 150 °C) and pressure limitations and can easily tear or jam in harder formations. They are also prone to bending during laydown on the rig floor and handling. Even a small kink or flexing during handling can create sufficient tensile and compressive force to cause core parting along bedding planes in thin-bedded intervals and core shear failure.

FIGURE 2.8 Fibreglass inner liner. *Courtesy of Halliburton.*

Aluminium liners are used for the majority of coring operations. The slick bore design of thin wall and thick wall aluminium solid liners, as shown by the examples in Fig. 2.9, allows for smooth core entry into the liner. The fluted design provides better fluid displacement via channels that allow gas to vent. The flutes also can ensure that stabilisation materials, such as foam and resin, are more evenly distributed around the core than in slick bore tubes. The effective ID of the fluted tubes is slightly less than slick wall tubes. Whereas this can offer better core support and can minimise incidences of wedging or displacement along natural fractures, it can make it more difficult to push the core out of the liner when it reaches the test laboratory.

Specialist liners have been designed to allow core inspection and sampling at wellsite as they can be opened to reveal the core, similar to opening a clam shell. Figure 2.10 provides an example. The liners are pre-cut longitudinally with a laser and are provided with key holes and ports that vent drilling fluid, decreasing the hydraulic resistance to the incoming core and allowing trapped gases to escape during tripping. A special tool is inserted into the key holes at surface to open the liners and examine the core, as indicated in Fig. 2.11. They have advantages where samples need to be preserved or sampled at wellsite, for example, in shale formations where the rock has to be preserved

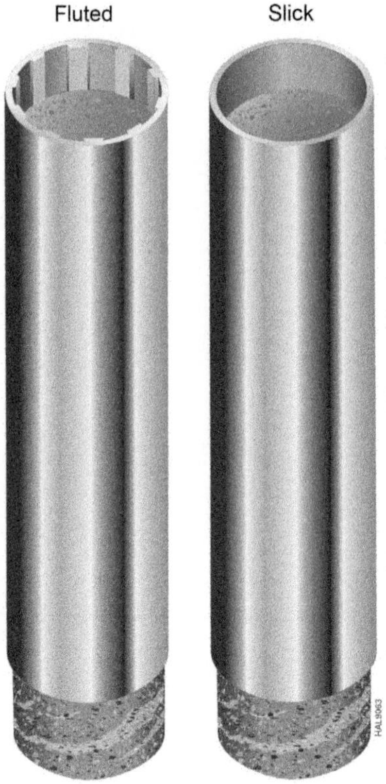

FIGURE 2.9 Solid (rigid) aluminium liners. *Courtesy of Halliburton.*

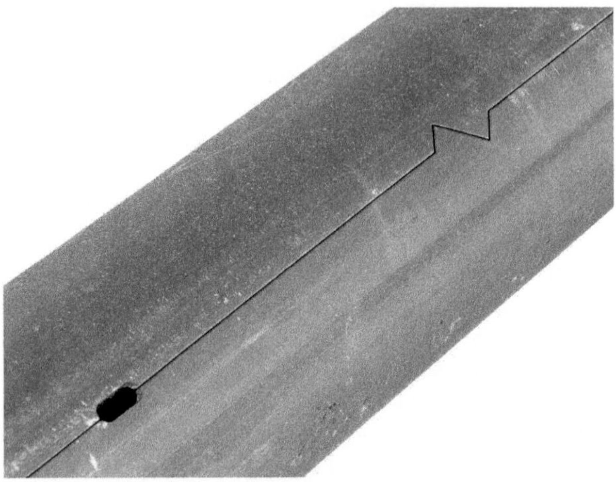

FIGURE 2.10 Aluminium "LaserCut" liner. *Image courtesy of Baker Hughes.*

FIGURE 2.11 Aluminium "LaserCut" liner opened to expose core. *Image courtesy of Baker Hughes.*

immediately, or for Dean-Stark extraction samples. However, in poorly consolidated formations where the core has been stabilised with foam or resin, the core can stick to the inside of the liner and can part along the length of the core when the liner is opened. There is little point in stabilising these type of liners in any event.

Selecting the appropriate liner for the conventional coring application depends on a number of factors including the degree of formation consolidation, the reservoir fluids and reservoir pressure and temperature. It is important to allow any fluid pressure to vent on the core barrel's journey out of the drill hole. Each system has advantages and disadvantages.

Fibreglass liners are inert and, unlike aluminium liners, are unaffected by mud pH and oxidation during storage. They have a lower thermal expansion coefficient compared to aluminium, and a long storage life if stored correctly. However, they require bonded steel couplings to achieve a reliable connection and do not thread easily for the inclusion of safety vents for gas release. The fibreglass degrades in strong sunlight with no means to evaluate the degree of change. Perhaps most importantly, compared to aluminium liners, fibreglass liners are significantly higher cost, have lower operating temperature limitations, are less resistant to core jamming and have lower pressure collapse resistance than aluminium. They are less well suited to long barrel applications. In addition, they can create harmful airborne dust when being sawed, suffer from bending and flexing during coring and offer a lower tolerance to poor transit and storage conditions. They can adopt a permanent 'bend' if stored unsupported in high ambient temperatures.

Aluminium liners are much more robust and thread easily and accurately for the inclusion of safety vents for gas release. The comparatively stiff tube is ideal for rig floor handling with a reduced risk of core damage from bending and flexing, and the higher pressure collapse resistance is better suited to long barrel applications than fibreglass.

However, aluminium liners can be affected by mud pH other than neutral when using water-based muds (WBMs) and can become oxidised due to local weather conditions during storage. Fluted tubes offer less contact resistance than slick bore tubes and provide gas vent channels along the complete length, but the flute restrictions can make it difficult to push out the core when it reaches the laboratory. The advantages and disadvantages of the principal liner systems are summarised in Table 2.2.

TABLE 2.2 Advantages and Disadvantages of Standard Liner Systems

Liner Type	Advantages	Disadvantages
Fibreglass	Unconsolidated formations Flexible Lower thermal expansion	Prone to bending/flexing Pressure and temperature limitations Difficult to install gas vents Higher cost than aluminium Dust when cut/slabbed Less resistant to jamming
Aluminium slick	Robust Lower cost than fibreglass More tolerant to poor transit Stiff tube resists bending Lower friction than fibreglass Better suited for long barrels	Can be affected by mud pH Oxidation Higher thermal expansion Gas vents restricted
Aluminium fluted	Less contact resistance Flutes allow gas venting More even distribution of stabilisation material	ID less than slick tubes can make it difficult to push out core
Aluminium clam shell	Allow inspection/ sampling at wellsite Key cuts and vents allow gas to escape	Not recommended for stabilisation. Foam can stick to liner and slab core when opened

2.1.2 Wireline-Retrievable Cores

In conventional coring operations, after the core sample has been taken, the drillstring is pulling out of hole and the rotary coring bit, barrel and catcher are removed. The drillbit is reattached, and drilling can commence again. Because coring requires the suspension of drilling, the process is quite expensive and usually only performed at the reservoir interval. At average 2014 day rates of $450,000 per day for deep-water operations, one coring run can easily cost around $300,000 just in rig time, plus the cost of the coring and core handling equipment and tools.

Wireline-retrievable cores offer significant cost and time advantages in that the core is cut using a conventional coring bottom-hole assembly (BHA), but the inner barrel is retrieved on wireline. Consequently, the driller can core or drill ahead without the need to trip the BHA. The wireline coring assembly contains a shock absorber that is connected to the outer tube of the core barrel, a wireline-retrievable core bit (Fig. 2.12) and a wireline-retrievable core barrel. The BHA can be converted to a drilling assembly by tripping in with a drill plug.

The core barrel assembly is forced down the inside of the drill pipe using drilling mud pressure. When the core barrel assembly reaches the lower end of the drill stem, a locking device holds the barrel in place. During coring operations, the circulating fluid passes between the core barrel assembly and the drill collar. After the core has been cut, the core barrel assembly with its core is retrieved by lowering an overshot through the drill pipe which is designed to engage the upper end of the core barrel. As the overshot is lowered over the upper end of the assembly, the locking devices are released, permitting removal of the entire assembly.

Historically, core diameters have been limited to around 2¾ in. However, as shale gas exploration and appraisal drilling have become more extensive so

FIGURE 2.12 Example wireline retrievable core bit. *Used with permission from NOV.*

the wireline-retrievable coring technology has evolved and improved. Systems have been developed that will deliver 3½ in. diameter core in up to 120-ft. intervals, for example, Corpro's Quickdrill 89™ and National Oilwell Varco's Corion Express®, amongst others.

However wireline systems may require non-standard drill pipe, and rapid retrieval can lead to gas expansion fracturing of the core which negates the cost advantages as core quality is poor and the recovered core may be unanalysable.

2.1.3 Gel Coring Systems

Gel coring systems were developed principally as an alternative to operator-intensive wellsite core preservation (Skopec et al., 1996). The encapsulated gel coring system was designed to minimise filtrate invasion and to prevent core damage during and after coring. Gel coring can help preserve *in situ* saturations and rock properties of the reservoir sample and can improve core recovery and reduce jamming. The gels are non-toxic polypropylene glycol (PPG)-based zero spurt-loss gels which are preloaded in the inner tube of the core barrel, as indicated in Fig. 2.13. The gel is distributed around the core by a core-activated floating piston valve after the core begins to enter the core head. As the core is cut and enters the inner tube, it displaces all but a substantial coating of gel which forms a water-impermeable latex-like film on the surface of the core. This gel coating can help mechanically stabilise weak core, provides a barrier to static drilling mud-filtrate invasion in the barrel as the core is tripped and also protects it while in transit to the lab. The gel coring process helps provide the core analyst with what is claimed to be unaltered reservoir rock sample, with formation fluid saturations that are largely undisturbed.

In the authors' experience, gel coring job results have been mixed—particularly in terms of gel coverage. There are also concerns over the effects of PPG additives on native core wettability alteration.

2.1.4 Liquid and Gas Retention Coring Systems

When a core is brought towards the surface and the hydrostatic mud pressure reduces, the hydrocarbon fluid will expand and, in an oil reservoir, gas will be liberated when the oil is brought below the bubble point pressure. Gas liberation or expansion provides a force which will cause displacement of both the native fluids and the invaded mud filtrate from the core into the drilling mud system. The use of specially designed liquid and gas retention core barrels provides a means to prevent loss of oil from the core on hydrocarbon expansion on core recovery by maintaining reservoir pressure throughout the tripping operation or by collecting the displaced fluids.

The principal applications of liquid retention systems are to determine oil-in-place in fractured and vuggy carbonates, and oil saturations in water-

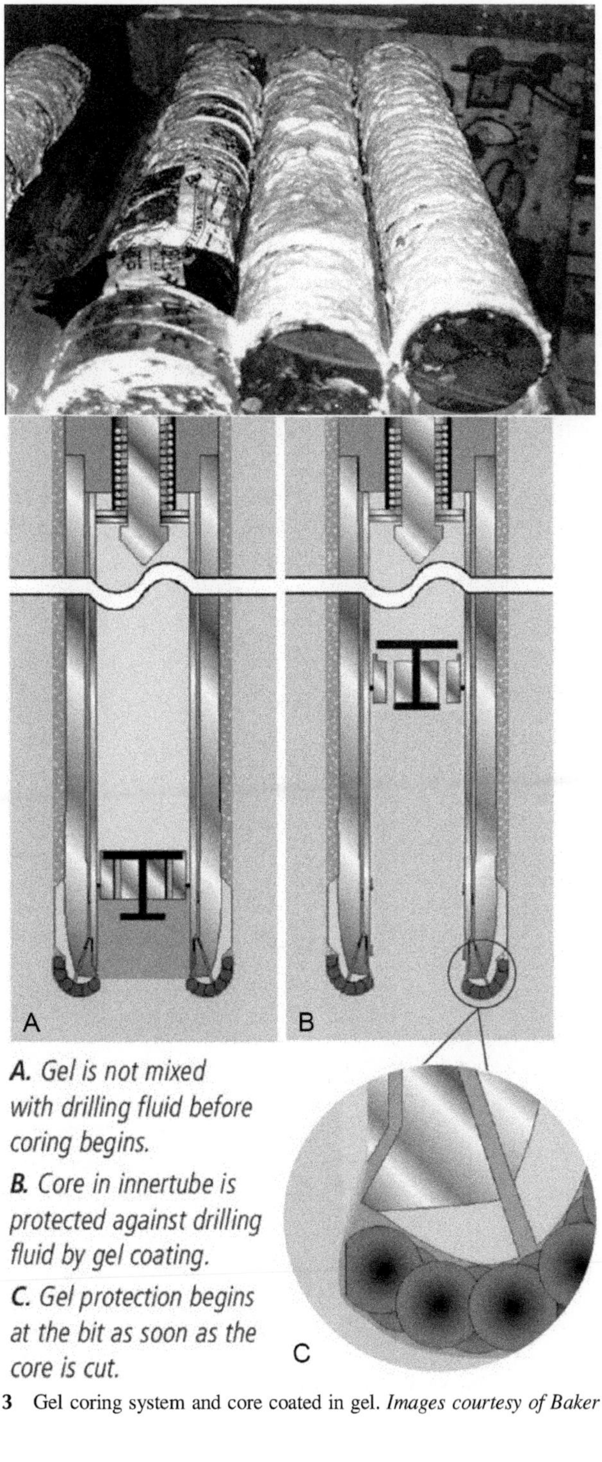

FIGURE 2.13 Gel coring system and core coated in gel. *Images courtesy of Baker Hughes.*

flooded zones or depleted zones prior to improved or enhanced oil recovery project evaluation. Pressure (gas retaining) barrels are now principally used for unconventional/shale gas coring operations.

2.1.4.1 Sponge Coring

Although the sponge core system can be modified to retain water, the conventional application is normally designed to trap oil expelled from the core. It involves the use of a polyurethane sponge liner in the annular space between the inside of the liner and the core (Durandeau et al., 1996). The sponge is chemically conditioned to be oil wet, so that it adsorbs oil that bleeds from the core and holds it opposite the formation from which it bled (Fig. 2.14). In this example, the aluminium liner is perforated with 1/16 in. holes to allow pre-saturation fluid to be displaced and gas to escape during the trip out. The fins support the sponge and core during coring and transport to laboratory.

Sponge core analysis utilises whole cores typically 3.5 in. diameter selected to be 12–24 in. long. The oil from the sponge (Vo_{sponge}) adjacent to the core section is extracted using solvents in a specially adapted soxhlet

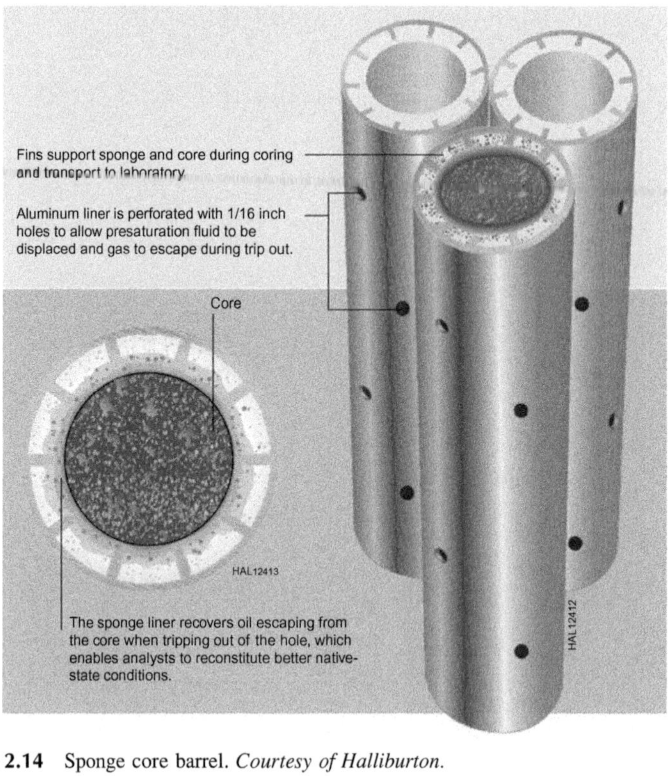

FIGURE 2.14 Sponge core barrel. *Courtesy of Halliburton.*

extraction apparatus. The core section itself is then extracted and the oil volume solvated is determined (Vo_{core}). Oil saturation (So) is calculated by summing the volume of oil collected in the sponge adjacent to the core (Vo_{sponge}) and the oil extracted from the core (Vo_{core}), divided by the cleaned core pore volume (Vp):

$$So = \frac{\left(Vo_{sponge} + Vo_{core}\right)}{Vp}$$

Figure 2.15 shows an example of the application of sponge coring in a flushed fractured and vuggy carbonate reservoir in Saudi Arabia (Funk et al., 2004). There is an excellent match between oil saturation from sponge core analysis and remaining oil saturation from nuclear magnetic resonance (NMR) spectroscopy.

Sponge coring has the advantage of being less expensive and safer than the pressure-retained coring systems, while providing an opportunity to improve the accuracy of the core-based oil saturation data. However, there are limitations related to operating temperatures and core diameters and lengths. Core diameter is typically 3½ in. (89 mm) and core lengths are normally limited to about 30 ft. (9.1 m) so that several trips may be required to recover longer core lengths compared to conventional systems. For unconsolidated formations, the foam insert in sponge cores may provide an obstruction as the core moves into the liner.

2.1.4.2 *Liquid Capture Systems*

These systems have been developed to capture liquids expelled from the core in core recovery, but allow gas pressure to vent during the trip to surface. The Liquid Trapper system (from ALS), for example, consists of a specifically designed liner assembly which, through an inflatable seal system, 'traps' oil and/or water escaping from the core. The liquid trapper inner tube system is pre-saturated with a selected fluid before it is run in hole. This barrel-saturating fluid is selected depending on the information targeted from the core. The saturating fluid can eventually be doped using a mud tracer and it is maintained inside the inner tube using an inner tube plug, which is made of a special ceramic plate that breaks as soon as the corehead tags bottom. Coring then proceeds as standard. During the pull-out trip, some of core fluids are expelled out of the core. The inner tube of the liquid trapper consists of stacked modules, 1 m in length each, that represent closed 'cells' in which expelled fluids would separate naturally based on their specific gravity. These cells, shown in Fig. 2.16, are sealed around the core, thanks to specifically designed seal joints that hold liquids yet still allow core gases to discharge into the mud system through a modified inner head. Within the closed cells, oil and water from the core are trapped in the annulus between inner and outer cell wall.

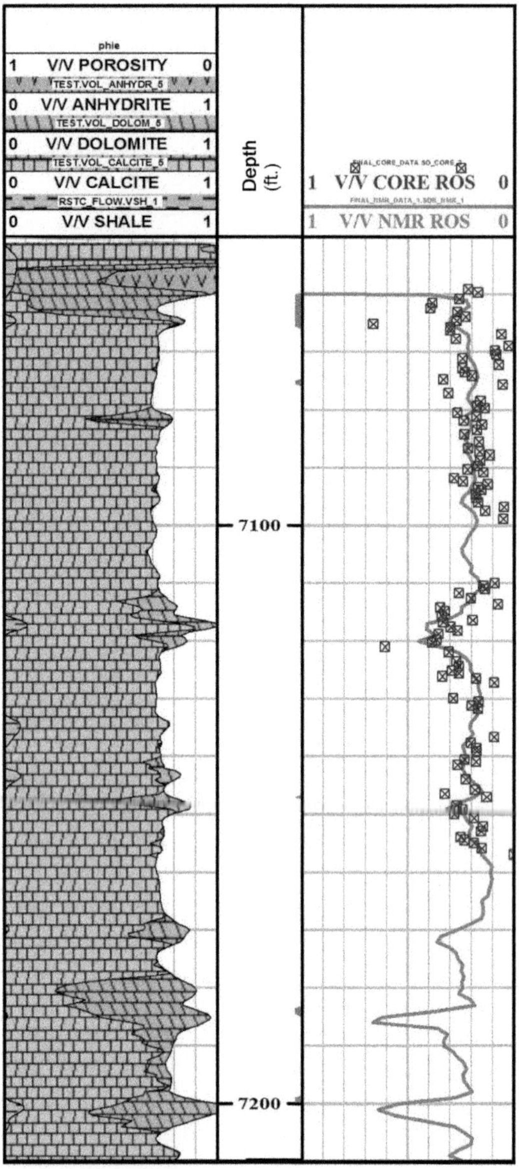

FIGURE 2.15 Example of sponge core oil saturation determination. *After Funk et al. (2004).*

On recovery, the core is laid down normally using core cradles. The liquid trapper cells are cut into 3 ft. or 1 m sections, sealed and stored vertically in special core containers. The natural separation of oil and water continues during storage and transport. In the lab, the fluids surrounding the core are collected. Whole-core analyses are conducted to determine remaining water and oil

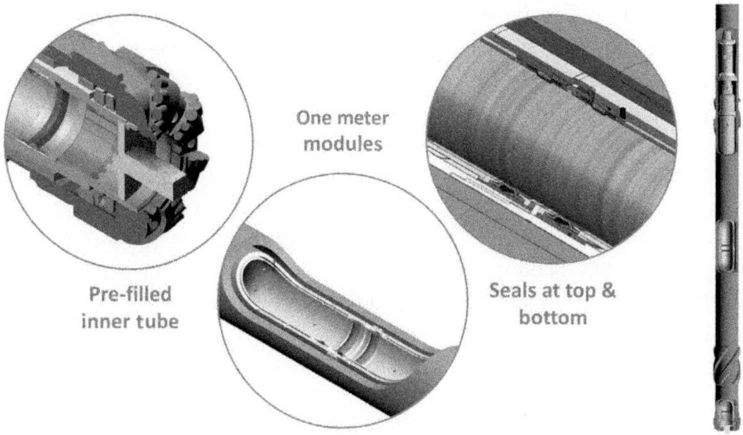

FIGURE 2.16 Liquid trapper closure cells. *Courtesy of ALS Oil and Gas.*

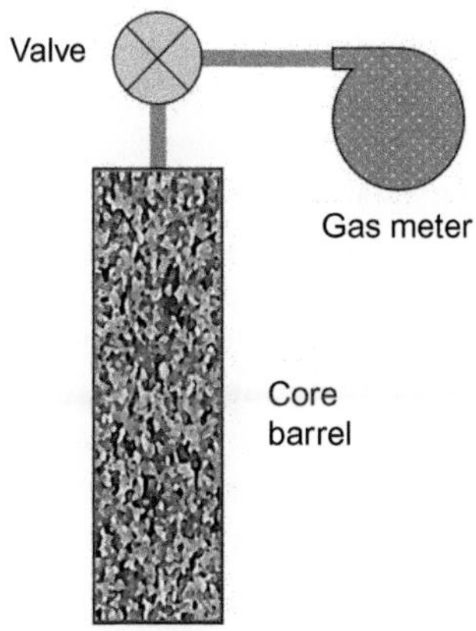

FIGURE 2.17 Pressure core barrel schematic.

volumes and the cleaned core pore volume, similar to sponge core analysis. The whole core and collected liquid volumes are recombined and corrected for fluid and pore volume expansion to determine the *in situ* saturations.

Like sponge coring, the liquid trapper has limitations on core diameter and length of core barrel that can be run: 3½ in. diameter by typically 10–50 ft. long.

2.1.4.3 Pressure-Retaining Systems

Pressure barrel coring systems are designed to maintain the reservoir pressure at surface so preventing gas, oil and water displacement from the core on depressurisation. A schematic example is shown in Fig. 2.17.

The procedure can be used for any project in which success is dependent upon accurate calculation of reserves in place (Eaton et al., 1991). Capturing and maintaining the core at or near bottom-hole pressure prevent gas expansion and fluid loss. Once the core is cut, the core chamber is sealed at hydrostatic pressure. When pulling the barrel out of the hole, a valve mechanism incorporated in the system will close the chamber to maintain bottom-hole pressure or, in some applications (for example, gas compositional analysis), allows gas to vent until a predefined hydrostatic pressure has been reached. On surface, the drilling mud is flushed from the annulus between the inner and outer barrel. Using an inert fluid under controlled backpressure, the core is then frozen for transport to the laboratory where controlled thawing allows accurate liquid saturation measurements.

Gas volumes can be determined by determining the volume of evolved gas using a gas meter. For oil cores, if the gas–oil ratio of the oil is known at surface temperature from PVT tests, then the volume of oil originally in the core can be determined from the volume of gas evolved. The core sample is then cleaned to remove residual liquids and the pore volume determined. The volume of oil determined from gas evolution and the pore volume are used to determine oil saturation. By combining porosity, fluid saturation and adsorbed gas, gas saturation can be quantified at reservoir temperature and pressure conditions. Gas samples can be collected and shipped to a geochemical or PVT lab for gas compositional analysis.

Historically, the principal drawback of pressure coring is the inherent safety concerns created by bringing high-pressure systems to the rig site surface. A combination of safety risks and poor reliability has historically limited the use of pressure-coring technologies. With the expansion of unconventional reservoir exploration and appraisal over the last few years, pressure coring has re-emerged. Pressure cores provide a means to obtain a direct measurement of total gas content in shale gas cores and to collect gas samples. Further, the tripping time required to vent pressure slowly enough to prevent core fracturing during pressure dissipation in ultra-low-permeability formations makes conventional coring and retrieval extremely expensive in rig time. Pressure coring allows cores to be vented at controlled depletion rates on a batch basis as the pressure coring drilling operation continues.

However, pressure cores are limited in diameter and length. Original systems recovered a 1.2″ diameter core which is difficult to analyse conventionally. Modern systems (e.g. Bjorum, 2013) allow larger core diameters (3–4 in.) but core lengths are restricted to 10 ft. per run and maximum operating pressures are restricted. The lack of sophisticated on-site facilities to service the barrel and handle the pressurised cores and the availability of laboratories with the necessary equipment to handle pressured cores have also constrained its application, but technology is improving to meet the challenges and risks.

2.1.5 Oriented Coring

Directionally (oriented) referenced core enables the determination of the dip and strike of inclined strata or fractures, stress field direction and directional permeabilities. Oriented coring refers to the combination of cutting a core which is marked with a narrow groove along its length and providing directional survey data referenced to this groove. As the core is cut, it is pushed through a shoe (located behind the corehead) which has three scribe blades pointing inwards. As the core moves through this shoe, the blades cut grooves on the surface of the core (Fig. 2.18).

By orienting a survey instrument to one of these blades (Fig. 2.19), the direction of the groove can be determined, referenced to magnetic north. The magnetic survey tool is locked within a non-magnetic collar to ensure no interference from the steel assembly. At surface, the depth, inclination and azimuth are recorded along the length of the core according to the survey density. Features on the core can then be oriented to the scribe line, and in turn oriented to grid or magnetic north, using a goniometer.

Hay et al. (1990) report several issues with oriented coring using older systems including instrument failures, differential sticking and core damage when taking check shots, spiralling scribe lines and poor reliability. To a large extent, new technology enabling continuous core orientation without the need for check shots, and electronic goniometers to measure planar and linear features have enhanced the reliability and improved accuracy.

FIGURE 2.18 Illustration of scribe lines cut into oriented core. *Used with permission from NOV.*

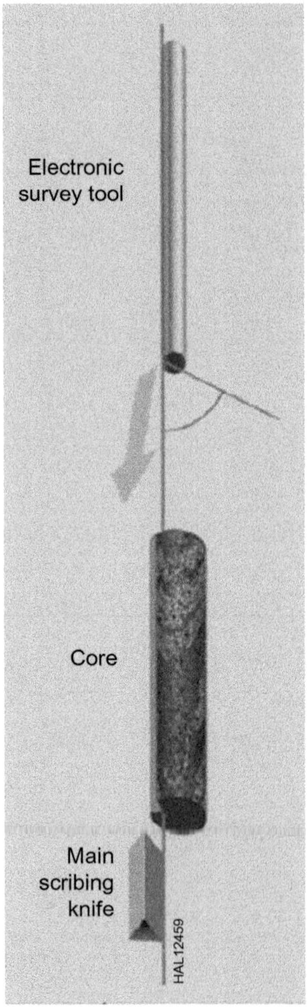

FIGURE 2.19 Electronic survey instruments are referenced with the primary line scribed into the oriented core. *Courtesy of Halliburton.*

If oriented features are important yet the core is un-oriented, then palaeomagnetic reorientation or matching formation micro-image log features with the core can provide viable alternatives. Palaeomagnetism is the study of the Earth's magnetic field variations through time. Magnetic grains in sediments may align with the magnetic field during or soon after deposition. This is known as viscous remnant magnetisation (VRM) and can remain unchanged in rocks for periods of hundreds of millions of years. The vector components (inclination and azimuth) of this VRM field can be recorded using a magnetometer on plugs taken at known orientations from the core. When this palaeomagnetic information from a sample is compared with known records of polar wandering and magnetic reversals throughout geological time, the original magnetic orientation of the grains can be determined and, when compared to the present-day magnetic field, the core (and its bedding or fracture features) can be oriented to Magnetic North. Potential applications of palaeomagnetic reorientation are provided by Hartley et al. (1991) and Hamilton et al. (1995), amongst others.

2.1.6 Sidewall Cores

Sidewall cores are taken to minimise coring costs or to obtain reservoir rock samples in an interval which has either been cored and core recovery lost, or in an interval which has not been cored conventionally. There are two main types of sidewall cores: percussion and rotary.

Percussion sidewall coring systems are essentially an adaptation of a wireline-conveyed perforation gun. Instead of firing perforation charges, the tool is designed to shoot a series of hollow, chisel-edged 'bullets' which are loaded with explosive charges. An example is shown in Fig. 2.20. The bullets are loaded into the tool which is run in on wireline to the interval of interest. The gun is fired and the explosive charge shoots the bullets into the formation. The core plugs cut by the bullets are retained within the bullets which are then brought back into the tool using a wire chain and the tool returned to surface. Typical percussion sidewall plug sizes are 0.75–1 in. diameter by 1 in. long and up to 66 plugs could be taken in a single trip. The principal advantage with this method is that it is cheap. There are few, if any, advantages. The driving force required for the bullets to penetrate the formation can cause stronger rock to fracture and weaker rocks to consolidate. Porosity and

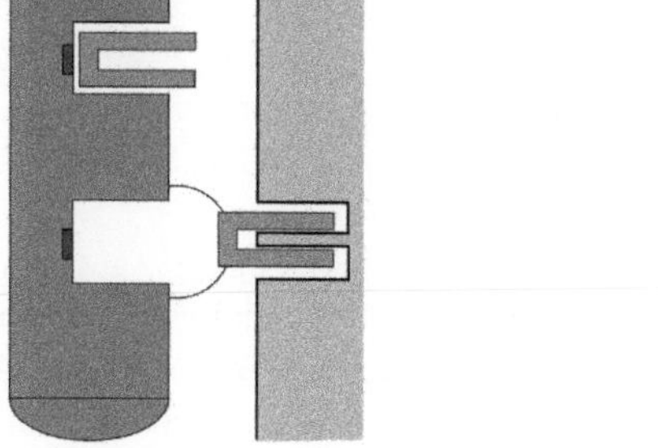

FIGURE 2.20 Percussion sidewall core tool schematic.

FIGURE 2.21 Rotary sidewall tool bit. *Image courtesy of Baker Hughes.*

permeability data measured on such samples will not be representative. The prime application of percussion sidewalls is therefore restricted to obtaining samples for lithological description, grain size and palynology and palaeontology, although grain density measurements and particle size analysis measurements should also be possible.

Rotary sidewall tools have now almost completely replaced percussion sidewalls. The tool is still run in on wireline, but instead of shooting hollow cylinders, it has a series of rotary coring bits which cut a core plug from the formation at the borehole wall. These are spaced according to client's sampling requirements. The plugs are typically 0.92 in. diameter by 1–1.5 in. long. The tool is lowered to the interval of interest, and the core plugging bit is extracted from the tool and pressed against the borehole wall. Either mud is circulated through the tool which causes the core bit to rotate and take the sample or an electric motor (Fig. 2.21) is used. On completion, the bit is retracted into the tool and the tool is taken back to the surface. The quality of the plugs obtained using rotary techniques is far superior to the percussion sidewall. Generally, reliable porosity and permeability data can be obtained on the plugs, but many plugs are frequently misshapen (Fig. 2.22) which makes them difficult to test in a standard coreholder for routine and special core analysis, and which can impact on certain porosity measurements.

Recently, Baker Hughes ('MaxCOR™') has developed sidewall coring systems (Fig. 2.23) that take much larger plugs—typically 1.5 in. diameter by up to 3 in. long, so plug dimensions and volumes are more comparable with typical SCAL plugs. Other examples include Schlumberger's 'XL Rock' system.

Rotary sidewall core samples are taken directly at the formation face at the wellbore wall. In formations where the bedding dip is not orthogonal to the wellbore axis, such as in steeply dipping beds exposed in a vertical well or in a deviated well, the plugs taken will not be parallel to bedding so will not sample the maximum permeability. The rotary plug can be oriented to

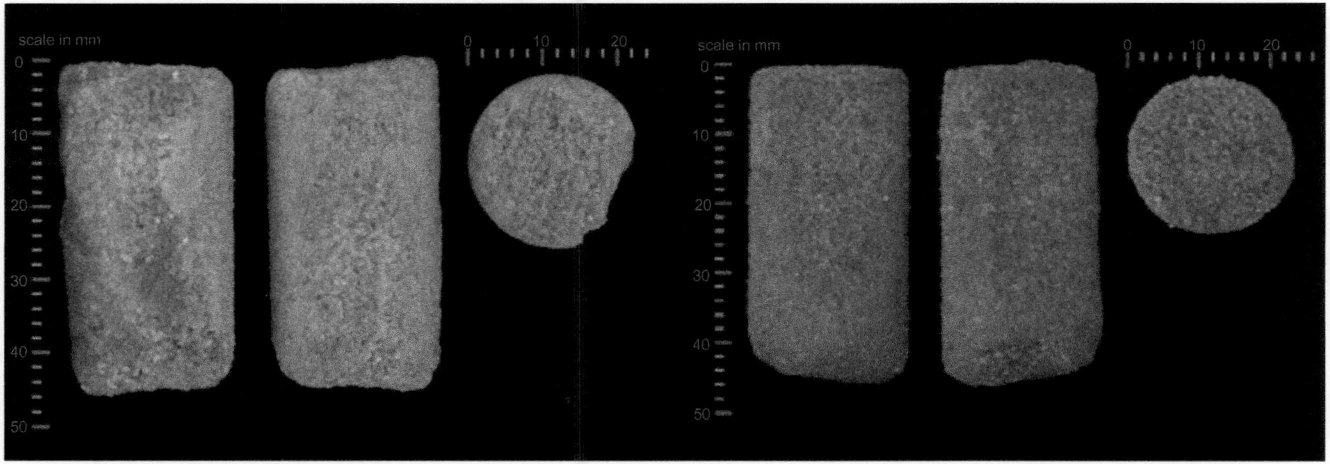

FIGURE 2.22 Example rotary sidewall plugs.

FIGURE 2.23 Baker Hughes MaxCOR large-diameter rotary sidewall plugs. *Image courtesy of Baker Hughes.*

allow much smaller 'mini' plugs to be taken from the sample in the appropriate bedding plane orientation, but this largely defeats the purpose of taking a larger plug in the first place.

In addition, the plugs are also taken from the formation face at the wellbore wall which may be invaded by mud solids and mud filtrate and which may be potentially weakened as a result of possible wellbore deformation on drilling. The authors have seen several examples where rock strength measured on sidewall cores is between 25% and 50% of the strength measured on standard plugs taken from conventional full-diameter core, and where sidewall plugs have anomalously high grain densities as a result of infiltration of mud solids weighting agents into the sample. External mud cakes can be trimmed off the wellbore face of the plug, but if they invade deeper into the sample, they are almost impossible to remove without destroying the sample. Any rotary sidewall cores used for RCA or SCAL should be thoroughly examined and checked for potential invasion and damage prior to testing.

2.2 CONVENTIONAL CORING OPERATIONS

The wellsite operations' personnel involved in core recovery and handling must ensure that the best quality core is received by the testing laboratory. Normally coring contractors will be following their own quality control procedures, but it is essential that wellsite personnel ensure that core liners are laid out in the correct order and marked up correctly, and ensure that the core is not damaged during pulling out of hole (POOH), wellsite handling and shipment.

In any case, each coring job and reservoir should be carefully examined prior to the design of a wellsite coring, handling and preservation programme. The typical sequence of events at wellsite is:

1. Coring
2. POOH (pulling out of hole)
3. Remove from barrel and lay out liners
4. 'Way-up' and depth marking
5. Wellsite screening/gamma-ray logging or sampling
6. Wellsite sampling and sample selection (if required)
7. Core preservation/stabilisation (if required)
8. Liner division into suitable lengths (if required)
9. Core transportation to laboratory

2.2.1 Health, Safety and Environmental Considerations

Core handling, by its nature, presents manual handling risks in the form of lifting, pinching and exposure to hydrocarbons and chemicals. Health, safety and environmental (HSE) protection during coring operations should be managed according to the current standards and procedures of the operating company and coring contractors, and according to local regulatory legislation. An HSE Organisation and Management System should be in force for the duration of the coring project—including the compliance standards expected from service contractors.

The operator should establish arrangements to ensure that contractors and service providers are properly selected, managed and monitored according to their HSE standards. All core handling and processing activities must be reviewed to make sure that it does not conflict with other activities. Pre-job briefings and training must be given to any staff that assists the core recovery team. General work permits must be obtained for any specialised procedures and equipment. Special permits must be issued for electrical equipment in the presence of potential hydrocarbon gases escaping the core. Core liners and core may require cutting using powered abrasion saws which must only be used by qualified operators with appropriate personal protective equipment.

The area between the drill pipe and the drill collar racking area must be kept clear of obstructions. Drill collars should be racked back using a racking arm. The elevator lift sub should be hoisted with an air winch to the elevators and latched into the elevators with the lift sub standing on the deck. When breaking out the safety joint, personnel must stand back from the rig tongs as the connection breaks. The safety joint should be laid down for unloading the inner barrels from the outer barrel. A tugger line should be attached to the lifter, taking the weight, so when the elevators are released the safety joint will not fall over.

The presence and concentration of hydrogen sulphide (H_2S) must be established. Any work related to core handling (retrieving, laying out, measuring, marking, gamma-ray logging, cutting and packing) must only be carried out after safe clearance is given. All personnel in direct contact or within the vicinity of the core material must have completed H_2S training. Full H_2S precautions must be implemented including the use of portable H_2S sensors and toxicity clips with the initial checks carried out using breathing apparatus. Core handling and preservation operations must be conducted in an open, well-ventilated area. If high levels of H_2S are anticipated, then consideration should be given to catching core in aluminium liners that enable rapid degassing of any trapped H_2S at wellsite. Time may be required to degas and confirm that H_2S levels are at zero prior to transporting to the core laboratory. This could restrict and constrain wellsite sampling (for example, Dean-Stark samples). Core shipment using vented containers and packaging will enable release of any further small accumulated amounts of hydrocarbon or toxic gases from individual core samples prior to reaching the core laboratory. The core laboratory must ensure any core arriving from the rig site is stored, unpacked and initially laid out in an open, well-ventilated area and not in a confined building or room. Laboratory staff should also implement full H_2S precautions.

2.2.2 Coring Team

Drilling and coring engineers, wellsite geologists and the core analysis laboratory personnel should hold a formal pre-coring wellsite meeting with rig crew and key personnel to highlight the importance of good coring and core handling and to promote teamwork. Ideally, the core analysis company must be involved in any pre-coring activities, and at wellsite as the core is recovered, as they have a vested interest in getting good quality core back to the laboratory. Key safety issues and coring risks should also be discussed.

The roles and responsibilities of the coring team are highlighted in Table 2.3.

2.2.3 Managing Coring Risks

There are risks associated with coring that can lead to damage to the core or loss of core recovery. With best practice procedures in place these can be effectively managed.

The option of using a core barrel with a full-closure core catcher or similar (dog or flapper/basket) may need to be considered if there is a risk of losing the core before retrieval to surface. Options should be discussed with the coring contractor.

Prior to running in with the core barrel, the bottom of hole should be clean, and the depths defined and agreed between drilling superintendent

TABLE 2.3 Typical Roles and Responsibilities of Coring Team

Team Member	Roles and Responsibilities
Drilling Supervisor (Company)	Responsible for overseeing all operations
Mud Engineer (Contractor)	Maintaining desired mud properties during coring operations
Wellsite/Operations Geologist (Company)	Responsible for overseeing marking core, identifying plugging points in the core, identifying preserved sample zones and supervising core preservation. Responsible for initiating and leading all wellsite meetings and conversations to ensure that best practice wellsite procedures are followed resulting in the receipt of core in a good structural condition at the analysis laboratory
Core Hand (Coring Contractor)	Responsible for coring operations from picking up the core barrel through to cutting the core and providing parameter adjustments throughout each run. Responsible for communicating coring-associated hazards and risks to rig personnel. Provision of all coring, lay down and processing equipment including end-caps, clips, pneumatic screwdriver, air saw spare blades in a safe and functional condition
Coring Hand (Coring Contractor)	Operates air saw and cuts inner liners to lengths and depths determined by Wellsite Geologist. Ensures that cut liners, once inspected and chips or core plugs taken from the cut faces, are sealed as soon as practical after the cuts are made
Core Handlers (Core Analysis Lab)	Responsible for core preservation, wellsite sampling and supervising or carrying out core stabilisation, plug labelling, preservation and transportation

and coring contractor. This prevents short runs and problems after the event. The rotary table should not be used to reverse the string in as these prevent safety joints backing out. Ideally, the first core run should commence with as near as possible to a full 90 ft. (28 m) on the stand so a connection to complete the coring run need not be made. This reduces the risk of shortening the core run. Picking up the top drive every 1000–1500 ft. and pumping through the core barrel reduce the risk of ingesting cuttings debris into the inner barrel. Tight spots should be pumped through using light weight on bit (WOB) and initially low rotation (RPM). This reduces the risk of the core barrel sticking.

Jamming or blocking the inner barrel is one of the more common problems of coring. Jamming is usually caused by a formation condition such as fractures, unconsolidated material or swelling shales. If the barrel jams in soft unconsolidated formation, the penetration rate may remain normal, but most likely will decrease. The stand pipe pressure will increase initially and then decrease as the core bit drills off. A change in torque, pump strokes, or a decrease in pump pressure will also indicate a jammed core barrel. An abrupt increase in stand pipe pressure may be caused by plugging of the core barrel from an accumulation of foreign particles in the mud system such as rubber, LCM or other junk. Most coring contractors offer anti-jamming technology. For example, the Baker Hughes JamBuster™ anti-jamming coring system overcomes up to two inner barrel core jams enabling the assembly to stay on bottom coring before ultimately being tripped on the third inner barrel core jam.

Inner core barrels should be fitted with shoes, catchers and inner tube stabilisers off the critical path of the well prior to pick up of the assembly. Sufficient spares should be on hand to prevent reuse of these items in the event of continued coring. The cores are back-loaded with these subs in place on the inner tubes, thus preventing any potential for core damage or loss at the rig site through their removal.

Flow rates when pumping the drop ball down the hole should be approximately 150 gallons per minute. This reduces the risk of collapsing the inner barrels. WOB during coring should not be allowed to drill off as this reduces the risk of washing or unnecessarily disturbing the core, shortening the run and damaging the core.

In weak formations, excessive WOB can cause the rock to fail ahead of the bit immediately before it is cored. In addition, the core barrel must not be completely filled as this can limit core compression damage. Shorter core barrels—30–60 ft. (10–20 m)—are recommended in poorly consolidated sands.

Circulation times should be limited in order to reduce the risk of core loss unless hole conditions dictate heavy circulation is needed. Well safety should always take precedence over core recovery. Setting the slips gently during the trip out of the hole reduces the risk of core loss.

It is advisable not to rotate the core inner barrel during coring and the drillstring while tripping out. Serious damage to the core will result if the inner barrel rotates with the outer barrel. Slips should be set softly to prevent jarring of the core. Tripping rates should be closely controlled and monitored to ensure that sufficient time is allowed to allow pore pressure to dissipate and to prevent damage to the core through fracturing. Under normal circumstances, the rotary table should not be used to help back out the drillstring. This reduces the risk of core loss and damage. Slow rotation, 40 RPM or less, is acceptable for a limited period.

2.3 CORING FLUIDS

2.3.1 Mud Types

The selection of coring fluids may often be dictated by drilling and logging requirements rather than by core analysis requirements. Generally, if the core is to be taken to determine residual fluid saturations, then WBM is used for residual oil saturation determination and OBM for residual water saturation. If sensitive clays or shales are expected, then OBMs, synthetic oil-based muds (SBMs) or inhibited WBMs are used.

The extent of any filtrate invasion depends on the core bit used, mud type and mud rheology, the coring overbalance pressure, formation wettability and formation absolute and relative permeabilities. Invasion can be minimised by selecting an appropriate low-invasion core bit and optimising the coring parameters, especially mud weight. In weak sands, high overbalance is often used to prevent the hole from collapsing and this can promote invasion. High coring rates (greater than 100 ft./h), high WOB (which must be balanced against potential damage in weak core) and high RPM can all help minimise excessive filtrate invasion. Low overbalance (especially in depleted zones) and high loading of properly sized calcium carbonate ($CaCO_3$) solids can help to promote filter cake formation and minimise spurt losses. As a rule of thumb, the size of the solids should be equal to or greater than 1/3 of the median pore throat size. OBM and SBM should have low fluid loss characteristics (around 2–4 ml). In addition to minimising filtrate loss, the mud rheology design should also consider reducing the yield point and viscosity to minimise core erosion in loose sands, and consider low water content and emulsifiers in OBM/SBM to minimise free water.

As far as possible, addition of surfactants to the mud should be minimised (or eliminated) to prevent wettability alteration. OBM often carries the risk of contaminating core wettability. WBM should have a salinity similar to formation water (to minimise salinity gradients in formations with sensitive clays) and pH adjusted to prevent clay dispersion and carbonate mobilisation. Coring contractors and mud vendors should be asked to advise on their most appropriate coring and fluid system for minimal invasion coring, especially in potentially friable, depleted or flushed sands.

2.3.2 Mud Tracers

Oil- or water-soluble tracers can be added to the mud system to determine the degree of oil or water-filtrate invasion from the coring muds. Subsequent fluid extraction measurements on recovered core samples, such as Dean-Stark analysis, can determine the amount of tracer per unit volume of fluid in the core and so quantify the amount of filtrate invasion. This information is extremely useful in determining *in situ* core saturations (Ringen et al.,

2001) and potential fluid and rock property alteration (Bruno, 2007). The most commonly used tracers in core analysis are chemical salts, stable isotopes and radioactive isotopes, but tracers containing magnetic nanoparticles and synthetic DNA (Fehling, 2013) are promising, but expensive, new developments.

The key features are that the tracers must not occur naturally in the formation or in the formation water and must be able to be detected easily and accurately in the lab at low concentrations. They must be soluble in the mud filtrate and must be biologically and chemically stable so that they do not react with, or precipitate within, the formation.

2.3.2.1 Water-Soluble Tracers

Water-soluble tracers are either chemical (salt compounds, dyes or alcohols) or nucleonic isotopes of hydrogen (deuterium and tritium).

Salt compounds include iodide, lithium and bromide salts. The advantages of these systems are that they are thermally and chemically stable and are relatively cheap and readily available. However, lithium reacts with clay minerals. The principal disadvantages are variable background levels in the formation; the detection limits are high, typically in the range from 1 to 10 ppm; and high dopant concentrations are required. Most importantly, these salts are non-distillable in Dean-Stark-extracted water which means that the salt compounds are precipitated and remain in the core after water extraction. The plugs therefore have to be carefully eluted to release the salts and this often requires crushing the core plugs to elute all of the tracer with no guarantee that all the salts will be released.

Dyes and alcohols are cheap and readily available but they can react with the rock, evaporate at low temperatures and less volatile alcohols can break down in contact with crude oil.

Deuterium oxide (D_2O) is a non-radioactive, non-toxic isotope with properties similar to water but D_2O occurs naturally and at variable levels in water and formation brine. Fjerstad et al. (1993) were amongst the first to successfully use D_2O for native water saturation estimation. Because of the relatively high background levels, it requires a large dosage—about ~35 kg/1000 bbls of mud which increases the cost substantially. D_2O can also interfere with NMR logging.

Tritium or HTO (Bruno, 2007) has the distinct advantage over D_2O in that it requires only a very low dosage (about 10 ml for 60 ft. of core) and the detection limit is very low—less than 1 ppb. It does not interfere with NMR logging. It is a beta-particle emitter so safety precautions are essential, but with proper training tritium can be managed (dosing and disposal) safely and securely. The concentration levels required are about 50% of the allowable limits for unregulated discharge. Tritium, provided it can be regulated/certified for use by local regulatory authorities, probably represents the most accurate water-soluble tracer system.

Tritium tracers should be added while the drilling mud is circulating about 100–200 ft. above the coring point. Bottoms up should be circulated several times. Samples of the traced mud should be taken before, during and after coring at 5–10 ft. or 30-min intervals depending on coring speed. These samples have to be depth-lag corrected. Although diffusion of tracers to the centre of the core can occur within 1–2 days, ambient temperature and permeability can influence diffusion rates, so it is advisable to sample the core as soon as possible after collection—ideally at wellsite. Wellsite sampling and plugging should ensure that no water is added to the core (e.g. from plug drillbits) and that the plug is sealed/preserved as soon as possible. Figure 2.24 shows an example of a preserved wellsite plug. It is wrapped in plastic film and then aluminium foil which is sealed with liquid plastic or wax. The core plug, once received in the lab, is extracted in a Dean-Stark extractor and the water (containing tritium) is collected and analysed using a liquid scintillation counter.

The principal application of tracers is in WBM to assess filtrate invasion. Water-soluble tracers can also be used to dope the water phase of OBMs if there are concerns about water-filtrate invasion increasing the 'native' water saturation. Analysing results from several wells in the North Sea, Ringen et al. (2001) found that there were no or insignificant amounts of water-phase invasion with OBMs with water content between 15% and 35%. Woodhouse (1998) found similar results in an extensive study of Prudhoe Bay cored wells.

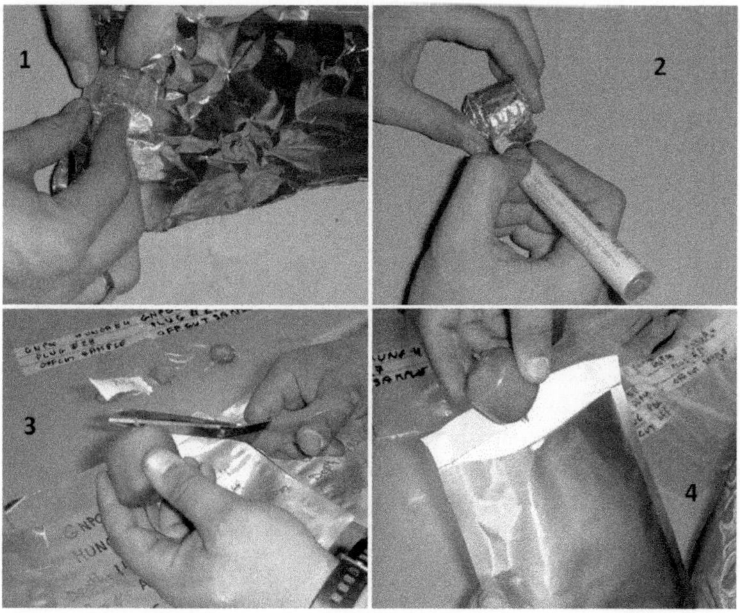

FIGURE 2.24 Sealing core plug sample at wellsite. *Courtesy of ALS Oil and Gas.*

However, there is some largely anecdotal evidence that SBMs, especially systems with high solids loading and higher water contents, can give up significant volumes of water to the formation.

2.3.2.2 *Oil-Soluble Tracers*

Oil-soluble tracers are used to dope the oil phase in OBM where determination of native oil saturation is important. Typical applications include remaining oil saturations in flushed zones for enhanced oil recovery project screening when WBM cannot be used for drilling reasons. The two most common oil-soluble tracers are based on deuterated hydrocarbons and halogenated paraffins. When the core reaches the laboratory, the OBM tracer is eluted in toluene and collected in the toluene/oil extract from the Dean-Stark extraction process. It should be easily detectable at low concentrations. If there is concern that OBMs give up the water phase to the formation, then it may be advisable to trace both the water and oil phases of the mud. This is expensive as it requires muds to be traced at the vendor base facility as they cannot be added 'on the fly' at wellsite.

2.4 CORE DAMAGE AND CORE FLUID/PETROPHYSICAL PROPERTY ALTERATION

During its journey from the reservoir to the wellsite or laboratory, the stress acting on the cored formation is released rapidly, the pore pressure depletes and the temperature reduces. These changes result in:

- alteration of the spatial and volumetric distribution of fluids within the reservoir pore space;
- alteration of the rock's pore texture and mineralogical properties;
- alteration of the rock's native wettability.

Understanding these damage mechanisms and the nature of the damage that can occur to a core sample are essential to our evaluation of the data acquired from the core and our understanding of relationships between core analysis data and other reservoir data sources.

2.4.1 Fluid Saturation Alteration

In an oil or gas reservoir, the fluid distributions (water and oil or gas saturation) will be controlled by capillary forces. During coring with either an OBM or WBM system, mud filtrate can enter the pore system of the rock before and as it is being cored, altering the initial volumetric and spatial fluid distribution. The degree of mud-filtrate invasion will depend upon coring bit design, drilling rate, the rheological properties of the mud and the rock properties such as porosity, capillary pressure, wettability and both the absolute and relative permeability. Figures 2.25–2.27 provide schematic representations of the changes in saturation that occur during coring for both oil and gas reservoirs.

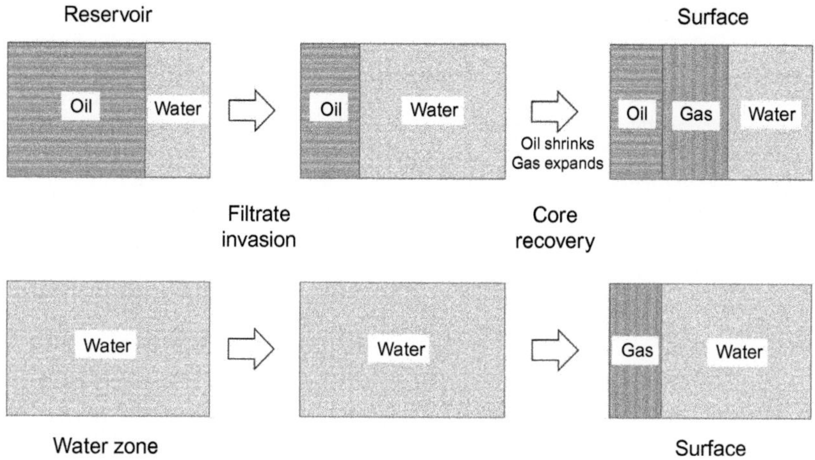

FIGURE 2.25 Saturation alteration—WBM-filtrate invasion in oil reservoir.

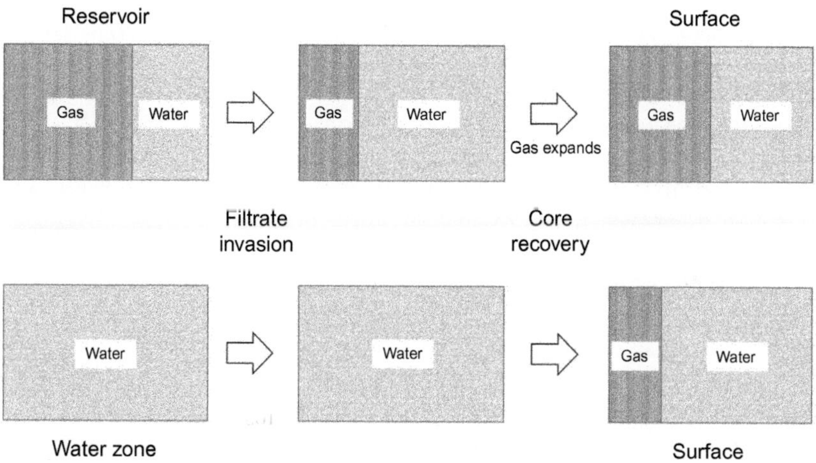

FIGURE 2.26 Saturation alteration—WBM-filtrate invasion in gas reservoir.

For example, when drilling with a WBM (Fig. 2.25) in the irreducible or immobile water zone in the oil or gas legs, the relative permeability to water filtrate at high oil or gas saturations will be low and so the amount of fluid flushing might be small. However, as the overbalance increases, more filtrate can enter the core pore space and the water saturation increases, so it is easier for WBM to invade deeper into the core as the relative permeability to water increases. This will result in oil being flushed from the core towards residual oil saturation. During core recovery below the bubble point depth, gas in the oil evolves, expands and displaces both unflushed oil and invaded WBM

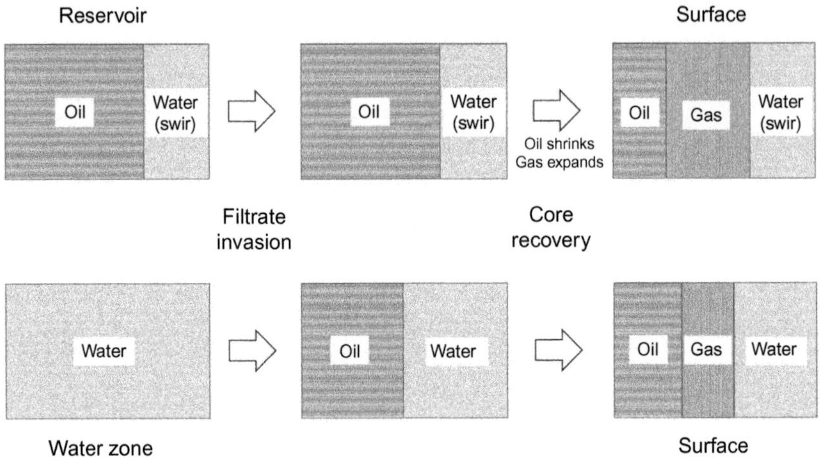

FIGURE 2.27 Saturation alteration—OBM-filtrate invasion in oil reservoir.

filtrate so the oil and water volumes may be reduced towards residual saturations. Due to capillary hysteresis between drainage and imbibition (water invasion) processes, the final water saturation in the core will be higher than the native or connate saturation. If the core is oil wet, WBM invasion may be restricted, especially in low-permeability formations. WBM filtrate flushing in the water leg may result in partial or complete replacement of the connate water with filtrate. There is always a limited amount of gas dissolved in the formation water, which will expand and drive off some water as a result of pressure evolution on core recovery.

Similar processes applied in gas reservoirs drilled with WBM (Fig. 2.26). In tight- or low-permeability gas sands especially, the wetting capillary forces are so strong that WBM filtrate is spontaneously imbibed, flushing the core towards residual gas saturation. Gas then expands on core recovery, displacing the WBM filtrate towards residual saturation.

When drilling with an OBM in the immobile water zone of a gas or oil reservoir (Fig. 2.27), the relative permeability to OBM filtrate will be high so that OBM may be able to thoroughly flush the core. However, if the water is at immobile saturation, the water saturation should remain unaltered, trapped by capillary forces and low relative permeability, unless the interfacial tension (IFT) between oil and water is reduced, for example, in contact with surfactants in the mud system. This would mobilise 'irreducible' water. In the absence of IFT effects and provided that the water saturation in the core is immobile; the core is not allowed to dry out; and the mud does not give up the water phase to the formation and the water saturation in the core, measured at surface, will be the same or very close to that in the reservoir. Even in capillary transition zones, the water remaining in the core may not be mobilised provided the relative permeability to water (and hence water mobility) is low.

In the water leg and towards the base of the transition zone where water becomes increasingly mobile, OBM filtrate invasion and gas expansion on core recovery might drive the water towards residual saturation. The result is that core saturation measurements at surface conditions may show little difference in saturations between oil-leg and water-leg samples. In the transition zone through the water zone, the relative permeability to oil will be low, so OBM filtrate invasion will be restricted. However, if the mud pressure is sufficiently high (overbalance drilling), filtrate will be forced into the pore system.

As the core is brought to the surface, gas liberation or expansion provides a force which will cause displacement of both the native fluids and the invaded mud filtrate. Subsequent extraction saturation measurements of water and/or oil saturation in an oil reservoir core will therefore indicate a gas saturation, even although there may be no gas cap in the reservoir. There will always be a 'gas' (air) saturation in the recovered core—even in the water leg. This is partly due to gas evolution on recovery (some gas is often dissolved in the formation water), and the differential expansion of the pore volume and water volume on depressurisation. The pore volume expansion is greater than water volume expansion, so additional pore space is created that is not filled with water.

At the surface at the wellsite, if core is exposed to air, the saturation in the core can be altered through evaporation. Preserving the core provides an opportunity to prevent further fluid loss as well as other forms of core damage. The use of liners can aid in preventing fluid loss as well as other forms of mechanical and physio-chemical core damage. End-caps are placed over each section of liner (Fig. 2.28), so that exposure of the core to air and fluid loss are minimised during wellsite operations and transportation to the laboratory.

FIGURE 2.28 Sealing end of liner section with cap. *Used with permission from NOV.*

If the core transportation logistics are difficult—for example, a long distance or time between the wellsite and the laboratory—it may be advisable to take core plugs for saturation determination from the ends of each liner section immediately or very soon after the core has been recovered, just in case the core dries out. These plugs can then be sealed and shipped to the laboratory for saturation analysis.

2.4.2 Stress Release

As the core bit approaches the target formation, the vertical stress reduces as overburden is removed with little or no change in the horizontal stresses (Holt, 1993). As the core is drilled free from the surrounding rock, the horizontal stresses are also released. Depending upon the shape of the core bit, there will be a zone of compression underneath the teeth of the bit and a zone of vertical tension around the external side of the core, just above the bit. In this stress condition, tensile failure of the core can occur (Maury et al., 1988). If the stress situation at some point exceeds the failure strength for the core, the rock may fail macroscopically, often with a disked appearance. In less brittle rocks, if the stresses exceed the yield envelope of the rock, permanent inelastic deformation can occur, without necessarily causing visible mechanical damage. Since it looks intact, the core will be used for petrophysical property measurements but the results of these analyses at low-confining stresses may be invalid. Minerals within rocks have different coefficients of thermal expansion. Depending upon the rock mineralogy, this can lead to microcrack formation on cooling as the core is brought back to the surface.

2.4.2.1 Pore Pressure Dissipation

The most common cause of core damage on core recovery is a consequence of differential pore pressure dissipation. In conventional coring systems, the pressure in the annulus between the outside of the core and the inside of the liner is hydrostatic. In high-permeability formations or in core lacking a filter cake, the pressure at the centre of the core dissipates at approximately the same rate as the pressure in the annulus as there is little or no barrier to fluid movement and pressure equalisation out of the core. However, in low-permeability formations or where there is an ultra-low permeability filter cake, the pressure in the centre of the core dissipates more slowly than in the annulus. This will create a pressure difference between the centre of the core and the outside of the core. As a result, the effective stress within the core may enter a tensile regime. If the effective tensile stress exceeds the tensile strength of the rock, which is about 10% or less of the unconfined compressive strength, then tensile fractures can be propagated, extending radially from the centre of the core towards the outside of the core. The highest risk of differential pore pressure damage occurs if there is a rapid release of formation gas as the core is POOH to surface. This is particularly important

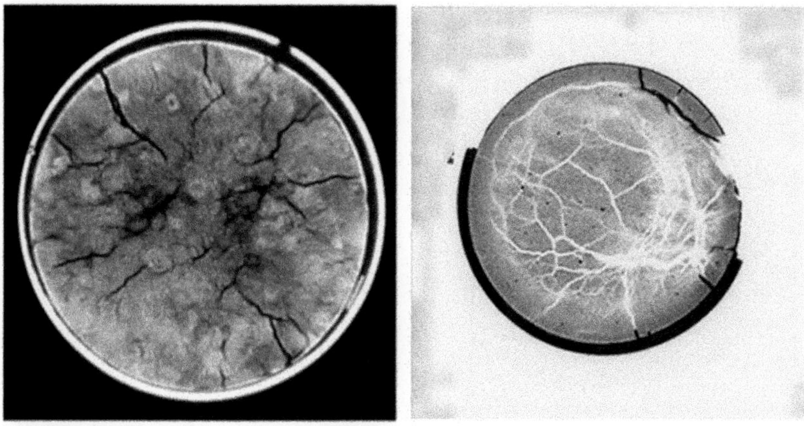

FIGURE 2.29 CT scan images of core that has been damaged by rapid gas expansion/depressurisation.

TABLE 2.4 Recommended Generic Tripping out of the Hole Rates

Depth Range (TVD)	Trip Rate
TD to *XYZ* m	1.5 min per stand
XYZ to 150 m	3.0 min per stand
150 m to surface	10.0 min per stand

during the last part of the trip if the pipe is pulled too quickly and core/liner annulus pressure is reduced too rapidly. An example of CT scans from a damaged core is shown in Fig. 2.29.

In order to reduce the risk of core mechanical damage, trip rates are provided for general guidance purposes based on coring contractors' experience (Table 2.4).

A stand is defined as three drill pipe lengths—3×30 ft. $= 90$ ft. The *XYZ* depth is the true vertical depth at which the hydrostatic (mud) pressure equals the bubble point for the hydrocarbon in the core. The time specified relates to the time spent pulling the stand, which excludes the time taken to set slips, etc.

Several companies have developed their own pulling out of hole rates. Examples for oil and gas reservoir cores are presented in Tables 2.5 and 2.6.

These core tripping out schedules/rates are often based on rules of thumb or experience and can lead to unjustifiably long tripping out operations, or core damage if the core is retrieved too quickly. Long tripping schedules

TABLE 2.5 Example Recommended Tripping out of the Hole Rates for a Water or Oil Core

Depth Range (TVD)	Trip Rate
TD to 300 m below rotary table	As per swab calculations
300–150 m below rotary table	3 min per stand
150 m to top of barrels	6 min per stand
Coring bottom-hole assembly (BHA)	10 min per joint

TABLE 2.6 Example Recommended Tripping out of the Hole Rates for a Gas Core

Depth Range (TVD)	Trip Rate
TD to 3000 m	4 min per stand
3000–2000 m	6 min per stand
2000–1000 m	7–8 min per stand
1000–200 m	9–10.5 min per stand
200 m–surface	12–13.5 min per stand

are hard to justify to drillers who face high daily rig costs and generally describe coring as 'too expensive'. Worse, drillers' key performance indicators are often based on maximising footage drilled, whereas for coring they should be based on quality of core recovered. As a result, there is a lack of clarity regarding tripping out schedules which impacts the integrity of the core and the economics of coring.

Software is available that can be used to optimise tripping out schedules, offering a design and diagnostic tool that helps identify and minimise core damage potential due to restricted bleed-off while pulling the core barrel. Zubizarreta et al. (2011) used computational fluid dynamics (CFD) numerical modelling software to predict the pressure distributions of fluids and the onset of tensile failure in the core during core retrieval. A typical workflow is shown in Fig. 2.30.

The model predicts the pressure contours (Fig. 2.31) within the core at any time during the trip out of hole and compares the difference in pressure between the centre of the core and the outside of the core (which is at hydrostatic pressure during the trip) with the tensile strength. The magnitude of the

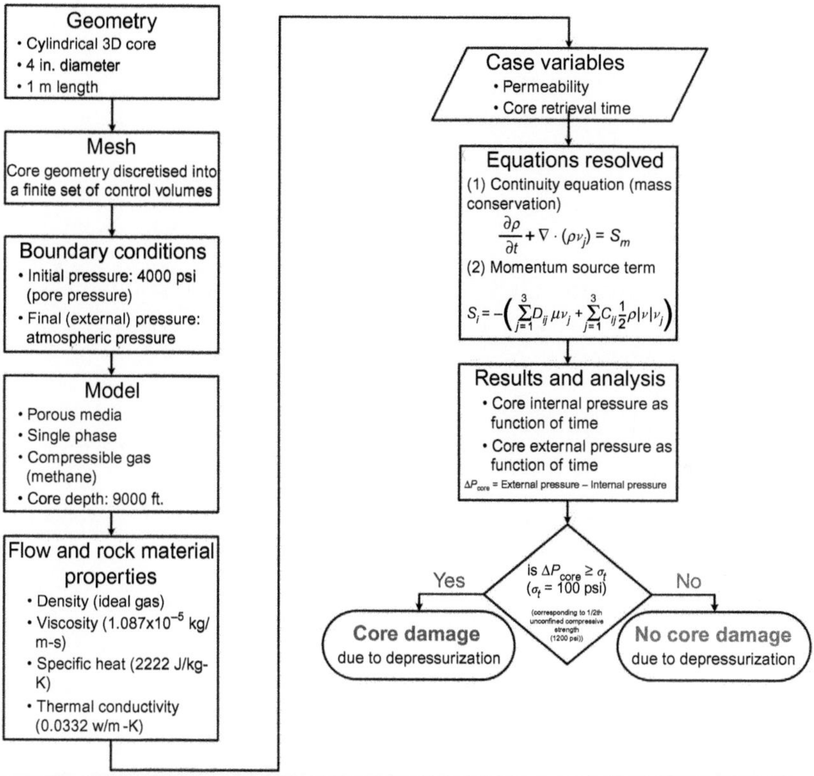

FIGURE 2.30 Core trip rate CFD model example workflow. *From Zubizarreta et al. (2011).*

pressure difference depends on trip rate, core strength, core permeability and mud cake permeability. Where the pressure difference exceeds the tensile strength of the rock, the core will fail in tension causing fractures to develop. Model results support and confirm empirical evidence that the pressure differentials created in a core during core retrieval are very low for relatively high-permeability rock (>300 mD). The model realisations demonstrate that the largest differential pressure and the highest risk of tensile failure are more likely to be reached during the final last stages of the trip (Fig. 2.32).

Low formation core permeability and/or the presence of an ultra-low permeability mud cake will restrict pressure dissipation and increase the time required for the core internal pressure to approach the external pressure increases, increasing equilibration times. The extreme case of a very low permeability rock of 0.0002 mD (Fig. 2.33) shows internal core pressures exceeding the critical internal pressure from around 1250 min. Simulations for shales show that critical pressures are exceeded throughout the maximum simulated trip time of 12 h, suggesting that pore pressure release damage is expected in shales during conventional tripping operations. This is why

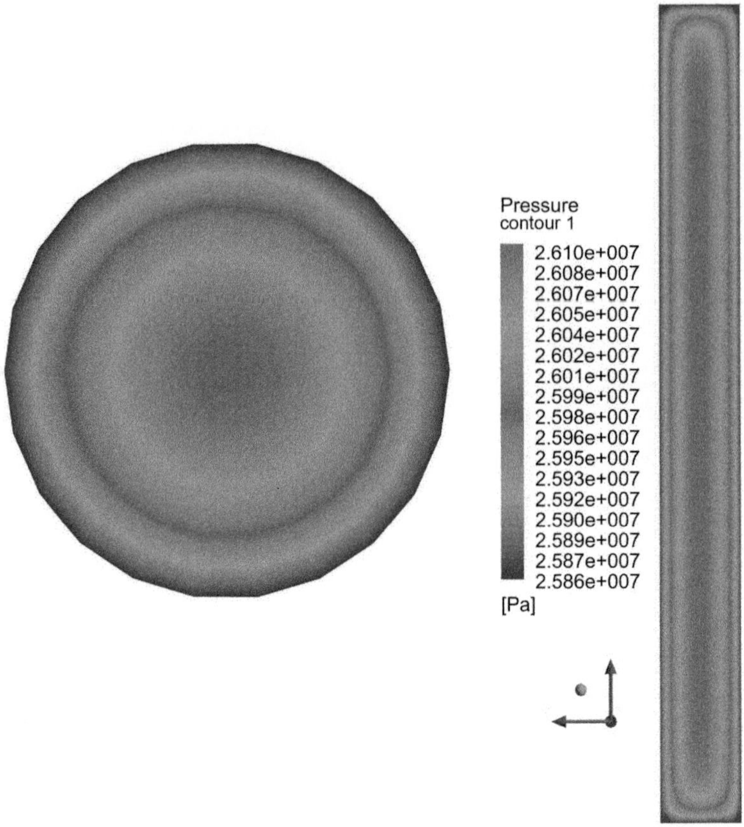

FIGURE 2.31 Core pressure distribution during tripping. *From Zubizarreta et al. (2011).*

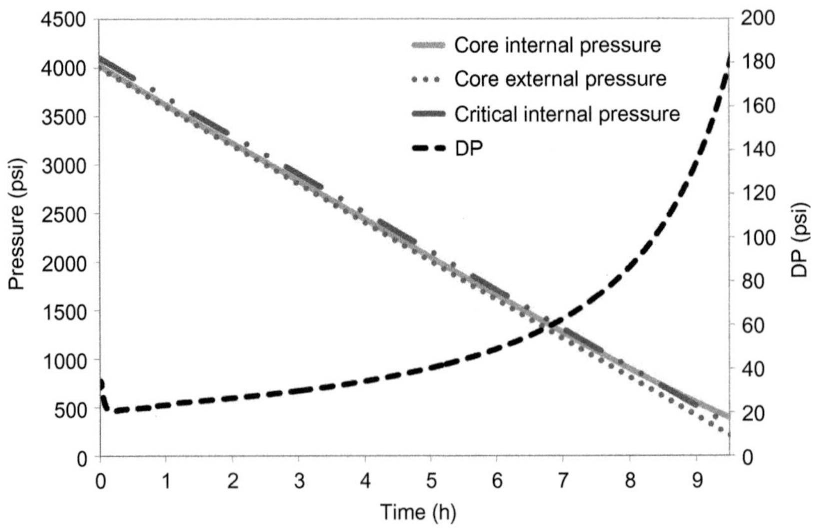

FIGURE 2.32 Core pressure evolution—0.001 mD core with 10-h trip time. Green (light grey in the print version) curve is differential pressure. *From Zubizarreta et al. (2011).*

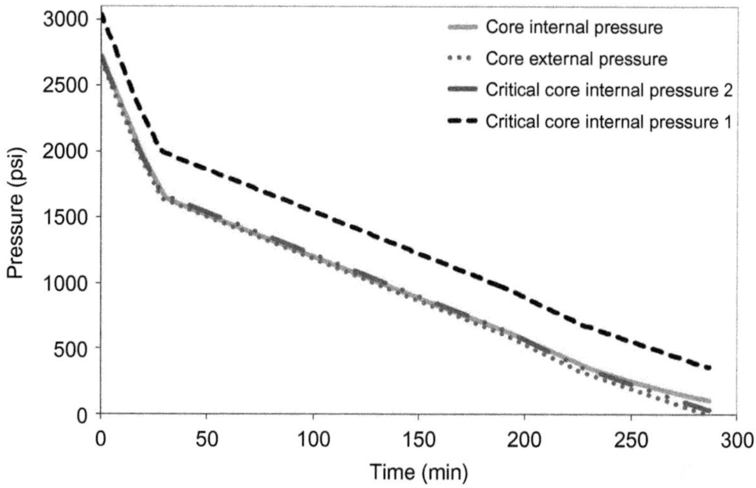

FIGURE 2.33 Core pressure evolution—0.0002 mD core with 0.00016 mD mud cake. *From Zubizarreta et al. (2011).*

pressure barrel coring in unconventional reservoirs has re-emerged as a valuable and viable method to control pressure reduction at surface rather than downhole.

2.4.2.2 Shear Failure and Compressive Damage

Shear failure in weak or poorly consolidated formations (Fig. 2.34) can result from excessive WOB ahead causing shear failure in the formation just ahead of the core barrel, or due to rotation of the core inside the liner during drilling (the core often shows concentric rings).

Compressive failure often results at the base of the core barrel due to the weight of the overlying core in the barrel (Fig. 2.35). In this case, the use of long core barrels should be avoided. The recommended core barrel length should be no more than 20 m (66 ft.), and in very weak rocks, 10 m (33 ft.). When coring fractured formations, short core lengths may also be beneficial to reduce the risk of jamming.

2.4.3 Wettability Alteration

Wettability is a fundamental rock and fluid property that controls fluid distributions within porous media (Anderson, 1986a). It is the result of a complex interaction of forces, which are related to both the reservoir fluid system type and the rock type, and has a major impact on capillary behaviour (Anderson, 1987a), core electrical measurements (Anderson, 1986b), relative permeability tests and residual saturations (Anderson, 1987b).

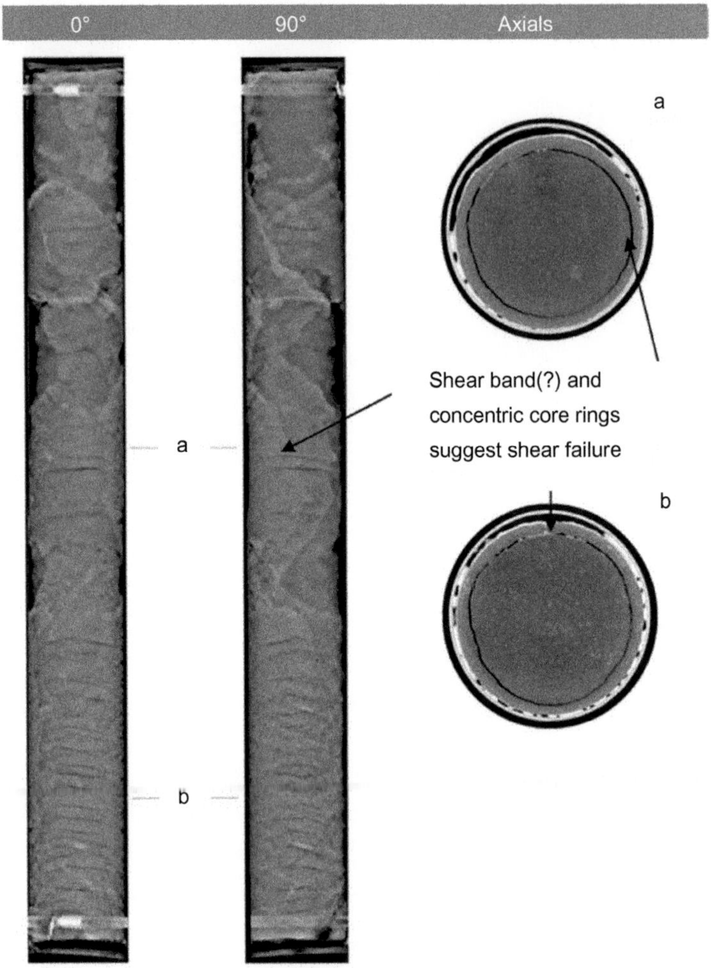

FIGURE 2.34 CT scan of core showing shear bands in weak sand.

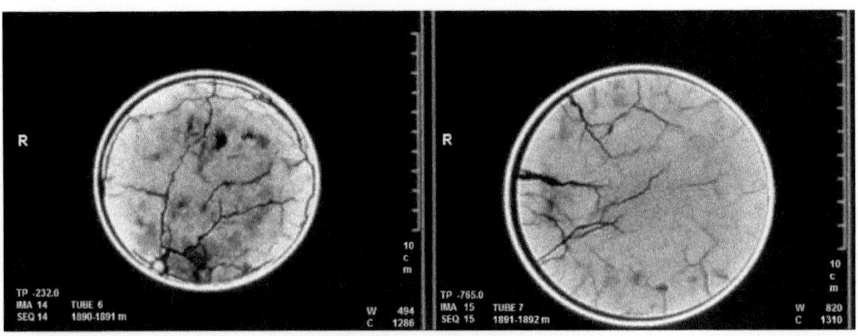

FIGURE 2.35 CT scan showing compressive failure in weak sands.

It is important to maintain the appropriate reservoir wettability conditions if special core analysis tests are to provide reliable results. For single-phase properties—such as porosity, absolute permeability, and formation resistivity factor—wettability plays no role. However, for resistivity-index tests, imbibition capillary pressure and effective and relative permeability measurements, maintenance or restoration on rock native wettability is essential for reliable results. Despite many studies on the effects of wettability on rock properties measured in core analysis experiments, appropriate wettability conditioning is frequently still ignored. Tests performed on 'fresh-state' or cleaned cores may be wholly inappropriate. This chapter provides an introduction to wettability controls and wettability alteration during coring, core retrieval and surface handling. Wettability tests are described in detail in Chapter 7.

2.4.3.1 Definition

Wettability is defined as the tendency for one fluid to spread or adhere to a solid surface in the presence of a second fluid. An example is shown in Fig. 2.36. When a drop of water is placed on a preferentially water-wet glass slide immersed in oil, it will displace oil. The droplet will form a contact angle, θ. By convention, this contact angle is measured through the denser phase. If the slide has a preference for water compared to oil, the water droplet will tend to spread so that the contact angle is low. This system is said to be preferentially water wet. The adhesion force (or tension, A_t) is a function of the IFT between the oil and water system and the contact angle, θ:

$$A_t = \sigma_{o-w} \cos\theta$$

Now suppose the glass slide is made to be oil wet, for example, by chemical treatment, oil tries to contact the glass slide but the water droplet gets in

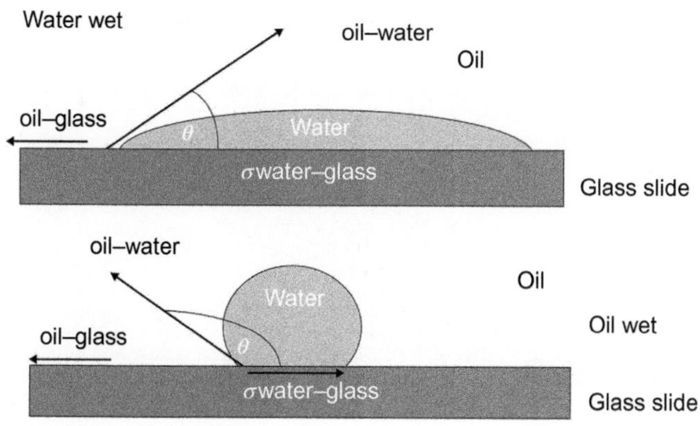

FIGURE 2.36 Wettability concepts.

the way. Consequently, since the adhesion force between the plate and water is very weak, water has no tendency to spread, the water droplet forms a ball and the contact angle exceeds 90°. In this case water is said to be non-wetting.

In oilfield applications, many wetting states are thought to occur. Some of these are:

Water wet where water coats the grains and may fully occupy the smaller pores. Oil would occupy only the centre of the larger pores.

Oil wet where oil contacts the grain surfaces and may fully occupy the smaller pores. Water would occupy only the centre of the larger pores.

Neutrally wet where the system has a uniform non-preference for oil or water.

Intermediate wet where the system has a slight but equal oil- or water-wetting state throughout the pore system.

Fractional wet where certain minerals have a strong wetting preference in one direction, while other minerals have a strong opposite preference.

Mixed wet A special case of fractional wettability in which oil-wet surfaces form a continuous path through the larger pores although water still occupies the smaller pores. This wetting state is probably unproven but does allow very low residual oil saturations, and some unusual relative permeability behaviour to be explained.

Most oil reservoirs are not uniformly water wet or oil wet throughout the pore system. Figure 2.37 provides an example of an environmental scanning electron microscope (SEM) photomicrograph of a sandstone at irreducible water saturation. The SEM shows droplets of brine (surrounded by oil) adhering to quartz grain overgrowths and on clays. The contact angles indicate that the quartz surface is water wet (low contact angle) and the clays possibly oil wet (high contact angle). This variation in wettability occurs over a distance of less than 20 μm in the pore system.

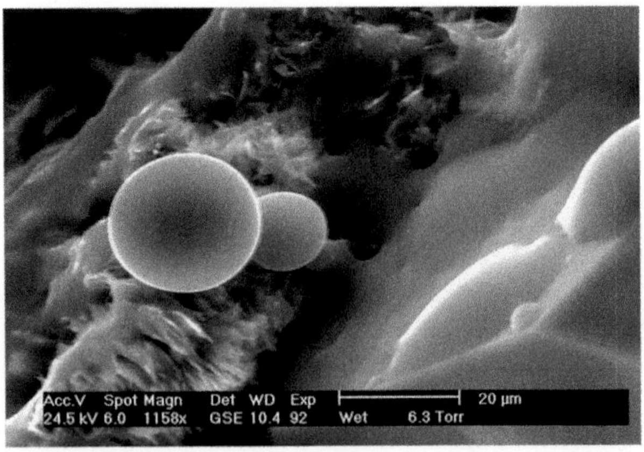

FIGURE 2.37 Environmental SEM photomicrograph for reservoir rock at irreducible water saturation.

2.4.3.2 Native Reservoir Wettability Controls

Before oil migrates into the reservoir the rock can only be water wet, by definition. It might also be expected that, in a virgin reservoir, depending upon the rock mineralogy and reservoir fluids, wettability might grade towards a less water-wet state through the transition zone, possibly becoming non-water wet in the irreducible water zone where oil saturation reaches a maximum and has more opportunity to contact the rock grains.

Treiber et al. (1972) and Chilingar and Yen (1983) studied the wettabilities of 30 sandstone reservoirs and over 50 carbonate formations. They classed wettability on the basis of contact angles, and the results were surprising for the time: the vast majority of the reservoirs studied were apparently oil wet. The results may have been biased, however, as many of the tests may have been performed on fresh-state core drilled with oil-based or on unpreserved core, both of which may indicate an oil-wetting tendency. It is generally recognised that most sandstone reservoirs, certainly in the North Sea, exhibit an overall intermediate to slightly oil-wetting behaviour and are not strongly water wet.

The literature on reservoir wettability is vast. Anderson (1986a) provides an extensive and excellent summary of reservoir wettability and its controlling factors. The principal controls can be summarised as below.

2.4.3.2.1 Oil Type/Composition

It is assumed that before oil migration, reservoir formations had to be water wet as the grain surfaces were only in contact with water. As oil migrates and fills much of the pore space, adsorption of polar compounds and surfactants, and deposition of organic matter from the crude oil, can induce oil-wetting (or at least weakly or non-water wetting) tendencies. The polar compounds in resins and asphaltenes both combine hydrophilic and hydrophobic characteristics. If the crude oil is a poor solvent for its own surfactants, it will have a greater propensity to change wettability than one that is a good solvent (Al-Maamari and Buckley, 2003). Pressure, temperature and crude oil composition all affect asphaltene stability.

2.4.3.2.2 Brine Chemistry

For components of an oil to alter wetting, the oil phase must displace a brine film which originally coats the surface of the grains (Hirasaki, 1991). If the film destabilises, polar components of the crude oil are attracted to the grain surface and render it more oil wet. High brine pH, and a high concentration of divalent cations (Ca^{2+}) in the formation brine, can cause water films to destabilise and promote surfactant adsorption on the rock surfaces.

2.4.3.2.3 Grain/Mineral Type

The range of pH causing water films to destabilise is different for carbonate and sandstones. Silica surfaces are negatively charged above a pH of 2, so positively charged ions (base chemical species) can adsorb. Conversely, calcite surfaces may be positively charged below pH 9.5, so negatively charged ions (acidic species) can adsorb (Buckley et al., 1998). Carbonate wettability is also controlled by specific interactions with carboxylic acids and by the reactivity of carbonate minerals (Thomas et al., 1993). Generally, under similar fluid conditions, adsorption of asphaltenes on silica surfaces which can lead to oil wetting can be an order of magnitude less than adsorption of asphaltenes on carbonates. As suggested by Fig. 2.37, evidence points to many clays being oil wet. For example, Jerauld and Rathmell (1994) established that in oil-saturated sandstones, kaolinites are preferentially wetted by oil, while illites and other minerals are essentially wetted by water. Bantignies et al. (1997) found that the wettability contrasts between illite and kaolinite are related to structural differences between the two clay types.

NMR is uniquely sensitive to surface wetting conditions of oil and water at the pore scale due to surface relaxation effects on NMR relaxation (T2), so the NMR tool response can provide a qualitative estimate of wettability from the reduction of the oil relaxation time away from its bulk value. Through detailed modelling of the NMR response, Looyestijn and Hofman (2005) developed a quantitative wettability index based on a model for the microscopic distribution of the crude oil and the wetting state of the rock at any given overall saturation which was verified against core wettability measurements. Similar modelling of the NMR wettability response at the pore scale in cores is reported by Talabi and Blunt (2010) and Al-Muthana et al. (2012). However, there is no convenient tool with which to quantitatively characterise wettability under downhole or *in situ* conditions. Unfortunately, native reservoir wettability estimates from reservoir core samples can be affected and altered by the process of taking a core in the first place! This arises as a result of interactions between the rock and fluids during coring and core recovery, including:

- contact with drilling mud,
- pressure and temperature losses on core recovery,
- oxidation of the oil through lack of appropriate core preservation.

All can alter the native wettability to different and varying degrees.

2.4.3.3 Wettability Alteration Mechanisms: Drilling Mud

Components of drilling muds can alter the wettability of portions of the formation with which they come into contact. Wettability is sensitive to surfactants, brine chemistry and pH, and many drilling muds and oilfield chemicals are known to alter wettability. The ideal mud should contain no surfactants and a minimum of additives. Wettability alteration is severe for OBM fluids because of the oil-wetting surfactants commonly used. For example, in many reservoirs, contact with OBMs (especially those using diesel as the base oil) is known to induce a strong oil-wetting tendency that can be difficult to remove, even by harsh cleaning. Core analysis tests on ‘fresh-state’ core drilled with

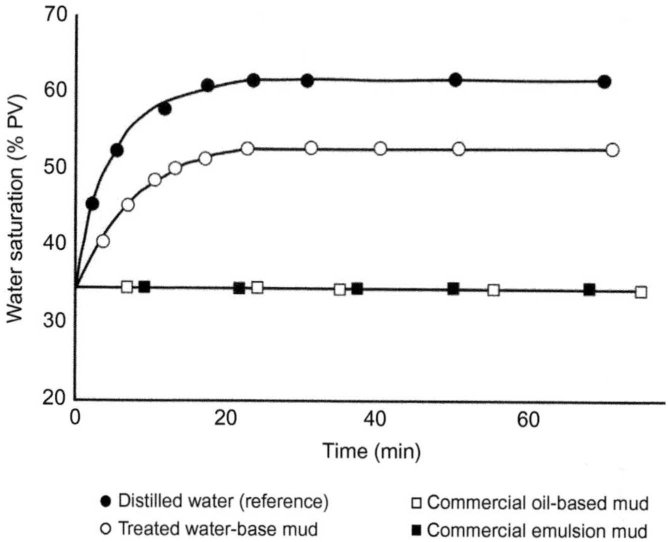

FIGURE 2.38 Effects of oil-based mud on spontaneous imbibition. *After Bobek et al. (1958).*

an OBM will therefore be unrepresentative—the wettability will be altered from its native condition. The only drilling fluid that is more or less guaranteed to preserve native wettability is unoxidised reservoir crude but, unsurprisingly, this is seldom, if ever, used.

Bobek et al. (1958) were amongst the first to investigate wettability alteration due to contact with drilling mud. Water-wet plug samples were flushed with a variety of mud filtrates and additives and the subsequent effect on brine imbibition rates was examined. The sample wettability was tested by measuring the spontaneous imbibition of water following treatment. The low imbibition rates following treatment with oil mud filtrate indicated little affinity for brine, suggesting that the samples had become oil wet (Fig. 2.38). Stiles and Hutfilz (1992) carried out a similar study on the Brent field and found a marked reduction in brine imbibition rates for samples treated with an OBM filtrate (Fig. 2.39). This presents a potential problem for special core analysis studies because OBM filtrate can alter the wettability of the reservoir rock. Figure 2.40 shows the results of experiments made on Berea outcrop samples which were contacted with both WBMs and OBMs simulating typical reservoir coring exposure times. Berea is naturally strongly water wet, and clearly contact with OBM has suppressed water imbibition.

Ballard and Dawe (1988) studied the influence of surfactants in OBM on the wettability of glass surfaces. They showed that even small concentrations of surfactants in mud filtrate could make rock surfaces become more oil wet. Pour et al. (2011) developed an equation-of-state compositional simulator which modelled the processes of mud-filtrate invasion and ensuing wettability

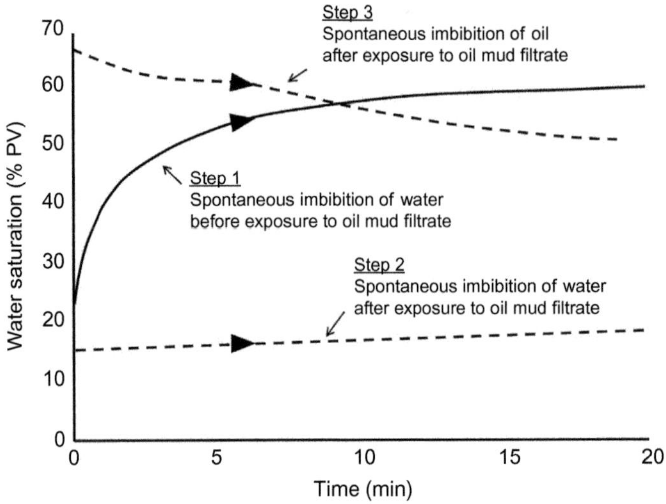

FIGURE 2.39 Effects of oil-based mud on spontaneous imbibition. *After Stiles and Hutfilz (1992).*

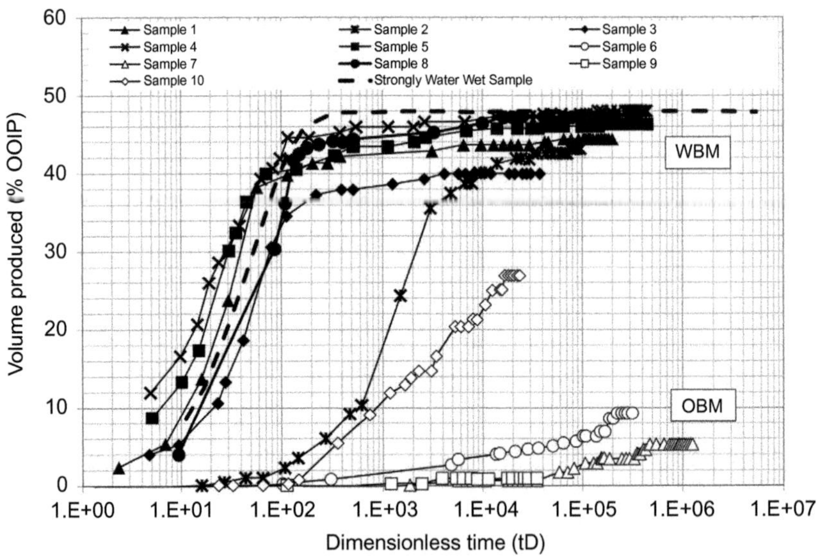

FIGURE 2.40 Imbibition curves for Berea sandstone core treated with different mud systems.

alterations once emulsifiers included in OBM filtrate made contact with grain surfaces. Simulation results confirm that irreducible water saturation decreases as grain surfaces make contact with emulsifiers in the mud.

Jia et al. (1994) showed (with some discrepancies) that all OBMs, filtrates and components evaluated increased oil wettability or reduced water

TABLE 2.7 Example 'Bland' Water-Based Mud Recipe

Basic System	Formation Brine or Seawater
Weighting material	Barite or $CaCO_3$
Filtration control	Bentonite+starch, CMC and XC polymers
pH control	$NaHCO_3$ or soda ash
Viscosifier	Xanthan gum
Bactericide	*para*-Aldehyde

wettability. However, WBMs and mud components had little effect on strongly water-wet rocks with the exception of lignosulphonate which reduced the water wetness. Sharma and Wunderlich (1985) found that rock surfaces with a strongly oil-wet condition become less oil wet after becoming in contact with WBM filtrate.

Arguably a 'bland' WBM, such as the recipe in Table 2.7, is required to obtain core material with 'representative' wettability. Such a mud contains no surfactants and a minimum of additives. The pH of the mud also needs to be maintained within a fairly narrow range of 7–8.5. It should not contain corrosion inhibitors, emulsion breakers or other chemicals which can alter wettability.

If possible, the following mud components should be avoided:

- deflocculants such as lignosulfonates, lignite, phosphates and quebracho,
- bactericides,
- hydroxyethyl cellulose (HEC) polymer (viscosifier),
- corrosion inhibitors,
- diesel,
- sodium hydroxide (alters pH).

In reality, it is not possible to prepare a truly bland mud that has no effect on native core wettability yet still delivers the required rheology. Certainly, WBMs are less likely to cause oil-wetting contamination than OBMs, but even the act of recovering the core to the surface and exposing it to air can cause significant wettability alteration.

2.4.3.4 Pressure and Temperature Reduction

Water-wet rock can be made more oil wet (or less water wet) during core retrieval as the core is brought to the surface. Pressure reduction results in the liberation of gas as the core pressure is reduced below the bubble point of the crude oil. This increases the relative concentration of the heavier end components and increases the surfactant concentration in the oil which can be adsorbed onto the rock surfaces. Temperature reduction reduces the solubility of the surfactants in the oil and increases the potential for adsorption.

Oxidation of the crude oil, through exposure of the core to air during handling at wellsite and in the laboratory, causes an increase in the asphaltene component and provides an opportunity to increase the oil wettability (McPhee, 1994). Correct core preservation, designed to eliminate or minimise core exposure to air, will prevent this form of wettability damage.

2.5 BEST PRACTICE IN WELLSITE HANDLING

2.5.1 Core Recovery on Rig Floor

Once the core reaches the rig floor, core handling procedures, core preservation and stabilisation must be designed to minimise physical and chemical alteration of the rock material and exposure to the elements.

A core-filled liner is lowered to the catwalk within the metal inner core barrel using a system of pulleys anchored to the end of the catwalk. The barrel should be lowered gently onto the catwalk using a rigid core cradle or shuttle ideally provided with rollers to prevent jarring. Examples are shown in Figs. 2.41 and 2.42.

FIGURE 2.41 Core laydown on cradle/shuttle. *Used with permission from NOV.*

FIGURE 2.42 Rigid core barrel shuttle to avoid mechanical core damage by flexing. *Used with permission from NOV.*

FIGURE 2.43 (a) Lifting liner without support causes core damage due to flexing. (b) Example of core damage due to liner flexing.

When the liner sections are removed from the barrel they must be supported during handling and lifting, otherwise the core liners—especially fibreglass—can flex and cause mechanical damage to the core (Fig. 2.43).

The core liner sections are now labelled and marked so that they can be correctly oriented and reassembled in the lab. As the cores should be protected from exposure to extreme heat, sunlight and rain, the core sections should be maintained, sealed in their liners and, if possible, should not be removed from the liners at wellsite. The exceptions would be to take wellsite samples (e.g. for Dean-Stark analysis) or preservation of core sections that cannot be allowed to dry out, such as shale sections. In the former case, it is better to drill samples from the ends of the liner sections rather than through the liners, and in the latter case, use of clam shell barrels will enable speedy removal of selected core sections. The liner can then be closed and sealed for transport.

The core recovery must be recorded and any under- or over-recovery reported. The convention is to assume that under- or over-recovery is assigned to the bottom of the core though exceptions must be noted. The top depth of the next core should be given the same depth as the bottom of the previous core. This means that in the case of over-recovery in the previous core, some sections will have the same depth in both cores. However, these cores will be distinguishable from each other because of their core box numbers.

As discussed above, unless circumstances and sampling protocols dictate, the core should not be extracted from the liner at wellsite—this is better left until it reaches the core laboratory, where it can be removed under more closely controlled conditions than is possible at wellsite. Even a few minutes exposure, depending on atmospheric conditions, can cause a significant loss of both water and light hydrocarbon fractions from cores. If required, samples can be taken at one end of each liner section. If clam shell liners are used, the liner can be opened to allow core description and sampling. This is not recommended for weak, friable core or for oil cores which are sensitive to oxidation.

In H_2S environments, the inner core barrels are laid out and allowed to degas before any further handling until such time that it is safe to do so.

2.5.2 Linear Layout and Marking

The liners are wiped down and cleaned of mud, and then marked with water-resistant colour wax crayons or marker pens for identification, correct orientation and depth. Typically, the liner is striped from top to bottom with two parallel red (light grey in the print version) and black lines—'tramlines'. Marking schemes may vary from company to company, but the key is to ensure consistency of marking—no matter which scheme is used it must be rigorously and consistently applied. The normal convention is that the red (light grey in the print version) line is on the right of the liner and the black line is on the left of the liner looking up the core.

A circular depth marker is then marked on the core liner at regular intervals with an arrow pointing towards the top of the core. The measured

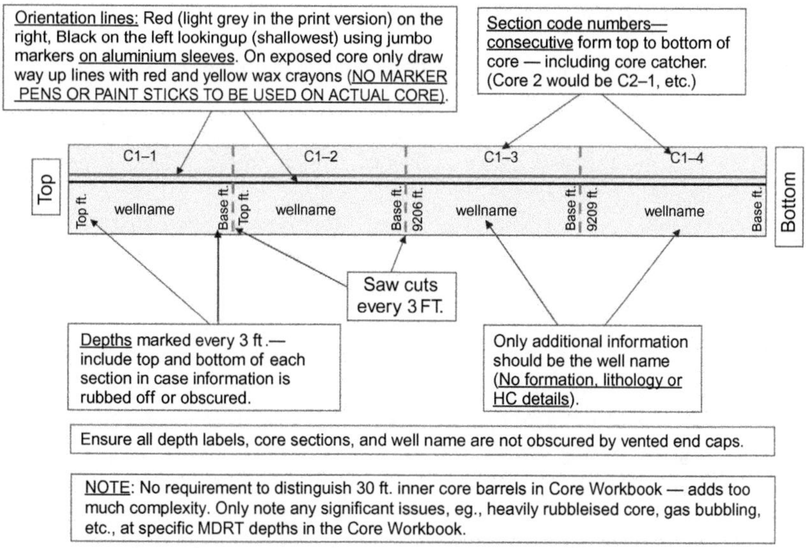

FIGURE 2.44 Example core linear markup. *Courtesy of Wintershall.*

FIGURE 2.45 Core liners marked up at wellsite. *Courtesy of ALS Oil and Gas.*

depths are written on the side of the liners. Depths must be marked on both tops and bottoms on each core section. All markings must allow fitting of end-caps to each liner section after cutting without obscuring the depth information.

A template for liner markup is shown in Fig. 2.44 and an example of well-site marking is shown in Fig. 2.45.

A detailed list documenting the well name, core number, cored interval, core recovery, number and depth interval of each core section, chip or plug sample, rubble section, etc., has to be maintained for all recovered core.

2.5.3 Wellsite Gamma-Ray Scanning

Prior to cutting liner sections, the core can be gamma ray scanned in the liners. The measurements are normally performed using portable or handheld devices (Fig. 2.46) with a sodium iodide crystal gamma-ray detector. The cores should be kept well away from casing or pipe racks as the metal can cause magnetic interference in the gamma logger.

These gamma-ray scans are not intended to replace core scanning in the laboratory but, by comparing lab and wellsite traces or logs, they can ensure that the core linear sections have been reassembled in the correct order in the lab. Wellsite logs generated using the portable gamma logger can be used for initial assessments of core recovery and core-to-log depth correlation; to differentiate sand/shale sequences, to select locations for on-site plugging (if required); to enable operators to make quick, real-time decisions of further coring, logging, testing, perforating and well-completion activities; and to ensure that the core is reassembled correctly in-depth order in the lab.

2.5.4 Wellsite Liner Cutting

Before the core sections leave the derrick, the barrel is cut at each joint using a shear plate. An example is shown in Fig. 2.47. The shear plate operates on a guillotine principle with the blade slowly advancing through the core itself driven by a mechanical screw or a hydraulic piston. As it closes, the blade supports the core and does not allow it to fall.

The core barrel joint and liner sections are around 30 ft. (9.1 m) long. The liner sections are moved around the rig floor on a rigid shuttle or cradle (Fig. 2.48) which is supported along its entire length during lifting. These

FIGURE 2.46 Handheld portable wellsite gamma-ray logger. *Courtesy of ALS Oil and Gas.*

FIGURE 2.47 Shear plate. *Courtesy of ALS Oil and Gas.*

FIGURE 2.48 Liner lifted on rigid supported cradle. *Courtesy of ALS Oil and Gas.*

FIGURE 2.49 Wellsite liner cutting system. *Used with permission from NOV.*

liner sections can then be shipped to the laboratory intact, but best practice now dictates that the liner sections are cut into 3-ft. (1 m)-long sections as shorter sections are less subject to damage due to flexing of the liner during lifting and transport. The liners are cut using an air or electric-powered circular saw. Care should be taken to avoid vibration and rotation of the core, and sparks generated by the cutting process. The liners should be cut at a small angle to ensure that core can be fitted back together in the lab. Figure 2.49 shows an example of a wellsite liner cutting system.

The liner sections can be topped up with the drilling mud and the tops and bases of each section capped or sealed (Fig. 2.28) to prevent evaporation/air exposure and fluid ingress. This is not recommended as a long-term preservation method, but will allow the core to be processed quickly and without special equipment, and will prevent the core from drying out on its journey to the laboratory.

2.5.5 Wellsite Core Stabilisation and Preservation

The core integrity and pore fluid content of the rock can be altered when bringing the core back to surface. These changes are largely unavoidable and in some cases are manageable through use of specialist coring systems and trip rates. However, it is important to avoid further changes to the core and the pore fluids when the core reaches the rig floor. This means that selected core sections or samples may have to be preserved at wellsite to prevent further fluid loss or exposure to air and, if the core is particularly weak or friable, the core liner sections may have to be stabilised for transport to the lab.

Where possible, core recovered in solid liners should not be extracted from the liners at wellsite as this can lead to core contamination and damage. If the transportation logistics are favourable, it is much better to remove and preserve core sections in the lab under controlled conditions, as described in Chapter 3. If required samples can be plugged from the ends of the liner section and then sealed, the liner section can be topped up with drilling fluid to provide temporary preservation, or the entire liner section can be sealed by dipping in wax. If using clam shell liners, core sections can be lifted out but they must be preserved quickly and securely.

Liner sections from friable, weak or unconsolidated formation core should be mechanically stabilised prior to shipment to the laboratory. In such formations, erosion of the core during cutting tends to increase the annular gap between the outside of the core and the inside of the liner. If this gap is not filled or stabilised, then the core is free to move and rotate during transportation which will ultimately damage the core.

The three main stabilisation methods are:

- core freezing,
- resin injection,
- gypsum and foam injection,

Weak cores caught in clam shell liners should not be stabilised by resin or foam/gypsum injection as the core section can become slabbed along its length when the liner is forced open in the lab.

2.5.5.1 Freezing

Core liner sections can be frozen at wellsite using:

1. Chest freezers: −30 °C
2. Dry ice (liquid CO_2): −78 °C
3. Liquid nitrogen: −196 °C

Freezing prevents further loss of volatiles and prevents alteration of the rock fabric on handling. However, core freezing is a procedure that nearly all operators use with a degree of unease. Freezing can cause significant evaporative losses through sublimation and cores may have to be protected by a thin layer of frozen brine to limit sublimation process. Freezing may also result in the migration or diffusion of fluids within the core structure and can cause atmospheric water to condense into the pore system in higher permeability formations.

On freezing, water expands so that ice crystals can disrupt the grain structure and damage the grain boundaries in weak sands. At best, this can increase porosity and permeability (Everett et al., 1987; Torsaeter and Beldring, 1987): at worst, frost heave can destroy the grain structure. The size and potential destructive power of ice crystals depend on the rate of freezing which affects the size of the ice crystals. Slow freezing (for example, in a chest freezer) results in the growth of larger ice crystals than faster freezing (in dry ice) and is potentially more damaging. However, rapid freezing can cause thermal shocks leading to fracturing along thin-bed boundaries and may disrupt grain-to-grain contacts. Freezing is inefficient if liquid saturations are low, or if the freezing points of liquids are below temperatures which can be easily be achieved at wellsite. The cost of storing large amounts of core at low temperatures can also be prohibitive.

In some cases, core is often frozen on receipt by the laboratory—not at wellsite—so damage may have already occurred on transport to the lab.

All plugs scheduled for routine analysis, SCAL, geomechanics and formation damage tests should be taken at the same time when the core is frozen as successive thaw (to allow inspection and plug selection) and freeze (to allow plugging) cycles will disrupt and weaken the core and preclude further plugs from being taken. It is important that plugs taken from frozen core should be stored frozen until they are required for analysis.

In view of the damage that freezing can create, core freezing should only be considered in extremely weak, effectively unconsolidated formations where there may be no annular gap between the core and the liner. In this case, it may be the only viable method to stabilise the core.

2.5.5.2 Resin Consolidation

Any mud in the annular space between the core and the inner liner of the core barrel is drained and then filled with self-polymerising liquid plastic or rapid hardening resin. The resin secures the core to the liner and preserves the morphology of beds and bed contacts. Worthington et al. (1987) claim that this method results in less disturbance of the rock fabric and sedimentary structures, and fewer uncertainties in measurement of petrophysical properties compared to freezing.

Although handling and storage costs are much reduced compared to freezing, chemical contamination of the core by resin components and invasion of the resin into the pore space in higher permeability, friable, fractured or vuggy formations are possible, especially if no mud cake is present on the core surface. Resin components can also alter the wettability of the stabilised rock.

Resins require premixing because they are usually composed of two-component systems in order to control the hardening time. If the resin sets too early, it can prevent full coverage of the core/liner annulus. It sets too late, and the resin can enter the pore system and then harden. Resination is an intrinsically hazardous operation as most resins contain isocyanates which require careful HS&E management. In addition, the use of resins involves using various items such as mixers, hosepipes and jugs which should be properly disposed of.

2.5.5.3 Foam and Gypsum Stabilisation

Foam (Fig. 2.50) or gypsum (Fig. 2.51) injection is designed to fill the core/liner annulus and encapsulate the core in a protective, non-invasive and

FIGURE 2.50 Foam injection into 1 meter liners.

FIGURE 2.51 Gypsum poured to top-up liners.

non-absorbent cushion. Holes are normally drilled along the liner to ensure that drilling mud is displaced by the foam or gypsum and to ensure full encapsulation.

In moderately cemented to consolidated core, the core liner sections are held horizontal during the complete stabilisation process. In weak sands, any movement while the core is in a horizontal position will cause damage to the core. In this case, the core is held vertical during cutting of the liner sections. After cutting, an end cap with an insert is placed at the bottom of the core section. This insert allows foam to be injected into the end cap, providing a cushion onto which the core section can rest.

2.5.5.4 *Gypsum Versus Foam: Advantages and Limitations*

Gypsum as a stabilisation product was initially developed by Reservoir Laboratories (now Weatherford Laboratories) to overcome the problems associated with resins, especially those related to chemical alteration of wettability (Hjelmeland et al., 1996). Although foam may successfully protect the core material, any organic compounds in the foam might potentially alter the wettability of the grain surfaces. Gypsum ($CaSO_4$) is an inorganic compound and therefore avoids the potential geochemical alteration problems associated with organic resins and some foams, and is also a cheap, non-toxic, benign alternative. Retardation and curing agents have been developed to control viscosity and setting times. According to Hjelmeland et al., invasion of the gypsum into the pore space is minimal, reaching only 1–2 mm within a synthetic high-permeability sandstone.

Gypsum is injected in liquid form and requires unrestricted access to the core/liner annulus to be successful. Any restrictions will block the passage of liquid, whereas foam will bypass any blockages. Mixing of the gypsum solution and management of the water phase during the injection processes are crucial to ensure even distribution of the gypsum, control premature setting and restrict adsorption of the water phase into the pore space. Gypsum requires experienced staff to know the correct quantities to produce a mixture that will set correctly within a given time frame. If the mixture is not made to the correct consistency, it may take an extended time to set, or not set at all. Gypsum can also absorb water from the core which can lead to the gypsum becoming structurally unstable. This possible absorption of water from the core could also affect latter fluid saturation tests, such as Dean Stark.

Compared to epoxy resins, polyurethane foam has a minimal penetration into the pore space as the sizes of the foam bubbles are greater than the pore diameters, and causes only minimal grain disturbance (Moss, 2014). Epoxy resin can partially penetrate the pore space and polyester resin can penetrate deeply into the pore space if there is no or partial mud cake on the core surface. Foam-based stabilisation materials should be able to be easily removed from the core surface with a scalpel blade, but polymer and epoxy resins cannot.

The injection of foam involves the utilisation of a self-contained pressurised canister and does not require any premixing of any of the components it contains. In addition, cured foams are classified as general waste, and after use the cylinders can be easily disposed according to certified waste management practices.

Foam has lower weight compared to resins, is simpler to process and has technical and environmental advantages over resins and gypsum (Baraka-Lokmane et al., 2010). However, foam setting is an exothermic process which might cause evaporation of water in the pore space.

2.5.6 Wellsite Sampling

2.5.6.1 Reasons for Sampling

In general, sampling of core recovered at the wellsite is not recommended if the core has to be removed from the liner or exposed to wellsite ambient conditions during the sampling process. If sampling of the core at wellsite is required, the exposure time of the core should be minimised.

There are a number of reasons why wellsite samples may be required. These include:

- samples for lithological description and/or mineralogical analysis,
- water saturation (Dean-Stark) and water resistivity analysis sample,
- fast-track measurement of basic rock properties,
- geomechanics testing,
- wettability studies,
- oil fluorescence observations,
- *in situ* stress estimation from strain recovery measurements,
- coal or shale gas methane desorption analysis.

However, reasons for sampling must be critically reviewed—in many cases it may be better to wait until the core arrives back in the laboratory if core atmospheric exposure causes issues or the sampling process complicates the logistics and economics of core handling at wellsite. Unless there are very strong technical and business cases, analysis is better performed in the laboratory rather than at wellsite.

2.5.6.2 Sampling Methods

All samples or core sections removed from the core must be recorded and locations identified using rigid spacers. Hammering to obtain samples should be avoided. Ideally, samples should be taken at natural breaks in the core or using a trimsaw and sealed/preserved pending analysis. In some cases, plug samples may be required to be taken and the core analysis contractor should have appropriate plugging equipment (and plugging fluids) on site. It is always better to take plugs at the end of liner cut sections rather than plugging through the liner. In this case, the plug holes must be filled with non-adsorbent filler and the holes in the liners sealed. The liners are then topped up with drilling mud, oil or kerosene and then sealed for shipment.

If transportation logistics dictate that plug samples for Dean-Stark water saturation analysis and extraction of formation water from core for water resistivity (Rw) and compositional analysis cannot be taken after core receipt in the lab, then plugs can be taken and sealed/preserved quickly at wellsite and then shipped back to the lab for analysis. Wellsite plugging is also recommended if the core is to be subsequently stabilised with foam as the exothermic reaction causing the foam to set can evaporate water in the pore space, or if the core is to be frozen (to prevent atmospheric water condensation).

FIGURE 2.52 Example of plug sampling at wellsite. *Courtesy of Vahid Tavakoli.*

Although sample site and lithology selection can be aided by wellsite core gamma logging, essentially wellsite plugs are taken 'blind'. Plugging 1″ or 1.5″ diameter samples from the centre and potentially least invaded part of the core, at the ends of each liner section, is a better option as there is no need to subsequently seal plug holes drilled through the liner. The plugs are cut parallel to the liner long axis, away from the halo of mud invasion, as indicated in Fig. 2.52. These plugs should not be used for permeability measurements as their bedding plane orientation will not be known at this stage. Plugs cut through the liner sections, perpendicular to the core long axis, may not be parallel to bedding if bedding is oblique to the core axes. In this case, permeability measurements could prove misleading.

For unstabilised core caught in clam shell liners, the liners can be opened to expose the plugging sites.

2.5.6.3 Dean-Stark and Water Analysis Plugs

For water analysis, plugs should be cut with mineral oil or air as the bit coolant, taking care to minimise air positive pressure (or cut using a vacuum system). The plugs should be sealed immediately (see Fig. 2.24). Core chilling may be considered to prevent loss of light volatiles.

In the laboratory, wellsite core Rw samples are placed in an ultracentrifuge and centrifuged at 600–800 psi equivalent capillary pressure under air to recover brine samples (Clinch et al., 2010). Normally, only a few drops of brine are recovered but these can be sufficient to obtain ambient condition measurements of water resistivity, and complete chemical analysis, if attention to sample handling and analysis protocols is strictly followed. Results can be interpreted in conjunction with tracer analysis for Sw.

2.5.6.4 Shale Preservation

It is normally much better to maintain core intact in a sealed liner and remove it in the laboratory under controlled conditions than to attempt to take and preserve whole core samples at wellsite. One exception is preservation of shale sections for cap rock analysis and geomechanical tests where core must not be allowed to dry out. This can cause shales to delaminate and can adversely impact the results of the tests, or can render sampling impossible.

If the core is caught in solid liners, then they should be cut into 1 m (or shorter) sections, the liner sections topped up with mineral oil or drilling fluid, the end-caps secured and the entire core liner section dipped and preserved in wax. If clam shell liners are used, core sections can be lifted out and placed in plastic tubes, the tube annulus filled with pure synthetic oil, and the ends of the tubes sealed. This procedure should be designed to ensure that the core does not dehydrate during shipping and storage. The plastic tubes can also confine the core to some extent which can help prevent further splitting along fissility planes. Alternatively, core sections can be sealed using a conventional film–foil–wax technique and then securely wrapped with tape to confine the core, but this is more laborious, and it is much more difficult to guarantee an air-tight seal.

2.5.7 Core Transportation

As for core handling on the rig, the liner sections must be supported during handling and lifting. Cutting liners into shorter lengths generally makes for easier handling and avoids flexing damage.

The core liner sections should be shipped in specialist core containers which are designed to avoid transfer of shock loads to the core sections either using spring-loaded devices or completely cushioning the liner sections using pre-shaped layers of hard or soft foam (Figs. 2.53 and 2.54). Shock tabs (Fig. 2.55) or accelerometer monitoring systems should be placed on or within containers before transportation to indicate whether the container has been subject to shock load (e.g. dropped) and potentially damaged during transport. This also sends a warning message to all people involved in core handling that

FIGURE 2.53 Core Care liner Transportation Service. *Image courtesy of Baker Hughes.*

FIGURE 2.54 Core liners in transport container awaiting shipment. *Used with permission from NOV.*

FIGURE 2.55 Shock tag mounted on transport container. *Image courtesy of Baker Hughes.*

potential damage may have occurred before the core begins to be processed by the laboratory. Garcia et al. (2010) carried out an extensive study to understand the shock distribution inside core containers during transportation. They found that the hardness and shape of foam inserts for storing liner sections was critical to preventing damage, and that more fragile samples should be stacked in the container starting from top to bottom and stowing them equally spaced within the container. Any empty space should be filled with foam pillows.

2.5.8 Coring Report

One of the difficulties faced by many geoscientists and engineers who are reviewing or auditing legacy core analysis data is that the reports rarely include details of the coring operation. It therefore becomes a forensic exercise to trace the provenance of the core and how it was cored, handled and transported. As core damage due to inappropriate handing or chemical alteration by drilling mud can have a significant impact on the core analysis results, it is essential that this information is readily available with the routine and/or SCAL report (for example, as an Appendix) and not 'buried' in an end of well report or daily drilling report. This information will be invaluable in qualifying the interpretation of the core analysis data. The information that needs to be recorded includes:

1. well identification and location,
2. type of well (vertical, deviated, sidetrack, etc.),
3. well inclination,
4. core orientation details (if run),
5. drilling mud details (type, water content if OBM, mud weight, YP and PV, pH)
6. core and bit type,
7. details of coring assembly including liners,
8. coring parameters (drilling and circulation rates, WOB, runs, etc.),
9. cored intervals and total core recovered,
10. core diameter,
11. core tripping out of the hole rates (per 90 ft. stand),
12. core liner cutting equipment and methodology,
13. core stabilisation method,
14. wellsite sampling methods and sample details,
15. well logs and mud logs (if available) over cored interval,
16. wellsite core gamma-ray log (if run),
17. difficulties encountered: jamming, lost recovery, mud losses to formation, etc.,
18. core transportation log,
19. any other relevant information.

2.6 SPECIAL HANDLING CONSIDERATIONS FOR DIFFICULT ROCK TYPES

Consolidated formations normally require no special treatment at wellsite. Specialist precautions for other formations and fluid types are summarised below.

2.6.1 Unconsolidated Core

Unconsolidated or poorly consolidated core containing light oil or gas is susceptible to significant fluid loss during handling at the surface. As with most rock types, by the time the core has been brought to the surface, it has undergone mechanical stress relief due to the removal of overburden pressure and has had varying amounts of gas expansion when internal pore pressure is lost. The degree to which these effects will alter the core depends on depth, reservoir pressure, oil gravity, fluid properties, sediment type and coring procedures. As discussed earlier in this chapter, care must be taken to ensure that sufficient time is allowed to prevent excessive pressure differentials during tripping out of hole especially if there is a thick and resilient mud cake. Cored intervals should be limited in length to prevent possible collapsing under its own weight, and liners must be cut into 1 m (3 ft.) sections. Specialised handling protocols and liner stabilisation will almost certainly be required at wellsite to minimise disturbance to the rock texture.

Heavy oil introduces additional challenges particularly due to the low mobility of gas which can result in very slow drainage. This can, in turn, cause the core to expand, filling the annulus between the core and the liner. Once the radial expansion is constrained by the liner, the core can only further expand in the axial direction and expansion may cause the core to extrude out of the end of the liner if there is no axial constraint. Rigid rather than rubber liner end-caps can help in this circumstance. The use of pre-perforated liners can also shorten gas evolution times. Freezing should only be used as a last resort as it inhibits gas drainage. Sufficient time should be allowed in the laboratory to ensure full gas dissipation before analysis of the core and the core may have to be restrained until fully equilibrated.

2.6.2 Carbonates

Generally, most carbonates are competent and do not often require special precautions. However, large vugs and fractures can affect core recovery and core disking can be an issue in vertical wells if the horizontal stresses are high enough and if the carbonate is brittle enough (Santarelli and Dusseault, 1991). Disking is thought to result from the unusual stress distributions underneath the advancing core disc which can result in extensional failure causing the core to part in a series of stacked disks. It is more common where the horizontal stresses are similar to or even exceed the vertical stress and the mud overbalance pressure is low. Disking is less pronounced in sands or shales as these tend to deform rather than fracture.

Resin, foam or gypsum used for stabilisation will enter and may fill larger vugs and natural fractures. In some carbonate lithologies, evaporites as either beds or vugs or fracture fills are quite common. If evaporites are expected, cores should be drilled with OBM or highly saline WBM, and the recovered core should never be exposed to freshwater.

2.6.3 Shaly Sands

The presence of clay minerals in sands can also cause problems. For example, smectite is a highly reactive clay that can swell in the presence of water leading to potential disruption of the rock textural framework. Chlorite is another clay mineral that is sensitive to acids. Illite can occur in a very delicate filamentous morphology that can be easily damaged as a result of the passage of an air/liquid interface on drying. Kaolinite is a more robust clay but clay platelets can migrate in the presence of water that is not in cationic equilibrium with the pore fluids.

If the clay content is high or reactive enough, OBM or highly inhibited WBM should be used as the coring fluid. Contact with freshwater should be avoided, especially when slabbing core sections prior to taking plugs, or when taking plugs. Smectitic sandstones may swell when confining stress is removed if free water is available, even if the free water has properties identical to the interstitial fluids. In weak sands with high smectite content, rehydration of the clay interlayer sites on exposure to water after drying can cause swelling forces sufficient to destroy the rock texture.

2.6.4 Shales

Shales are a mixture of organic and inorganic components with and have high kerogen and clay contents, low porosity and ultra-low permeability. While shale grain size and content can vary substantially: from dark, fissile organic shales to siltstones, in general the tightness of the rock and abundance of clay minerals and kerogen can cause a number of challenges in core recovery and core handling.

In shale gas core, the intergranular porosity is predominantly water saturated, while pores in organic material may be gas saturated. This can promote imbibition of both oil-based and WBM filtrates during coring.

The ultra-low shale permeability means that pore pressure dissipation is extremely slow. The Zubizarreta et al. (2011) CFD model demonstrated a high risk of pore pressure release-induced tensile fracturing in shale gas core at economic trip rates. Pressure-retaining core barrels allow controlled, offline depressurisation to help prevent this form of core fracturing.

One of the biggest problems in shales is that a large number of partings or splitting are induced by stress unloading and relaxation during coring and core recovery. To some extent, this is largely unavoidable but can be exacerbated by inappropriate core handling and preservation once it gets to the surface.

Chenevert and Amanullah (2001) showed that shales must be preserved at their native water content if accurate physical measurements are to be made. Swelling data show that shales that are altered during handling, either via dehydration (drying out) or via rehydration (exposed to water), do not respond properly even when restored to their native hydration conditions. They tend to experience excessive swelling compared to cores kept at their native water content. Any level of desaturation during handling or storage can cause an irreversible change in the shale matrix and thus the ultimate behaviour of the shale is preventing delamination or splitting along fissile bedding planes. Dehydration in dark fissile shales can occur rapidly and can cause parting or splitting along bedding planes. Once a fissile shale core has split, it may be impossible to obtain specimens large enough for core analysis or geomechanics tests.

Best practice to preserve shale properties and to prevent further core damage therefore dictates that excessive handling or movement of the core must be avoided, any excess water from the core surface must be removed and shale core sections should be preserved immediately at wellsite.

The water content in most shales is high and therefore preserved shale sections should never be frozen, and large temperature fluctuations should be avoided to prevent micro-fissuring and changes in water content.

2.6.5 Coal

Coal cores are normally taken for coal-bed methane or *in situ* combustion projects. For coal-bed methane *in situ* gas content, gas sorption behaviour, permeability, relative permeability, cleat and fracture analysis, core composition and mechanical behaviour are the major interests.

Coal formations vary greatly between different coal seams and areas. As a result, it is not appropriate to core all coal-bed methane wells with one specific fluid. The drilling mud should not alter the coal structure and should be tailored to each specific area relative to the composition of the coal, the produced water, whether it is a wet or dry reservoir, as well as the reactivity of the formations above and below the coal seam. As for shales and unconventional gas reservoirs, time may be required to permit full gas dissipation so pressure-retained coring techniques may be required if conventional coring is time constrained.

Careful handling is essential and coal can be very brittle and shatter when the core liner is flexed, dropped or struck with a sharp blow.

The low-temperature oxidation of coal is a very complex process that can dramatically influence and alter the inherent coal properties. Oxidation can render the coal less hydrophobic and more hydrophilic, makes the coal more friable, increases the moisture content and decreases the volatile oil content (de Korte, 2001). As for shales, exposure to air should be minimised and freezing is not recommended. Special containers may be required to preserve coal core sections that enable flushing of the cylinders with inert gas or methane.

REFERENCES

Al-Maamari, R.S.H., Buckley, J.S., 2003. Asphaltene precipitation and alteration of wetting: the potential for wettability changes during oil production. SPE Reserv. Eval. Eng. 6 (4), 210–214.

Al-Muthana, A.S., Hursan, G.G., Ma, S.M., Valori, A., Nicot, B., Singer, P.M., 2012. Wettability as a function of pore size by NMR. In: International Symposium of the Society of Core Analysts held in Aberdeen, Scotland, UK, 27–30 August 2012, Paper SCA2012-31.

Anderson, W.G., 1986a. Wettability literature survey—part 1: rock/oil/brine interactions and the effects of core handling on wettability. J. Pet. Technol. 38 (10), 1125–1144.

Anderson, W.G., 1986b. Wettability literature survey—part 3: the effects of wettability on the electrical properties of porous media. J. Pet. Technol. 38 (12), 1371–1378.

Anderson, W.G., 1987a. Wettability literature survey—part 4: the effects of wettability on capillary pressure. J. Pet. Technol. 39 (10), 1183–1300.

Anderson, W.G., 1987b. Wettability literature survey—part 5: the effects of wettability on relative permeability. J. Pet. Technol. 39 (11), 1453–1468.

Ballard, T.J., Dawe, R.A., 1988. Wettability alteration induced by oil-based drilling fluid. In: SPE Formation Damage Control Symposium, Bakersfield, California, 8–9 February, Paper SPE 17160.

Bantignies, J.L., Dit Moulin, C.C., Dexpert, H., 1997. Wettability contrasts in kaolinite and illite clays: characterization by infrared and X-ray absorption spectroscopies. Clay Clay Miner. 45 (2), 184–193.

Baraka-Lokmane, S., Wigness, M., Smith, M., Schipper, B., 2010. Preservation of core integrity: a comparative study of core stabilization products. In: International Symposium of the Society of Core Analysts, Halifax, Nova Scotia, Canada, 4–7 October, Paper SCA2010-07.

Bjorum, M., 2013. A new coring technology to quantify hydrocarbon content and saturation. In: International Symposium of the Society of Core Analysts held in Napa Valley, California, USA, 16–19 September 2013, SCA Paper 2013-094.

Bobek, J.E., Mattax, C.C., Denekas, M.O., 1958. Reservoir rock wettability—its evaluation and significance. Pet. Trans. AIME 213, 155–160.

Bruno, D., 2007. Drilling mud tracers for formation evaluation and reservoir engineering applications. Houston Geol. Soc. Bull. 50 (4), 15.

Buckley, J.S., Liu, Y., Monsterleet, S., 1998. Mechanisms of wetting alteration by crude oils. SPE J. 3 (1), 54–61.

Chenevert, M.E., Amanullah, M., 2001. Shale preservation and testing techniques for borehole-stability studies. SPE Drill. Complet. 16 (3), 146–149.

Chilingar, G.V., Yen, T.F., 1983. Some notes on wettability and relative permeability of carbonate reservoir rocks. Energy Sources 7 (1), 67–75.

Clinch, S., Wei, W., Lasswell, P., Shafer, J., 2010. Determining formation water salinity in the oil leg using cores and logs. In: SPWLA 51st Annual Logging Symposium, 19–23 June, Paper XXX.

de Korte, G.J., 2001. Dewatering and drying of fine coal. Presented at Coaltech 2020.

Durandeau, M., El-Emam, M., Anis, A.H., 1996. Successful field evaluation of the efficiency of a gas gravity drainage process by applying recent developments in sponge coring technology. In: International Symposium of the Society of Core Analysts, Montpellier, France, 8–10 September, Paper SCA-9605.

Eaton, N.G., Radford, S.R., Segrest, C., 1991. Pressure coring: better cores for better analysis. In: Eurocas Symposium, SCA Paper 1991-05.

Everett, R.V., Herron, M., Pirie, G., Schweitzer, J., Edmundson, H., 1987. Faja case study results on a single well. SPE Form. Eval. 2 (3), 361–375.

Fehling, D., 2013. Tracing the culprit if fracking pollutes water supplies. http://stateimpact.npr.org/texas/2013/01/22/scientists-close-to-testing-tracers-for-drilling-fluid/ (accessed 30 July, 2014).

Fjerstad, P.A., Thereby, H., Pallatt, N., Stockden, I., 1993. Application of deuterium tracer in estimating native water saturation in the Gyda Field. In: SPE Asia Pacific Oil and Gas Conference and Exhibition, Singapore, February, Paper SPE 25376.

Funk, J., Al-Khan, M., Eyvazzadeh, R., Dudenas, P., Jones, D., Wilson, B., 2004. Revitalizing the use of sponge core for the determination of remaining oil saturations in carbonate reservoirs. In: International Symposium of the Society of Core Analysts, Abu Dhabi UAE, 5–9 October, SCA Paper SCA2004-31.

Garcia, J.-V., Rousseau, J., Dorel, D., 2010. Well site core stabilization and packaging—the first step in acquiring undisturbed core. In: International Symposium of the Society of Core Analysts, Halifax, Nova Scotia, Canada, 4–7 October, Paper SCA2010-35.

Hamilton, W.O., Van Alstine, D.R., Butterworth, J.E., Raham, G., 1995. In: Annual Technical Meeting of the Petroleum Society of Canada, Calgary, Alberta, 7–9 June, Paper PETSOC-95-56.

Hartley, A.J., Turner, P., Paleo, 1991. Paleomagnetic studies of Carboniferous and Permian in the U.K. Southern North Sea: core orientation, paleocurrent analysis, and diagenetic application. Presented at AAPG Annual Convention, Dallas, Texas, April 7–10.

Hay, W., Knott, I., Shiells, G., 1990. Development of coring and electronic data logging systems enhance data quality and quantity facilitating improved oriented core analysis. http://www.scaweb.org/assets/papers/1990_papers/1-SCA1990-08EURO.pdf (accessed 4 August 2104).

Hirasaki, G.J., 1991. Wettability: fundamentals and surface forces. SPE Form. Eval. 6 (3), 217–226.

Hjelmeland, O., Tjetland, B.B., Ardo, B.A., Bolle, L., Scheie, A., Venturini, C., 1996. A new method for stabilization of friable and unconsolidated core samples at well-site. In: International Symposium of the Society of Core Analysts, Montpellier, France, 8–10 September, Paper SCA-9603.

Holt, R., 1993. Effects of coring on petrophysical measurements. http://scaweb.org/assets/papers/1994_papers/1-SCA1994-07.pdf (accessed 30 July 2014).

Jerauld, G.R., Rathmell, J.J., 1994. Wettability and relative permeability of Prudhoe Bay: a case study in mixed-wet reservoirs. In: Proceedings of 3rd International Symposium on Evaluation of Reservoir Wettability and Its Effect on Oil Recovery; Laramie, Wyoming.

Jia, D., Buckley, J.S., Morrow, N.R., 1994. Alteration of wettability by drilling mud filtrates. In: SCA Symposium, Stavanger, 1994. International Symposium of the Society of Core Analysts held in Stavanger, Norway, 2012, Paper SCA 9408.

Looyestijn, W.J., Hofman, J., 2005. Wettability index determination from NMR logs. In: SPE Middle East Oil and Gas Show and Conference, 12–15 March, Kingdom of Bahrain, Paper SPE 93624.

Maury, V., Santarelli, F.J., Henry, J.P., 1988. Core discing: a review. In: Proceedings of the ANGORM Symposium on Rock Mechanics in Africa, pp. 221–231.

McPhee, C.A., 1994. Effects of strong wetting on end point relative permeability. Dialog. London Petrophysical Society.

Moss, 2014. Stabilisation of unconsolidated reservoir rocks for core analysis. http://www.lps.org.uk/docs/unconsolidated_cores.pdf (accessed 30 July 2014).

Pour, R.H., Torres-Verdín, C., Sepehrnoori, K., 2011. Quantifying wettability alteration during oil-base mud-filtrate invasion and corresponding effects on resistivity logs. In: SPWLA 52nd Annual Logging Symposium, Colorado Springs, Colorado, United States, 14–18 May.

Ringen, J.K., Halverson, C., Lehne, K.A., Rueslatten, H., Holand, H., 2001. Reservoir water saturation measured on cores: case histories and recommendations. In: Proceedings of the 6th Nordic Symposium on Petrophysics, Trondheim, Norway, 15–16 May 2001.

Santarelli, F.J., Dusseault, M.B., 1991. Core quality control in petroleum engineering. In: Roegiers, J.-C. (Ed.), 32nd US Symposium on Rock Mechanics, 10–12 July, Norman, Oklahoma. Paper AARMA-91-111.

Sharma, M.M., Wunderlich, R.W., 1985. The alteration of rock properties due to interactions with drilling fluid components. In: SPE Annual Technical Conference and Exhibition, Las Vegas, Nevada, USA, 22–26 September, Paper SPE 14302.

Skopec, R.A., Collee, P., Tyler, P., 1996. Recent advances in coring technology: new techniques to enhance reservoir evaluation and improve coring economics. In: Society of Core Analysts Annual Meeting, Montpellier, France, 8–10 September, SCA Paper 9604.

Stiles, J., Hutfilz, J.M., 1992. The use of routine and special core analysis in characterising Brent Group reservoirs, UK North Sea. J. Pet. Technol. 44 (6), 704–713.

Talabi, O., Blunt, M.J., 2010. Pore-scale network simulation of NMR response in two-phase flow. J. Pet. Sci. Eng. 72, 1–9.

Thomas, M.M., Clouse, J.A., Longo, J.M., 1993. Adsorption of organic compounds on carbonate minerals: 1. Model compounds and their influence on mineral wettability. Chem. Geol. 109 (1–4), 201–203.

Torsaeter, O., Beldring, B., 1987. The effect of freezing on slightly unconsolidated cores. SPE Form. Eval. 2 (3), 357–360.

Treiber, L.E., Owens, W.W, 1972. A laboratory evaluation of the wettability of fifty oil producing reservoirs. SPE J. 12 (6), 531–540.

Woodhouse, R., 1998. Accurate reservoir water saturations from oil-mud cores: questions and answers from Prudhoe Bay and beyond. Soc. Professional Well Log Anal. J. 39 (3), A1.

Worthington, A.E., Gidman, J., Newman, G.H., 1987. Reservoir petrophysics of poorly consolidated rocks, Part I: wellsite procedures and laboratory methods. In: Society of Core Analysts Symposium, Dallas, Texas, Paper 8704.

Zubizarreta, I., Byrne, M., Jiminez, M.A., Rojas, E., Sorrentino, Y., Velazco, M.A., 2011. Pore pressure evolution and core damage: a computational fluid dynamics approach. In: Society of Core Analysts Annual Meeting, Austin, Texas. 18–21 September. SCA Paper 2011-A073.

RECOMMENDED READING

AAPG Wiki. Conventional coring. http://wiki.aapg.org/Conventional_coring (accessed 31 July 2014).

AAPG Wiki. Core handling. http://wiki.aapg.org/Core_handling (accessed 31 July 2014).

AAPG Wiki. Core alteration and preservation. http://wiki.aapg.org/Core_alteration_and_preservation (accessed 31 July 2014).

American Petroleum Institute (API), 1988. Recommended Practice RP 40. Recommended Practices for Core Analysis, second ed. API Publishing Services, Washington, DC.

Carillo, E.R.P., Pachon, Z., Gomez, J.J., Marin, M.P., 2014. Integrated methodology for laboratory evaluation of shale plays cores. In: SPE Latin America and Caribbean Petroleum Engineering Conference, 21–23 May, Maracaibo, Venezuela, Paper SPE-169301.

Farese, T., Ahmed, H., Mohanna, A., 2013. A new standard in wireline coring: recovering large diameter wireline core through standard drill pipe and custom large bore jar. In: SPE/IADC Drilling Conference, 5–7 March, Amsterdam, The Netherlands, Paper SPE 163507.

Gidman, J., Worthington, A.E., Newman, G.H., 1987. Well-site handling of poorly consolidated core, Inglewood Field, Los Angeles Basin. Society of Exploration Geophysicists. In: 1987 SEG Annual Meeting, 11–15 October, New Orleans, Louisiana, Paper SEG-1987-0278.

Guarisco, P., Meyer, J.R., Thomson, I.J., Robichaux, J., Young, C., Lina, E., 2011. Maximizing core recovery in Lower Tertiary through drilling optimization service and intelligent core bit design. In: SPE/IADC Drilling Conference and Exhibition, 1–3 March, Amsterdam, The Netherlands, Paper SPE 140070.

Lamb, C.F., Ruth, D.W., 1991. Laboratory program design for unconsolidated heavy oil reservoirs. In: Society of Core Analysts International Symposium, San Antonio, Texas, Paper SCA 9104.

Mohamed, S.S.E.D., Kalam, M.Z., Dernaika, M., Brahmakulam, J.V., 2011. Remaining oil saturation study in a giant oil field: integration of reservoir core scale and field scale measurements. In: SPE Reservoir Characterisation and Simulation Conference and Exhibition, 9–11 October, Abu Dhabi, UAE, Paper SPE 148288.

Rathmell, J.J., Gremley, R.B., Tibbitts, G.A., 1993. Field applications of low invasion coring. In: SPE Annual Technical Conference, Houston, 3–5 October, SPE Paper 20413.
Shale, L., Radford, S., Uhelenberg, T., Rylance, J., Kvinnesland, A., Rengel, C., 2014. New sponge liner coring system records step-change improvement in core acquisition and accurate fluid recovery. In: SPE/EAGE European Unconventional Resources Conference and Exhibition, 25–27 February, Vienna, Austria, Paper SPE 167705.
Siddiqui, A., Al-Algam, W., Nuic, I., 2013. Core acquisition from a new sidewall rotary coring tool: overview and experience from Saudi Arabia. In: SPE Saudi Arabia Section Technical Symposium and Exhibition, 19–22 May, Al-Khobar, Saudi Arabia, Paper SPE 168068.
Sincock, J.J., Lombos, L., MacGregor, K.W., 1990. Major advance in sampling and preserving unconsolidated core. In: Proceedings of the First Society of Core Analysts European Symposium, London, 21–23 May.
Skopec, R.A., 1994. Proper coring and wellsite core handling procedures: the first step towards reliable core analysis. SPE J. Pet. Technol. 46 (4), 280.
Unalmiser, S., 1988. Handling unconsolidated cores to preserve wettability and pore structure. In: International Symposium of the Society of Core Analysts, Houston, Texas, Paper SCA 1988-03.
Whitebay, L.E., 1986. Improved coring and core-handling procedures for the unconsolidated sands of the Green Canyon Area, Gulf of Mexico. In: SPE Annual Technical Conference and Exhibition, 5–8 October, New Orleans, Louisiana, Paper SPE 15385.

Chapter 3

Core Laboratory Processing and Screening

Chapter Outline

3.1 Introduction 89
3.2 Core Receipt and Cutting 91
3.3 CT Scanning 91
3.4 Gamma Ray Logging 95
3.5 Removal from Liners 100
3.6 Core Viewing and Sample Selection 102
3.7 Sample Preservation 104
3.7.1 Dry Preservation 104
3.7.2 Wet Preservation 106
3.8 Core Plugging 106
3.8.1 Plug Samples 106
3.8.2 Drill Press and Plugging Fluids 108
3.8.3 Plug Orientation 110
3.8.4 Plug Allocation 112
3.8.5 Plug Trimming 116
3.9 Core Slabbing 117
3.10 Core Resination 118
3.11 Core Photography and Imaging 120
3.11.1 Conventional Imaging 121
3.11.2 360° Core Imaging 124
3.12 Weak or Unconsolidated Core Processing 125
3.12.1 Core Receipt and Cutting 125
3.12.2 CT Scanning 125
3.12.3 Core Gamma Ray Scanning 129
3.12.4 Core Slabbing 129
3.12.5 Core Viewing and Sample Selection 129
3.12.6 Core Plugging and Plug Protection 130
3.12.7 Core Photography 133
3.12.8 Core Preservation 133
References 134
Recommended Reading 134

3.1 INTRODUCTION

When the core is shipped from the wellsite and arrives in the core analysis laboratory, it is processed, screened and sampled. Although there are important exceptions depending on the nature, type, quality and quantity of the core, and the type of analyses required, this is done more or less in the following order.

- Receipt (and cutting)
- Computer tomography (CT) scanning (in liners)
- Gamma ray logging (either with core in or removed from the liners)
- Removal from liners

Developments in Petroleum Science, Vol. 64. http://dx.doi.org/10.1016/B978-0-444-63533-4.00003-2

- Layout and marking
- Sample selection and sample preservation
- Plugging for core analysis testing
- Slabbing
- Photography and digital imaging
- Preservation and presentation.

Figure 3.1 provides a flow chart for a typical core analysis processing and screening operation. The key points of the methods and procedures, the issues

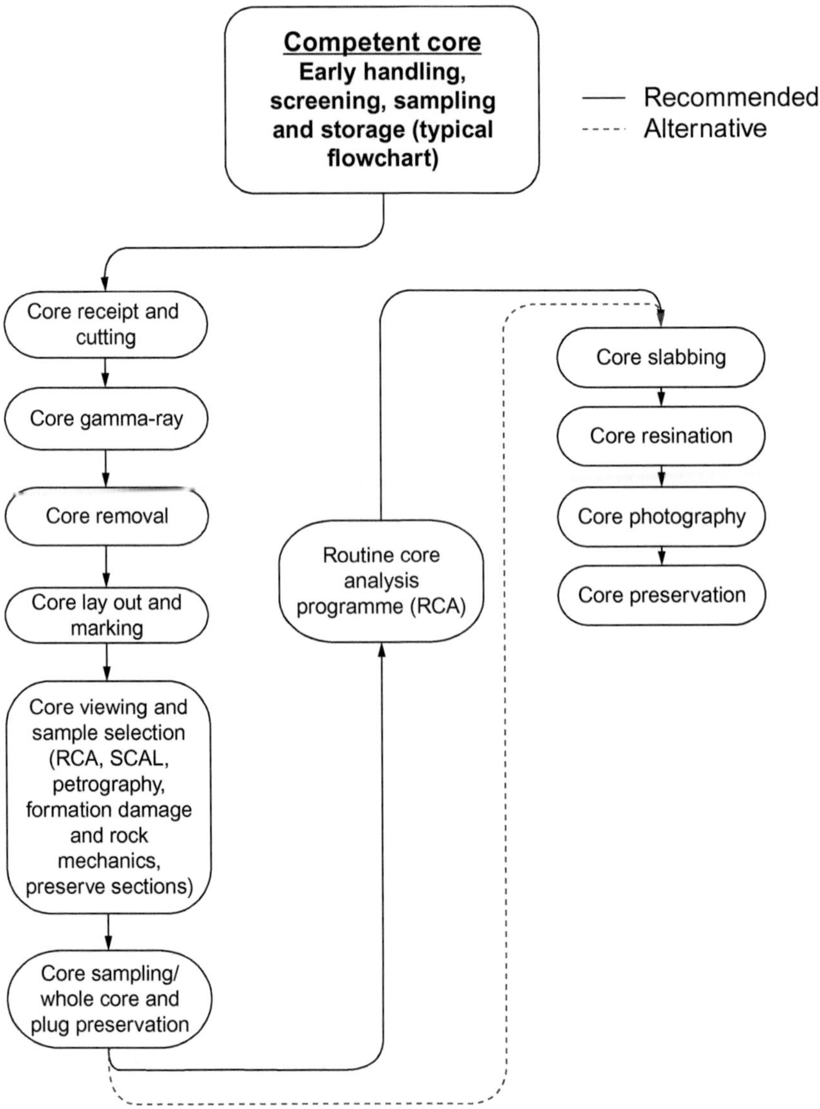

FIGURE 3.1 Flow chart for typical handling, screening, sampling and preservation processes for core.

and challenges involved and recommended best practice for both competent and weak rocks are described below.

3.2 CORE RECEIPT AND CUTTING

The core is received in the laboratory where it is checked to ensure that it corresponds with the coring reports, transportation box log and shipment details from wellsite. It is checked for obvious damage to the liners.

The received core, the correspondence from the client, and the Core Catcher's logs should be cross-referenced and checked. Particular attention should be paid to:

- Well number.
- Core catcher's box log and information written on, or attached to, the individual core liner sections.
- The top and bottom depths of the core (driller's report) and those given on the top and bottom liners of the received core.
- Total number of liner sections as indicated on the shipper's documents and the client's despatch fax/email, and the number received.

If the liners have not been cut into 1 m (3 ft.) sections at wellsite, they can be cut in the laboratory though this is never recommended for weak or friable cores.

3.3 CT SCANNING

Linear X-ray and CT scanning of core samples can be applied to scan the core normally in (or sometimes removed from) its liner, prior to plugging. These are hospital scanners which are adapted to investigate rock rather than bone and soft tissue. Allan Cormack and Godfrey Hounsfield shared the Nobel Prize in Physiology or Medicine in 1979 for their contributions to its development. This technology is also used to scan preserved samples and plugs prior to analysis. Some laboratories have CT scanning facilities in house, but many transport the core sections to use scanners at local hospitals or clinics. A typical CT scanner is shown in Fig. 3.2. The X-ray scanner rotates around the core sample which is laid out on a tube. In this set-up, around 500 X-ray slices will be taken per foot although in conventional applications this may only be one or two slices per foot.

The core is scanned by a highly collimated X-ray source. Detectors on the opposite side of the core measure the intensity of the transmitted beam and measure the average linear attenuation coefficient which reflects the degree to which X-ray intensity is reduced by the material it passes through. The X-ray tube and detectors rotate and translate around the core. For each direction of the X-ray beam, the scanner effectively records the X-ray adsorption by the core. A computer program then computes the brightness

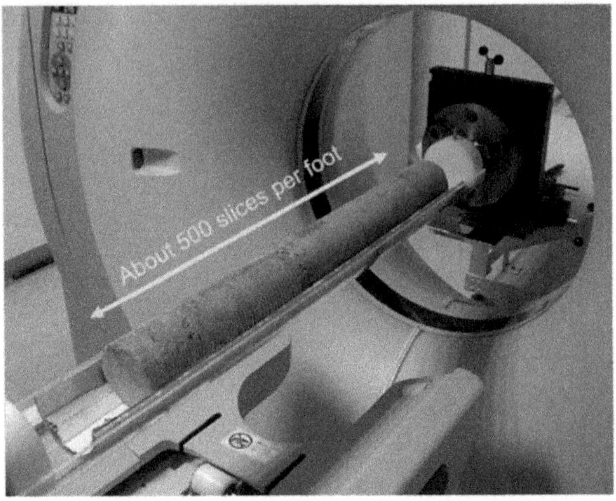

FIGURE 3.2 Example of whole-core CT scanner. *Courtesy of Ingrain.*

of each pixel from all of these separate recordings. Since the CT scan requires many X-ray exposures, the amount of radiation used to make a CT scan is typically greater than that used to make a traditional X-ray. A CT image is composed of pixels, whose brightness corresponds to the adsorption of X-rays in a thin rectangular slab of the cross section, which is called a 'voxel'. In conventional CT scans, dense material adsorbs a lot of X-rays and appears white. Less dense material which does not adsorb X-rays appears darker grey. Consequently, changes in bulk density across and along the core, caused by bedding, mineralogical and lithological changes, invaded high-density mud solids, nodules and fractures, can be identified by CT scan before the core is removed from the liner. This aids core orientation when taking plug samples and rapidly and easily evaluates core damage and core heterogeneity.

The spatial resolution in a CT image is determined principally by the size and number of detector elements, the size of the X-ray focal spot, the source–object–detector distances and the size of the pixel array used to reconstruct the image. Conventional medical CT instruments provide resolutions of the order of 1–2 mm. This is normally sufficient for screening purposes on full diameter core. Higher resolution instruments provide resolution of the order of 100–200 μm for decimetre-scale to centimetre-scale core, and 10–20 μm scale for centimetre-scale to millimetre-scale core plugs. With micro-tomography, which is performed using dedicated beamlines at synchrotron facilities, it is possible to achieve micron-scale resolution within objects at the millimetre to sub-millimetre scale. This technology forms the basis of digital core analysis.

Measurements in the medical sector are taken in Hounsfield units (HU) calibrated universally with air at −1000 HU and water at 0 HU. Typical values for muscle and bone are 40 and 1000 HU, respectively. One of the issues faced when using medical CT scanning facilities is that the units need to be correctly calibrated and adjusted for denser rock rather than soft tissue and bone. A typical medical standard CT setting may obscure detail in the core sections where the Hounsfield units are typically greater than those for bone. In the example shown in Fig. 3.3, the standard setting masks the density variation detail evident in the core.

It is incorrect to assume that CT scanning only shows up variations in rock density. CT scanning demonstrates the ability of a given mineral to attenuate X-rays, and in general, the denser a mineral is, the more it attenuates X-rays. However, several elements can attenuate X-rays to a very high degree, such as iodine and iron, and minerals containing these elements appear very dense on a CT scan. Silt, clay and carbonaceous materials generally show dark contrasts, whereas pure limestones/dolomites show pale contrasts. Grain size differences also tend to affect contrast, with very fine micrite showing dark contrast and sparite showing pale contrast. The elemental composition of the micrite and sparite will have an impact if one or other is contaminated by an iron component.

An example of a CT-scanned core section is shown in Fig. 3.4. In this representation, high-density material, such as heavy minerals and barite from invaded mud solids in the fractures, shows up light grey/white, whereas

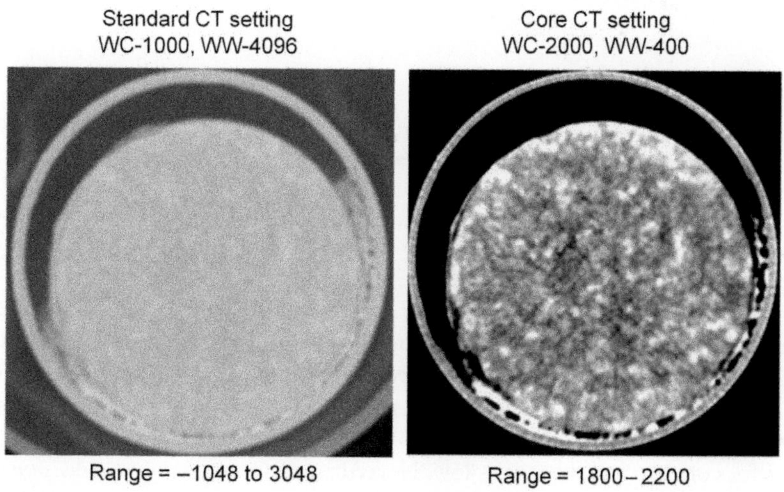

FIGURE 3.3 CT images of core showing the effects of CT settings on image quality. Standard medical settings are WC (Centre HU setting) = 1000 and WW (width HU setting) = 4096. For core WC = 2000 and WW = 400.

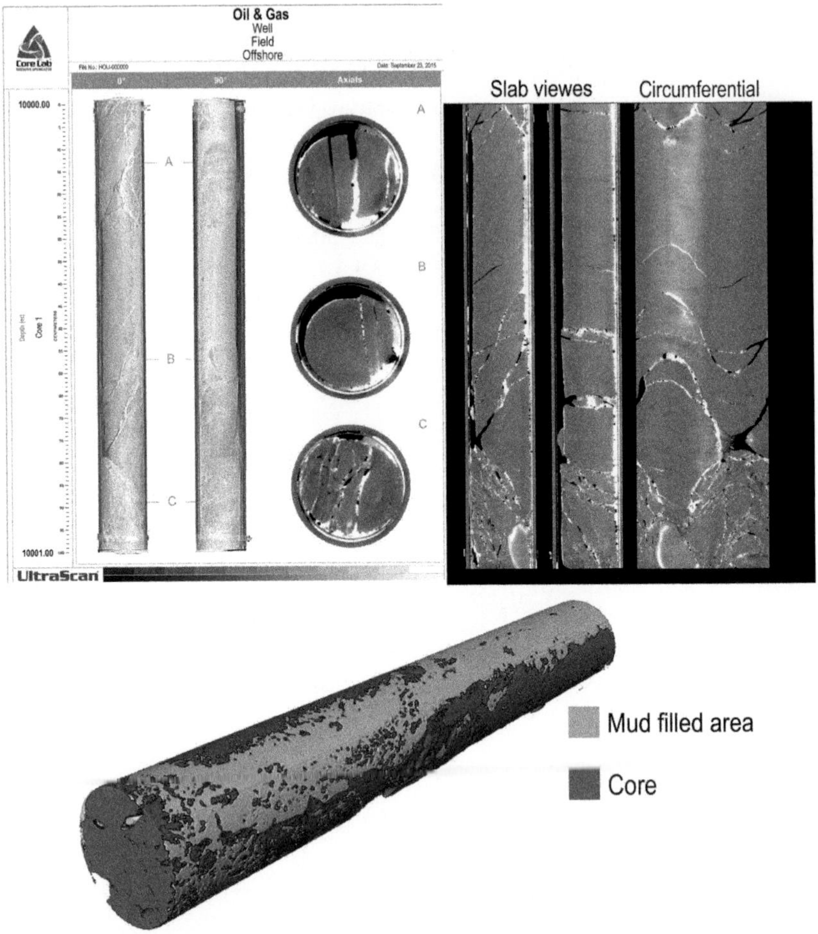

FIGURE 3.4 CT scan example. *Courtesy of Core Laboratories and its Petroleum Services and Core Lab Instruments divisions. All Other Rights Reserved.*

low-density material (lower density minerals and many open fractures) appears dark grey or black.

CT scanning with the core in its liner is used to orient the liner with respect to bedding planes, to evaluate damage and to select representative plug locations avoiding heterogeneous, invaded or fractured intervals. However, as the costs of CT scanning and imaging of long core sections can be onerous, they are not normally run for competent sandstone core as the potential for core textural damage is low. Nevertheless, CT scanning should always be considered for heterogeneous carbonates, heterogeneous/laminated sandstones and naturally fractured formations and is essential for weak or poorly consolidated formations. CT scanning can aid in plugging orientation where

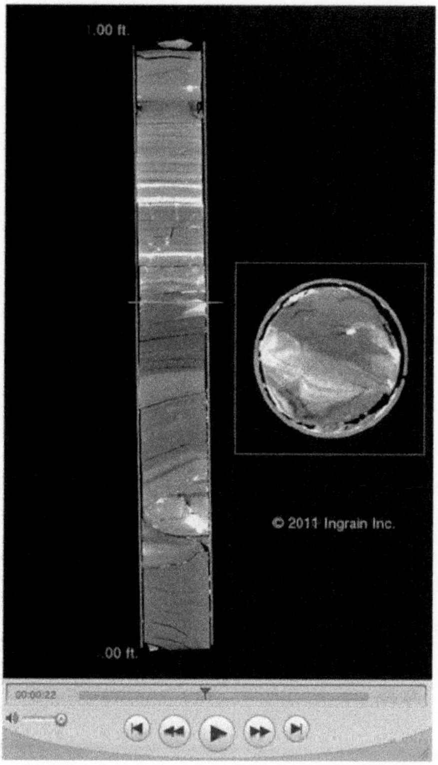

FIGURE 3.5 Continuous CT scan core viewer. *Courtesy of Ingrain.*

visual inspection of the core does not provide much information or the core cannot be removed from the liner.

By combining sequential CT scan slices at closely spaced intervals, it is possible to provide slab images and three-dimensional images. Advanced digital imaging and image capture techniques allow CT images to be run as a continuous moving image along the core (Fig. 3.5).

3.4 GAMMA RAY LOGGING

The principal purpose of gamma ray logging is to match the logging depth to the core depth or driller's depth by comparing log and core natural gamma ray traces. Commonly drillers' depths are based on pipe tallies, while log depths are based on cable or wire measurements corrected for stretch. The two depths differ but not normally by more than a few feet. In addition, spectral core gamma logs provide information on the spectral decomposition of the natural gamma ray energy which can identify the relative concentration of naturally occurring gamma ray emitters such as uranium, thorium and potassium-40.

The spectral gamma ray can be used to differentiate shale in sandstone cores with large amounts of potassium feldspar and mica. Spectral gamma ray data can also be used to assess sediment provenance in sedimentological studies.

As most cores are now recovered in fibreglass and aluminium liners, and as these are relatively 'invisible' to gamma radiation, the core gamma ray log is normally run with the core intact in the liners. Of course, the gamma ray scan can be run, or rerun, when the core has been removed from the liners and has been reassembled.

Typical equipment (Fig. 3.6) comprises:

- A time or speed controlled conveyor belt for moving core.
- A tunnel containing a lead-shielded gamma ray detector is designed such that the core can be moved at a fixed speed underneath the detector.
- A computer data acquisition system that can display and plot total and spectral gamma counts versus depth.

The gamma ray scintillation crystal is normally sodium iodide, often doped with thallium. Other crystals such as caesium iodide can be used. As the naturally emitted gamma radiation from a core sample is much lower than from a downhole formation of similar thickness, the signal/ratio noise ratio of core gamma detectors is high but can be optimised by using a large, lead-shielded, gain-stabilised detector which is typically 4″ diameter by 3″ long.

The core or core liners are placed in depth order on the conveyor belt. The core gamma ray response is normally logged from bottom to top depth, consistent with the sequence used in downhole logs. The logging operation is conducted at constant speed without interruption, in the correct order and spacing, with matching core breaks fitted together. The precision of the measurement varies with the square root of the count rate, so there has to be a compromise between precision and scanning time and costs. Typical scan speeds

FIGURE 3.6 Core gamma ray logging equipment. *With permission from Ofite Website (http://www.ofite.com/).*

are around 0.5 ft./min for total gamma scanning of cores in aluminium liners and for spectral gamma, to 1 ft/min for total gamma scans of exposed core. A duplicate section should be run near the bottom of the core to ensure repeatability.

The signals from the detectors are processed by the acquisition computer, and the gamma ray counts are displayed and/or plotted in downhole log format at a scale of typically 1:200 to allow direct comparison with downhole logs. An example output from a spectral gamma log run is shown in Fig. 3.7. Total gamma is shown as counts per second (cps), and potassium and uranium/thorium contents in per cent and ppm, respectively.

The instruments should be calibrated against pure samples of potassium, uranium and thorium of known activities and concentrations contained in tubes. Total gamma radiation can be calibrated in downhole log API units if known samples of API primary standards are available. A blank calibration tube should be measured to assess the extent to which background gamma rays are interfering with the apparatus. Calibrations are sensitive to the size of the core, the gamma ray energy range and the magnitude of background radiation. In Aberdeen (Scotland) for example, background radiation can be high due to granite bedrock.

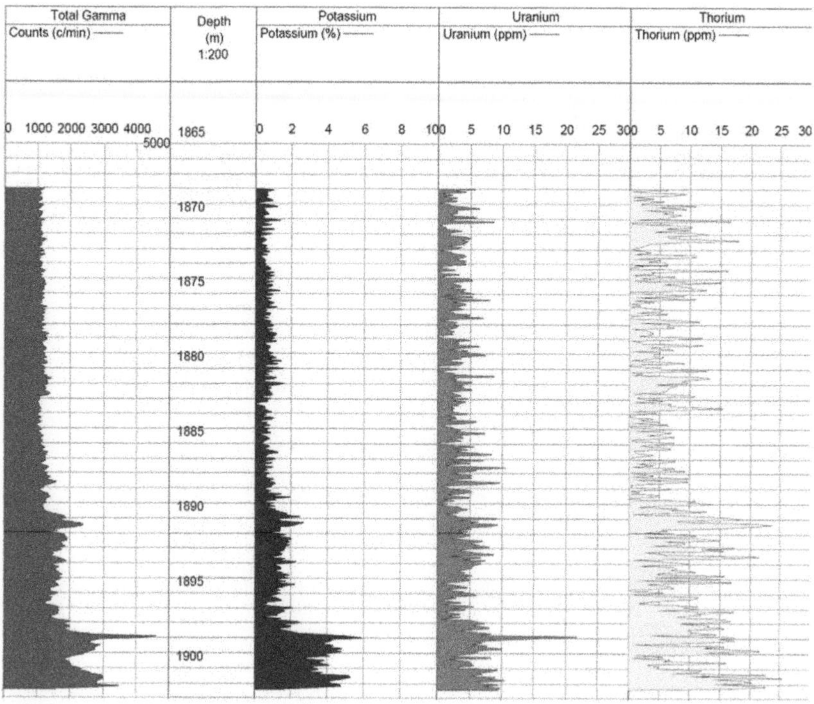

FIGURE 3.7 Example of total and spectral core gamma ray logs run in the lab.

Figure 3.8 plots an example of core gamma ray (core GR in counts per minute) and log gamma ray (LOGGAMMARAY) traces over a reservoir interval. The core gamma ray trace is shifted to best match the log trace in Track 2. Core depths are always shifted to the log depth—never the other way round. For 100% core recovery, the same block or constant depth shift

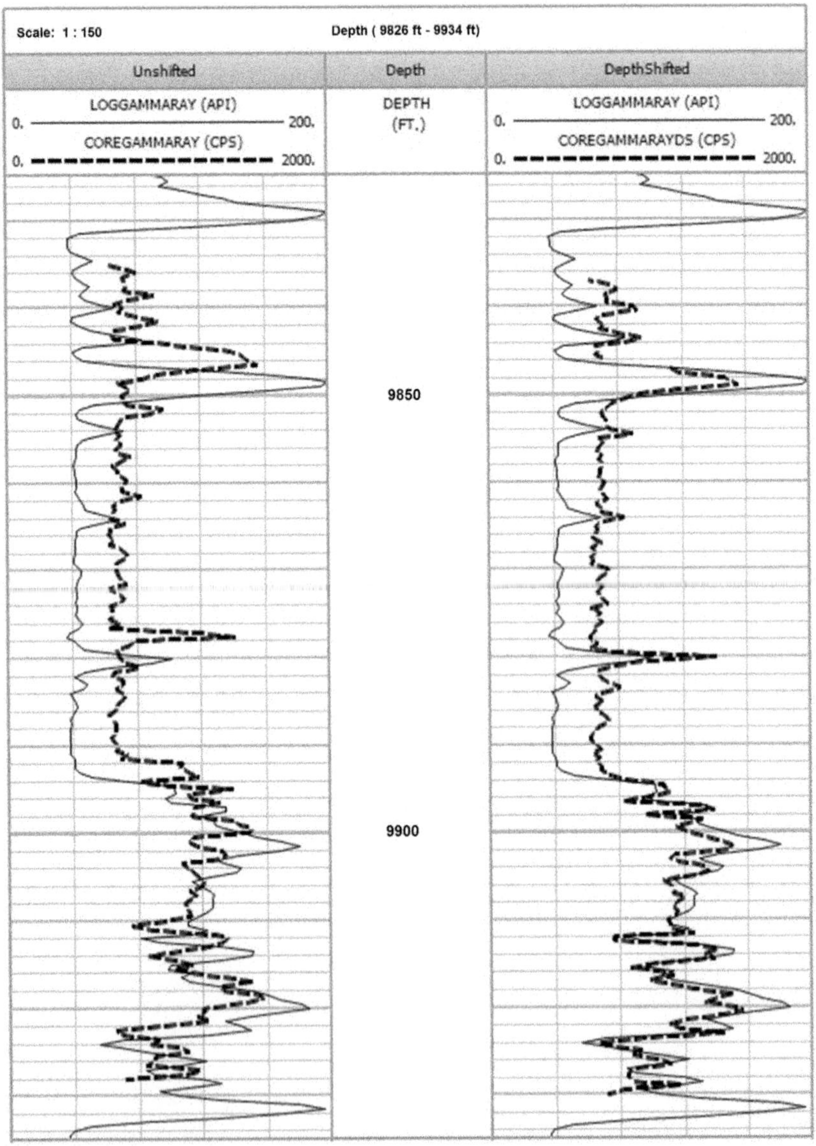

FIGURE 3.8 Core and log gamma ray traces over reservoir interval. Track 1 is unshifted and Track 2 is depth shifted.

should be applied to all data within a given core length unless there is evidence that the core is fragmented, in which case stretch shifts (variable fitting depths) may be attempted.

Two of the issues involved in the application of core gamma ray logging data in core to log depth shifting are that the core gamma ray can have a very different character to the log gamma ray response due to the difference in gamma ray energy detected; and the core gamma ray has a higher vertical resolution. In addition, artefacts occur at the gaps in the core. In clean sands, there may be very little gamma ray activity or curve character so that depth matching becomes impossible or highly uncertain. In this case, comparison of core porosity with an approximate density porosity is often used to improve or determine depth shifts as demonstrated in Fig. 3.9.

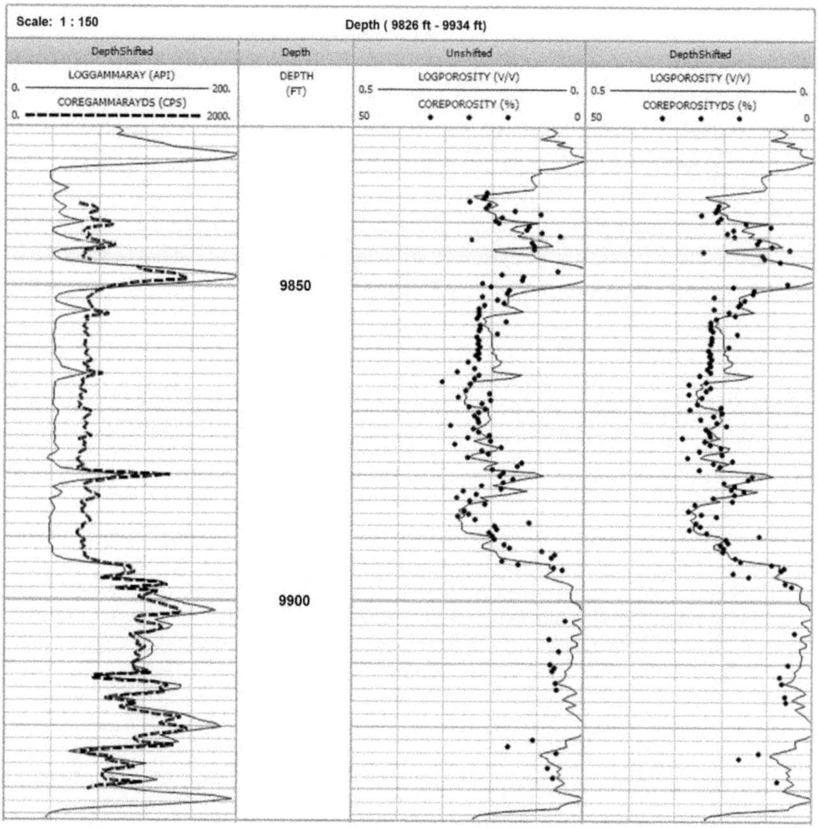

FIGURE 3.9 Core and density porosity versus depth. Original depth in Track 2 and shifted depth in Track 3.

3.5 REMOVAL FROM LINERS

The laboratory is required to carefully piece all the core back together in the correct sequence. Every break in the core should be categorised by the laboratory (e.g. goodness of fit) to aid in depth matching core and log data and to identify loss of core recovery, natural or induced fractures and core damage. The overriding objective is to prevent damage to the core when it is removed from the liners.

Core recovered in clam shell liners (Fig. 3.10) can normally be lifted out without any need to push it out. These liners should not be used for unconsolidated core that requires stabilisation, as opening the liner can create sufficient force to cause the core to slab along its length.

For core recovered in solid liners then, provided the core is consolidated and competent, the core can be removed by gently lifting one end of the liner and allowing the core to slide out under gravity. An example is shown in Fig. 3.11. In some cases, gently tapping the liner with a hammer can help move the core but excessive force must be avoided. If this fails, it should be pumped out of the barrel using a piston that ideally prevents fluids from directly contacting and contaminating the core. The pumping fluid should be inert and not react or contaminate the core and should be suitable for the analysis proposed on the core. For example, formation water should not be used if Dean-Stark (DS) samples are required, and freshwater should never be used if the rock contains sensitive clays. Finally, the core catcher end of the liner should be cut off with an aluminium saw (Fig. 3.12) and the core material from this end collected and laid out in depth sequence.

The core is then laid out on a viewing table (Fig. 3.13). The lab should check the amount of core recovered in the coring report to ensure that no core material has been lost during transportation. Any reported damage or alteration of the core during wellsite handling and transportation should also be

FIGURE 3.10 "Laser Cut" liner opened to expose core. *Image Courtesy of Baker Hughes.*

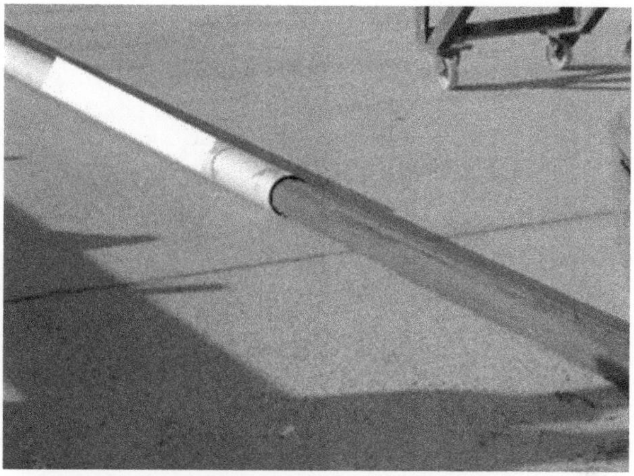

FIGURE 3.11 Lifting end of core liner to enable core to slide out.

FIGURE 3.12 Cutting core catcher end of liner section.

noted. The top and bottom depths of the core (driller's/coring report) must match those given on the top and bottom liners of the received core. Any mismatch should be noted and reported.

Speed is essential in removing, laying out, labelling and preserving the core to minimise any alteration due to atmospheric exposure. However, this should not compromise the correct sequencing of the core.

When laying out the core, the following guidelines are applicable:

- The bottom of the core should be placed on the viewing table first with each succeeding piece being placed closer to the top.
- The proper sequence and orientation of the core must be respected. Comparison of wellsite and lab core gamma ray traces can help ensure that individual core segments are not positioned out of place or upside down.

FIGURE 3.13 Core laid out on viewing table.

The individual core pieces should be fitted together to give continuous runs. The laboratory technicians should use their judgement to assess and record the continuity of adjacent sections: e.g. definite continuity, possible and impossible. Any broken core should be put in thick plastic bags and placed in its correct sequence.

- The total recovery can then be measured. If any errors are noted, the lab should contact the client so that some agreed corrective action can be taken.
- Any excess fluid or excessive mud cake should be wiped off the surface. The core should never be washed with water.

The core is then marked up with tramlines, depth marks and orientation arrows in a similar fashion to core liner sections at wellsite. Marking schemes varies from company to company, but should be consistent within a company. The convention is that the core is striped from top to bottom with parallel and continuous red and black lines ensuring that red is on the right when looking up the core. The core depth is marked on the core liner at regular depth intervals (every 1 ft. for example) with a continuous band around the circumference of the core and marked with depth and sometimes 'way up' arrows. Figure 3.14 provides an example.

3.6 CORE VIEWING AND SAMPLE SELECTION

Immediately after the core has been laid out and marked, it is wrapped with plastic film to avoid fluid evaporation and further core deterioration. Speed is of the essence and the core viewing area should be protected from temperature extremes, rain and direct sunlight. If plugs for DS analysis have not been taken at wellsite, they should be taken now, and plugs and whole core sections must be preserved as soon as possible.

The objectives of core viewing are to assess the quality of the core material and to select plugging sites and preserved sample locations. At this stage, different

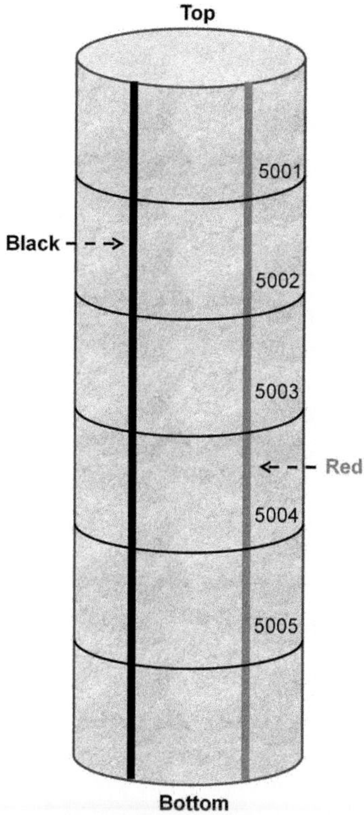

FIGURE 3.14 Core orientation markings.

parties generally will contribute to sample selection. This normally includes geologists, petrophysicists, sedimentologists, reservoir engineers, lab core analysis technicians and, in some cases, rock mechanics and formation damage specialists.

It is recommended that a comprehensive core analysis programme is designed and agreed internally among geologists, reservoir engineers and petrophysicists and with the service lab, prior to the core been taken. Testing and final sampling requirements will depend on the objectives of the core acquisition programme and the final conditions/characteristics of the recovered core. Consequently, the pre-designed core analysis programme can be modified if necessary and agreement reached to proceed to sample selection. Final sample selection is based on a number of factors including:

- Petrophysical sampling requirements.
- Well logs from cored interval.
- Core gamma ray and CT scans.
- Core lithology and condition.

Where a formation contains widely varying lithology and heterogeneous porosity types, such as conglomerates, cherts, fractures, laminae, vuggy carbonates and inter-laminated shales and sands, the proper selection of representative samples requires greater care.

3.7 SAMPLE PRESERVATION

Whole core sections can be temporarily preserved by sealing the ends of the core liner sections. This is not recommended as a longer term preservation method but will allow the core to be processed quickly and without special equipment.

The aim of longer term sample preservation is to prevent fluid ingress/egress from the core and, for wettability maintenance, to prevent exposure to air to minimise oxidation reactions and evaporation. Both whole core samples and plugs can be preserved. Selected whole core sections for preservation (approximately 10–12″ long) are carefully cut from the core and the sample depths measured.

Sample preservation methods are either 'wet' or 'dry'. Dry methods encapsulate the core in a material intended to provide a vapour and water barrier and should prevent exposure to air during handling and storage. Wet methods involve immersing the core in liquid in a deoxygenated container.

3.7.1 Dry Preservation

The two most common methods of dry core preservation are sealing in barrier foil laminates and hot wax or strippable plastics.

3.7.1.1 Barrier Foil Laminates

The most common barrier foil laminate is ProtecCore™, similar to that shown in Fig. 3.15. This consists of six layers of different materials. The first and innermost layer next to the core is Barex™ plastic film, followed by biaxial nylon, aluminium foil, two layers of low-density polyethylene and a layer of polyester. The aluminium foil is the principal liquid and oxygen barrier. The outer layers provide strength and rigidity.

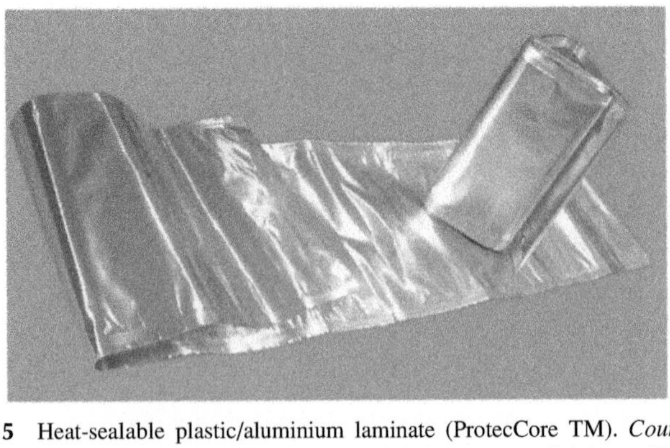

FIGURE 3.15 Heat-sealable plastic/aluminium laminate (ProtecCore TM). *Courtesy of Core Laboratories and its Petroleum Services and Core Lab Instruments divisions. All Other Rights Reserved.*

The core sections are wrapped in two to four layers of Barex film to prevent tearing or rupturing of the laminated material. The wrapped core is then inserted into the tubular laminate and the ends of the laminate heat sealed. Heat sealing is crucial as any discontinuity in the seal will negate the barrier properties of the material. To reduce any air space around the core, a small hole can be left in a corner and a vacuum applied. The vacuum holes are then heat sealed. The method provides a good vapour barrier which is claimed to be superior to the hot wax or strippable plastic method (Auman, 1989; Hunt and Cobb, 1988).

3.7.1.2 Hot Wax or Strippable Plastics

This technique, often referred to as the 'seal peal' method, involves wrapping the core in plastic film and then aluminium foil, and this core package is then dipped or coated in hot wax or molten (strippable) plastic to protect the aluminium (Fig. 3.16). The plastic film prevents fluids in the core from coming into contact with, and potentially corroding, the aluminium foil. Saran Wrap® is often used but can degrade with certain oils, solvents and oil phase drilling muds. Barex ™ is relatively inert to hydrocarbons but is less flexible and more difficult to wrap around the core. The aluminium foil provides the principal vapour barrier. The core package is multiple dipped in melted wax or

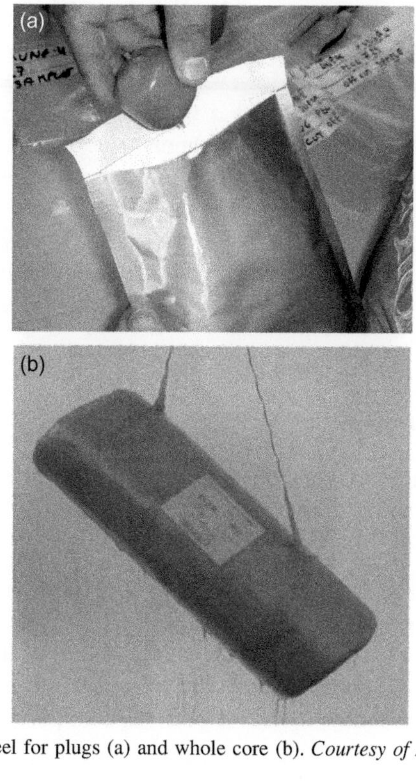

FIGURE 3.16 Seal peel for plugs (a) and whole core (b). *Courtesy of ALS Oil and Gas.*

plastic to provide a 4-mm to 7-mm thick coating. Each coat is allowed to harden before applying the next coat. API RP 40 (1988) recommends tying a wire around the core to hold it during dipping, but this can rip the aluminium foil. String is an alternative, but the string should be cut off and the end sealed in wax to prevent it acting like a wick and drawing fluids from the core. Wax and plastic are permeable to oxygen and water vapour, but Core Seal™ and several common polymers are not.

The seal peal method is also used to preserve core plugs pending SCAL or other tests.

Tests have shown that with the right sealant, water loss can be reduced to practically zero with this method (Auman, 1989), and that very long-term preservation of core properties is possible.

3.7.2 Wet Preservation

The objective of wet preservation is to immerse the core sample in an appropriate liquid to prevent oxidation, evaporation and drying. A bactericide, such as formaldehyde, is added to prevent bacterial growth during storage. The jars are sealed and any air removed by evacuation or sparging/flushing with inert gas (helium or nitrogen).

The most common containers are glass or PVC, but PVC allows diffusion of oxygen and water (Basan et al., 1988). Glass anaerobic jars (Cornwall, 1990) have a sealable lid with a valve system that allows air evacuation or inert gas sparging (Fig. 3.17). However, glass can easily break on handling and transportation, and storage and maintenance costs (regular sparging is required) are high compared to dry methods.

Wet preservation methods tend to be used for short-term preservation of wettability rather than longer term storage, so the choice of immersion liquid is crucial. Synthetic formation brine, depolarised kerosene or mineral oil are commonly used as these have little influence on wettability, but imbibition of brine in water-wet rock and oil in oil-wet rock will result in alteration of the fluid saturations.

3.8 CORE PLUGGING

3.8.1 Plug Samples

Most routine and special core analysis tests are performed on plug samples cut from the full diameter core. The plug samples are right cylinders typically 1″ or 1.5″ in diameter and 1–3″ long. The rule of thumb is that the plugs for RCA or SCAL should have a length equivalent to one to two times the plug diameter. Some typical plugs cut from competent rock are shown in Fig. 3.18. In friable or 'difficult' core material plugs are not such perfect cylinders, may be wrapped in polytetrafluoroethylene (PTFE) tape to prevent collapse and are

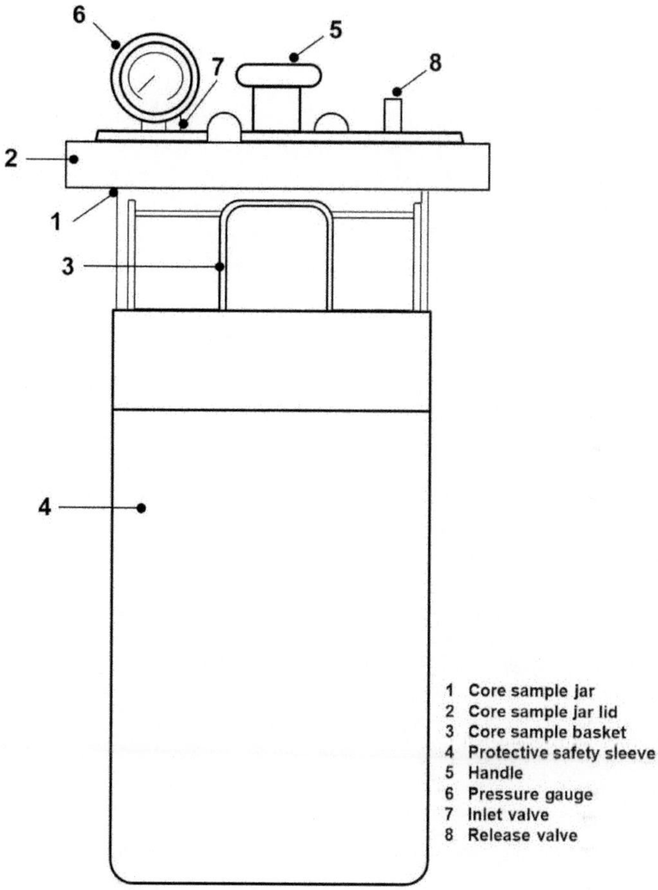

FIGURE 3.17 Example anaerobic jar for core preservation. *From Cornwall (1990).*

of variable length. Figure 3.19 provides some typical examples from one core. Plug shape and surface topology can have a significant influence on porosity and capillary pressure measurements, and plug length completely dominates two phase flow behaviour in relative permeability tests.

The use of larger plugs is preferred for both RCA and SCAL since the errors involved in porosity and saturation measurements on small plugs can be large and can have a significant impact on the data. For example, the pore volume of a 'standard' 1.5″ plug is about four times greater than the pore volume of a typical 1″ plug for the same porosity. As cutting larger plugs can lead to excessive removal of material from the core, some companies compromise on 1″ plugs for RCA and 1.5″ plugs for SCAL. Ultimately, the plugging requirements depend on the final rock type/quality as recovered in the core and the core diameter. In lithologies where plug measurements are not considered to be

FIGURE 3.18 Typical core plugs (competent material).

FIGURE 3.19 Typical core plugs (friable material).

representative—such as fractured or vuggy carbonates, conglomeratic sands and where the scale of the formation heterogeneity is greater than the plug sample—the analyses may be performed on the whole, full diameter core.

3.8.2 Drill Press and Plugging Fluids

Core plugs are taken using a diamond-tipped, hollow cylindrical, rotary core bit mounted on a drill press, similar to that shown in Fig. 3.20. A variety of bit lubricants are used, depending upon the fluid content, core drilling mud and lithology/mineralogy. Typical plugging fluids are:

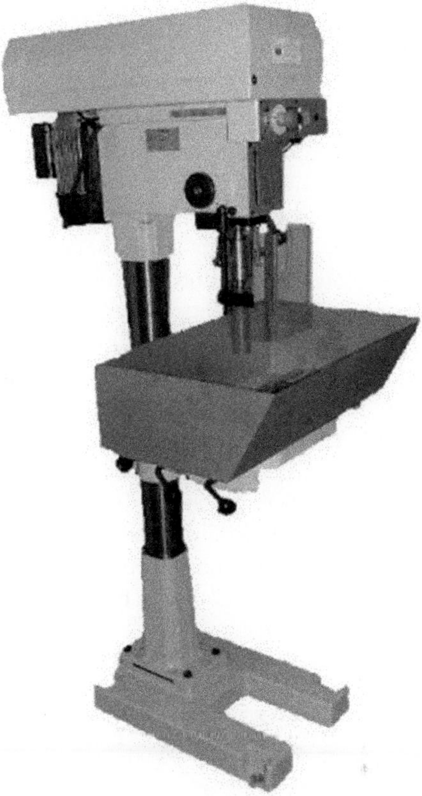

FIGURE 3.20 Core plug drilling press. *Courtesy of Vinci Technologies.*

- Brine or synthetic formation water, which is made up to the same composition as the formation water. This should not be used for taking plug samples for Rw analysis or DS analysis. Plugs should never be taken using tap water as this can result in sample alteration if the core contains reactive clays or soluble minerals.
- Depolarised kerosene, base oil, or mineral oil (e.g. Blandol) are used where brine–rock incompatibility is expected, where the formation brine composition is unknown or where cores are cut with oil-based mud and scheduled for DS or fresh-state analyses.
- Compressed air is used where the invasion of plugging fluid may change the plug saturations.
- Liquid nitrogen or chilled air is often used where the core is too weak to survive conventional plugging due to erosion by liquids.

If there is any doubt about potential problems arising from water formation incompatibility, then depolarised kerosene or mineral oil should be used.

3.8.3 Plug Orientation

For most core analysis, formation damage or geomechanics test applications, plugs are taken either in the 'horizontal' or in the 'vertical' direction with respect to bedding planes on the core. The principal exception is plugs taken for DS analysis which should be taken from the centre of the core and parallel to the core long axis irrespective of the bedding orientation. This minimises the chances of taking the plug within the halo of mud filtrate invasion. The plugs should be cut as close as practicable to the location of the horizontal plugs to avoid excessive core damage and to allow for comparative data sets.

Figure 3.21 shows typical plug orientations that are cut from core—in this case from a vertical well with relatively steeply dipping beds.

Horizontal or bedding plane parallel plugs are cut parallel to the apparent bedding plane features. The objective is to sample the maximum permeability in the formation which is normally parallel to bedding. If the bedding forms an angle with the core long axis, then plugs cut perpendicular to the long axis of the core will be oblique to bedding and will sample neither the minimum nor the maximum permeability. For horizontal plugs, the samples should be ideally taken from the centre of the core to minimise the influence of mud solids and filtrate invasion (Fig. 3.22), which is particularly important for

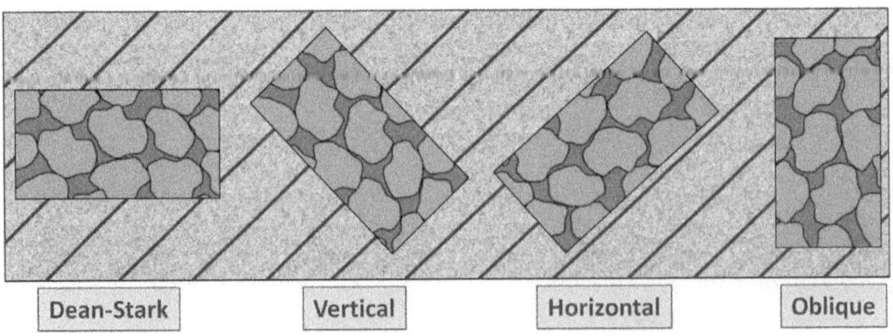

FIGURE 3.21 Example plug orientations with respect to core bedding.

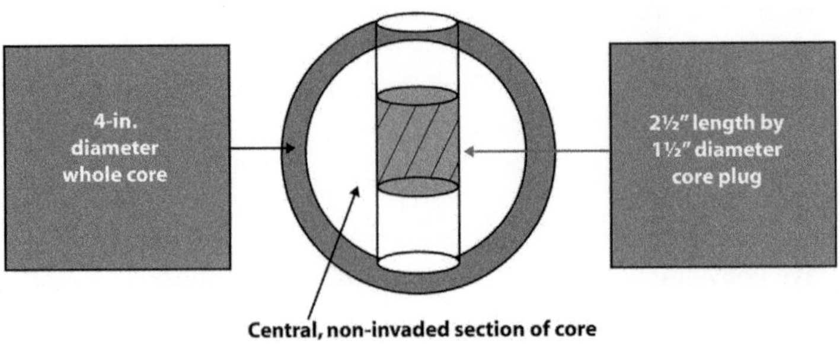

FIGURE 3.22 Schematic of core plug taken from the centre of the core.

SCAL and formation damage test plugs. The depth of penetration of the bit in a thinly bedded or laminated core section should be the entire core width (drill right through the core) to prevent damage when releasing the plug.

Horizontal plugs can be taken parallel to bedding in two directions: one where the long axis of the plug is parallel to bedding plane ('x'-direction) and the other taken orthogonally, at 90° to the x-direction ('y'-direction). For consistency, all horizontal samples should be taken in the same direction (either x or y) unless plugs are specifically required for permeability or resistivity horizontal anisotropy analysis.

Vertical or bedding plane normal plugs are typically used for rock mechanics tests and routine core analysis tests. They are taken perpendicular to the maximum dip (bedding plane normal or perpendicular), not parallel to the long axis of the core. These sample the minimum permeability direction. These plugs should be cut as close as practicable to the location of the horizontal plugs to allow for comparative data sets.

Plugs can be taken either before or after core slabbing. There are advantages and disadvantages of both methods, as listed in Table 3.1.

TABLE 3.1 Advantages and Disadvantages of Plugging Before or After Slabbing

Plugging Before Slabbing	Plugging After Slabbing
Bedding can be difficult to distinguish if the core surface is obscured by rock flour or mud cake CT scans can be used to orient the core prior to removal from the liners	Bedding can be easily distinguished. Assumes that core is slabbed to expose the maximum bedding dip
Correct plug orientation can rely heavily on the expertise and experience of the lab technicians	Slabbing the core to expose the maximum dip will enable horizontal plugs to be taken easily—at 90° (normal) to the slabbed surface
No or restricted fluid contamination or alteration before plugging	Water is often used for core slabbing which could have an adverse affect on saturations and sensitive formation clays. Crude oil may dry out pending slabbing and may make cleaning much more difficult
No restriction on horizontal plug length other than core diameter and depth of mud filtrate/solids invasion	Horizontal plug lengths are more restricted. The maximum possible plug length on 2/3 slabbed 4″ diameter core will be less than 2.6″ compared to over 3″ if plug is cut from unslabbed core

3.8.4 Plug Allocation

Figure 3.23 provides a general schematic guide of the plugging requirements for core analysis, rock mechanics and formation damage testing.

3.8.4.1 Routine Core Analysis Plugs (R)

1.5″ or 1″ diameter horizontal plugs (bedding plane parallel) are cut for routine poroperm analysis at approximately 1 per ft. (1 per 25 cm) in reservoir quality intervals. For the test results from these plugs to be statistically rigorous, a constant spacing should be observed. In reality, technicians often shift the plug locations to avoid friable, difficult to plug or difficult to test intervals, thereby introducing bias in the data sets. Small-scale profile permeability measurements on a slabbed surface can be used to optimise the horizontal plug spacings: for example, fewer plugs would be required in a thick homogeneous interval compared to a thin-bedded, heterogeneous interval. The spacing for vertical (bedding plane perpendicular plugs) depends on the lithology—closer spacings are adopted for heterogeneous or thin-bedded lithologies.

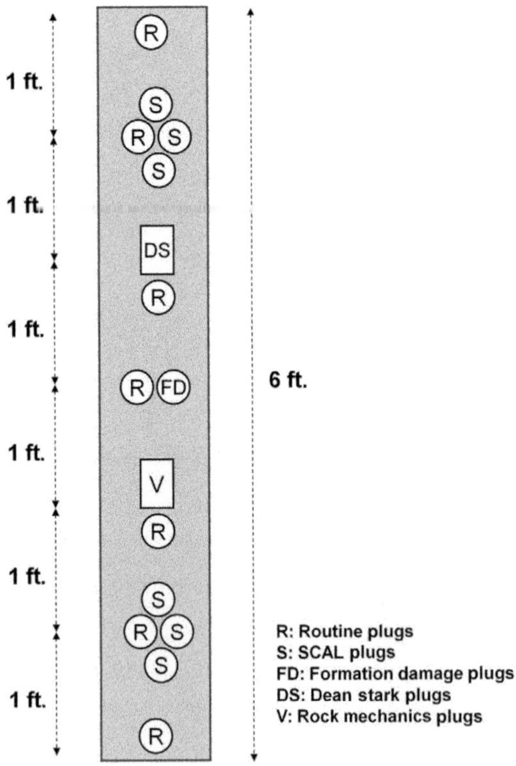

FIGURE 3.23 Schematic guide for plug sampling.

3.8.4.2 Special Core Analysis (S) and Formation Damage (FD) Test Plugs

A separate set of horizontal plugs (bedding plane parallel) are taken for most SCAL and formation damage tests. The larger the plug sample, the lesser the potential uncertainty in pore volume and saturation estimates. For this reason, 1.5″ diameter plugs have become the standard for SCAL tests.

Though there are the occasional exceptions, most SCAL and formation damage tests are carried out on horizontal (bedding plane parallel) plugs.

SCAL plugs taken from preserved core sections offer the advantage that the core should have been protected from evaporation, drying out and exposure to oxygen, but has the principal disadvantage that the selection of preserved sections is often biased and concentrated on the best quality intervals. In addition, the porosity and permeability of preserved sections are difficult to estimate precisely other than from routine core analysis plugs taken immediately above and below the section. SCAL plugs can also be taken (and preserved) adjacent and parallel to RCA plug locations on the unslabbed or slabbed core, as indicated in Fig. 3.24. This has the advantage that the porosity and permeability of the proposed SCAL plugs are reasonably well estimated from the RCA plugs but core properties—especially wettability—may have continued to alter, or the core may have deteriorated further, by the time the plugs can be taken. In some cases, SCAL tests can be performed on RCA plugs provided that the plug preparation and subsequent measurement procedures do not compromise the subsequent SCAL data such as core cleaning and drying damage, and potential damage caused by stress cycling.

Plug sampling frequency depends on the sampling frequency of preserved sections, the core condition, core geometry and formation lithology, but typically ranges from 1 per 3 ft. (1 m) to 1 per 6 ft. (2 m) for horizontal SCAL plugs. These plugs are taken as close as possible to a routine plug site and are often preserved pending analysis.

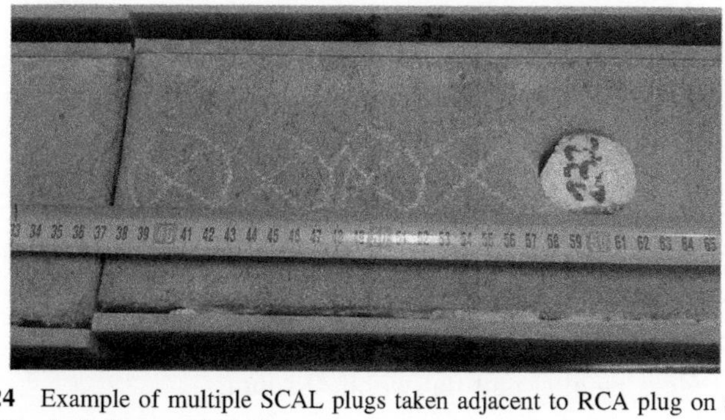

FIGURE 3.24 Example of multiple SCAL plugs taken adjacent to RCA plug on slabbed core. *Courtesy of Woodside Energy Limited.*

3.8.4.3 DS Plugs

Vertical (parallel to core axis) plugs for DS are spaced at about 1 per 3 ft. (1 m) and are taken from the centre of the core to minimise the chances of taking a mud filtrate-invaded sample. The 1.5″ diameter plugs should be taken as close as possible to a routine horizontal plug. Smaller 1″ diameter plugs could be acceptable if the plug size is constrained, but saturation uncertainties involved with samples with smaller pore volumes should be borne in mind.

Permeability measurements on DS plugs are misleading unless the plugs have been taken parallel (in a horizontal well) or perpendicular (vertical well) to bedding.

3.8.4.4 Rock Mechanics Test Plugs (V)

Unconfined compressive strength (UCS), single-stage (SST) and multi-stage triaxial, pore volume compressibility and acoustic travel time tests are normally performed on vertical plugs cut perpendicular to the bedding planes. The exception is plugs cut from lithologies such as shales which can exhibit bedding plane strength anisotropy. In this case, a series of plugs are often cut from 0° to 90° to bedding. Thick-wall cylinder (TWC) tests are normally performed on horizontal (bedding plane parallel) plugs. Where UCS and TWC plugs are required at the same (or similar) depths, a horizontal plug and a vertical plug must be cut at closely adjacent locations. Where a set of three or four plugs are required for an SST test set, the plugs must be cut as parallel plug 'triplets' or 'quadruplets' immediately adjacent to one another. 1.5″ diameter samples are preferred, especially in coarse-grained material where the grain diameter/plug diameter ratio must respect ISRM (1983) recommendations (Brown, 1981). Due to core slabbing constraints and the available core geometry and condition, 1″ plugs may have to be cut.

TWC tests are performed on a 1″ or 1.5″ diameter hollow cylinders, with a standard outside to inside diameter (OD:ID) ratio of 3:1. The example shown in Fig. 3.25 is for a 1.5″ diameter plug sample. The internal hole must be centred with a maximum diameter variation of 0.2 mm. To ensure a concentric hole geometry, it is advisable to drill or core the centre hole first (0.33″ or 0.5″ diameter) and thereafter overcore the centre hole with a 1″ or 1.5″ bit to produce the TWC sample (Fig. 3.26). If the hole is eccentric, then the sample will fail prematurely and the result is a significant underestimation of the rock strength.

Plugs cut for geomechanics tests require much stricter control on plug geometry and end face allowance than either RCA or SCAL plugs. The length to diameter ratio must be 2:1 or greater and the ends must be lathed or cut to produce a sample that meets ISRM (1983) tolerances:

- End surfaces parallel and flat within 0.1 mm.
- End surfaces perpendicular to sample axis within 0.5°.

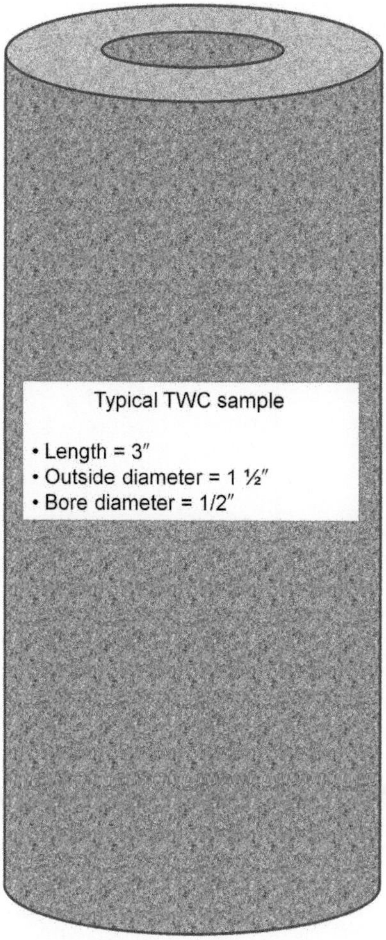

FIGURE 3.25 Standard 1.5″ diameter TWC test sample.

Failure to ensure flat end faces can lead to tensile splitting in UCS tests (Fig. 3.27) as the load is not evenly distributed over the cross-sectional area. When drilling the plugs, it is important to apply a constant weight on bit to prevent plug irregularities, as evidenced by the plug sample shown in Fig. 3.28.

For these reasons, plugs for rock mechanics tests are better taken by experienced rock mechanics lab technicians rather than by the routine core analysis laboratory. If this is not possible due to local regulations, or the logistics or cost of transporting or exporting core material to the rock mechanics lab, then the local lab can cut plug 'sticks' with an L:D ratio greater than 2. The sticks can then be sent to the rock mechanics test lab for final sample end face preparation. An example plug stick is shown in Fig. 3.29.

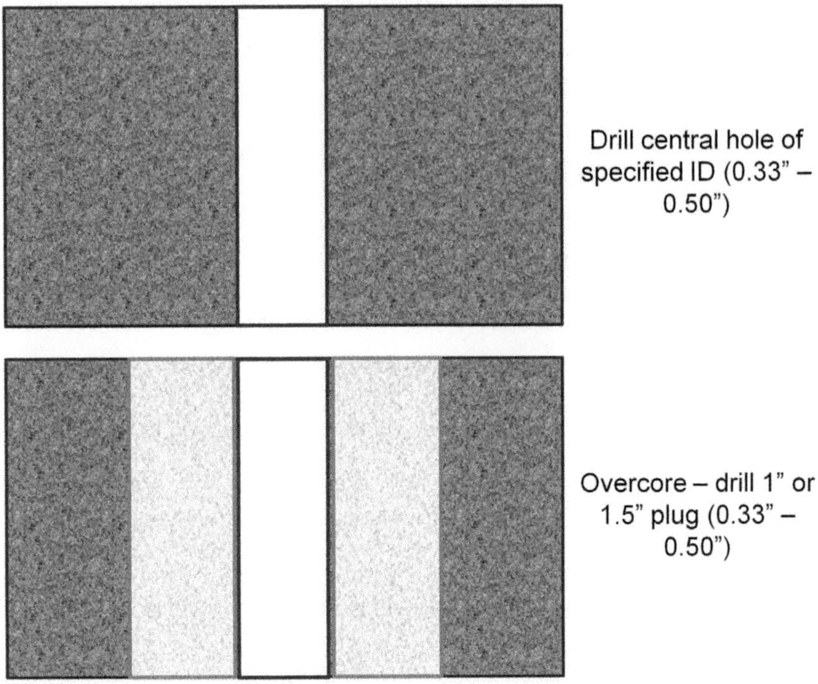

FIGURE 3.26 Diagram of over-coring procedure for TWC test samples.

3.8.5 Plug Trimming

For RCA, SCAL and formation damage test plugs, the samples are trimmed using a saw capable of performing a cut at 90° to the plug long axis. The saw blade cutters should ensure a smooth cut with minimal heat transfer. The trimming saw fluids should be the same as those used for plugging.

Once trimmed, the plug end faces are lightly brushed under the trimming liquid to remove the moveable mud and rock flour. For formation damage tests, particular care is required as rock flour produced on the end faces can be misinterpreted as formation fines. In this case, acetate peel techniques should be considered. A sheet of plastic film is placed on the end face and softened by a solvent which then evaporates. The plastic conforms in minute detail to the end face and is peeled off to remove the embedded rock flour.

Offcuts or end trims from, and the remains of, the plug samples (post-testing) should be securely stored, labelled and retained for potential mineralogical/petrographic examination.

At least three measurements of final plug length and diameter are made using callipers. These dimensions can be used to give an initial estimate of plug bulk volume.

FIGURE 3.27 Example of tensile splitting of UCS test plug.

3.9 CORE SLABBING

Core slabbing exposes the sedimentological, lithological and bedding features of the core and allows preparation of a clean surface for core photography and for further geological examination. Slabbing is essential for adequate reservoir description, as it allows detailed observation of sedimentary structures poorly displayed in the rough/mud-invaded outer surface of the core.

Cores are slabbed parallel to the maximum apparent dip of any bedding or dominant fracture trend (with the aid of the CT scans if available), to show maximum apparent dip from top left to bottom right of the cut face for photography and to enable plugs cut parallel to bedding. Once the orientation has been selected, it is essential to maintain it throughout the job. Where no actual bedding is seen, orientation should follow the direction of maximum dip extrapolated from adjacent sections.

A slabbing saw for competent core slabbing is shown in Fig. 3.30. Specialist band saws are used to slab weak or poorly consolidated core. A trained slabbing operator is essential, as mis-orientation of the slab can be confusing and lead to plug mis-orientation, and saw marks may obscure 'real' core features.

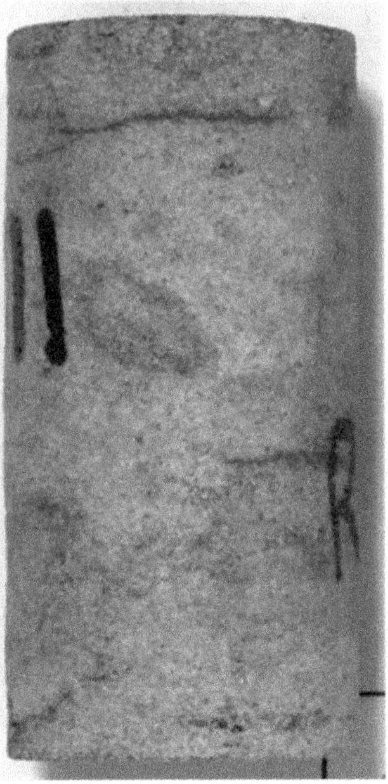

FIGURE 3.28 UCS plug with irregular sides due to variable weight on bit when plugging.

The slab cuts depend on the client and the local regulatory authorities. Normally, once the routine core analyses are completed the lab, with authorisation from the client, will perform a 2/3 to 1/3 cut (or similar) longitudinally along the length of the core or core liner, as shown in Fig. 3.31.

3.10 CORE RESINATION

Core slab resination is carried out to protect one section of the core from further deterioration and to present a clean core face for inspection/photography and archiving. This biscuit or sliver slab is around 1″ (25 mm) thick. An example is shown in Fig. 3.32. To create the resination slab, the 2/3 slabbed core is placed flat side down in plastic trays, which are labelled with the company, well name, core number, box numbers, tops and depths and measuring scale. Sample locations can also be added. A transparent epoxy resin–hardener mixture is subsequently poured around the core. After the resin has hardened, the top section of the core is removed by slabbing the core parallel to the resin tray. This leaves a thin biscuit slab or sliver. The core biscuit provides a much easier way to

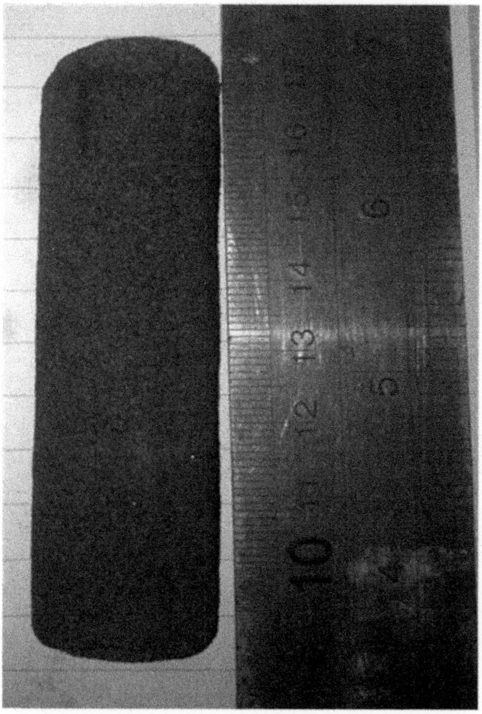

FIGURE 3.29 Example of plug stick without end trimming.

FIGURE 3.30 Core slabbing saws—conventional (left) and band saw for weak core (right). *Courtesy of Core Laboratories and its Petroleum Services and Core Lab Instruments divisions. All Other Rights Reserved.*

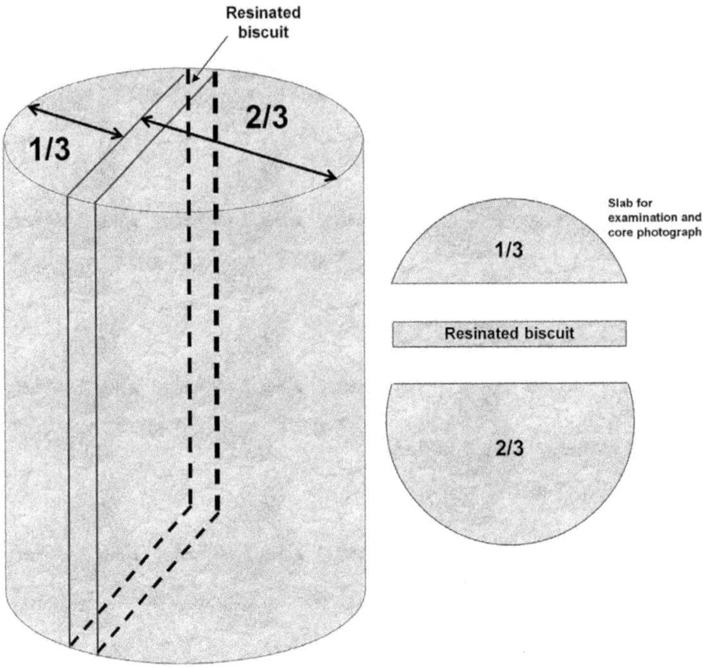

FIGURE 3.31 Illustration of 1/3 to 2/3 core slab cuts.

access the information displayed within the core, and it is particularly useful for core viewing, photography and detailed geological examination. An example of a recommended resinated slab presentation is shown in Fig. 3.33, where petrophysical data from plug locations on the core section are also shown. The thin slabs are easily stored for archiving (Fig. 3.34).

3.11 CORE PHOTOGRAPHY AND IMAGING

A recorded photographic or digital image of the core is essential. This record will provide information that can be used remotely if viewing of the core is not possible, and may include visual images of the surface features of the core and visual representations of the internal structures of the core such as CT tomographs. Any record should include information on core depth along with a scale or legend indicating the intensity range of the recorded image. Core images provide a visual record of the core that can be used as the basis for routine core analysis sampling, facies analysis, reservoir quality studies, net to gross estimation, core analysis data interpretation, supplementary core analysis sample selection, and comparison with downhole log interpretations.

Nowadays, virtually all core images are captured digitally rather than the large format photographic plates that have been used historically. Not only

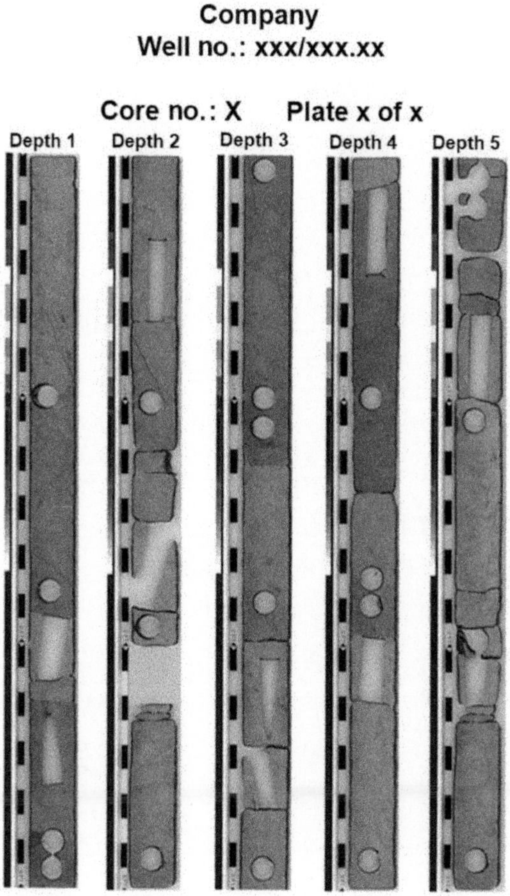

FIGURE 3.32 Example of a resinated biscuit slab.

does this make it easier to store, manipulate and interpret the image data, high-resolution digital imaging means that it is possible to view sedimentary or other features in very fine detail.

3.11.1 Conventional Imaging

Core is normally photographed under both natural and ultraviolet (UV) light. Natural (or white) light images show lithology and sedimentary structures and allow specific features recorded in the core description to be examined such as styolites and fractures.

UV exposure causes most oils to fluoresce. UV light excites susceptible electrons within the atomic structure of the crude oil molecule. These excited electrons temporarily jump up to a higher orbital within the atomic structure.

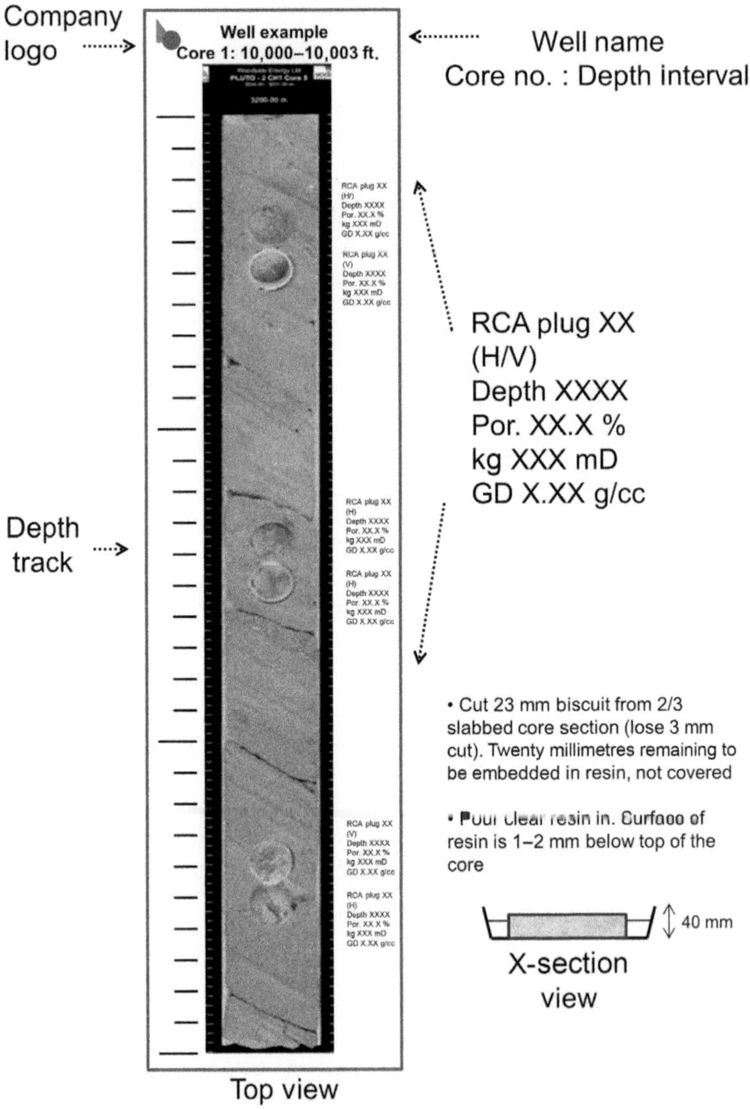

FIGURE 3.33 Recommended resinated slab presentation example. *Courtesy of Woodside Energy Limited.*

When those electrons fall back down to their original orbital, a small amount of energy is released in the form of light, known as fluorescence. Fluorescence colours range from orange-brown for heavy oils, to bright yellow for high-gravity (low density) oils. Condensates can appear as a very light white to blue-white colour. Non-hydrocarbon-bearing zones will generally appear as dark purple, but some common minerals in reservoir rocks can also show

FIGURE 3.34 Storage of resinated biscuit slab trays. *Courtesy of Woodside Energy Limited.*

fluorescence. Calcite has been known to fluoresce red, blue, white, pink, green and orange.

UV imaging should be carried out as soon as possible following slabbing as light hydrocarbons can often dissipate over relatively short times. Oil-based mud filtrate can also produce a dull yellow fluorescence, so a sample of the mud filtrate should be included in the image to differentiate between invaded filtrate and native crudes. Slabbed core surfaces must be free from rock flour and uniformly dry, although some features within the rock are clearer when the rock is wetted with a glycerol solution. UV imaging should be carried out on a dry core surface. Care should be taken to ensure that there is no surface contamination from cleaning rags which may cause erroneous fluorescence.

White light and UV digital images from a cored interval are shown in Fig. 3.35. The original RCA (1″ diameter) and SCAL (1.5″ diameter) plug

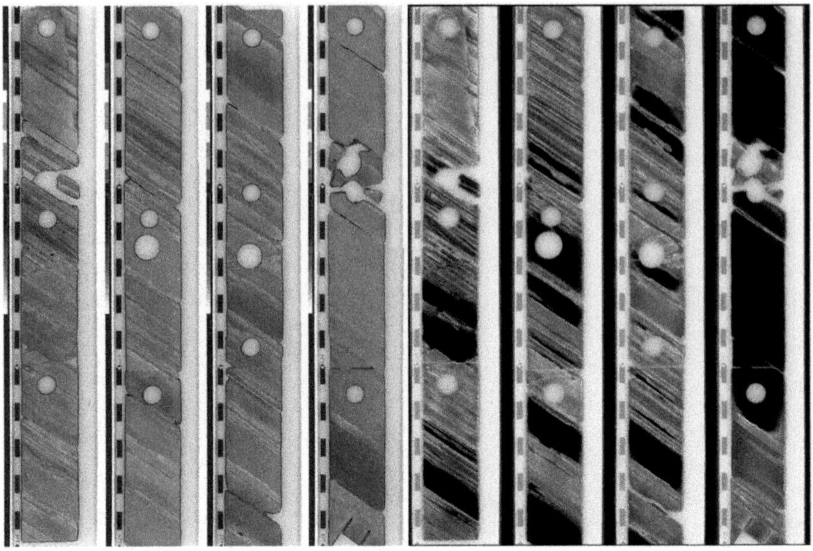

FIGURE 3.35 Digital image of core under white light (left) and UV light (right).

holes are clearly evident. Note that the RCA plug spacing is not strictly regular. The yellow fluorescence is indicative of oil and in this case was used to improve estimates of net to gross ratio (net represents the vertical length of the fluorescent intervals) in this thin-bedded sand where the bedding plane thickness was below the level of log resolution.

High-resolution core images can be imported into petrophysical analysis packages (e.g. Interactive Petrophysics™) and oilfield data management systems (e.g. ODM™) to link core analysis, logs and reservoir geology information.

3.11.2 360° Core Imaging

A recent development in core imaging is the 360° core digital scanning system (Fig. 3.36) which enables high-resolution core images to be taken around the entire full core circumference of the core, or on slabbed surfaces. Slabbed core can be temporarily 'glued' back together to enable circumferential imaging.

Scans are normally taken over a core box length across the interval of interest and then they are added and fitted together, creating larger aligned core intervals. Figure 3.37 provides an example of a scan which has been 'flattened' out to simulate the kind of image from borehole image logs. The scanned core image features such as fractures and bedding dip can be matched and oriented to similar distinctive features on the borehole images.

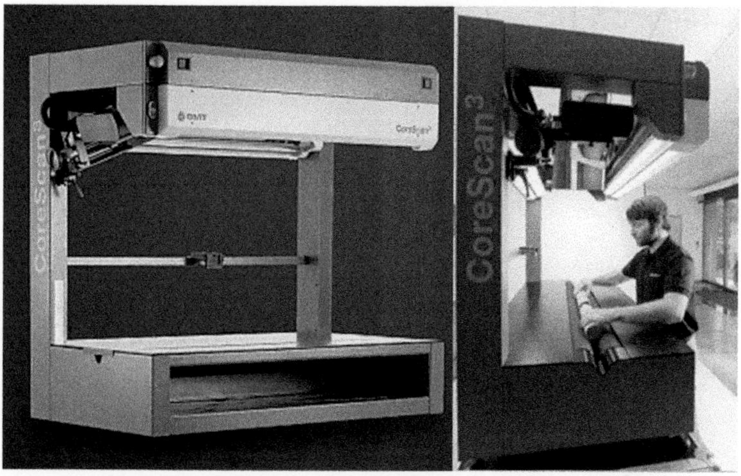

FIGURE 3.36 360° digital core scanner. *http://www.dmt.de*.

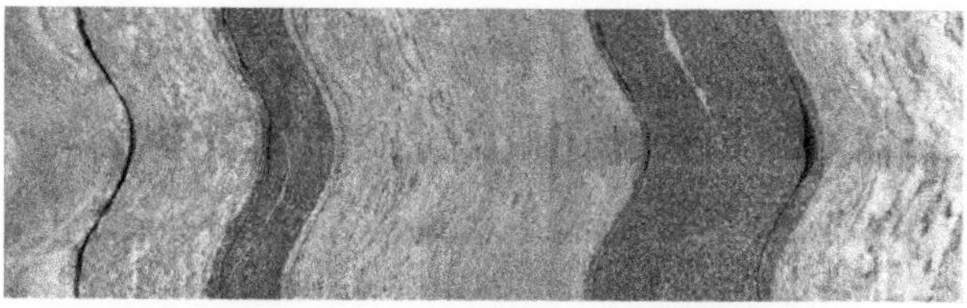

FIGURE 3.37 Example of 360° digital core image. *http://www.dmt.de*.

3.12 WEAK OR UNCONSOLIDATED CORE PROCESSING

In general, the processes and procedures for both competent and weak core are broadly similar but special precautions and procedures are required for processing weak, friable and unconsolidated core to prevent core damage. Figure 3.38 provides a flow chart for a typical core analysis process for a weak core.

3.12.1 Core Receipt and Cutting

If the cores are received frozen or chilled, then they should remain so during the whole handling, cutting and storage processes. If the core thaws out, it may then be difficult to further process the core (cutting plugs for example) without causing damage to the core.

3.12.2 CT Scanning

CT scanning of received core intact in its liner is recommended best practice for all weak and friable core. Any core damage can be assessed and specialist

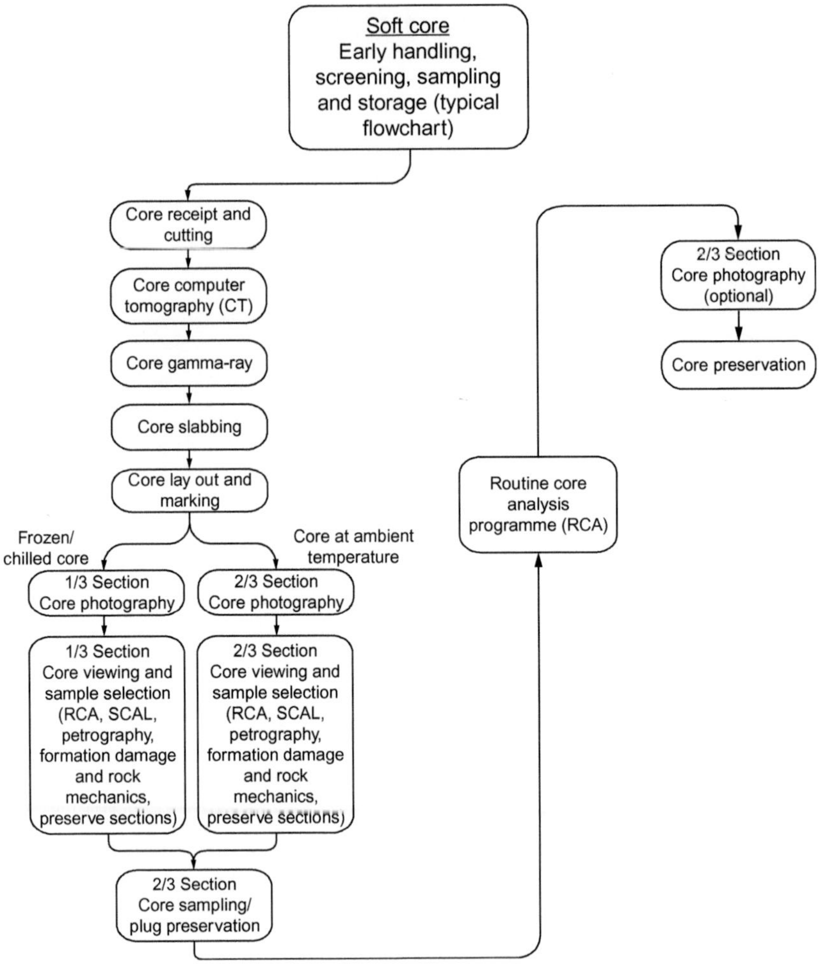

FIGURE 3.38 Example flow chart for processing, screening and sampling of weak, friable or unconsolidated core.

precautions or remedial actions can be planned without having to remove the core from the liners, risking further damage.

The value of CT scanning is demonstrated by the following examples. Figure 3.39 is a CT scan of a weak sandstone core where the classic radially emanating fractures have been propagated by rapid pore pressure dissipation caused by pulling out of the hole too fast. These fractures have been extended by subsequent foam injection during the stabilisation process. Figure 3.40 shows false colour image CT scans of two sections of weak core where two concentric 'rings' of core are clearly evident. The 'outside' ring,

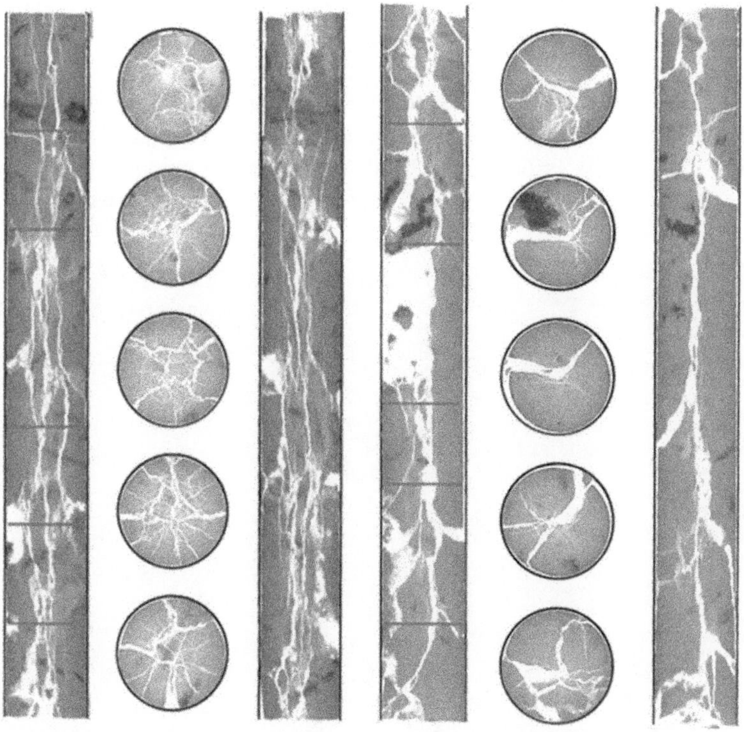

FIGURE 3.39 CT scan images of weak sandstone core showing radial fractures caused by rapid pore pressure dissipation.

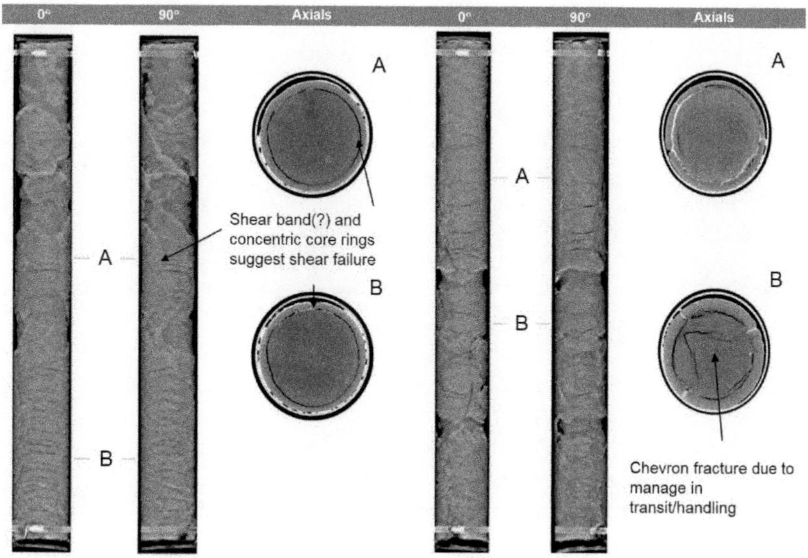

FIGURE 3.40 CT scan images of weak sandstone core exhibiting (left) possible spiralling shear failure bands and concentric core rings, and chevron fracture (right).

closer to the liner, could be confused with a layer of caked mud, but the presence of spiralling bands in the longitudinal image on the right may be more indicative of shear failure caused by compressive loading, possibly due to excessive torque during drilling or on surface core handling. The image on the left shows a clear 'chevron' fracture which can be a result of dropping the core section during handling or transportation. The CT scans illustrated in Fig. 3.41 show shear failure of weak sand towards the bottom

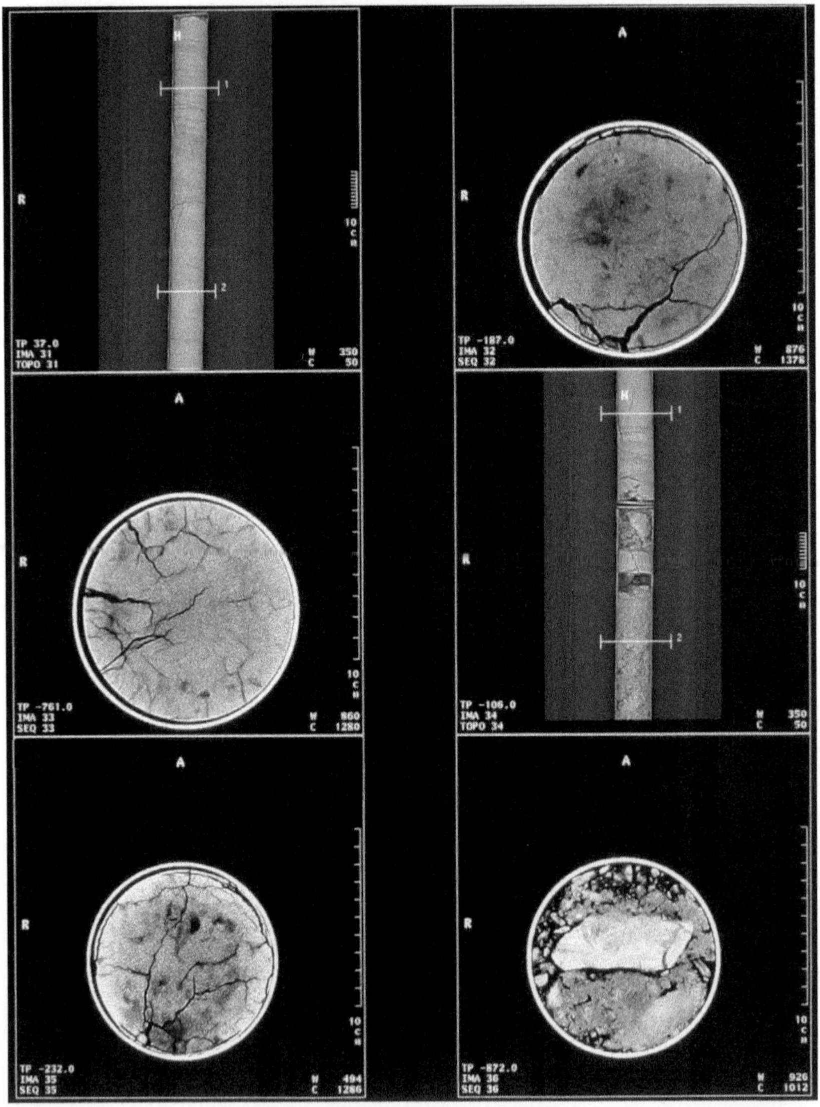

FIGURE 3.41 CT scan images of compressive shear failure from the weight of overlying core.

of the core due to the compressive force from the weight of the overlying core in the barrel. In each case, it was not possible to take test plugs from the damaged core.

At least one longitudinal image and three to five (or more) cross-section images per 3 ft. liner section are normally required. If the core has been frozen, core liners should be kept frozen in dry ice during transportation from the lab to the CT scanning facility and vice versa, as well as while waiting to be processed. Exposure to ambient temperature should be kept to a minimum during the CT scanning process, allowing only for the time the core has to spend in the scanning equipment.

3.12.3 Core Gamma Ray Scanning

The core gamma ray scan is carried out with the core intact in its liners. The scanning speed should be adjusted to compensate for the presence of aluminium liners, in which case the maximum scan speed will be 0.5 ft./min. Again, exposure of frozen core to ambient temperature should be minimised.

3.12.4 Core Slabbing

For unconsolidated materials, the standard procedure is to slab the core before any plugging takes place. This procedure makes the plugging operation easier and avoids excessive damage by having to extract the core from the liner. In some cases, plugs can be taken through the liners themselves provided the liners have been oriented with respect to bedding by CT scanning.

The core should not be removed from its liners, especially if it has been stabilised with foam, gypsum or resin, as this is a recipe for destruction of the core. Opening stabilised clam shell liners run the risk of uncontrolled slabbing of the core along its length if the stabilisation material sticks to the core.

The soft core in the liners is slabbed using a special band saw (Fig. 3.30). If the core is frozen, then the 2/3 slab is kept frozen and later used for sampling/plugging. The 1/3 slab is wrapped in plastic film, to avoid condensation of ambient humidity onto the core surface, and is allowed to thaw out. Once the core has reached ambient temperature, it is used for digital imaging (white and UV light), core viewing/sample selection and surface examination (e.g. sedimentology and stratigraphy).

3.12.5 Core Viewing and Sample Selection

As it is not feasible to mark the surface of a soft core, the lines are marked up on the liner. For frozen core, as soon as possible after the core has been slabbed, the 1/3 section is allowed to thaw and is photographed. Plugs are taken from the 2/3 slab which must be maintained frozen, if required.

For soft cores, it is generally not recommended to take preserved whole sections as the potential for damage is great. Freezing and storing whole core sections for preservation can also be expensive. Taking and preserving a series of additional plugs for SCAL can help minimise core disturbance. Plugs are marked with the sample number and depth and then wrapped in plastic film and aluminium foil. The sample must be sealed securely so that fluids cannot escape from, nor air enters the sample. They are sealed inside a vacuum foil bag and marked again on the outside. Vacuum should be applied gently to avoid damaging the plugs. These plugs should be maintained frozen if taken from frozen core.

3.12.6 Core Plugging and Plug Protection

3.12.6.1 Plugging

Conventional plugging in weak, unconsolidated or friable formations is difficult. Using oil or water as core bit coolants will lead to erosion of material from the plug. Chilled air is often used if the rock is friable but not particularly weak.

Liquid nitrogen (LN2) drilling is used where the core is too weak to survive conventional plugging or if the core has been frozen. Low vibration drill presses (Fig. 3.42) are used with the addition of a special container to keep the core frozen with dry ice during this operation. If LN2 is used to plug unfrozen

FIGURE 3.42 Liquid nitrogen plugging.

core, contact with liquid nitrogen might cause damage to the plug and grain rearrangement, as a result of ice formation and thermal shock.

Plugs can be cut in soft sediments using plunge cutters (Moss, 2014). This involves pushing a special core bit with a 'chisel' edge into the core and then extracting the sample by carefully twisting the bit. Provided the cutting edge of the plunge-cutting bit is designed to deflect the most of the force of penetration to the sides, away from the downward direction, the force on the part of the core which will form the plug is minimised. Plunge cutting is more successful in very poorly consolidated cores than in better consolidated (though still weak) material which can fracture more easily. It avoids thermal shock damage from liquid nitrogen but can cause better consolidated core to fracture if excessive force is used to take the plug.

More recent developments include a specialised weak sand plugging system that uses a non-rotating inner barrel/liner tube to contain and support the core plug during drilling. This prevents damage to the core during bit rotation and is similar to field coring systems.

3.12.6.2 Plug Protection/Mounting

The plug samples from unconsolidated cores must be protected against further disturbance during core preparation and testing.

To minimise handling and sample degradation, the plugs are best confined in individual coreholders throughout the core cleaning, drying and measurement procedures. Frozen plugs are firstly loaded into individual coreholders and then confined under nominal stress, typically 200 psi or less. The imposition of confining stress acts to reinforce and strengthen the core plug when it is allowed to thaw. The samples are then cleaned *in situ* in the coreholders by flushing (flowing) with solvents and then are dried by lenient air flow. Pore volume and gas permeability are then measured with helium and nitrogen, respectively. The plugs are only removed for final grain volume measurements.

Very few commercial laboratories have access to, or can routinely provide, such capability as it requires substantial investment in customised coreholder/cleaning systems. If this is not available, the plugs are protected within a sleeve or jacket and transferred from instrument to instrument through the cleaning, drying and measurement phases like a conventional, competent plug. The most common protective solution is to mount the plugs in thin metal (nickel or tin) or PTFE jackets/sleeves.

Heat-shrinkable PTFE tubing has the advantages that, once heated, it shrinks around the core plug providing some nominal effective stress reinforcement to the plugs and is chemically resistant to most common solvents. Typical operating temperature ranges from −70 to +260 °C. Using a heat gun to shrink PTFE can have a tendency to cause the material to overheat in small areas, while under-heating other areas, causing uneven shrinking of the material. The most reliable method for shrinking PTFE Teflon is the use of a controlled temperature oven, but this may be impractical if the core contains temperature-sensitive minerals like clays.

Metal sleeves cannot be used if the plugs are scheduled for electrical properties testing, and can easily wrinkle (Fig. 3.43) which may allow by-pass during permeability measurements (Rosen et al., 2007) and cause overestimation of pore volume in porosity measurements.

The unconsolidated plug mounting procedure is illustrated in Fig. 3.44. The plug sample is wrapped in PTFE tape and then inserted into a sheath of PTFE heatshrink tubing or metal sleeve. Fine (100 mesh) and then coarse (16 mesh) steel gauzes or screens are placed at each end of the plug inside

FIGURE 3.43 PTFE sleeve (left) and metal sleeve (right). *From Rosen et al. (2007).*

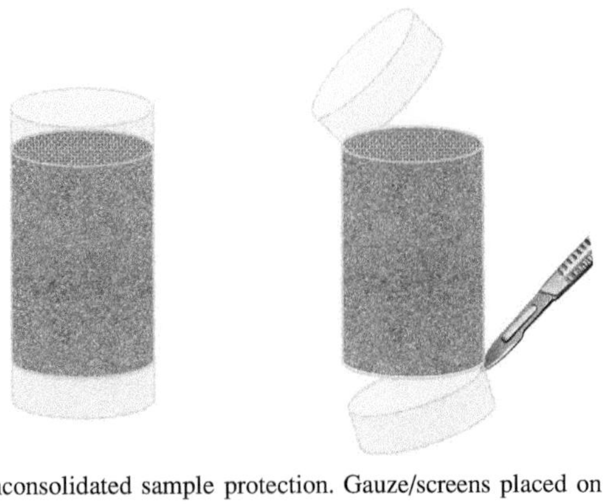

FIGURE 3.44 Unconsolidated sample protection. Gauze/screens placed on plug ends and into PTFE heatshrink tube (left) and trim excess PTFE (right).

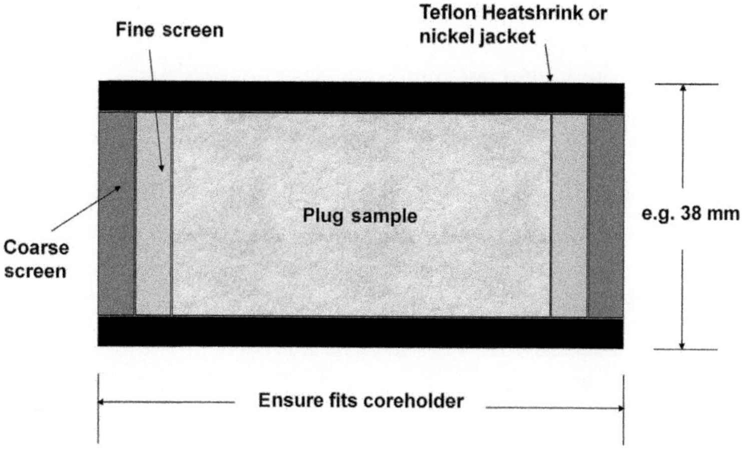

FIGURE 3.45 Example unconsolidated plug/jacket assembly for standard porosity and permeability measurements.

the PTFE heatshrink or metal sleeve. These protect the end faces and prevent grain loss on subsequent handling. The assembly is heat shrunk and excess material is then trimmed off and weighed. The assembly is loaded into a coreholder (Fig. 3.45) and confined at a nominal pressure. This improves conformance between the metal and screens and the plug surface.

It is essential that the weights, densities, both solid and pore volumes of PTFE tape, heatshrink/metal sleeve and the screens and the weight of the excess trimmed material are recorded, as these must be accounted for in subsequent porosity and grain density measurements.

3.12.7 Core Photography

For frozen core, photographs are initially acquired on the 1/3 slab section once thawed and marked. It is recommended that UV imaging is carried out as soon as possible following slabbing, as light hydrocarbons can often dissipate over relatively short times.

For unfrozen core, core photographs are initially acquired on the 2/3 slab. As an option, core photographs will also be acquired on the 2/3 slab once the routine core analysis programme is completed and the client authorises the section to be thawed out on completion of the programme. The objective of this second photography set is to have a visual proof of the sampling locations for future reference during data analysis and interpretation.

3.12.8 Core Preservation

In the case of unconsolidated material, core will inevitably suffer a rapid deterioration process with time, especially if the core has been frozen and it is left to thaw. Preservation will only help to minimise the process but will not be able to avoid it.

Two options for long-term storage are generally available:

- film, foil and wax;
- barrier laminates.

The slabs and plugs are then dispatched to the operator's core store and/or to partners and regulatory authorities where they are archived and stored.

REFERENCES

American Petroleum Institute (API), 1988. Recommended Practice RP 40. Recommended Practices for Core Analysis, second ed. API Publishing Services, Washington, DC.

Auman, J.B., 1989. A laboratory evaluation of core preservation materials. SPE Form. Eval. 3 (4), 691–695.

Basan, P., Hook, J.R., Hughes, K., 1988. Measuring porosity, saturation, and permeability from cores. Schlumberger Tech. Rev. 36 (4), 22–36.

Brown, E.T. (Ed.), 1981. Rock Characterization, Testing and Monitoring: ISRM Suggested Methods. Pergamon Press, Oxford.

Cornwall, C.K., 1990. Core preservation—an alternative approach. In: Proceedings of the First Society of Core Analysts European Symposium, London, 21–23 May.

Hunt, P.K., Cobb, S.L., 1988. Core preservation with a laminated, heat-sealed package. SPE Form. Eval. 3 (4), 691–695.

International Society for Rock Mechanics, 1983. Suggested method for determining the strength of rock materials in triaxial compression: revised version. Int. J. Rock Mech. Min. Sci. Geomech. Abstr. 20, 283–290.

Moss, A., 2014. Stabilisation of unconsolidated reservoir rocks for core analysis. Petrophysical Society, London. http://www.lps.org.uk/docs/unconsolidated_cores.pdf (accessed 30 July 2014).

Rosen, R., Mickelson, B., Fry, J., Hill, G., Knabe, B., Sharf-Aldin, M., 2007. Recent experience with unconsolidated core analysis. In: Society of Core Analysts International Symposium, Calgary, Canada, 10–12 September, Paper SCA 2007-41.

RECOMMENDED READING

AAPG Wiki. Core alteration and preservation. http://wiki.aapg.org/Core_alteration_and_preservation (accessed 31 July 2014).

Bajsarowicz, C.J., 1992. Core alteration and preservation: part 3 Wellsite methods. In: Dev. Geol. Refer. Manual, pp. 127–130 (AAPG special volumes).

Lamb, C.F., Ruth, D.W., 1991. Laboratory program design for unconsolidated heavy oil reservoirs. In: International Symposium of the Society of Core Analysts, San Antonio, Texas, Paper 9104.

LaTorraca, G.A., Carlo, M.R., 1989. Combined resistivity, porosity, brine saturation, and capillary pressure measurements on poorly consolidated samples. In: International Symposium of Core Analysts, New Orleans, Louisiana, Paper SCA 8904.

Unalmiser, S., 1988. Handling unconsolidated sandstone cores to preserve wettability and pore structure. In: International Symposium of Core Analysts, Houston, Texas, Paper SCA 8803.

Worthington, A.E., Gidman, J., Newman, G.H., 1987. Reservoir petrophysics of poorly consolidated rocks: 1. Wellsite procedures and laboratory methods. In: SPWLA 28th Annual Logging Symposium, 29 June–2 July, London, England, Paper SPWLA-BB.

Chapter 4

Core Sample Preparation

Chapter Outline

4.1 Introduction **135**
4.2 Cleaning **136**
4.2.1 Solvents 136
4.2.2 Common Cleaning Methods 138
4.2.3 Core Cleaning Methods: Advantages and Drawbacks 146
4.3 Core Drying **146**
4.3.1 Conventional (and Vacuum) Oven Drying 146
4.3.2 Humidity Oven Drying 146
4.3.3 Critical Point Drying 150
4.3.4 Flow-Through Drying 152
4.3.5 Core Drying Methods: Advantages and Drawbacks 153
4.4 Quality Control Issues, Checks and Diagnostics **153**
4.5 Clays and Clay Damage Mechanisms **156**
4.5.1 Clay Mineral Structures 158
4.5.2 Clay Types 160
4.5.3 Cation Exchange Capacity 161
4.5.4 Clay Morphology and Rock Property Controls 163
4.5.5 Clay Damage Mechanisms 165
4.5.6 Testing Without Drying 168
4.6 Core Conditioning for Porosity Measurements **170**
4.6.1 Porosity Evaluation 170
4.6.2 Total and Effective Porosity Concepts 170
4.6.3 Log and Core Porosity 172
4.7 Special Considerations in Core Preparation **175**
4.7.1 Wettability 175
4.7.2 Carbonates and Chalks 176
4.7.3 Halite 176
References **178**
Recommended Reading/Viewing **179**

4.1 INTRODUCTION

For most RCA and SCAL tests, the plug samples must be initially prepared by cleaning with solvents followed by drying to remove oil and water as well as evaporated salts, mud filtrate and other contaminants. There are a number of methods and techniques used to clean and dry plug samples.

API RP40 (1988) lists the following methods for cleaning core and plug samples:

1. Flushing by centrifuge

Developments in Petroleum Science, Vol. 64. http://dx.doi.org/10.1016/B978-0-444-63533-4.00004-4

2. Gas-driven solvent extraction
3. Liquefied gas extraction (pressurised Soxhlet extraction)
4. Distillation–extraction (Soxhlet extraction)
5. Solvent flush cleaning by direct pressure.

The first three methods have largely fallen out of favour in most commercial labs.

Gas-driven solvent extraction involves filling the gas-filled pore space remaining after core recovery with a solvent containing dissolved CO_2. The solvent is pressurised and mixes with oil. Subsequent depressurisation removes some of the oil. Although once common for cleaning large, whole core samples, the method has effectively been judged by the industry to be largely inefficient, expensive and potentially hazardous. Flushing by centrifuge involves the use of a centrifuge to spray warm solvent against the core samples as they are rotated. The centrifugal force causes solvent to flow-through samples displacing and extracting the oil and water. The method was found to be very inefficient and notoriously dangerous—the flash point of toluene is only 6 °C and a small electrical spark from the motor caused many centrifuges to burst into flames. Liquefied gas extraction uses a pressurised Soxhlet extractor and a condensed, low-boiling-point, polar solvent. It is rarely used and only for special applications such as cleaning core containing heat-sensitive minerals such as gypsum.

The two most commonly used methods in most commercial labs nowadays, are the distillation–extraction method (and its variants) and the solvent flush clean method.

API RP40 (1988) lists the following methods for drying core and plug samples:

1. Conventional oven
2. Vacuum oven
3. Humidity oven.

These methods are still in common use today and other specialised drying technologies, such as critical point drying (CPD), have been developed for sensitive cores.

Selection of the most appropriate method depends on the lithology, clay content, potential wettability contamination of the core, as well as the objectives of the analysis for both RCA and SCAL.

4.2 CLEANING

4.2.1 Solvents

Core plugs need to be cleaned and dried to:

1. remove native oil, mud filtrates, evaporated salts and connate water prior to porosity and permeability analysis and single-phase SCAL experiments (e.g. formation factor and brine permeability);

2. condition or render the core water wet prior to certain two-phase SCAL experiments (primary drainage capillary pressure) and prior to wettability conditioning (e.g. restoration).

Effective cleaning of oilfield core can only reliably be achieved using at least two solvents:

(a) a non-polar solvent to dissolve the lighter ends of the oil and evaporate water in high-temperature extraction and
(b) a polar solvent to dissolve the heavier ends of the oil; to miscibly solvate water in lower temperature flush cleaning methods; and to leach/dissolve any precipitated salts from the core remaining after evaporation or removal of formation brine.

Non-polar (aromatic) solvents include:

- alkanes (e.g. hexane, heptane, decane, cyclohexane)
- alkenes, alkynes and aromatics (e.g. toluene, xylene)
- halogenated (e.g. chloroform).

The most common non-polar solvents in commercial core analysis lab applications are toluene, xylene and chloroform.

Polar solvents include:

- ketones, alcohols and carboxylic acids (e.g. acetone, methanol, ethanol and acetic acid);
- halogenated (e.g. dichloromethane).

The more common polar solvents used by labs are methanol, ethanol and dichloromethane.

A variety of solvents are therefore available depending upon the application. The usual sequence is to begin with a non-polar solvent (e.g. toluene) followed by the polar solvent (such as methanol). If the core is difficult to clean, more aggressive and potentially more damaging solvents are often used.

An azeotrope is a mixture of two or more solvents whose proportions cannot be altered by simple distillation. This happens because when an azeotrope is boiled, the vapour has the same proportions of constituents as the unboiled mixture. The boiling point of a heteroazeotropic mixture is lower than its individual components. Common core cleaning azeotropes include:

- chloroform/methanol,
- chloroform/methanol/acetone,
- methanol/methyl chloride.

The combination of both polar and non-polar solvents in the same mixture is often used for convenience, and in lower temperature cleaning systems, such as total immersion cleaning. At least two repeat cleaning cycles are recommended where core contamination is suspected, such as core drilled with oil-based muds or containing particularly difficult crude oils. For gas reservoirs, normally only methanol cleaning is employed to dissolve water and precipitated salts, although non-polar solvents may be used if the core was cut with OBM or was plugged with mineral oil.

Naturally occurring halite cements can be removed by solvent cleaning (especially when methanol is used) and this can pose problems if halite is part of the rock matrix.

4.2.2 Common Cleaning Methods

The three principal methods that are commonly used to clean cores nowadays are:

1. Soxhlet extraction (hot and cool),
2. constant or total immersion (submerged) Soxhlet extraction and
3. solvent flush cleaning (mild, miscible cleaning).

4.2.2.1 Soxhlet Extraction

The Soxhlet extractor (Soxhlet, 1879) was originally designed for the extraction of a lipid from a solid material but is used extensively in core analysis to remove water and oil from core samples. A schematic of a Soxhlet extractor assembly is shown in Fig. 4.1.

The principal components are:

- a round bottomed flask or extraction pot (2) filled with solvent
- a stirrer bar or anti-bumping granules (1)
- a side arm (3)
- a soxhlet thimble containing the core samples (5)
- a siphon arm inlet (6) and outlet (7)
- a reduction adaptor (8)
- a condensor unit which can be circulated with chilled water via ports 9 and 10

A typical apparatus set-up in a core analysis lab is shown in Fig. 4.2.

The apparatus (Fig. 4.1) can accommodate a large number of core plugs depending on the core sample chamber size. The plugs are placed in the soxhlet thimble which is fitted with a reflux (siphon) tube (6 and 7). The round-bottomed flask (2) sits on a heating mantle and is part filled with solvent. In most applications, this will be toluene initially. The heating mantle is switched on. When the solvent reaches its boiling point it evaporates. Table 4.1 provides the boiling points of most common core cleaning solvents. The solvent vapour travels up the side arm (3), bypassing the sample chamber itself, to the condenser (9) which is continually circulated with chilled or cold water. The solvent vapour condenses and falls down into the soxhlet thimble containing the plugs. As the hot solvent fills the chamber, it diffuses into the pores of the core and leaches oil. As the temperature of toluene in the

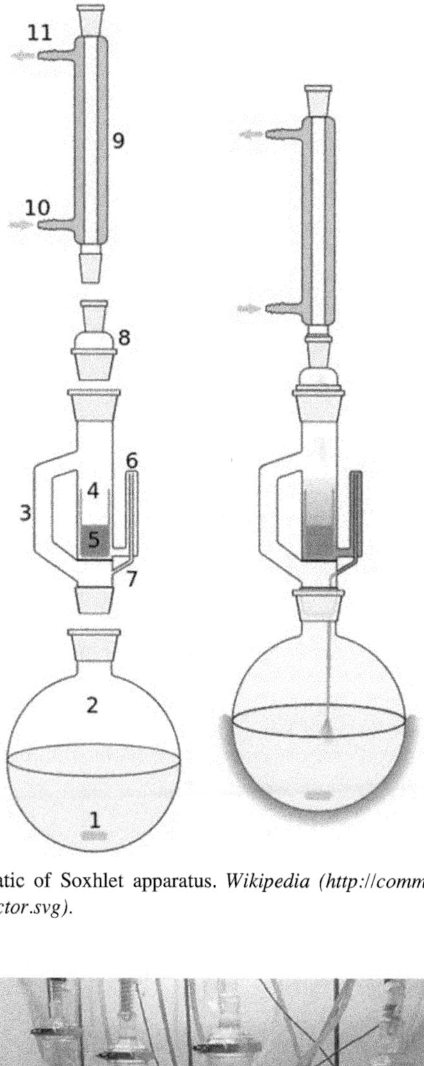

FIGURE 4.1 Schematic of Soxhlet apparatus. *Wikipedia (http://commons.wikimedia.org/wiki/File%3ASoxhlet_extractor.svg).*

FIGURE 4.2 Typical hot Soxhlet apparatus set-up in the core analysis laboratory.

TABLE 4.1 Boiling Points and Water Solubility of Core Cleaning Solvents

Solvent	Boiling Point (°C)	Solubility in Water (g/100 g)
Acetone	56	Miscible
Chloroform	61	0.795
Ethanol	78	Miscible
Isopropanol	83	Miscible
Methanol	65	Miscible
Tetrahydrofuran	65	30
Toluene	111	0.867
Xylene	138–144	Insoluble
Chloroform/methanol azeotrope (65/35)	54	–
Water	100	–

chamber is around 100 °C or greater, it causes water from the core to evaporate. The distilled solvent slowly fills the sample chamber, eventually submerging the core samples. When the level of solvent reaches the siphon point (6) the solvent and oil (coloured green) siphon or reflux from the thimble (via the syphon arm outlet (7)) into the bottom flask exposing the core plugs in the chamber. The bottom flask will now contain solvent and oil extracted from the core. As the oil/solvent mixture that has refluxed into the flask continues to be heated, fresh solvent vapour is produced, condensed, falls into the sample chamber, extracts oil, refluxes and the cycle repeats. This whole cycle is repeated many times. The solvent accumulating in the sample chamber or refluxed into the flask is regularly checked for discolouration. Normally, the solvent is replaced by fresh solvent near the endpoint of cleaning to assess any further discolouration. In addition, plugs can be removed and checked under UV light for oil fluorescence.

At the end of the toluene cleaning stage, the core samples will contain salts from the formation water, some residual water and perhaps some heavier ends of residual oil. To complete the cleaning process, the samples are then extracted (leached) in the Soxhlet apparatus with methanol. Again the distillation–extraction–reflux process cycles are continued until the core is cleaned of any remaining fluids and salts. Near the endpoint of cleaning, samples of methanol extract in the bottom flask are tested with a 10% $AgNO_3$ (silver nitrate) solution. This tests for the presence of chloride ions in the extract. These cause the solution to change colour and signals that salts still remain in the cores.

As this method uses high-boiling point extraction solvents such as toluene and methanol, it is referred to as hot Soxhlet extraction. Cleaning times depend on the permeability of the core, the viscosity and properties of the oil in the pores, formation water salinity, the presence of mud filtrates and the temperature and viscosity of the solvents. It can take from a few days to several weeks to complete the cleaning process.

Hot Soxhlet extraction has become the standard method to process plugs and cores scheduled for routine core analysis and some SCAL tests because it is fast, simple, cheap and convenient. It can be operated with minimal supervision. It is commonly considered to be a harsh cleaning method, since the plugs in the chamber are subject to high temperatures (greater than 100 °C for toluene and xylene, for example) and are subjected to repeated cycles of hot liquid immersion/hot vapour soak and reflux. The core exposure to air in the sample chamber following reflux creates a hot and evaporative environment that removes water from the pores and from water-sensitive minerals like clays, and causes a solvent/air interface to move through the sample pore system as the solvent evaporates. The interfacial tension (IFT) forces associated with this interfacial movement may be sufficient to cause damage to delicate mineral structures like illite clays, for example.

The use of solvents with lower boiling points (e.g. chloroform/methanol azeotrope) will not cause pore or clay water vapourisation but will still involve evaporative drying during reflux, so will not entirely eliminate interfacial movements.

4.2.2.2 Cool Soxhlet Extractor

This variation on the standard hot Soxhlet extraction method increases the solvent residence time in the apparatus after condensation. Often referred to as the 'cool' Soxhlet extractor, it is still possible to use high-boiling point solvents such as toluene, but the temperature of the solvent entering the core chamber is much lower. However, as the solvent is cooler, it is less effective and it takes significantly longer to clean cores than in a hot Soxhlet apparatus. It is commonly considered to be a benign method, since plugs are subject to cooler liquid immersion, but evaporative phase changes and interfacial movement still occur during the cleaning process as the core chamber drains. Once a relatively popular method for lower temperature cleaning, it has nowadays been largely replaced by the total immersion Soxhlet.

4.2.2.3 *Total Immersion Soxhlet*

Total (or constant) immersion cleaning combines the non-damaging benefits of cool static cleaning and the efficiency of hot Soxhlet cleaning. It is commonly described as a benign method, since plugs are subject to warm liquid immersion (preventing water vapourisation) and are not exposed to potentially damaging evaporative cycles during the cleaning process as samples remain submerged in the solvents throughout the cleaning cycle.

The immersion method is a variant of hot Soxhlet. A schematic is shown in Fig. 4.3 and the equipment assembly is indicated in Fig. 4.4. Samples are placed inside a core sample chamber under solvent which is connected to the solvent chamber of a Soxhlet apparatus. The solvent is boiled by heating the solvent flask and vapour rises and is directed into a condenser which sits above the core sample chamber. Core samples are placed on a perforated shelf which sits a few inches off the bottom of the sample chamber. The chamber is filled with solvent to the level of the return tube back to the solvent flask which is a higher level than the top of the cores. Condensed solvent drips down into the sample chamber but, unlike the standard Soxhlet, the solvent is directed to the base of the chamber via a funnel and percolates up from the base through the core samples. The chamber is heated and can include a stir bar which agitates the incoming solvent. As the fresh solvent enters the chamber, it displaces the solvent and oil extracts from the cores and these run off through the return tube back into the solvent chamber. The set-up allows for continuous re-distillation and circulation of solvent. In practice, fresh batches of clean solvent are introduced periodically. The system design means that the cores are continually submerged under solvent

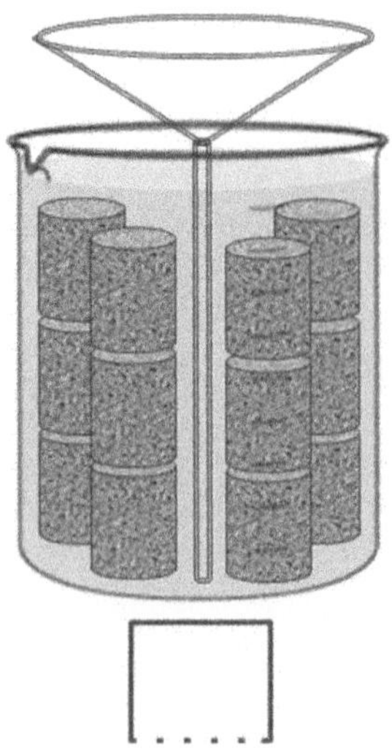

FIGURE 4.3 Total (constant) immersion Soxhlet equipment schematic.

FIGURE 4.4 Typical constant immersion cleaning assembly. *Courtesy of Corex.*

throughout the cleaning process and are not exposed to air during the reflux cycle in the standard solvent. Normally, a chloroform/methanol azeotrope is used as the cleaning solvent, as the boiling point (54 °C) is significantly less than toluene (111 °C). Throughout the process, the core samples are in contact with warm clean liquid solvent (thus enabling efficient cleaning) and are not exposed to the cyclical drainage/evaporation. As the cleaning temperature is lower than standard Soxhlet and the interfacial forces are minimised, delicate clay structures can be preserved.

4.2.2.4 Flush (Flow) Cleaning

Flow-through cleaning or flush cleaning involves the sequential flow of solvents through a core sample to solvate and displace resident hydrocarbons and aqueous fluids. Two variations of flow-through cleaning are commonly used: one utilising miscible solvents (mild, miscible cleaning) and the other immiscible solvents. Both methods involve flowing solvents or solvent mixtures through a confined core sample. Unconsolidated or friable samples should be mounted in heat-shrinkable Teflon, or similar solvent-resistant elastomer, before loading into the coreholder.

Figure 4.5 provides a schematic of the flush clean coreholder and Fig. 4.6 shows a typical system set-up. The coreholder (normally constructed from steel or aluminium) is a pressure vessel with a rubber core sleeve (or 'boot') which is often referred to as a Hassler cell. The core sample is loaded in the rubber sleeve which is confined by applying hydraulic or pneumatic pressure

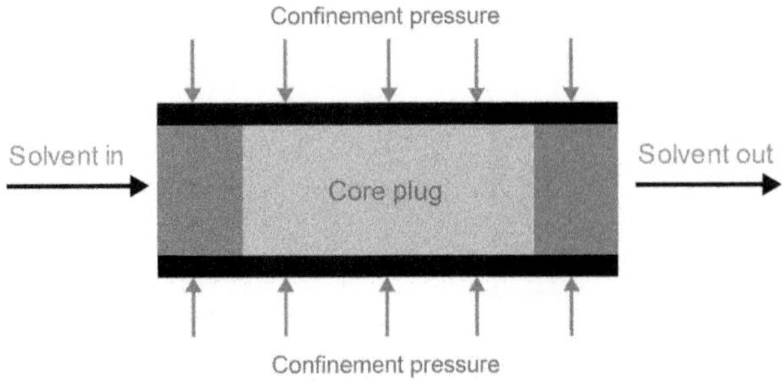

FIGURE 4.5 Schematic of solvent flushing coreholder.

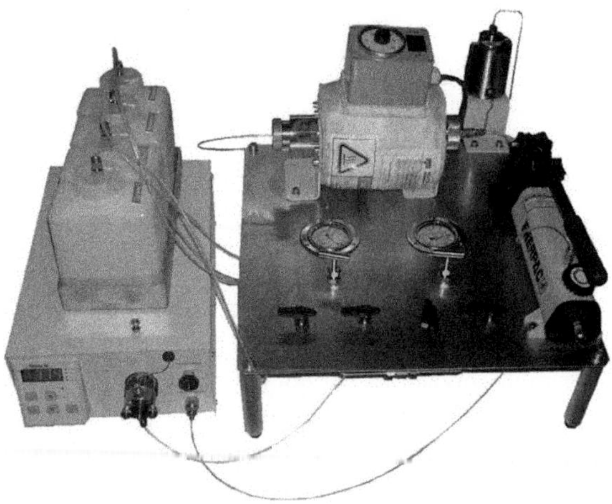

FIGURE 4.6 Flow-through core cleaner. *Courtesy of Vinci Technologies.*

to the outside of the sleeve. Inlet and outlet platens are used to allow flow into and out of the core samples. The core sample must necessarily be cylindrical in shape—this method cannot be used for irregularly shaped samples. The solvent is injected into the core either at constant rate using a pump or at constant pressure, using a transfer cell under air or nitrogen drive. Constant flow rate systems are preferred as it is much easier to control flow velocities through the core. Nominal backpressure (slightly elevated pore pressure) is used to assist in dissolving any air in the test samples. Confining pressure can be applied up to the limits of the coreholder and solvents can be injected at pressures up to the limits of the pump and back pressure regulators. In stress-sensitive formations, a nominal confining pressure is used sufficient to seal the plug in the sleeve. The coreholders can be located within an air bath or water bath, so the core sample can be cleaned at elevated temperatures.

Flow-through cleaning is also used to clean unconsolidated cores where handling must be kept to a minimum. The cores can be dried and pore volume and permeability measured with the core intact in its individual coreholder. Only one plug sample can be cleaned in the test coreholder at any one time so simultaneous cleaning of multiple plugs requires multiple coreholders and sophisticated control and pumping equipment. Provision of multiple core flush systems is extremely expensive, and few laboratories have the equipment or resources to routinely process the large number of plugs required for RCA. Many labs will cannibalise core flooding equipment rather than have a dedicated flush cleaning resource, so throughput can be extremely limited. Consequently, flush cleaning is normally restricted to preparing limited number of SCAL plugs or unconsolidated samples.

Several 'soak' or stasis periods—in which flow is temporarily stopped—can improve cleaning efficiency. The dynamic displacement process, at elevated temperatures, can clean an individual plug in a fraction of the time normally required for Soxhlet extraction. However, a Soxhlet apparatus can process many more plugs simultaneously so the total time for cleaning a large plug suite can be significant, depending on the lab resources.

As the solvent temperature can be controlled, high temperatures can be avoided and water evaporation can be eliminated. As the cores are continually saturated in solvents, there are no evaporative cycles so interfacial movement can be minimised. The process is especially useful for cleaning cores containing clays, unconsolidated plugs and samples where wettability contamination is suspected. However, injection pressures are largely constrained by the requirement to maintain solvent flow rates below critical velocities to prevent fines migration.

Flush cleaning uses similar solvents to the Soxhlet extraction technique. However, to prevent interface movements through the core, a miscible sequence of cleaning solvents is often required, especially if the core contains delicate clays or the core is to be critically point dried. Miscible cleaning implies that the solvent used is miscible with the preceding and succeeding solvents in the sequence. Typical flush clean sequences for cores containing oil include:

- toluene → isopropanol → methanol;
- toluene → chloroform/methanol azeotrope → methanol.

For gas cores drilled with water-based muds only methanol or ethanol cleaning may be required as both solvents are miscible with water. However, in cores where the formation water is highly saline, a sequence of increasing formation water (FW)/solvent ratios is used to prevent salt precipitation and pore plugging. For example,

$$100\% : 0\% \, \mathrm{FW/methanol} \rightarrow 75\% : 25\% \, \mathrm{FW/methanol} \rightarrow 50\% : 50\% \, \mathrm{FW/methanol} \rightarrow 25\% : 75\% \, \mathrm{FW/methanol} \rightarrow 100\% \, \mathrm{methanol}.$$

4.2.3 Core Cleaning Methods: Advantages and Drawbacks

At least three different cleaning methods and a variety of solvents are available to clean the core. The selection will depend on the nature and conditions of the core, its residual liquids and the nature of the analysis to be performed on the plugs. There is no single 'perfect' system.

Summaries of the advantages and drawbacks of the Soxhlet extraction systems and the flush cleaning systems are provided in Tables 4.2–4.4.

4.3 CORE DRYING

In addition to the conventional and vacuum oven-drying and the humidity oven-drying (HOD) methods listed in API RP40 (1988), two other drying methods are sometimes employed:

- critical point drying—CPD;
- flow-through (lenient air) drying.

4.3.1 Conventional (and Vacuum) Oven Drying

Plugs are placed in an oven (Fig. 4.7) and dried at a constant temperature in an unhumidified environment to constant weight. Vacuum is often used to speed up the drying process. Typical drying temperatures can range from 40 °C to over 100 °C. Higher temperatures are required for clay-rich samples (e.g. 'shaly' sands) if plugs are to be conditioned for total porosity measurements and to ensure all water is removed from the plugs for mercury injection capillary pressure tests. Plug weights are monitored regularly by the lab. The cessation of drying criterion is that plug weight should be constant within ±0.01 g or better, for three or four successive measurements over a couple of days. Plugs are processed on a batch basis so many plugs can be dried at the same time.

Conventional oven drying used in combination with hot Soxhlet extraction is considered to be a harsh preparation method as it can result in clay destruction due to dehydration of water held within or on the clay mineral structures.

4.3.2 Humidity Oven Drying

HOD is used where excessive temperature and anhydrous conditions associated with conventional oven drying can cause damage to clays. HOD evaporates the water in the large pores but is supposed to retain two molecular layers of water on the surface of clays (Keelan, 1982). This can prevent the clays from collapsing as they would under unhumidified drying.

TABLE 4.2 Standard Soxhlet Extraction Systems: Advantages and Drawbacks

Advantages	Drawbacks and Issues
• It is the cheapest, quickest, and in terms of processing a large number of core plugs (for example for RCA), the most effective cleaning method for most conventional oil and gas reservoir core samples. • Several samples can be cleaned simultaneously in the same batch. • Minimal supervision is required. • Normally used where core total porosity measurements are required to calibrate density logs as clay-bound water can be evaporated allowing helium to enter the pore space. • Cool Soxhlet apparatus is designed to prevent core damage associated with high-temperature cleaning.	• High-boiling point solvents are commonly used. These can potentially cause significant damage to delicate minerals such as halites and clays contained in the porous media. Removal of clay-bound water can increase porosity and permeability. • The method involves many rapid drying cycles as samples are subjected to repeated sequences of liquid immersion/hot vapour soak creating interfacial movement through the pores. IFT forces can cause collapse of delicate minerals such as fibrous clays, altering the pore media structure, and increasing porosity and permeability. • Hot Soxhlet extraction can induce an artificial oil-wetting condition in cores drilled with OBM or containing crude oil with high asphaltene content. High temperatures (>100 °C) associated with boiling toluene or xylene can vapourise the thin water film coating the grains and filling smaller pores, and helping promote the adsorption of surfactant molecules in the oil onto the rock grain surfaces. Hot Soxhlet-cleaned cores may not therefore be rendered water wet. • Samples must be protected from erosion by the drip of condensed solvent. • Cool Soxhlet cleaning efficiency is limited and cleaning times are long. • Samples in cool Soxhlets are still subject to repeated sequences of evaporative drying. • Cool Soxhlet extraction is inefficient at removing wettability contaminants associated with OBM.

TABLE 4.3 Total Immersion (Submerged Cleaning) Soxhlet: Advantages and Drawbacks

Advantages	Drawbacks and Issues
• The system is designed to prevent core damage associated with high temperatures and evaporative drying cycles (interfacial tension forces). As samples are always submerged under solvent, there is no evaporative cycle as occurs during solvent drainage from the core chamber in hot or cool Soxhlet extraction. • Recommended method for SCAL tests where clay damage must be prevented. • Several samples can be cleaned simultaneously in the same batch. • Minimal supervision is required. • More expensive than hot Soxhlet, but still relatively cheap.	• Cleaning efficiency is more limited and cleaning times are longer than hot Soxhlet. • Constant immersion extraction can be inefficient at removing certain wettability contaminants associated with oil-based muds and some reservoir crudes.

TABLE 4.4 Flush (Flow Through) Cleaning: Advantages and Drawbacks

Advantages	Drawbacks and Issues
• Low-temperature cleaning and lack of interfacial movement (in mild miscible cleaning) prevent damage to delicate clays. • Can be more rapid and efficient than Soxhlet extraction on an individual core plug basis. • Most reliable method to clean SCAL plugs where wettability contamination (from oil or OBM) is suspected. • Ideal method to clean unconsolidated sand plugs as the samples can remain under confining stress which reinforces weak core and holds the grains together during thawing and cleaning.	• Fines migration is possible at high rates, resulting in fines blocking and occluding the pore space. • Only one sample per coreholder is possible. • Very expensive (for multiple plugs) compared to the other cleaning methods. • Operator intensive. • Its application in routine core analysis is limited as many labs do not have the equipment resources to process a large number of plugs. It is generally used in SCAL where fewer plugs are usually tested. • If heavy oil or waxy oil is to be removed from test plugs, then higher temperatures will be required to reduce oil viscosity and optimise cleaning. • Not recommended for low-permeability core due to limitations on flow rate and injection pressure and confining stress.

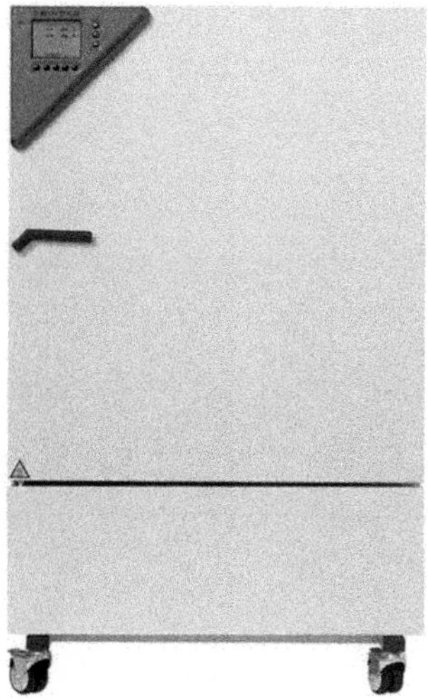

FIGURE 4.7 Vacuum oven. *Courtesy of Vinci Technologies.*

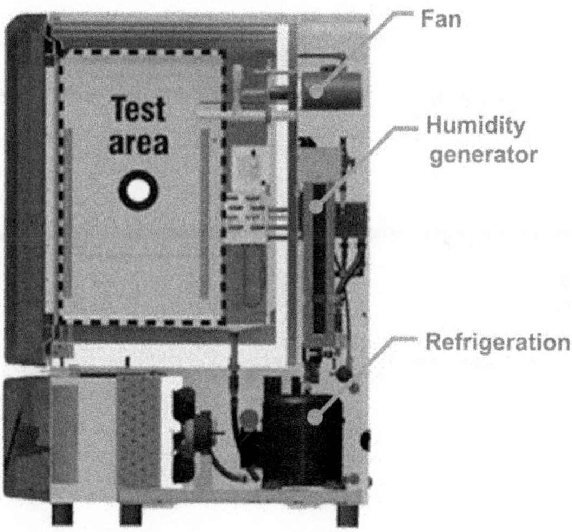

FIGURE 4.8 Humidity chamber details.

There are two types of HOD:

1. The direct injection environmental test chamber (Fig. 4.8). These ovens are composed of an air bath, an electronic temperature regulator, a water vapour injection system, a capacitance probe to control humidity level and a timer. A portion of the air returning from the conditioned area is directed through a water spray area with a closely controlled temperature. The water spray is adjusted to maintain the dew point temperature of the targeted condition. As the air passes through the water spray, it is cooled down to the temperature of the water and approaches saturation.

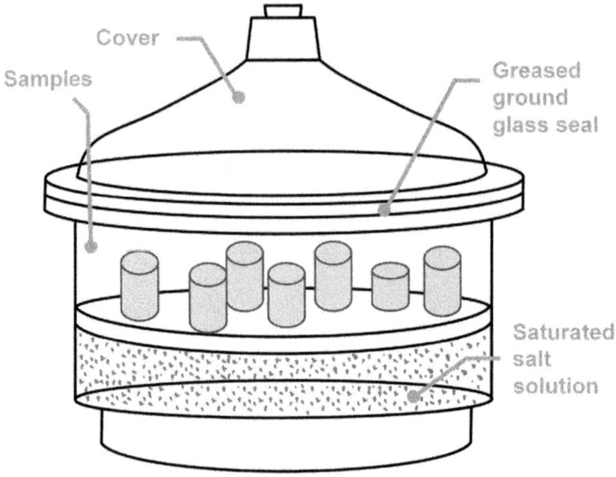

FIGURE 4.9 Dessicator used for humidity-conditioned drying over saturated salt solution.

2. A much cheaper solution is to use a dessicator vessel inside a conventional oven. Plugs are placed on a stand in the vessel over a saturated salt solution (Fig. 4.9). The salt solution is selected such that vapour pressure in the dessicator is sufficient to provide the necessary relative humidity (RH) at the desired temperature.

The conditions for HOD are remarkably consistent in labs throughout the world: 60 °C and 40–45% RH. There appears to be no great science attached to the derivation of these parameters—this was apparently the only humidity oven available to Bush and Jenkins (1977) who were amongst the first to advocate HOD in core analysis of shaly sands.

4.3.3 Critical Point Drying

CPD is a well-established method of dehydrating biological tissue prior to examination in the scanning electron microscope (SEM). The technique was first described for core analysis by McHardy et al. (1982).

As discussed for both conventional Soxhlet and both conventional and humidity oven drying, evaporative drying of rock specimens containing delicate clays such as 'hairy' or filamentous illite can cause severe deformation and collapse of the clay structure due to the passage of a liquid/vapour interface through the pore systems. The clay structures are subjected to considerable IFT forces, which are present at the phase boundary as the liquid evaporates. IFT can be reduced by substitution of a liquefied gas with a lower surface tension. If the temperature of liquefied gas is increased, the IFT becomes very small and the liquid surface becomes very unsteady and ultimately disappears. When this ‘critical point’ is reached, it is possible to pass from liquid to gas without any abrupt change in state.

The CPD technique prevents the development and advancement of a gas/fluid or fluid/fluid interface within the rock by raising the fluid within its pores above its critical point. Figure 4.10 shows the phase diagram for CO_2. Consider the phase envelope separating the liquid from the gas phase. At pressures and temperatures above this line, CO_2 exists as a liquid. At pressures and temperatures below this line, it exists as a gas. If at a certain temperature below the critical point, *C*, the pressure is reduced, the pressure/temperature locus crosses the phase envelope. Thus, there will be a phase change from liquid to vapour and an interface will be created in the sample. This interface can cause damage to delicate clay structures, such as fibrous or filamentous illite. If the pressure and temperature can be adjusted such that the locus is navigated around the critical point, the sample can be dried without involving a phase change or creating an interface. When the pressure and temperature of the pore fluid are raised above its critical point, the fluid is brought into a super-critical state in which no phase transition exists between the gas and the liquid. The gas is then vented from the core isothermally.

CO_2 remains the most common medium for the CPD procedure as the critical pressure and temperature (32 °C and 1072 psi) can be easily achieved in practice in the lab. However, CO_2 is not miscible with water and therefore water (and oil) has to be replaced in the specimen with another fluid which is miscible with CO_2, and this is termed the ‘intermediate fluid’. Consequently, samples scheduled for CPD have to undergo a rigorous cleaning

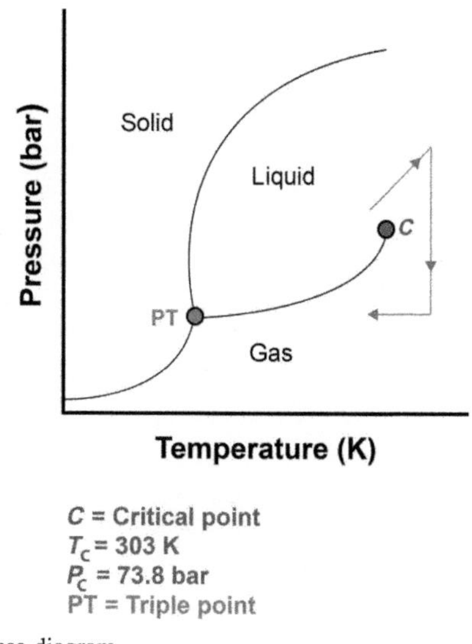

FIGURE 4.10 CO_2 phase diagram.

procedure which involves the miscible displacement or original pore fluids by the intermediate fluid via a sequence of mutually miscible fluids, such as:

$$\text{Toluene} \rightarrow \text{isopropanol} \rightarrow \text{methanol} \rightarrow \text{ethanol (intermediate fluid)}$$
$$\rightarrow \text{liquid } CO_2 \text{ (transition fluid)}.$$

Figure 4.11 shows a typical CPD apparatus developed principally to prepare biological specimens for SEM analysis though specialist systems have been developed for core drying. CPD reached its zenith in core analysis in the late 1980s and early 1990s, but since then, CPD has only rarely been used for limited SCAL plug processing due to cost and, in many cases with more robust clays, the lack of any significant advantage over total immersion cleaning and HOD.

4.3.4 Flow-Through Drying

Often referred to as 'lenient air' drying this method is generally used to dry unconsolidated core plugs as it means that the core remains in the confined coreholder following flush cleaning and thereby minimises sample handling. Following the final cleaning solvent (typically methanol), warm nitrogen gas is then flowed gently through the samples until no more liquid is displaced and the plugs are judged to be dry. Weighing the samples *in situ* in the coreholder to check for constant weight is not sufficiently accurate and removing the plugs to verify weight stability could lead to sample

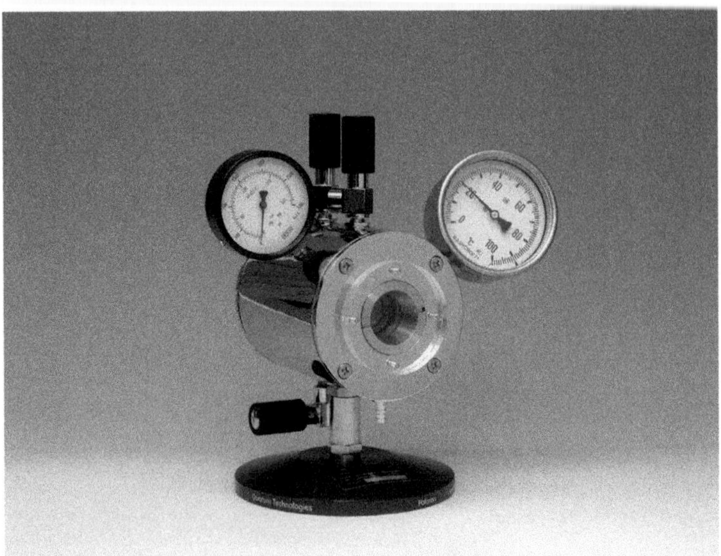

FIGURE 4.11 Example CPD apparatus. *Courtesy of Quorum Technologies.*

damage. Consequently, there can be some uncertainty about how dry the samples actually are.

4.3.5 Core Drying Methods: Advantages and Drawbacks

Summaries of the advantages and drawbacks of the principal drying methods are provided in Tables 4.5–4.8.

4.4 QUALITY CONTROL ISSUES, CHECKS AND DIAGNOSTICS

In this section, we provide a synopsis of the key quality control issues, checks and diagnostic tools associated with core cleaning and drying, so the interpreter is better placed to judge the viability of the subsequent core analysis measurements:

1. SEM and XRD analysis should be considered to identify the presence of sensitive minerals and delicate clays in the core so that the cleaning and the subsequent drying processes can be selected to minimise potential damage due to water evaporation and IFT movements.
2. In some cases, a cleaning pre-study should be undertaken, depending on the test objectives. For example, for wettability restoration in SCAL, the

TABLE 4.5 Conventional Oven Drying: Advantages and Drawbacks

Advantages	Drawbacks and Issues
• Rapid drying at relatively high temperatures (typically up to 105 °C). • It is the cheapest drying method. • High-temperature drying is required to remove clay-bound water for total porosity measurements and water in microporosity systems prior to mercury injection capillary pressure tests. • The method is acceptable for RCA and SCAL tests on clean sands, clean carbonates, or on sands which contain robust clays with low CBW contents (e.g. kaolinite).	• Drying at high temperatures (e.g. 105 °C) can lead to destruction of the clay framework through removal of CBW and is not an acceptable technique for reliable SCAL tests on shaly sands. • The interfacial tension forces between air and solvent as the solvent meniscus passes through the sample on evaporative drying can be sufficient to collapse very delicate illites, so that they are forced back against the host grains, thereby artificially increasing effective pore throat diameters (Worthington et al., 1988). High-temperature evaporative excessive drying normally has a much greater impact on sample permeability and resistivity that on porosity.

TABLE 4.6 Humidity Oven Drying: Advantages and Drawbacks

Advantages	Drawbacks and Issues
• It is a convenient method to help preserve clay structures for SCAL measurements. • In general, HOD is around twice as expensive as conventional (or vacuum) oven drying, though still relatively cheap and much cheaper than critical point drying.	• Humidity drying is much slower than conventional drying. • In shaly sands, HOD may remove some but not all clay-bound water, so the resultant porosity measured on the test plug lies somewhere between total and effective porosity and has no unique relationship with either log-derived effective or total porosity. • Many direct injection humidity ovens cause atmospheric water to condense on the samples, which has the misleading effect of reducing porosity and increasing 'dry' weight. The apparent removal of 'clay' water on unhumidified drying may therefore have nothing to do with clays—but may result only from evaporation of condensed water on the samples. • Despite the humidified condition, drying is still achieved using evaporation. The interfacial tension forces between air and solvent as the solvent menisci passes through the sample on evaporative drying can be sufficient to collapse very delicate illites, so that they are forced back against the host grains, thereby artificially increasing effective pore throat diameters. • There is little point in employing HOD to preserve clays if damage (excess temperature/interfacial movement) has occurred during the cleaning phase (e.g. hot Soxhlet extraction).

core must be rendered water wet prior to ageing in oil and this should be tested for the different cleaning methods and solvents. Pre- and post-test SEM (and if possible porosity and permeability) can be used to identify any clay damage mechanisms caused by the core preparation methods and their impact on the post-cleaning petrophysical properties.

TABLE 4.7 Critical Point Drying: Advantages and Drawbacks

Advantages	Drawbacks and Issues
Method has been designed and adapted to prevent the collapse of delicate illitic clays by eliminating the interfacial tension forces between air and solvent during 'drying' and the transition from a liquid to a vapour phase.	• Very expensive method. More suited to SEM biological sample preparation than for core plugs for RCA or SCAL. • The time required to dry an individual sample can vary depending on size and permeability. • Core has to be flush cleaned before drying. • CPD has only rarely been used since the early 1990s due to cost and the lack of any significant advantage over HOD in samples where clays are more robust. • It is not commonly available. Only some of the major commercial laboratories offer this service at their principal technical centres. • Other and simpler methods (e.g. flush cleaning followed by brine saturation) can be used to mitigate evaporative drying damage (at least for certain SCAL tests).

TABLE 4.8 Flow-Through Drying: Advantages and Drawbacks

Advantages	Drawbacks and Issues
Method is designed to help prevent mechanical damage to unconsolidated plugs during conventional drying.	• Method is much less efficient than conventional or humidity drying. • Difficult to determine when samples have been fully dried, and have reached weight stability, without removing the plug from the coreholder. • Only one sample per coreholder can be used. • Very expensive (for multiple plugs) compared to oven drying. • Its application in routine core analysis is limited as many labs do not have the equipment resources to process a large number of plugs. If used at all, it is generally used in SCAL where fewer plugs are usually tested.

3. During cleaning, the lab must check the non-polar solvent effluent for solvent discolouration and oil fluorescence in core (presence of oil), and polar solvent effluent for presence of chlorides (silver nitrate solution test) to ensure that the cleaning process has been completed.
4. During drying, plugs should be dried until a constant weight is achieved: ±0.01 g over three or four successive measurements over 2–3 days.
5. Grain density is a powerful QC tool. Anomalously, low grain density values could result from insufficient cleaning or drying. Liquid might remain in the pore system and block helium. As liquid is less dense than rock grains, this would suppress the grain density measurements. The presence of liquid would also reduce both porosity and permeability. For example, Fig. 4.12 compares porosity and grain density measurements on the same plugs. Porosity was measured after 'hot shot' preparation. Hot shot analysis is designed to give fast results, so plugs are normally only cleaned for 8–12 h and dried for 8–12 h before porosity and permeability measurements. After testing, the plugs are re-cleaned and re-dried using conventional protocols and timings. Clearly, the grain densities and porosities of the hot shot samples are significantly lower than the fully cleaned plugs indicating the cores probably still contain liquids as they have not been fully cleaned or dried.
6. It is worth checking with the laboratory which type of humidity oven they use: direct injection or saturated salt solutions, and if the latter, the salt solution used or proposed.
7. In CPD, the transition liquid should fill 50% of the specimen pressure chamber. This will ensure that specimens are not uncovered during initial flushing stages and in addition, this should enable critical constants of temperature, pressure and density to be achieved relatively simultaneously without excessive pressure or evaporation conditions occurring. The temperature should be maintained somewhat above the critical temperature during decompression. This will avoid the possibility of gas re-condensing. It is also important to control the decompression rate itself as there is evidence that rapid pressure equalisation can cause specimen damage.

4.5 CLAYS AND CLAY DAMAGE MECHANISMS

Clay minerals are hydrous alumino-silicates with a sheet-like structure (phyllosilicates) which adsorb water on their surface. They form in the presence of water, normally from feldspars and micas, and may be allogenic, deposited with the sediments that form the sand, or authigenic created by diagenesis.

Authigenic clay minerals form within the pore spaces of sediments after deposition and during cementation, and their presence usually reduces permeability. Further, they are in contact with the pore fluids in the reservoir and

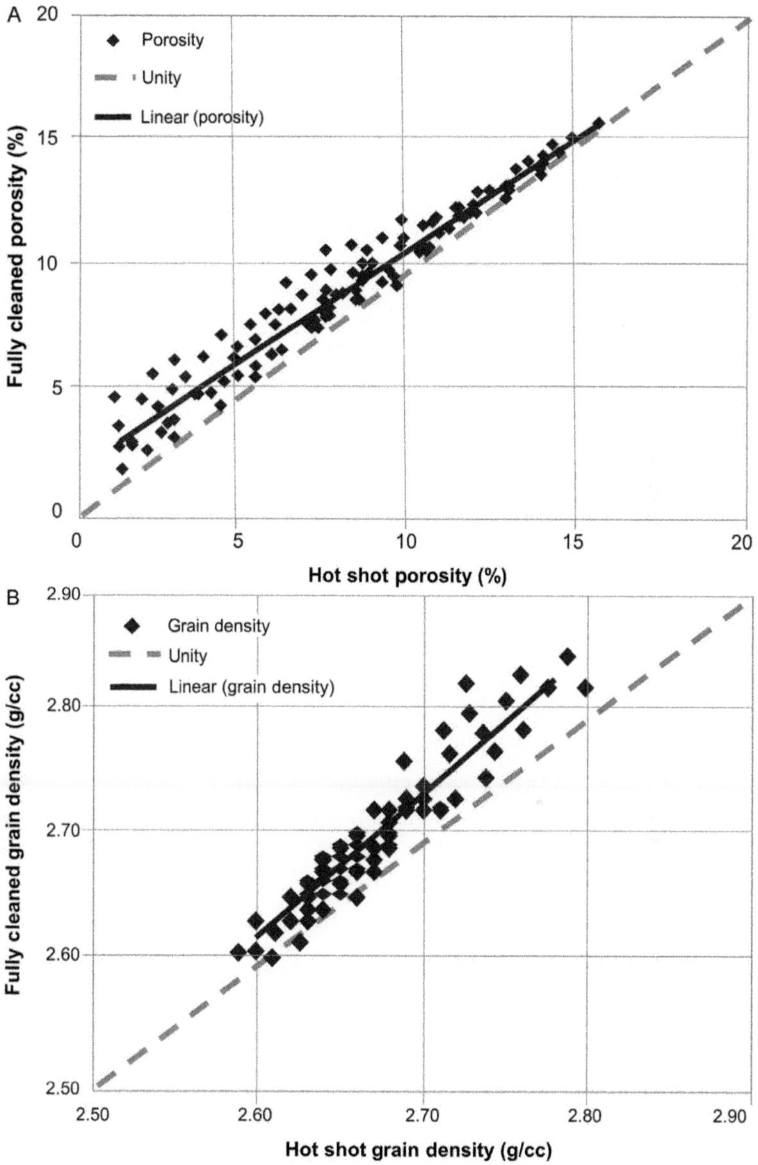

FIGURE 4.12 Comparison of 'hot shot' and fully cleaned porosity (A) and grain density (B) measurements.

can react with these, as well as drilling and production fluids, as they flow through the formation. They have an effect on the saturation distributions and production characteristics of the reservoir which far outweighs the percentage distribution which they occupy in the reservoir pore system.

Clays can form in discrete beds and in structural or laminated habits (Spooner, 2014), but in core analysis of conventional reservoirs, it is primarily dispersed clays that are of major importance (Fig. 4.13). Clay primarily occupies pore spaces between the sand grains and reduces porosity and permeability. Sand permeability depends on the clay type, volume and morphology (discrete, pore lining or pore bridging).

Whether or not clays will significantly impact formation evaluation depends on a number of factors including:

- mode of distribution of the clay
- type of clay mineral present and their relative volume
- clay morphology within the pore system
- formation water salinity and resistivity

4.5.1 Clay Mineral Structures

Clay minerals are described as hydrous alumino-silicates. They have crystal structures that consist of combinations of sheets of silica and alumina held together by water molecules or positive ions:

- Silica (SiO_2). Si^{4+} ions (often partly replaced by Al^{3+}) are tetrahedrally coordinated with oxygen atoms.
- Alumina (Al_2O_2). Al^{3+} ions are octahedrally coordinated with oxygen and hydroxyl anions. Often some Al^{3+} ions are replaced by Mg^{2+} and Fe^{2+} ions. They can also be replaced by K or Ca ions.

The silica and alumina sheets extend indefinitely in the *a* and *b* directions and are stacked in the *c* direction in composite layers of either two or three sheets held together by exchangeable cations (positive ions) and/or water molecules,

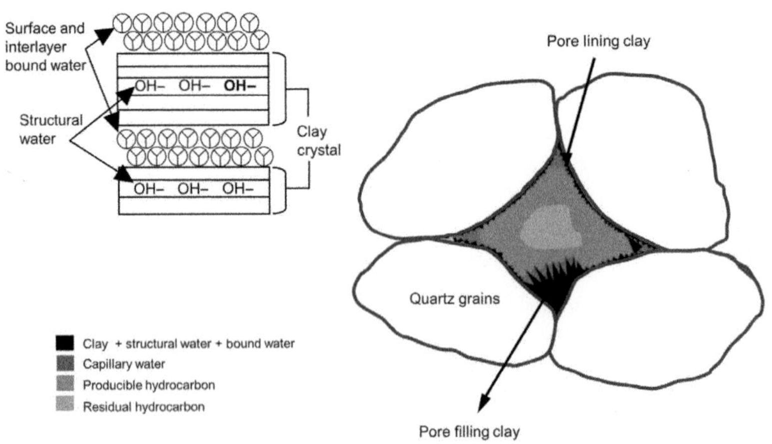

FIGURE 4.13 Schematic diagram of authigenic clay structures and morphology.

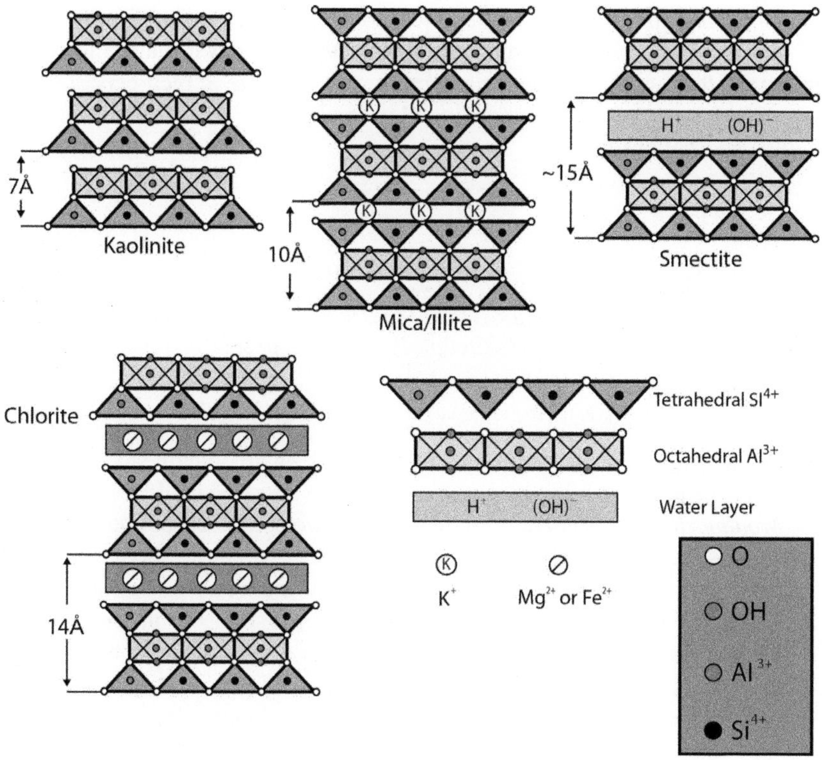

FIGURE 4.14 Schematic of common clay mineral structures.

as illustrated in Fig. 4.14. Different arrangements of composite layers along the *c*-axis lead to a variety of crystal structures and clays with a range of properties. Clays having two sheets per composite layer include kaolinite (dickite and nactite) and those having three sheets per composite layer include illite, glauconite, smectite (montmorillonite) and chlorite.

Substitution of Si^{4+} by Al^{3+} in the tetrahedron sheet and replacement of Al^{3+} with Mg^{2+} or Fe^{3+} in the octahedral sheet produce a net negative charge on the clay. This negative charge is balanced by cations attracted to the surface of each flat platelet. The cations on the clay surface are termed exchangeable since they can be readily removed by other cations in their environment.

The exchange layers are associated with layers of water molecules which, to varying degrees, are bound to the negative sites on the clay platelet surface. This is often referred to as structural water or water of hydration. The number and location of these layers of water molecules, which are considered to be an integral part of the clay mineral structure and not 'free' pore water, depend on the mineralogy and nature of the clay present. Because of the similarity in structures of the three-sheet clay minerals, mixed-layer clay minerals often develop amongst illites, smectites, vermiculites and chlorites. The addition or loss of water from clay minerals which have interlayer water molecules as part of their structure causes an expansion or collapse of their crystal structure along the *c*-axis. The expanding clay minerals include smectites, vermiculites and halloysites.

4.5.2 Clay Types

The major clay groups are:

- montmorillonite—$(Na,Ca)0.33(Al,Mg)2Si_4O_{10}(OH)_2{\cdot}n(H_2O)$
- illite—$(K,H_3O)(Al,Mg,Fe)_2(Si,Al)_4O_{10}[(OH)_2,H_2O]$
- kaolinite—$Al_2Si_2O_5(OH)_4$
- chlorite—$(Ca,Na,K)(Mg,Fe,Al)_9(Si,Al)_8O_{20}(OH)_{10}{\cdot}n(H_2O)$.

These clays are classified according to their molecular structure and comprise relatively simple 'building blocks' arranged in a variety of different ways.

Montmorillonite (smectite) group minerals are composed of two silica tetrahedral sheets with a central alumina octahedral sheet. Some 80% of exchange cations (commonly Na^+ or Ca^{2+}) occur between the silicate layers, with the remainder associated with the external surfaces of the particle. Figure 4.15 is a SEM photomicrograph of pore-filling smectite. The smectite itself has quite a high porosity but the pores are small (less than 1 μm in diameter) and, unless the hydrocarbon column is extremely high, these pore spaces will normally always be filled with water.

Illite (mica) group minerals comprise a layer composed of two silica tetrahedral sheets with a central octahedral sheet. The structure is similar to smectite, with the exception that the charge deficiency at the interlayer exchange is balanced by potassium cations, which are stable and not easily replaced. This clay usually occurs in a pore-lining or pore-bridging morphology, often in

FIGURE 4.15 SEM photomicrograph of pore-filling smectite in shaly sand.

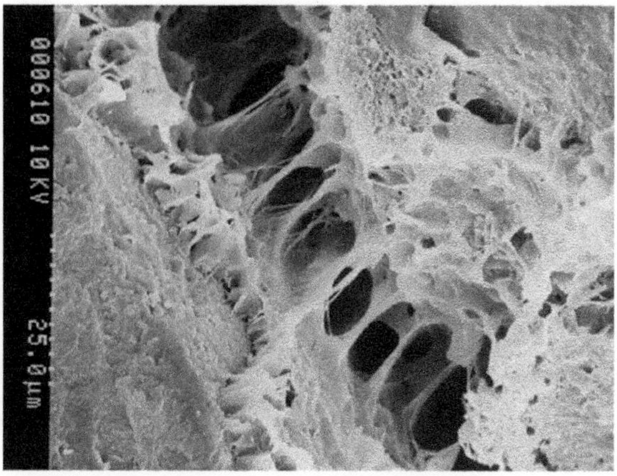

FIGURE 4.16 SEM photomicrograph of fibrous illite bridging across pore.

filamentous form (or 'hairy' illite) which is easily damaged during core preparation. Figure 4.16 provides a SEM photomicrograph of fibrous illite in a gas sand in its pore-bridging habit.

Kaolinite minerals are composed of a single silica tetrahedral sheet and an alumina octahedral sheet combined in a unit such that the tips of the silica tetrahedrons and one of the layers of the octahedral sheet form a common layer. Electrical charges within the structural unit are balanced and there is very little substitution within the lattice. Kaolinite commonly occurs as pseudohexagonal 'booklets' which are loosely attached to the host grains, frequently in discrete patches as shown in Fig. 4.17. This loose attachment may result in mobilisation during hydrocarbon production and core flow testing and can result in pore blockage. It has a low cation exchange capacity (CEC) and its main effect on rock properties is simple pore volume reduction.

Chlorite group minerals consist of alternate mica- and brucite-like layers. The mica layer is unbalanced by substitution of Al^{3+} for Si^{4+} and the charge deficiency is balanced by an excess charge in the brucite layer, as a consequence of substitution of Al^{3+} for Mg^{2+}. Chlorite can be very stable in its well-crystallised form but is easily degraded by acids to an unstable form. It typically occurs with a 'cornflake' texture as clay coatings (e.g. Fig. 4.18).

4.5.3 Cation Exchange Capacity

The ability of clay materials to sorb or attract certain cations or anions and retain them in an exchangeable state is one of their most important properties. The ions are exchangeable for other anions or cations by treatment with such ions in a water solution.

Two measures of the concentration of cations are available: the CEC which is measured relative to mass; and Qv, which is measured relative to

FIGURE 4.17 SEM photomicrograph of kaolinite booklets loosely adhering to quartz grains (discrete habit).

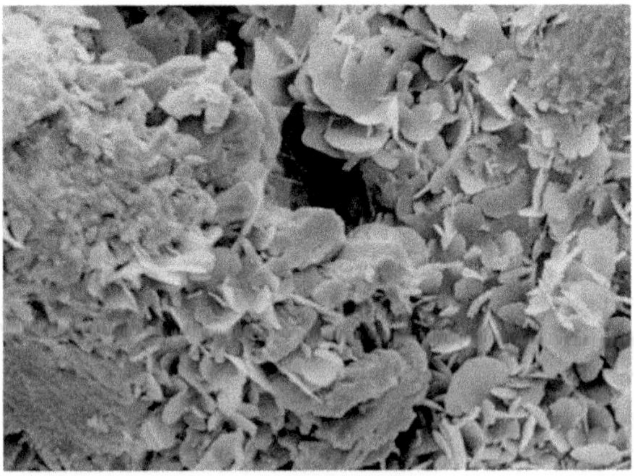

FIGURE 4.18 SEM photomicrograph of pore-lining chlorite.

pore volume. CEC is expressed in milli-equivalents of exchangeable ions per 100 g of dry clay or rock (meq/100 g). Qv is expressed in milli-equivalent per millilitre of pore volume (meq/ml). Table 4.9 provides a summary of the CEC of common authigenic clays in reservoir sands.

In smectites, some 80% of the exchangeable cations occur on the base of plane surfaces with the remainder on the edges. In illite and chlorite, most of the cations are on the edges, between crystals. With kaolinite, the exchangeable cations occur only on the base of the silica tetrahedral units, and the CEC depends on the thickness of the kaolinite particles. Thus, in general, kaolinite has the lowest CEC increasing through illite and chlorite, to smectite, which has a CEC about 20–30 times that of kaolinite. The exchange cations in these clays are associated with an integral number of water layers.

TABLE 4.9 Cation Exchange Capacities of Common Authigenic Clays

Clay	CEC (meq/100 g dry clay)
Montmorillonite (smectite)	70–130
Illite (hydrous mica)	25–40
Kaolinite	3–15
Chlorite	10–40

In smectite, two layers of water molecules occur between the silicate layers (Fig. 4.19). Interlayer water molecules are coordinated around the sodium cations constituting an inner primary hydration cell referred to as Type I water. Type II water forms an outer secondary coordination sphere, being indirectly linked to the cations and the inner Type I layer via weak ion–dipole bonds, and is more mobile than Type 1 water. Exchange reactions in kaolinite, illite and chlorite generally occur as the result of broken bonds at the edges of the crystal lattices, and consequently, exchange sites occur at interparticle (that is, between crystals) rather than at interlayer (within crystals) sites.

In petrophysical terms, the water molecules' negative poles are attracted to the cations and results in the formation of a diffuse double layer of water molecules attached to the surfaces of the clays, the first layer immediately adjacent to the clay surface and the second held by the hydrated cations. The positive ions in the clay-bound water (CBW) layer are mobile and therefore provide an additional path for electrical current (Fig. 4.20). The resistivity of the rock is thus reduced.

4.5.4 Clay Morphology and Rock Property Controls

In his classic paper, Neasham (1977) identified three clay morphologies from SEM analysis of shaly sands and examined the impact on porosity–permeability relationships (Fig. 4.21).

- Discrete
- Pore lining
- Pore bridging (or filling).

This does not mean to say that all sands with 'discrete' or 'pore-lining' or 'pore-bridging' clays will have identical relationships. Rather Fig. 4.21 serves to illustrate the impact on permeability as more complex clay morphologies decrease effective pore throat diameters and increase pore tortuosity.

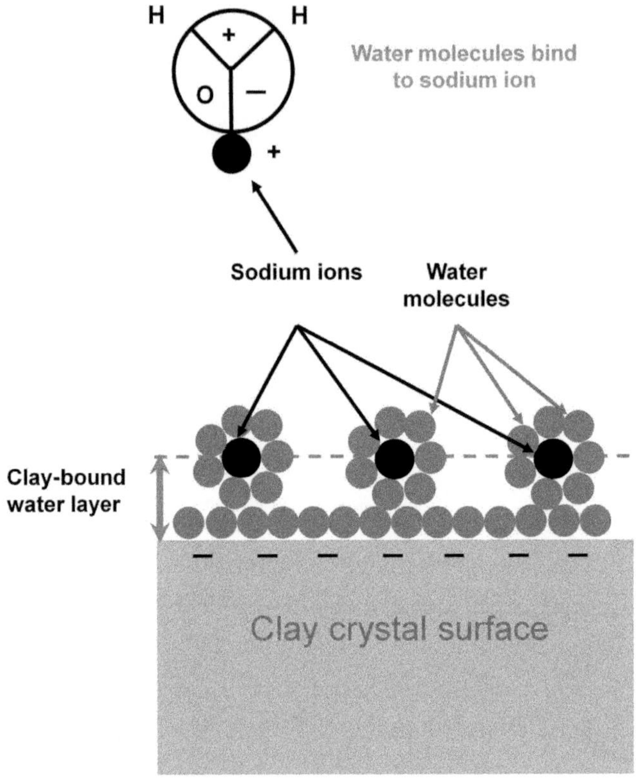

FIGURE 4.19 Diffuse double layer of water molecules in smectite.

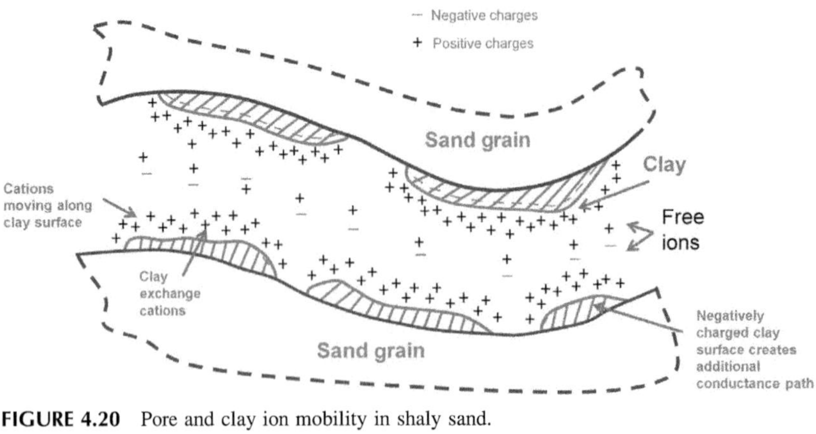

FIGURE 4.20 Pore and clay ion mobility in shaly sand.

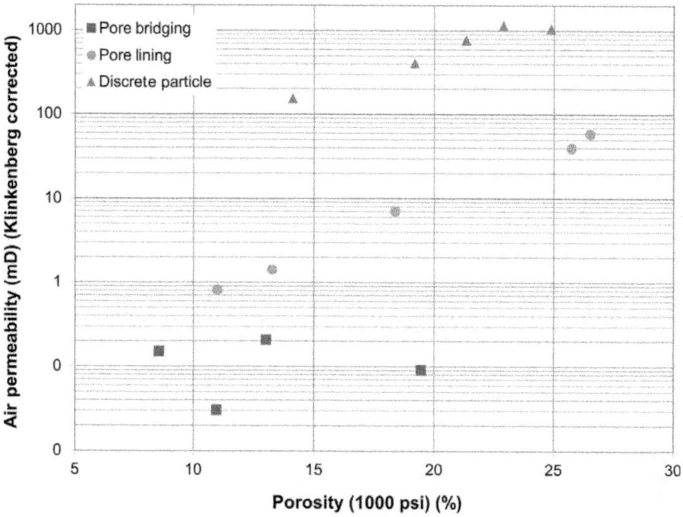

FIGURE 4.21 Example poroperm relationships for different clay morphologies in sandstones. *After Neasham (1977).*

Discrete clays, such as the kaolinite shown in Fig. 4.17, are generally located attached to grain surfaces, and because their surface area to volume ratio (which directly correlates with CEC) is low, they do not occupy too much of the remnant pore space and normally have a minimal control on porosity and permeability. Reasonable to good reservoir quality can be maintained.

The data set for the pore-lining clays (for example, the chlorite shown in Fig. 4.18) can still show good total porosity, but this porosity may not be well connected and the permeability is reduced by one or two orders of magnitude compared to discrete kaolinite. Pore throat channels have become significantly reduced by a relatively small amount of clay. This pattern of permeability reduction is continued further with the intergrown pore-bridging clays, as exemplified by the hairy illite shown in Fig. 4.16, and pore-filling clays (the smectite in Fig. 4.15) where the permeability is reduced by a further two orders of magnitude.

These results clearly show that if the clay morphology is altered in any way during core and plug preparation or during testing, then the impact on the derived core petrophysical data may be very significant.

4.5.5 Clay Damage Mechanisms

4.5.5.1 Core Cleaning and Drying

Ill-considered and/or inappropriate core cleaning and drying can lead to significant damage to delicate clays. Damage must be avoided or minimised if reliable petrophysical data are required. The exceptions are:

- The special case of core porosity measurements where the samples have to be conditioned in a total porosity system so that the measurements are consistent with porosity derived from density logs.
- Sample preparation for mercury injection capillary pressure tests where all water has to be removed from the sample pore system so that it does not act as a wetting phase.

The use of high-boiling point solvents, for example, xylene or toluene, in Soxhlet extraction, and evaporative drying conditions in both Soxhlet extraction and conventional unhumidified drying, can be sufficient to cause significant dehydration of CBW in pore-lining and -filling clays. This opens up the microporosity to helium in porosity measurements but can cause the clays to collapse. Porosity increases from effective to total, but permeability increases significantly as the pore throat diameters increase and tortuosity reduces.

Similarly, evaporative drying can cause destruction of delicate pore-bridging clays like fibrous illite. Heaviside et al. (1983) describes a classic example of clay damage on cores from the Magnus field in the UKCS. At that time, the field development plan was to drill deviated injectors into the water leg sands. A pilot well was drilled, cored and tested to ensure sufficient formation injectivity for matrix injection. In this case, routinely measured plug air permeabilities on core from the water leg showed a particularly poor correlation with well test-derived injectivity index—core permeabilities were, on average, 30 times greater than the well-test permeability. Subsequent investigations showed that filamentous, diagenetic illite was being severely altered by routine core preparation and measurements were actually performed on prepared samples with a texture that was very different to that occurring in the reservoir. It was found that the IFT forces between air and solvent as the solvent menisci passed through the pore space on evaporative drying were sufficient to collapse the very delicate illites so that they were forced back against the host grains, thereby increasing effective pore throat diameters, reducing tortuosity and improving permeability. CPD using carbon dioxide was found to preserve these delicate illites as the technique does not result in the formation of a liquid/vapour phase interface.

The Magnus work may have been an isolated example as the illite fibres were very thin (around 2–4 μm) and thicker illite filaments may not suffer the same degree of damage since they are much stronger. However, the risk of evaporative drying-induced illite clay damage still exists (Fig. 4.22 from Byrne and Patey, 2004). Nevertheless, the use of lower temperature flush cleaning and total immersion cleaning techniques followed by humidified oven drying can normally prevent most forms of clay damage.

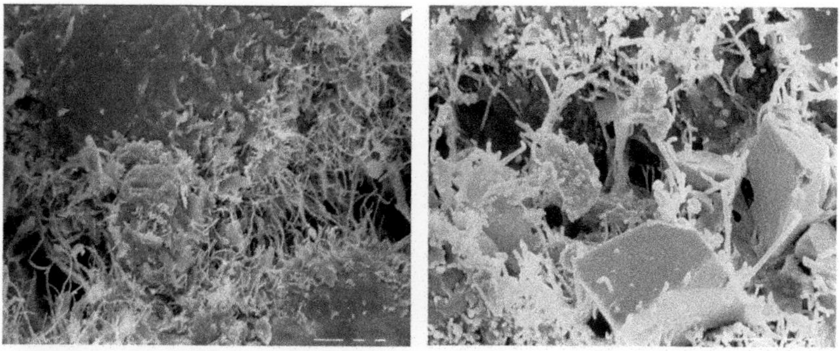

FIGURE 4.22 SEM photomicrograph indicating damage to illite on evaporative drying. *From Byrne and Patey (2004).*

4.5.5.2 Clay Swelling and Migration

Other forms of clay-induced petrophysical and rock textural damage include clay swelling and clay migration and dispersion on saturation with water, or as a result of excessive flow rates on flow-through cleaning or SCAL testing.

Drying the core sample or exposure of the core sample to lower salinity brine than the formation water will disturb the often delicate chemical equilibrium at the exchange sites. On resaturation with formation brine, for example, prior to SCAL testing, or on exposure to lower salinity brine or fresh water (for example when cutting plugs in tap water), the existing cations on the clays will hydrate. The hydration forces will depend upon the original cation at the exchange sites. Sodium cations are the most readily exchangeable and can promote adsorption of up to 32 layers of water molecules in smectite interlayers, causing clay volume expansion and flocculation. This expansion of the swelling clays will block pore throats and cause severe permeability reduction. In some cases, if the rock is inherently weak, the expansion force on rehydration with formation water containing Na^+ cations can be so powerful that it can cause disintegration of the test plugs. Figure 4.23 shows photographs of two plugs of weak reservoir sands from SE Asia which have been exposed to formation water after drying. The sands have a clay content of 20–30% composed predominantly of smectite.

In lower exchange capacity clays such as kaolinite, illite and chlorite, hydration does not cause swelling but can generate sufficient osmotic pressure to cause separation of individual clay platelets which, under the influence of flowing liquid, are dispersed into the pore network. Clay dispersion, subsequent migration and flocculation can cause plugging or bridging of pore throats, leading to severe permeability damage. The clay structural and morphological characteristics are permanently changed.

FIGURE 4.23 High smectite content core plug exposed to formation water after drying.

Similar mechanisms can occur during high-rate cleaning by solvent flushing or high-rate flooding in SCAL tests. If the rate exceeds the critical velocity, illite and kaolinite can become detached from the host grains and are free to migrate through the pore network (like logs floating downstream) and can eventually block pore throat restrictions. Again this causes permanent petrophysical damage. One way to test for this critical velocity is to determine plug liquid permeability to the cleaning or SCAL test fluid at a base or reference low rate (normally less than 1 ml/min) and then increase the flow rate in a series of steps, each time determining permeability at the higher rate and again after returning to the base low rate. Analysis of the permeability response to increasing rates can be used to set a maximum flow rate for SCAL tests or cleaning procedures for the particular formation.

4.5.6 Testing Without Drying

The specialist cleaning and drying techniques described above are available to help prevent petrophysical damage in particularly susceptible clay-rich samples, but there is always a suspicion that some damage could still occur. In this case, a pre-cleaning study is essential to determine the optimum cleaning method for SCAL tests. In this example from a gas reservoir in the Southern North Sea, SEM analysis indicated that the primary intergranular porosity has been significantly reduced by the presence of thick grain coating and pore-bridging illite, as indicated in Fig. 4.24.

The plugs were cut from preserved core using synthetic formation water (SFW) and then were saturated and flushed with brine, and permeability to brine (k_w) was determined without drying the samples. The plugs were then flush cleaned (benign cleaning) in a sequential brine/methanol sequence of decreasing brine—increasing methanol concentrations, as direct methanol contact with brine would have led to salt precipitation. After flowing with

100% methanol, the samples were dried under standard humidified conditions (HOD) and air permeability (k_{air}) and porosity were determined. The samples were then harshly cleaned using hot Soxhlet extraction in methanol and dried at 60 °C in a conventional vacuum oven (COD). The permeability results for three samples are plotted in Fig. 4.25. Although brine permeabilities and air permeabilities are not equivalent (due to rock/fluid interaction and Klinkenberg effects), the data do show an increase in permeability by a factor of 8–10 times. There is little difference in the poroperm properties between benignly cleaned (flush clean-HOD) and harshly cleaned (Soxhlet-COD)

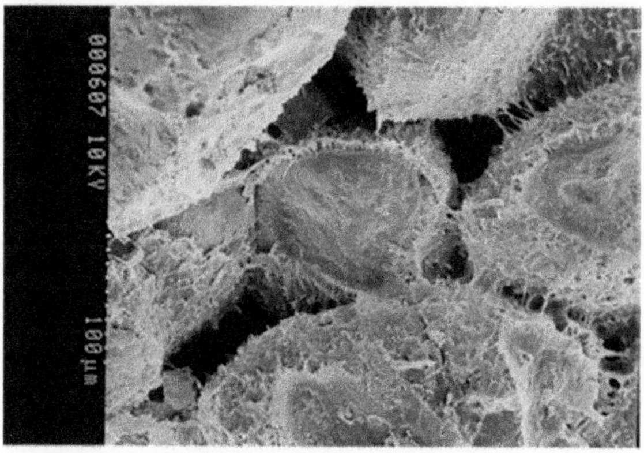

FIGURE 4.24 SEM photomicrograph—Southern North Sea gas reservoir illitic sands.

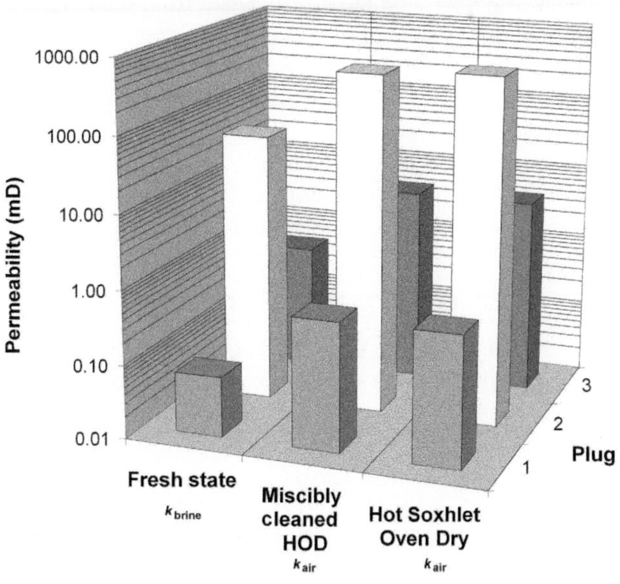

FIGURE 4.25 Permeability enhancement as a result of drying on three illitic sand samples.

samples. The petrophysical damage can primarily be attributed to evaporative drying causing partial collapse of the illitic clays which have a powerful control on the rock permeability.

Based on the results of this pre-cleaning study, it was decided not to risk permeability damage so the samples scheduled for SCAL were tested without initially drying the plugs. This involved cutting plugs from preserved core with SFW, flush cleaning to methanol and then saturation with formation brine under elevated pore pressure to remove any air. SCAL tests—including brine permeability, formation resistivity factor, resistivity index and air-brine porous plate capillary pressure—were performed directly on the saturated samples without the need to dry. Only at the end of the test programme were the plugs flush cleaned and humidity dried.

This procedure is also sometimes used for SCAL testing of unconsolidated or weak plugs, but it can compromise basic quality control of saturation calculation data since it is difficult to obtain an initial sample pore volume without resorting to the use of flushing with tracers such as nitrated water. Pore volume can be measured on completion of the tests with the sample finally flush cleaned and dried. As porosity is less affected by clay damage than permeability, the pore volume determined after drying is normally sufficiently accurate for porosity and saturation calculations.

4.6 CORE CONDITIONING FOR POROSITY MEASUREMENTS

4.6.1 Porosity Evaluation

Sound assessment of porosity on the basis of data from both cores and logs is essential if reliable and meaningful evaluation of reserves are to be made. Assessment of porosity data must be compatible with the petrophysical model employed and with all other porosity-related parameters which are ultimately fed into a petrophysical model.

The basic underlying reason for potential uncertainty is usually the different porosity concepts and related petrophysical models employed in the evaluations—principally the difference between total and effective porosity.

4.6.2 Total and Effective Porosity Concepts

Total porosity (ϕ_t) relates to the total pore volume in the formation or core that is filled with fluids. In some reservoirs, such as vuggy limestones, there are a number of isolated pores which contain water—and contribute to the overall pore volume—but are not well interconnected so that they do not contribute to the hydrocarbon storage capacity of the rock. In sandstones, however, the relative volume of isolated pores is small and normally considered to be negligible. The interconnected pores that are available for hydrocarbon storage make up the *effective porosity*, which is numerically less than total porosity. If clays are present, CBW contributes to the total porosity but not the effective porosity. In petrophysics, debate still rages over whether petrophysical evaluation should be based on total or effective porosity in shaly sands. One of the principal problems is that the various effective porosity definitions, concepts and determination techniques yield different effective porosities. Total porosity, on the other hand, is much less contentious.

Shales are fine-grained sediments, consisting of clay-size particles. The particles are a mixture of clay and other minerals and shales have porosity but no permeability. The porosity is made up of the very small pore spaces (microporosity) between the clay minerals and other particles and the water bound to the surfaces of the clay mineral. This CBW is often considered to be an integral part of the clay minerals. The volume occupied by CBW is not available for storage of hydrocarbon. Thus, while clays and shales can have a significant total porosity, effective porosity and permeability are effectively zero.

For a formation which is 100% sand (without isolated pores), the rock is made up of only two components: the sand matrix (grains) and the pore space which will be filled partly with oil and partly with water, the relative saturations of each depending on the pore throat size distribution and the height of the formation above the free water level. In this system, the total and effective porosities will be equal. As the clay content in a reservoir rock increases so the amount of dry shale contribution to the matrix increases, the amount of CBW in the total pore space increases, and the effective porosity and permeability reduce until, at a certain clay content, the 'sand' will be considered to be non-reservoir quality.

Efforts over the last 50 years to integrate the effective and total porosity concepts in shaly sands with core and log analysis involve either models that use the wetted clay volume, V_{cl}, or models that use the properties of chemically bound (double layer) clay water ('Qv' models).

4.6.2.1 V_{cl} Models

It is argued that the CBW associated with clays occupies part of the total porosity rendering that component unavailable for hydrocarbon storage. Hence, the total porosity of shaly sand (ϕ_t) should be discounted by the porosity of the clay present in it (ϕ_{cl}), and the resultant effective porosity (ϕ_e) becomes:

$$\phi_e = \phi_t - V_{cl}\phi_{cl}$$

V_{cl} is the ratio of the volume of wetted shale to the volume of the rock and includes the clay-bound and structural water. There are a number of methods available in petrophysics to determine V_{cl}. ϕ_{cl} is the total porosity of the clay whose pore volume only comprises CBW and structural water.

The traditional V_{cl} models are mainly log interpretation oriented and it can be argued that these models are impossible to integrate with core analysis data since V_{cl} cannot be measured on cores. In fact, V_{cl} is derived using parameters derived from shale laminae in the formation and does not necessarily correspond to the properties of dispersed authigenic clays lining the pores and in direct contact with the pore fluids.

4.6.2.2 *Clay Water Models*

The common feature of the two principal clay water models (Clavier et al., 1984; Waxman and Smits, 1968) is the recognition that it is not the amount of clay, but the CEC (related to Qv), that is the critical parameter for the quantification of both clay conductivity for Sw calculations and CBW for porosity calculations.

The amount of CBW defined via Qv is less than that determined from $\phi_{cl}V_{cl}$ because only the water adsorbed on the clay mineral surfaces (proportional to CEC) and the water required for the hydration of the clay counterions (CEC) are included—capillary-bound 'far' water in the shales is excluded.

In this approach, the effective porosity is the total porosity excluding 'anion-free CBW' and, according to Hill et al. (1979), can be defined as

$$\phi_e = \phi_t\left(1 - 0.6425S^{-0.5} + 0.22\right)\mathrm{Qv}$$

where S is the formation water salinity (in g/l NaCl) and Qv is the CEC normalised by pore volume (meq/ml).

These models have the advantage that both ϕ_t and Qv can be measured on cores, facilitating the integration of all porosity-related input and output parameters with core data, and the validity of the models can be judged on the basis of the scientific considerations and supporting laboratory experiments on which they are based.

4.6.3 Log and Core Porosity

Figure 4.26 shows one interpretation of the concepts of the log- and core-derived porosity associated with a shaly sand. It is assumed that the volume of isolated pores is insignificant.

The neutron porosity logging tool employs a neutron source to measure the hydrogen index in a reservoir. As hydrogen atoms are present in both water- and oil-filled reservoirs, measurement of the hydrogen index allows estimation of the amount of liquid-filled porosity. The measurement therefore samples all hydrocarbons and all water in the formation and cannot discriminate between the pore water, CBW in the clay microporosity (interlayers) and the structural water (OH^-) within the clay minerals themselves. If samples are dried in a furnace at temperatures higher than 450 °C, then the pore water, CBW and structural water are lost and the clay mineral structures become amorphous. If porosity is measured on a furnace-dried sample, then this would represent the same porosity system as sensed by the neutron tool.

In the density logging measurement, gamma rays interact with electrons in the formation and are scattered. The number of scattered gamma rays that reach the detector is related to the formation's electron density which is

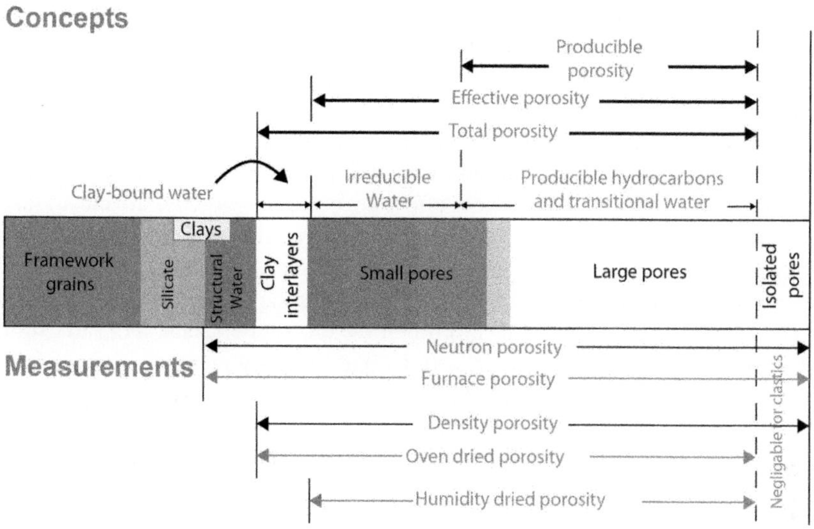

FIGURE 4.26 Porosity concepts in shaly sand. *After Eslinger and Pevear (1988).*

related to the formation's bulk density. Knowing bulk density, matrix or grain density and the density of the fluid filling the pore space, it is possible to determine porosity. The density tool responds to the pore spaces filled with CBW and pore water—not the structural water—so the porosity from a density tool is less than that from a neutron tool in a shaly sand. In this definition, *total porosity* is classified as the summation of the pore space associated with large pores plus the small pores filled with irreducible water trapped by capillary forces plus the CBW microporosity. This is argued to be equivalent to the total porosity pore volume 'seen' by the density log.

Juhasz (1988) suggested that drying cores at 105 °C in a vacuum oven exactly removes all the CBW without removing structural water. Consequential porosity measurements will force helium into the microporosity previously occupied by water, so core porosity will be the same total porosity as seen by the density tool. However, drying core plugs at 105 °C may not remove all pore fluid that would be seen by the logging tools as total porosity. Experiments reported by Peveraro and Hook (1988) showed that only in smectites would clay water loss be likely to occur at 105 °C: samples containing mixed-layer illite/smectite did not lose clay water until at least 150 °C.

In the definition illustrated by Fig. 4.26, the *effective porosity* (ϕ_e) is equal to the total porosity ϕ_t minus the CBW, since the CBW is not free to move. Humidity drying is a convenient method to preserve clay structures so that permeability and grain density (for density log interpretation) could be better preserved. Therefore, it is assumed that shaly sand cores dried using humidity drying (HOD) would condition a core to represent the effective porosity: that is HOD would remove the pore water, the capillary-bound water in small pores, but retain the CBW in the clay interlayers. In reality however, this is not true and a certain amount of CBW (correlatable with CEC) will be removed by HOD so the amount of clay-associated water retained by this drying technique is too small. Therefore, the resulting HOD core porosity lies somewhere between the total and effective porosity and has no unique relationship with either log-derived effective or total porosity. In clean sands and carbonates without isolated pores, humidity-dried and harshly dried plug porosities will be the same, or at least within the measurement repeatability.

The requirement for total porosity conditioning creates a major dilemma when measuring petrophysical properties on shaly sands. For all SCAL measurements—permeability, resistivity and capillary pressure—it is essential to preserve clay structures so that core measurements are reliable and as representative of the intact reservoir formation as possible. Consequently, utmost care must be taken to ensure that the optimum plug preparation method is selected that preserves the clay structures and does not alter porosity or permeability. However, if routine core porosity measurements in RCA are used to calibrate density logs, then they need to be prepared in such a way as to remove the interlayer CBW so that helium can access the CBW-vacated space and bring core porosities closer to total porosity. Whereas the core porosity may be appropriate for density log calibration and integration, the permeabilities measured on these plugs may not. Harsh Soxhlet cleaning and high-temperature conventional oven drying can lead to clay destruction such that RCA permeabilities are significantly enhanced, and the samples become no longer representative of the reservoir for SCAL measurements. Using benign preparation methods and humidity drying may preserve the petrophysical properties, but the core porosities are no longer representative of a total or an effective porosity system.

Figure 4.27 provides an example of air permeabilities measured initially after humidity (HOD) drying and after conventional (COD) drying on shaly sand samples. On average, the COD permeabilities are one to two orders of magnitude higher than HOD permeabilities. Faced with this quandary, the petrophysicist may consider preparing plugs for RCA porosity and permeability by different methods. That is, porosity and permeability are determined on separate suites of parallel pair plugs with one set harshly cleaned and dried (COD) and the other set prepared using benign methods (HOD). Alternatively, permeability can be determined after humidity drying, and porosity is measured on the plugs after harsh cleaning and/or high-temperature oven drying. In the petrophysics interpretation, permeability prediction models are often based on density porosity so the correct core–log integration procedure would be to plot harshly dried (total) porosity versus benignly dried permeability. Otherwise, the predicted permeability from density logs would be much higher than in the reservoir, even after stress and saturation corrections have been applied.

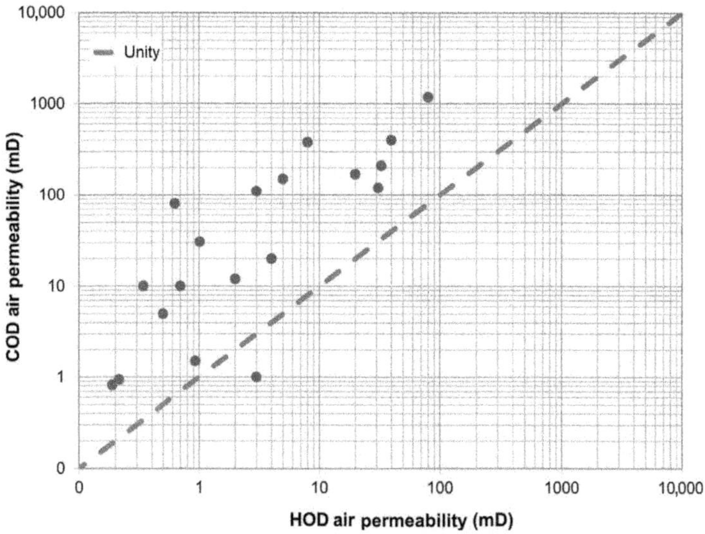

FIGURE 4.27 Comparison of air permeabilities measured after humidity oven drying (HOD) and after conventional oven drying (COD).

4.7 SPECIAL CONSIDERATIONS IN CORE PREPARATION

4.7.1 Wettability

Core cleaning methods are designed to remove oil and water (plus other drilling mud contaminants) from a plug sample. Invariably therefore, core cleaning should render the core samples water wet. In fact, core cleaning to condition cores to an initial water-wet condition is a prerequisite prior to wettability restoration processes. The cleaning methods and solvents are selected primarily on this basis. Anderson (1986) provides an excellent description of the methods and techniques used. Depending upon the oil and rock types, extremely harsh cleaning methods and solvents may be required, so the requirements for wettability restoration must be balanced against the necessity to preserve delicate minerals. Wettability and wettability testing are discussed further in Chapter 7.

Recent, but mostly unpublished, evidence shows that hot Soxhlet extraction in toluene can promote a non-water-wetting condition, so that the cleaned core is not wholly water wet. When aged in crude oil in an attempt to restore native wettability, the Soxhlet-cleaned samples can become quite oil wet. Hot Soxhlet extraction in toluene removes both water surrounding the grains and the non-polar fraction of the crude oil, thus permitting the heavier (polar) oil fraction which contains surfactants and asphaltenes to come into direct contact with mineral surfaces and thereby create a mechanism to alter grain wettability. The phenomenon may be exacerbated by low initial water saturation and single cycle cleaning. A second cleaning cycle in toluene can help to dissolve any asphaltene precipitates remaining after first cycle cleaning and aid in restoring water wetness. Solvent flush cleaning at elevated temperatures has proven to be quite effective in removing wettability contaminants and, as water is not vapourised, should not induce a non-water-wetting condition.

A cleaning pre-study may be required to assess the optimum wettability decontamination cleaning method and to evaluate any petrophysical damage to the core.

4.7.2 Carbonates and Chalks

As these carbonates rarely contain sensitive or delicate clays, the normal core preparation method is hot Soxhlet cleaning with methanol and toluene as the most used cleaning liquids. Due to the low permeability of chalk extended cleaning periods are often required and Soxhlet cleaning may not clean the chalk plugs completely. However, the effect on chalk grain density is normally within measurement error.

Drying is normally done at 105–110 °C to remove adsorbed water and obtain the most accurate porosity and grain density figures.

4.7.3 Halite

Halite forms an important cement in certain formations—especially those with a high formation water salt concentration. The challenge in cleaning samples containing halite is to remove salt from the pore water and yet not dissolve the natural halite cement.

In conventional hot Soxhlet cleaning with toluene and methanol, the water content will evaporate rapidly in contact with toluene, precipitating its salt content into the pore volume. Removal of precipitated salt from the pore space using methanol will simultaneously remove native halite cements and alter the porosity and permeability properties of the core.

The selected cleaning method must remove the formation water and its salts without significantly affecting the native halite cement content. Methanol will dissolve both, so it is recommended to use iso-propyl alcohol (IPA) as the solvent. It is impossible to avoid some dissolution of the native halite, but the relatively large surface area of the precipitated salts will cause them to be dissolved more rapidly than cementing halite, which has a relatively small exposure surface. As direct application of water-miscible solvents (e.g. methanol and IPA) with saline pore brine can often cause precipitation, the initial stage of cleaning should therefore attempt to remove a significant volume of water prior to application of IPA.

Before attempting to prepare samples containing halite, a cleaning study should be performed to assess any damage to the core and the optimum cleaning method (Fig. 4.28). This would normally involve flush cleaning plugs with toluene at elevated temperatures (70–80 °C), then IPA, using cyclic

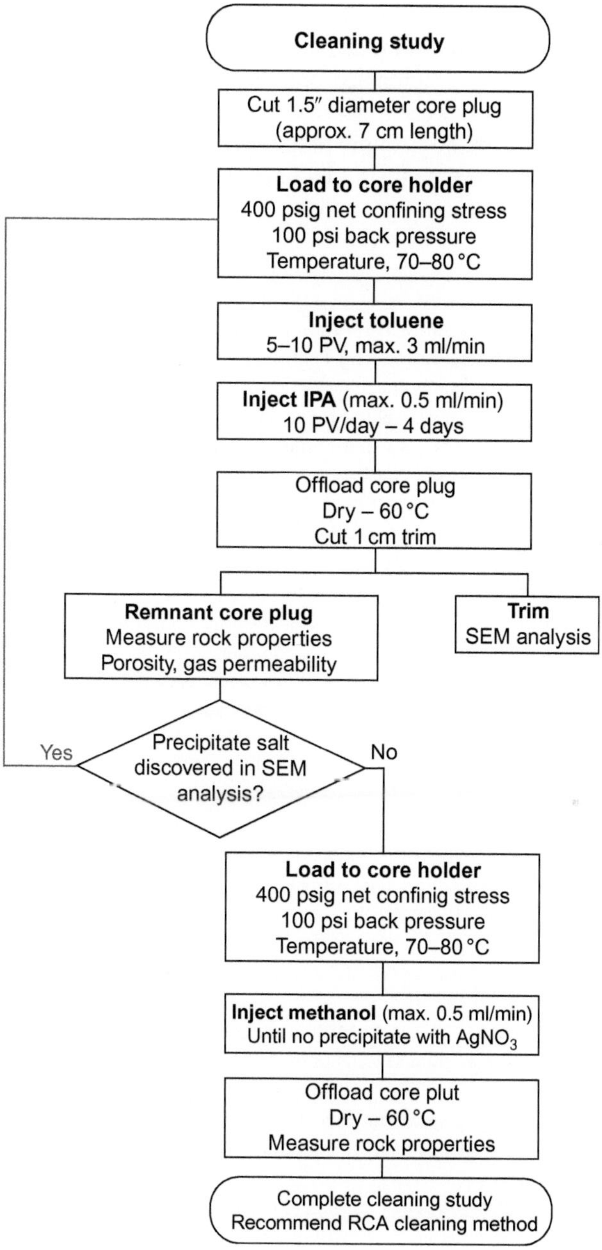

FIGURE 4.28 Example flow chart for cleaning pre-study for halite-cemented samples.

periods of injection and stasis (to allow diffusion). A thin end trim is sliced off of one end of the plug and examined under the SEM. If this shows salt precipitate, the IPA cleaning is continued until no precipitated salts remain. When the salt precipitates have been cleared, the plugs are flushed with methanol until the effluent shows no precipitation to silver nitrate. Measuring porosities after each cleaning treatment, in conjunction with SEM analysis, can help define the optimum cleaning method.

REFERENCES

American Petroleum Institute (API), 1988. Recommended Practice RP 40. Recommended Practices for Core Analysis, second ed. API Publishing Services, Washington, DC.

Anderson, W.G., 1986. Wettability literature survey—part 1: rock/oil/brine interactions and the effects of core handling on wettability. J. Pet. Technol. 38 (10), 1125–1144.

Bush, D.C., Jenkins, R.E., 1977. CEC determinations by correlations with adsorbed water. In: Society of Petrophysicists and Well Log Analysts 18th Annual Logging Symposium, 5–8 June, Houston, Texas, Paper SPWLA-1977-H.

Byrne, M., Patey, I., 2004. Core sample preparation—an insight into new procedures. In: International Symposium of the Society of Core Analysts, Abu Dhabi, UAE, 5–9 October, Paper SCA2004-50.

Clavier, C., Coates, G., Dumanoir, J., 1984. Theoretical and experimental bases for dual-water model for interpretation of shaly sands. Soc. Pet. Eng. J. 24 (2), 153–168.

Eslinger, E., Pevear, D., 1988. Clay minerals for petroleum geologists and engineers. Society of Economic Paleontologists and Mineralogists Short Course Notes No. 22, Tulsa, Oklahoma.

Heaviside, J., Langley, G.O., Pallat, N., 1983. Permeability characteristics of Magnus reservoir rock. In: 8th European Formation Evaluation Symposium. Society of Petrophysicists and Well Log Analysts, London, pp. 1–29.

Hill, H.J., Klein, G.E., Shirley, O.J., Thomas, E.C., Waxman, W.H., 1979. Bound water in shaly sands—its relation to Qv and other formation properties. Log Anal. 20 (3), 3–19.

Juhasz, I., 1988. Porosity systems and petrophysical models used in formation evaluation. In: Society of Petrophysicists and Well Log Analysts, London Chapter, Porosity Seminar, 26th April.

Keelan, D.K., 1982. Core analysis for aid in reservoir description. J. Pet. Technol. 34 (11), 2483–2491.

McHardy, W.J., Wilson, M.J., Tait, J.M., 1982. Electron microscope and X-ray diffraction studies of filamentous illitic clay from sandstones of the Magnus field. Clay Miner. 17, 23–39.

Neasham, J.W., 1977. The morphology of dispersed clay in sandstone reservoirs and its effect on sandstone shaliness, pore space and fluid flow properties. Paper SPE 6858. In: Society of Petroleum Engineers Annual Fall Technical Conference and Exhibition, 9–12 October. Denver, Colorado.

Peveraro, R., Hook, J., 1988. The determination of effective porosity—the experimental basis and comparison with log data. In: Society of Petrophysicists and Well Log Analysts, London Chapter, Porosity Seminar, 26th April.

Soxhlet, F., 1879. Die gewichtsanalytische Bestimmung des Milchfettes. Dingler's Polytechnisches J. 232, 461–465, (in German).

Spooner, P., 2014. Lifting the fog on confusion surrounding clay and shale in petrophysics. In: Society of Petrophysicists and Well Log Analysts 55th Annual Logging Symposium, 18–22 May, Abu Dhabi, United Arab Emirates.

Waxman, M.H., Smits, L.J.M., 1968. Electrical conductivities in oil-bearing shaly sands. Soc. Pet. Eng. J. 8 (2), 107–122.

Worthington, P.F., Toussaint-Jackson, J.E., Pallatt, N., 1988a. Effect of sample preparation upon saturation exponent in the Magnus Field, U.K. North Sea. Log Anal. 29 (1), 48–53.

RECOMMENDED READING/VIEWING

Bil, K.J., De Waal, J.A., Dicker, A.I.M., Kantorowicz, J.D., 1988. Petrophysical core analysis of sandstones containing delicate illite. Log Anal. 29, 05.

Dacy, J.M., Lerner, D.B., Raible, C.J., Rathmell, J.J., Swanson, G., Walls, J.D., 1990. SCA guidelines for sample preparation and porosity measurement of electrical resistivity samples: part II—sample preparation and porosity measurement. Log Anal. 31, 02.

McCrone, A.W., 1963. Quick preparation of peel prints for sedimentary petrography. J. Sediment. Petrol. 33 (1), 228–230.

Petrowiki, 2014. Formation Damage from Fines Migration. http://petrowiki.org/Formation_damage_from_fines_migration (accessed 9 September 2014).

Ruhovets, N., Fertl, W.H., 1982. Digital shaly sand analysis based on Waxman-Smits model and log derived clay typing. Log Anal. 23 (03), 7–22.

Shaw, H.F., 1980. Clay minerals in sediments and sedimentary rocks. In: Hobson, G.D. (Ed.), Developments in Petroleum Geology. Allied Sciences Publication Ltd., London, pp. 53–85.

Thomas, E.C., Stieber, S.J., 1975. The distribution of shale in sandstones and its effect upon porosity. In: Society of Petrophysicists and Well Log Analysts 16th Annual Logging Symposium, 4–7 June, New Orleans, Louisiana, Paper SPWLA-1975-T.

Waxman, M.H., Thomas, E.C., 1974. Electrical conductivities in shaly sands—I. The relation between hydrocarbon saturation and resistivity index. II. The temperature coefficient of electrical conductivity. J. Pet. Technol. 26 (02), 213–225.

Worthington, P.F., Toussaint-Jackson, J.E., Pallatt, N., 1988b. Effect of sample preparation upon saturation exponent in the Magnus Field, U.K. North Sea. Log Anal. 29 (1), 48–53.

Worthington, P.F., 1985. The evolution of shaly-sand concepts in reservoir evaluation. Log Anal. 26 (01), 23–40.

Yaalon, D.H., 1962. Mineral composition of the average shale. Clay Miner. 5, 31–36.

http:/www.youtube.com/watch?v=XBZkVEWxlnU. Soxhlet apparatus.

http:/www.youtube.com/watch?v=2L6MDOW88zc. Soxhlet apparatus.

Chapter 5

Routine Core Analysis

Chapter Outline

5.1 Introduction 181
5.2 Fluid Saturation Measurements 182
5.2.1 Retort Method 183
5.2.2 Dean–Stark Analysis 187
5.3 Porosity Measurements 195
5.3.1 Helium Grain Volume and Grain Density 199
5.3.2 Helium Pore Volume 204
5.3.3 Bulk Volume 210
5.3.4 Liquid Saturation Porosity 216
5.3.5 Accuracy and Repeatability of Porosity Measurements 219
5.4 Permeability Measurements 221
5.4.1 Definitions 221
5.4.2 Darcy's Law 221
5.4.3 Non-Darcy Flow: Klinkenberg Effects 224
5.4.4 Non-Darcy Flow: Forchheimer Effect 228
5.4.5 Steady-State Permeability Measurements 229
5.4.6 Unsteady-State Permeability Measurements 239
5.4.7 Steady-State Liquid (Absolute) Permeability Measurements 247
5.4.8 Probe or Profile Permeability Measurements 252
5.5 Whole Core Analysis Measurements 262
5.5.1 Sample Preparation 262
5.5.2 Fluid Saturations 263
5.5.3 Porosity 263
5.5.4 Gas Permeability 263
References 266
Recommended Reading/Viewing 267

5.1 INTRODUCTION

There is no strict discrimination between routine core analysis (RCA), which is often referred to as basic or conventional core analysis (CCA), and special core analysis (SCAL). One lab's RCA capabilities might reflect another lab's SCAL capabilities.

Generally, RCA is accepted to be lower cost and faster turnaround than SCAL. It involves fluid saturation measurements and petrophysical measurements on dry samples normally at ambient or nominal stress and temperature conditions, whereas most SCAL measurements are made on plugs which have

Developments in Petroleum Science, Vol. 64. http://dx.doi.org/10.1016/B978-0-444-63533-4.00005-6

been conditioned and tested to reflect reservoir-appropriate saturations and often reservoir conditions.

A typical RCA programme involves the following measurements on plugs and core samples:

- Fluid saturations
- Porosity
- Air (nitrogen) and Klinkenberg permeability
- Probe (or profile) permeability.

The test plugs used for RCA are typically 1″ or 1.5″ diameter, with larger samples being preferred for saturation and porosity measurements, in particular.

Fluid saturations, porosity and permeability can be measured on full diameter or whole core samples. However, the time and costs involved mean that whole core analysis is normally only performed where plug measurements would be considered to be unrepresentative. Examples include sandstone conglomerates and vuggy or naturally fractured carbonates where the scale of the heterogeneity is larger than the plug scale.

5.2 FLUID SATURATION MEASUREMENTS

The two principal methods used to determine fluid saturations on core by extraction of the fluids are:

- The retort method (also referred to as the summation of fluids porosity method).
- The Dean–Stark extraction method.

Historically, the retort was the more common method. It provides direct measurements of water, oil and gas saturation, is rapid and is cheaper than Dean–Stark. The principal disadvantages are that gas and liquid saturations are measured on different samples, possible oil volume errors introduced by hydrocarbon cracking, the release of clay and hydrated water at higher temperatures and its poorer accuracy (compared to Dean–Stark). For these reasons, retort fluid saturations are not now considered to be a best practice method and the method is not recommended for core fluid saturation determination on conventional reservoir core other than in qualitative terms. However, as the technique is seeing a renaissance in shale core analysis, and as petrophysicists and engineers may be faced with interpreting and integrating data from legacy retort data, the method and its advantages and drawbacks are described below.

The Dean–Stark method involves vaporisation and distillation of water from the sample using toluene or xylene extraction. The water volume is collected and measured, and salt correction factors applied depending on the formation water salinity. Saturations and porosity are measured on the same sample and provide total rather than effective porosity and water saturation.

5.2.1 Retort Method

This was, and in some labs still is, a relatively common method, used by many laboratories. It provides estimates of water, oil and gas saturation.

5.2.1.1 Sample Preparation

The technique does not rely on testing a regularly shaped plug sample. Liquid (oil and water) and gas saturations are measured on separate, though adjacent, rock samples chipped from the core, normally close to a RCA plug location.

In some applications, a core plug is used, but since this is a destructive test, the plug cannot then be used for poroperm measurements. In this case, gas saturation is determined from initial helium expansion measurements on the core which is partly saturated with water and oil. This provides an apparent grain volume, which includes the volume of the grains plus the volume of oil and water. Subsequent retort extraction measurements can be used to determine oil and water saturation.

5.2.1.2 Test Equipment

Typical apparatus is shown in Fig. 5.1, and a schematic is provided as Fig. 5.2.

The core scheduled for oil and water volume measurements is placed inside a stainless steel retort which can be sealed with a cap. A long condensing tube is fitted to the base of the tube which is fitted with a screen to prevent grain loss from the retort vessel. The retort tubes are placed inside a temperature controlled oven. The condensing tubes are cooled by circulating chilled water in a water bath, and a graduated tube is used to collect condensed fluids.

A mercury pump is often used for gas volume measurement. It consists of a valve-sealed pressure vessel to accommodate the sample, a pressure gauge and a pump with Vernier scale.

FIGURE 5.1 Retort apparatus.

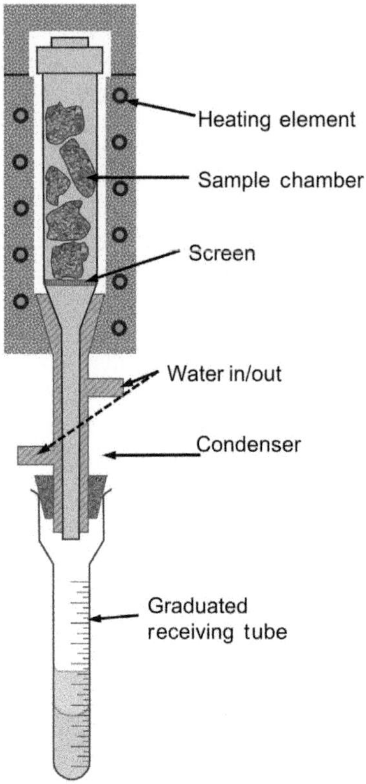

FIGURE 5.2 Retort apparatus schematic.

5.2.1.3 Test Procedures

Separate samples, from the same core sections, are used to determine water and oil volumes and gas volumes in the core.

Gas saturation is determined on around 10–15 g of core chips which are placed in the pressure cell of the mercury pump. The bulk volume of the sample is determined by immersion in mercury. The cell valve is closed and the mercury pressurised to 750–1000 psi to compress the air in the sample. At 1000 psi, the air will be compressed to around 2% of its volume at atmospheric conditions, while it is assumed that the volume of water and oil in the sample remains essentially constant as liquid is much less compressible. Consequently, the volume of mercury injected (corrected for pump expansion) is taken to represent the gas volume. In some labs, gas volume is determined from initial helium expansion measurements on the core chips.

For water and oil volume measurement, 100–200 g core chipped from the same core location as the gas sample is crushed, screened to remove any fines, and the crushed rock is poured into individual retorts. The retort end caps are sealed and placed in the oven, connected to the condenser tubes, and the oven temperature is raised.

The oven temperature is initially set at 350 °F (Hensel, 1982) to 400 °F for times that vary between 20 min and an hour. Some laboratories select a lower temperature, just above the boiling point of water, and some using slower heating periods. The temperature is selected to vapourise water from the core samples which is then condensed and measured in the graduated tube. The water includes pore water, adsorbed water, clay-bound water, water of hydration of common rock forming minerals, but not the structural water of clays. Cores with significant evaporate or smectite content are therefore not suitable for this analysis. The condensed water volumes are monitored until production volumes stabilise and are recorded. An example water production trend is shown in Fig. 5.3.

The temperature is then raised to 1000–1200 °F which vapourises the oil. The vapour then condenses and, when no more liquid is given up, the final volumes of liquid in the graduated tube are recorded. Oil being less dense than water sits on top of the water in the collection tube. Any water recovered during high-temperature heating is assumed to be crystalline from clay or salt minerals and should not be included in the computed water saturation as the water is not part of the total porosity. Both the water volume and oil volume are corrected for volume changes due to distillation (water) and cracking (oil), on the basis of the formation water brine salinity and an experimentally derived oil retort curve. The sample bulk volume may be determined by mercury immersion of the sample, or it may be computed from grain density data from an adjacent plug.

To speed up the process, it was common in some labs to set the initial retort temperature at the maximum value (1200 °F) so that water is given up first as the core heats up then, at higher temperatures, oil. In this case, it is difficult to distinguish between pore water, clay-bound water, water of hydration and the structural or crystalline water production trends, so the total water volume recovered cannot be corrected for crystalline water.

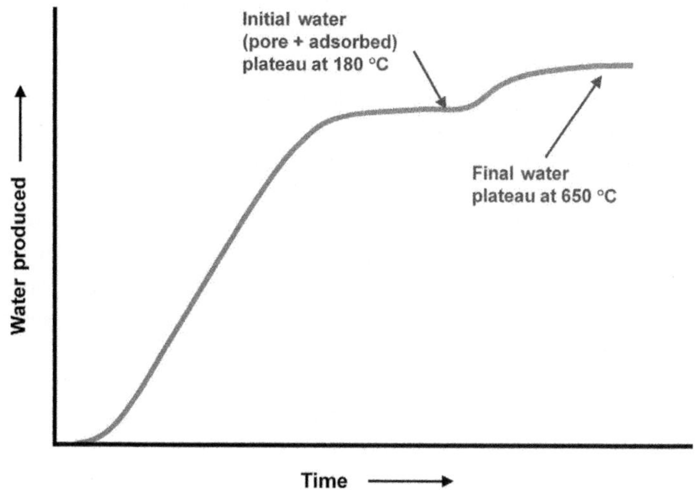

FIGURE 5.3 Example water production trend during retort heating.

The test is destructive, so samples cannot be used for any other core analysis tests.

5.2.1.4 Saturation and Porosity Calculations

The gas (Sg), oil (So) and water saturations (Sw) are determined from the corrected volumes of oil (Vo) and water (Vw) recovered, the bulk volume of the retort sample (Vb_{ret}), the volume of mercury injected (V_{Hg}) in the gas saturation sample and the bulk volume of this sample (Vb_{Hg}):

$$Sg_{(b)} = \frac{V_{Hg}}{Vb_{Hg}} \text{(fraction bulk volume) and Sg} = \frac{Sg_{(b)}}{\left(Sg_{(b)} + So_{(b)} + Sw_{(b)}\right)} \text{(fraction pore volume)}$$

$$So_{(b)} = \frac{Vo}{Vb_{ret}} \text{(fraction bulk volume) and So} = \frac{So_{(b)}}{\left(Sg_{(b)} + So_{(b)} + Sw_{(b)}\right)} \text{(fraction pore volume)}$$

$$Sw_{(b)} = \frac{Vw}{Vb_{ret}} \text{(fraction bulk volume) and Sw} = \frac{Sw_{(b)}}{\left(Sg_{(b)} + So_{(b)} + Sw_{(b)}\right)} \text{(fraction pore volume)}$$

The volumetric and saturation data can be used to compute a summation of fluids porosity, ϕ_{sum}, for the combined retort and gas samples with:

$$\phi_{sum} = \frac{Vo + Vw + Vg}{Vb}$$

The gas volume from the separate mercury injection sample is adjusted on the difference in weights between the gas volume and oil/water retort samples assuming a constant bulk density.

Figure 5.4 plots the volume of water produced from a clean sand compared to a shaly sand as a function of time, at a temperature of 400 °F. Water production (pore- and capillary-bound water) from the clean sand rapidly becomes stable, whereas water loss from clays in this case is continuous. Since no plateau can be seen, it is difficult to determine the relative proportions of clay and pore water. As a result, in shaly sands, the water volume may be too high, the water saturation will be too high and the calculated summation of fluids porosity will be systematically higher than helium porosity measured on adjacent samples. Typical data are shown in Fig. 5.5 from a slightly shaly sand. The summation of fluid porosities is systematically higher (by around 4 porosity units (p.u.) on average) than the partner plug helium porosities.

5.2.1.5 Advantages and Drawbacks

The main advantages, drawbacks and issues associated with the retort method are summarised in Table 5.1.

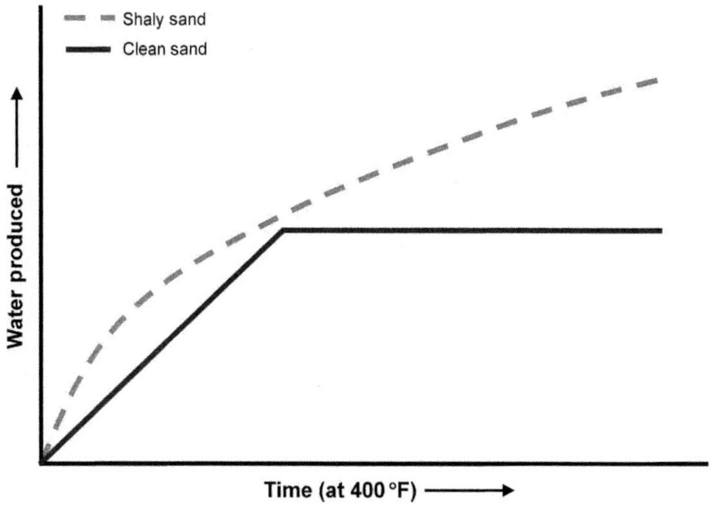

FIGURE 5.4 Example retort water production curve for shaly sand.

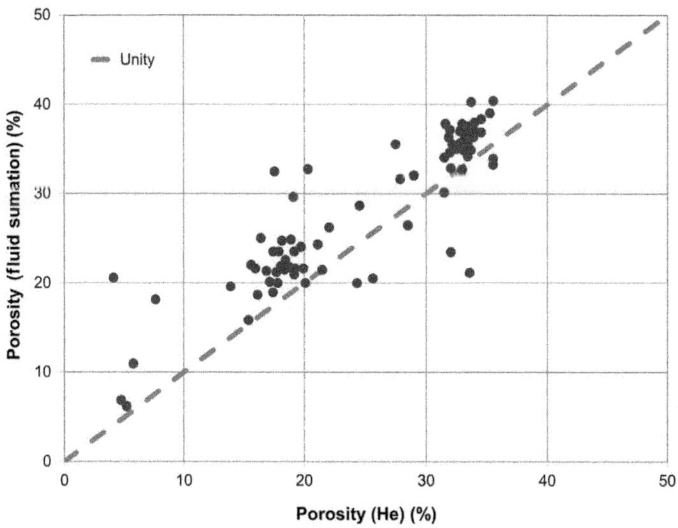

FIGURE 5.5 Comparison of retort fluid summation porosities and helium plug porosities in a shaly sand.

5.2.2 Dean–Stark Analysis

Dean–Stark extraction (Dean and Stark, 1920) is an extraction method that has been adapted to determine the volume of water (Sw) that can be vapourised from a core plug. It is the recommended method for core saturation determination. Unlike the retort, the method is not destructive as it allows the subsequent measurement of pore volume, grain density and permeability

TABLE 5.1 Fluid Saturation by Retort: Advantages and Drawbacks

Advantages	Drawbacks and Issues
• Cheap and rapid. • Large sample volume may be more representative than small plugs. • Minimal supervision is required. • Grain loss errors on saturation calculations are potentially less than Dean–Stark extraction. • Becoming more common for shale gas/oil core analysis.	• Difficult to differentiate between pore water, clay water (bound water) and crystalline water leading to overestimation of water saturation and summation of fluid porosity. • Not recommended for cores with high evaporate or smectite contents. • Organic compounds in core (as in some shales) can break down at high temperatures and released and can be mistaken for crude oil. This will increase oil volumes and saturations. • Oil volume correction curves for the specific field crude may not have been measured so correction may rely on generic curves. • Porosity 'measurement' is not repeatable. • Oil and water volumes and gas volumes are obtained on different samples so samples must be lithologically similar.

on the plug. The pore volume and grain volume data are used to calculate fluid saturations.

The technique is particularly suited to the case where the cores have been drilled with an oil-based mud, or where the core water saturation has not been affected by mud-filtrate flushing, or fluid flushing on pressure expansion during core recovery. In these cases, it is assumed that the core in the laboratory may be assumed to be at the same irreducible water saturation as in the reservoir (Woodhouse, 1998).

In the case of cores that have been drilled using a water-based mud, Dean–Stark analyses require the aqueous mud to have been doped during drilling with a tracer, with the objective of quantifying mud filtration during lab testing (Fjerstad et al., 1993). Details of suitable tracers are discussed in Chapter 2. Additionally, depth-lagged mud samples need to be taken over the reservoir interval during coring (every 3–5 m), so that the actual tracer concentration in the mud system at a particular depth can be determined either directly or by extrapolating from nearby points. Dean–Stark analysis equipment and procedures are the same as for samples drilled with an oil-based mud. However, the extracted waters are analysed on completion of the extraction to determine tracer content. The aqueous phase of the mud system is also extracted and analysed for tracer content. The amount of aqueous mud-filtrate invasion that a sample has exposed to is calculated from the tracer content of the extracted core water and the mud from a similar depth. The aqueous filtrate invasion is subtracted from the measured core sample water volume, and saturation is corrected for mud invasion.

5.2.2.1 Sample Preparation

Samples are selected and plugged as soon as possible after coring to prevent dehydration of the samples. In some cases, especially if it is logistically or politically difficult to transport the core to the test lab in a reasonable time, the plugging can be carried out at wellsite, as discussed in Chapter 2. Dean–Stark plugs can be taken by drilling through the liners. The plugs are preserved in film, foil and wax for shipment, and the plug holes in the liners are sealed. If the core has been caught in liners and the liners have been capped, Dean–Stark plugging can be done immediately after the core is removed from the liners in the laboratory. Dean–Stark extraction should also be performed immediately.

Plugs should be drilled from the centre of the core parallel to the long core axis, where possible, to avoid the mud invasion zone extending from the outside of the core. Horizontal plugs run the risk of having some water invasion from WBM. If the Dean–Stark plug is taken in the centre of the core along the core long axis, then permeability subsequently measured on Dean–Stark plugs may not represent the permeability perpendicular to bedding. Permeability in this case is used only as a reference value and does not necessarily correspond to vertical permeability.

If Dean–Stark extraction measurements are to be performed on cores cut with OBM or WBM, then the plugs must never be cut in water as this will invalidate the results. Plugs are cut in kerosene or mineral oil.

For SCAL, Dean–Stark extraction should also be used as an independent check on calculated water saturation on completion of the test sequence, where the test programme allows. Examples include final water saturations on completion of drainage or imbibition capillary pressure, relative permeability corefloods, Amott wettability and resistivity index tests.

5.2.2.2 Test Equipment

A Dean–Stark apparatus schematic is shown in Fig. 5.6, and a typical laboratory set-up is shown in Fig. 5.7. The extraction system is similar to that described for hot Soxhlet cleaning (Chapter 4). The sample is placed in the sample chamber (3), often in a cellulose tare to prevent grain losses), but the condensed water is now collected in a graduated receiving tube (8) rather than being returned to the solvent flask. Toluene or xylene is used to extract water from the sample. Xylene, which has a higher boiling point, is preferred if the formation brine contains significant amounts of salts, particularly calcium carbonate, since these will increase the boiling point of the water.

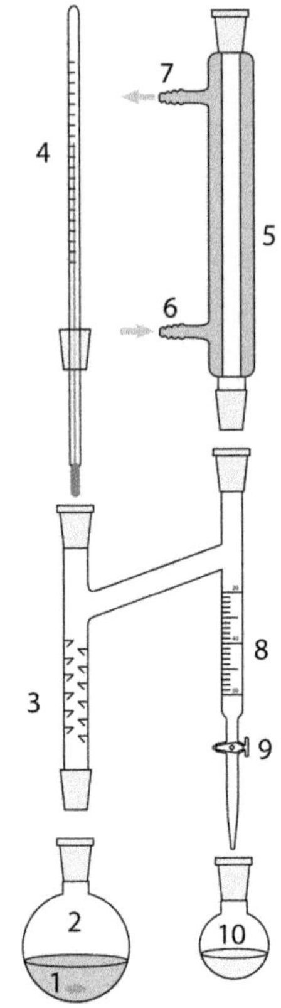

FIGURE 5.6 Dean–Stark schematic. *Original PNG by Quantockgoblin, SVG adaptation by Slashme (http://en.wikipedia.org/wiki/Dean-Stark_apparatus).*

The solvent should be Analar grade. If industrial grade solvent is used, it should be dewatered prior to use by distillation of the solvent in the Dean–Stark test apparatus before loading the test sample.

The Dean–Stark apparatus must include a guard tube containing dry silica gel to minimise condensation of atmospheric moisture into the receiving tube caused by chilled water circulating in the condenser unit.

5.2.2.3 Test Procedures

The sample is weighed prior to the Dean–Stark extraction process. The selection of a suitable capacity graduated side arm should be made following estimation of the volume of water in the core based on the sample weight and an approximate estimate for sample porosity. The sample is loaded in the test apparatus and the solvent is heated up.

As the solvent boils, its vapour rises up through the sample, which is held in the chamber immediately above the solvent flask, and vapourises the water in the sample. The solvent/water vapour then rises up through the condenser

FIGURE 5.7 Typical Dean–Stark set-up in a laboratory. *By Rifleman 82 via Wikipedia Commons. http://commons.wikimedia.org/wiki/File%3ADean_Stark_apparatus.jpg.*

unit where it is cooled by circulating chilled water. It then condenses and falls down into the graduated side arm, where it is collected and measured. The water and solvent settle under gravity with the water settling in the base of the receiving tube, and the less dense solvent above it. As the liquid level rises in the tube, solvent will overflow and drip back down through the sample chamber into the solvent flask, where it is re-distilled. Care should be taken to drain off collected water before water reaches the top of the receiving tube; otherwise, water will be lost back into the solvent flask.

The collected water volume is recorded over time until a stable value is achieved when recorded over a period of at least 6 h. Normally, the minimum extraction period is 48 h but may be much longer in lower permeability samples. On completion of the test, any water retained on the inside wall of the condenser should be flushed down into the side arm using a xylene or toluene wash bottle.

Once extraction is complete, the sample is removed. At this stage, the sample contains precipitated formation water salts remaining after water extraction. Provided water-soluble tracer salts are not to be eluted, the sample is weighed then extracted in methanol in a Soxhlet extractor to dissolve them. The silver nitrate test (Chapter 4) is used to determine the end point of removal of salts.

Following conventional oven drying at 100–105 °C, sample grain volume (Vg), bulk volume (Vb) and base air or nitrogen permeability are measured.

5.2.2.4 Saturation Calculation

The water collected in the Dean–Stark side arm is distilled water, not formation water (brine). The volume of water (Vw) must therefore be converted to an equivalent brine volume (V_{FW}) using a salt correction factor

$$V_{FW} = \mathrm{Vw\,SCF}$$

The SCF value depends on the formation water (ρ_{FW}) and distilled water density (ρ_W) at 20 °C, and formation water salinity (weight percent salt, $W\%_{salt}$):

$$\mathrm{SCF} = \frac{\rho_W}{\rho_{FW}} \times \frac{100}{100 - W\%_{salt}}$$

If brine concentration is provided in TDS (in ppm), then:

$$\mathrm{SCF} = \frac{998{,}000}{\rho_{FW}\left(10^6 - \mathrm{TDS}\right)}$$

Accurate estimation of water saturation demands that the properties of the formation brine originally saturating the samples are known. These should be provided by the client.

Pore volume (Vp) and water saturation (Sw) are calculated from:

$$\mathrm{Vp} = \mathrm{Vb} - \mathrm{Vg}$$

$$\mathrm{Sw} = \frac{V_{FW}}{\mathrm{Vp}}$$

Oil and/or gas (air) volumes in the sample can be determined by weighing it before and after extraction in the Dean–Stark, and after extraction in methanol in a Soxhlet apparatus following conventional oven drying. However, as these measurements can be affected by grain loss, they should be viewed as qualitative. The whole point of Dean–Stark extraction is to determine water saturation.

The Dean–Stark brine volume determined at ambient laboratory conditions will be greater than that at reservoir pressure and temperature conditions due to compressibility effects. The water compressibility (c_w) can be estimated by:

$$c_w = \frac{1}{V_{FWS}}\left(\frac{\partial \mathrm{Vw}}{\delta P}\right)$$

where V_{FWS} represents the water volume at reservoir conditions. Thus, the increase in water volume (δVw) can be calculated for a reduction in pressure (δP) from reservoir to ambient conditions

$$\partial \mathrm{Vw} = \delta P V_{FWS} c_w$$

The rock will also relax when the reservoir stress is removed by coring so the pore volume will expand. The amount of pore volume expansion depends on the pore volume compressibility (c_p) of the rock which is estimated from:

$$c_p = \frac{1}{Vp}\left(\frac{\partial Vp}{\delta P}\right)$$

The increase in pore volume (δVp) can also be calculated for a reduction in pressure (δP) from reservoir to ambient conditions:

$$\partial Vp = \delta P Vp_s c_w$$

Typical formation water compressibilities range from around 2 to 8×10^{-6} psi^{-1}. In relatively stiff formations, the pore volume and water compressibilities have a similar range, so the pore volume and water volume expansions effectively cancel one another out. Thus, the ambient condition Sw is close to the reservoir saturation. Only in highly compressible formations or with very low compressibility formation waters will the difference be significant enough to warrant correction.

One popular, but rigorously incorrect, method to correct Dean–Stark water saturations (Sw_{amb}) to reservoir conditions (Sw_{stress}) is based on the ambient sample porosity (ϕ_{amb}) and the ratio of porosity at reservoir stress to ambient porosity—the porosity compaction factor (PCF):

$$Sw_{stressed} = Sw_{amb}\left[\frac{[1/PCF] - \phi_{amb}}{1 - \phi_{amb}}\right]$$

However, this ignores water compressibility effects, so will over-correct the reservoir saturation—Sw_{stress} will be too high.

Dean–Stark saturations from the water leg rarely achieve 100% saturation. This is due to differential expansion of water and pore volume on core recovery and the presence of dissolved gas in the water-leg formation water.

In the case of cores drilled with a water-based mud doped with tracers, the water extracted from the core needs to be analysed to determine tracer content, if a distillable tracer, or eluted from the core, if a non-distillable salt tracer (Fjerstad et al., 1993). The aqueous phase of the mud system is also extracted and analysed for tracer content. The amount of aqueous mud-filtrate invasion that a sample has suffered is calculated from the tracer content of the extracted core water and the mud from a similar depth. The aqueous invasion is subtracted from the measured core sample water volume, and saturation is corrected for mud invasion.

5.2.2.5 Data Reporting Requirements

It is essential that the test laboratory provides the data to allow the interpreter to quality control the test, to check and verify the lab calculations and, if necessary, to reinterpret the results. Example data requirements are provided in Table 5.2.

TABLE 5.2 Data Requirements Checklist for Dean–Stark Extraction Measurements

Data	Comments
Brief description of procedures and experimental apparatus	Including extraction time and solvent
Plug depth and orientation	Horizontal, vertical or parallel/perpendicular to core
Plug length and diameter	
Distilled water density and temperature	
Measured distilled water volume Vw	Final, stable value
Formation water density and temperature	Normally supplied by client
Formation water salinity	Normally supplied by client
Calculated salt correction factor (SCF)	
Sample grain volume	Test method for Vg
Sample bulk volume	Test method for Vb
Calculated sample pore volume	Vp
Calculated water saturation, Sw	Third decimal place (fraction)
Calculated sample porosity, ϕ	Third decimal place (fraction)
Sample air or Klinkenberg permeability	At net confining stress
Water analysis and tracer concentration from core samples	If mud doped with tracers
Water analysis and tracer concentration from mud samples	If mud doped with tracers
Calculated mud water volume invasion volume	If mud doped with tracers
Mud invasion-corrected water saturation Sw_{corr}	If mud doped with tracers
Water compressibility if used in saturation calculations	Measured or generic
Pore volume compressibility if used in saturation calculations	Measured or generic
Porosity compaction factor if used in saturation calculations	Measured or generic

5.2.2.6 Advantages and Drawbacks/Issues

Summaries of the advantages and drawbacks of the Dean–Stark method are provided in Table 5.3.

5.2.2.7 Dean–Stark Quality Control Issues, Checks and Diagnostics

The water saturation from Dean–Stark extraction essentially represents water saturation in the Dean–Stark samples more or less at the time they were cut and preserved. Mud-filtrate invasion and saturation hysteresis on coring and core recovery (Chapter 2), or water evaporation from the core before plugging, could mean that the Dean–Stark Sw is not representative of the reservoir Sw.

Not all commercial labs are capable of measuring water saturation when core is drilled with doped water-based mud filtrate. This requires the accurate elution and/or detection of chemical or nucleonic tracers from core, distilled water and mud samples. Always check and audit the lab's resources experience and expertise in tracer analysis and interpretation before awarding or commissioning the tests.

If unconsolidated, sleeved samples are not able to maintain their shape following extraction, then measurement of the pore volume by helium expansion might involve a relatively high error, generally overestimating the pore volume, and hence underestimating Sw. In this case, the option is to measure bulk volume of the un-sleeved sample prior to running the test so pore volume can be calculated from grain volume and bulk volume. Potential grain loss errors in friable formations mean that lab-reported oil and gas saturations may not be accurate.

When irreducible water saturation is plotted against porosity for a single rock type, saturation should increase with decreasing porosity, as shown in Fig. 5.8. Different rock types can also be distinguished if enough data are available. The plot can also help to differentiate between intervals where water is mobile and intervals at irreducible water saturation. When dispersed data are obtained, as shown in Fig. 5.9, it could mean samples are either not at irreducible water saturation or have suffered from water invasion during coring or handling.

If reliable capillary pressure data and log-derived water saturation curves are available, the endpoint Pc Sw or formation Sw can be checked against Dean–Stark values. There are several reasons why laboratory Dean–Stark Sw may be differ from water saturation from logs. Some of these are listed in Table 5.4 in terms of Sw being higher or lower than log Sw. In this case, it is assumed that the log interpretation parameters in clean formations are accurate and that there are no uncertainties introduced by bad hole conditions, thin beds, low-resistivity contrast pay, etc.

5.3 POROSITY MEASUREMENTS

Porosity is a measure of the space available for storage of fluids in a rock. As discussed in Chapter 4, the total porosity is the ratio of all pore space to bulk

TABLE 5.3 Dean–Stark Extraction: Advantages and Drawbacks

Advantages	Drawbacks and Issues
• Relatively cheap and accurate method for native water saturation in core samples. • Recommended method for Sw measurements on cores. • In oil-based mud drilled cores, Sw from Dean–Stark can represent reservoir Sw. • Can be used in water-based mud drilled cores provided water filtrate doped with tracers. • Accurate Dean–Stark results can be used in verify Archie log interpreted Sw and saturation height model Sw. • Can be used, in conjunction with formation factor tests, log-derived formation resistivity data to determine Archie saturation exponent at reservoir conditions. • Provides independent verification of laboratory-calculated final water saturations in SCAL tests.	• Care needs to be taken to use dewatered solvents prior to the test, and to prevent water loss through leakage, or water gain through atmospheric condensation. A silica gel water trap should be used to prevent atmospheric water entering via the condenser tube. • Water volume and pore volume errors in measurements on small plugs (reduced pore volume) can be large and can have a significant impact on the data. • As the use of high-temperature extraction and plug drying will remove clay-bound water, the porosity and water saturation of Dean–Stark plugs represent a total porosity/Sw system. • The use of the lower boiling point toluene in higher salinity brines may not efficiently vapourise the water if it contains divalent cations. Xylene is often used for carbonates as it has a higher boiling point (138 °C) than toluene (112 °C). • Not suited for cores drilled with a water-based mud, unless suitable tracers are used during coring. This is normally expensive and complicated in terms of logistics and HSE on wellsite and lab. Interpretation is also required to determine the degree of mud invasion. • If the OBM contains surfactants that interact with the reservoir water, the Dean–Stark Sw values may not be representative of the reservoir as reduction of interfacial tension between oil and water could result in mobilisation of immobile water. • Some synthetic oil-based muds are suspected to give up some of the water phase to the formation. This will increase Sw. • If samples are taken in a zone of high water mobility (transition zone or aquifer), Dean–Stark values will probably not produce representative results due to displacement of mobile water by OBM filtrate. The preservation of Sw is a key concept in the reconciliation of water saturation data. • Unconsolidated/friable samples are subject to error from grain loss. Also, rapid freezing can cause increase Sw due to atmospheric water vapour condensation in the core.

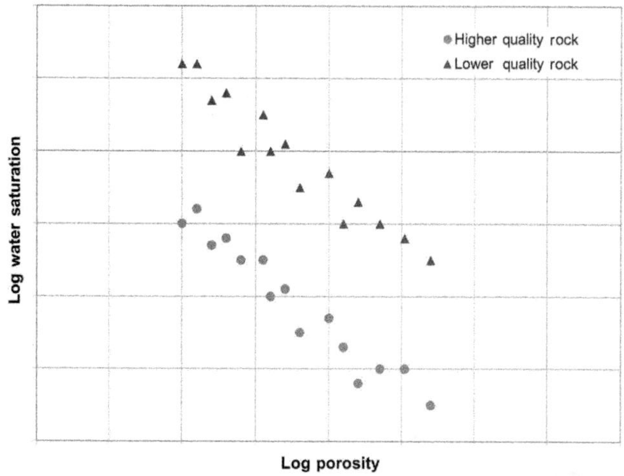

FIGURE 5.8 Hypothetical example of well-behaved water saturation data.

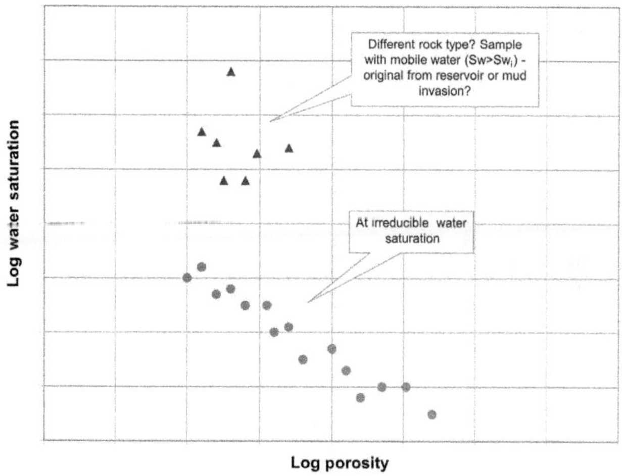

FIGURE 5.9 Comparison of well-behaved water saturation data versus samples not at irreducible water saturation.

volume. It includes all pores that are filled with water or hydrocarbon, regardless of their size and degree of connectivity, and therefore includes clay microporosity (CBW) as well as isolated pores (e.g. vuggy systems in carbonates). Effective porosity is the ratio of interconnected pore volume to the bulk volume and has many determinants.

Assessment of porosity data must be compatible with the petrophysical model employed and with all other porosity-related parameters which are ultimately fed into a petrophysical model. In practice, there is a dilemma when

TABLE 5.4 Dean–Stark Sw Differences from Log-Derived Sw

Reason	Dean–Stark Sw
Core drilled with WBM (no tracers used to discriminate filtrate)	Higher
Core drilled with OBM or SBM with high water content (assumes mud gives up the water phase to the formation—doping of water phase recommended)	Higher
Mud tanks uncovered and exposed to sea spray or rain	Higher
Toluene not dewatered in laboratory	Higher
No equipment control on humidity condensation (water trap)	Higher
Core sections plugged with water	Higher
Core sections frozen and condense atmospheric water	Higher
Plugging performed after slabbing (most slabbing uses water as lubricant)	Higher
Water saturation mobile ($>Sw_{critical}$) and part displaced by OBM filtrate	Lower
Equipment joints not tight (water vapour escapes)	Lower
Insufficient distillation time	Lower
Dean–Stark plugs at wellsite not preserved—water evaporates	Lower
Dean–Stark plugs taken in lab from unpreserved core	Lower
Stress correction not applied	Lower
Salt salinity correction not applied	Lower

measuring petrophysical properties on shaly sands. If RCA core porosity measurements are used to calibrate density logs, then they need to be harshly prepared to allow helium to access the CBW space and bring core porosity closer to total porosity. This will lead to clay destruction such that RCA permeabilities are significantly enhanced, and the samples are no longer representative of the reservoir for SCAL measurements. Benign preparation methods may preserve the petrophysical properties, but the core porosities are no longer representative of a total (or indeed an effective) porosity system.

In vuggy and similar carbonates, total porosity (including isolated pores) is determined by measuring porosity conventionally on intact plugs, then crushing the rock and determining the grain volume and bulk volume of the crushed rock sample.

In the laboratory (irrespective of whether the plugs have been harshly prepared or not), porosity, ϕ, is determined from:

$$\phi = \frac{Vp}{Vb} = \frac{Vb - Vg}{Vb} = \frac{Vp}{Vp + Vg}$$

where Vb is the sample bulk volume, Vp the pore volume and Vg the grain volume. Consequently, at least two independent measurements of the three volume contributors, bulk volume (Vb), pore volume (Vp) and grain volume (Vg), are needed to determine porosity. The methods used often differ between and within RCA labs and SCAL labs. The different techniques must measure any two of the three volume contributors.

RCA programmes are designed to be cheap and with a fast turnaround, both of which limit the techniques that are used to estimate porosity. Much RCA core porosity data are derived from independent measurements of helium grain volume and mercury bulk volume. Some are based on a combination of helium grain volume and helium pore volume at low confining stress. The awareness that mercury vapour is an extremely hazardous substance has caused some laboratories to move away from immersion bulk volume measurements to the potentially unsatisfactory helium grain volume/ helium pore volume combination.

The measurement of porosity in SCAL programmes generally involves determining the re-saturation porosity which requires the samples to be saturated in brine. This is a time-consuming and more expensive method, but potentially more accurate and, in certain cases, more representative than the RCA techniques. Some labs use brine re-saturation methods to determine porosities as part of their RCA procedures, but this is relatively rare.

5.3.1 Helium Grain Volume and Grain Density

5.3.1.1 Sample Preparation

Samples to be used for helium grain volume and grain density tests must be thoroughly cleaned of fluids and dried using a method consistent with the application of the porosity data, as described in Chapter 4. The dry weight (W_d) of sample is measured before testing as this is used to determine grain density.

The shape of the test sample is largely irrelevant: tests can be performed on a cylindrical plug, irregular rock chip or even loose sand. For RCA and SCAL tests, plugs must be used and larger plugs (1″ versus 1.5″) are preferred since the errors involved in grain volume or pore volume measurements on small plugs can have a significant impact on the calculated data. The impact of measurement errors becomes much greater, the smaller the pore volume. The pore volume of a 1″ plug is about four times smaller than the pore volume of a typical 1.5″ plug of the same porosity. The potential for error in core measurements should therefore be considered in these circumstances.

5.3.1.2 Test Equipment

Grain volume is determined using a twin-cell helium porosimeter or pycnometer, which operates on the principal of Boyle's law (Fig. 5.10). This states that, for an ideal gas assuming constant temperature, the product of pressure and volume in a closed system remains constant. That is $P_1V_1 = P_2V_2$. Helium is used as the gas in all grain volume experiments because it is an ideal gas at ambient conditions, is inert and has a very small molecular size so that it can penetrate all the accessible pores in a rock. Provided care is taken, the method is accurate and reproducible.

Consider the application of Boyle's law to grain volume measurements, shown schematically in Fig. 5.11. In this design, which is more or less standard for most porosimeters, helium is allowed into a reference volume of volume V_r, at a pressure P_r. Valves V_1 and V_2 isolate the reference volume. A matrix cup, with volume V_s, is used as the expansion chamber. Valves V_2 and V_3 isolate this expansion volume. As valve V_2 is opened, gas at pressure P_r in volume, V_r, expands to fill both V_r and V_s. The final expanded pressure, P_x, will depend on the ratio of V_r to V_s; thus,

$$P_rV_r = P_x(V_r + V_s)$$

A core sample is now placed in the matrix cup (Fig. 5.12). The volume of the matrix cup (V_s) will be reduced by an amount, Vg, equivalent to the matrix, grain or solids volume of the sample. Any pores in the sample will fill with helium on expansion. If the expansion sequence is repeated with the plug in the matrix cup, with helium at pressure, P_r, in reference volume, V_r, then Boyle's law becomes:

$$P_rV_r = P_x(V_r + V_s - Vg)$$

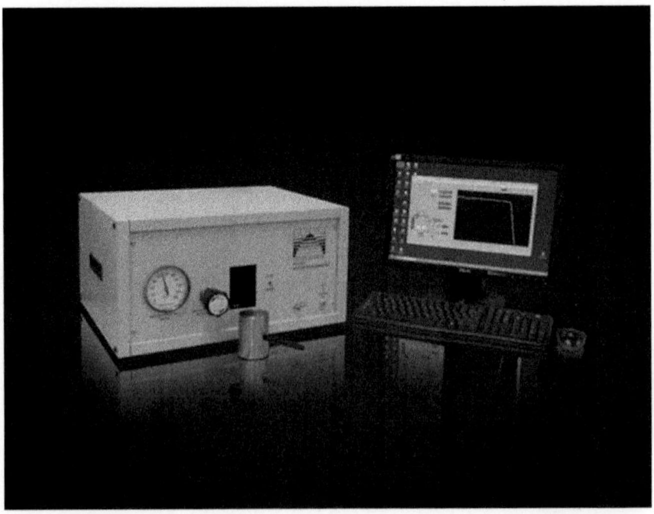

FIGURE 5.10 Grain volume porosimeter equipment. *Courtesy of Coretest Systems, Inc.*

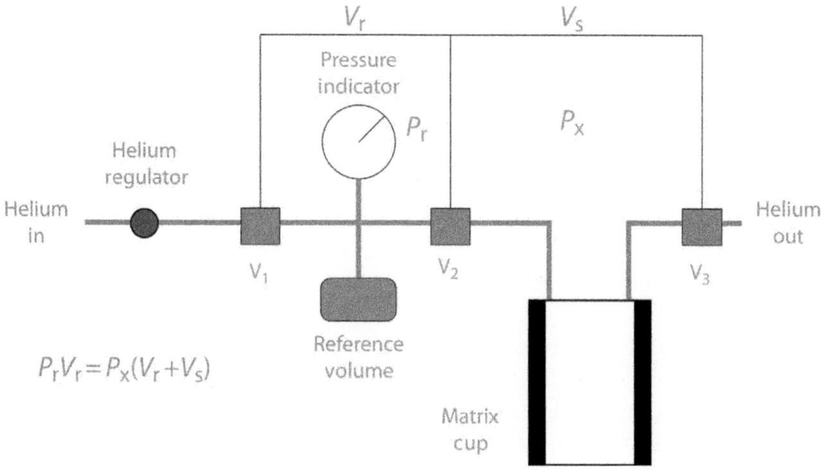

FIGURE 5.11 Helium grain volume instrument schematic (matrix cup empty).

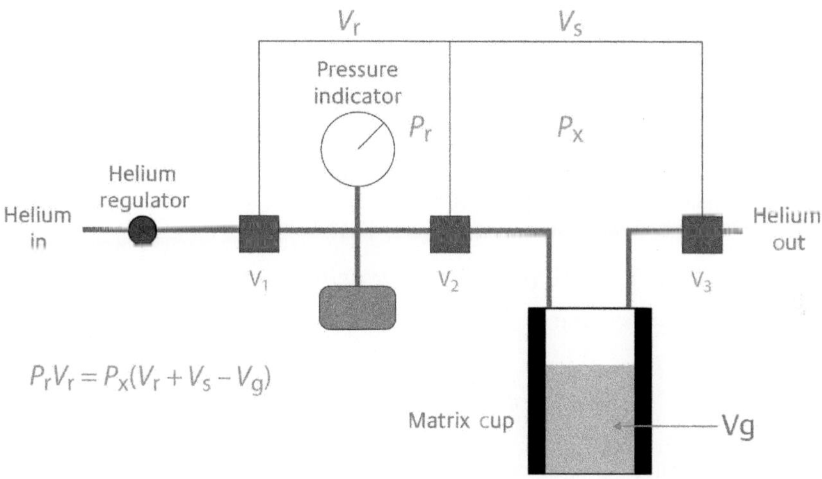

FIGURE 5.12 Helium grain volume instrument schematic (matrix cup filled with plug).

Rearranging, and solving for Vg:

$$\text{Vg} = \frac{P_x(V_r + V_s) - P_rV_r}{P_x}$$

Therefore, if we know values of V_r and V_s, and can measure initial reference volume pressure and expanded pressure, then the grain volume (Vg) can be calculated. V_r and V_s are found from initial calibration experiments, in which a series of solid (non-porous) steel or metal blocks (billets) of known bulk volume (Vb = Vg) are placed in the matrix cup.

5.3.1.3 *Test Procedures*

The sample is placed in the matrix cup which is selected on the basis of the core dimensions. To optimise accuracy and repeatability, if the plug is too short, billets of known grain (bulk) volume are used to fill up the cup. Any air in the system and sample is removed by flushing the system with helium either by going through the expansion process or by drawing a vacuum on the matrix cup and then flushing with helium.

The matrix cup is vented to atmosphere, and valves V_2 and V_3 are closed. Valve V_1 is opened, and helium fills the reference volume and is regulated to around 100 psi. With valves V_3 and V_1 now closed, and when the reference volume pressure is stable, helium is allowed to expand into the matrix cup (V_2 open) and the expanded pressure monitored until stable. In higher permeability samples, this can be seconds, but in lower permeability samples, it can take several minutes so the operator must be sure that the instrument is leak tight.

The lab grain volume measurements should also be made at the same temperature (within ±1 °C) of the instrument calibration temperature; otherwise, inaccuracies will result. Although, in theory, temperature variations on gas volumes can be accommodated in Charles's law (*PV*/*T* is constant), the instrument pipework and reference volumes can expand or contract to different degrees depending on temperature variation and pipe material.

The instrument calibration should be checked using billets or reference check plugs before, during and after the suite of plug measurements. If necessary, the instrument is recalibrated.

5.3.1.4 *Grain Volume and Grain Density Calculation*

Grain volume is determined from:

$$\mathrm{Vg} = \frac{P_{\mathrm{X}}(V_{\mathrm{r}} + V_{\mathrm{s}}) - P_{\mathrm{r}}V_{\mathrm{r}}}{P_{\mathrm{X}}}$$

Grain density, ρ_g, is determined from

$$\rho_{\mathrm{g}} = \frac{W_{\mathrm{d}}}{\mathrm{Vg}}$$

5.3.1.5 *Data Reporting Requirements*

Data reporting requirements from the laboratory are listed in Table 5.5.

5.3.1.6 *Advantages and Drawbacks/Issues*

Summaries of the advantages and drawbacks of the principal drying methods are provided in Table 5.6.

TABLE 5.5 Data Requirements Checklist for Helium Grain Volume and Grain Density Measurements

Data	Comments
Brief description of procedures and experimental apparatus	
Description of cleaning and drying procedures	For core–log calibration
Plug depth and orientation	
Plug length and diameter	
Plug/sample dry weight	
Unconsolidated plug mounting materials	State whether used
Calculated grain volume	Based on final, stable P_x
Calculated grain density (g/cc)	
Results of repeat tests on plug samples	Confirm measurement accuracy

TABLE 5.6 Helium Grain Volume: Advantages and Drawbacks

Advantages	Drawbacks and Issues
• Fairly rapid for majority of samples. • Relatively cheap method. • Under close control, measurements are accurate and repeatable. • Common measurement for all RCA porosity determination methods. All commercial core analysis labs offer this fundamental measurement. • The method is non-destructive so samples are used for further testing. • Any sample shape can be tested.	• Could be slow for low permeability samples so sufficient time must be allowed for pressure stabilisation. • Errors involved in measurements on small samples are likely to have larger errors than measurements on larger samples. • Inefficient drying of free fluids can suppress grain density.

5.3.1.7 Helium Grain Volume and Density Quality Control Issues, Checks and Diagnostics

Grain volume and grain density are easy and rapid to measure and, if the helium pycnometer and weigh balance have been calibrated and maintained correctly, it is potentially very accurate.

The calculated grain density should match the expected lithology. For example, samples with grain densities outside the range 2.71 ± 0.02 g/cc in chalks, or 2.65 ± 0.02 g/cc in clean sands, should be inspected for unusual mineralogy or possible laboratory errors such as insufficient cleaning and/or drying, or fundamental grain volume measurement errors.

It is recommended that grain density repeatability tests are carried out on at least every 10th plug by different operators within the laboratories, and the results of the tests made available in the RCA report.

5.3.2 Helium Pore Volume

5.3.2.1 Sample Preparation

Samples to be used for pore volume tests must be thoroughly cleaned of fluids and dried using a method consistent with the application of the porosity data.

Unlike grain density, the shape of the test sample is crucial to the accuracy and reliability of the pore volume measurement. It must be as close to a perfect right cylinder as possible. The method is very inaccurate for plugs with irregular or rounded ends or with fractures or vugs on the surface that are not sealed during the measurement.

Again, larger plugs are preferred to minimise measurement uncertainty on the porosity data.

A combination of helium grain volume and helium pore volume is a common method to determine porosity on unconsolidated or weak sands and vuggy carbonates where either mercury is considered to be too hazardous to use or is considered might potentially intrude into the pore system.

5.3.2.2 Test Equipment

Helium pore volume is determined using a modified version of the twin-cell helium porosimeter, which is used to determine grain volume. Figure 5.13

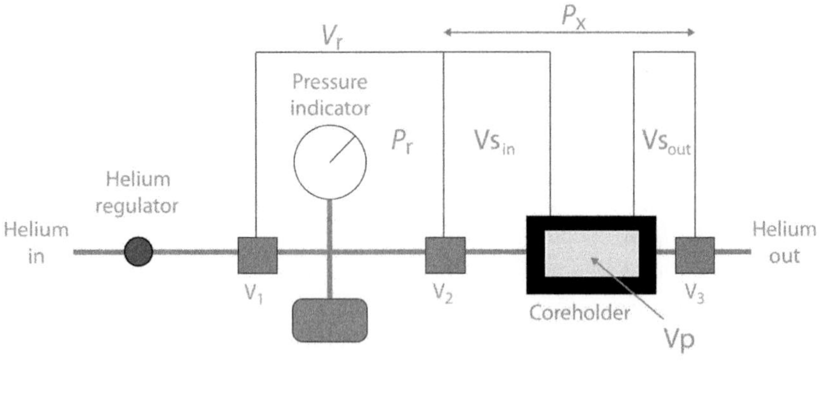

$$P_r V_r = P_x (V_r + Vs_{in} + Vs_{out} + Vp)$$

FIGURE 5.13 Helium pore volume porosimeter schematic.

provides a schematic of typical equipment. The principal difference is that the sample is held within a rubber sleeve within a suitable Hassler-type coreholder which replaces the grain volume matrix cup. A moderate confining pressure is required to seal the sleeve around the surface of the plug (surface conformance) although the test can be performed at elevated stress simulating the reservoir stress. Unlike the porosity determined from helium grain volume and mercury bulk volume, pore volume measurements depend on the level of confining stress applied and this should be considered when comparing results from different systems or labs.

The CMS-300™ system (Core Laboratories), and most automated porosity equipment, base their porosity calculations on the pore volume/grain volume combination.

For the arrangement shown in Fig. 5.13, Boyle's law becomes:

$$P_r V_r = P_x (V_r + \mathrm{Vs_{in}} + \mathrm{Vs_{out}} + \mathrm{Vp})$$

Rearranging, and solving for Vp:

$$\mathrm{Vp} = \frac{P_r V_r - P_x (V_r + \mathrm{Vs_{in}} + \mathrm{Vs_{out}})}{P_x}$$

The calibration procedures to calculate the reference volume, V_r, and the inlet, $\mathrm{Vs_{in}}$, and outlet, $\mathrm{Vs_{out}}$, are similar for grain volume, except that hollow metal cylinders of known pore volume are used.

Caution must be exercised in interpretation of pore volume measurement. The technique provides a measure of all empty space between the inlet and outlet platens of the coreholder. This includes platen dead volume and any empty annulus space occupied between the core sample and the internal sleeve of the coreholder.

5.3.2.3 *Test Procedures*

The sample is placed in the coreholder which is then confined at the specified confining stress.

Any air in the system and sample is removed by flushing with helium either by going through a 'dummy' expansion process or by drawing a vacuum then flushing with helium.

The reference volume is filled with helium to around 100 psi or 200 psi and, when stable, helium is allowed to expand into the coreholder and the pressure monitored until stable. Again, in lower permeability samples this can take several minutes.

As with grain volume, the lab pore volume measurements should also be made at the same temperature (within ±1 °C) of the instrument calibration temperature. The instrument calibration should be checked using hollow billets or reference check plugs before, during and after the plug measurements. If necessary, the instrument is recalibrated.

5.3.2.4 *Pore Volume and Porosity Calculation*

Pore volume is determined from:

$$\mathrm{Vp} = \frac{P_r V_r - P_x (V_r + \mathrm{Vs_{in}} + \mathrm{Vs_{out}})}{P_x}$$

If helium pore volume is then used in combination with grain volume to determine porosity, then:

$$\phi = \frac{\mathrm{Vp}}{(\mathrm{Vp} + \mathrm{Vg})}$$

5.3.2.5 *Data Reporting Requirements*

Data reporting requirements from the laboratory are listed in Table 5.7.

5.3.2.6 *Advantages and Drawbacks/Issues*

Summaries of the advantages and drawbacks of the helium pore volume measurement are provided in Table 5.8.

TABLE 5.7 **Data Requirements Checklist for Helium Pore Volume Measurements**

Data	Comments
Brief description of procedures and experimental apparatus	
Description of cleaning and drying procedures	For core–log calibration
Plug depth and orientation	
Plug length and diameter	Lab should state whether plug shape is a regular cylinder and/or provide photographs of the plug
Plug/sample dry weight	
Unconsolidated plug mounting materials	State whether used
Confining stress used	
Calculated pore volume	Based on final, stable P_x
Calculated porosity (if based on Vp and Vg)	As a fraction to 3 D
Results of repeat tests on plug samples	Confirm measurement accuracy

TABLE 5.8 Helium Pore Volume: Advantages and Drawbacks

Advantages	Drawbacks and Issues
• Recommended porosity determination method for unconsolidated samples and vuggy carbonates with large surface pores (in conjunction with helium grain volume). • Fairly rapid for majority of samples. • Relatively cheap method. • Can be run quickly and conveniently at elevated confining stresses to determine porosity as a function of stress. • The method is not destructive (unless samples are stress sensitive) so samples can be used for further testing.	• Could be slow for low permeability samples so sufficient time must be allowed for pressure stabilisation. • Errors involved in measurements on small samples are likely to have larger errors than measurements on larger samples. • Inefficient drying of free fluids can suppress pore volume. • The method is plug shape dependent and can produce pore volume values that are too high if the sample is not a regular shape, has rounded ends or unsealed fractures or vugs on the surface. • For unconsolidated samples, the dead volume between the plug and the protective jacket, and the pore volume of the end screens must be accounted for.

5.3.2.7 Helium Pore Volume Quality Control Issues, Checks and Diagnostics

The principal concern about helium pore volume centres around the regularity of the plug shape and its surface condition or topology. When helium is expanded, it will enter and fill all empty space between the end platens of the coreholder including the pore volume and the false pore volume created by irregular plug ends or a poor seal in the sleeve/plug annulus. Measurements made on plugs with irregular or rounded ends (Fig. 5.14), or with fractures or vugs in the surface, will always overestimate the true pore volume, since it is almost certainly impossible to cut a truly perfect, right cylindrical plug. There is little that can be done in this case, though the potential for overestimating porosity must be addressed in the petrophysical evaluation.

The impact of these false volume measurements can be substantial.

In the example shown in Fig. 5.15, the 'RCA Plug Porosity' measurements were made on the plugs using a combination of helium grain volume (which is accurate) and helium pore volume (which can be inaccurate). Subsequent measurements were made on the same, selected samples prior to SCAL tests (SCAL Plug Porosity). These used helium grain volume and mercury immersion bulk volume. Since the pore volume was determined at 800 psi confining

FIGURE 5.14 Example of irregularly shaped plugs with rounded end faces.

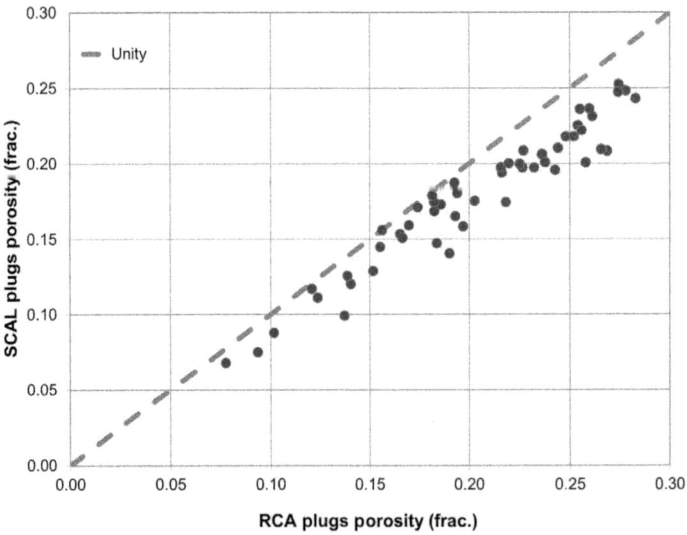

FIGURE 5.15 Comparison of 'RCA' (Vg and Vp) and 'SCAL' (Vg and Vb) porosities on same plugs.

stress, then it is expected that the RCA porosity should be slightly lower than the SCAL plug measurements made at unconfined conditions. In fact, the opposite is true: there is a significant difference between the measurements caused by the plug irregularity, especially at higher porosities (up to 6 p.u.) as the plugs became more friable and less cylindrical in shape. The use of the initial RCA plug data to calibrate density logs eventually resulted in an overestimation of around 7% of GIIP.

Figure 5.16 provides a similar example, this time on a strong competent sand. On average, the porosities at 800 psi from confined helium pore volume and unconfined ambient helium grain volume are higher than the ambient (unconfined) porosities from helium grain volume and mercury bulk volume on the same samples, whereas they should be lower.

Similar issues are found with weak or poorly consolidated samples which have been mounted in protective jackets and screens. Figure 5.17 plots the normalised PCF, which is the ratio of porosity at stress to a base porosity (in this case 800 psi confinement) as a function of the base porosity for the same samples. The porosities in both cases were calculated from helium pore volume and helium grain volume. The samples which have been sleeved for protection (mounted) have significantly lower stressed porosities in general than the unmounted samples. This is because the false dead volume between the rock, protecting jacket and screens is filled with helium at low stress (800 psi) but reduces significantly due to better conformance at higher stresses.

If the helium pore volume method is used, selected plugs should be photographed to have a qualitative idea of how regular the samples actually are.

Unless the samples are stress sensitive, it is recommended that pore volume repeatability tests are carried out on at least every 10th plug by different operators within the laboratory, and the results of the tests made available in the RCA report.

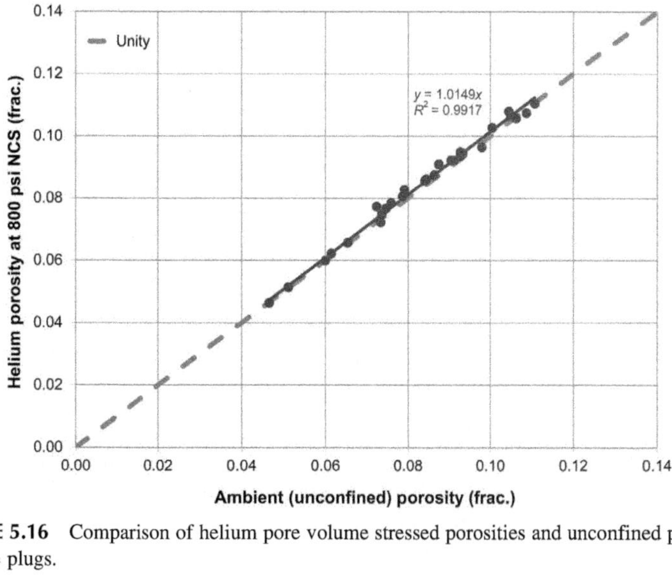

FIGURE 5.16 Comparison of helium pore volume stressed porosities and unconfined porosities on same plugs.

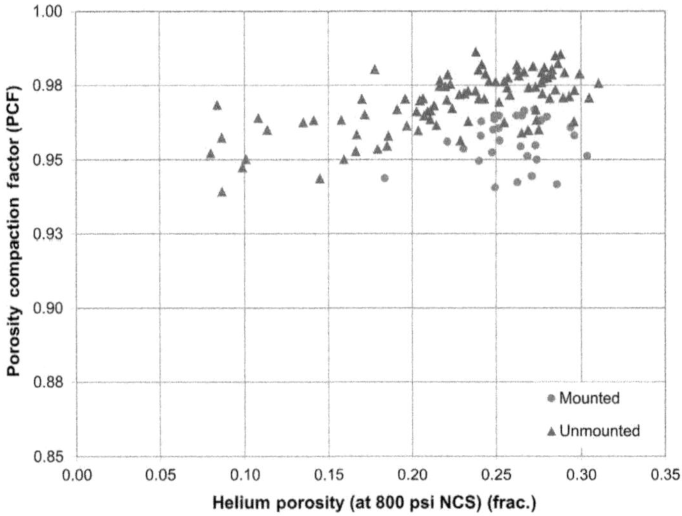

FIGURE 5.17 Normalised helium porosity at stress for unmounted and mounted samples.

For consolidated and non-vuggy samples, porosities should be checked against measurements of porosity from helium grain volume and mercury immersion bulk volume on at least every fourth plug.

5.3.3 Bulk Volume

In RCA, three methods are commonly used to determine bulk volume (Vb):

1. Mercury pycnometer.
2. Mercury immersion system.
3. Calliper system.

Bulk volume calculated from measurements of the length and diameter of the sample using a calliper will always overestimate the bulk volume of the plugs since real plugs are rarely perfect right cylinders (see Fig. 5.14). The only circumstance when this method should be used is for ambient condition measurements on samples where mercury may enter the pore space of the sample, for example, in vuggy carbonates. In this case, the sample should be as large as possible to minimise the impact of measurement inaccuracy (e.g. whole core samples).

Mercury bulk volume relies on the fact that mercury is a high-density, non-wetting liquid and will only enter the pores of the plug under applied pressure—it will not spontaneously imbibe. Since the pore volume is not intruded with mercury, then the air in the pore volume plus the grain volume will displace an equivalent volume (or weight) of mercury. If mercury enters the true pore system then this will result in an underestimation of bulk volume. If mercury enters the sample pore system, some of it may be trapped on withdrawal which renders the samples unsuitable (and hazardous) for further testing. To prevent mercury entering the sample pores, the mercury must be free of rock dust or similar contaminants which can result in surface wetting, and the largest pore throat radius of the true pore system has to be sufficiently large to prevent mercury penetration. Samples with surface vugs, fractures, very coarse grain fabrics or sleeved in metal/Teflon (unconsolidated samples) should not be measured, as mercury will penetrate the samples resulting in a lower bulk volume value. In the latter case, bulk volume may be better determined by calliper measurements or by total saturation of the pore space by a fluid of known density.

In addition, the depth of immersion in mercury should be minimised. The pore throat radius, r in cm, penetrated by mercury at a given immersion depth, H in cm, can be determined from

$$r = \frac{c2\sigma\cos\theta}{\left(\rho_{Hg} - \rho_{air}\right)gH}$$

where ρ_{Hg} is the mercury density (g/cc), ρ_{air} is air density, θ is the mercury contact angle (°), σ is the air–mercury interfacial tension (dyne/cm), g is the gravitational constant and c is a units constant. For a mercury density of 13.6 g/cc, then the pressure at the base of a 1.5″ diameter by 3″ long plug immersed vertically in mercury will be about 1.5 psi. This pressure would cause mercury to fill pore bodies connected to pore throats larger than 70 μm radius. Therefore, provided the largest pore throat radii of the true rock pores are less than 70 μm, mercury will not invade the pore system. If the core plug is held horizontally under mercury, the minimum pore throat radius criterion increases to around 140 μm.

5.3.3.1 Sample Preparation

Precautions must be taken when using mercury in core analysis. Mercury vapour is recognised as a highly toxic substance and therefore has been banned from some core analysis labs. Most labs are increasingly reluctant to use it. Managed properly, carefully and correctly, with mercury vapour monitoring systems, suitable extraction systems, and employee health monitoring and surveillance protocols in place, there is no reason why mercury immersion methods cannot be used regularly. Ideally mercury equipment should be contained within and isolated and chilled room or fume cupboard.

Like grain volume, the shape of the test sample is not crucial to the accuracy and reliability of the measurement, although regular right cylindrical plugs are preferred. Again, larger plugs are preferred to minimise measurement uncertainty on the porosity data.

The method is unsuitable for samples with surface vugs, fractures, very coarse-grained rock fabrics or as mercury can enter the pore volume which will result in an underestimation of bulk volume. The measurement is valid only if the mercury does not enter the pores of the sample. As it is often difficult to differentiate between false volume and pore volume for samples sleeved in metal or PTFE, this method is not recommended.

The sample dry weight is measured before (Wd_i) and after (Wd_f) completion of the measurement to check for mercury contamination.

5.3.3.2 Test Equipment—Mercury Pycnometer

All pycnometers (Fig. 5.18) have two key features in common (Fig. 5.19):

- a pressure chamber which accommodates the sample;
- a mercury pump, of known bore and piston diameter and with digital or analogue (scale plus Vernier) readout to determine the volume of mercury injected into the sample chamber.

The sample chamber is provided with a small hole at the top, so that when mercury fills the chamber, a small bead of mercury emerges and touches a needle datum. Steel forks or prongs are often used in the top of the sample chamber to hold the plug under the surface of the mercury, to counteract

FIGURE 5.18 Mercury pycnometer. *Courtesy of Vinci Technologies.*

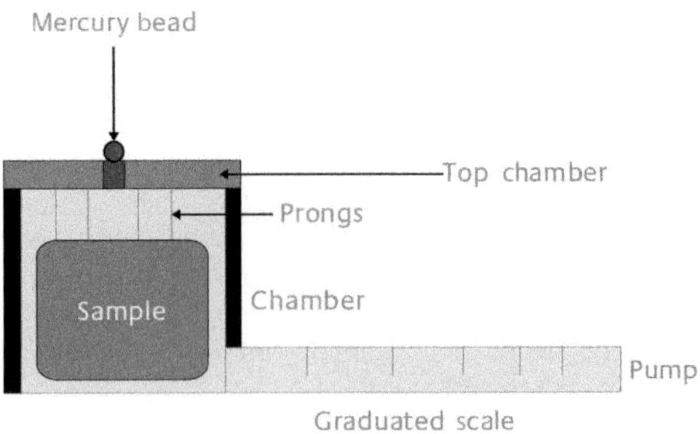

FIGURE 5.19 Typical mercury pycnometer schematic.

the buoyancy force of the mercury acting on the sample and to prevent the sample from blocking the exit hole.

5.3.3.3 Test Procedures—Mercury Pycnometer

In operation, the pump is used to inject mercury into the empty sample chamber until the mercury just appears at the hole in the top of the chamber. The pump is zeroed. The mercury is then withdrawn, and the sample placed into the chamber. Mercury is then pumped into the chamber until it just appears at the top hole. The volume required to reintroduce mercury to exactly the same level is equivalent to the sample bulk volume (Vb).

5.3.3.4 Test Equipment—Mercury Immersion System

In general, mercury immersion systems (Fig. 5.20) have the following key features in common:

- a plastic or metal mercury bath which accommodates the sample;
- an electronic balance with tare (zero) facility on which the mercury bath is placed;
- a sample cradle which holds the sample, together with an immersion mechanism (screw type) to lower the sample into the mercury bath;
- an indicator system, usually a needle connected to a DC battery and bulb, which is activated when the needle touches the mercury surface;
- a temperature probe to measure the mercury temperature.

5.3.3.5 Test Procedures—Mercury Immersion System

The cradle is lowered into the mercury to the reference mark, and the balance is tared (zeroed). The cradle is withdrawn and the sample is placed in the

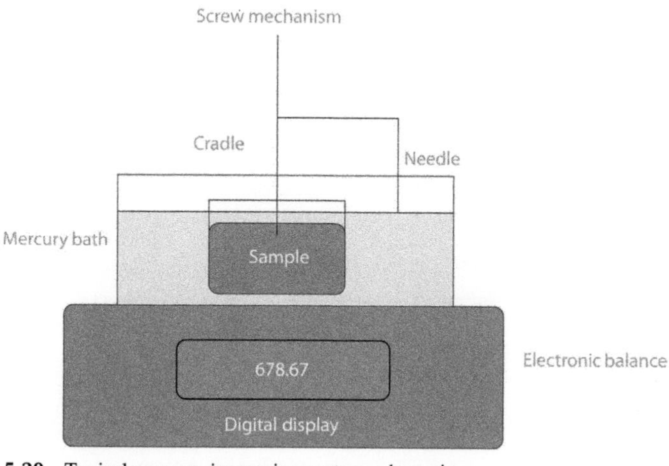

FIGURE 5.20 Typical mercury immersion system schematic.

cradle then re-immersed to the same reference mark. The increase in weight on the balance represents the immersed weight, W_{imm}.

The sample bulk volume is found from:

$$\mathrm{Vb} = \frac{W_{imm}}{\rho_{Hg}}$$

where ρ_{Hg} is the mercury density at test temperature (g/cc).

5.3.3.6 *Data Reporting Requirements*

Data reporting requirements from the laboratory are listed in Table 5.9.

5.3.3.7 *Advantages and Drawbacks/Issues*

Summaries of the advantages and drawbacks of the helium pore volume measurement are provided in Table 5.10.

5.3.3.8 *Mercury Bulk Volume Quality Control Issues, Checks and Diagnostics*

For the majority of test plugs, this is the recommended method for bulk volume determination and, in conjunction with ambient helium grain volume, the preferred method for porosity determination in RCA:

TABLE 5.9 Data Requirements Checklist for Mercury Immersion Bulk Volume Measurements

Data	Comments
Brief description of procedures and experimental apparatus	Pycnometer or immersion system
Description of cleaning and drying procedures	For core–log calibration
Plug depth and orientation	
Plug length and diameter	
Plug/sample dry weight before mercury immersion	
Plug sample dry weight after mercury immersion	Check for Hg invasion
Mercury density at test temperature	
Calculated bulk volume	
Calculated porosity (if based on Vb and Vg)	As a fraction (to 3 D)

TABLE 5.10 Mercury Bulk Volume: Advantages and Drawbacks

Advantages	Drawbacks and Issues
• Fairly rapid for majority of samples. • Relatively cheap method. • The depth of immersion of the sample in the mercury is minimised if the system design allows the samples to be immersed horizontally. This reduces the potential for mercury to enter the sample pores since the pressure of mercury acting on the base of the sample is reduced. • Mercury immersion system is potentially the most accurate method to determine sample bulk volume provided the sample is suitable.	• Unsuitable for samples with surface vugs, fractures, very coarse-grained fabric or sleeved in metal/Teflon (unconsolidated samples). Mercury can enter the pore volume which will result in an underestimation of bulk volume. • It is important to use the correct mercury density for the measurement temperature in the calculations. A temperature variation of 5 °C will induce a systematic error of 0.02% in the bulk volume. • Air bubbles can adhere to the sample, resulting in bulk volume overestimation. • Errors involved in measurements on small samples are likely to have larger errors than measurements on larger samples. • The mercury has to be maintained scrupulously clean as dirt or surface contamination can reduce the air/mercury interfacial tension and make it easier to enter the sample pores. The surface of the mercury is often cleaned with acetic acid. • Could potentially result in mercury contamination of the sample, invalidating further testing. For this reason, this method is not recommended for samples scheduled for electrical property measurements.

$$\phi = \frac{(\mathrm{Vb} - \mathrm{Vg})}{\mathrm{Vb}}$$

The key issue is whether mercury will enter the sample and underestimate bulk volume and also potentially contaminate the sample. For these reasons, it is not suitable for samples with large surface pores or unconsolidated sands. To prevent the possibility of any mercury vapour affecting test equipment, permeability is often measured on the plug sample after permeability measurement. That is: the mercury bulk volume is the final measurement in the RCA porosity and permeability test sequence.

Sample dry weight must be measured before and after testing, to check for potential mercury contamination of the sample.

Some laboratories now prefer to use helium pore volume and helium grain volume as this eliminates an H&SE risk. However, this increases the uncertainty in the porosity measurements. The selected lab should be audited to make sure that they adopt all necessary mercury H&SE procedures to manage the risks correctly.

5.3.4 Liquid Saturation Porosity

This method is used to determine the porosity of a sample saturated and immersed in a liquid of known density. The method is not generally used during RCA programmes as it is relatively more expensive and time consuming. Saturating the sample in toluene was once a popular method, but this has been superseded by helium porosity methods. Many SCAL tests (such as electrical resistivity, capillary pressure, wettability and relative permeability tests) involve initially saturating the sample in formation water. This provides an opportunity to determine sample re-saturation porosity, independent from helium grain volume and mercury bulk volume measurements. In fact, in electrical property measurements where any mercury contamination could influence the measurements, re-saturation porosity is the preferred method.

The method relies on the sample's pore space being completely saturated in brine of known density, and that the brine volume in the pore space plus the grain volume displaces an equivalent volume of brine when immersed in brine of the same density.

5.3.4.1 Sample Preparation

Samples must be thoroughly extracted from fluids and properly dried, following cleaning and drying procedures described in Chapter 3. Porosity can be initially measured following any of the methods provided in the previous section. This provides a check on the re-saturation porosity measurement. Care should be taken with mercury bulk volume due to the potential for sample contamination.

Synthetic formation water (SFW) is prepared in accordance with the recommendations in Chapter 6, and density (ρ_{FW}) is measured at the test temperature.

5.3.4.2 Test Equipment

Pore volume is determined by weight difference between the saturated and the dry weight of the plug using an electronic balance. Bulk volume is determined following the Archimedes immersion principle, using an experimental set-up similar to that shown in Fig. 5.21. The electronic balance should have a weigh hook underneath so that the core cradle can be immersed in the same SFW.

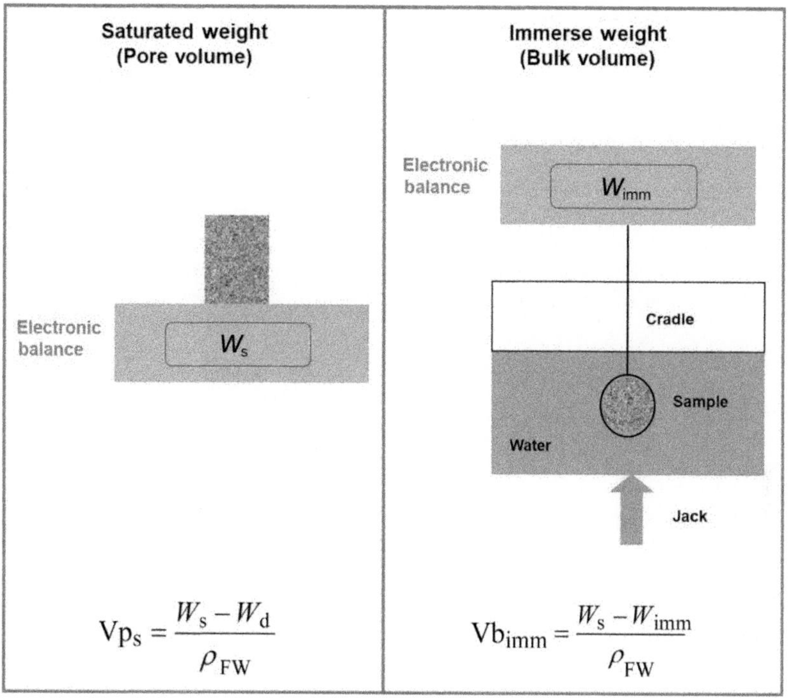

FIGURE 5.21 Archimedes brine immersion system.

5.3.4.3 *Test Procedures*

The sample dry weight is measured (W_d) and the plug is placed in a vacuum dessicator and saturated in deaired SFW. In lower permeability samples, the plugs are then loaded into a pressure vessel and pressure saturated in SFW at approximately 2000 psi for 2–3 days to fully saturate the pore space. Sample saturated weight is measured (W_s) on top of the balance.

Bulk volume is determined by suspending the plug under the weigh balance so that it is immersed in the same brine using a cradle attached to the underside of the balance. The saturated weight of the plug is firstly determined on top of the balance. The core cradle is lowered (or the brine container jacked up) until a reference mark on the wire is reached. The balance is tared to compensate for the cradle displacement and the cradle is withdrawn. The plug is then placed in the cradle and re-immersed in SFW to the same reference mark on the wire and the immersed weight (W_{imm}) is recorded.

5.3.4.4 *Porosity Calculation*

The re-saturation pore volume (Vp_s) is obtained from:

$$Vp_s = \frac{W_s - W_d}{\rho_{FW}}$$

and the sample immersed bulk volume (Vb_{imm}) is determined from:

$$Vb_{imm} = \frac{W_s - W_{imm}}{\rho_{FW}}$$

The re-saturation porosity is calculated from:

$$\phi = \frac{Vp_s}{Vb_{imm}}$$

If helium porosity measurements were initially performed on the samples at ambient conditions, then the data can be cross-checked.

5.3.4.5 Data Reporting Requirements

Data reporting requirements from the laboratory are listed in Table 5.11.

TABLE 5.11 Data Requirements Checklist for Re-saturation Pore Volume Measurements

Data	Comments
Brief description of procedures and experimental apparatus	
Description of cleaning and drying procedures	For core–log calibration
Plug depth and orientation	
Plug length and diameter	
Plug/sample dry weight before saturation	W_d
Brine ionic composition and TDS (mg/l or ppm)	
Brine density at test temperature	ρ_{FW}
Sample saturated weight (in air on top of balance)	W_s
Sample immersed weight in SFW	W_{imm}
Calculated pore volume	Vp_s
Calculated bulk volume	Vb_{imm}
Calculated saturated porosity (if based on Vb and Vg)	As a fraction (to 3 D)
Helium pore volume	If measured
Helium grain volume	If measured
Mercury or other bulk volume	If measured
Helium porosity	As a fraction (to 3 D)

5.3.4.6 Advantages and Drawbacks/Issues

Summaries of the advantages and drawbacks of the helium pore volume measurement are provided in Table 5.12.

5.3.4.7 Re-Saturation Porosity Quality Control Issues, Checks and Diagnostics

The re-saturation porosity should be within ±0.2 p.u. of the helium porosity. If the re-saturation porosity is too low in comparison with helium porosity then not all of the pore space may be saturated in SFW. If it is too high, then the plug may have suffered from grain losses.

5.3.5 Accuracy and Repeatability of Porosity Measurements

The API RP 40 (1988) 'standard' for reproducibility of helium porosity measurements between laboratories is ±0.5 p.u. Most reputable laboratories can achieve a much more stringent repeatability of ±0.25 p.u. (Amabeoku

TABLE 5.12 Re-saturation Porosity: Advantages and Drawbacks

Advantages	Drawbacks and Issues
• Since the test principle is based on measurement of weights rather than volumes, the potential measurement accuracy can be greater than the other porosity determination methods, provided there is no grain loss. • The method is not destructive so samples can be used for further testing. • Saturation in formation water is the first stage in most SCAL experiments.	• It is important to use the correct brine density for the measurement temperature in the calculations. A temperature variation of 5 °C will induce a systematic error of 0.02% in the bulk volume. • Relatively more expensive and significantly more time consuming that other porosity methods. • Samples need to be fully saturated in brine. • Air bubbles can adhere to the sample surface, resulting in bulk volume overestimation. • Measurements on small samples are likely to have larger errors than measurements on larger samples. • Errors can occur in more friable samples due to grain loss during the experiment. • Difficult to use on unconsolidated/sleeved samples, unless weights and volumes of sleeve material and screens are fully accounted for.

and BinNasser, 2012; Thomas and Pugh, 1989) especially if porosity is determined from helium grain volume and mercury immersion bulk volume. Measurements on smaller plugs are likely to have larger errors than measurements on larger plugs, and measurements using helium pore volume or bulk volume measured by callipers can have much larger errors if the plugs are not perfect right cylinders.

Grain density or matrix density (ρ_m) is an essential measurement to calibrate wireline density logs, where porosity is determined from the environmentally corrected bulk density (ρ_b) and fluid density (ρ_f):

$$\phi = \frac{(\rho_m - \rho_b)}{(\rho_m - \rho_f)}$$

Any errors in grain density will obviously translate to errors in estimated log porosity. For example, Fig. 5.22 plots error in porosity as a function of error in grain density. The lines shown correspond to a clean sandstone (true grain density = 2.65 g/cc) saturated with brine with a fluid density of 1 g/cc for four different porosities. For low porosity formations, an error in grain density of 0.03 g/cc corresponds to an error in calculated porosity of ~2 p.u. The errors are less for higher porosity formations and decrease with decreasing error in grain density. For errors of around 0.01 g/cc (API standard), the errors in porosity for almost all ranges of reservoir quality rock are less than

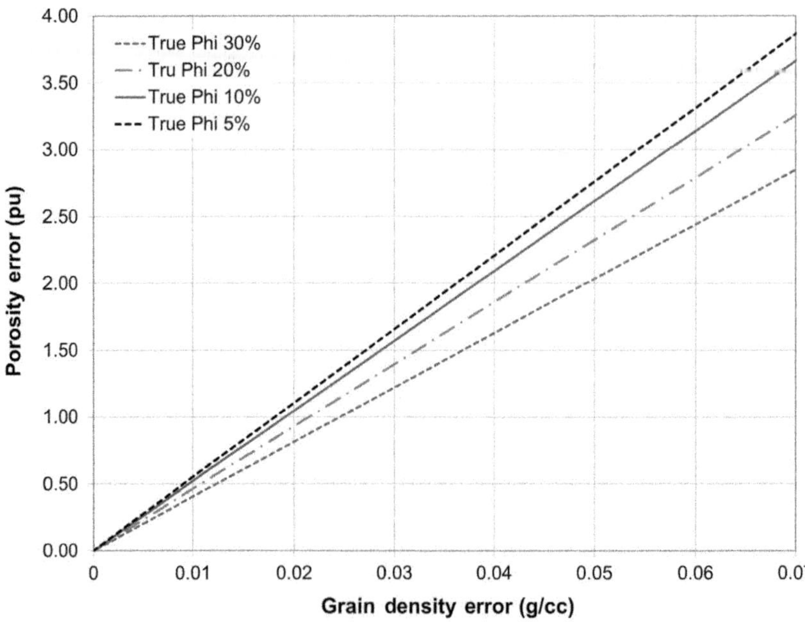

FIGURE 5.22 Grain density errors in density porosity calculation.

0.5 p.u. Adopting a stricter standard of ±0.25 p.u. will translate to a grain density standard of ±0.005 g/cc.

Core measurements made at ambient conditions in RCA programmes cannot be reconciled directly with log measurements for a number of reasons. For example, logs respond to the bulk densities of formations at reservoir *in situ* stress conditions, so that the effects of stress on core porosity must be determined as part of the SCAL programme.

5.4 PERMEABILITY MEASUREMENTS

5.4.1 Definitions

There are three principle definitions of permeability in core analysis:

1. Absolute permeability: The permeability to a single phase where only that phase fills the pore space (e.g. air, oil or brine).
2. Effective permeability: The permeability to one phase when two or three phases are present in the pore space. For example, oil permeability (Ko) and water permeability (Kw) at some specific water saturation.
3. Relative permeability: The ratio of effective permeability to a base permeability which for SCAL laboratories is normally an endpoint effective permeability (Ko′ at Swir or Kg′ at Swir) or a single-phase permeability (air or Klinkenberg permeability) in reservoir modelling.

For historical reasons, which have very little to do with the reality of reservoir evaluation, RCA programmes utilise ambient condition, absolute permeability measurements to air or nitrogen. These are cheap, fast and convenient. Unfortunately, air permeability at ambient conditions bears little relationship to the reservoir condition permeability, though it does highlight permeability trends and distributions. SCAL programmes provide an opportunity to determine permeability at conditions which better approximate the reservoir situation.

The core plug samples are usually cleaned and dried prior to analysis, to remove oil, brine and drilling mud contaminants, as described in Chapter 4. Cleaning and drying, especially in shaly sands, provides an opportunity for damage to the rock texture that must be addressed if reliable data are required.

5.4.2 Darcy's Law

In 1856, Henry Darcy, a hydraulic engineer working in Dijon, France, published the first definition of the fluid conductivity of a porous medium. His experiments were simple, yet elegant, and, with some modification, are essentially the same as used in every core analysis laboratory today. All his experiments were carried out by flowing water through tubes filled with sorted sand—from which he derived his final equation describing fluid flow. By considering a cylinder of porous media (Fig. 5.23), Darcy's equation for horizontal, rectilinear, steady-state and incompressible flow can be defined as:

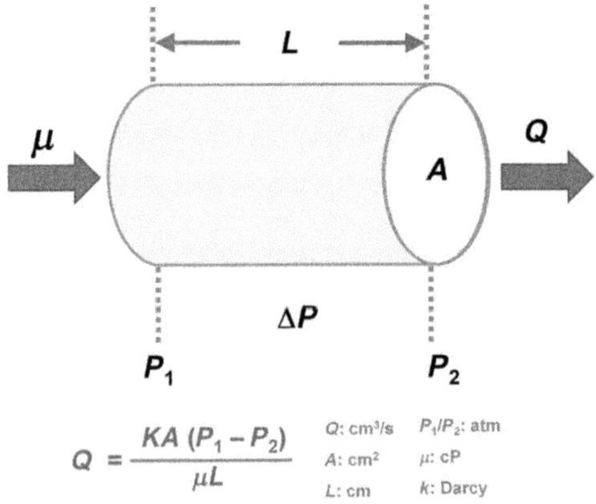

FIGURE 5.23 Diagram explaining parameters in the Darcy equation for incompressible liquid flow through a core plug.

$$Q = \frac{kA(P_1 - P_2)}{\mu L}$$

In SI units, the unit of permeability is m^2. In the petroleum industry, the Darcy (D) is more convenient ($1\ D \sim 1 \times 10^{-12}\ m^2$) and is the standard field unit for permeability (k). It represents 1 cm^3 of fluid with a viscosity (μ) of 1 cP flowing through a 1 cm^2 cross-sectional area (A) of rock in 1 s under a pressure difference ($P_1 - P_2 = \delta P$) of 1 atm. per 1 cm length (L) in the direction of flow. It is a linear relationship: if the flow rate (Q) is doubled so the pressure drop (δP) doubles. Thus, a plot of flow velocity (Q/A) against the pressure drop per unit length ($\delta P/L$) on a core sample will be a straight line with slope equal to k/μ.

In comparison to water, gas is compressible so corrections have to be made for gas compression and expansion across the core sample. For example, if the pressure at the core inlet is twice the pressure at the core outlet, then, in accordance with Boyle's law, the volumetric flow rate of gas at the inlet will be half the rate at the outlet. Thus, if the flow rate is measured at the core outlet, it has to be corrected for volumetric expansion. One convenient method to correct for gas compressibility, which is used by almost all laboratories for the solution of the Darcy equation for the steady-state flow of gas yet is not rigorously correct (API RP 40), is to correct the flow rate measured at the either core inlet or outlet to the mean flow rate and the mean pressure in the sample.

Rearranging the Darcy equation for incompressible flow in terms of permeability, k, then:

$$k = \frac{Q_m \mu L}{A(P_1 - P_2)}$$

where the mean flow rate of incompressible liquid, Q_m, is constant along the length of the plug.

If the flow rate is measured at the core outlet at atmospheric pressure, P_a, it is corrected to the mean flow rate, Q_m, and mean pressure using:

$$Q_m = \frac{Q_a P_a}{P_m}$$

where the mean pressure is:

$$P_m = \frac{P_1 + P_2}{2}$$

Therefore, the gas permeability, k_g, in Darcy units (where 1 D = 1000 mD) can be estimated from

$$k_g = \frac{2 Q_a P_a \mu_g L}{A\left(P_1^2 - P_2^2\right)}$$

The parameters in the gas permeability equation and the data sources are provided in Table 5.13.

TABLE 5.13 Parameters in Darcy Equation for Gas Permeability

Term	Parameter	Units	Source
Q_a	Atmospheric flow rate	cm^3/s	From downstream manometer or mass flow metre
μ_g	Gas viscosity	cP	At flow pressure and temperature. Normally obtained from PVT correlations such as Tables 6.4 (for air) and 6.5 (for nitrogen) in API RP 40 (1988)
L	Plug length	cm	From average of 2 or 3 calliper measurements
A	Cross section area	cm	From average of 2 or 3 calliper measurements of plug diameter $A = \frac{\pi D^2}{4}$ or from mercury bulk volume, Vb, and length, L $A = \frac{Vb}{L}$
P_1	Upstream pressure	atm.	Pressure transducer or gauge at inlet face
P_2	Downstream pressure	atm.	Pressure transducer or gauge at outlet face. If test carried out with no backpressure, then $P_2 = P_a$. If test performed with backpressure then $P_2 > P_a$

This is the standard laboratory equation used for steady-gas permeability measurements. It assumes that the mean gas compressibility factor is unity and that the mean temperature of the flowing gas is equal to the gas temperature at atmospheric pressure, conditions which are approximately true for nitrogen at typical conventional core analysis test conditions.

Darcy's law involves four other key assumptions that need to be taken into account:

1. The saturating fluid is inert and does not interact with the porous medium. This is true for nitrogen or helium.
2. The flow through the sample is laminar. This is untrue for gas under certain flow rate conditions.
3. Fluid flow is single phase and the fluid should completely saturate the porous medium. This is true for gas in clean and dry cores.
4. The permeability is constant and does not vary with the nature of the fluid, flow rate nor pressure. This is not true for gas due to non-Darcy effects, principally the Klinkenberg and Forchheimer effects.

5.4.3 Non-Darcy Flow: Klinkenberg Effects

When a liquid is flowing through a capillary (analogous to a pore), there is a zero velocity layer at the pore walls. When gas is flowing, the layer of gas next to the surface of the wall is in motion with respect to the solid surface. Interactions between the gas molecules and the pore walls help move the gas in the direction of flow. The degree of interaction is governed by the equation:

$$\frac{4c\lambda}{r} = \frac{b}{P_m}$$

where c is a constant, λ is the mean free path of a gas which is related to the inverse of the molecular weight of the gas, r is the mean pore (capillary) radius and P_m is the mean pressure. As the pore radius approaches the mean free path of the gas, the velocity of individual gas molecules accelerates or 'slips' when contacting the pore surfaces. The b factor is a measure of slippage.

At lower pressures (P_m), the molecules will collide less frequently and the gas slippage effect (b) is enhanced—the gas molecules are so far apart that they slip through the pore spaces with little friction loss. Hence, gas permeability is increased. At higher mean pressures, the gas molecules are closer together and experience a friction drag at the side of the pore walls. This increases as the mean pressure increases, with the gas acting more and more like a liquid, and measured gas permeability decreases accordingly. As b is an inverse function of r then, in low permeability samples, as the pore radii get smaller, so the slippage effects increase. In very low permeability formations, the gas slippage effects therefore make a very significant contribution to gas flow. If the mean pressure is increased to an infinite value, the mean free path reduces to zero, and the gas molecules behave as liquids. In high permeability samples with larger pore radii, the slippage effect becomes much less significant.

Klinkenberg (1941) argued that if measurements could be made at infinite gas pressure, then the gas permeability measured at infinite pressure would be equivalent to the permeability to a non-reactive liquid, irrespective of the gas type (and molecular weight). This means that no matter the type of gas or the mean pressure used (laboratories can use different gases and measure gas permeability at different mean pressures), the Klinkenberg permeability at infinite mean pressure will always be the same. This is one reason why Klinkenberg permeability is often used to describe permeability in many static reservoir models in preference to nitrogen or helium permeability—it is considered to be somehow 'definitive' and independent of gas type and mean pressure.

However, this measurement at infinite pressure is impossible, so Klinkenberg devised an experiment to extrapolate the equivalent liquid permeability. Measurements of steady-state gas permeability are made at a number of mean pressures (P_m), normally by adjusting the system backpressure. Gas permeability data (k_g) are plotted against the corresponding inverse mean pressure data $\left(P_m^{-1}\right)$. Ideally, these should fall on a straight line as indicated in Fig. 5.24. The Klinkenberg permeability for each plug is then calculated from the multi-pressure gas permeability data using the equation:

$$k_g = k_l\left(1 + \frac{b}{P_m}\right)$$

where k_l is the Klinkenberg permeability (mD), often referred to as k_∞.

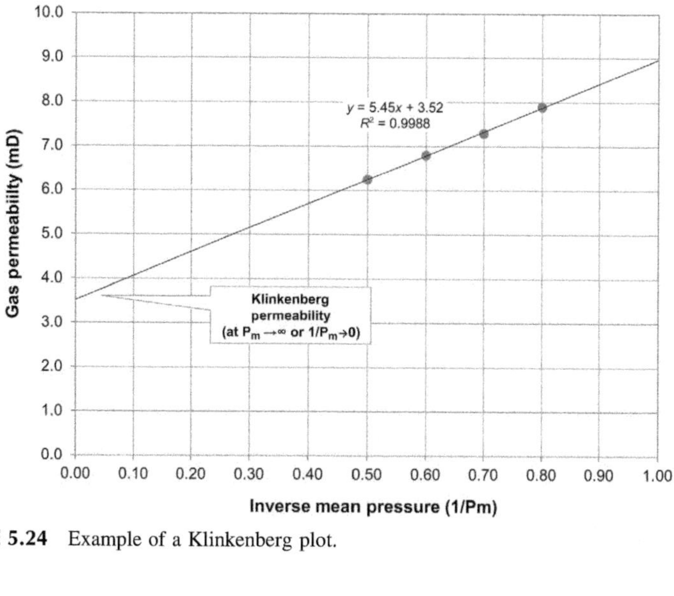

FIGURE 5.24 Example of a Klinkenberg plot.

Values of Klinkenberg permeability and slippage factor can be determined from linear regression analysis of the gas permeability and inverse mean pressure values. k_l corresponds to the intercept for $P_m^{-1}=0$ ($P_m \rightarrow \infty$), and b, at a specified mean pressure, can be calculated from the slope, m:

$$b=\frac{mP_m}{k_l}$$

In the laboratory determining Klinkenberg (slippage-corrected) permeability under steady-state flow conditions requires the measurement of gas permeability at three or four mean pressures. McPhee and Arthur (1991) demonstrated that Klinkenberg permeabilities determined by conventional steady-state methods were sensitive to the methods and procedures used to acquire and analyse the data and, in low permeability samples where the Klinkenberg effect is more pronounced, the test procedures were tedious and protracted. Steady-state measurements of Klinkenberg permeability in RCA programmes are too time consuming and are rarely performed in most commercial labs unless specifically requested. Most lab-reported Klinkenberg permeabilities are not 'measurements' at all but are derived from correlations. An example for a North Sea field is shown in Fig. 5.25. An alternative is to use a correlation between slippage factor (b) and Klinkenberg permeability of the form

$$b=ak_l^{-x}$$

as shown in Fig. 5.26. This is substituted into the Klinkenberg equation, which is rearranged so that Klinkenberg permeability can be estimated from individual single gas permeability measurements at a known mean pressure:

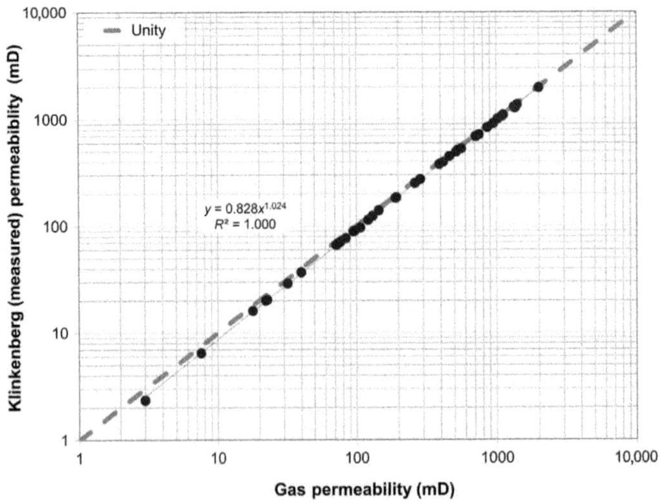

FIGURE 5.25 Klinkenberg versus gas permeability relationship for a North Sea field.

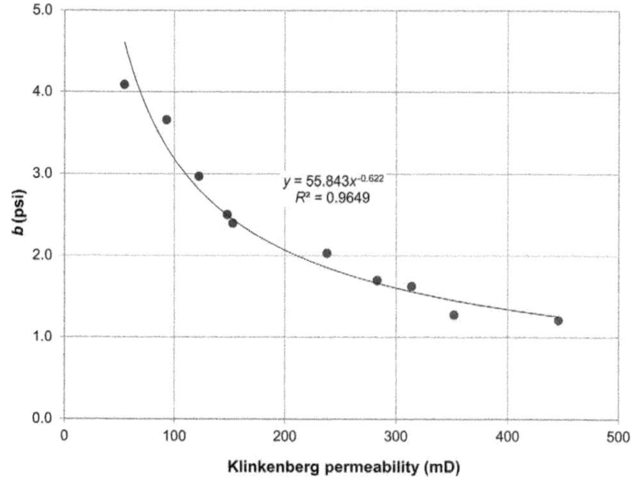

FIGURE 5.26 Example of Klinkenberg permeability versus slippage factor plot.

$$k_l = \left[\frac{k_g}{1 + (a k_l^{-x} / P_m)} \right]$$

As this is a non-linear equation, an iterative method (e.g. Newton–Raphson) is used to solve for k_l given, b, P_m and k_g.

The issue with using correlations is that they may be valid only for a particular field or lithology and only if all gas permeability measurements were made at the same, or very similar, mean pressures. It may not be representative of other fields or lithologies, or for gas permeability measurements made at different mean pressures. For example, many laboratories still use the Klinkenberg permeability correlation reported by Heid et al. (1950) which was developed from a very limited dataset. The correlations assume that the slip factor is constant and does not vary with mean pressure. In fact, Klinkenberg (1941) comments that 'the value of the constant, b, increases with increasing pressure': a most significant statement which has been overlooked by many analysts.

Klinkenberg found that the measured permeability of the same samples to inert liquid (isooctane) and the slippage-corrected gas permeability agreed 'within experimental error'. The Klinkenberg permeability is based on fluid flow through capillaries and takes no account of rock–fluid interaction between real liquids and real rock pore systems. For example, Klinkenberg permeabilities are found to be routinely higher (by factors of 2 or 3 in clay-rich rocks) than the absolute permeability to brine (Fig. 5.27) or the effective permeability to oil. Reasons for this discrepancy are rehydration, flocculation and rearrangement of clays on exposure to brine following harsh core drying; and undersaturation in low permeability samples—the presence of air in the pore system will suppress the apparent permeability to brine.

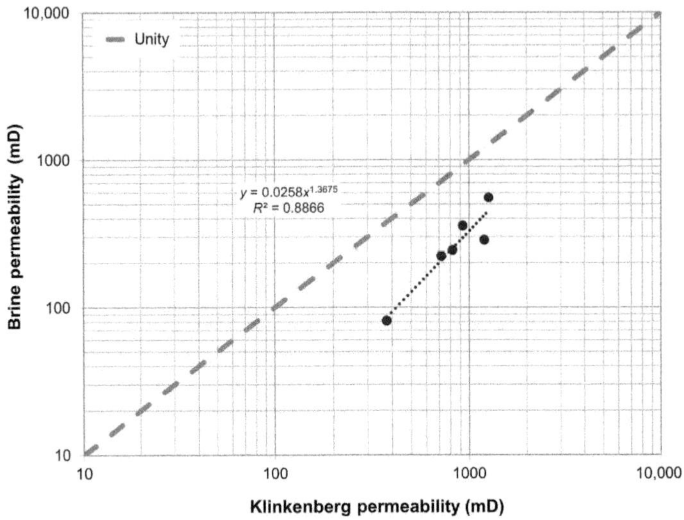

FIGURE 5.27 Relationship between Klinkenberg and brine permeability (clay-rich samples).

5.4.4 Non-Darcy Flow: Forchheimer Effect

If the flow rate used in steady-state gas permeability measurements is too high, then the flow regime may no longer be viscous, but turbulent (inertial). This turbulent flow is described by the Forchheimer (1901) equation and will create an additional non-Darcy pressure drop across the core, in addition to the Darcy viscous pressure differential:

$$\frac{dP}{dL} = \frac{dP}{dL_{\text{Darcy}}} + \frac{dP}{dL_{\text{non-Darcy}}}$$

The non-Darcy flow pressure differential for linear flow across as sample with permeability k is described by:

$$-\frac{dP}{dL_{\text{non-Darcy}}} = \frac{\mu v}{k} + \beta \rho v^2$$

The β term is referred to as the Forchheimer inertial resistance coefficient with units of inverse length. ρ is the gas density and v is the gas velocity (Q/A). There is a strong inverse correlation between β and permeability (Jones, 1987). If $\beta \rho v^2$ approaches zero, Darcy's law will adequately describe flow. In most laboratory cases, measurements of RCA gas permeability are normally within a Darcy flow regime, unless very high flow velocities (v) are used, or where measurements are made on very low permeability samples (high β). If non-Darcy flow is not recognised, then applying Darcy's law to flow data that are not in a viscous flow regime will underestimate permeability due to the additional pressure drop.

Combining both Klinkenberg and Forchheimer corrections to Darcy's law, the governing equation for gas permeability becomes:

$$-\frac{dP}{dL} = \frac{\mu v}{k_l(1 + (b/P_m))} + 3.238 \times 10^{-8} \beta \rho v^2$$

where β has the units of ft.$^{-1}$, ρ is gas density (g/cc) and v is gas velocity (cm/s).

McPhee and Arthur used an iterative technique, based on non-linear optimisation of the parameters k_l, b and β, to solve for non-Darcy flow from steady-state experiments. Ruth and Kenny (1987) contest that if inertial effects are not present, or can only be barely captured, yet an analytical model allows for them, then any minor experimental errors (such as pressure transducer inaccuracy) can cause calculated values of β to arise which have nothing to do with inertial effects. They suggest that derived values for b and β should only be accepted if they lead to experimental Klinkenberg numbers (b/P_m) greater than 0.1.

5.4.5 Steady-State Permeability Measurements

5.4.5.1 *Sample Preparation*

Samples must be thoroughly extracted from fluids and properly dried, following cleaning and drying procedures described in Chapter 3.

5.4.5.2 *Test Equipment*

Figure 5.28 provides a schematic of a typical steady-state gas permeability apparatus. All gas permeameters have the following key features in common:

- a coreholder, or pressure cell, provided with a rubber sleeve (or 'boot') which is used to confine the test sample (Fig. 5.29). This is often referred to as a Hassler cell. Inlet and outlet end platens allow gas to enter and

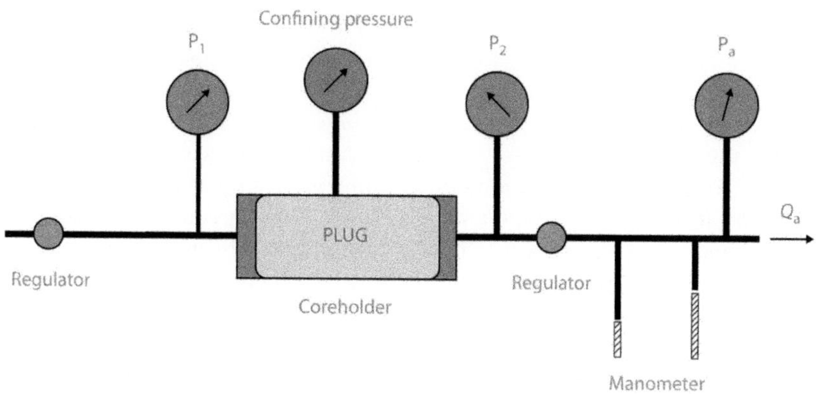

FIGURE 5.28 Gas permeameter schematic.

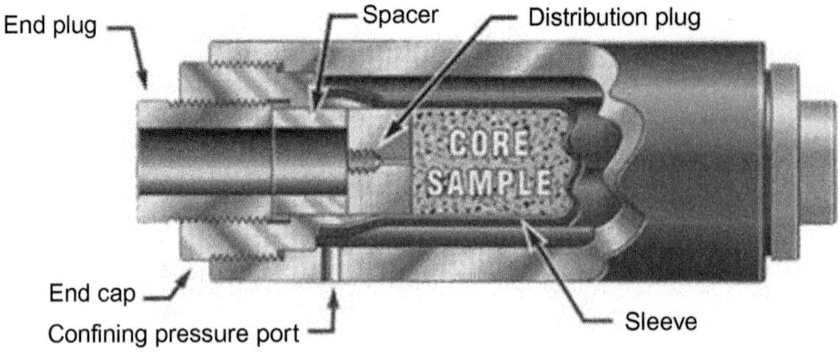

FIGURE 5.29 Hassler coreholder for gas permeability measurements. *Courtesy of Core Laboratories and its Petroleum Services and Core Lab Instruments divisions. All Other Rights Reserved.*

leave the plug and are also sealed by the rubber boot. The end platens are often provided with a 'spider web' groove pattern to promote even flow distribution into the plug. In most coreholder designs, the end platens are free to move against the plug when a pneumatic or hydraulic pressure is applied to the outside of the rubber sleeve. Hence, the axial stress applied along the sample is equal to the radial stress around the plug. This is referred to as isostatic loading conditions. The stress applied has to be sufficient such that the rubber sleeve conforms to the surface of the sample, sealing the sample to prevent gas flowing along the outside of the plug (gas bypass). Measurements can also be made at a confining stress that attempts to replicate the equivalent effective isostatic reservoir stress.

- a system to apply pressure to the sleeve, usually hydraulic pressure (high stress) or pneumatic pressure (low stress), provided with a pressure indication gauge or electronic transducer.
- a method to control the pressure of gas entering the sample (injection pressure) usually a gas regulator, upstream of the sample. The maximum injection pressure is usually limited to about 100 psi.
- a pressure indicator gauges sited upstream (P_1) and downstream (P_2) of the sample. Alternatively, some permeameter designs utilise a gauge transducer upstream of the sample and a differential pressure transducer to measure the pressure drop ($P_1 - P_2$) across the sample. For high permeability samples, the flow rate is normally increased to optimise the accuracy of the differential pressure measurement, as errors in transducer calibration at low-pressure differentials have a significant impact on permeability calculations. High rates can induce non-laminar flow in both the core sample and the inlet and outlet pipework which will result in a significant underestimation of permeability by as much as 50% in plugs with permeabilities greater than 5000 mD. In this case, it is recommended that the coreholder end platens are provided with separate pressure taps so that P_1 and P_2 are measured directly at the core end faces on non-flowing lines. This will eliminate non-laminar flow effects in the pipework on the pressure measurements.

- a method to control the outlet pressure in the sample (typically a backpressure regulator), which is used for flow control in high permeability samples and in Klinkenberg measurements.
- a system to measure gas flow rate for a wide range of sample permeabilities, generally up to about 2000 ml/min. In many modern permeameters, gas flow is controlled and measured by a mass flow meter upstream of the core plug (at P_1). In this case, the measured flow is not at atmospheric conditions (Q_a) but at inlet pressure conditions (Q_1) and is corrected to mean pressure conditions by $Q_m = (Q_1 P_1)/P_m$. In the case shown in Fig. 5.27, gas flow from the sample is directed through a selected capillary tube downstream of the backpressure regulator at atmospheric pressure, P_a. The pressure drop caused by gas flowing through the tube is directly related to the flow rate through the tube and measured using a manometer. The manometer response is precalibrated using wet gas meters and/or soap film meters.
- a thermocouple to measure flowing gas temperature (to allow prediction of gas viscosity).
- a barometer to measure atmospheric pressure (P_a) if flow rate is measured at atmospheric conditions. That is $Q_m = (Q_a P_a)/P_m$.

5.4.5.3 *Test Procedures*

At least three measurements of sample length and diameter should be made. If the bulk volume has been determined during porosity measurements, the cross-sectional area, A, can be found from:

$$A - \frac{\mathrm{Vb}}{L}$$

The plug is loaded in the coreholder and a confining stress is applied. The confining stress must be sufficient to avoid nitrogen passing through the annular space between the sample and the confining sleeve.

Oxygen-free nitrogen is flowed through the sample at a constant flow rate or pressure drop. Low injection pressures are preferred. For higher permeability plugs, flow under nominal backpressure generally provides a more stable rate. If the injection pressure has to be increased significantly (e.g. in low permeability samples), the confining stress is adjusted accordingly to maintain the net confining stress (NCS) constant. The NCS is the difference between the total applied stress (TCS) and the pore pressure (p_p): $\mathrm{NCS} = \mathrm{TCS} - p_p$. In other words, if the specified NCS is 300 psi and the mean pore pressure is 50 psi, then the TCS must be set to 350 psi.

There is no definitive guide to the selection of gas flow rates for permeability measurements but as a rule of thumb the maximum rate for high permeability samples flow is 2000 ml/min.

Once pressure drop and flow rates reach stability (steady-state conditions), flow rate, differential pressure, temperature and atmospheric pressure are recorded. The outlet pressure (P_2) is normally equivalent to atmospheric pressure (P_a) if no backpressure is used. If using backpressure, P_2 will be greater than P_a. The use of back pressure is highly recommended as it provides improved control of gas flow rate and core differential pressure, assisting in maintaining Darcy's viscous flow at higher mean pressures.

If Klinkenberg permeability is required, the mean pressure in the sample is increased either by increasing the inlet pressure (P_1) or flow rate or, via the backpressure regulator, by increasing the downstream pressure, P_2. Once pressure drop and flow rates reach stability again flow rate, differential pressure, temperature and atmospheric pressure are recorded. This is repeated a further one or two times at the same NCS until three or four gas permeability measurements at increasing mean pressures are acquired.

5.4.5.4 Gas Permeability and Klinkenberg Permeability Calculations

The gas (nitrogen) permeability is determined from:

$$k_g = \frac{2Q_a P_a \mu_g L}{A\left(P_1^2 - P_2^2\right)}$$

if flow (Q_a) is measured at atmospheric conditions, or

$$k_g = \frac{2Q_1 P_1 \mu_g L}{A\left(P_1^2 - P_2^2\right)}$$

if Q_1 is measured at inlet (upstream) pressure (P_1) conditions.

Klinkenberg permeability, k_l, is determined from linear regression analysis of at least 3 (and ideally more) k_g and $1/P_m$ pairs.

5.4.5.5 Evaluation of Klinkenberg and Non-Darcy Effects in Steady-State Flow

Three different graphical analysis techniques can be used to evaluate the steady-state measurements for non-Darcy flow:

1. *Klinkenberg plot*. The gas slippage effect can be assessed using the conventional Klinkenberg plot of the inverse mean pressure against the apparent gas permeability (Fig. 5.30) where Klinkenberg-corrected permeability can be estimated as explained above. Non-Darcy effects may also be identified as deviations from the straight line trend to lower permeability at high mean pressures (low P_m^{-1}), which can be attributed to the onset of inertial effects.
2. *Slippage-corrected pressure plot*. The plot shown in Fig. 5.31 is used to correct for slippage effects. The slippage-corrected Darcy pressure drop is plotted against mass flow rate. This plot should yield a line drawn through the origin, with slope equal to the inverse of Klinkenberg-corrected permeability (K_l), as shown in the following analytical expression (Rushing et al., 2004):

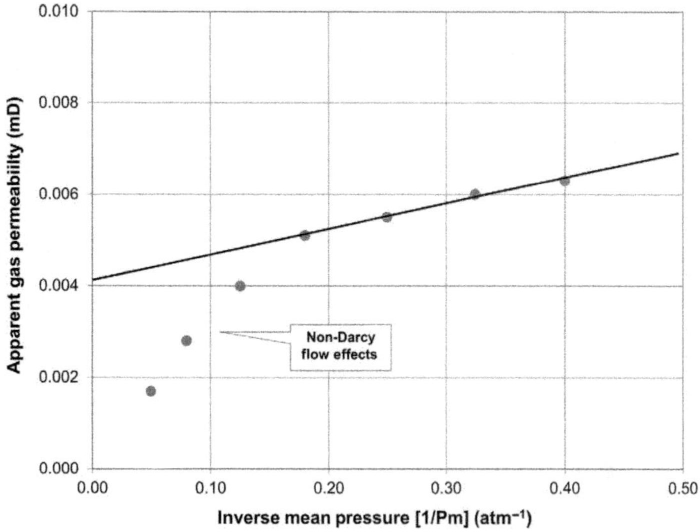

FIGURE 5.30 Hypothetical Klinkenberg plot showing non-Darcy flow behaviour. *Modified from Rushing et al. (2004).*

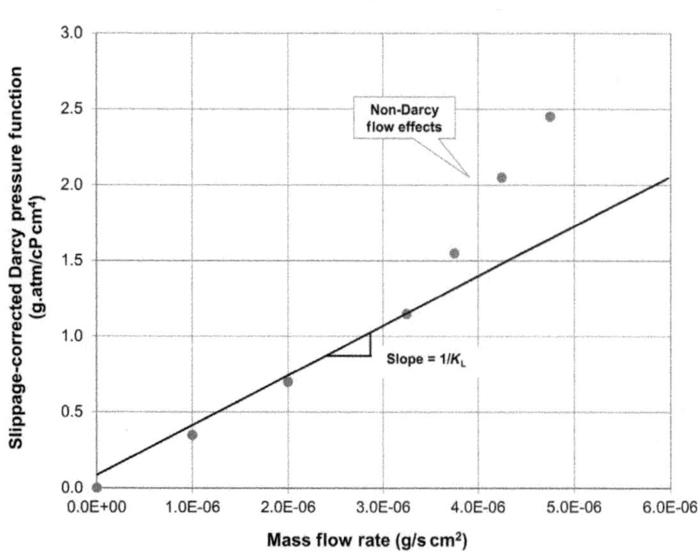

FIGURE 5.31 Hypothetical slippage-corrected Darcy plot showing non-Darcy flow behaviour. *Modified from Rushing et al. (2004).*

$$\frac{M(1+b/P_{\rm m})(P_1^2-P_2^2)}{2\mu zRTL}=\frac{\dot{m}}{k_1}$$

where:

M gas molecular weight (g/mol)
b gas slippage factor (atm.)
P_m mean pressure (atm.)
P_1 inlet pressure (atm.)
P_2 outlet pressure (atm.)
z average gas deviation factor (dimensionless)
R universal gas constant = 82.057 atm cm^3/g mol K
L core length (cm)
T absolute temperature (K)
$\dot{m}$ mass flow rate = ρv (g/cm^2 s)
ρ gas density (g/cm^3)
v interstitial gas velocity (cm/s)
μ average gas viscosity (cP)
k_1 Klinkenberg permeability (D)

Non-Darcy flow effects are indicated by a trend which deviates upward from the line.

3. *Inertial flow-corrected pressure plot*. The plot shown in Fig. 5.32 is used to correct for inertial effects (Rushing et al., 2004). The slippage-corrected Forchheimer pressure change is plotted against the slippage-corrected mass flow rate term. This plot should yield a line with slope equal to the inertial

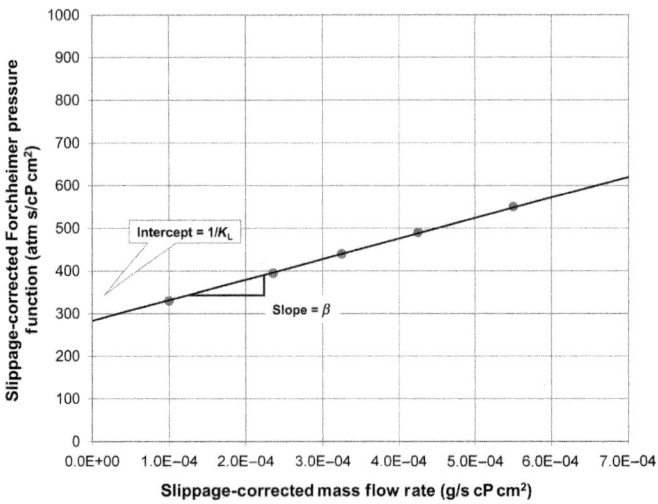

FIGURE 5.32 Hypothetical inertial flow and slippage-corrected Forchheimer plot. *Modified from Rushing et al. (2004).*

resistance coefficient (β) and an intercept equal to the inverse of Klinkenberg-corrected permeability (k_l), as shown in the following analytical expression:

$$\frac{M(1+b/P_m)(P_1^2-P_1^2)}{2\mu zRTL\dot{m}}=\frac{\dot{m}(1+b/P_m)\beta}{\mu}+\frac{1}{k_l}$$

where

β Forchheimer inertial resistance coefficient (cm^{-1})

5.4.5.6 Data Reporting Requirements

Data reporting requirements from the laboratory are listed in Table 5.14 for both gas and Klinkenberg steady-state permeability measurements

5.4.5.7 Advantages and Drawbacks/Issues

Summaries of the advantages and drawbacks of the steady-state gas permeability measurement systems are provided in Table 5.15.

5.4.5.8 Gas and Klinkenberg Permeability Quality Control Issues, Checks and Diagnostics

Repeatability tests should be carried on at least every 10th plug by different operators at the same NCS and ideally using different permeameters. Thomas and Pugh (1989) were among the first to publish definitive repeatability data for steady-state gas permeability measurements. The statistical analysis was based on 1900 check measurements made on plugs prepared by Amoco Research by 65 laboratories representing 35 service companies, over more than 30 years. They expressed the results in terms of experience-based deviation, as indicated in Table 5.16. The Thomas and Pugh results are compared with repeatability criteria for three major European commercial laboratories using more modern equipment than tested by Thomas and Pugh. The permeability ranges quoted are not the same as used by Thomas and Pugh but the data do indicate that the optimum permeability range for steady-state measurements is from 0.1 to 1000 mD where repeatability is expected to in the 6–15% range. These results are in line with those provided by Amabeoku and BinNasser (2012) who reported repeatability of 0.4–15% for steady-state gas permeability in the range from 10 mD to 1000 mD. The repeatability tends to be generally poorer for lower permeability samples (<0.1 mD) which Thomas and Pugh ascribed principally to insufficient stabilisation time allowed to reach steady-state conditions. Neither Thomas and Pugh nor Lab 1 and Lab 2 specifically report data for permeabilities greater than 1 Darcy but, as seen with Lab 3, the error increases significantly due to pressure measurement inaccuracy at low differential pressures (gas permeability is an inverse function of pressure squared) and potential non-linear flow conditions in the permeameter tubing. Before embarking on a major RCA

TABLE 5.14 Data Reporting Requirements Checklist for Steady-State Gas and Klinkenberg Permeability Test Measurements

Data	Comments
Brief description of procedures and experimental apparatus	Steady state or unsteady state
Description of cleaning and drying procedures	Evaluate potential clay dehydration effects
Plug depth and orientation	Horizontal or vertical permeability
Plug length and diameter	
Net confining stress (or stresses)	
Gas used	Air, nitrogen, helium or other
Gas temperature	
Gas viscosity (measured or correlation)	State correlation used
Gas flow rate and measurement pressure	Q at P_1 or P_a
Inlet pressure	P_1
Outlet pressure	P_2
Calculated mean pressure	$P_m = (P_1 + P_2)/2$
Calculated gas permeability	Specify mD or D
k_g versus $1/P_m$ data and chart	Klinkenberg permeability
Slippage factor at specified inverse mean pressure	b
Klinkenberg permeability from correlation	State correlation used and provide details

programme, the test laboratory should be audited by testing a set of 'standard' samples and comparing results.

Standardisation of confining stress is important for repeatable measurements between and within laboratories. The pressure should be sufficient to prevent gas bypass through the annular space between the sample and the rubber sleeve. In certain circumstances, especially in 1″ samples where the surface area is much smaller than 1.5″ samples, or with vuggy or rugose plug surfaces, low confinement stresses (<150–200 psi) may not be sufficient to cause a seal between the rubber sleeve and the core surface. As a result, gas may be able to pass along annular space left between the plug and the sleeve (gas bypass) which will result in an overestimation of gas permeability. The effect is enhanced in low permeability samples where the injection pressure

TABLE 5.15 Steady-State Gas Permeability: Advantages and Drawbacks

Advantages	Drawbacks and Issues
• Relatively rapid and cheap for all but low permeability samples. • Wide measurement range from 0.01 mD to over 20,000 mD depending on permeameter equipment. • Gas permeability is measured directly in the sample (not inverted from Klinkenberg permeability as in unsteady-state methods). • Klinkenberg permeability can be calculated by extrapolating P_m^{-1} vs. k_g measurements to $P_m \rightarrow \infty$. • The method is not destructive so samples can be used for further testing. • Using gas instead of brine in permeability measurements has the advantage of being non-reactive (a crucial parameter for clay-rich samples) and less prone than liquid to mobilise fines. It is also considerably cheaper and faster. • To a large extent, steady-state honours original Darcy law.	• Samples need to have a cylindrical shape. • If the plug permeability is low (<0.1 mD), the measurement takes a long time to reach steady state for each mean pressure. Also, artefacts can occur in long-term experiments (condensation, leaks, etc.). In this case, the unsteady-state method produces faster and arguably more reproducible results. • The effects of non-Darcy flow in any Klinkenberg measurement must be recognised and corrected for, especially in tight samples. Most laboratories assume negligible non-Darcy flow effects in their gas permeability and Klinkenberg permeability calculations. • Results are not representative of reservoir condition permeability and are sample preparation dependent (preservation, cleaning and drying). • Net confining stress must prevent gas bypass through the annular space between the sample and the rubber sleeve. • In carbonate formations, there is a higher frequency of induced fractures and stylolites. These features can cause unrealistically high permeability values if the confining pressure is insufficient. • Failure to maintain constant net effective pressure during the test may result in non-constant permeability as a result of progressive pore volume expansion, or gas bypass due to poor sleeve conformance with the sample's surface.

is increased to achieve a measurable flow rate yet the total confining stress is not increased to compensate for the increase in pore pressure. As a result the NCS is reduced. This tends to push the sleeve away from the plug if the NCS is not sufficient.

Until the early 1990s, very few RCA labs reported the confining stress used for the 'ambient' condition measurements. Typically this was in the range from 150 to 300 psi, though in some systems such as those using Fancher-type coreholders (API RP 40, 1988), it was not possible to even

TABLE 5.16 Permeability Reproducibility Data

Permeability Range (mD)	Thomas and Pugh (%)	Europe Lab 1 (%)	Europe Lab 2 (%)	Europe Lab 3 (%)
<0.02				±34
0.02–0.05				±13
0.05–0.1				±6
0.01–0.1	±30	±12	±30	
0.1–1.0	±25	±8		±4
1.0–10	±15	±8	±15	±4
1.0–50	±15	±8	±15	
10–100				±6
50–500	±15	±8	±5	
500–1000				±8
50–1000	±15	±8	±10	
1000–5000				±25
5000–10,000				±50

determine the stress used as the plugs were simply squeezed in a rubber compression ring. The lack of standardisation of systems between different laboratories leads to serious inconsistencies in data. Depending upon the confining stress used, one laboratory provided gas permeabilities that were at least five times higher than another laboratory using a slightly higher confining stress on the same plugs. More recently, 400 psi has been adopted as an industry 'ambient' standard, at least in the United Kingdom. In Norway, values of 20 bar (300 psi) or 30 bar (450 psi) are commonly used.

Prior to embarking on an RCA programme, the legacy RCA permeability data from the field should be reviewed. If the 'ambient' confining stress is found to vary, then permeability measurements should be made as a function of these stresses so that different databases can be converted to a common ambient confining stress so the data sets are comparable. For example, if historical values have been measured at 400 psi, and the RCA lab proposes a higher value (e.g. 800 psi), then measurements should be made at both confining stresses on a representative subset of samples. Of course, it is also possible to test gas and Klinkenberg permeability at confining stresses which are selected to represent the reservoir confining stress.

Some labs do not calculate Klinkenberg permeability from multi-pressure measurements but use empirical correlations. It is recommended to check which correlation was used.

If data are available, plot the diagnostic slippage- and inertial flow-corrected charts to detect gas slippage and inertial effects in Klinkenberg measurements, as discussed above. If non-Darcy flow is not recognised, then applying Darcy's law to flow data that are not in a viscous flow regime will underestimate permeability.

5.4.6 Unsteady-State Permeability Measurements

The pressure decay method is an unsteady-state technique which was initially introduced for very low permeability rocks (<0.1 mD) where attainment of steady-state conditions can take anywhere from minutes to hours. Typically, the time required to reach steady-state conditions roughly varies with the square of the sample length and is inversely proportional to the intrinsic permeability when constant pressures applied at the upstream and downstream faces of the sample are considered.

In the unsteady-state system, the core sample is saturated in an inert (ideal) gas and a pressure transient is induced across the sample. The differential pressure across the sample is recorded as a function of time and analysed to solve for permeability. As the pressure pulse can induce changes in mean pressure and can induce non-Darcy flow, the solution should account for both Klinkenberg and Forchheimer effects.

5.4.6.1 Test Equipment

There are two types of apparatus which operate on similar principles. The first is often referred as the pressure drawdown method, and the second is the pulse decay system.

Figure 5.33 provides a simple schematic of a typical unsteady-state gas permeability pressure drawdown apparatus. It consists of a calibrated tank and pressure transducer that can be pressurised with gas. A coreholder is attached to the tank, separated by a quick opening valve. In the pulse decay method (Fig. 5.34), the downstream end of the core is connected to a tank of known volume. It has been found that estimation accuracy is improved when the upstream and downstream tank volumes are similar and close to the pore volume of the sample when porosity is to be determined.

In the drawdown equipment, the downstream tank is removed and the gas flows through the sample directly to atmosphere. This is a special case of the pulse decay technique and can be treated with the same set of equations with the downstream reservoir volume taken as infinite featuring a constant pressure boundary condition at the plug outlet.

One of the most common commercially available drawdown systems is the Core Laboratories CMS-300® equipment, shown in Fig. 5.35. Permeability is

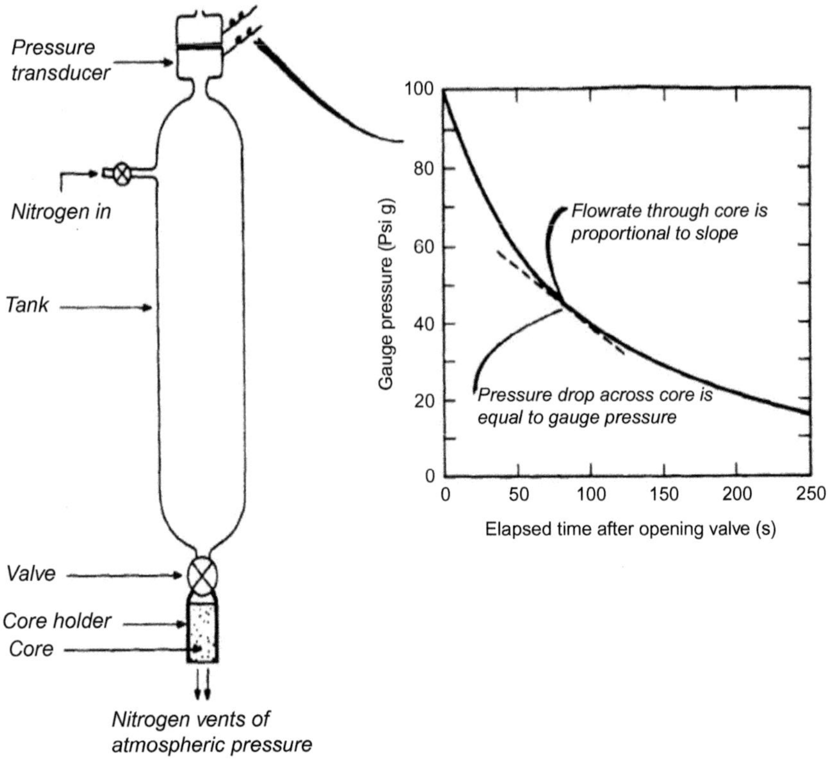

FIGURE 5.33 Example of unsteady-state pressure drawdown permeameter. *From Jones (1972).*

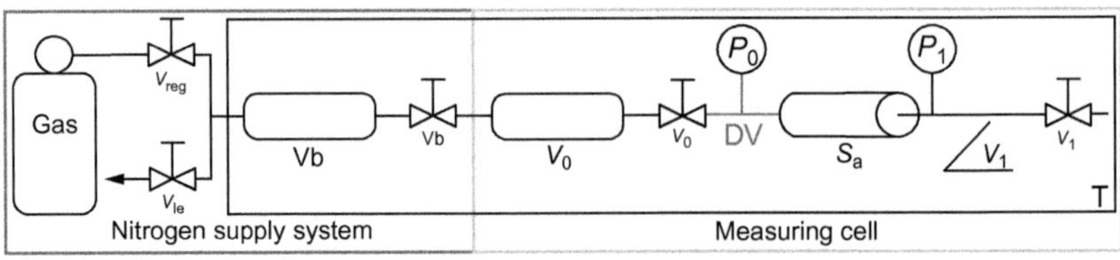

FIGURE 5.34 Example pulse decay apparatus. *From Lasseux et al. (2012).*

commonly determined to helium (although the option of nitrogen is also available), as pore volume (as part of porosity) is measured in the same loading operation. In the permeability measurement, helium is initially contained within one of five reservoirs of accurately known volume. A valve isolates the coreholder from the helium tank. For porosity measurement, a valve downstream of the plug sample is closed while, for permeability measurement, it is opened as helium flows from the reference tank though the sample. The core plug is loaded automatically via a carousel into the test coreholder

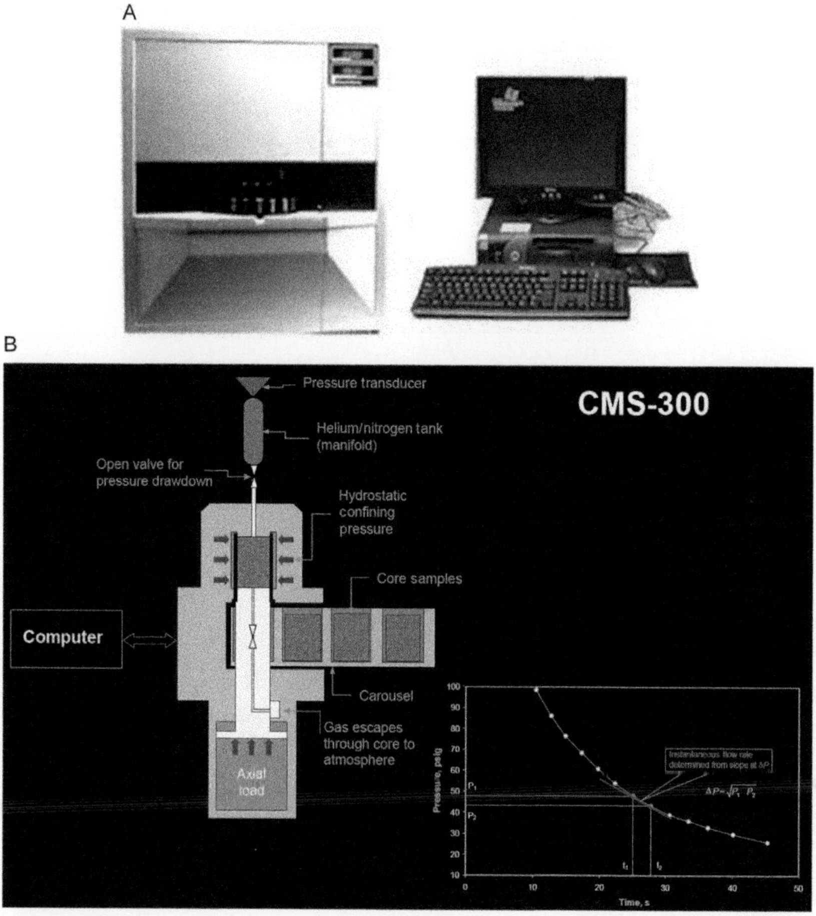

FIGURE 5.35 Core Laboratories CMS-300 unsteady-state permeameter/porosimeter. *Courtesy of Core Laboratories and its Petroleum Services and Core Lab Instruments divisions. All Other Rights Reserved.*

which is designed for confining stresses of up to 10,000 psi. The upstream pressure in the reference tank, which is initially about 240 psi is monitored as a function of time by a microcomputer as the gas flows through the core. Instantaneous flow rates are calculated from the known reservoir volume and the rate of pressure decay. Instantaneous pressure differences across the plug are numerically equal to the upstream pressure readings.

5.4.6.2 Test Procedures and Permeability Calculation

The core plug is loaded into the test coreholder, confined at the specified confining stress, and connected to the upstream and (if used) downstream reference volumes which are initially isolated by valves. The source chamber is filled with gas (helium or nitrogen). The upstream (and/or downstream) valve is opened at time zero and gas expands through the core sample either to atmosphere (CMS 300), or vacuum, or into the downstream tank. The upstream and/or differential pressure across the core is recorded as a function of time until a predetermined pressure or pressure stability criterion has been reached.

Since the mean pressure in the core is continually declining and the initial flow rates are rapid, the analytical solution must take account of both Klinkenberg (viscous) and Forchheimer (inertial) effects. That is:

$$-\frac{dP}{dL}=\frac{\mu v}{k_1\left(1+\frac{b}{P_m}\right)}+3.238E-08\beta\rho v^2$$

As only differential pressure is measured and reduces with time, data interpretation is considerably more complex than for steady-state. In the Core Laboratories CMS 300 instrument, the data analysis procedures are based on the work of Jones (1972). This method relies on solving for the mass flow rate out of the core, Q_{out}:

$$\mathrm{Vt}=\frac{d\rho}{dt}=Q_{out}$$

where Vt is the source tank volume, ρ is the gas density and t is time.

The instantaneous volumetric flow rate of helium leaving the source chamber is approximated by:

$$Q_{out}=\mathrm{Vt}f_1\left(\frac{-d\ln[p_o]}{dt}\right)\sim\frac{\mathrm{Vt}p_g f_1}{p_g+p_a}\left\{\frac{\ln\left[\frac{p_1}{p_2}\right]}{t_2-t_1}\right\}$$

where:

$$f_1=\left[1-\frac{p_o}{z_o}\left(\frac{dz_o}{dp_o}\right)\right]$$

p_o pressure
z_o gas deviation factor
p_1 pressure at selected time, t_1
p_2 pressure at selected time, t_2
p_a atmospheric pressure
p_g geometric mean pressure between p_1 and p_2, $p_g=\sqrt{(p_1p_2)}$

and is evaluated between two adjacent points (p_1/t_1 and p_2/t_2) on the pressure decay curve (Fig. 5.36).

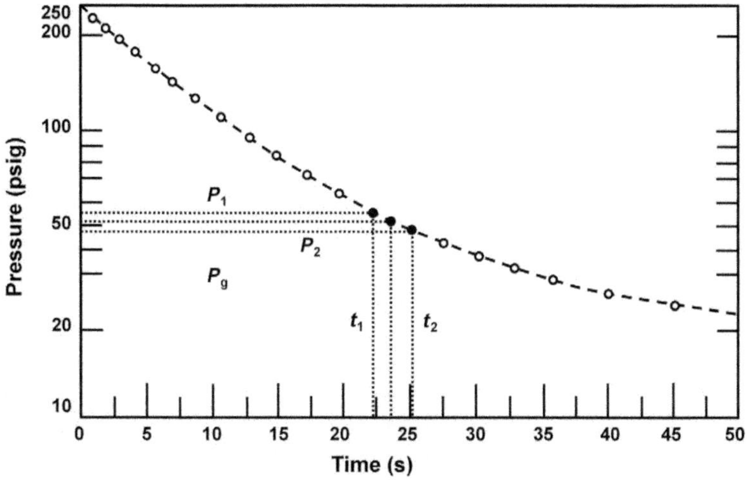

FIGURE 5.36 Example pressure decay curve from CMS 300. *Courtesy of Core Laboratories and its Petroleum Services and Core Lab Instruments divisions. All Other Rights Reserved.*

The superficial (Darcy) velocity at any point in the core plug is related to the instantaneous flow. The analysis accounts for non-constant mass flow rate through the length of the core plug which is a complex function of pore volume, tank volume and pressure. Essentially, the solution returns an equation of the form:

$$Y(n) = A_0 + A_1X_1(n) + A_2X_2(n)$$

for n number of p/t datapoint pairs as indicated above. The A_0 term is a function of the slip factor, b; the A_1 term is a function of Klinkenberg permeability, k_1; and the A_2 term is a function of the Forchheimer factor, β. A least squares fitting technique is applied to the n datasets and is used to find values for A_0, A_1 and A_2, and hence b, k_1 and β.

The Jones method is considered to be an approximate solution and is based on a number of assumptions including that the gas is ideal and the mass flow rate through the core is essentially constant (although the correction method is provided for non-constant mass flow). To solve for the three unknowns: Klinkenberg permeability, b factor and Forchheimer β, using essentially two equations requires an iterative scheme which may introduce numerical errors. In addition, Ruth and Kenny (1987) have shown that experimental errors may cause inaccurate measurements, especially above 1000 mD. Ruth and Kenny present a numerical solution to simultaneously solve for k_1, b and β using a finite difference scheme where the core sample is divided into a number of equal sized grid blocks. When fluid expands rapidly its temperature decreases. As the analytical solutions assume isothermal conditions, it is essential to maintain these to prevent significant changes in density and viscosity. Thus, rapid gas Joules–Thomson expansion in high permeability samples may invalidate isothermal assumptions so many labs impose a maximum permeability cut-off of around 1000 mD for unsteady-state tests.

5.4.6.3 Data Reporting

The CMS 300 is the most commonly used unsteady-state equipment and an example of the CMS 300 data output is presented in Table 5.17. This is explained below.

- *Net confining stress*: In this case, measurements were made on the same plugs at 800 and 1500 psi NCS. The minimum NCS value in the CMS 300 is 800 psi. This is because the instrument was designed for confining stresses up to 10,000 psi which requires a thick rubber sleeve to avoid puncturing. Therefore, 800 psi represents the confining stress required to guarantee a seal between the plug and the sleeve.
- *Porosity*: This is determined from helium pore volume and (prior) helium grain volume measurements immediately before the unsteady-state pressure pulse experiment. Helium gas is allowed to expand from the source chamber through the plug with the outlet valve closed. As discussed earlier, the pore volume is dependent on plug shape and surface topology and can be overestimated at lower stresses. Some Core Laboratories labs do not use the CMS 300 for porosity determination, preferring conventional helium grain volume and mercury bulk volume measurements where possible.
- *Klinkenberg permeability*: This is determined directly from the solution to the pressure decay equation, based on Jones's (1972) method. Note that at permeabilities higher than 1000 mD, air permeability is determined using a steady-state method (SS Ka). This represents the cut-off to prevent possible cooling effects invalidating the assumptions in the permeability solution.
- *Air permeability*: Air permeability is not measured but is back-calculated from the Klinkenberg permeability (k_l) and slip factor (b) derived from the approximate solution to the pressure decay. The slip factor (for helium) is scaled to that for nitrogen based on the ratio of molecular weights (and mean free paths) of the gases, and the Klinkenberg equation is used to solve for k_{air} at a given inverse mean pressure:

$$k_{air} = k_l\left(1 + \frac{b}{P_m}\right)$$

  The mean pressure used for the air permeability calculation is $\sim$0.84 atm.$^{-1}$.
- *Slip factor b*: This is determined directly from the solution to the pressure decay equation. In the example above, the b factor for helium is converted to an equivalent b for air.
- *Forchheimer factor β*: This is determined directly from the solution to the pressure decay equation.

TABLE 5.17 Example Core Laboratories CMS 300 Output

Sample Number	Depth (ft.)	Net Confining Stress (psi g)	Porosity (%)	Permeability		*b* (Air; psi)	*β* (ft.$^{-1}$)
				Klinkenberg (md)	k_{air} *(md)*		
1001	4874.60	800	16.98	0.007	0.015	99.72	2.33E+14
1001	4874.60	1500	16.60	0.005	0.012	106.85	3.99E+14
1002	4875.20	800	13.45	0.052	0.058	9.81	2.25E+11
1002	4875.20	1500	13.20	0.018	0.036	77.59	3.44E+13
1003	4879.30	800	13.61	0.158	0.248	14.04	7.02E+11
1003	4879.30	1500	13.42	0.124	0.205	16.25	6.83E+11
1007	4893.75	800	20.96	0.989	1.19	3.65	3.69E+12
1007	4893.75	1500	20.68	0.939	1.12	4.26	3.23E+12
1009	4898.65	800	14.24	0.191	0.247	7.10	4.48E+10
1009	4898.65	1500	14.06	0.147	0.198	8.69	1.54E+12
1013	4906.75	800	16.96	0.727	0.858	4.00	1.38E+12
1013	4906.75	1500	16.61	0.661	0.788	4.31	1.72E+12
1014	4916.50	800	29.70	118	128	1.44	1.29E+08
1014	4916.50	1500	29.41	115	124	1.42	1.40E+08
1018	4926.10	800	34.79	SS Ka	1100	n/a	n/a
1018	4926.10	1500	34.70	SS Ka	1080	n/a	n/a
1021	4938.80	800	30.63	241	255	0.92	9.20E+06
1021	4938.80	1500	30.45	238	252	0.93	9.28E+06

5.4.6.4 Data Reporting Requirements

Data reporting requirements for unsteady-state measurements from the laboratory are listed in Table 5.18.

5.4.6.5 Advantages and Drawbacks/Issues

Summaries of the advantages and drawbacks of the unsteady-state gas permeability measurement systems are provided in Table 5.19.

5.4.6.6 Unsteady-State Permeability Quality Control Issues, Checks and Diagnostics

Repeatability tests should be carried on at least every 10th plug by different operators at the same NCS and ideally using different permeameters (if this is possible). Repeatability should be broadly in line with repeatability for steady-state measurements (Table 5.16) with two possible exceptions:

1. Repeatability should be better than steady state for low permeability samples (<0.1 mD)
2. Steady-state methods should be used if permeability is greater than 1000 mD. For these samples, k_{air} is approximately equal to k_l.

TABLE 5.18 Data Reporting Requirements Checklist for Unsteady-State Permeability Test Measurements

Data	Comments
Brief description of procedures and experimental apparatus	Pressure drawdown or pulse decay
Description of cleaning and drying procedures	
Plug depth and orientation	Horizontal or vertical permeability
Plug length and diameter	
Net confining stress (or stresses)	
Gas used	Air, nitrogen, helium or other
Source chamber initial pressure	
Final pore pressure at achieved stability criterion	
Klinkenberg permeability, k_l	
Gas slippage factor, b	
Forchheimer inertial resistance coefficient, β	
Reference mean pressure for air permeability	
Calculated air permeability at reference, P_m	Specify mD or D

TABLE 5.19 Unsteady-State Gas Permeability: Advantages and Drawbacks

Advantages	Drawbacks and Issues
• Much faster than steady-state methods for low permeability samples and for Klinkenberg permeability. • No flow meter calibration required. • Klinkenberg permeability (k_l), slippage coefficient (b) and Forchheimer inertial resistance factor (β) can be calculated in a single experiment. • Measurement time and accuracy can be optimised by selection of appropriate tank volume. • Using gas instead of brine in permeability measurements has the advantage of using a non-reactive fluid (a crucial parameter for clay-rich samples) and less prone than liquid to mobilise fines. • The method is not destructive so samples can be used for further testing.	• Samples need to have a cylindrical shape. • Results are sample preparation dependent (preserving, cleaning and drying). • Gas permeability is not measured directly during the test. It is back-calculated from the Klinkenberg permeability and slippage factor at a certain reference inverse mean pressure. • Minimum confining stress in CMS-300 is 800 psi which is significantly higher than the 'standard' confining stress in most steady-state systems. In stress-sensitive formations, the higher NCS could result in lower permeability so data may be inconsistent with legacy data. • Measurement of Klinkenberg coefficients becomes progressively difficult for samples over 1000 mD. For these samples, the use of the vacuum at the exit of the core is recommended. Alternatively, these samples can be more conveniently measured by steady-state methods (at the same confining stress) as the Klinkenberg effect will be negligible. • Adequate thermal insulation must be provided to minimise temperature changes induced from ambient changes.

5.4.7 Steady-State Liquid (Absolute) Permeability Measurements

Although not normally part of an RCA programme, many laboratories now offer liquid permeability measurements on a routine basis. The measurement equipment is relatively simple but equipment resources required to handle a large number of plugs, and the time that may be required to prepare the samples, make the tests more involved, slower and more expensive than gas permeability measurements. Consequently, liquid permeability tests are usually performed as part of a SCAL programme. Compared to gas, the tests measure permeability that better approximates the reservoir fluids. Oil (normally mineral oil) is preferred over brine in clay-rich formations as it is much less reactive than brine and can prevent rehydration and reflocculation of the clays.

The preparation and test procedures presented below are for absolute, single-phase liquid permeability measurements using a steady-state method.

5.4.7.1 Sample Preparation

Samples to be tested for liquid permeability are normally prepared following the procedures described in Chapter 4. In sensitive lithologies, liquid permeability (to formation water) can be determined after miscibly flush cleaning the samples to brine without drying the plugs. This will help prevent clay damage and damage to unconsolidated cores.

5.4.7.2 Saturation Procedures

Where plug preparation allows, samples should be saturated in deaired brine or mineral oil at a pressure of 2000 psi for a minimum period of 3 days. Although initial vacuum saturation can be used to assist the saturation process, the maximum time under vacuum should be limited to 20 min for brines, to prevent excessive water evaporation increasing the brine density and conductivity. Brine or oil viscosity must be measured at the test conditions.

If helium porosity data are available, the saturated weight of the core can be checked to ensure full saturation. If any plug appears to be significantly undersaturated, it should be replaced in the saturator, and pressure saturation continued.

5.4.7.3 Test Procedures and Permeability Calculation

The plugs are then loaded into a coreholder under brine or oil to ensure no air gets into the samples, and then confined at the desired NCS.

Liquid injection can be controlled either using a constant pressure system (e.g. regulating nitrogen pressure in a transfer cylinder) or using constant rate pumps (e.g. HPLC pumps). The latter is normally preferred. A backpressure (downstream) regulator is recommended as application of even nominal backpressures (e.g. 200 psi) can assist in compressing and flushing trapped air and can minimise pressure and flow fluctuations.

The initial flow rate should be no more than 0.5 or 1 ml/min to prevent possible fines migration. When the flow rate (Q) and pressure drop ($P_1 - P_2$) are stable after flowing the equivalent of at least two pore volumes (PV) of oil or brine injected, liquid permeability (k) is determined from Darcy's law:

$$k = \frac{Q\mu L}{A(P_1 - P_2)}$$

Viscosity, μ, of the test liquid is measured at the lab temperature.

As oil and brine viscosity are much larger than gas, pressure differentials during liquid flow can be much higher, and it is important to keep the NCS as constant as possible throughout the experiment. In low permeability samples where the pressure differential across the plug is large, the NCS at the outlet can be much lower than the NCS at the inlet. This can lead to differential pore volume compaction across the core which will affect the permeability measurement. In this case, tests should be carried out at same rate/differential pressure but using much higher total confining stress as this can minimise differential NCS.

The accuracy of the measurement, especially in higher permeability samples, depends on the pressure measurement system accuracy and resolution. Modern digital pressure transducers are designed for different pressure ranges, and their accuracy depends on regular calibration (using a dead weight tester) at zero and maximum pressure; and inherent error (expressed as percentage of full transducer range) termed the full-scale deflection (FSD). Differences in pressure accuracy and responses from different transducers are expected: gauge transducers are typically less accurate than differential transducers although this depends on transducer type, calibration and FSD error. Small transducer errors (non-linearity, pressure close to FSD range, zero shift, etc.) can lead to significant errors in calculated permeability if permeability is measured at a single flow rate. In this case, a multipoint technique can be used. The sample pressure differential ($\delta p = P_1 - P_2$) is measured at three or four incremental flow rates, trying to ensure that the flow velocity does not exceed the critical velocity for fines mobilisation: for example: 0.5, 1.0, 1.5 and 2.0 ml/min.

The Darcy equation is a linear function:

$$\frac{Q}{A} = \frac{k}{\mu}\frac{\delta p}{L}$$

so a plot of Q/A versus $\delta p/L$ for each flow rate should fall on a straight line

$$y = mx + c$$

with slope equal to k/μ as shown in Fig. 5.37. Permeability can therefore be determined from linear regression of the Q/A versus $\delta p/L$ datapoints. A non-zero intercept for $Q = 0$ on the $\delta p/L$ axis indicates transducer gauge error (zero shift). Deviation from linearity could indicate transducer FSD error but is normally associated with fines movement: an increasing slope indicates permeability increasing and fines being flushed from the core; a decreasing slope indicates permeability reducing and fines blocking pore throats. To test for this, it is recommended to reverse flow through the plug after stability at each rate and monitor the pressure response: a short-lived pressure transient can be indicative of dynamic fines movement. On completion of the differential pressure measurement at the highest flow rate, the rate should then be returned to the initial low (or base rate) to assess pressure repeatability. Fines migration plugging internal

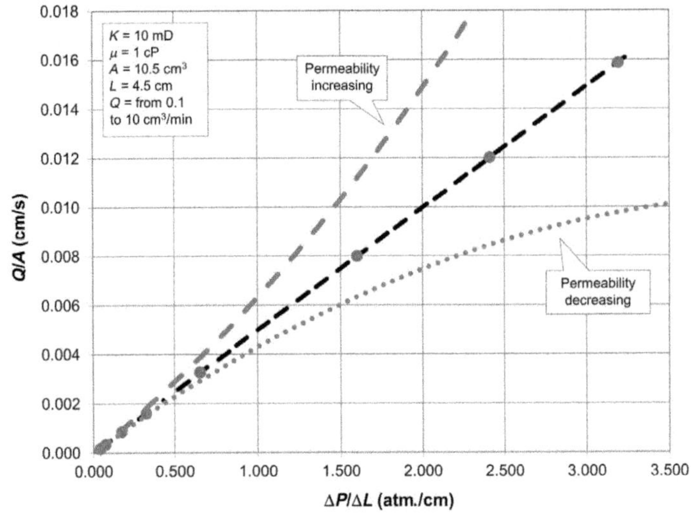

FIGURE 5.37 Example of Darcy plot for liquid flow.

pore throats will cause the differential pressure at the base rate to be higher, whereas fines flushing from the plug will result in a lower differential pressure.

5.4.7.4 Data Reporting Requirements

Data reporting requirements from the laboratory are listed in Table 5.20 for steady-state liquid permeability measurements.

5.4.7.5 Advantages and Drawbacks/Issues

Summaries of the advantages and drawbacks of the steady-state liquid permeability measurement systems are provided in Table 5.21.

5.4.7.6 Steady-State Liquid Permeability Quality Control Issues, Checks and Diagnostics

Multi-rate tests are preferred as this minimises impacts of pressure transducer error on permeability derived from single rate analysis but are obviously more expensive to perform. A plot of Q/A versus $\delta p/L$ for each flow rate is used to assess linearity and to determine permeability from regression analysis. The multi-rate tests can also be useful to assess the critical velocity for fines migration which can be used to constrain other dynamic coreflood test flow rates. XRD and SEM analysis of the plug end trims and solids produced in the flow effluent can be used to assess formation sensitivity, if required.

Compare liquid permeability (k) with air or Klinkenberg permeability (k_l) on the same samples at the same NCS. The k/k_l ratio is expected to be around 0.8–1.0 for all but low permeability samples (less than 0.1 mD or so) or clay-

TABLE 5.20 **Data Reporting Requirements Checklist for Steady-State Liquid Permeability Test Measurements**

Data	Comments
Brief description of procedures and experimental apparatus	Steady state
Description of cleaning and drying procedures	
Plug depth and orientation	Horizontal or vertical permeability
Plug length and diameter	
Type of liquid	Formation water or oil
Sample dry weight prior to saturation (if applicable)	
initial helium porosity, grain volume and bulk volume (if applicable)	
Pressure and time allowed for saturation	
Sample saturated weight after saturation	
Sample immersed weight after saturation	
Liquid density used to calculate Archimedes bulk volume	
Brine salinity	Formation water test
Re-saturation pore volume	From liquid-filled pore volume and Archimedes bulk volume
Net confining stress (or stresses)	
Pore pressure	Backpressure
Test temperature	
Liquid viscosity (measured or correlation)	At test temperature
Flow rate(s)	Q_1, Q_2, etc.
Differential pressure(s)	δP_1 at Q_1, δP_2 at Q_2, etc.
Calculated liquid permeability (single rate tests)	From Darcy equation
Q/A versus $\delta P/L$ multi-rate data and chart (multi-rate test)	
Calculated liquid permeability (multi-rate tests)	From linear regression Q/A versus $\delta P/L$

TABLE 5.21 Steady-State Absolute Liquid Permeability: Advantages and Drawbacks

Advantages	Drawbacks and Issues
• Permeability more representative of reservoir fluids than air or Klinkenberg permeability. • Brine permeability can be conveniently added to certain cleaned-state SCAL test protocols such as permeability prior to capillary desaturation; relative permeability (cleaned-state); formation factor measurements. • Oil in permeability measurements is non-reactive in the presence of clays. • Multi-rate measurements can be used to assess and correct for pressure transducer errors.	• Samples need to have a cylindrical shape. • Results are sample preparation dependent (preserving, cleaning and drying). • Much slower and more expensive than steady-state or unsteady-state gas permeability measurements. • Sample must be fully saturated with test liquid. Any air in the sample will suppress liquid permeability due to relative permeability effects. • Formation water re-saturation can promote adverse clay reactions in sensitive lithologies. • Higher NCS may be required in low permeability samples to prevent differential compaction. • High rates can results in fines migration if rate is above critical velocity. • In tests on chalk or certain carbonates, the formation water should match the *in situ* formation water concentration and be balanced with respect to bicarbonate. This may require equilibrating the formation water with crushed chalk to ensure equilibrium.

rich sandstones. A low brine to air permeability ratio can result from adverse clay reactions and/or fines migration, or as a result of undersaturation (air in sample). A k/k_l ratio greater than 1 might indicate measurement errors or fines production from the plug sample.

5.4.8 Probe or Profile Permeability Measurements

Probe (or profile) permeability has been used to provide highly resolved measurements of permeability, especially in formations where conventional plug analysis cannot adequately capture the degree of heterogeneity. First described by Dykstra and Parsons (1950), and developed further by Eijpe and Weber (1971), the use of probe permeability has greatly improved log–core correlations, inter-well permeability correlations and net to gross determination. The ability to take a large number of highly resolved measurements enables a more detailed description of reservoir heterogeneity and its relationship to sediment type and architecture, and is used increasingly in upscaling small-scale measurements to reservoir grid block size.

The small diameter of the injection tip seal means that permeability spatial resolution is about 4 mm—enabling permeability to be determined on individual laminae, clasts and fractured carbonate matrices, for example. The probe permeameter allows much higher resolution measurements of permeability than is possible with plugs and is designed to complement and, in some cases where plug measurements are unrepresentative or impossible, supplement plug measurements.

The measurement of probe permeability on 'un-cleaned' core will reflect the permeability to gas at some undefined liquid saturation. Relative permeability effects will therefore reduce the permeability compared to the absolute value. Tests on gas core suffer less from these effects but precipitated salt can suppress permeability.

Both steady-state and unsteady-state instruments are available. The former requires flow and pressure stability while injecting under steady-state, constant rate of constant pressure conditions. Analysis is based on the steady-state solution for Darcy's law and can include for Klinkenberg effects, normally via a correlation algorithm. In the unsteady-state method, gas is expanded from a source tank through the probe tip and into the core sample. Measurement times in low permeability cores are much faster than steady state. The pressure decay response is recorded and analysed using similar techniques to those for the unsteady-state plug permeameter. The method returns simultaneous values of k_l, b and β.

5.4.8.1 Sample Preparation

Measurements can be performed on whole cores, resinated biscuits, plug samples or outcrop rock faces but, most commonly, on slabbed core surfaces that have been allowed to dry.

A certain degree of surface flatness is required as the probe needs to form a leak-tight seal on the rock surface during the measurement. The core surface should be treated to remove dust and rock flour after slabbing.

5.4.8.2 Steady-State Equipment and Test Procedures

The injection probe comprises a small diameter (4 mm or so) probe which is sealed against a rock surface by pneumatic pressure, using a rubber 'O' ring seal (Fig. 5.38). The seal size is different for different lithologies—a thicker seal is used to reduce stress loading for unconsolidated sands—and should be changed by the lab operator accordingly. Nitrogen is used as the test fluid.

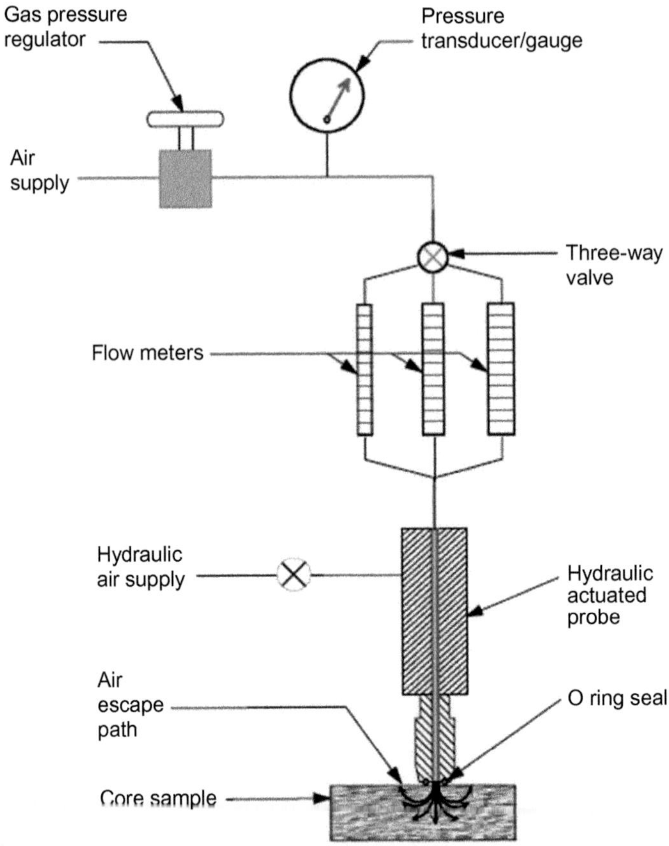

FIGURE 5.38 Probe permeameter schematic. *Courtesy of Core Laboratories and its Petroleum Services and Core Lab Instruments divisions. All Other Rights Reserved.*

The probe is positioned on the core surface manually or under computer control. Figure 5.39 shows semi-automated equipment. The slabbed core is laid out on the permeameter table, and the probe moves along the slab taking steady-state measurements at pre-defined intervals. Some modern equipment is fully automated and computer controlled and incorporates laser technology to identify the exact location of the measurement and to detect and avoid plug holes, breaks in the core and fractures.

Nitrogen gas is injected through the probe into the rock, and the injection pressure, P_i, and flow rate, Q_i, are recorded. The time to reach steady state in each measurement varies from only few seconds to a few minutes, depending upon permeability. Once stable conditions are attained, the data are recorded either manually or automatically via a computer, and the probe is released and moved to the next location.

Measurement frequency depends upon the level and scale of core heterogeneity, core length, budget and required resolution, but a spacing of at least

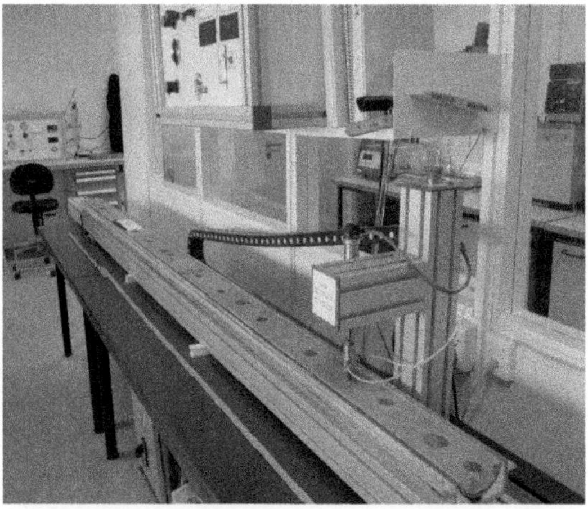

FIGURE 5.39 Steady-state probe permeameter. © *2015 Weatherford.*

every 1–2 in. (3–5 cm) is recommended. In some cases, measurements can be taken over a grid pattern both along and across the slabbed core surface.

5.4.8.3 Steady-State Permeability Calculation

The gas flow pattern from the injection probe is assumed to be broadly hemispherical. Goggin et al. (1988) performed numerical calculations to produce dimensionless geometrical flow factors that can be used for permeability calculation and which account for the geometry of the flow path, the probe tip and the sample being tested. These dimensionless factors were developed for test surfaces with a plane upper surface and sufficient lateral extent and depth (around 2 cm or so) such that exterior boundaries have no influence on permeability measurements. Goggin et al. found that the geometric factors can be used provided that they have a depth at least four times the interior radius (r_i) of the probe seal, and that the closest lateral boundary is at least $4r_i$ from the axis of the probe. For a typical 4 mm diameter probe tip, this means that the test sample should have a depth of 8 mm (0.3″) and the tip should be placed at least 8 mm from a boundary. This means that it is theoretically possible to test the probe permeameter on the end faces of a 1″ or 1.5″ diameter plug with a 4-mm diameter probe, and provided the probe seal is gas tight and the radius of the sample is at least $12r_i$ (24 mm or 1″) on a cylindrical surface of a sample. Probes should not be placed close to the edge of a slab or end of a core plug, or on a very thin slab or sample; otherwise, gas will escape at the boundary and affect the permeability measurement.

The injection flow rate responds to permeability of the rock underneath the tip seal. For a constant injection pressure, Q_i, flow rate increases in proportion to an increase in permeability and reduces with lower permeability. Quantification of the flow rate response in terms of permeability is determined using the numerical solution proposed by Goggin et al. (1988) or, and most commonly, by means of calibration of the probe response on core plugs of known gas permeability (Halvorsen and Hurst, 1990).

Compare the simplified form of Darcy's law for linear gas flow through porous media, for the cases of core plug measurements and probe measurements (Fig. 5.40):

$$\text{Core plugs:}\quad k_a = \frac{2000 Q_a \mu_a P_a L}{\left(P_i^2 - P_a^2\right) A}$$

$$\text{Probe:}\quad k_a = \left(\frac{2000 Q_i \mu_i P_i}{\left(P_i^2 - P_a^2\right)}\right) F$$

for air permeability, k_a, in mD. The flow path geometrical factor—length/area (L/A)—in the plug measurement is known, but for the probe it is unknown. However, the geometric flow term in the plug equation is replaced by a dimensionally consistent term, F or G—the geometric factor—in the probe permeability equation. In this case, probe measurements are made on homogeneous plugs of known permeability, k_a, and the probe injection pressure, P_i, and flow rate, Q_i, are recorded for each measurement. Linear regression analysis of plug permeability, k_a, as a function of $\left(\frac{2000 Q_i \mu_a P_i}{P_i^2 - P_a^2}\right)$ is used to determine the geometric factor, as illustrated in Fig. 5.41, where F is found from the slope of the regression line.

5.4.8.4 *Unsteady-State Equipment and Test Procedures*

One of the most commonly available instruments is the Core Laboratories PDKP-400™ pressure decay profile permeameter as shown in Fig. 5.42. The profile permeameter equipment comprises the gas source volume tank,

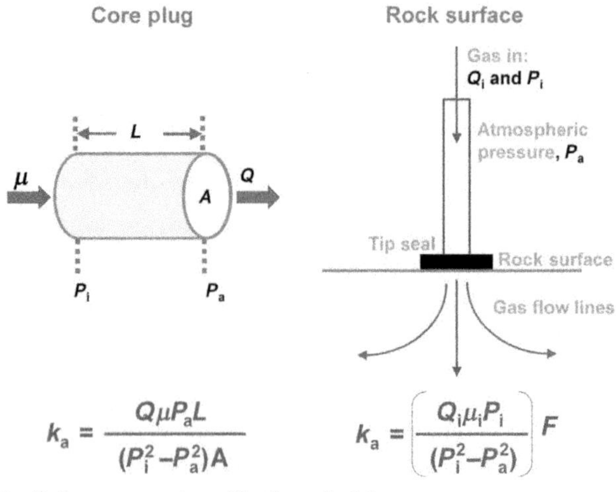

FIGURE 5.40 Probe permeameter calibration principle.

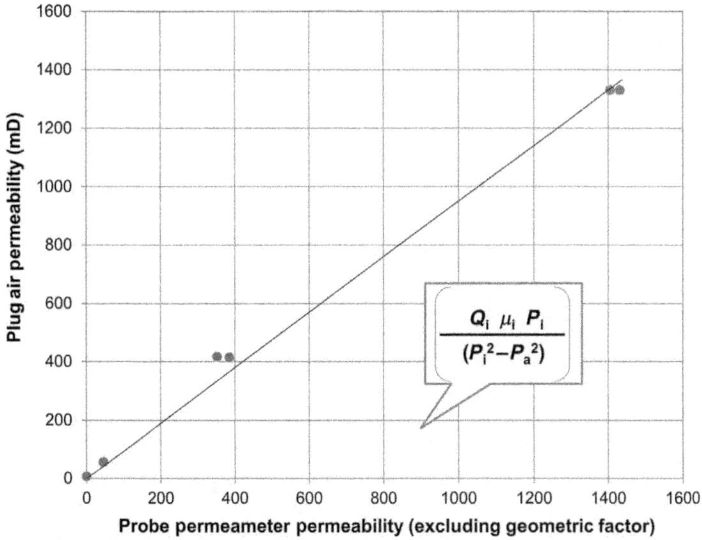

FIGURE 5.41 Probe permeameter calibration data.

a digital pressure transducer, valves and a probe-on pressure gauge to measure the seal force. The probe tip is similar to the steady-state version. Nitrogen is used as the saturating fluid. It has a claimed permeability measurement range from 0.001 mD to 30,000 mD.

The measurement principle (Jones, 1994) is similar to that described for the CMS 300™ pressure decay permeameter. In this set-up, a probe-on pressure is usually maintained at a certain value with the seal being formed by a rubber gasket seal. Permeability is computed monitoring the pressure decay of nitrogen from calibrated reservoir volumes measured against time. Geometric flow factors are used in the calculations based on sample and probe tip geometries. The monitored pressure decay curve allows calculation of Klinkenberg permeability as well as the Klinkenberg slippage term, b, and the inertial resistance factor β.

5.4.8.5 *Data Reporting Requirements*

Data reporting requirements from the laboratory are listed in Table 5.22 for probe permeability measurements.

5.4.8.6 *Advantages and Drawbacks/Issues*

Summaries of the advantages and drawbacks of the probe permeability measurement systems are provided in Table 5.23.

5.4.8.7 *Probe Permeability Quality Control Issues, Checks and Diagnostics*

The probe seal pressure must be applied with a constant force and a constant orientation to the sample. The probe tip should be made of material that does

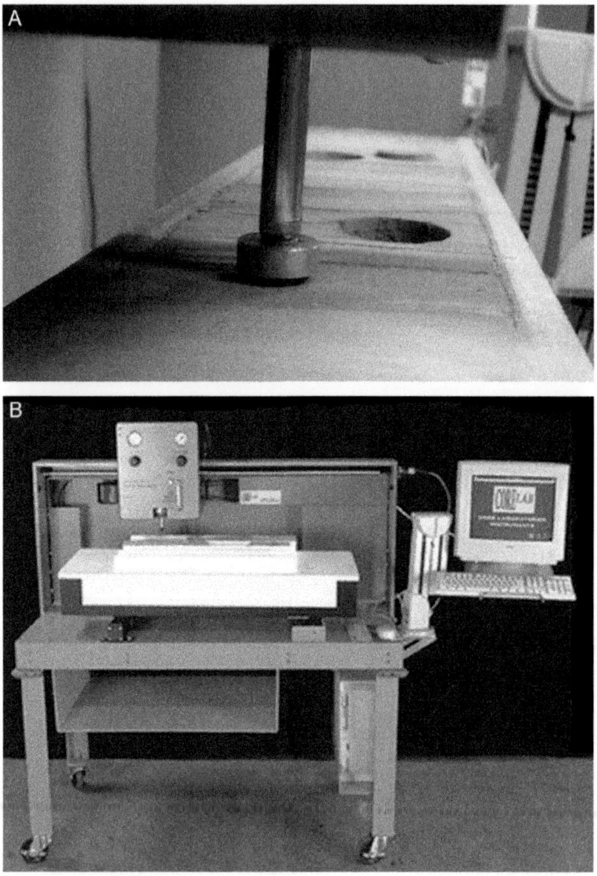

FIGURE 5.42 Unsteady-state probe permeameter equipment (PDPK-400™). *Courtesy of Core Laboratories and its Petroleum Services and Core Lab Instruments divisions. All Other Rights Reserved.*

not deform. Probe tips should be checked, replaced and recalibrated at regular intervals.

The probe must be placed with a high degree of precision while being able to record the position of each point. This is important for repeat measurements and assessing the influence small-scale geological features can have on the measured permeability.

The plugs used for probe permeameter calibration must be homogeneous and cover a wide range of permeabilities. The degree of heterogeneity within most core plug samples often gives a poor correlation between permeability and flow rate. A much better correlation is obtained if the plugs are reasonably homogeneous. Halvorsen and Hurst (1990) suggest that at least three measurements should be made on each end of the plug to get a statistically robust estimate of the heterogeneity of the 'homogeneous' plug. They contend

TABLE 5.22 Data Reporting Requirements Checklist for Probe Permeability Test Measurements

Data	Comments
Brief description of procedures and experimental apparatus	Steady state or unsteady state (USS)
Description of test surface	Whole core, slabbed core, resinated biscuits, plug samples or outcrop rock faces. Saturation conditions (core cleaned and/or dried; fresh state)
Measurement depth	
Probe diameter	
Probe seal pressure	
Type of gas	Air, nitrogen or helium
Inverse mean pressure	Steady state only
Gas flow rate, Q_i, and gas injection pressure, P_i	Steady state only
Atmospheric pressure, P_a	Steady state only
Details of steady-state calibration on core plugs	As Fig. 5.40 (steady state)
Test temperature	
Gas viscosity (measured or correlation)	At test temperature
Air permeability	Steady-state only
Klinkenberg permeability	Direct (USS) or correlation (SS)
Gas slippage factor, b	Unsteady-state only
Forchheimer inertial resistance factor β (ft.$^{-1}$)	Unsteady-state only
Depth chart plotting permeability as a function of depth along the core	1:200 or as otherwise specified

that similarity between arithmetic, geometric and harmonic mean values for both ends of the core plugs is a good indicator of sample homogeneity.

Compare measured core permeabilities with values obtained from core plugs (clean and dry) on the same interval/depth (if available). Note that depending on the saturation conditions of the core, results could differ, as an effective permeability could be measured using the probe permeameter. In addition, probe measurements are not made exactly at the positions where core plugs are taken, thus making direct comparison of the measurements of uncertain value. This increases proportionally with respect to the degree of

TABLE 5.23 Probe Permeability: Advantages and Drawbacks

Advantages	Drawbacks and Issues
• Fast, cheap and non-destructive measurement. • The method provides highly resolved (to about 4 mm) measurements of permeability, especially in formations where conventional plug analysis cannot adequately capture the degree of heterogeneity (e.g. laminated reservoirs). • Allows statistically significant volumes of permeability data to be collected, without excessive cost. • The data improve the understanding of permeability distribution and allow more confident assignment of net to gross values and zonation of reservoirs. • Designed to complement and, in some cases where plug measurements are unrepresentative or impossible, supplement plug measurements. • Test samples can have any shape, as long as the dimensions of the sample with respect to the location of the probe tip do not violate the geometrical constraints described by Goggin et al. (1988).	• The measurement of probe permeability on 'un-cleaned' or 'un-dried' core will reflect the permeability to gas at some undefined liquid saturation. Relative permeability effects will therefore reduce the permeability compared to the absolute value. • Core slabs should be surface dried to at least 2 cm from the surface as this represents the probable depth of investigation of the instrument. Salt, solids and heavy crude oil precipitation in this zone can suppress permeability. • Probe permeameter measurements may be a problem in unconsolidated cores as the stress generated by the probe seal may damage the core. Larger seals spread the force over a wider area but will have a different calibration factor (*F* or *G*) than thinner seals. • Measurement points need to avoid plug holes, fractured or damaged core/sample sections, as the measurements will be unrepresentative. This is particularly important in fractured cores and vuggy carbonates. Boundary effects practically impose a constraint that measurements should not be within 1–2 cm of plug holes, fractures or vugs. • Measurements should not be performed on thin, resinated biscuit slabs nor near the edges of the core slab, as the flow pattern from the tip may be affected by infinite or impermeable (resin) boundaries. • Measurements are dependent on the selection of probe geometry, accurate control of probe application force and the use of an appropriate seal on the probe tip.

heterogeneity of the material. An example of some typical steady-state probe permeameter results is shown in Fig. 5.43. In this case, the lab doing the probe measurements had no prior knowledge of the plug data and precalibrated the probe against their own internal calibration plugs. Nevertheless, there is an excellent agreement between plug and probe data over a reasonable

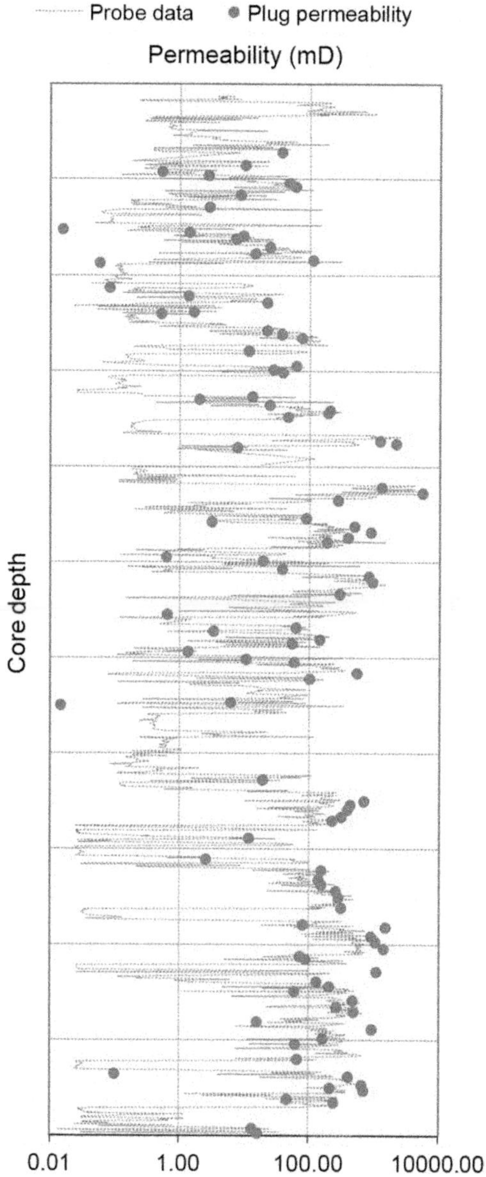

FIGURE 5.43 Example of probe permeability and plug measurements on same cored interval.

permeability range. The probe data highlight a sequence of very low permeability 'baffles' that were not captured by the plugs but helped explain some of the production history from the reservoir. However, the data show that the probe can often produce optimistic results in low permeability intervals due to seal leakage at higher injection pressures and pessimistic permeability in high permeability intervals due to non-Darcy flow effects at high flow velocities near the tip.

5.5 WHOLE CORE ANALYSIS MEASUREMENTS

Whole core analysis is essential for characterising directional permeability in heterogeneous, fractured and/or anisotropic rock where the scale of the heterogeneity is larger than a standard RCA or SCAL plug. However, the time and costs involved in preparing and testing the samples mean that whole core analysis is only performed where plug measurements would be considered to be completely unrepresentative, or where directional permeability (permeability anisotropy) is judged to be important. Examples include sandstone conglomerates, and vuggy or naturally fractured carbonates.

Only a limited number of commercial laboratories have the experience and resources to perform RCA measurements on whole cores up to 4″ in diameter and up to 12″ long, especially at elevated stresses. Even fewer commercial laboratories have the capabilities to perform anything other than basic SCAL measurements.

5.5.1 Sample Preparation

All whole core samples should be photographed and CT scanned to evaluate heterogeneity. Post-test plug measurements and/or probe permeameter measurements on cleaned surface should also be considered to assess heterogeneity.

A whole core sample may not be cut exactly normal or parallel to the rock's bedding planes as this depends on the well deviation and formation dip. Therefore, measurements of 'horizontal' and 'vertical' permeability on the whole core may not be corresponding exactly with longitudinal and transverse measurements on plugs.

Sandblasting the outside of the core to remove attached mud solids may be required. In soft carbonates, or where the mud cake on the outside of the core is difficult to remove undercutting cores to a smaller diameter is often recommended.

Most laboratories use large capacity soxhlet apparatus (Chapter 4), but Honarpour et al. (2005) argue that this can be inefficient in whole core samples. Gas-driven CO_2/toluene extraction on whole cores (API RP 40, 1988) is a common cleaning method for whole core analysis but is not recommended for carbonates due to the risk of rock dissolution as a result of CO_2/water reactions.

5.5.2 Fluid Saturations

Dean–Stark measurements are possible in whole core samples provided the lab have the necessary large capacity apparatus. However, the extraction process take considerably longer than plugs, equipment availability may mean that only one or two samples might be processed at a time and the whole core samples must be sealed and preserved at wellsite.

5.5.3 Porosity

The most common porosity measurement method involves determination of helium expansion pore volume (Vp) into a whole core coreholder at stress, and ambient condition calliper bulk volume (Vb). Porosity is better measured at reservoir NCS especially in vuggy/fractured and soft carbonates and chalks. Length and diameter volume measurement inaccuracies are less significant in larger samples than they are in plug samples. The large scale of the sample is much closer to the porosity logging tool scale than plug measurements.

5.5.4 Gas Permeability

Whole core analysis uses a similar system to steady-state core plug tests, but the coreholder is designed to allow flow both across the sample (radial transverse flow) and along the axis of the core (linear flow). Figure 5.44 provides a schematic of a typical whole core system. If the core is cut exactly orthogonal to bedding planes, then the axial (transverse) measurements represent the 'horizontal' permeability and the along axis measurement represents the 'vertical' permeability of the core. Axial permeability measurements are made in

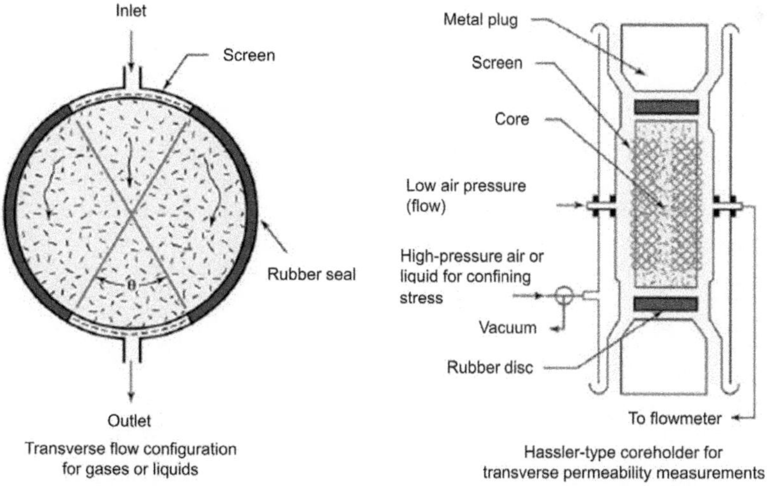

FIGURE 5.44 Example apparatus for whole core transverse permeability measurements. *From Honarpour et al. (2005).*

TABLE 5.24 Results of Repeat Measurements on Whole Core Samples

	No. of Tests	Average	Maximum	Minimum	Standard Deviation (SD)	SD/True % Error	"True" Value
Sample #25							
k_{max}	13	93.7	118	51.0	15.9	17.1	93
k_{90}	13	63.5	79.5	51.9	7.6	12.5	61
k_v	11	134.5	171	49.5	36.5	25.3	144
Porosity	13	17.2	17.6	16.9	0.21	1.2	17.2
Grain density	13	2.66	2.67	2.65	0.01	0.3	2.66
Sample #34							
k_{max}	14	0.060	0.480	0.005	0.120	601	0.020
k_{90}	14	0.048	0.220	0.005	0.065	325	0.020
k_v	9	0.011	0.020	0.005	0.005	49.3	0.011
Porosity	13	4.66	5.60	3.20	0.76	15.4	4.90
Grain density	13	2.65	2.67	2.63	0.01	0.5	2.65
Sample #43							
k_{max}	12	0.035	0.260	0.006	0.068	486	0.014
k_{90}	11	0.047	0.260	0.004	0.087	871	0.010

k_v	8	0.012	0.020	0.006	0.005	41.3	0.011
Porosity	13	1.71	2.02	0.90	0.32	17.7	1.78
Grain density	13	2.69	2.71	2.68	0.01	0.3	2.70
Sample #52							
k_{max}	15	4.49	5.84	3.15	0.69	15.1	4.59
k_{90}	15	4.09	5.19	2.85	0.69	16.0	4.34
k_v	13	1.08	1.28	0.81	0.16	14.3	1.15
Porosity	13	13.2	14.2	12.4	0.49	3.8	13.1
Grain density	13	2.69	2.71	2.68	0.01	0.3	2.69

From Honarpour et al. (2005).

a similar fashion to plug measurements. For horizontal permeability measurements, gas is flowed across the core. The coreholder is designed with rubber sleeve containing screens which allow gas in and out of the core. The sleeve must not penetrate the screens or it will restrict flow. In transverse flow mode, the end platens inlet and outlets are plugged. Gas flows from the inlet screen, which covers a subtended angle, θ, on the core diameter and the entire length of the sample, across the sample and into a similar outlet screen placed diametrically across from the inlet screen. The flow pattern is complex but Collins (1952) computed a dimensionless geometrical factor that is a function of the angle subtended by the screens. This geometric factor is used to correct the standard Darcy equation for steady-state (axial) flow in core plugs to transverse whole core measurements. In the transverse direction, the length, L, is the diameter of the core. The cross-sectional area A is the area of the core surface covered by the screen. If the coreholder uses screens that subtend a 90° angle on the core circumference, then the geometric correction factor is 1.

After measuring transverse permeability in one direction, further permeability measurements are often taken with the core rotated by 45° (4 measurements) or by 60° (3 measurements) each time. These data can be used to assess permeability anisotropy in the horizontal direction.

Table 5.24 (from Honarpour et al., 2005) lists the results of repeat porosity, permeability and grain density measurements on four whole core samples conduced at four commercial laboratories over a 5-year period. The agreement between grain density and porosity measurements is acceptable, but relative error increases for permeability measurements in lower permeability samples.

REFERENCES

Amabeoku, M., BinNasser, R.H., 2012. Quality control/quality assurance assessments of core analysis data from multiple commercial laboratories. In: SPE Saudi Arabia Section Technical Symposium and Exhibition, April, 2012, SPE 160868.

American Petroleum Institute (API), 1988. Recommended Practice RP 40. Recommended Practices for Core Analysis, second ed. API Publishing Services, Washington, DC.

Collins, R.E., 1952. Determination of the transverse permeabilities of large core samples from petroleum reservoirs. J. Appl. Phys. 23 (6), 681–684.

Darcy, H., 1856. Les fontaines publiques de la ville de Dijon: exposition et application. Victor Dalmont, Paris.

Dean, E.W., Stark, D.D., 1920. A convenient method for the determination of water in petroleum and other organic emulsions. J. Ind. Eng. Chem. 12 (5), 486–490.

Dykstra, H., Parsons, R.L., 1950. The prediction of oil recovery by waterflood. In: Secondary Recovery of Oil in the United States. American Petroleum Institute, Dallas, pp. 160–174.

Eijpe, R., Weber, K.J., 1971. Mini-permeameters for consolidated rock and unconsolidated sand. Am. Assoc. Pet. Geol. Bull. 55 (2), 307–309.

Fjerstad, P.A., Thereby, H., Pallatt, N., Stockden, I., 1993. Application of deuterium tracer in estimating native water saturation in the Gyda field. In: SPE Asia Pacific Oil and Gas Conference and Exhibition, Singapore, February, Paper SPE 25376.

Forchheimer, P., 1901. Wasserbewegung durch Boden. Z. Ver. Dtsch. Ing. 45, 1731.

Goggin, D.J., Thrasher, R.L., Lake, L.W., 1988. A theoretical and experimental analysis of minipermeameter response including gas slippage and high velocity flow effects. In Situ 12 (1 and 2), 79–116.

Halvorsen, C., Hurst, A., 1990. Principles, practice and applications of laboratory minipermeametry. In: Worthington, P.F. (Ed.), Advances in Core Evaluation: Accuracy and Precision in Reserves Estimation. Gordon and Breach Science Publishers, London, pp. 521–549.

Heid, J.G., McMahon, J.J., Neilson, R.R., Yuster, S.T., 1950. Study of the permeability of rocks to homogeneous fluids. In: Drilling and Production Practice. American Petroleum Institute, pp. 230–246.

Hensel, W.M., 1982. An improved summation-of-fluids porosity technique. Soc. Pet. Eng. J. 22 (2), 193–201.

Honarpour, M., Djabbarah, N., Sampath, K., 2005. Whole-core analysis—experience and challenges. SPE Reserv. Eval. Eng. 9 (6), 460–469.

Jones, S.C., 1972. A rapid accurate unsteady-state Klinkenberg permeameter. Soc. Pet. Eng. J. 12 (5), 383–397.

Jones, S.C., 1987. Using the inertial coefficient, β, to characterise heterogeneity in reservoir rock. In: 62nd Annual Technical Conference of Society of Petroleum Engineers, New Orleans, Louisiana, Paper SPE 16949.

Jones, S.C., 1994. A new, fast, accurate pressure-decay probe permeameter. SPE Form. Eval. 9 (3), 193–199.

Klinkenberg, L.J., 1941. The permeability of porous media to liquids and gases. In: Drilling and Production Practice. American Petroleum Institute, New York, New York, pp. 200–213.

Lasseux, D., Janoor, Y., Profice, S., Mallet, M., Hamon, G., 2012. The "step decay": a new transient method for the simultaneous determination of intrinsic permeability, Klinkenberg coefficient and porosity on very tight rocks. In: International Symposium of the Society of Core Analysts, Aberdeen, Scotland, August 27–30, Paper SCA2012-25.

McPhee, C.A., Arthur, K.G., 1991. Klinkenberg permeability measurements: problems and practical solutions. In: Worthington, P.F., Longeron, D. (Eds.), Advances in Core Evaluation II: Reservoir Appraisal. CRC Press, Amsterdam, pp. 495–519.

Rushing, J.A., Newsham, K.E., Lasswell, P.M., Cox, J.C., Blasingame, T.A., 2004. Klinkenberg-corrected permeability measurements in tight gas sands: steady-state versus unsteady-state techniques. In: SPE Annual Technical Conference and Exhibition Held, Houston, Texas, 26–29 September, Paper SPE 89867.

Ruth, D.W., Kenny, J., 1987. The unsteady-state permeameter. J. Can. Pet. Technol. 28 (3), Paper PETSOV-89-03-06.

Thomas, D.C., Pugh, V.J., 1989. A statistical analysis of the accuracy and reproducibility of standard core analysis. Log Anal. 30 (2).

Woodhouse, R., 1998. Accurate reservoir water saturations from oil-mud cores: questions and answers from Prudhoe Bay and beyond. Log Anal. 39 (3).

RECOMMENDED READING/VIEWING

Ahmed, U., Crary, S.F., Coates, G.R., 1989. Permeability estimation: the various sources and their interrelationship. J. Pet. Technol. 43 (5), 578–587.

Bravo, M.C., 2007. Effect of transition from slip to free molecular flow on gas transport in porous media. J. Appl. Phys. 102, 074905.

Buckles, R.S., 1965. Correlating and averaging connate water saturation data. J. Can. Pet. Technol. 9 (1), 42–52.

Corbett, P.W.M., Jensen, J.L., 1992. Variation of reservoir statistics according to sample spacing and measurement type for some intervals in the Lower Brent group. Log Anal. 33 (1), 22–41.

Goggin, D.J., Chandler, M.A., Kocarek, G., Lake, L.W., 1998. Patterns of permeability in eolian deposits: page sandstone (Jurassic), Northeastern Arizona. SPE Form. Eval. 3 (2), 297–306.

Hurst, A., 1992. Sedimentary flow units in hydrocarbon reservoirs: some shortcomings and a case for high resolution permeability data. In: Bryant, I.D., Flint, S.S. (Eds.), The Geological Modelling of Hydrocarbon Reservoirs and Outcrop Analogues. Special Publication International Association of Sedimentologists, Blackwell Publishing Ltd., Oxford, UK, pp. 191–204.

Jannot, Y., Lasseux, D., Vize, G., Hamon, G., 2007. A detailed analysis of permeability and Klinkenberg coefficient estimation from unsteady-state pulse-decay or drawdown experiments. In: Society of Core Analysts International Symposium, Calgary, Canada, 10–12 September, Paper SCA2007-08.

Jensen, J.L., 1990. A model for small scale permeability measurements with application to reservoir characterization. In: Society of Petroleum Engineers/Department of Energy Symposium on Enhanced Oil Recovery, Tulsa, Oklahoma, Paper SPE/DOE 20265.

McPhee, C.A., Robertson, G.M., 1990. High resolution probe permeability: an aid to reservoir description. In: Worthington, P.F. (Ed.), Advances in Core Evaluation: Accuracy and Precision in Reserves Estimation. Gordon and Breach Science Publishers, London, pp. 495–519.

O'Sullivan, T., Ananthakrishnan, B., Zittel, R.J., Wheaton, S., Beliveau, D., Warner, H., Woodhouse, R., 2009. Evidence and verification of very low water saturations within the Fatehgarh Sandstone, Barmer Basin, India. Petrophysics 50 (03).

Rathmell, J., Atkins, K., Kralik, J., 1999. Application of low invasion coring and outcrop studies to reservoir development planning for the Villano field. In: Latin American and Caribbean Petroleum Engineering Conference, 21–23 April, Caracas, Venezuela, Paper SPE 53718.

Ringen, J.K., Halverson, C., Lehne, K.A., Rueslatten, H., Holand, H., 2001. Reservoir water saturation measured on cores: case histories and recommendations. In: Proceedings of the 6th Nordic Symposium on Petrophysics, Trondheim, Norway, May.

Stiles, J.H., Hutfiltz, J.M., 1992. The use of routine core analysis in characterising Brent group reservoirs, UK North Sea. J. Pet. Technol. 44 (6), 704–713.

Tiab, D., Donaldson, E.C., 1996. Petrophysics: Theory and Practice of Measuring Reservoir Rock and Fluid Transport Properties. Gulf Publishing Company, Houston.

http://petrowiki.org/Corrections_to_core_measurements_of_permeability (accessed 20 October 2014).

http://wiki.aapg.org/Overview_of_routine_core_analysis#Whole_core_analysis (accessed 20 October 2014).

Dean–Stark process. http://www.youtube.com/watch?v=A7s_UN4tHKs (accessed 20 October 2014).

Chapter 6

Preparation for Special Core Analysis

Chapter Outline

6.1 Fluid Preparation and Characterisation 269
- 6.1.1 Formation Water Samples and Brine Preparation 269
- 6.1.2 Oil 276
- 6.1.3 Gas 279
- 6.1.4 Data Requirements 280

6.2 Interfacial Tension 281
- 6.2.1 Methods to Determine IFT 282

6.3 Sample Selection for SCAL 286
- 6.3.1 Sample Provenance 286
- 6.3.2 CT Scanning and Photography 286
- 6.3.3 Petrophysical Properties 292

6.4 Reservoir Stress Estimation 300
- 6.4.1 Test Cell Design 300
- 6.4.2 Effective Stress Concepts 300
- 6.4.3 Net Confining Stress Loading Conditions 301
- 6.4.4 Reservoir Stress Data Sources 304

References 311

Recommended Reading 311

6.1 FLUID PREPARATION AND CHARACTERISATION

The proper preparation of brine solutions is critical to data that are to be used in field applications, especially for electrical property tests.

6.1.1 Formation Water Samples and Brine Preparation

The test brines are normally synthesised on the basis of the ionic compositions of formation brine samples. An example of a water analysis (from a wireline test sample) is shown in Table 6.1. Obtaining a sample of undisturbed formation water is not trivial. Wireline test tool sample collection volume is restricted such that the collected water may only be produced from the flushed and invaded zone, so the sample may be contaminated by drilling mud filtrate. Corrosion of these sample chambers is also a problem. Mud filtrate invasion can be identified if the mud is doped with tracers (see Chapter 2). If the lab analysis indicates a relatively high proportion of iron (as indicated in Table 6.1), this is almost certainly nothing to do with the formation brine,

Developments in Petroleum Science, Vol. 64. http://dx.doi.org/10.1016/B978-0-444-63533-4.00006-8

TABLE 6.1 Example Water Sample Analysis From RFT Sample

Appearance before filtration	{Sample from tool} Clear very pale yellow
Appearance after filtration	Clear colourless water
Total dissolved solids	51,470
Specific gravity at 60 °F	1.037
Resistivity, Ω m at 60 °F (determined)	0.182
Hydrogen sulphide	None detected
pH	7.03

Constituents		mg/L	meq/L
Cations			
	Sodium	16,470	716.45
	Potassium	120	3.07
	Calcium	2010	100.30
	Magnesium	885	72.80
	Barium	45	0.66
	Strontium	135	3.08
	Total iron	15	–
	Dissolved iron	2.9	0.1
Anions			
	Chloride	31,730	895.10
	Sulphate	L 0.5	L 0.01
	Bicarbonate	71	1.16
	Carbonate	Nil	Nil
	Hydroxide	Nil	Nil
Additional elements			
	Boron	15	
	Aluminium	L 0.5	
	Silica	4.1	
	Phosphorus	2.0	

but is a corrosion product. The best water samples are taken during production tests where significant volumes of formation water are produced. Micro-analysis of core water samples spun out of fresh-state core using a centrifuge (see Chapter 2) can be used for obtaining an ionic composition, but these could also be contaminated with mud filtrate.

The special core analysis laboratory (SCAL) then prepares a formation water sample (synthetic formation water (SFW)) which is synthesised based on the ionic composition and prepared from reagent grade salts and distilled and deionised water that is free of readily filtered solids. An example of an SFW composition calculated from a water sample ionic composition is shown in Table 6.2.

The proper preparation of brine and its resistivity measurement are critical to the acquisition of data. Worthington et al. (1990) provide detailed guidelines for the preparation of brine for resistivity measurements. The key points are:

TABLE 6.2 Example of SFW Composition

Ion	mg/L	Salt Weight (g/L)	Salt Weight (g/5 L)	Salt
Na(Cl res.)	7866.00	19.550	97.750	NaCl
Na(Na res.)		19.551	97.757	
Chloride balance (%)		100.007		
Ca^{2+}	3567.00	13.082	65.410	$CaCl_2{\cdot}2H_2O$
Mg^{2+}	38.50	0.322	1.610	$MgCl_2{\cdot}6H_2O$
K^+	331.40	0.632	3.159	KCl
Sr^{2+}	0.00	0.000	0.000	$SrCl_2{\cdot}2H_2O$
Ba^{2+}	0.00	0.000	0.000	$BaCl_2{\cdot}2H_2O$
Cl^-	18,583.00			
SO_4^-	65.40	0.219	1.096	$Na_2SO_4{\cdot}10H_2O$
HCO_3^-	381.36	0.525	2.626	$NaHCO_3$
CO_3^{2-}	0.00	0.000	0.000	Na_2CO_3
Fe^{3+}	0.00	0.000	0.000	$FeCl_3{\cdot}6H_2O$
TDS	30,832.66	34.330	269.407	

- Salts should be reagent grade and NaCl should be oven dried at 450–550 °C prior to use if in a high humidity environment or when using older, open stock.
- When using hydrated salts that contain water as part of their crystal structure, the most stable hydrate should be used when different hydrates are available, and salt weight must account for the water content.
- High-purity distilled and/or deionised water should be used. High-purity water without dissolved carbon dioxide should have a resistivity of approximately 100,000 Ω m. Deionised water may have a lower pH than distilled water and could cause negative interactions with the rock. This is especially important in carbonate cores. If this is the case, the pH should be adjusted.
- Masses of salts and water should be determined to better than ±0.1% and a buoyancy correction may be required. This correction arises from the difference between the density of the substance being weighed and the density of the balance weights.
- It is not usually necessary to degas the water or the brine in order to obtain accurate concentrations by either volumetric or gravimetric methods. However, brines may require degassing during core saturation.
- Some composite brines may form a precipitate during mixing or after standing due to reaction between the components, or from formation of insoluble carbonates by reaction with carbon dioxide dissolved from the atmosphere. All brines used for saturation of cores should be very carefully filtered immediately prior to contact with the core. Normally, a 0.45-μm filter is used.

In the absence of an ionic composition for the formation water, an estimate of the resistivity of the formation water can be determined from petrophysical analysis, such as a Pickett plot in a clean, water-bearing sand. For example:

$$R_w = R_0 \cdot \varphi^m$$

where R_w is the formation water resistivity (Ω m), R_0 is the true formation resistivity in clean, water-filled sand (Ω m), φ is the formation porosity (fraction) and m is the Archie cementation exponent (core measurements normally required).

In this case, the SFW sample is made up to replicate the formation water resistivity (R_1 estimated at reservoir temperature, T_1), at lab temperature conditions (R_2 at T_2) using the Arps equation (Arps, 1953):

$$R_2 = R_1 \cdot \frac{(T_1 + 21.5)}{(T_2 + 21.5)} \text{(for temperature in °C for NaCl brines)}$$

The salt content of the brine is normally made up of 80% NaCl with the addition of 20% KCl (to minimise potential clay reactivity effects).

6.1.1.1 Resistivity

Resistivity of the formation water SFW sample is measured as part of formation factor, resistivity index and Waxman–Smits parameter tests.

Accurate measurements require optimisation of cell design so that the solution resistance is far smaller than the impedance arising from the capacitive coupling between the electrodes at the measurement frequency, and far larger than the electrical impedance due to electrode polarisation. Ideally, the calibration factor should be constant over the useful range of solution concentrations.

Typical resistivity cells consist of two platinum black electrodes (~1 cm diameter, rigidly mounted 5–10 cm apart) and contained within a glass tube. An example is shown in Fig. 6.1. The silver electrodes are provided with a platinum black coating to increase the electrode surface area and to reduce polarisation effects to maximise the performance of the system. Stoppered sidearms, connected to the main cell body behind the electrodes, are provided for washing. The electrodes, of known cross-section area (A), are separated by a known distance (L) in the cell.

The cell factor (A/L), used to convert resistance measurements to resistivity, is determined by pre-calibrating the cell against a range of standard brine solutions whose resistivity is known. Both potassium chloride and sodium chloride solutions can be used as their electrical conductivity as a function of concentration is well known and tabulated. Each cell should be calibrated over its useful range. Some cells may not be very ruggedly constructed so that the cell constant may change with handling; therefore, repeat calibrations may be necessary.

The cell should be maintained at constant temperature by immersion in a constant temperature bath that uses a non-conducting fluid, such as oil or air. Precise control and measurement of temperature are a necessary requirement for maintaining accuracy during electrical resistivity measurements of brines or brine-saturated rocks. Best practise is to determine brine resistivity, R_w, from a number of measurements made over a range of temperatures or at a specified test temperature (if testing will be performed at elevated temperature). R_w data should be plotted against temperature and regressed to obtain the R_w equation as a function of temperature. This will allow use of actual R_w at the measured test temperature.

If R_w cannot be performed at temperature, measurements should be performed at several temperatures close to ambient temperature. Subsequent brine and rock resistivity measurements are thus typically related to a standard temperature, using the Arps relationship:

$$R_{wr} = R_{wl} \cdot \frac{(T_l + 21.5)}{(T_r + 21.5)}$$

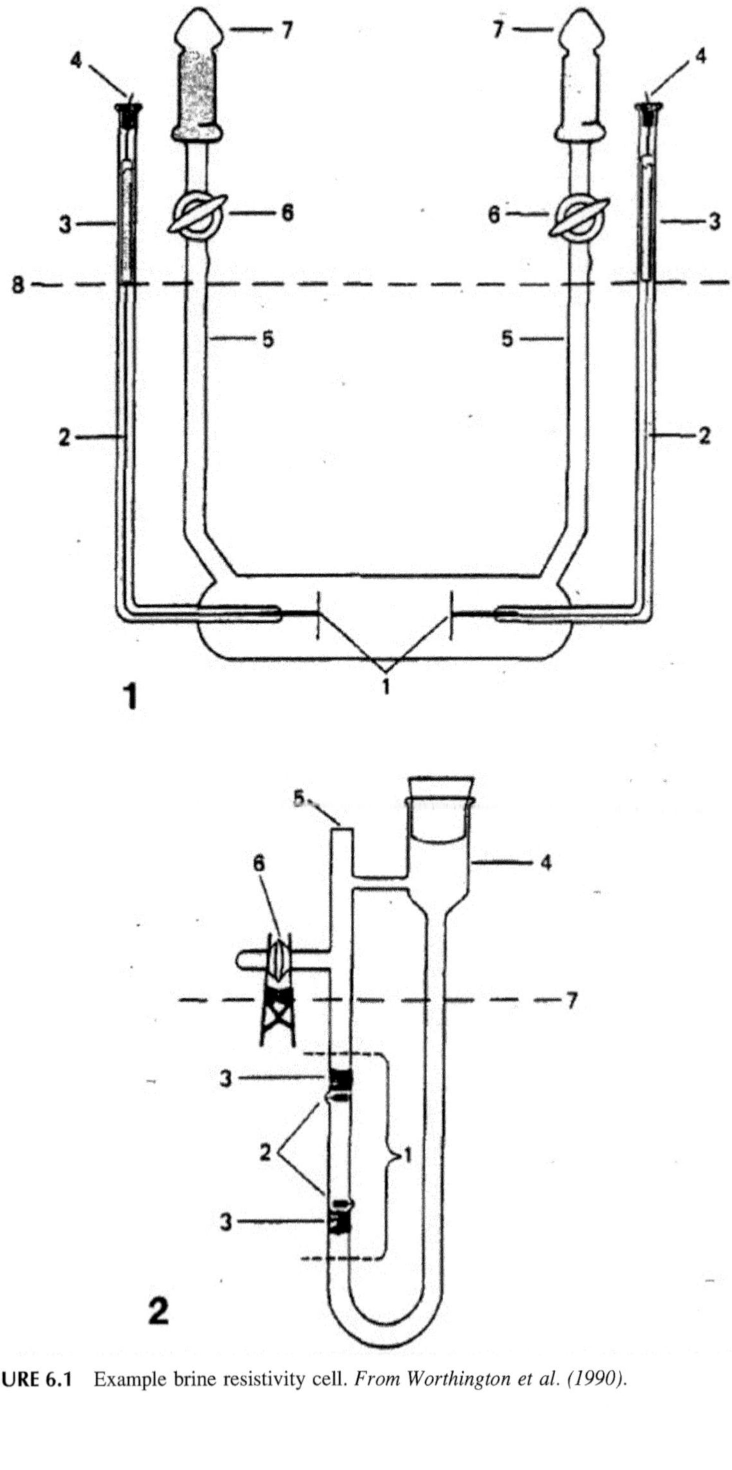

FIGURE 6.1 Example brine resistivity cell. *From Worthington et al. (1990).*

where T_l is the lab measurement temperature (°C), T_r is the reference temperature (typically 25 °C or at specified test pressure), R_{wl} is the brine resistivity at laboratory conditions (Ω m) and R_{wr} is the brine resistivity at reference conditions (Ω m).

Because this formula represents an empirical fit between temperature and resistivity for NaCl solutions, resistivity corrections of non-NaCl brines with the Arps equation should be verified, and, as it is approximate, measurements of resistivity should be made as close as possible to the specified temperature.

The measurement AC frequency should be high enough that platinising of the electrodes effectively eliminates contributions to the measured impedance and low enough that stray capacitance and/or inductive coupling between measurement leads and between the leads and the surroundings are easy to eliminate. Until quite recently, the 'industry-standard' AC measurement frequency for brine (and rock) resistivity measurements was 1000 Hz. However, at lower frequencies phase angle shifts (a measure of the lag between voltage and current sine waves) occur due to electrode polarisation (and contact resistance effects when testing in rocks). This creates an extra impedance which consists of a resistive component and a reactive component. If the measurement is made at a frequency which minimises the phase angle, then the reactive component will be negligible and the measured impedance is equal to the resistance. A multi-frequency analyser should be used to measure cell impedance. These usually contain circuitry for automatically correcting the observed impedance for the phase shift induced by the electrode capacitance. Phase angles of less than 0.5° should be possible with properly prepared cells. Increasingly therefore, measurements are made over a range of frequencies to determine the minimum phase angle, and measurements are made at that frequency.

6.1.1.2 Viscosity and Density

Brine density is determined either using a pycnometer or a digital densitometer. Viscosity is determined using a suitable viscometer.

As for resistivity, viscosity and density are temperature dependent so measurements are made over a range of three or four temperatures to cover the typical laboratory range. An example is shown in Table 6.3. For ambient temperature condition measurements, all resistivity measurements should be as close to the specified temperature as possible or normalised to a standard temperature (e.g. 20 or 25 °C). Density and viscosity data as a function of temperature can also be regressed to produce equations to allow accurate calculation of these parameters at recorded test temperatures.

6.1.1.3 Special Considerations for Brine Preparation: Carbonates

Carbonate samples (especially chalks) may require additional calcium carbonate to permit stability between the core samples and brine. Instability may cause dissolution of sample components into the brine. A preferred option is

TABLE 6.3 Example SCAL Reported SFW Characterisation Results

Salt	Salt Weight (g/L)
NaCl	97.750
$CaCl_2 \cdot 2H_2O$	65.410
$MgCl_2 \cdot 6H_2O$	1.610
$SrCl_2 \cdot 2H_2O$	0.000
KCl	3.159

Temperature (°C)	Density (g/cc)	Viscosity (cp)
20	1.046	1.115
25	1.045	0.988
30	1.043	0.892

to crush a 'sacrificial' core plug (or large plug trim) and place into the water prior to addition of other salts. The other option is to add laboratory grade calcium carbonate prior to addition of other salts. This first step should ensure a remnant precipitate to ensure super-saturation of calcium carbonate.

6.1.2 Oil

SCAL tests (e.g. resistivity index, capillary pressure and relative permeability) will use live reservoir oil, dead crude oil and/or mineral oil as the oil phase in experiments. Dead oil is a sample of the reservoir oil obtained at ambient conditions (e.g. stock tank oil [STO] or separator oil sample). Most of the lighter (gaseous) hydrocarbon components will have evaporated during depressurisation to ambient conditions. Live oil is the reservoir oil with all hydrocarbon components under reservoir temperature and pressure; obtained either as an acquired reservoir fluid sample (bottom-hole sample or MDT fluid sample) or by recombination (e.g. mixing separator oil and gas together under reservoir pressure and temperature). Mineral oils are refined, paraffinic oils.

6.1.2.1 Mineral Oil

Different mineral oils are used by different laboratories for different analyses, based on closest available supplier and upon the properties of the oil. Mineral oil tends to be easier to work with in the lab, since there are lower HSE risks, elevated temperature is not necessary and there is less chance of emulsion formation, to name a few potential benefits. However, mineral oils cannot be used for wettability restoration—which requires a sample of reservoir oil (either live or dead). If using a mineral oil in relative permeability testing, it is usual to ensure that viscosity be selected such that the ratio of oil to water viscosities at laboratory test conditions is the same as the reservoir viscosity ratio at reservoir conditions.

Before use, mineral oil should be filtered through a 0.45-μm filter and allowed to flow through a depolarising column (e.g. containing silica crystals), to remove any remnant polar components from the refining process (this is to ensure that the mineral oil has no wettability altering capabilities).

Mineral oil properties shall be measured—density and viscosity. Mineral oil density is determined using either a 50-ml pycnometer (in a water or air bath) and a top-loading electronic balance or a suitable digital density metre. Mineral oil viscosity is measured using a suitable viscometer cell (e.g. Cannon-Fenske Routine viscometer), contained within a water bath.

Measurements of density and viscosity are made at a range of temperatures, allowing at least 1 h to achieve temperature stability at each test temperature. Properties should be plotted against temperature and regressions of each curve performed to obtain the equation of each property as a function of temperature. The equations allow determination of density and viscosity at test temperatures.

6.1.2.2 Reservoir Crude Oil

Reservoir oil is a complex of millions of different hydrocarbon components. Not all hydrocarbon components are compatible with one another and are held in solution because of the presence (or abundance) of other components at reservoir temperature and pressure: alteration of temperature and pressure has the potential of causing an imbalance and precipitating the incompatible components.

Reservoir oil components are standardly divided into four major hydrocarbon categories, dependent upon carbon bonding, polarity and inorganic (non-hydrocarbon) content: saturates, aromatics, resins and asphaltenes (SARA). Saturates are simple to long-chain, saturated hydrocarbons, contain no inorganic content and are non-polar. Aromatics are unsaturated (i.e. carbon bonds exhibit electron sharing in the form of one or more benzene rings), contain no inorganic content and are non-polar. Resins are more simple hydrocarbons units containing a single inorganic group (nitrogen, oxygen or sulphur) and are polar components. Asphaltenes are highly complex hydrocarbons containing multiple bonded unsaturated hydrocarbons (benzene rings) with various saturate and resin appendages. Due to polar differences, saturates are incompatible with polar resins and asphaltenes. Aromatics are considered to be the solvent, allowing the mix to stabilise in the saturate solution.

It is recommended that SARA analysis as well as acid and base content should be measured on uncontaminated reservoir oil samples for characterising the oil, to aid in deciding whether live and dead oil should be used.

These analyses provide information about the polar content of the oil, which may give early indications of the potential wetting ability of the oil.

In addition, an asphaltene stability analysis will provide vital information about the asphaltic content of the oil and its propensity to precipitate by contact with saturate hydrocarbons (e.g. hydrocarbon gases and/or laboratory mineral oil). Certain SCAL processes will require interchanging between reservoir oil and laboratory oil, during which certain oils may precipitate asphaltenes, artificially altering wettability and effective permeability. This can be moderated by using either decalin or a saturate/aromatic mixture (e.g. decane/xylene or mineral oil/toluene). A saturate/aromatic mixture ratio is best created to simulate the reservoir oil's original saturate/aromatic ratio, otherwise an 80:20 ratio should be used.

NB. SARA analysis is performed on dead crude oil (not live oil), where many gaseous saturate components may have already been lost during depressurisation. For live ratios, the PVT separation analyses are required to correct to the live composition.

6.1.2.3 Live Reservoir Oil

The decision must be made whether to use live reservoir oil or dead crude oil for SCAL tests: either 'live' (or recombined) crude oil or dead (gas removed) crude oil can be used. Of the two, dead crude oil is easier for laboratories to use, since lower pressure and temperature might be employed, while live oil is obviously more representative of reservoir properties since reservoir temperature and pressure are employed, but becomes more difficult for laboratories to use: longer test times are required to apply and stabilise at the higher temperature and pressure, processes must ensure against loss of gaseous components, etc. Fewer labs have capability to perform live oil conditions, since the equipment requires greater experience and financial investment: analyses are subsequently more time consuming and expensive.

Dead crude oil (often called STO) may differ from live oil due to loss of gaseous components during depressurisation and drop in temperature from reservoir to surface. The amount of difference will depend upon the content (molar weight percent) of those lighter (gaseous) hydrocarbons, lost during trip to surface. Gas–oil ratio (GOR) measurements, the volume ratio of reservoir gas to oil evolved during depressurisation, are recommended on the live oil. A reservoir with high GOR may be significantly altered during depressurisation and the resulting dead crude oil may not be representative of the true reservoir oil: it will not have the same viscosity and density properties, it may have different wetting properties. It is therefore recommended that a GOR limit be employed to determine what constitutes a representative dead oil sample. Certain companies suggest a GOR less than 200 scf/bbl ($\sim$40 m^3/m^3) may indicate that the dead oil can be used to represent the live oil.

In theory, the use of 'live oil' is preferred to STO, especially for oils with high GOR content or where asphaltene or wax deposition can obstruct sample permeability. In practise, this is limited by the quality and availability of reservoir oil samples: either bottom-hole samples or dead crude oil to recombine with gas, the viability of PVT data, and by the associated, normally high, cost. Crudes containing H_2S have HSE issues and even fewer labs are able to conduct tests using them.

Certain waxy crudes have high wax appearance temperatures or pour points which means they can become solid at room temperature. In this case, it is important to ensure that elevated temperature is maintained during characterisation and throughout testing with such crudes. As such, it is preferable that laboratory processes and equipment have been designed to minimise temperature variation/loss during transfer between equipment.

All reservoir oil (whether live or dead) must be pre-filtered/treated to remove any contaminants (such as sand particles and water droplets) prior to characterisation. Filter size is important to consider, especially for low-permeability plug samples. Standard filter sizes are 10, 3 and 0.45 μm. If mean pore throat size is smaller than these filter sizes, the plug sample will act as a further filter during testing, exhibited as gradually decreasing permeability. For such material, a sacrificial, low-permeability plug (or several, since each plug may gradually lose injectivity as large waxy and/or asphaltenic components are removed) is recommended to use as an inline filter during injection studies.

Density and viscosity of the crude oil are measured using similar methods and equipment to those for mineral oil. For dead crudes, a range of temperatures around test temperature should be measured and regressed to obtain the equation of properties as a function of temperature. For live oils, reservoir temperature should be used.

6.1.3 Gas

Gas–liquid systems (drainage and imbibition) are prone to disequilibria owing to gas compressibility and dissolution in pressurised liquids. Disequilibria effects (mass transfer) will entail one or more of the following artefacts: dissolution of gas, evolution of gas, evaporation of liquid and condensation of liquid. During injection studies, it is important to control gas–liquid equilibrium and limit disequilibria effects.

The majority of SCAL tests on gas–brine systems (e.g. porous plate and centrifuge capillary pressure) use oxygen-free nitrogen as the gas phase. Nitrogen viscosity is normally determined from correlations at the test pressure and temperature.

Most laboratories perform gas–brine analyses at atmospheric pressure, while some will apply a nominal pore (line) pressure to reduce gas compressibility issues. To reduce mass transfer between gas and water during core testing (evaporating water from the core), the test nitrogen is humidified by bubbling it through water. For pressurised systems, gas and brine are usually equilibrated by high-rate, co-current flooding in a closed loop, by-pass system for several hours (i.e. by-passing the core plug).

Nitrogen is also used for gas–oil analyses, e.g., gas–oil relative permeability tests. The gas and oil are equilibrated by flooding gas through oil to reduce mass transfer between phases. For gas–oil systems, it is usual to have an irreducible water saturation, when it is then important to ensure all phases are in equilibrium (oil, water and gas) to prevent reduction of the initial water saturation. Nitrogen is not often miscible with crude oil and where miscibility and thin film drainage plays a part in gas flood oil recovery, live fluids should be used. Gas can be taken from a separator sample or recombined from hydrocarbon gases based on a PVT composition. Gas viscosity and density are normally determined from correlations at the test pressure and temperature, though live hydrocarbon gas samples may often have viscosity measured as part of a PVT study.

Disequilibrium is inescapable during gas–liquid flooding experiments. During liquid–gas imbibition, additional gas will be dissolved into the injected liquid, thus artificially reducing residual gas. During gas–liquid drainage, liquid molecules may be evaporated into the gas, thus artificially reducing residual liquid saturation. To attempt to reduce these effects, it is recommended that differential pressure should be controlled to less than 2% of pore pressure; also attempt to minimise extended flooding periods/volumes, thus reducing the time disequilibria can affect the process.

6.1.4 Data Requirements

Data Requirement	Comments
• Brief description of procedures and experimental apparatus	
• Formation water ionic composition and SFW composition	
• Viscosity cell calibration data and procedures	
• Brine resistivity cell calibration data and procedures	
• Resistivity cell factor	
• Brine resistance and phase angle as a function of time at each test temperature and as a function of frequency	
• Stabilised brine viscosity, density and resistivity data as a function of temperature	Including plot of properties versus temperature and the subsequent equations for the properties as a function of temperature

Data Requirement	Comments
• Stabilised oil viscosity and density data as a function of temperature	Including plot of properties versus temperature and the subsequent equations for the properties as a function of temperature
• Gas viscosity and density at test temperature and pressure	Including plot of properties versus temperature and the subsequent equations for the properties as a function of temperature
• Reservoir oil (crude or live) equilibration procedures	
• PVT data and compositional analysis for recombined live gas and oil	
• SARA analysis	
• Acid number and base number of oil	
• Asphaltene study results	
• Wax appearance temperature	

6.2 INTERFACIAL TENSION

Interfacial tension (IFT) is essential for conversion of laboratory data to reservoir. Laboratory analyses, both static and dynamic, are most often performed using synthetic fluids (SFW, mineral oil, nitrogen, etc.) at laboratory conditions. IFT between these fluids at these conditions can be significantly different than reservoir IFT. Failure to correct for this may induce large uncertainty.

For example, most laboratories report IFT of standard laboratory fluids (SFW and mineral oil) as being 48 mN/m (48 dyne/cm). However, Fig. 6.2 indicates that IFT varies as a function of salt concentration (from a little over 30–45 mN/m). For dynamic testing, where *in situ* saturation monitoring requires doped mineral oil, the iododecane used to dope oil can reduce laboratory IFT to approximately 16 mN/m at elevated temperatures.

Capillary pressure and relative permeability must be corrected for IFT. Correction of laboratory data to reservoir representative incorporates an IFT factor:

$$\text{IFT factor} = \frac{\sigma\cos\theta_{\text{res}}}{\sigma\cos\theta_{\text{lab}}}$$

If $\sigma_{\text{res}} = 24$ mN/m, then using a standard IFT value (48 mN/m) produces an IFT factor = 0.5, while if a measured value showed IFT = 35 mN/m, IFT factor = 0.686. For the case of iododecane at elevated temperature (16 mN/m), the IFT factor may be 1.5.

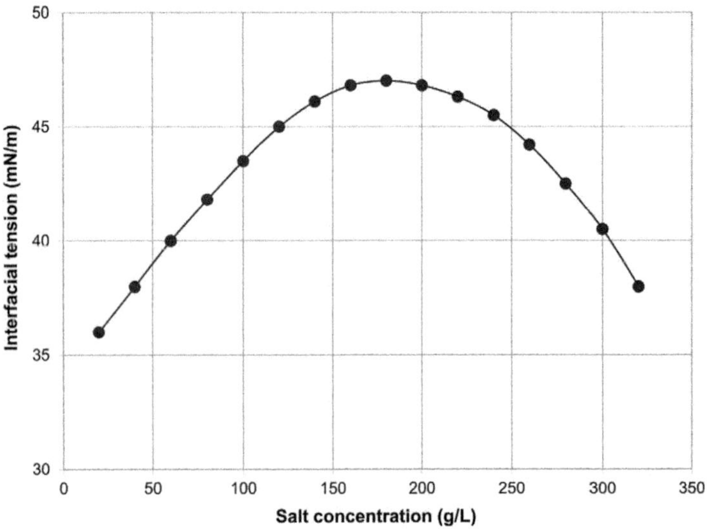

FIGURE 6.2 SFW-mineral oil IFT versus salt concentration.

This is a possible error >200% in the conversion factor of laboratory to reservoir data, since most laboratories have not measured IFT of water and iododecane doped oil and would assume it to be standard.

It is strongly recommended that IFT be measured for the test fluids, at test conditions (all test conditions to be used) and of the reservoir fluids under reservoir conditions.

6.2.1 Methods to Determine IFT

Care should be taken when using/reviewing IFT data. IFT changes with time from first contact of the fluids until a stable/equilibrated fluid system is obtained—even if fluids have been brought into equilibrium outside of the test equipment. This may be due to differences in dynamic and static IFT. As an example, an equilibrated oil–water system may produce IFT in the order 35–40 mN/m within the first few seconds of test contact, but decreases (as a decay curve) to 35–30 mN/m over the course of a few hours. Laboratories should be asked to provide information about their procedures.

6.2.1.1 From Compositional Data

IFT can be estimated from fluid compositional data, e.g.:

$$\sigma = \left[6.95 \frac{\Delta\rho}{T_r^{0.3125}}\right]^4$$

where $T_r = \frac{T}{T_c}$, in which T is the reservoir temperature (°R), T_c is the mixture critical temperature (°R) and $\Delta\rho$ is the water–oil density difference at reservoir temperature (lb/ft.^{3}).

6.2.1.2 The Pendant Drop Method

Experimentally, the pendant drop technique is the most commonly used method for the measurement of IFT in the petroleum industry. Drop shape analysis is a convenient way to measure IFT. The principal assumptions are:

- the drop is symmetric about a central vertical axis: this means that it is irrelevant from which direction the drop is viewed
- the drop is not in motion in the sense that viscosity or inertia are playing a role in determining its shape: this means that surface (or interfacial) tension and gravity are the only forces shaping the drop

Surface tension is determined by fitting the shape of the drop (in a captured video image) to the Young–Laplace equation, which relates IFT to drop shape (Fig. 6.3). Normally, software does this automatically. Once the dimensions of the droplet are measured, they are used along with density data of the liquid and vapour phases to calculate the IFT.

The two principal practical advantages of the technique are:

- calibration is straightforward in that only optical magnification is needed. This can be measured with high accuracy (density must be known);
- solid surfaces of the apparatus involved need not have any special cleanliness because their wettability, per se, does not affect the result. This is a significant advantage over such techniques as the Wilhelmy plate, where cleanliness is required;
- the tests can be run at pressure (17,000 psi) and temperature (200 °C);
- digitised image processing allows measurements of IFTs down to 10–2 mN/m, important for EOR and gas condensate processes.

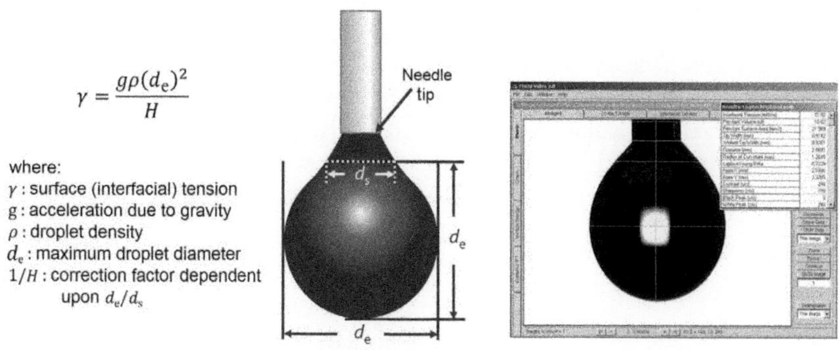

FIGURE 6.3 Illustration of pendant drop method. (Right: *Courtesy of First Ten Angstroms.*)

Certain laboratories will provide only the first-contact IFT (average of 5–10 drops), while others will provide only equilibrated IFT (average of 5–10 drops). It is recommended that every drop is recorded as a function of time to provide both average first contact and average equilibrated data. IFT versus time curves can be used as an additional quality control tool to ascertain variance and the potential of interference from impurities in the system.

6.2.1.3 Force Tensiometry

The measurement of surface and IFT, as performed by a force tensiometer, is based on force measurements of the interaction of a probe with the surface interface of two fluids. Measurements are generally available only at ambient pressure and slightly elevated temperatures.

With this technique, IFT measurements can be performed just by insuring that the bulk of the probe is submersed in the light phase prior to beginning the experiment.

In these experiments, a probe is hung on a balance and brought into contact with the liquid interface to be tested. The forces experienced by the balance as the probe interacts with the surface of the liquid can be used to calculate surface tension. The forces present in this situation depend on the following factors: size and shape of the probe, contact angle of the liquid/solid interaction and surface tension of the liquid. The size and shape of the probe are easily controlled. The contact angle is fixed to zero (complete wetting). This is achieved by using probes with high energy surfaces. The probes are generally made of a platinum/iridium alloy which insures complete wetting and easy cleaning.

The mathematical interpretation of the force measurements depends on the shape of the probe used. Two types of probes are commonly used, the Du Noüy ring and the Wilhelmy plate.

6.2.1.4 Du Noüy Ring

This method utilises the interaction of a platinum ring with the surface being tested. The ring is submerged below the interface and subsequently raised upwards. As the ring moves upwards, it raises a meniscus of the liquid. Eventually, the meniscus tears from the ring and returns to its original position. Prior to this event, the volume, and thus the force exerted, of the meniscus passes through a maximum value and begins to diminish prior to the actually tearing event. The process and ring are shown in Fig. 6.4. The calculation of surface or IFT by this technique is based on the measurement of this maximum force:

$$\sigma = \frac{F_{max} - F_v}{L\cos\theta}$$

where σ is the surface (interfacial) tension, F_{max} is the maximum force, F_v is the weight of volume of liquid lifted, L is the wetted length and θ the is contact angle.

FIGURE 6.4 Illustration of Du Noüy ring method. *https://en.wikipedia.org/wiki/Tensiometer_(surface_tension).*

Contact angle (θ) decreases as the extension increases and has the value 0° at the point of maximum force; this means that the term cos θ has the value 1. Unfortunately with this method, due to the volume of the liquid hanging from the ring as it is pulled above the surface, a correction needs to be applied (F_v). Various correction methods are available.

6.2.1.5 Wilhelmy Plate

This method utilises the interaction of a platinum plate with the surface being tested. The calculations are based on the geometry of a fully wetted plate in contact with, but not submerged in, the heavy phase. As the surface is brought into contact with the probe, the tensiometer notices this event by the change in forces it experiences. It will register the height at which this occurs as the 'zero depth of immersion'. The plate will then be wetted to a set depth to insure that there is indeed complete wetting of the plate (zero contact angle). When the plate is later returned to the zero depth of immersion, the force it registers can be used to calculate surface tension. The tension is calculated using the following equation:

$$\sigma = \frac{F}{L\cos\theta}$$

where σ is the surface (interfacial) tension, F is the force acting on the balance, L is the wetted length and θ is the contact angle.

The plate is made of roughened platinum and is optimally wetted so that contact angle (θ) is virtually 0°. This means that the term cos θ has a value of approximately 1, so that only the measured force and the length of the plate need to be taken into consideration (Fig. 6.5). Unlike the ring method, correction calculations are not necessary with the plate method.

A summary of the advantages and disadvantages of the different IFT methods is given in Table 6.4.

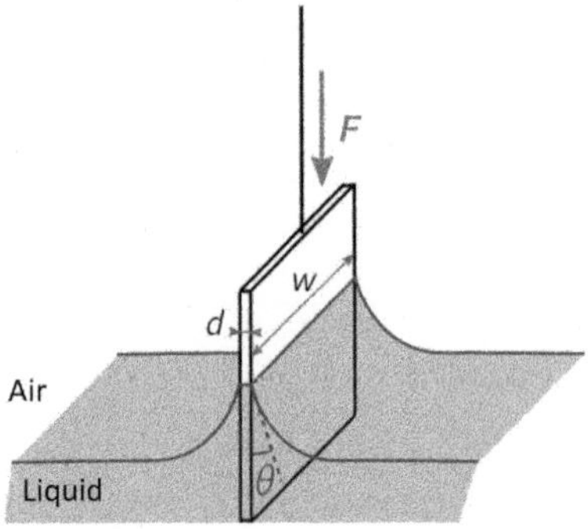

FIGURE 6.5 Illustration of Wilhelmy plate method. *https://en.wikipedia.org/wiki/Wilhelmy_plate*

6.3 SAMPLE SELECTION FOR SCAL

6.3.1 Sample Provenance

Plugs used for SCAL tests normally originate from four sources:

- Plugs cut from slabbed core
- Plugs cut from preserved core
- Preserved plugs (preserved during routine core analysis (RCA) programme)
- Plugs previously used in the RCA test programme

The advantages and disadvantages of the different sample types are listed in Table 6.5.

6.3.2 CT Scanning and Photography

CT scanning should be performed on:

1. Selected preserved whole-core sections
2. Preserved plugs
3. Ex-RCA plugs or plugs cut from preserved whole-core sections

CT scanning can help identify any anomalous material, sample damage and/or significant heterogeneity in the samples of plugs.

Figure 6.6 shows an example of CT scans on preserved whole-core sections. The upper sample is damaged and unsuitable for plugging. The lower sample shows extreme heterogeneity which would make the sample unsuitable for plugging.

Figure 6.7 shows examples of plug CT scans (including a brief description) and Table 6.6 shows an example of a plug selection chart traffic light system following CT scanning of preserved plug samples and assessment of

TABLE 6.4 Summary Table—Interfacial Tension Measurement Methods

Method	Advantages	Disadvantages
Pendant drop	• Calibration is straightforward in that only optical magnification is needed. This can be measured with high accuracy (density must be known) • Solid surfaces of the apparatus involved need not have any special cleanliness because their wettability, per se, does not affect the result. This is a significant advantage over such techniques as the Wilhelmy plate and where cleanliness is required • The tests can be run at pressure (17,000 psi) and temperature (200 °C)	• It is a relatively recent methodology and then is not the best option for comparison purposes with legacy data • It is not as widely available as any of the traditional tensiometry methodologies
Du Noüy ring	• Many values in the literature have been obtained with the ring method. This means that in many cases, the ring method should be preferred for comparison purposes • The wetted length of the ring exceeds that of the plate by the factor 3. This leads to a higher force on the balance and accordingly to a better accuracy. This effect does not influence the results of surface tension measurements, but small interfacial tensions can be carried out more accurately with the ring method • Some substances, e.g., cationic surfactants, show poor wetting properties on platinum. In such cases, the surface line between a ring and the liquid is more even than that of a plate • Equipment requires thorough cleanliness as results may be affected	• The method requires mathematical corrections to account for the weight of the volume of liquid lifted • The method requires the density of the fluids to be known • In an interfacial tension measurement, the surface is pressed into/pulled out of the other phase. This may cause the phases become mixed • With the ring method, the surface or interface is renewed permanently due to the movement of the ring. If the ring is moving with high velocity, or if solutions of large molecules or with high viscosities are to be measured, the maximum force is obtained when the diffusion equilibrium at the surface or interface is still not reached. This may produce erroneous results • Measurements are generally available only at ambient pressure and slightly elevated temperatures

Continued

TABLE 6.4 Summary Table—Interfacial Tension Measurement Methods—Cont'd

Method	Advantages	Disadvantages
		• The method measures static interfacial tension, which is not the case in reactive systems, such as those involved in microbial improved oil recovery
Wilhelmy plate	• Unlike the ring method, no correction is required for measurement values obtained by the plate method • With the plate method, the densities of the liquids do not have to be known as they have to be with the ring method • In an interfacial tension measurement, the surface is only touched and not pressed into/pulled out of the other phase. This avoids the phases becoming mixed • The measurement failure caused by ring movement does not occur with the plate method. The plate method is a static measurement; i.e., the plate does not move after the surface or interface has been detected • The method is specially suited to check surface tension over long time intervals	• Some substances, e.g., cationic surfactants, show poor wetting properties on platinum. In such cases, the surface line between a ring and the liquid is more even than that of a plate • Less accurate than the ring method for small interfacial tension measurements as the wetted length of the plate is less than the ring by the factor 3. This leads to a lower force on the balance and accordingly to a poorer accuracy • Measurements are generally available only at ambient pressure and slightly elevated temperatures • The method measures static interfacial tension, which is not the case in reactive systems, such as those involved in microbial improved oil recovery

the images. Plug sample images should be graded: the following provides a suggested grading system:

1. Homogeneous plug
2. Mostly homogeneous but some heterogeneity
3. Questionable heterogeneity, use understanding there is a high risk of failure
4. Heterogeneous and should not be used
5. Damaged, fractured or other physical property that negates its use

TABLE 6.5 Advantage and Disadvantages of SCAL Sample Types

Sample Type	Advantages	Disadvantages
Plugs from preserved core	Normally, preferred option as core is full diameter and protected from dehydration and alteration	Ensure integrity of preservation before plugging Preserved core *must* be CT-scanned before plugging to evaluate any damage, heterogeneity, and to orient plugs Porosity and permeability are essentially unknown but can be estimated from closest adjacent RCA plug Preserved core selection is often subjective and may be biased (especially towards better quality intervals)
Preserved plugs	Plug prevented from dehydration and alteration Porosity and permeability estimated from parallel RCA plug (assuming RCA and SCAL plugs cut as parallel pair)	Ensure integrity of preservation before selection Preserved plug *must* be CT-scanned before plugging to evaluate any damage and heterogeneity
Plugs from slabbed core	Plugs can be selected at any location on core surface with heterogeneity clearly visible	Plug size can be constrained by slab geometry and dimensions Core may have dried out, and in oil cores, residual crude can cause wettability alteration Slabbing fluids and subsequent drying can cause damage to delicate clays
Ex-RCA plugs	Porosity and permeability are known Acceptable for some SCAL tests (e.g. mercury injection)	Plug preparation method must be consistent with aims of SCAL test. Clays may have been destroyed during RCA preparation, or sample may not have been rendered water wet Evaluate plug history (including preparation method and any stress cycling) to ensure the RCA procedures have not damaged the plug

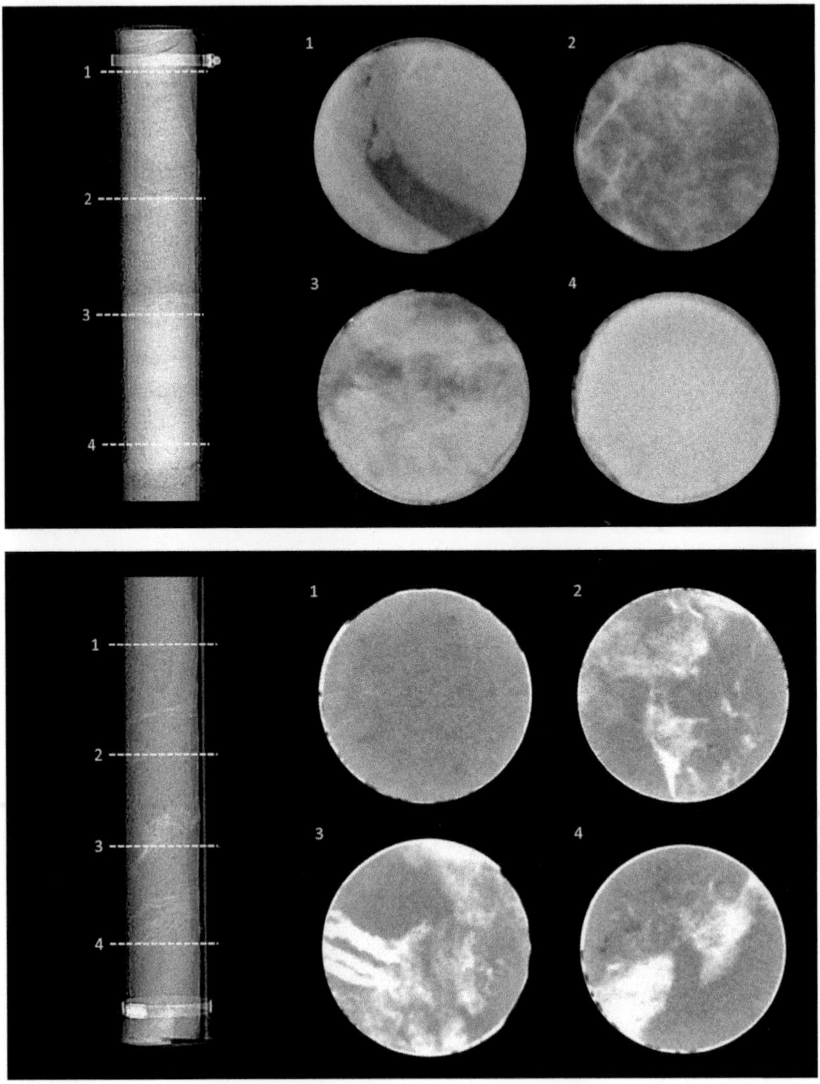

FIGURE 6.6 Example of CT scans on preserved cores.

It is recognised that CT image grading can be somewhat subjective; thus, it is suggested that more than one person assesses the samples and a discussion should ensue where different grades have been given. When selecting plugs, grade 1 samples should be used before grade 2 samples (where possible). Please note that plug samples denoted grade 3 are those that are heterogeneous and would not usually be recommended, but may be required should there be insufficient usable material: however, use of such plug samples must

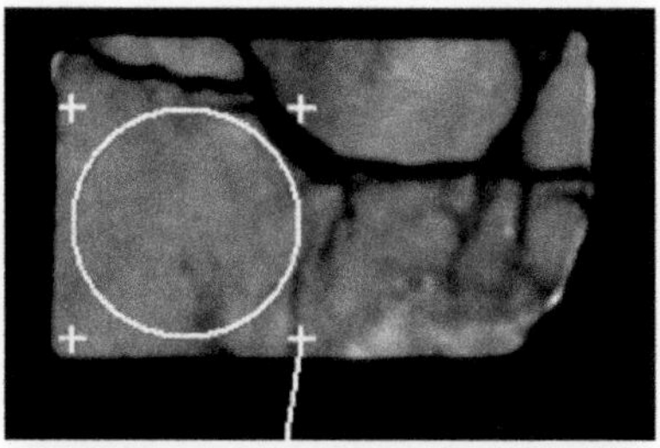

Fractures. The CT scan image shows that the core plug is generally massive but contains several fractures

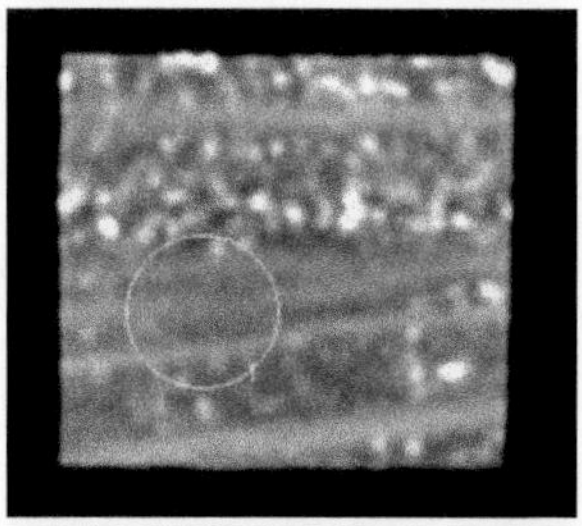

Split lithology. The CT scan image shows that the core plug has two lithologies; a parallel-laminated lower section and a generally massive upper section containing abundant small X-ray dense patches

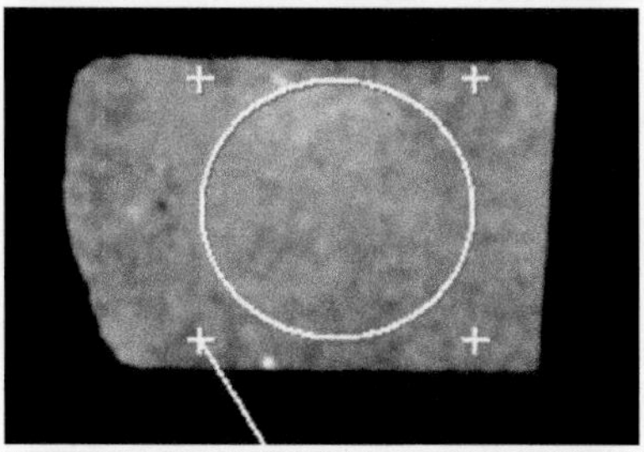

Generally massive. The CT scan image shows that the core plug is generally massive

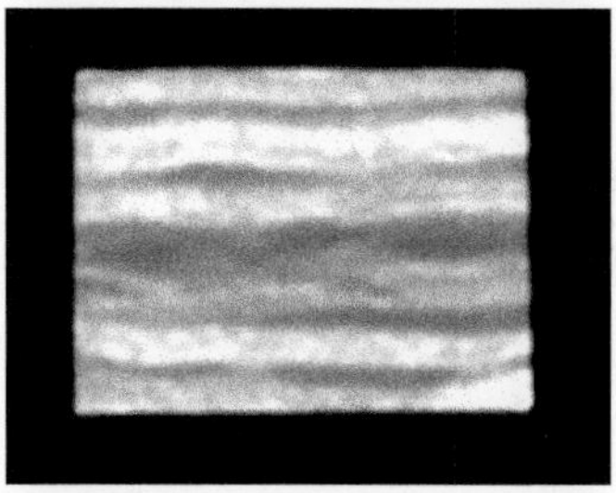

Thick laminations. The CT scan image shows thick X-ray dense continuous, parallel but slightly wavy laminations, running parallel relative to the inlet and outlet faces

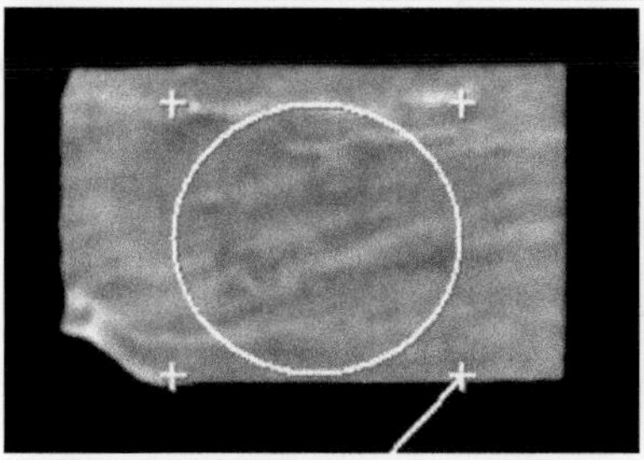

Faint, discontinuous laminations. The CT scan image shows faint, wispy, sub-parallel, discontinuous laminations running slightly oblique relative to the inlet and outlet faces

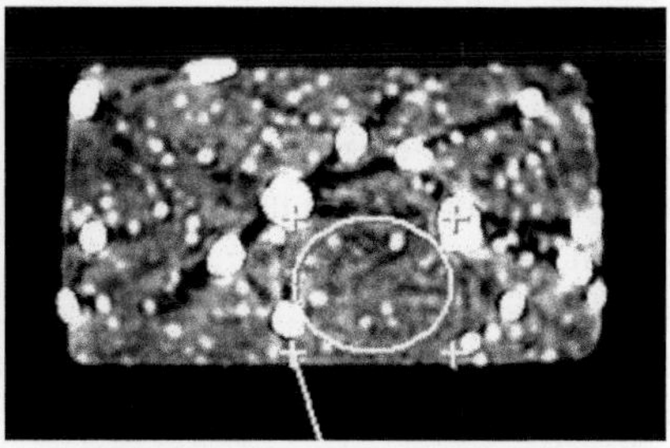

Extremely dense patches & beam hardening (not fractures). The CT scan image shows that the core plug is generally massive but contains abundant X-ray dense patches which are causing severe 'cross-shaped' beam hardening

FIGURE 6.7 Example of plug CT scans with a brief description.

consider the benefits of obtaining data and consequent costs involved, versus the risk that the data may be defective. Grades 4 and 5 denote plug samples that are either too heterogeneous (unsuitable for use) or in some way damaged (unable to use), respectively. Plugs which should not be used (grade 4 or 5) are shown shaded in dark-grey and light font and those shaded in light grey

TABLE 6.6 Example of Plug Selection Chart After CT Scanning

Sample ID	Core Depth (m)	Received Plug State	Graded Assessment	Comments
1	1081.02	Unclean	3	Some heterogeneity
2	1081.07	Unclean	3	Some heterogeneity
3	1081.44	Unclean	4	Non-horizontal laminae
4	1081.49	Unclean	4	Non-horizontal laminae
5	1082.04	Unclean	3	Some heterogeneity
6	1082.08	Unclean	1	OK
7	1082.13	Unclean	1	OK
8	1082.18	Unclean	1	OK
9	1082.42	Unclean	4	Non-horizontal laminae
10	1085.05	Unclean	5	Fractured
11	1085.12	Unclean	4	Laminated, high-density nodules
12	1088.23	Unclean	2	Fairly homogeneous, OK
13	1088.23	Unclean	4	Heterogeneous
14	1088.28	Unclean	3	Some heterogeneity
15	1088.28	Unclean	2	Fairly homogeneous, OK
16	1088.33	Unclean	2	Fairly homogeneous, OK
17	1088.33	Unclean	5	Damaged

should only be used with caution (grade 3—pending plug inspection and petrophysical measurements).

Once the final plug selection process had been completed, the plugs test should also be photographed pre- and post-testing. Fig. 6.8 provides a typical presentation from a SCAL report. In this case, a plug history chart with CT scans and white light digital images is provided for each SCAL plug so that the end user can evaluate the SCAL results in the context of the plug condition, lithology, homogeneity and shape. For example, Fig. 6.9 shows plug photographs and CT scans which exhibit sub-parallel bedding and may be subject to permeability anisotropy.

6.3.3 Petrophysical Properties

The SCAL sample set must reflect the full range of rock types and poroperm properties in the reservoir and should cover all the rock types comprising the significant reservoir flow units. Consequently, the final SCAL sample set should not be selected until the results of the RCA programme (and any petrophysical interpretation) become available.

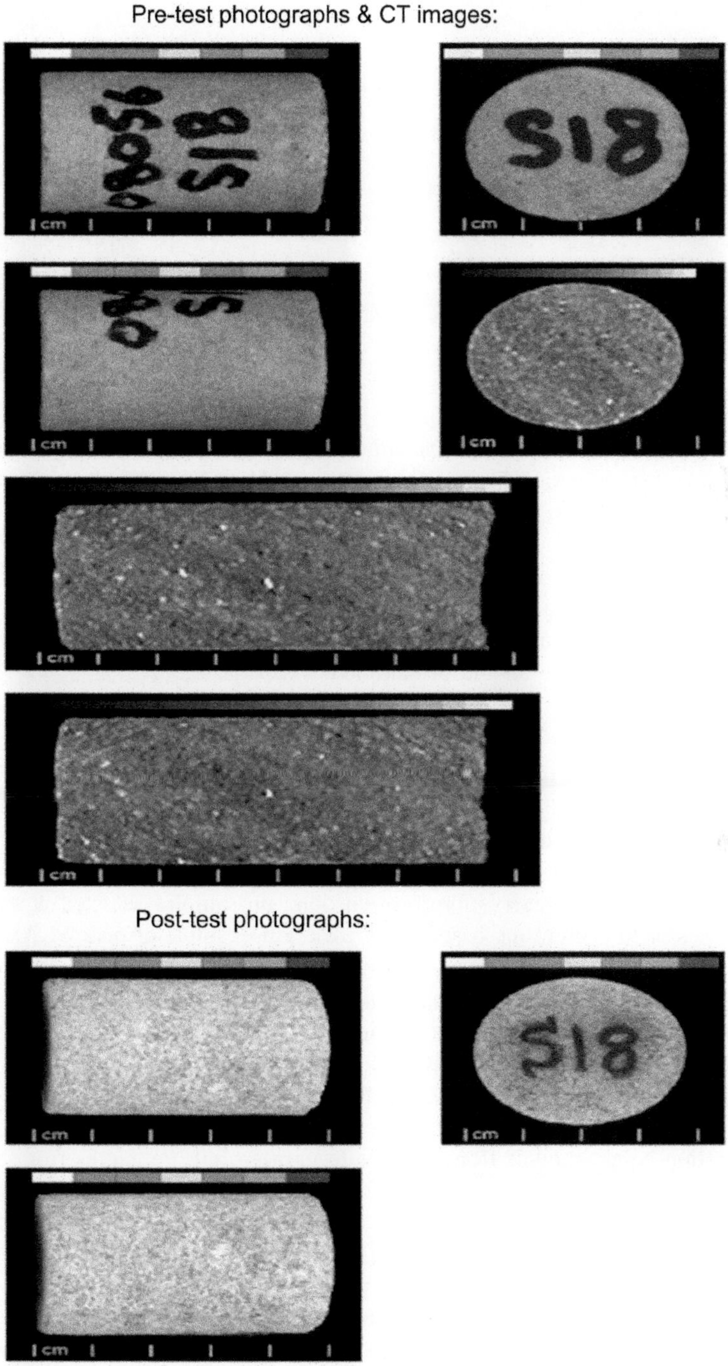

FIGURE 6.8 Example of plug photography presentation. *Courtesy of Woodside Energy Limited.*

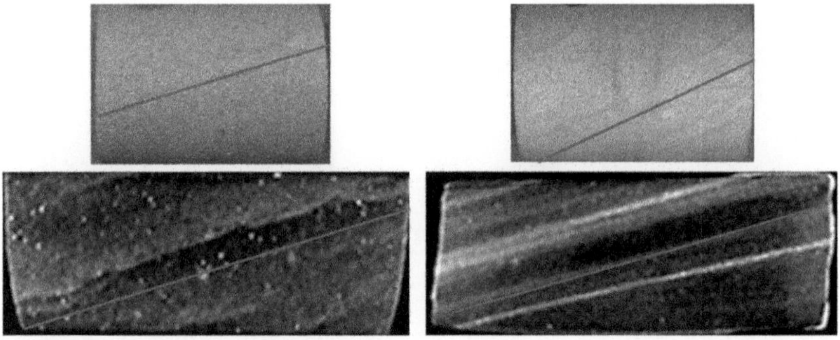

FIGURE 6.9 Example of unsuitable plugs.

Normally, 10 samples per test suite are considered to be sufficient to provide a statistically significant SCAL data set. However, the number of samples will depend on heterogeneity in the reservoir, and samples should be grouped by zone, facies, porosity class or hydraulic flow units. Normally, the range of core porosities and permeabilities used in SCAL tests should reflect the porosity and permeability distribution of the RCA database. This is especially important for electrical properties and capillary pressure, where the entire net interval should be sampled.

6.3.3.1 Preserved Sample Selection Bias

Despite adopting a rigorous approach to sample selection, bias can be introduced by core availability and preserved sample selection. For example, if tests are to be carried out on 'fresh-state' core, or where it can be shown that de-preservation may alter key rock properties, samples are selected from preserved core. Core preservation is often done on samples selected at regular intervals or to capture the best quality rock. An example of poor SCAL sample selection (in this case for capillary pressure) is shown in Fig. 6.10 where the poroperm relationships of the RCA plugs and the proposed SCAL samples (green (grey colour in the print version) dots) are shown. The net cut-off is 10% porosity so all the cored interval is net, yet the SCAL sampling is biased towards the best quality rock. Even the lowest porosity SCAL sample is an outlier. Poor or biased sample selection is one of the reasons why saturation-height models from core Pc tests often fail. Pc curves from better quality samples must be extrapolated to poorer intervals with no constraint.

In the example shown in Fig. 6.11, sample selection covers a range of permeability and porosity but samples below 12% porosity/2 mD are excluded as they are considered to be non-net.

It is essential to ensure that selected whole-core sections for preservation are representative and not biased. Despite being taken at regular intervals, they are selected before the RCA data become available and therefore may

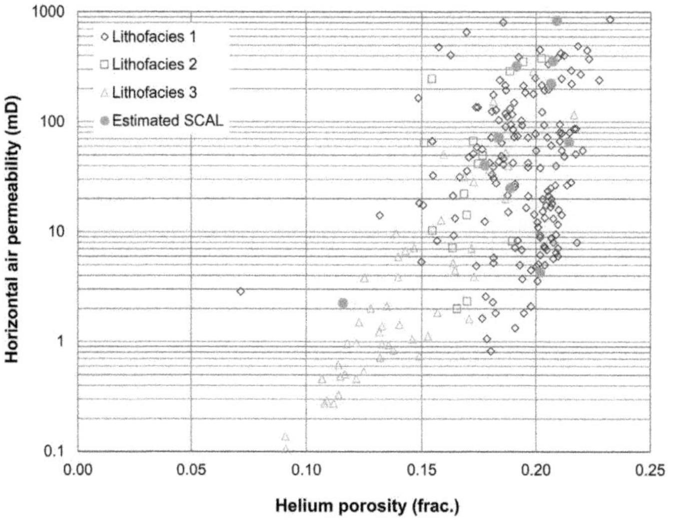

FIGURE 6.10 Example of poor SCAL sample selection (green (grey in the print version) dots are SCAL samples).

not be representative of the poroperm properties of the principal lithologies in the cored intervals. Consequently, it is recommended that parallel SCAL plugs are cut and preserved at every two or three RCA plug locations. This provides a better chance of obtaining representative samples than relying on whole-core preservation alonc.

In addition, core recovery is often poor in high-permeability zones (especially in unconsolidated sands) so may not be sampled by RCA or SCAL. Samples displaying heterogeneity are best avoided, since data analysis is ultimately dependent on a 1D flow regime and upscaling of results may not be particularly reliable.

6.3.3.2 Sample Preservation and Size

Sample preservation prevents fluid loss, wettability alteration and rock deterioration during storage. However, some SCAL tests can also be performed on samples which have been used in the RCA programme provided these are suitable, and neither the preparation nor test method has damaged the sample properties.

The samples themselves should be homogeneous, although they need not be perfect right cylindrical plugs. 1.5″ diameter plugs are preferred over 1″ samples as the larger the sample, the larger the potential pore volume, and the less will be the impact of saturation measurement errors. It should be noted that automatic mercury injection capillary pressure apparatus, which are designed for pore size distribution measurement, is often used for capillary pressure tests. The sample size capacity of these instruments is small (of the

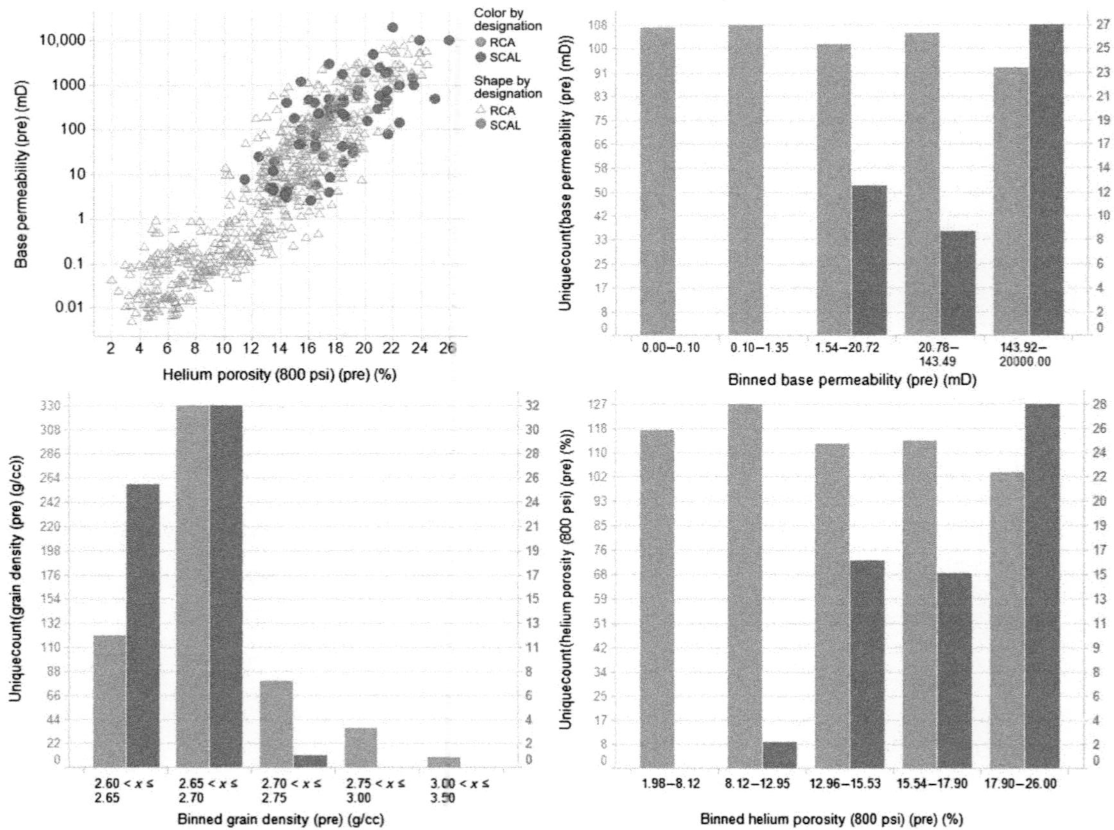

FIGURE 6.11 Example of SCAL sample selection.

order of 8 cc or less), and saturation estimates of these low pore volume samples can be subject to considerable error.

In addition, offcuts or end trims from the selected samples should be characterised for pore size distribution (mercury injection test) and for petrography/mineralogy (thin section, SEM and XRD).

6.3.3.3 Rock Quality Index

One useful tool in selection of SCAL samples, or assessing their representativeness, is by means of their rock quality index (RQI) and flow zone indicator (FZI) distribution (Amaefule et al, 1988).

FZI is a unique parameter that incorporates the geological attributes of texture and mineralogy in the discrimination of distinct pore geometrical hydraulic units. Capillary pressure, electrical properties and relative permeabilities are often clearly distinguishable on the basis of defined ranges of FZI values (hydraulic units).

FZI values are calculated from:

$$\mathrm{RQI} = 0.0314\sqrt{\frac{k}{\varphi}}$$

$$\varphi_z = \frac{\varphi_e}{(1-\varphi_e)}$$

$$\mathrm{FZI} = \frac{\mathrm{RQI}}{\varphi_z}$$

where k is the horizontal air (or Klinkenberg) permeability, mD, φ_e is the core helium porosity, fraction and RQI is the Rock (Reservoir) Quality Index, μm.

On a log–log plot of RQI versus φ_z, core samples with similar hydraulic characteristics and hence similar petrophysical properties will lie on a unit slope straight line. The intercept of the unit slope with the y-axis at φ_z equal to one defines the FZI for that group of samples. As discussed previously, if the SCAL data set is to be as descriptive of the petrophysical properties of the reservoir, the distribution of FZI values for the Pc plugs should be similar to the net intervals from the RCA data set, within core availability constraints.

Figure 6.12 plots the permeability–porosity relationships for a North Sea field based on RCA data. The corresponding RQI–φ_z relationships are shown in Fig. 6.13. The majority of the data (apart from data from one well below the oil–water contact) fall within a narrow range of FZI values, so the majority of the SCAL samples should be selected on the basis of FZI values between 7 and 10 μm.

6.3.3.4 Dynamic Test Sample Selection Procedures

Mohammed and Corbett (2003) studied the number of relative permeability curves required for a specific reservoir, based on geological analysis, rock typing, heterogeneity analysis and flow simulation, which they considered

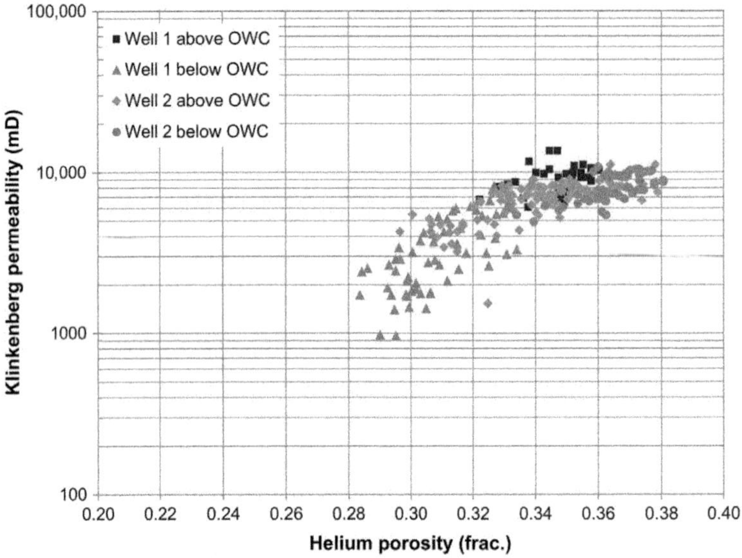

FIGURE 6.12 Porosity–permeability distributions North Sea reservoir.

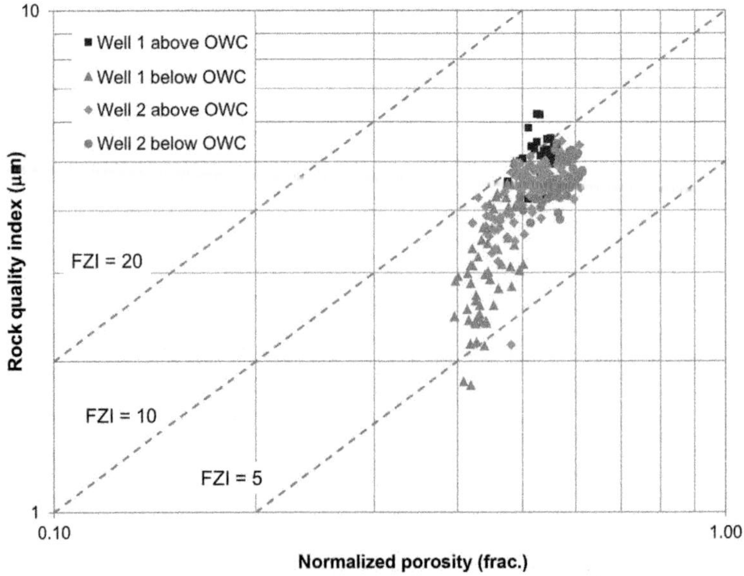

FIGURE 6.13 RQI-normalised porosity distributions for North Sea reservoir.

most appropriately represents a framework for the selection of the appropriate material for relative permeability tests. The selection of relative permeability curves from the more important rock types, defined as those making up the dominant transmissive and storage units in the reservoir, can be sufficient.

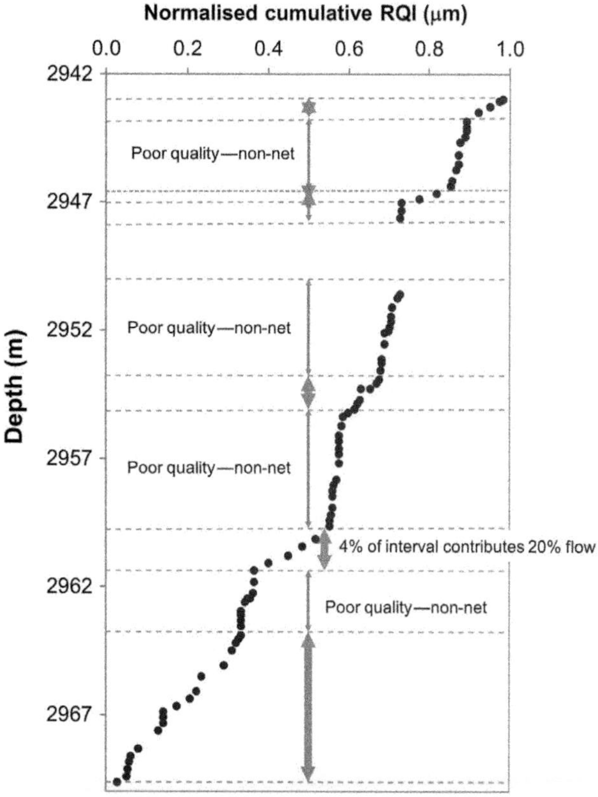

FIGURE 6.14 Example of normalised cumulative RQI plot.

A simpler approach is to plot normalised cumulative RQI data from core in a well. This might also include probe permeameter measurements if available. Both RQI and cumulative RQI are calculated for each core point and normalised by dividing the RQI at each point by the cumulative RQI for the section of interest, i.e., $RQI/\sum(RQI)$. This is effectively a pseudo-PLT, as indicated in Fig. 6.14 where the green (grey in the print version) arrows pick out the units which will provide the majority of the contribution to flow. In addition, to ensuring that representative rock types are employed in testing, the authors recommend that critical velocity and water sensitivity analyses are performed - to assess the impact of water on the core samples. These can be performed on lower grade samples that would not normally be selected for dynamic analyses.

Also, miscible dispersion analyses are recommended on all selected core plug samples, in order to assess the homogeneity of the selected samples during a displacement process. Samples requiring more than 3 pore volumes injected to fully displace are not recommended for relative permeability.

6.4 RESERVOIR STRESS ESTIMATION

The acquisition of SCAL data at representative reservoir conditions firstly demands that the tests are carried out in the laboratory under a net confining stress that is representative of the reservoir effective stress conditions. Whilst applying this effective confining stress, consideration must be made for corrections that are necessary owing to additional surface volumes created between the confining equipment and the plug sample (for example, Nieto et al., 1990)

6.4.1 Test Cell Design

There are various types of laboratory overburden test systems that can be used to determine SCAL properties at overburden stress.

In general, the test sample is enclosed in a rubber sleeve within a pressure vessel (coreholder) (Fig. 6.15) which may be constructed from steel, aluminium or carbon composite. The test plug is contacted on the end faces by end platens fitted with inlet and outlet fluid lines to allow pore fluid into and out of the sample. Many systems incorporate electrical end platens so that electrical properties such as formation factor at overburden can be determined in the same experiment, or additional pressure tappings on the platens.

A confining pump is used to pressurise the sample in the rubber sleeve with hydraulic oil or water. The cell is designed such that the pressure acting radially around the sample is equal to the pressure acting on the end faces of the sample (hydrostatic or isostatic stress). Pore fluid is usually maintained at ambient pressure, although some systems allow for both fluid and confining pressures to be controlled independently of one another.

6.4.2 Effective Stress Concepts

In SCAL testing, the tests are normally performed at an effective stress that simulates the effective reservoir stress. The effective stress (σ') is simply the total stress (σ) minus the pore pressure (p_p)

$$\sigma' = \sigma - \alpha p_p$$

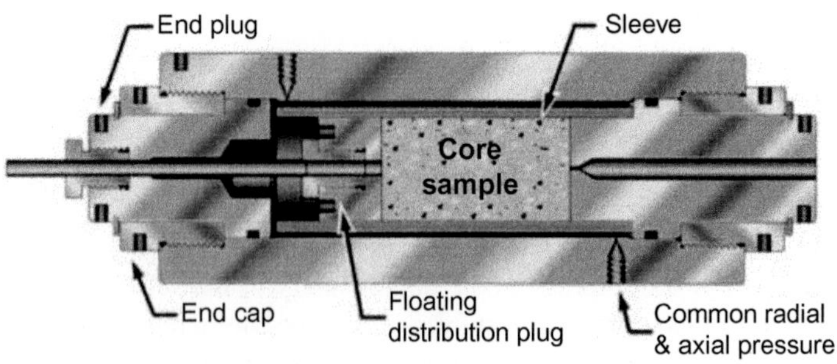

FIGURE 6.15 Example isostatic test cell. *Courtesy of Core Laboratories and its Petroleum Services and Core Lab Instruments divisions. All Other Rights Reserved.*

The α multiplier to pore pressure is often referred to as Biot's poro-elastic constant which is a parameter which describes the 'efficiency' of the fluid pressure in counteracting the total stress. Its value is dependent upon the pore geometry and the physical constituents of the solid system, thus:

$$\alpha = 1 - \frac{K_b}{K_{ma}}$$

where K_b is the rock bulk modulus ($=\frac{1}{\beta}$), K_{ma} is the matrix (skeletal) bulk modulus ($=\frac{1}{\beta_s}$) for which β is the bulk (dry) rock compressibility and β_s is the compressibility of individual grains (matrix).

The larger the value of β, the easier it is to compress the rock. When the compressibility of the dry rock is much greater than the compressibility of the solid grains ($\beta \gg \beta_s$), which is normally the case for sedimentary reservoir rocks, then α approaches 1. In the case of very low porosity rocks (such as igneous rocks), the values of β and β_s are very close, thus α approaches zero, which means that pore pressure has minimal effects on the stress carried by the rock matrix.

Values of α of between 0.5 and 1 have been measured for a range of reservoir rocks, although values of between 0.9 and 1.0 are normally assumed for SCAL testing. In the absence of any data, α is normally taken as 1. In practice α is difficult to measure experimentally but the rock and matrix bulk moduli or compressibilities can be derived from DSI log data:

$$K_b = 1.34 \times 10^{10} \cdot \rho_b \left(\frac{1}{\Delta t_c^2} - \frac{4}{3\Delta t_s^2} \right) K_{ma} = 1.34 \times 10^{10} \cdot \rho_{ma} \left(\frac{1}{\Delta t_c^2} - \frac{4}{3\Delta t_s^2} \right)$$

where ρ_b is the rock bulk density (g/cm^3) and ρ_{ma} is the rock matrix (grain) density (g/cm^3), but these must be scaled to static values from rock mechanics tests (Chapter 12).

Typical matrix bulk moduli data are:

- Sandstones: 37.5 GPa
- Limestones: 70.1 GPa
- Dolomites: 83.0 GPa

6.4.3 Net Confining Stress Loading Conditions

On stress release on coring, microcracks can develop in the core fabric. These can be closed by reimposition of a net confining stress. In the SCAL test therefore, the effective stress is increased simply by increasing the total confining stress since the pore pressure is normally maintained at ambient conditions. Alternatively, both the pore pressure and confining stress can be raised to the initial reservoir lithostatic and pore pressure values. The confining stress is held constant and the pore pressure is allowed to reduce to simulate the increase in effective pressure. This simulates reservoir depletion and is the method recommended to determine pore volume compressibility, porosity and permeability under simulated depletion conditions. This is discussed further in Chapter 12. For SCAL measurements where the net confining stress is simply reimposed, the former method (effective stress method) is preferred.

There are two principal methodologies which are used to determine the equivalent reservoir stress in SCAL testing:

- Net overburden stress method
- Equivalent effective isostatic stress method

6.4.3.1 Net Overburden Stress Method

The basis of this method attempts to correct for or compensate for the different loading mechanisms between the lab (isostatic loading) and the reservoir (uniaxial loading).

Geertsma (1957) considered that the reservoir may be often represented as being laid down between very stiff lateral boundaries. In this case, the vertical movement of the reservoir, as a response to lithostatic loading, is much greater than its lateral movement. As a result, any strain will be entirely vertical and the lateral strain will be zero. This is referred to as a uniaxial loading condition because strain is only in one direction (Fig. 6.16).

Biaxial systems (mistakenly called 'triaxial') allow independent application of axial and radial loading to the plug (simulating different vertical and horizontal stress regimes). These cells (instrumented for strain measurements) are normally only available in rock mechanics test labs. In the SCAL lab, the sample is loaded isostatically so strains are in both horizontal (radial) and vertical (axial) directions. Consequently, if isostatic coreholders are used to replicate a uniaxial stress regime, corrections have to be made to convert compressibilities measured under isostatic loading to a uniaxial loading (reservoir) conditions.

In the method developed by Teeuw (1971), the SCAL measurement (porosity in this case) is made at the same effective vertical stress as in the reservoir:

$$\sigma'_{\mathrm{NOB}} = \sigma_{\mathrm{v}} - \alpha p_{\mathrm{p}}$$

where σ'_{NOB} is the net (effective) overburden stress, σ_{v} is the total vertical stress, α is the Biot's poro-elastic factor and p_{p} is the initial reservoir pressure.

Porosity data are acquired at ambient conditions and at the net vertical (overburden) stress under isostatic loading. The stressed porosity measurements (at the selected net overburden stress) are converted to uniaxial-equivalent conditions, to account for differences between isostatic laboratory and uniaxial reservoir loading conditions, using:

$$\varphi_{\mathrm{u}} = \frac{\varphi_0(1-\varphi_{\mathrm{i}}) - A(\varphi_0 - \varphi_{\mathrm{i}})}{(1-\varphi_{\mathrm{i}}) - A(\varphi_0 - \varphi_{\mathrm{i}})} \quad A = \frac{(1+v)}{3(1-v)}$$

where φ_{u} is the porosity under uniaxial stress, φ_0 is the porosity under unstressed (ambient) conditions, φ_{i} is porosity under isostatic stress ($\sigma' = \sigma - \alpha p_{\mathrm{p}}$) and v is Poisson's ratio (if $v = 0.30$, $A = 0.62$).

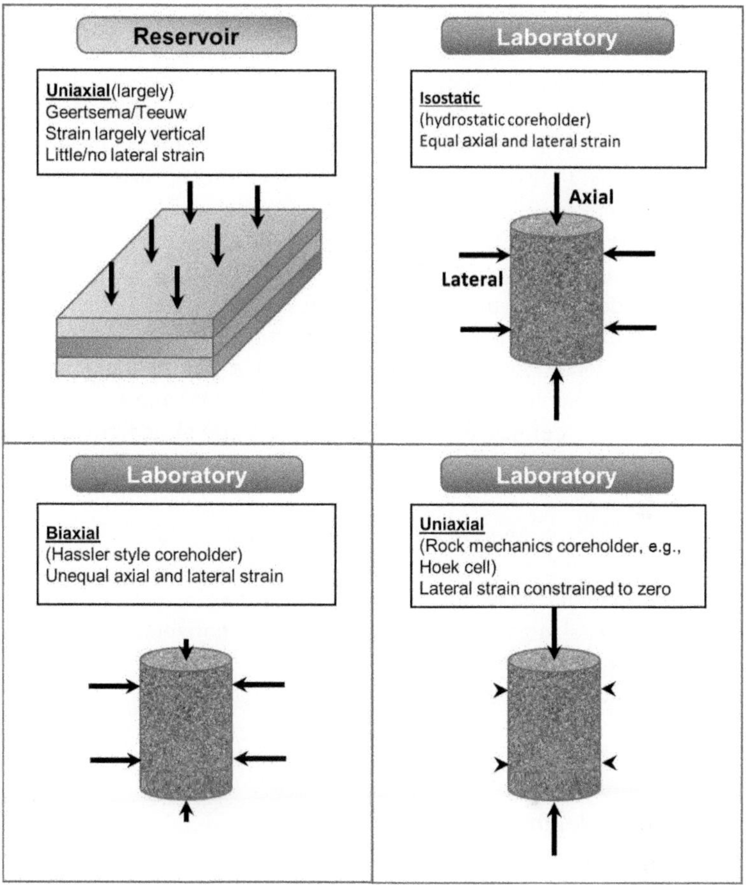

FIGURE 6.16 Reservoir and lab loading.

The uniaxial correction protocol assumes that only vertical (not lateral) strain occurs during sediment burial, and that the two horizontal stresses (σ_H and σ_h) are equal in magnitude. This implies a reservoir whose height (thickness) is much less than its width and length and has stiff lateral restraints. The uniaxial horizontal stress condition is only valid if there is no imposed tectonic stress. In tectonically active basins, the horizontal stress component is increased, horizontal stress anisotropy is manifest (often signalled by borehole breakouts) and strain is not entirely vertical.

6.4.3.2 *Equivalent Effective Isostatic Stress*

Teeuw's method was specifically derived to determine pore volume compressibility under simulated depletion conditions—not just to reimpose a representative overburden stress. Unfortunately, many laboratories impose the effective vertical stress and determine porosity at this stress without making the subsequent correction to uniaxial conditions. As the method ignores the influence of the two horizontal stresses, the imposed confining stress is higher than methods which include the horizontal stresses in the stress calculations. It also assumes that Poisson's ratio is constant under loading which it is not.

Worthington et al. (1997) and Nieto et al. (1994) comment that the uniaxial correction method is valid for linear-elastic conditions only—a requirement that is defied by most porosity–stress curves—and state that the uniaxial correction factor is not valid for correcting core porosity from ambient to reservoir conditions. Holt et al. (2001) assert that the overburden corrected porosity needs to be evaluated by reloading the core to the effective mean *in situ* stress defined by the mean of the vertical and two horizontal effective reservoir stresses. That is:

$$\sigma'_{\text{iso}} = \left(\frac{\sigma_{\text{v}} + \sigma_{\text{H}} + \sigma_{\text{h}}}{3}\right) - \alpha p_{\text{p}}$$

where σ'_{iso} is the equivalent effective isostatic reservoir stress, σ_{H} is the total maximum horizontal stress, σ_{h} is the total minimum horizontal stress and p_{p} is the initial reservoir pressure.

This is the preferred methodology for SCAL testing at overburden.

6.4.4 Reservoir Stress Data Sources

6.4.4.1 Stress Regime

The three principal (far field) reservoir stresses are:

- Vertical stress: σ_{v}
- Maximum horizontal stress, σ_{H}
- Minimum horizontal stress, σ_{h}

The magnitude of, and the relationship between, these stresses depends on the local stress regime (Fig. 6.17).

In a normal stress regime, the greatest principal stress, σ_1, is vertical, while the least principal stress, σ_3, is horizontal and will be aligned 90° to the fault plane strike. The maximum horizontal stress (σ_2) is aligned parallel to the strike. The stress magnitudes for a normal fault regime are:

$$\sigma_{\text{v}} > \sigma_{\text{H}} > \sigma_{\text{h}}$$

Normal stress regimes are associated with passive basins with extensional faults. Where tectonic activity has been or is active, then the horizontal stress components can be increased significantly and can exceed the vertical stress.

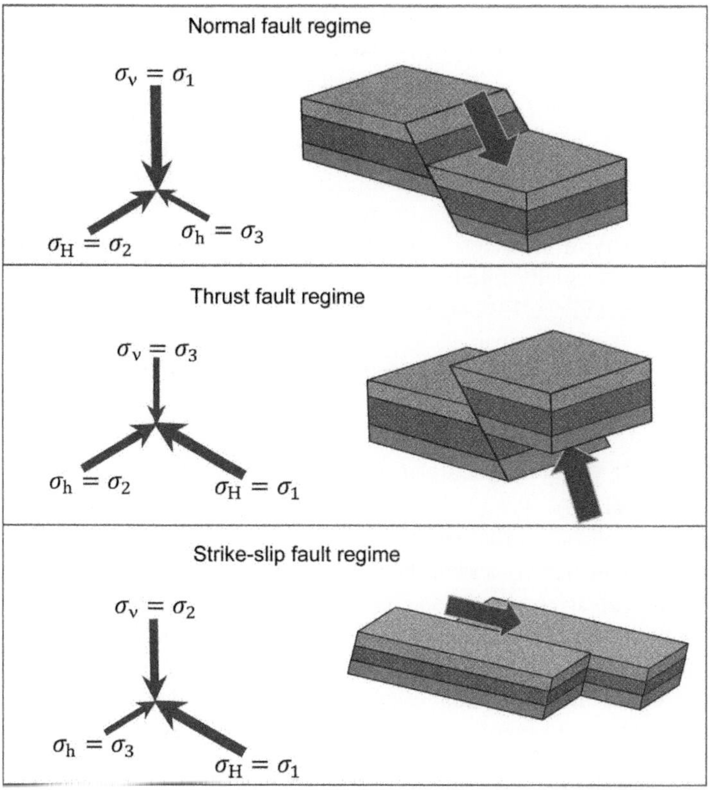

FIGURE 6.17 States of stress and fault regimes.

In regions where reverse faulting (or thrust faulting) is taking place, the least principal stress is vertical, while the greatest stress is horizontal and oriented perpendicular to the strike of the fault plane. That is:

$$\sigma_H > \sigma_h > \sigma_v$$

In the case of transcurrent (strike-slip) faulting, both the greatest and least principal stresses are horizontal, with the intermediate stress, σ_2, equal to the vertical stress. The stress magnitudes for a strike-slip fault regime are:

$$\sigma_H > \sigma_v > \sigma_h$$

6.4.4.2 *Total Vertical Stress*

In normally stressed reservoirs, the total vertical stress is assumed to be equivalent to the maximum principal *in situ* stress (σ_1) and is determined from density logs over the entire overburden and reservoir sections.

Density logs are integrated to determine total vertical stress at any depth; thus:

$$\sigma_v = \sum(\rho_{bi} \cdot h_i \cdot 0.4335)$$

where ρ_{bi} is the formation bulk density obtained from RHOB data (g/cc) and h_i is the vertical thickness of each formation, including seawater depth (ft.)—0.4335 is a unit conversion factor from g ft./cc to psi.

The density of seawater is assumed in the range from 1.01 to 1.03 g/cc.

A model is usually fitted to the density data from wells with the best density coverage (Fig. 6.18) and is integrated to calculate the overburden stress gradient.

Vertical stress models (e.g. Fig. 6.19) generally have the form:

$$\sigma_{v(grad)} = aZ^2 + bZ + c$$

where $\sigma_{v(grad)}$ is the total vertical stress gradient (psi/ft.) and Z is the vertical depth (ft. TVDSS).

Where shallow overburden density data are limited, sonic log data can be used to estimate density and/or the model can be extrapolated to determine the stress gradients to the mud line.

Normally, the mid-point reservoir depth or the OWC/GWC is chosen as the reservoir datum for stress calculations.

The common assumption of a vertical (overburden) stress gradient of 1 psi/ft. is not valid for shallow reservoirs or reservoirs in deep water fields.

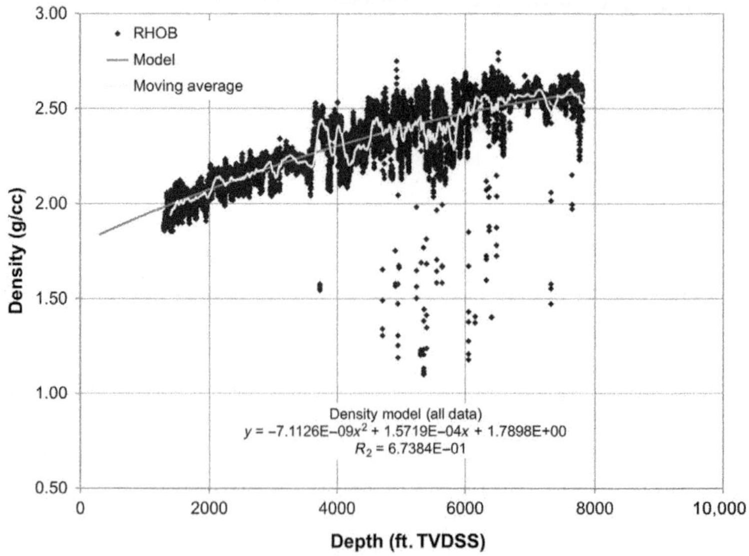

FIGURE 6.18 Density versus vertical depth example.

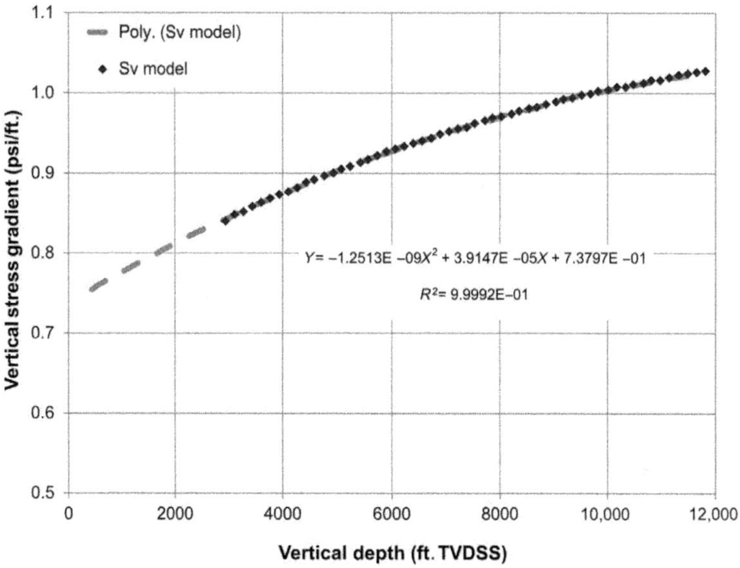

FIGURE 6.19 Vertical stress gradient model.

6.4.4.3 *Total Minimum Horizontal Stress*

The magnitude and orientation of the minimum horizontal stress can be determined from either theoretical analysis (elastic modelling), by direct measurements, or correlations.

Theoretical elastic models for horizontal stress estimation in passive basins are usually based on the assumptions that only vertical (not lateral) strain occurs during sediment burial (uniaxial condition), and that the two horizontal stresses (σ_H and σ_h) are equal in magnitude. This implies a reservoir whose height (thickness) is much less than its width and length and has stiff lateral restraints. For passive basins, the industry-standard (Anderson et al., 1973) elastic solution for minimum horizontal stress is:

$$\sigma_h = \frac{\nu}{1-\nu}\cdot\left(\sigma_v - \alpha p_p\right) + \alpha p_p$$

where σ_h is the total minimum horizontal stress, σ_v is the total vertical stress, ν is the Poisson's ratio and α is the Biot's poro-elastic factor.

In non-passive, tectonically active basins, the horizontal stress component is increased, horizontal stress anisotropy is manifest (often signalled by borehole breakouts) and strain is not entirely vertical. The increase in horizontal stress introduced by tectonic forces can be significant (e.g. Addis et al., 1996; Blanton and Olson, 1999).

Direct estimation methods include:

- Breakout analysis from multi-arm calliper logs—e.g. HDT and SHDT—and image log analysis—e.g. UBI and FMS. These can provide data on stress orientation but not magnitude but can be used in maximum horizontal stress estimation.
- Differential strain analysis tests on oriented core samples. These can provide both stress orientation and magnitude data, but the success rate of these tests can be limited.
- Hydraulic fracturing tests where the open-hole is isolated by a single or dual straddle packer (stress magnitude).
- Mini-fracs and/or step rate tests performed in cased-and-perforated hole (stress magnitude).
- Leak-off tests (LOTs) and extended LOTs in open-hole which, if reliably performed, provide data on the magnitude of the minimum horizontal stress.

Determination of horizontal stress from LOTs relies on the generation of macro-fractures in the rock surrounding the wellbore. The methods utilise the near-wellbore stress equations and the bottom-hole pressures at breakdown of the well, at fracture growth and at shut-in, to provide the relative magnitudes of the stresses acting in the cross-section of the wellbore. Since the bottom-hole pressure must be maintained slightly in excess of the *in situ* minimum stress in order for the fracture to propagate outwards from the well, the complete pressure history during the fracturing process (Fig. 6.20) enables

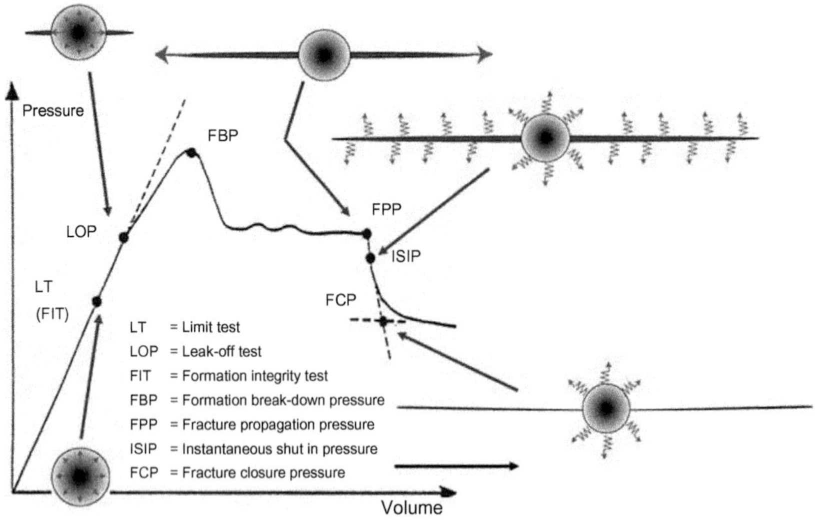

FIGURE 6.20 Idealised LOT chart. *After Gaarenstroom et al. (1993).*

the principal *in situ* stresses acting in the rock to be computed. The fracture propagation pressure, σ_3, is assumed to be equivalent to the minimum horizontal stress, σ_h.

LOT interpretation requires that the actual LOT pressure versus time (or volume) charts are reviewed to ensure that the frac extension pressure has been attained. Horizontal stress interpretation must also recognise that most LOTs are carried out in shales which are inelastic materials. Few tests are normally performed in sandstones, which are broadly elastic, and would be expected to yield different estimates of σ_h.

LOT interpretation is further muddied by the confusion in the minds of engineers when interpreting the meaning of a fracture gradient. To a production/fracturing engineer, the fracture gradient represents the pressure required to propagate a fracture a certain distance into the formation, after the formation has already broken down. For practical purposes, the fracture propagation pressure is equal to the total minimum *in situ* horizontal stress. To a drilling engineer, the fracture gradient is often taken to be the pressure at which the formation actually breaks down. This is a complex function of the stress distribution at the borehole wall, the permeability/condition of the mudcake and the pore pressure in the near-wellbore region. This pressure can be significantly higher than the minimum horizontal stress and, in some cases, breakdown is not actually achieved at all.

Most LOTs are actually FIT tests which mean that neither the formation breakdown pressure nor the fracture propagation pressure has been attained. The primary purpose of the FIT test is to evaluate the ability of the formation to withstand the pressure exerted by the drilling mud without fracturing the formation.

Most LOT and FIT data are, at best, measures of formation integrity which cannot be related to the *in situ* stress in any meaningful way. Extended LOTs which yield a clear fracture propagation pressure are required to derive horizontal stress magnitudes. Note that the fracture gradient must represent a maximum possible upper bound for minimum horizontal stress.

Using a large database of LOT and fracture test data, Breckels and Van Eekelen (1982) developed a relationship between σ_h (psi) and depth (Z, in ft. TVD) for data from the US Gulf Coast, Venezuela and the North Sea:

Normally pressured

$$\sigma_{hmin}(\text{psi}) = 0.197Z^{1.145}, \quad \text{for } Z < 11{,}500\,\text{ft.}$$

$$\sigma_{hmin}(\text{psi}) = 1.167Z^{-4596}, \quad \text{for } Z > 11{,}500\,\text{ft.}$$

Abnormally pressured

$$\sigma_{hmin}(\text{psi}) = 0.197Z^{1.145} + 0.46(p_p - p_{pn}), \quad \text{for } Z < 11{,}500\,\text{ft.}$$

$$\sigma_{hmin}(\text{psi}) = 1.167Z^{-4596} + 0.46(p_p - p_{pn}), \quad \text{for } Z > 11{,}500\,\text{ft.}$$

where p_p is the actual reservoir pore pressure and p_{pn} is the normal pore pressure corresponding to a gradient of 0.45 psi/ft.

6.4.4.4 Maximum Horizontal Stress

Maximum horizontal stress (σ_H) determination is problematic as there are no techniques available to measure this in the field. It is possible to obtain an estimate of maximum horizontal stress from modelling and back analysis of breakout in vertical wells, but these techniques rely on simplifying assumptions that may not be valid for a particular field or reservoir, a reliable estimate of rock failure characteristics (cohesive strength and friction angle) and a positive identification of breakout that should not be confused with other borehole failure features such as key seating.

Sedimentary rocks can exhibit some degree of anisotropy in their sonic response. This may arise from intrinsic structural effects, such as aligned fractures or layering, or from unequal stresses. Sonic waves travel fastest when the polarisation direction is aligned with the formation's stiffest direction. If the formation is anisotropic, shear wave polarisation aligned in the stiff direction will travel faster than one aligned in more compliant direction. As a result, the shear wave splits into two components: one polarised along the stiff (fast) direction and the other along the compliant (slow direction). With two orthogonal dipole transmitters and multiple receiver pairs aligned in orthogonal directions, shear sonic imaging tools can measure the components of shear slowness in any direction in a plane perpendicular to the borehole axis. Techniques have been developed to extract information on stress anisotropy and stress magnitudes from these "fast and slow shear sonic" data.

In the absence of any breakout in vertical wells, a broadly isotropic stress regime can be reasonably assumed, i.e.:

$$\sigma_H \sim 1.05 \text{ to } 1.10\sigma_h$$

If there is significant breakout in a vertical well, then an anisotropic horizontal stress regime exists. For this scenario in a normal stress regime, the maximum horizontal stress cannot exceed the vertical stress and can be estimated by

$$\sigma_H = \frac{\sigma_v + \sigma_h}{2}$$

which is considered a reasonable default for deeper, normally pressured reservoirs.

6.4.4.5 Pore Pressure

Pore pressures (for effective stress calculations) can be determined from wireline formation test or welltest data.

REFERENCES

Addis, M.A., Last, N.C., Yassir, N.A., 1996. Estimation of horizontal stresses at depth in faulted regions and their relationship to pore pressure variations. Soc. Pet. Eng. http://dx.doi.org/10.2118/28140-PA.

Amaefule, J.O., Kersey, D.G., Marshall, D.M., Powell, J.D., Valencia, L.E., Keelan, D.K., 1988. Reservoir description: A practical synergistic engineering and geological approach based on analysis of core data. Soc. Pet. Eng. http://dx.doi.org/10.2118/18167-MS.

Anderson, R.A., Ingram, D.S., Zanier, A.M., 1973. Determining fracture pressure gradients from well logs. Soc. Pet. Eng. http://dx.doi.org/10.2118/4135-PA.

Arps, J.J., 1953. The effect of temperature on the density and electrical resistivity of sodium chloride solutions. Soc. Pet. Eng. http://dx.doi.org/10.2118/953327-G.

Blanton, T.L., Olson, J.E., 1999. Stress magnitudes from logs: effects of tectonic strains and temperature. SPE Res. Eval. Eng. 2, 62–68.

Breckels, I.M., Van Eekelen, H.A.M., 1982. Relationship between horizontal stress and depth in sedimentary basins. Soc. Pet. Eng. JPT 34, 2191–2199.

Geertsma, J., 1957. The effect of fluid pressure decline on volumetric changes of porous rocks. Trans. AIME 210, 331–340.

Holt, R.M., Lehr, B.C., Kenter, C.J., Spits, P., 2001. In-situ porosity from cores: the rock mechanics approach to overburden correction. In: International Symposium of Special Core Analysts Edinburgh. SCA2001-09.

Mohammed, K., Corbett, P., 2003. How many relative permeability measurements do you need? A case study from a North African reservoir. Petrophysics 44 (4), 262–270.

Nieto, J.A., et al., 1990. Core compaction correction—a different approach. In: Proceedings of the First Society of Core Analysts European Symposium, London, pp. 21–23.

Nieto, J.A., Yale, D.P., Evans, R.J., 1994. Improved methods for correcting core porosity to reservoir conditions. Log Anal. 35, 21 30.

Teeuw, D., 1971. Prediction of formation compaction from laboratory compressibility data. Soc. Pet. Eng. J. 11, 263–271.

Worthington, P.F., Daines, J.M., Bratli, R.K., Nicolaysen, R., 1997. Comparative evaluation of the core compaction corrections for clastic reservoirs. Log Anal. 38 (5), 19–29.

Worthington, A.E., Hedges, J.H., Pallatt, N., 1990. SCA guidelines for sample preparation and porosity measurement of electrical resistivity samples. Log Anal. 31 (2), 20–28.

RECOMMENDED READING

Ahmed, U., et al., 1991. Enhanced in-situ stress profiling with microfracture, core and sonic logging data. SPE Form. Eval. 6, 243.

Archie, G.E., 1942. The electrical resistivity log as an aid in determining some reservoir characteristics. Trans. AIMI 146, 54–62.

Birch, F., 1966. Compressibility, elastic constants. In: Clark, S.P. (Ed.), Handbook of Physical Constants, In: Geological Society of America, Memoir 97, 97–173.

Gatens, J.M., et al., 1990. In-situ stress tests and acoustic logs determine mechanical properties in the Devonian Shales. SPE Form. Eval. 5, 248–254.

Hillier, I., 2000. Prediction of rock elastic and strength properties to determine optimum mud weights and suitability of drilling methods in mechanically weak formations. In: Paper Presented at PESGB PETEX 2000 Conference, Nov., 2000.

Iverson, W.P., 1996. Log-derived stress in anisotropic formations. Log Anal. 37 (5), 33–40. Discussion in 1997 by K.W. Katahara: Log Anal. 38 (5), 37–38.

Kaye, L.N., Hacksma, J.D., Knudson, S., Smith, J.E., 1990. Special core analysis loading conditions for a friable stress sensitive sandstone. In: Worthington, P.F. (Ed.), Advances in Core Evaluation: Accuracy and Precision in Reserves Estimation. Gordon and Breach Science Publishers, London, ISBN: 2-88124-766-0.

Klimentos, T., et al., 1996. Experimental determination of the Biot elastic constant: applications in formation evaluation (sonic porosity, rock strength, earth stresses, and sanding predictions). SPE Res. Eval. Eng. 1, 57–63.

Maute, R.E., et al., 1992. Improved data analysis method determines Archie parameters from core data. JPT, 103–107.

McPhee, C.A., 2012. The core analysis elephant in the formation evaluation room. In: SPE Annual Technical Conference and Exhibition, 8–10 October, San Antonio, Texas, USA. Soc. Pet. Eng., SPE Paper SPE-158087.

Nur, A., Byerlee, J.D., 1971. An exact effective stress law for elastic deformation of rock with fluids. J. Geophys. Res. 76, 6414–6419.

Sprunt, E.S., Maute, R.E., Rackers, C.L., 1990. An Interpretation of the SCA Electrical Resistivity Survey. The Log Analyst, (March–April). 76–88.

Worthington, A.E., 1975. Errors in the laboratory measurement of formation resistivity factor. In: 16th Annual SPWLA Logging Symposium, 4–7 June, New Orleans, Louisiana.

Chapter 7

Wettability and Wettability Tests

Chapter Outline

7.1 Introduction **313**
7.2 Contact Angle Method **315**
7.2.1 Sample Preparation 316
7.2.2 Test Equipment 317
7.2.3 Test Procedures 319
7.2.4 Results 320
7.2.5 Data Reporting Requirements 320
7.2.6 Contact Angle Summary 320
7.3 Amott (Amott–Harvey) Method **322**
7.3.1 Sample Preparation 323
7.3.2 Test Conditions 325
7.3.3 Test Equipment 327
7.3.4 Test Procedures 327
7.3.5 Amott–Harvey Wettability Index Calculation 329
7.3.6 Data Reporting Requirements 330
7.3.7 Amott (Amott–Harvey) Summary 331
7.4 USBM Method **332**
7.4.1 Sample Preparation 334
7.4.2 Test Equipment 334
7.4.3 Key Processes 334
7.4.4 Test Procedures 334
7.4.5 USBM Index Calculation 336
7.4.6 Data Reporting Requirements 337
7.4.7 USBM Summary 338
7.5 Combined Amott–USBM (Combination) Method **339**
7.5.1 Sample Preparation 340
7.5.2 Test Equipment 340
7.5.3 Test Procedures 340
7.5.4 Data Reporting Requirements 341
7.5.5 USBM and Amott–Harvey Index Calculation 341
7.5.6 Combined Amott–USBM Summary 342
References **344**
Recommended Reading **345**

7.1 INTRODUCTION

Wettability is a fundamental property of the interactions between the reservoir rock components and the pore filling fluids. Wettability is a controlling factor on the location of the reservoir fluids within its porous structure and upon the flow and subsequent distribution of fluids during the dynamic process of production. It is therefore an essential factor during laboratory experimentation, since it will have a controlling influence on all laboratory tests

Developments in Petroleum Science, Vol. 64. http://dx.doi.org/10.1016/B978-0-444-63533-4.00007-X

attempting to imitate reservoir characteristics and processes, especially capillary pressure and relative permeability. It is necessary to attempt to ensure that core material has reservoir representative wettability during laboratory analyses; else the resultant data could be meaningless. The authors recommend reading the 6 part literature survey by Anderson (Anderson 1986a,b, c, 1987a,b,c) as a good introduction to the basics of wettability and its effect of core analysis.

This section discusses the methods available to measure wettability in the laboratory. However, since no well-established techniques exist at present for downhole measurement of the wettability of reservoir material, it is not possible to state categorically whether a certain wettability is representative in a given case. The wettability measured in the lab is the wettability of the sample at the time it is measured. Its wetting state may not be representative of the reservoir wettability condition, (Bobek et al., 1958) as alterations can occur:

- on contact with drilling mud (Jia et al., 1994; Sharma & Wunderlich, 1985);
- during pressure and temperature reductions on core recovery;
- on oxidation of the oil through lack of appropriate core preservation;
- on core cleaning (Gant & Anderson, 1988);
- on ageing the core;
- during core testing.

All can alter the native wettability to different degrees. Wettability alteration is discussed in detail in Chapter 2.

In a rock/oil/brine system, many wetting states are thought to occur. Some of these are:

- water wet: where water coats the grains and may fully occupy the smaller pores, oil would occupy only the centre of the larger pores;
- oil wet: where the reverse of the above is the case, and oil coats the grains and occupies the smaller pore spaces;
- neutrally wet: where the system has a uniform non-preference for oil or water. Two sub-classes of "neutral" wettability are thought to exist:
  - fractional wettability: where certain areas of the rock (minerals) have a strong preference in one direction, while other areas (minerals) have a strong opposite preference (e.g. calcite-cemented quartz sandstone or clays lining smaller pores in a sandstone);
  - mixed wettability: a special case of fractional wettability, in which oil-wet surfaces form a continuous path through the larger pores although water continues to occupy the smaller pores. This wetting state is probably unproven, but does allow very low residual oil saturations, and some unusual relative permeability behaviour to be explained.

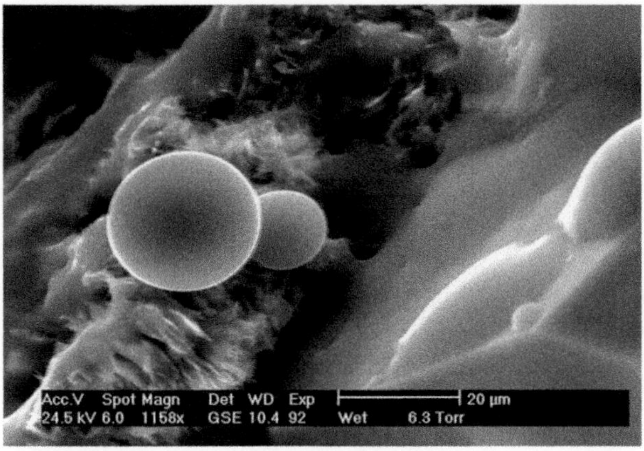

FIGURE 7.1 Environmental SEM of water droplets at Swir.

Both fractional and mixed wettability imply that wettability can vary at the pore (micron) scale. Figure 7.1 is an environmental SEM microphotograph of a sandstone core sample containing illitic pore lining clays which has been reduced to irreducible water saturation in the presence of oil. The water droplets on the clays (left) form a high contact angle with the illitic clays—implying oil wetting—while the contact angles on the quartz crystal (right) show a low contact angle—implying strong water wetting. This wettability in this sample can vary over a distance of ~20 μm.

There are a number of methods for quantitatively determining the wettability of the rock/fluid system. The most common methods are:

- contact angle;
- Amott method (often called Amott–Harvey);
- U.S. Bureau of Mines (USBM).

The contact angle method measures the wettability of a specific mineral surface and therefore may have little relevance to the reservoir formation, whereas the Amott and USBM methods measure the average (bulk) wettability of a core sample.

7.2 CONTACT ANGLE METHOD

The contact angle method has the distinct advantage over core tests that there is no possibility of mud filtrate components, surfactants or other compounds altering the wettability of the mineral surface. This method is also used to determine whether a crude oil can alter wettability and to examine the effects of temperature, pressure and brine chemistry on wettability. However, some difficulties are involved in applying contact angle methods to the reservoir situation.

The methods that are generally used in the oil industry are the sessile drop method and a modified form of the sessile method.

Purpose of test	• To obtain a measure of the wetting preference of a solid by a liquid
Sample requirements	• Two flat polished mineral surfaces of quartz (sandstone) or calcite (carbonate) crystals
Required input data for lab	• Synthetic formation water (SFW) composition and reservoir oil sample

7.2.1 Sample Preparation

The sessile drop method uses a single flat polished mineral surface (Fig. 7.2). The modified sessile drop method uses two flat polished mineral surfaces that are mounted parallel to one another on adjustable posts, as shown in Fig. 7.3. Quartz or calcite crystals are normally used to simulate sandstones and carbonates, respectively.

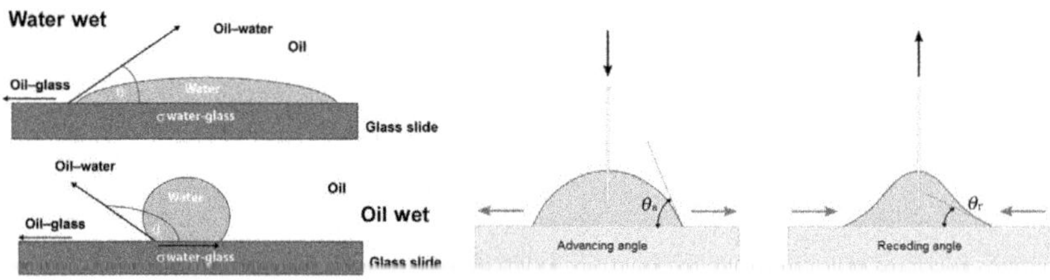

FIGURE 7.2 Sessile drop method. (Right: *https://commons.wikimedia.org/wiki/File%3ADynamic_contact_angle_measurement.svg*).

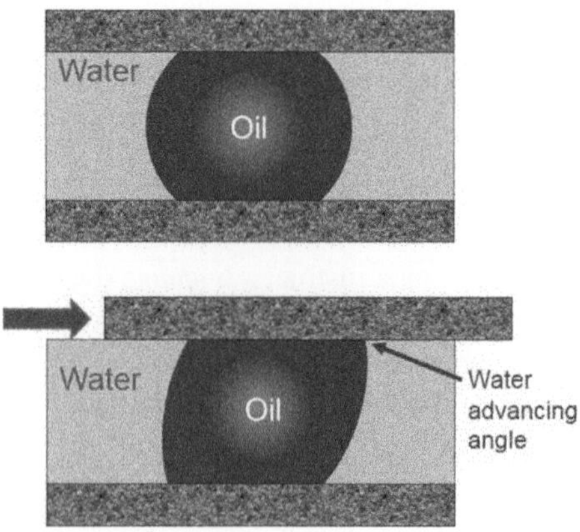

FIGURE 7.3 Modified sessile drop method.

7.2.2 Test Equipment

The sessile drop method is measured by a contact angle goniometer (Fig. 7.4) using an optical subsystem to capture the profile of a pure liquid on a solid substrate. Older systems used a microscope optical system with a back light. Current-generation systems employ high-resolution cameras and software to capture and analyse the contact angle (Fig. 7.5). The method uses mineral crystal surfaces that are mounted in a test cell composed entirely of inert materials to prevent contamination. The droplet is deposited by a syringe

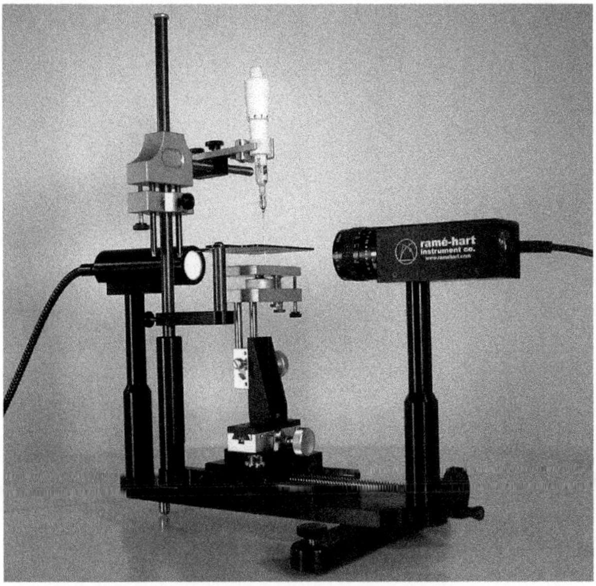

FIGURE 7.4 Contact angle goniometer. *By Ramehart (Own work) [CC BY 3.0 (http://creativecommons.org/licenses/by/3.0)], via Wikimedia Commons.*

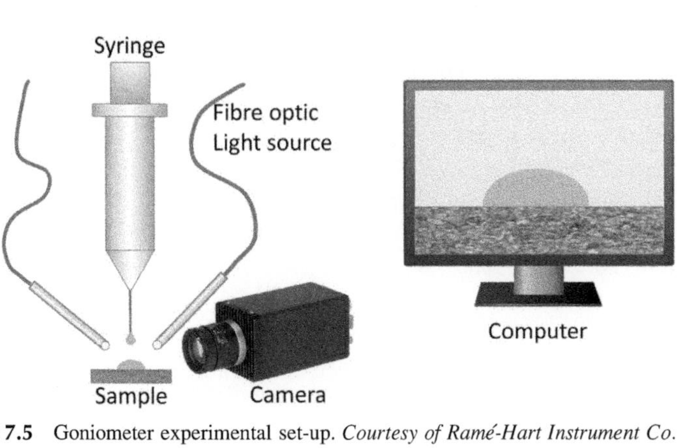

FIGURE 7.5 Goniometer experimental set-up. *Courtesy of Ramé-Hart Instrument Co.*

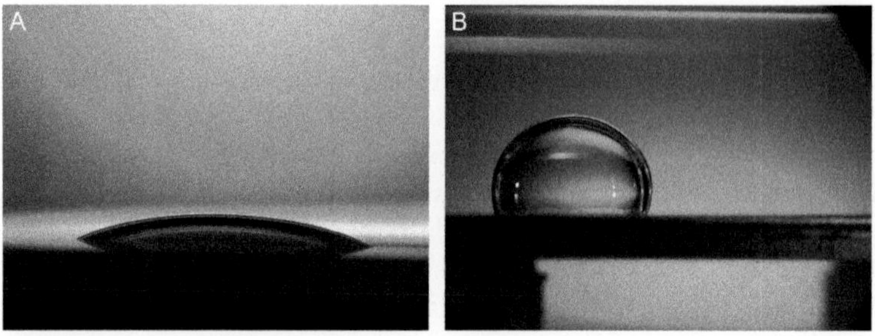

FIGURE 7.6 (A) "Water droplet in oil on glass surface" by Guro Aspenes, SINTEF Petroleum Research—Research at SINTEF Petroleum Research, Bergen, Norway. Licensed under CC BY-SA 3.0 via Wikimedia Commons—https://commons.wikimedia.org/wiki/File:Water_droplet_in_oil_on_glass_surface.JPG#/media/File:Water_droplet_in_oil_on_glass_surface.JPG. (B) "Water droplet in oil on brass surface" by Guro Aspenes, SINTEF Petroleum Research—Research at SINTEF Petroleum Research, Bergen, Norway. Licensed under CC BY-SA 3.0 via Wikimedia Commons—https://commons.wikimedia.org/wiki/File:Water_droplet_in_oil_on_brass_surface.JPG#/media/File:Water_droplet_in_oil_on_brass_surface.JPG.

pointed vertically down onto the sample surface, and a high-resolution camera captures the image (e.g. Fig. 7.6), which can then be analysed either by eye (with a protractor) or using image analysis software. The size of the droplet can be increased gradually so that it grows proportionally, and the contact angle remains congruent. By taking pictures incrementally as the droplet grows, the user can acquire a set of data to get a good average. If necessary, the receding contact angle can also be measured by depositing a droplet via syringe and recording images of the droplet being gradually sucked back.

The sessile and modified sessile drop methods allow determination of advancing and receding contact angles. In drainage, oil displaces water, whereas in imbibition water displaces oil. In general, these two displacement regimes result in different contact angles: *receding* contact angle, θ_r, in *drainage* and *advancing* contact angle, θ_a, in *imbibition*. Experiments have shown that advancing contact angle can be much larger than the receding contact angle (Ma et al., 1996; Morrow, 1975). This is partly due to domains on the crystal surface which present barriers to the motion of the contact line. For the case of chemical heterogeneity, these domains represent areas with different contact angles than the surrounding surface. For example, when wetting with water, hydrophobic domains will pin the motion of the contact line as the liquid advances, thus increasing the contact angles. When the water recedes, the hydrophilic domains will hold back the draining motion of the contact line, thus decreasing the contact angle. From this analysis, it can be seen that, when testing with water, advancing angles will be sensitive to the hydrophobic domains and receding angles will characterise the hydrophilic domains on the surface.

7.2.3 Test Procedures

The apparatus is thoroughly cleaned because even trace amounts of contaminants can alter the contact angle. Values reproducible to better than a few degrees are generally only obtained under laboratory conditions with purified liquids and very clean solid surfaces. The presence of an oxide layer, or contaminants, on the solid surface can significantly increase the contact angle.

The mineral surfaces are placed in a cell and this is then filled with deoxygenated synthetic formation brine.

In the sessile drop method, a droplet of crude oil is formed at the end of a fine capillary tube and brought into contact with the flat mineral surface (Fig. 7.2). The droplet is allowed to age on the surface. Advancing and receding contact angles can be measured by using the capillary tube to expand and contract the volume of the crude oil drop.

For the modified sessile method, an oil drop is then placed between the two crystals so that it contacts a large area of each crystal (Fig. 7.3). After the oil/crystal interface has been allowed to stabilise (aged) for 2–3 days, the two crystals are displaced parallel to one another. This shifts the oil droplet and allows the brine to move over a surface previously covered with oil. The contact angles measured with the goniometer in this fashion are termed water-advancing contact angles. A non-equilibrium angle is observed immediately after the drop is moved. This angle decreases for a day or two until a constant value is obtained.

When crude oil contains natural surfactants, the water-advancing contact angle increases as the oil/crystal interface ages, approaching a limiting value as adsorption equilibrium is reached. This may require hundreds or even thousands of hours of interface ageing time as shown in Fig. 7.7.

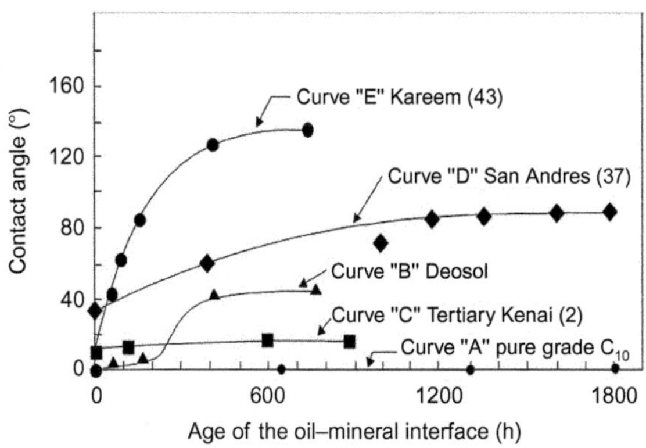

FIGURE 7.7 Contact angle equilibrium. *After Treiber et al. (1972).*

TABLE 7.1 Contact Angle Wettability Classification

	Water Wet	Neutrally Wet	Oil Wet
Minimum	0°	60–75°	105–120°
Maximum	60–75°	105–120°	180°

7.2.4 Results

The results of contact angle experiments are normally represented by an angle (θ). These can be related to wetting conditions as indicated in Table 7.1, providing cut-off angles similar to those described by Treiber et al., 1972.

7.2.5 Data Reporting Requirements

- brief description of procedures and experimental apparatus;
- mineral type (quartz, calcite or other);
- test temperature and pressure;
- oil API gravity and density at test conditions;
- water composition and density at test conditions;
- oil/water interfacial tension at test conditions (if available);
- measured contact angle (advancing and receding for modified sessile drop);
- contact angle as a function of time (age) to establish equilibrium (see Fig. 7.7).

7.2.6 Contact Angle Summary

Table 7.2 provides a summary of contact angle measurements: which fluid systems and lithologies are applicable, the advantages and disadvantages of measurements.

QC checks

- Analysis of the results should start with a thorough review of the laboratory report to check that acceptable procedures have been followed. This should be carefully evaluated.
- This method tests only the fluids and cannot reproduce the complexity of real formation pore systems. While the method is useful to establish ageing requirements, it should not be used for formation wettability estimates.

Suitable Laboratories

- Tests are most commonly done in PVT labs
- Some university petroleum engineering departments

TABLE 7.2 Summary of Contact Angle Measurements

Advantages	Drawbacks and issues
• It is a relatively low-cost technique. • It is the best wettability measurement method for pure fluids and pure rock crystals because there is no possibility of surfactants or other compounds altering the wettability. Therefore, the method is more suited to most research and development studies, as ideal conditions are required. • Method is useful to establish potential ageing times and ageing conditions for wettability restoration on clean, water-wet core samples.	• A significant length of time may be needed for a contact angle to achieve equilibrium if crude oil is used, due to the presence of surfactants (wettability altering compounds) in the crude oil. Early measurements can show that the system is water wet even though it may become oil wet at equilibrium (after a long time), as shown in Fig. 7.7. • The difference between the advancing and receding angle is contact angle hysteresis and can be greater than 60°. Hysteresis is caused by surface roughness, surface composition heterogeneity and surface immobility on a macromolecular scale. • The method cannot take into account the roughness, compositional heterogeneity and complex geometry of real reservoir rocks which can cause wettability variation at the pore scale. The test is performed on carefully prepared, pure mineral surfaces. For an idealised, clean, plane, non-deformable solid surface, there is only one equilibrium contact angle. In real situations, however, it is commonly found that for a given liquid–solid system, a number of deviations from idealised behaviour can be observed. These phenomena are important since they mean that the observed wetting behaviour can differ significantly from that which might be expected for a given liquid resting on an idealised solid. They also illustrate the practical uncertainty which should be attached to estimates of contact angles in a number of core analysis and reservoir engineering calculations. • Surface roughness of the test crystal can cause deviation from idealised behaviour. For all surfaces, the true contact angle at a liquid–solid interface is determined by the balance of surface energy terms. For roughened surfaces, however, the effect of surface roughness is to make the observed contact angle (which is measured relative to the average horizontal surface seen in a goniometer) different from the true contact angle (which is measured relative to the actual portion of the surface with which the liquid is in contact). Two effects of surface roughness are shown in Fig. 7.8. For the case where the true contact angle is less than 90°, the effect of surface roughness is to make the observed contact angle less than the "true" contact angle. Conversely, where the "true" contact angle is greater than 90°, the effect of surface roughness is to make the observed contact angle greater than the true contact angle. Anderson 1987a commented that a contact angle of 50° on flat surface translates to apparent angle of 0° on rough surface. • Even though it is possible, with great care, to get exact and reproducible contact angle measurements, these results are not representative of the wettability of reservoir core.

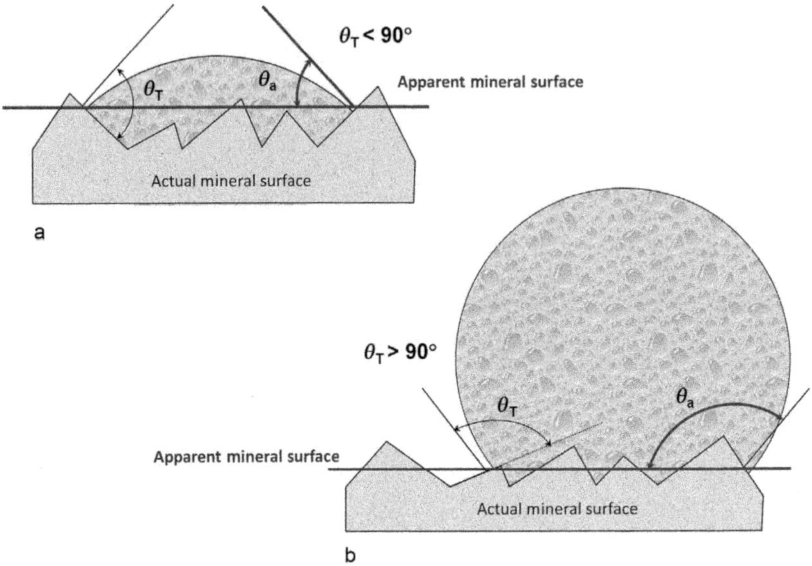

θ_T is the contact angle measured on a smooth, flat surface. (a) The droplet is the preferential wetting fluid. (b) The droplet us the non-wetting fluid

FIGURE 7.8 Surface roughness effects on measured contact angle. *After Popiel (1978).*

7.3 AMOTT (AMOTT–HARVEY) METHOD

The Amott, 1959 method combines imbibition (Amott cell) and forced displacement (centrifuge or coreholder) to measure the "average" (or bulk or complex) wettability of a core. Both reservoir samples and fluids can be used in the test. The method is based on the fact that the wetting fluid will generally imbibe spontaneously into the core, displacing the non-wetting fluid.

In strict terms, drainage refers to a reduction in the wetting phase saturation, whereas imbibition is the increase in the wetting phase saturation. Thus for a water-wet rock, drainage is water saturation decreasing and imbibition is water saturation increasing. However, for an oil-wet rock, drainage is oil saturation decreasing, while imbibition is oil saturation increasing. As this can be confusing, the accepted convention is that water saturation decreasing is always drainage and water saturation increasing is always imbibition, irrespective of wettability.

In the original Amott method, the plug saturated in oil is centrifuged under brine until residual oil saturation (Sro) is reached. In the modified method, known as Amott–Harvey (Boneau & Clampitt, 1977), the procedure has an additional step in the core preparation before the test is run: the core is centrifuged first under brine and then under oil to reduce the plug to initial water saturation (Swir). In conventional Amott tests, the forced imbibition stage is carried out using a centrifuge. In some Amott tests, forced imbibition is achieved by dynamic displacement in a sample coreholder. A centrifuge is preferred as the data can be also used to determine the wettability using the USBM methodology (Section 7.4; Table 7.3).

TABLE 7.3 Overview of Amott Wettability Method

Purpose of test	• To determine wettability preference of core plugs
Sample requirements	• Cylindrical shaped plug sample of 1″ or 1½″ diameter. The larger the sample, the larger the potential pore volume, and the less will be the impact of saturation measurement errors. • *Ideally*—cleaned (water-wet) sample, a sample taken from preserved core or a preserved plug, or plug after wettability restoration. *Alternative*—plug previously used in routine core analysis programme, provided that plug has not been pre-stressed and/or that plug preparation methods have not caused damage to delicate clay minerals, or have altered wettability.
Required input data for lab	• Formation water (FW) salinity and ionic composition, as close as practically possible to the formation water in the reservoir or FW resistivity at reservoir and lab temperature. Lab will prepare SFW accordingly. • Reservoir crude oil sample—STO or recombined crude oil (for ageing). Mineral oil for testing. • Required maximum capillary pressure (if forced displacement by centrifuge) or overburden pressure (if forced displacement in coreholder). • Specific equilibration time criteria to complete spontaneous phases. • Time required to spin sample (normally 24 or 48 h per pressure, depending on sample permeability) if forced displacements performed in centrifuge.

7.3.1 Sample Preparation

Wettability Amott and USBM tests can be carried out on core plugs under various initial conditions (states):

- "*Fresh state*" refers to core tested "as received" in the laboratory. This should not be confused with "native state", which is used for any core that was obtained and stored by methods that strictly "preserve the wettability of the reservoir". Thus, the term "fresh state" does not usually differentiate between oil-based and water-based muds. For "fresh-state" samples drilled with a water-based mud, the oil saturation may be close to residual saturation if filtrate has invaded the core. In this case, the samples are usually flushed in a coreholder or centrifuged with SFW towards Sro and the spontaneous imbibition of oil (secondary drainage) stage commenced from that point (Amott). For "fresh-state" samples drilled with an oil-based mud, the water saturation will be close to irreducible saturation (Swir). In this case, the samples are usually flushed in a coreholder or centrifuged with oil towards Swir and the spontaneous imbibition of water (primary imbibition) stage commenced from that point (Amott–Harvey). Two to three pore volumes of decalin are normally flushed before oil is injected attempt to prevent asphaltene deposition from contact with stock tank oil (STO).

- "*Cleaned state*" refers to the core material where attempts are made to remove all the fluids and adsorbed organic material by flushing or refluxing in solvents. The objective of cleaning is to return the cores to a water-wet state—the hypothesised condition in the reservoir prior to hydrocarbon accumulation. For cleaned-state samples, they are initially saturated in water and must be desaturated to Swir. This can be achieved by either dynamic displacement of brine with the test oil or capillary desaturation. Porous plate is the preferred method for establishing Swi, since it is the most likely to produce evenly distributed saturation throughout the sample. Dynamic displacement desaturation results in a higher and less uniform initial water saturation than porous plate or centrifuge desaturation. Capillary pressures used in porous plate or centrifuge desaturation must be sufficient to define Swir for the particular reservoir height and rock quality. Air or oil is used as the desaturation phase. If air is used for desaturating to Swi in an oil reservoir, then the resulting air space in the core at Swir should be displaced with oil by flowing in a coreholder under backpressure.
- "*Restored state*" refers to the core material where attempts are made to reestablish the reservoir rock native wettability condition. The core is initially cleaned to return the rock to a water-wet state, saturated in brine, established to Swi conditions using STO or recombined crude oil to simulate hydrocarbon accumulation and then aged at reservoir temperature and pressure for up to 40 days, which represents the industry "standard". Tests on "restored-state" samples normally commence from Swir, since this is the sample condition after restoration. In theory, the use of "live oil" opposed to "dead crude oil" is preferred, especially for oils were asphaltenes can precipitate in the pores. As a rule of thumb, ageing in STO (dead crude) can be used if the gas oil ratio is less than 400–500 Mcf/bbl. In practice, this is limited by the quality and availability of dead crude oil to recombine with gas and by the associated, normally high, costs. Wettability restoration using "live oil" is also prone to suffer from leaks, meaning that the ageing can be terminated before the required time is reached. Crudes containing H_2S have associated HSE issues and few labs would be able to conduct tests using them.

Core cleaning as part of wettability restoration is designed to induce a strong wetting state in the test samples as this would represent the reservoir formation prior to oil accumulation. The cleaning methods and solvents are selected primarily to ensure complete removal of pore contaminants (possible mud components, altered reservoir hydrocarbon, altered reservoir formation water, precipitated salts, etc.) and induce water-wet conditions. When the core contains oil with a significant asphaltene content and/or is drilled with some oil-based muds, it is essential to remove wettability contaminants from the pore space prior to restoration. Otherwise, the core may be too oil wet after restoration.

Recent (mostly unpublished) evidence shows that hot Soxhlet extraction in toluene and methanol often fails to remove oil-wetting contaminants so that the cleaned core is not wholly water wet. Hot Soxhlet extraction in toluene (the most usual industry practice) removes both water surrounding the grains (and where clays are present, water is not only removed from the surface but may also be partly removed from clay interlayers causing possible damage) and certain fractions of the crude oil, though may not be miscible with all fractions, thus possibly leaving remnant oil components in contact directly with mineral surfaces, altering wettability. The phenomenon may be exacerbated by low initial water saturation and long-term core storage and exposure to air. Lower temperature total immersion cleaning or mild miscible flush cleaning (Chapter 4) using toluene–isopropanol–methanol, for example, is normally considered to be better methods to remove contaminants and render the core water wet. In any wettability study, the cleaning method should be verified by performing wettability tests on cleaned-state core.

If forced imbibition or forced drainage is performed in the centrifuge, the lab should evaluate the selected plugs to ensure they can withstand the rigours of testing at elevated spin speeds. If necessary, the maximum spin speed (Pc) may be reduced.

7.3.2 Test Conditions

The conditions (temperature, pressure and type of oil) under which wettability is evaluated can potentially have an important influence on the measurement results; nevertheless, contradictory results on this subject are found in the literature. Cuiec (1995) reported that there is a tendency to increase the affinity to water as temperature increases. However, other results show the opposite tendency: water wettability diminishes when temperature increases. With regard to the effect of pressure on wettability, little is found in the literature. Most studies show that, in general, pressure has little or no influence on wettability. In practice, commercial labs are not equipped with a satisfactory apparatus to evaluate average reservoir wettability at full reservoir conditions. This can only be performed using the contact angle method, but it has its own limitations as previously described in Section 7.2.6. Commercial labs offer spontaneous imbibition at room conditions or nominal elevated temperature (Amott cell Fig. 7.10) and forced displacements under limited reservoir conditions (centrifuge or coreholder). However, it is not possible to carry all

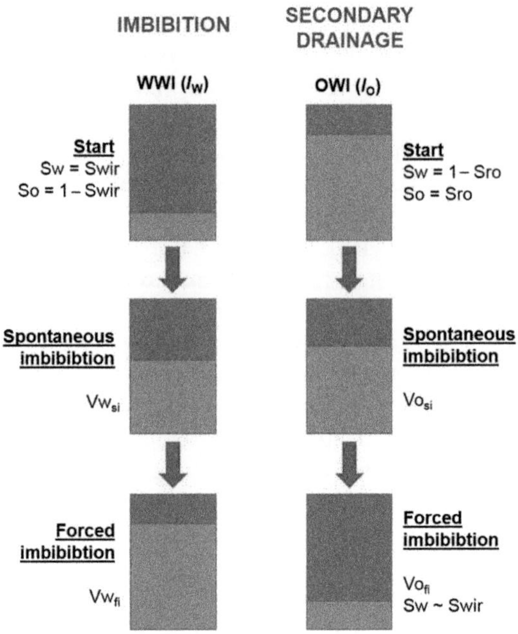

FIGURE 7.9 Schematic of Amott test.

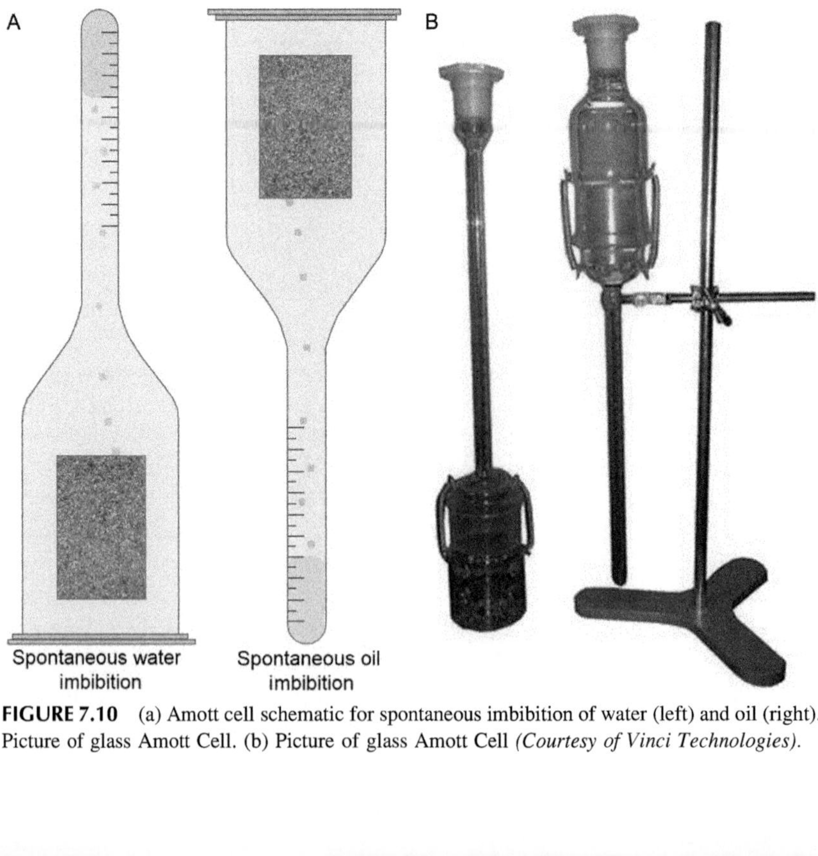

FIGURE 7.10 (a) Amott cell schematic for spontaneous imbibition of water (left) and oil (right). Picture of glass Amott Cell. (b) Picture of glass Amott Cell *(Courtesy of Vinci Technologies).*

four Amott stages (spontaneous and forced imbibition and drainage) while maintaining reservoir conditions using these systems. Pressure is released and temperature is usually reduced when moving from one apparatus to the other.

Commercial equipment is not designed to carry out wettability tests using "live oil" or at full reservoir conditions. Consequently, while crude oil may be used for ageing the samples, mineral oil, which is often easier to work with and has relatively constant properties, is more normally used for Amott tests.

Cuiec concluded that for a carbonate reservoir, it was observed that thermodynamic conditions have no significant influence in wettability and for sandstone reservoirs, when temperature was the sole or principal parameter modified, the trend most often observed was that water wettability increased with temperature, but the reverse was also observed. His results nonetheless confirm that each reservoir is unique and it was difficult to predict the influence of a change in thermodynamic conditions (T, P, kind of oil) on the wettability of a rock/fluid system.

7.3.3 Test Equipment

Spontaneous imbibition: the standard test equipment employs an Amott cell (Fig. 7.10) for spontaneous imbibition measurements. It has a graduated section to determine imbibition volumes. Spontaneous imbibition can be performed on unconsolidated cores using a coreholder, graduated separator system and a pump in a closed circulating system (Fig. 7.11). This may also be employed for any core, but incurs higher costs.

Forced imbibition: it is carried out in a high-speed centrifuge, similar to that described in Chapter 9 (preferred method), or by flooding in a coreholder. Note that when flooding, imbibition and drainage should employ the same displacement force (i.e. same constant differential pressure), but true residual saturations are most likely not achieved, which could skew results.

7.3.4 Test Procedures

Both methods, original Amott and modified Amott–Harvey, involve four stages. The order depends on the initial sample saturation stage, with the test plug at Swir or at Sro. The following assumes that the sample is reduced to initial water saturation, Swir, and each stage is shown schematically in Fig. 7.9.

(1) *Spontaneous brine imbibition*: (sometimes referred as Amott A) brine (SFW) is allowed to spontaneously imbibe into the core, which is initially at Swir in a graduated Amott cell (Fig. 7.10). As brine is imbibed, oil is displaced from the core and rises to the top of the graduated tube where volumes are monitored. The volume of oil displaced (equivalent to the volume of brine imbibed) is monitored as a function of time.

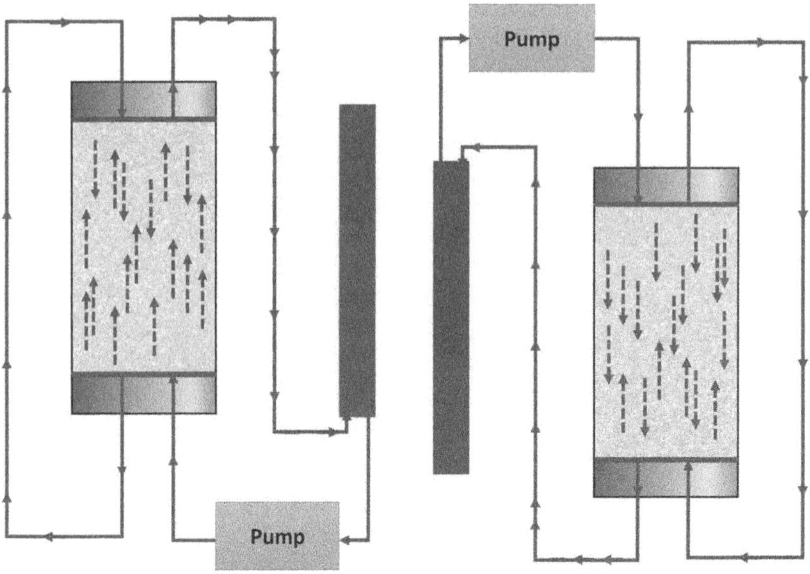

FIGURE 7.11 Spontaneous imbibition schematic for unconsolidated material (using coreholder, separator and pump).

The original Amott technique involved an arbitrary "cut-off" of 20 h for spontaneous imbibition. This is now recognised to be too short since, in neutral-wet rocks, imbibition can be slow yet still significant. More commonly, periods of up to 480 h are now allowed. Production volumes should be monitored as a function of log time or square root of time to assess when the cycle should be stopped. If the core is water wet, a significant volume of brine is spontaneously imbibed—if it is not water wet, spontaneous brine imbibition will be small or even zero. If core material is unconsolidated, the best approach is to perform spontaneous imbibition in a coreholder (Fig. 7.11). Fluid production is monitored in the graduated separator and plotted as a function of square root time or logarithm time.

(2) *Forced brine imbibition*: (sometimes referred as Amott B) brine is then forcibly imbibed into the core by applying sufficient pressure to force the displaced phase towards residual oil saturation (Sro). This is performed by dynamic displacement either by centrifuge (preferred) or in a coreholder. Viscous instability and capillary end effects in dynamic flooding will impact on measured forced imbibition volumes. Most often, displacement by flooding will not achieve true Sro; hence, total volumes and wetting index may be significantly in error. Because of capillary end effects, it is recommended that displacement by flooding is performed by the constant pressure method (see Chapter 10) rather than constant rate.

The volume of fluid displaced (brine imbibed) is determined by monitoring the displaced oil volume.

(3) *Spontaneous oil imbibition*: (sometimes referred as Amott C) oil is allowed to spontaneously imbibe into the core at residual oil saturation (Sro) in the Amott cell which, for this stage, is inverted. The volume of brine displaced (which is equivalent to the volume of oil imbibed) is monitored as a function of time. Again sufficient time should be allowed to achieve fluid production equilibrium. If the core is oil wet, a significant volume of oil is spontaneously imbibed—if it is not oil wet, spontaneous oil imbibition will be small or even zero.

(4) *Forced oil imbibition*: (sometimes referred as Amott D) the displacing phase (oil) is then forcibly imbibed into the core by applying the equivalent force (to that used during forced brine imbibition) upon the displaced phase towards irreducible saturation (Swir). This is performed by dynamic displacement either by centrifuge (preferred) or by flooding in a coreholder. It is important that the displacement force, used during forced oil imbibition, should be equal to the force used during forced water imbibition; this is to ensure comparability between the two displacement processes. The volume of fluid displaced (brine imbibed) is determined by monitoring the displaced oil volume.

7.3.5 Amott–Harvey Wettability Index Calculation

The displacement by brine ratio, δw or water-wetting index Iw, is determined from:

$$\delta w = \frac{Vwsi}{Vwsi + Vwfi},$$

where Vwfi is the volume of brine spontaneously imbibed (cc) and Vwfi is the volume of brine forcibly imbibed (cc). For a water-wet rock, Iw will be close to or equal to 1, since almost all the brine imbibed will be imbibed spontaneously.

When oil is the displacing fluid, the displacement by oil ratio, δo or oil-wetting index, Io, is determined from:

$$\delta o = \frac{Vo_{si}}{Vo_{si} + Vo_{fi}},$$

where Vo_{si} is the volume of oil spontaneously imbibed (cc) and Vo_{fi} is the volume of oil forcibly imbibed (cc). For an oil-wet rock, Io will be close to or equal to 1, since almost all the oil imbibed will be imbibed spontaneously.

The Amott–Harvey wetting index is then determined from:

$$AHWI = I - Io$$

TABLE 7.4 Amott Wetting Indices

Index	Water Wet	Neutral Wet	Oil Wet
δw	Positive	0	0
δo	0	0	Positive
AHWI	+0.3 to +1.0	−0.3 to +0.3	−0.3 to −1.0

This combines the two ratios into a single wettability index that varies from +1 for complete water wetness, to −1 for complete oil wetness. The larger the absolute value of WI, the greater the wetting preference. Table 7.4 shows index values and their relative wettability definitions for the Amott test.

On completion and provided the sample is not scheduled for other tests, Dean-Stark extraction is performed to provide an independent verification for end-point water saturation. The core is then leached in methanol and dried, and final pore volume, bulk volume and air permeability are measured.

In addition to the calculation of wetting index, it is useful to plot the spontaneous production as a function of dimensionless time, per Zhang et al. (1996). The shape of the plotted curves gives useful qualitative information about wettability.

7.3.6 Data Reporting Requirements

- brief description of procedures and experimental apparatus;
- SFW composition;
- SFW density and viscosity at test temperature (g/cc);
- type of oil used in test (STO or mineral oil);
- oil density and viscosity at test temperature (g/cc);
- initial sample state (clean, fresh or restored state);
- cleaning and restoration methods;
- base permeability (mD) pre- and/or post-test;
- base porosity (fraction) pre- and/or post-test;
- sample dimensions (cm);
- initial sample dry weight (g) if cleaned or restored plug;
- initial sample saturated weight (g);
- initial sample saturation (Swir or Sro);
- fluid production versus time for spontaneous imbibition;
- capillary pressure applied (if forced displacement by centrifuge) or overburden pressure and flood differential pressure (if forced displacement in coreholder);
- volume of brine spontaneously imbibed, Vw_{si} (cc);
- volume of brine forcibly imbibed, Vw_{fi} (cc);
- volume of oil spontaneously imbibed, Vw_{fi} (cc);
- volume of oil forcibly imbibed, Vo_{si} (cc);

Continued

- calculated displacement by brine ratio, δw;
- calculated displacement by oil ratio, δo;
- calculated Amott wetting index;
- plot spontaneous production versus dimensionless time;
- final sample dry weight (g);
- sample grain loss correction procedures (if applied), including typical worked example for one sample from sample database. Raw data for calculation.

7.3.7 Amott (Amott–Harvey) Summary

A summary of the Amott–Harvey wettability method is given here:

Advantages	Drawbacks and issues
• A major advantage of the Amott method over the USBM test is that it can often be more sensitive than USBM for neutral wettability. It will have positive displacement by water (Iw) and displacement by oil (Io) ratios, indicating that the system is non-uniformly wetted. • It is the best method for unconsolidated samples if forced displacement is performed in a coreholder. Centrifuge could potentially damage friable/unconsolidated plugs. • The method measures the average wettability of the core, while the contact angle measurement is at localised scale.	• Results can be dependent on duration of the spontaneous imbibition phase. • The maximum pressure applied for the forced stages should be the same for both drainage (+Pc) and imbibition (−Pc). • The spontaneous imbibition of brine is dependent on the amount and distribution of interstitial water present after primary drainage, as the surface available for wetting alteration increases as Swir decreases. • During the spontaneous phase, the adhesion of droplets of the expelled phase to the surface of the core can affect the results. The measured volume of the imbibed phase will be lower. • Dynamic flooding can result in viscous instability and capillary end effects which can impact on the forced imbibition volumes. • Centrifuge forced imbibition is preferred but can create an unusual stress regime, where fracturing is possible, especially at higher spin velocities. Therefore, the method using the centrifuge for the forced displacements is not suitable for samples that cannot withstand the forces applied during rotation at high speed (e.g. poorly consolidated core). In this case, tests could be carried out in centrifuge at nominal

Continued

Advantages	Drawbacks and issues
	confinement. Additionally, stopping and restarting can lead to sample fracturing. • The method has no validity as an absolute measurement, but it is industry standard for comparing the wettability of various core plugs.

7.3.7.1 QC checks & Suitable Laboratories

- Analysis of the results should start with a thorough review of the laboratory report to check that acceptable procedures have been followed. This should be carefully evaluated.
- Check the method used for cleaning and drying and potential impacts on wettability. Ensure care has been taken to avoid damage and preservation of the clay structures which can have a subtle, yet significant influence on wettability.
- Results should be consistent with test state and core condition. For example, a cleaned-state (MMC), clean sandstone or a fresh-state sample taken from a water zone with no residual hydrocarbon saturation is expected to be water wet. OBM can result in contamination and may induce stronger oil wetting. In some reservoirs, a wettability gradient is common so that fresh-state wettabilities might become less water wet with height above the FWL. Gas reservoirs are considered to be water wet, but the presence of residual oil in some gas reservoirs (representing a previous hydrocarbon saturation) or contamination with OBM filtrate can induce an oil-wetting tendency.
- Where possible, Amott tests should always be carried out in conjunction with USBM tests as it is possible to calculate USBM and Amott wettability indices and cross-compare.
- Be careful if results show a strong oil-wetting tendency. Very few clastic and carbonate reservoirs are entirely oil wet. Sample contamination must be suspected.
- Most commercial SCAL labs. Beware of labs who do not have automated data acquisition in centrifuge systems.

7.4 USBM METHOD

This method also measures the average wettability of the core. The test compares the work necessary for one fluid to displace the other. Because of the favourable free-energy change, the work required to displace the non-wetting fluid from the core is less than the work required for the opposite displacement.

The work (energy) is proportional to the area under the drainage and imbibition capillary pressure curves. The method comprises only the forced primary imbibition and forced secondary drainage stages.

The original (USBM) (Donaldson, 1981; Donaldson et al., 1969, 1980) criteria specified a maximum capillary pressure of 10 psi but, in most cases, this might not define Sro or Swir in poorer reservoir quality samples; thus, higher pressures must be used, depending on the core permeability. Irrespective of the maximum Pc pressure used, both drainage and imbibition Pc should be equal.

In the USBM tests, spontaneous imbibition is not specifically measured, but it occurs during initial centrifugation at low pressures. Consequently, the combined Amott/USBM (detailed in Section 7.5) method has largely replaced the USBM test as this allows spontaneous imbibitions to be determined in an Amott cell. This combination test also provides both Amott and USBM wetting indices (Table 7.5).

TABLE 7.5 Overview of USBM Wettability Method

Purpose of test	• To determine wettability preference of core plugs
Sample requirements	• Cylindrical shaped plug sample of 1″ or 1½″ diameter. The larger the sample, the larger the potential pore volume, and the less will be the impact of saturation measurement errors. • *Ideally*—cleaned (water-wet) sample, sample taken from preserved core or preserved plug, or plug after wettability restoration. *Alternative*—plug previously used in routine core analysis programme, provided that plug has not been pre-stressed and/or that plug preparation methods have not caused damage to delicate clay minerals, or have altered wettability.
Required input data for lab	• FW salinity and ionic composition, as close as practically possible to the formation water in the reservoir or FW resistivity at reservoir and lab temperature. Lab will prepare SFW accordingly. • Reservoir crude oil sample—STO or recombined crude oil (for ageing). Mineral oil for testing. • Number (normally 6–8) and estimated Pc for each test pressure station. Data from mercury injection capillary pressure tests (Chapter 9) can be used as an aid in test pressure design. Knowledge of the height above the free water level and fluid density difference can also be used as an aid to determine the highest Pc required, especially in the case poorer quality samples. • Time required to spin sample per each pressure step (normally 24 to 48 h, depending on sample permeability). However, times may be longer and production should have ceased at each step before continuing.

7.4.1 Sample Preparation

The test will normally start in samples at irreducible water saturation (Swir), in fresh-state (drilled with an oil-based mud), cleaned-state or restored-state. The test can commence at Sro in fresh-state samples drilled with a water-based mud.

7.4.2 Test Equipment

Forced primary imbibition and secondary drainage are carried out in a high-speed centrifuge, similar to that described in Chapter 9.

7.4.3 Key Processes

1. Prepare the sample to the required initial state: fresh, cleaned or restored. Determine base permeability and pore and bulk volume (porosity) for cleaned and restored plugs.
2. Spin samples under brine in a high-speed centrifuge at pre-designated pressure steps (rotational rates). Monitor oil production volume (equal to volume of water imbibed). Calculate saturation as a function of pressure (capillary pressure curve—see Chapter 9).
3. Spin samples under oil in a high speed centrifuge at pre-designated pressure steps (rotational rates). The final capillary pressure may either be the "standard" 10 psi or defined by the user to ensure Swir or Sro can be defined for the specific rock quality.
4. Determine average (not end-face) water saturation and plot $Sw_{average}$ versus Pc for drainage (+Pc) and imbibition (−Pc) curves.
5. Determine wettability index from areas below the primary imbibition and secondary drainage curves.
6. Carry out Dean-Stark extraction as an independent verification for end-point water saturation.

7.4.4 Test Procedures

The USBM method consists of two stages: primary forced imbibition and secondary forced drainage. In this example, it is assumed that the samples are previously prepared to initial water saturation (Swir).

(1) During primary forced imbibition, the plug at initial water saturation (Swir) is placed in brine and centrifuged at incrementally increasing speeds (equivalent to increasing pressure—calculated) until a maximum capillary pressure is reached. The first capillary pressure is normally as low as practicable (~±0.5 psi or ±1 psi) to simulate spontaneous imbibition (as closely as possible) and limit forced imbibition. During this stage, oil is produced from the sample, replaced by brine. Oil is produced into the centrifuge receiving cup. Oil production should be recorded and plotted as a function of logarithmic time or square root time to determine when production has ceased. At each incremental pressure step, the average saturation of the plug (not the end-face) is calculated from the volume of expelled oil. This is shown as Curve A_2 in Figs. 7.12–7.14. At the end of this stage, the sample is assumed to be at (or close to) irreducible oil saturation (Sro).

(2) During secondary forced drainage, the plug at residual oil saturation (Sro) is placed under oil and centrifuged at incrementally increasing speeds,

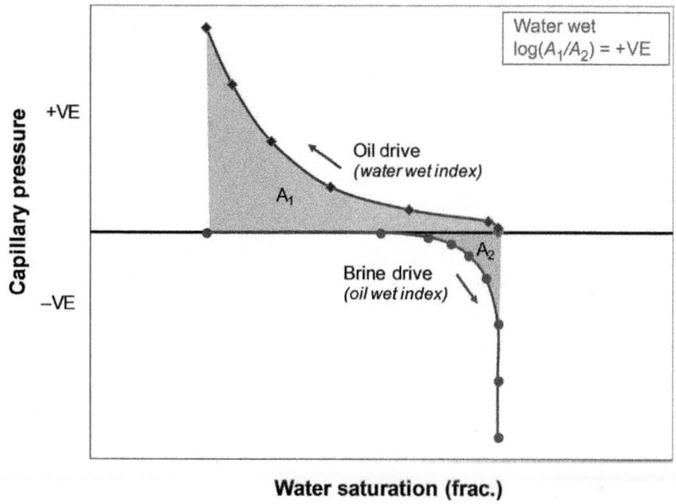

FIGURE 7.12 USBM example—water-wet sample. *After Donaldson et al. (1969).*

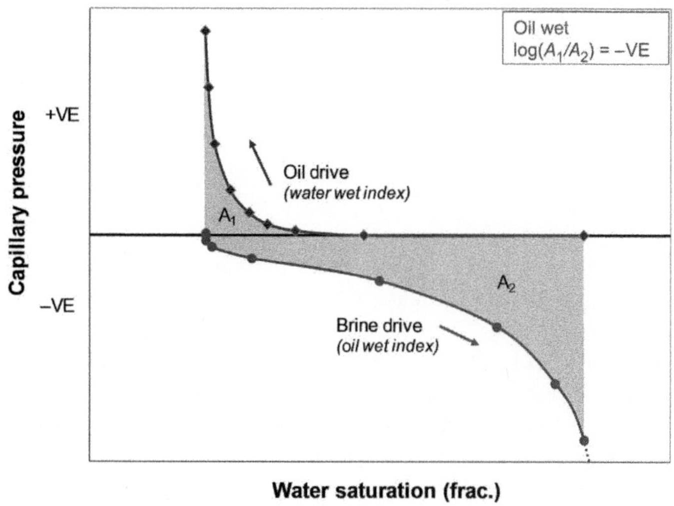

FIGURE 7.13 USBM example—oil-wet sample. *After Donaldson et al. (1969).*

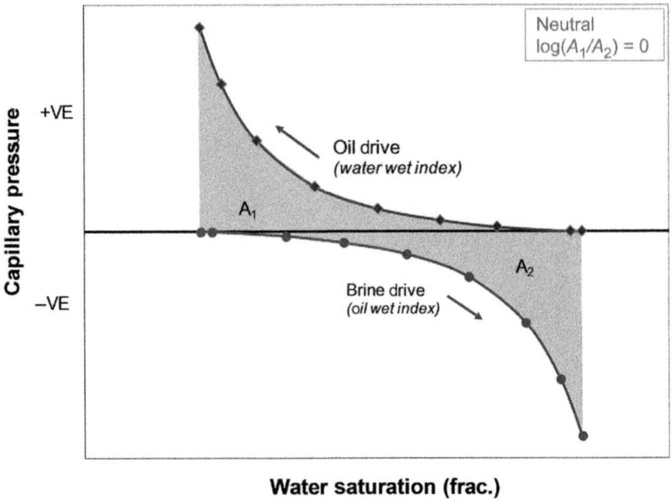

FIGURE 7.14 USBM example—neutral-wet sample. *After Donaldson et al. (1969).*

while fluid production is measured. The speeds should be selected to match the calculated pressures of the primary forced imbibition stage. The first pressure is normally as low as practicable (~±0.5 psi or ±1 psi) to simulate spontaneous secondary drainage, as closely as possible. During this stage, brine is produced from the sample, replaced by oil. The brine is produced into the centrifuge receiving cup. Brine production should be recorded and plotted as a function of logarithmic time or square root time to determine when production has ceased. The results of this secondary forced drainage stage are plotted as Curve A_1 in Figs. 7.12–7.14.

7.4.5 USBM Index Calculation

The USBM method uses the ratio of the areas under the two capillary pressure curves to calculate the wettability index:

$$\mathrm{USBM} = \log\left(\frac{A_1}{A_2}\right),$$

where A_1 is the area under the secondary drainage (oil drive) capillary pressure curve and A_2 is the area under the primary imbibition (brine drive) capillary pressure curve.

Figures 7.12–7.14 provide examples of typical curves for oil-wet, water-wet and neutral-wet samples. A positive USBM value indicates water wetting, whereas a negative value indicates oil wetting. A value of near zero indicates neutral wettability. The larger the absolute value, the greater the wetting preference. For the case of a water-wet sample (Fig. 7.11), the area below the secondary drainage curve (A_1) is much larger than the area under

TABLE 7.6 USB Wetting Indices

	Water Wet	Neutral Wet	Oil Wet
USBM	Positive	0	Negative

the primary imbibition curve (A_2), yielding a positive wettability index. For the case of an oil-wet sample (Fig. 7.12), the area below the secondary drainage curve (A_1) is much smaller than the area under the primary imbibition curve (A_2), yielding a negative wettability index. For the case of a neutrally wet sample (Fig. 7.13), both areas are equal, making the USBM wettability index = 0.

Table 7.6 shows index values and their relative wetting states for the USBM test.

On completion, and provided the sample is not scheduled for other tests, Dean-Stark extraction is performed (Chapter 5) to provide an independent verification for end-point water saturation. The core is then leached in methanol and dried (Chapter 4), and final pore volume, bulk volume and air permeability are measured (Chapter 5).

7.4.6 Data Reporting Requirements

- brief description of procedures and experimental apparatus;
- SFW composition;
- SFW density and viscosity at test temperature (g/cc);
- type of oil used in test (STO or mineral oil);
- oil density and viscosity at test temperature (g/cc);
- initial sample state (clean, fresh or restored state);
- cleaning and restoration methods;
- base permeability (mD) pre- and/or post-test;
- base porosity (fractional) pre- and/or post-test
- sample dimensions (cm);
- initial sample dry weight (g) if cleaned or restored plug;
- initial sample saturated weight (g);
- initial sample saturation (Swir or Sro);
- fluid production vs. time for each spin applied;
- capillary pressure steps applied (psi);
- average plug water saturation vs. Pc for drainage and imbibition;
- volume of brine forcibly imbibed, Vw_{fi} (cc);
- volume of oil forcibly imbibed, Vo_{fi} (cc);
- calculated area under brine forcibly imbibed curve (cm^2);
- calculated area under oil forcibly imbibed curve (cm^2);
- calculated USBM wetting index;
- final sample dry weight (g);
- sample grain loss correction procedures (if applied), including typical worked example for one sample from sample database. Raw data for calculation.

7.4.7 USBM Summary

A summary of the USBM wettability method is given here in Table 7.7.

7.4.7.1 QC checks & Suitable Laboratories

- Analysis of the results should start with a thorough review of the laboratory report to check that acceptable procedures have been followed. This should be carefully evaluated.
- Check the method used for cleaning and drying and potential impacts on wettability. Ensure care has been taken to avoid damage and preservation of the clay structures which can have a subtle, yet significant influence on wettability.
- Results should be consistent with test state and core condition. For example, a cleaned-state (MMC), clean sandstone or a fresh-state sample taken from a water zone with no residual hydrocarbon saturation is expected to be water wet. OBM can result in contamination and may induce stronger oil wetting. In some reservoirs, a wettability gradient is common so that fresh-state wettabilities might become less water wet with height above the FWL. Gas reservoirs are considered to be water wet, but the presence

TABLE 7.7 Advantages and Disadvantages of USBM Method

Advantages	Drawbacks and issues
• It is a relatively rapid, requiring only a few days, as no spontaneous imbibition is carried out. • The method measures the average wettability of the core, while the contact angle measurement is at localised scale. • Method can provide additional information on capillary pressure (especially imbibition) curves. For the USBM method, pressure versus saturation plots are based on the average saturation, whereas Pc curves should be based on the inlet face saturation.	• The Amott method tends to be more sensitive at neutral wettabilities. • Drainage and imbibition maximum Pc must be identical as this ensures consistency in USBM index calculation. • Centrifuge forced imbibition can create an unusual stress regime, where fracturing is possible, especially at higher spin velocities. Therefore, the method is not suitable for samples that cannot withstand the forces applied during rotation at high speed. Additionally, stopping and restarting the centrifuge can lead to sample fracturing. This is unavoidable when changing from imbibition to drainage. • The method has no validity as an absolute measurement, but it is industry standard for comparing the wettability of various core plugs.

of residual oil in some gas reservoirs (representing a previous hydrocarbon saturation) can induce an oil-wetting tendency.

- Where possible, USBM results should always be carried out in conjunction with Amott tests as it is possible to calculate USBM and Amott wettability indices and cross-compare (Fig. 7.16).
- Be careful if results show a strong oil-wetting tendency. Very few clastic and carbonate reservoirs are entirely oil wet. Sample contamination must be suspected.
- Most SCAL labs. Beware labs who do not have automated data acquisition for centrifuge tests.

7.5 COMBINED AMOTT–USBM (COMBINATION) METHOD

A modification of the Amott and USBM methods allows the calculation of both the Amott and USBM wettability indices in a single test. In this test, spontaneous imbibition is carried out in an Amott cell and forced imbibition in a high-speed centrifuge using the same, multiple speed method as used for the USBM test (Table 7.8).

TABLE 7.8 Overview of the Combined Amott–USBM Method

Purpose of test	• To determine wettability preference of core plugs using two analytical methods (Amott and USBM indices)
Sample requirements	• Cylindrical shaped plug sample of 1″ or 1½″ diameter. The larger the sample, the larger the potential pore volume, and the less will be the impact of saturation measurement errors. • *Ideally*—cleaned (water-wet) sample, sample taken from preserved core or preserved plug, or plug after wettability restoration. *Alternative*—plug previously used in routine core analysis programme, provided that plug has not been pre-stressed and/or that plug preparation methods have not caused damage to delicate clay minerals, or have altered wettability.
Required input data for lab	• FW salinity and ionic composition, as close as practically possible to the formation water in the reservoir or FW resistivity at reservoir and lab temperature. Lab will prepare SFW accordingly. • Reservoir crude oil sample—STO or recombined crude oil (for ageing). Mineral oil for testing. • Specific equilibration time criteria to complete spontaneous phases. • Number (normally 6–8) and estimated Pc for each test pressure station. Data from Mercury injection capillary pressure tests (Chapter 9) can be used as an aid in test pressure design. Knowledge of the height above the free water level and fluid density difference can also be used as an aid to determine the highest Pc required, especially in the case poorer quality samples. • Time required to spin sample per each pressure step (normally 24–48 h, depending on sample permeability).

7.5.1 Sample Preparation

The test will normally start with samples preferably at irreducible water saturation (Swir), or residual oil saturation (Sro)—for fresh-state samples drilled with WBM, either in fresh-state, cleaned-state or restored-state cores (see Section 7.3.1).

7.5.2 Test Equipment

- Spontaneous stages: Amott cell (see Fig. 7.10 and Section 7.3.3).
- Forced stages: high-speed centrifuge, similar to that described in Chapter 9.

7.5.3 Test Procedures

Like the Amott method, the combined method involves four stages (Fig. 7.15). It is most usual to initiate tests from Swir, though fresh-state plug samples drilled in WBM (exhibiting high invasion) may be initiated from Sro. The test order thus depends on the initial sample saturation conditions (Swir or Sro), but the following assumes initial water saturation, Swir:

(1) Spontaneous brine imbibition (Amott): brine is allowed to spontaneously imbibe as described in Section 7.3.4. The process is performed either in an Amott cell, for consolidated material (Fig. 7.10) or a coreholder, for unconsolidated material (Fig. 7.11). Further details are explained in Section 7.3.4.

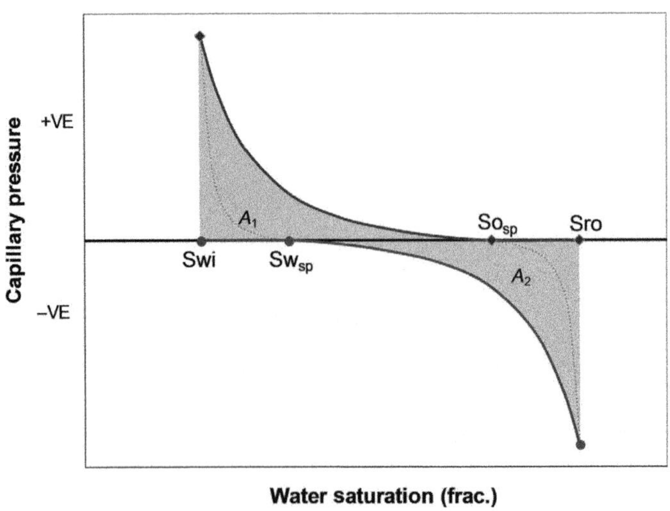

FIGURE 7.15 Combined Amott–USBM curve example.

(2) Forced brine imbibition (Amott+USBM): brine displaces oil by applying pressure under centrifugal force—driving the displaced phase (oil) towards residual oil saturation (Sro). The plug sample is placed under brine in a centrifuge at incrementally increasing speeds (incrementally increasing pressure), and fluid production is measured at each speed step, as in Section 7.4.4. Total fluid production measured at the final (highest) spin speed is used as the Amott forced fluid volume (Vw_{fi}). Amott water-wet index, Iw, is thus determined as explained in Section 7.3.5. Production at each pressure (each spin speed) is determined and the area under the curve is used to calculate the USBM index (A_2), as detailed in Section 7.4.5.
(3) Spontaneous oil imbibition (Amott): oil is allowed to spontaneously imbibe into the core which is initially at residual oil saturation (Sro). The spontaneous imbibition stage is carried out using an inverted Amott cell (or coreholder—unconsolidated). Further details are explained in Section 7.3.4.
(4) Forced oil imbibition (Amott+USBM): oil displaces brine by applying pressure under centrifugal force—driving the displaced phase (brine) towards irreducible water saturation (Swi). The plug sample is placed under oil in a centrifuge at incrementally increasing speeds (incrementally increasing pressure) and fluid production measured at each speed step. Total fluid production measured at the final (highest) spin speed is used as the Amott forced fluid volume (Vo_{fi}). Amott oil-wetting index, Io, is thus determined as explained in Section 7.3.5. Production at each pressure (each spin speed) is determined, and the area under the curve is used to calculate the USBM index (A_1), as detailed in Section 7.4.5.

7.5.4 Data Reporting Requirements

The reporting requirements are the same as those specified for Amott (Section 7.3.6) and USBM (Section 7.4.6) methods.

Figure 7.16 provides an example of the results of an Amott–USBM test. In this summary, both the drainage and imbibition curves, USBM index and Amott Iw and Io indices are provided. In this example, the test was carried out on fresh core and was commenced from Swir.

7.5.5 USBM and Amott–Harvey Index Calculation

The Amott–Harvey and USBM indices are calculated independently of one another, using the same equations as detailed in Sections 7.3.5 and 7.4.5, respectively. For the Amott–Harvey index, the total cumulative production volume at the highest centrifuge speed is the volume to use as the forced part of the index equation (Vw_{fi} and Vo_{fi}).

The AHWI and USBM indices should be compared for each sample.

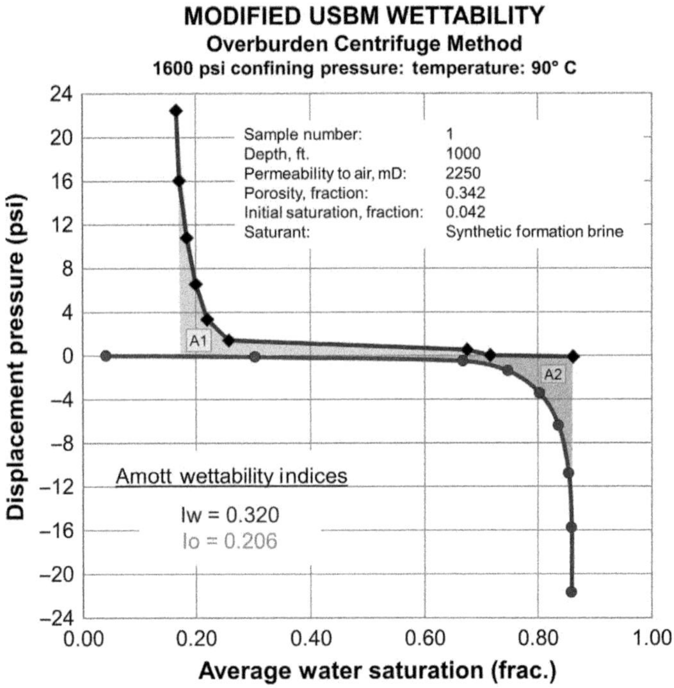

FIGURE 7.16 Typical Amott–USBM lab presentation.

7.5.6 Combined Amott–USBM Summary

A summary of the combined Amott–USBM method is provided in Table 7.9.

7.5.6.1 QC checks & Suitable Laboratories

- Analysis of the results should start with a thorough review of the laboratory report to check that acceptable procedures have been followed. This should be carefully evaluated.
- Check the method used for cleaning and drying and potential impacts on wettability. Ensure care has been taken to avoid damage and preservation of the clay structures which can have a subtle, yet significant influence on wettability.
- Results should be consistent with test state and core condition. For example, a cleaned-state (MMC), clean sandstone or a fresh-state sample taken from a water zone with no residual hydrocarbon saturation is expected to be water wet. OBM can result in contamination and may induce stronger oil wetting. In some reservoirs, a wettability gradient is common so that fresh-state wettabilities might become less water wet with height above

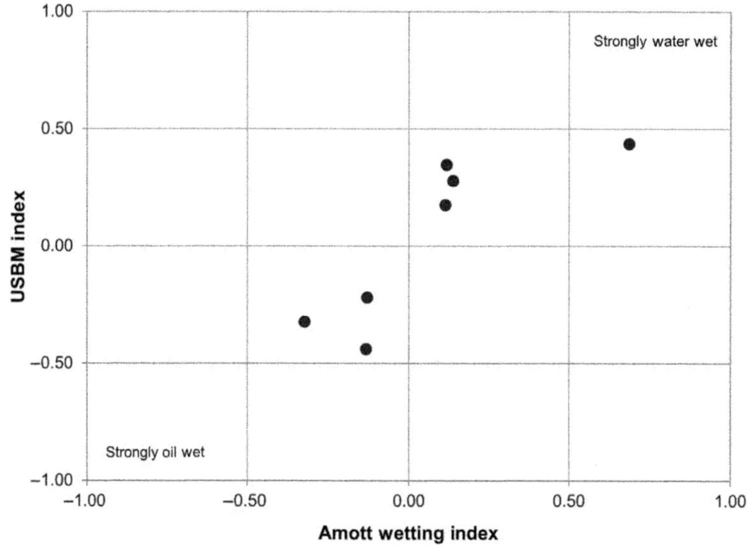

FIGURE 7.17 Combined Amott/USBM cross-plot.

TABLE 7.9 Combined Amott–USBM Summary

Advantages	Drawbacks and issues
• The method has all the advantages of the USBM method plus the additional resolution of spontaneous wetting provided by the Amott method. • Amott and USBM indices can be measured in a single experiment on the same core plug. The results can be cross-plotted as indicated in Fig. 7.17. This presentation can highlight outliers as both analytical methods should yield similar results (especially for strong wetting conditions). • Preferred method for wettability tests on core samples.	• Drainage and imbibition maximum Pc must be identical as this ensures consistency in USBM index calculation. • Method is not suitable for samples that cannot withstand the forces applied during rotation at high speed in the centrifuge. • The method has no validity as an absolute measurement, but it is industry standard for comparing the wettability of various core plugs.

the FWL. Gas reservoirs are considered to be water wet, but the presence of residual oil in some gas reservoirs (representing a previous hydrocarbon saturation) can induce an oil-wetting tendency.

- USBM and Amott wettability indices must be compared on the same samples (Fig. 7.16).

- Be careful if results show a strong oil-wetting tendency. Very few clastic and carbonate reservoirs are entirely oil wet. Sample contamination must be suspected.
- Most SCAL labs. Beware labs who do not have automated data acquisition for centrifuge tests.

REFERENCES

Amott, E., 1959. Observations relating to the wettability of porous rock. Trans. AIME 216, 156–162.

Anderson, W.G., 1986a. Wettability literature survey—part 1: rock/oil/brine interactions and the effects of core handling on wettability. SPE JPT 38, 1125–1144.

Anderson, W.G., 1986b. Wettability literature survey—part 2: wettability measurement. SPE JPT 38, 1246–1262.

Anderson, W.G., 1986c. Wettability literature survey—part 3: the effects of wettability on the electrical properties of porous media. SPE 38, 1371–1378.

Anderson, W.G., 1987a. Wettability literature survey—part 4: the effects of wettability on capillary pressure. SPE JPT 39, 1283–1300.

Anderson, W.G., 1987b. Wettability literature survey—part 5: the effects of wettability on relative permeability. SPE JPT 39, 1453–1468.

Anderson, W.G., 1987c. Wettability literature survey—part 6: the effects of wettability on waterflooding. SPE JPT 39, 1605–1622.

Bobek, J.E., et al., 1958. Reservoir rock wettability—its evaluation and significance. Trans AIME 213, 155–160.

Boneau, D.F., Clampitt, R.L., 1977. A surfactant system for the oil-wet sandstone of the North Burbank Unit. JPT 29, 501–506.

Cuioo, L., 1995. Wettability laboratory evaluation under reservoir conditions: a new apparatus In: Society of Core Analysts, Paper Number 9529.

Donaldson, E.C., 1981. Oil-water rock wettability measurement. In: Proceedings of the American Chemical Society, Division of Petroleum Chemistry, 26, March 29–April 3.

Donaldson, E.C., Thomas, R.D., Lorenz, P.B., 1969. Wettability determination and its effect on recovery efficiency. SPEJ 9, 13–20.

Donaldson, E.C., et al., 1980. Equipment and procedures for fluid flow and wettability tests of geological materials, Bartlesville energy technology center, report DOE/BETC/IC 79/5, U.S. DOE (May 1980).

Gant, P.L., Anderson, W.G., 1988. Core cleaning for restoration of native wettability. SPE Form. Eval. 3 (1), 131–138.

Jia, D., Buckley, J.S., Morrow, N.R., 1994. Alteration of wettability by drilling mud filtrates. In: Paper SCA 9408, SCA Symposium, Stavanger.

Ma, S., Mason, G., Morrow, N.R., 1996. Effect of contact angle on drainage and imbibition in regular polygonal tubes. J. Colloid Interface Sci. 117, 273.

Morrow, N.R., 1975. The effect of surface roughness on contact angle with special reference to petroleum recovery. J. Can. Pet. Technol. 14, 42–53.

Popiel, W.J., 1978. Introduction to Colloid Science. Exposition Press, Hickville, NY.

Sharma, M.N., Wunderlich, R.W., 1985. The alteration of rock properties due to interactions with drilling fluid components. In: SPE 14302, September.

Treiber, L.E., et al., 1972. A laboratory evaluation of the wettability of fifty oil producing reservoirs. SPEJ, 531–540.

Zhang, X., Morrow, N.R., Ma, S., 1996. Experimental verification of a modified scaling group for spontaneous imbibition, SPE Reservoir Engineering, 280–285.

RECOMMENDED READING

Craig, F.F., Jr, Society of Petroleum Engineers, 1971. The reservoir engineering aspects of waterflooding. H. L. Doherty Memorial Fund of AIME, New York.

Hjelmeland, O.S., Larrondo, L.E., 1986. Experimental investigation of the effects of temperature, pressure and crude oil composition on interfacial properties. SPE RE July, 321–328.

McCaffery, F.G., Mungan, N., 1970. Contact angle and interfacial tension studies of some hydrocarbon-water-solid systems. J. Can. Pet. Technol. 9 (3), 185–196.

Morrow, N.R., Lim, H.T., Ward, J.S., 1986. Effect of crude-oil-induced wettability changes on oil recovery. SPE Form. Eval. 1, 89–95.

Neumann, A.W., Good, R.J., 1979. Techniques of measuring contact angles. In: Good, R.J., Stromberg, R.R. (Eds.), In: Surface and Colloid Science, vol. 11. Plenum Press, New York City, pp. 31–91.

Wolfrom, R., Chander, S., Hogg, R., 2002. Evaluation of capillary rise methods for determining wettability of powders. Miner. Metall. Process. 19, 198–202.

Chapter 8

Electrical Property Tests

Chapter Outline

8.1 Introduction **348**
8.1.1 Resistivity Logs 348
8.1.2 Water Saturation Interpretation—Archie's Law 349
8.1.3 Water Saturation Interpretation—Shaly Sands 352
8.1.4 Water Saturation Interpretation Parameters from Core 353
8.2 FRF Tests **354**
8.2.1 Sample Selection and Preparation 354
8.2.2 Test Equipment 356
8.2.3 Test Procedures—Ambient (Base) Conditions 359
8.2.4 Test Procedures—Stress (Overburden) Conditions 360
8.2.5 Data Utilisation 366
8.2.6 Data Reporting Requirements 369
8.2.7 Advantages and Drawbacks/Issues 369
8.2.8 FRF Test Quality Control Issues, Checks and Diagnostics 371
8.3 Resistivity Index Tests **378**
8.3.1 Test Equipment 379
8.3.2 Test Procedures 382
8.3.3 Data Assignment and Utilisation 387
8.3.4 Saturation Exponent: Laboratory Artefacts 390
8.3.5 Saturation Exponent: Rock Type and Pore/Fluid System Controls 398
8.3.6 Data Reporting Requirements 406
8.3.7 Advantages and Drawbacks/Issues 409
8.3.8 RI Test Quality Control Issues, Checks and Diagnostics 409
8.4 Waxman–Smits Parameters **416**
8.4.1 Shaly Sand Models 416
8.4.2 Waxman–Smits Equation 420
8.4.3 CEC and Qv Tests 424
8.4.4 Data Utilisation and Assignment 430
8.4.5 Data Reporting Requirements 433
8.4.6 Advantages and Drawbacks/Issues 433
8.4.7 Qv, m^* and n^* Test Quality Control Issues, Checks and Diagnostics 433
8.5 Alternative Method for Saturation Exponent Determination **439**
References **444**
Recommended Reading **447**

Developments in Petroleum Science, Vol. 64. http://dx.doi.org/10.1016/B978-0-444-63533-4.00008-1

8.1 INTRODUCTION

Like porosity, water saturation, Sw (and by difference, hydrocarbon saturation, Sh), has a crucial role to play in reserves assessment. Fluid saturations can be determined by direct measurements on core (such as Dean-Stark extraction and capillary pressure tests) and estimated by indirect measurements from resistivity logs. Indirect water saturation estimates must be calibrated by core and, ideally, reconciled with core data. This chapter concentrates on the electrical properties of rock and how SCAL measurements are used to calibrate and interpret resistivity log measurements through clean sand and shaly sand water interpretation models.

8.1.1 Resistivity Logs

Resistivity logs do not measure water saturation directly—all respond to water saturation, albeit each responds to different physical characteristics of the formation and the interstitial fluids. More importantly, resistivity log responses must be calibrated by (and reconciled with) core measurements, if confidence is to be placed on the log interpretations.

A detailed description of logging tools and data interpretation is outwith the scope of this book, although the following points should be recognised.

8.1.1.1 Induction Logs

Focused induction logs are typically used in wells drilled with oil-based mud. The short normal log has a shallow depth of investigation and is designed to measure the resistivity of the invaded zone. The short normal records good resistivities in beds down to about 4-ft. thick. The size of the borehole and mud resistivity influences the tool response to some extent. The dual induction-focused log is designed to provide the resistivity data required to estimate the effects of invasion so that more reliable values of true formation resistivity can be obtained. The measurement devices are made up of a deep investigation, a medium investigation and a shallow investigation, guard-type, focused device. The separation between the curves produced by each device can be used to estimate the extent of invasion and whether variations in resistivity are due to a change in water saturation or porosity. The amount of separation between the curves depends on the depth of investigation and the ratio of the resistivity of the mud filtrate, Rmf, to the formation resistivity, Rw. Corrections are made to the induction logs for borehole/tool geometry, borehole conductivity, bed thickness and invasion effects.

8.1.1.2 Laterolog

The laterolog is designed to produce reliable resistivity measurements in boreholes containing saline drilling muds. A drop in potential is caused by the flow of current through the surrounding formation to a remote current return electrode. This potential difference is related to the resistivity of the formation. The induction log, in comparison, measures conductivity and sees the different zones as parallel circuits of resistors. The laterolog is a superior tool in high-resistivity formations and will obtain good values of resistivity in beds thicker than 2 ft. In highly resistive formations drilled with saline mud, laterologs with different depths of investigations are often recorded to obtain the true resistivity, Rt. As with the induction log, it is necessary to make environmental corrections for borehole and invasion effects.

8.1.2 Water Saturation Interpretation—Archie's Law

Before Gus Archie published his seminal work on rock resistivity (Archie, 1942), the resistivity log was purely qualitative. For example, for a constant water resistivity, an increase in rock resistivity, Rt, could result from a reduction in porosity or an increase in hydrocarbon saturation. Archie, working with samples from high-porosity clean sands (where the grains are assumed to be non-conductive), defined a fundamental set of equations which establishes the quantitative relationships between porosity, electrical resistivity and hydrocarbon saturation of reservoir rocks.

Archie's experiments showed that the resistivity of a clean formation is proportional to the resistivity of the brine saturating the rocks. The constant of proportionality is termed the formation resistivity factor or formation factor (F or FF or FRF):

$$F = \frac{\mathrm{Ro}}{\mathrm{Rw}}$$

Ro is the resistivity of a rock 100% saturated in brine with a resistivity, Rw.

The value of F, for a given rock sample, remains constant for a wide range of Rw values, provided the rock matrix is non-conductive. Thus, if we can measure the resistivity of the rock saturated in brine, and the brine resistivity, we can determine F.

Consider, for example, a rock with 20% porosity. If the rock's porosity is considered to consist of a parallel capillaries equal in length to the length of the rock, then the ratio of the total cross-section of the area of the capillaries to the total cross-sectional area of the rock sample would be equivalent to the porosity. The measured resistance of the rock sample, r, would be five times the resistance of a water sample of the same area, A, and length, L, since:

$$r = \frac{\mathrm{Rw}L}{\phi A}$$

In this case, the formation factor would be equal to 5. However, Archie's work showed that for a real rock with 20% porosity, F would be about 20—four times higher than that predicted from the capillary bundle model. The reason for this is the nature of the pore pathways in real rock, as opposed to capillaries. The porosity of a real rock is distributed through intergranular channels that are effectively longer than the length of the rock. Thus current must take a more tortuous flow path, so that the potential difference between the inlet and outlet of the path must be greater, so that resistance (and resistivity) must increase. The net result of the tortuosity effect is to cause the resistivity of the rock to be considerably greater than the product of Rw and $(1/\phi)$.

The results of Archie's experiments included an empirical relationship between formation factor and porosity:

$$F = \frac{1}{\phi^m}$$

The cementation exponent (often referred to as the porosity exponent) is determined by the type and degree of cementation holding the rock grains together and may numerically vary from about 1.3 in loose, unconsolidated sands to as high as 3.0 in carbonates though values of around 2.0 are considered typical.

A more general form of the Archie equation is:

$$F = \frac{a}{\phi^m}$$

where a is an empirical constant.

A value of a other than 1 is often used to compensate for variation in compaction, pore structure and grain size in the F and porosity inter-relationship. The numerical value generally falls between 0.6 and 1 and is often referred to as the tortuosity factor or cementation intercept.

Other relationships, based on the general form of the Archie equation, include:

$$F = \frac{0.62}{\phi^{2.15}}$$

Humble formula (Winsauer et al., 1952) and;

$$F = \frac{0.81}{\phi^2} \quad \text{Tixier formula}$$

These formulas are often employed in quick look interpretations until SCAL data are acquired.

Since hydrocarbons increase resistivity, (Archie, 1947) postulated that the rock resistivity, Rt, at some partial saturation, Sw, will increase. The increase in hydrocarbon saturation will reduce the pore volume filled by water, ϕ_w so that, $\phi_w < \phi$. Consequently, resistivity measurements will provide an apparent formation factor, F_a:

$$F_a = \frac{Rt}{Rw}$$

Therefore,

$$F_a = \frac{a}{\phi_w^m}$$

Water saturation can be expressed as:

$$Sw = \frac{\phi_w}{\phi}$$

If both sides of the equation are squared:

$$Sw^2 = \frac{\phi_w^2}{\phi^2}$$

the appropriate terms from the basic Archie equation relating porosity to Rw may be substituted:

$$\phi_w^2 = \frac{Rw}{Rt} \text{ and } \phi^2 = \frac{Rw}{Ro}$$

so

$$Sw^2 = \frac{Rw/Rt}{Rw/Ro}$$

which reduces to:

$$Sw^2 = \frac{Ro}{Rt}$$

The inverse ratio Rt/Ro is often referred to as the resistivity index, I or RI:

$$RI = \frac{Rt}{Ro}$$

so that

$$I = \frac{1}{Sw^2}$$

A more general form of the Archie saturation equation results from replacing the exponent 2, with n, referred to as the saturation exponent, so:

$$Sw^n = F\frac{Rw}{Rt}$$

Incorporating Archie's generalised equation for formation factor, the water saturation equation can be rewritten as:

$$Sw = \sqrt[n]{\frac{aRw}{\phi^m Rt}}$$

which is the fundamental equation on which water saturation interpretation from resistivity logs in clean (non-conductive matrix) sands is based.

8.1.3 Water Saturation Interpretation—Shaly Sands

Archie's equations explicitly assume that the rock matrix is non-conductive. However, clay minerals, by virtue of their structure and cation exchange capacity (CEC), allow current to pass through them. The presence of conductive minerals in the matrix therefore invalidates Archie's equations which implicitly assumes that the grains are non-conductive and ion transport can only occur through the pores. The presence of clays in a rock can permit some of the ions to be transported through the clay mineral/clay-bound water (CBW) structures. In this case, a modified approach is used that introduces a shale (or clay) conductivity term into Archie's Sw relationship. This additional term has been variously interpreted in terms of the physical properties of the clays, but the two most common groupings are models that use the wetted shale volume, Vsh, and models that use the properties of chemically bound (double-layer) clay water.

The basis of the modification is mathematically described by:

$$\frac{1}{\mathrm{Rt}}=\frac{1}{F\mathrm{Rw}}\mathrm{Sw}^{n}+X\mathrm{Sw}^{p}$$

where X is the extra conductivity introduced by the clays and p is the Archie saturation exponent modified for the clay effects.

8.1.3.1 Vshale Models

It is argued that shales are such fine-grained sediments that they have porosity but no permeability to allow hydrocarbons to enter or to contribute to fluid flow. Hence, the total porosity (ϕ_t) of a shaly sand should be discounted by the porosity of the shale present in it, so the effective porosity is:

$$\phi_e=\phi_t-\mathrm{Vsh}\phi\mathrm{Sh}$$

Vsh is the ratio of the volume of wetted shale to the volume of the rock and includes the CBW. In the generalised generic form of the shaly sand equation, the term X is replaced by

$$X=\frac{\mathrm{Vsh}}{\mathrm{Rsh}}$$

where Rsh is the resistivity of the shale.

There are many Vsh equations that have been used in shaly sand evaluation. Some common implementations are:

$$\frac{1}{\mathrm{Rt}}=\frac{1}{F\mathrm{Rw}}\mathrm{Swe}^{n}+\frac{\mathrm{Vsh}}{\mathrm{Rsh}}$$

Simandoux (1963) equation

$$\frac{1}{\text{Rt}} = \frac{1}{F\text{Rw}}\text{Swe}^n + \frac{\text{Vsh}}{\text{Rsh}} \cdot \text{Swe} \quad \text{Modified Simandoux equation}$$

$$\frac{1}{\sqrt{\text{Rt}}} = \frac{1}{\sqrt{F\text{Rw}}}\text{Swe}^{n/2} + \left(\frac{\text{Vsh}}{\sqrt{\text{Rsh}}}\right)^{1-(\text{Vsh}/2)}\text{Swe}^{n/2}$$

Indonesian equation (Poupon and Leveaux, 1971)

where Swe is the effective water saturation.

However, Vsh methods implicitly assume that the electrical properties of dispersed clays within the pore systems of sandstones are the same as the detrital clays in layers adjacent to the sandstones. There are no scientific justification and experimental support for this supposition, and there are many log clay indicators that give different values of Vsh (Spooner, 2014).

8.1.3.2 Double-Layer Models

The common feature of all these models is the recognition that it is not the amount of shale, but the CEC (related to Qv), that is, the critical parameter for the quantification of both clay conductivity (for Sw calculation) and CBW.

The amount of CBW defined via Qv is less than ϕ_tVsh because only the water adsorbed on the clay mineral surfaces (proportional to CEC), and the water required for the hydration of the clay counterions (CEC) are included—capillary-bound 'far' water in the shales is excluded.

The two principal equations used in double-layer models are:

$$\frac{1}{\text{Rt}} = \frac{1}{F^*\text{Rw}}\text{Swt}^{n^*} + \frac{B\text{Qv}}{F^*}\text{Swt}^{n^*-1}$$

Waxman – Smits equation (Waxman and Smits, 1968)

$$\frac{1}{\text{Rt}} = \frac{1}{F_o\text{Rw}}\text{Swt}^n + \frac{(\text{Cbw} - \text{Cw})V_q\text{QvSwt}^{n-1}}{F_o}$$

Dual – Water equation (Clavier et al., 1977)

where Swt is the total water saturation. The term n^* is the intrinsic saturation exponent, and F^* and F_o the intrinsic formation factor appropriate to shaly sands. In the Waxman–Smits model, Qv is the CEC per unit pore volume and B is the counterion equivalent conductance of sodium clay exchange cations (a measure of cation mobility which is related to clay water conductivity). The equivalent terms in the Clavier et al. dual water model are the clay-bound water conductivity, Cbw, and V_q, the volume of clay associated with 1 meq of clay counterions.

8.1.4 Water Saturation Interpretation Parameters from Core

Core analysis data provide essential input to the quantitative interpretation of water saturation from resistivity logs in both clean and shaly sands using the

Archie and Waxman–Smits shaly sand models. In the following sections, the measurements, interpretation and quality control of experimental data are described in relation to acquisition of the following parameters:

Archie model

Formation factor, F, FF or FRF, and cementation/porosity exponent, m
Resistivity index, RI or I, and saturation exponent, n

Waxman–Smits model

Qv (wet chemistry and multiple salinity tests)
Intrinsic formation factor, F^*, and intrinsic cementation/porosity exponent, m^*
Intrinsic resistivity index, RI* or I^*, and saturation exponent, n^*

8.2 FRF TESTS

The purpose of the FRF test is to determine the formation water resistivity and the resistivity and porosity of the core sample saturated fully in formation water. As ambient condition data are adversely affected by plug artefacts, cementation exponent must be obtained at representative stress conditions.

The combined porosity/FRF test is the recommended method to determine porosity compaction response to the imposition of reservoir stress in the cementation exponent calculations. Some labs determine porosity from helium pore volume on dry samples before saturating the samples and measuring FRF. Tests on dry samples using helium pore volume for porosity in FRF tests are not representative of the reservoir saturation; the helium porosity is adversely influenced by plug shape and surface topology; and stress cycling on the same samples (dry measurements then saturated FRF) can produce incompatible results. Only porosity and resistivity from saturated sample tests should be used in petrophysical evaluation of formation factor and cementation exponent.

8.2.1 Sample Selection and Preparation

Core analysis samples selected for measurement should be representative of the lithologies in the reservoir. This is particularly important for cementation exponent determination. The range of core porosities used in SCAL testing should reflect the porosity distribution of the RCA database. The derived cementation exponent could be affected if only a narrow range of porosities are covered. Ten samples per facies/reservoir rock type are normally considered to be the minimum required for statistical validity.

Tests are usually performed on plug samples that have been cleaned and dried. Care should be taken in the selection and application of cleaning and drying procedures to avoid damage to the samples and to preserve the clay structures. These methods are discussed in Chapter 4. After cleaning and drying, base air or Klinkenberg permeability at nominal confining pressure and porosity are measured, as described in Chapter 5. Helium grain volume is measured using Boyle's law procedure. To avoid any possible contamination from highly conductive mercury, mercury immersion bulk volume should not be performed at this stage. Initial porosity is determined from helium grain volume and brine immersion bulk volume, or for sleeved or mounted samples, from helium grain volume and helium pore volume. A final measurement of mercury bulk volume can be made at the end of the test sequence, if required.

If clays are particularly sensitive, then it may not be advisable to dry the samples: the plugs can be flush cleaned to methanol which is then miscibly displaced with formation water. The advantage of this approach is that clays may be better preserved: the principal disadvantage is that the pore volume and porosity of the test sample are difficult to determine in advance and can only normally be obtained or verified at the end of the experiment when the sample is dried.

Test formation water (synthetic formation water, SFW) is synthesised and characterised (density, viscosity, resistivity) as described in Chapter 6. SFW resistivity (Rw) is determined as a function of three to four temperatures covering the range of typical laboratory variation using a two-platinum electrode, pre-calibrated glass cell contained within a temperature-controlled air bath (incubator) or oil bath. The cell calibration factor used to calculate resistivity must be reported. Electrical measurements are made over a range of AC frequencies (typically 10–50 kHz). Rw is determined from data measured at a frequency which produces the minimum phase angle and is normalised to a standard temperature (20 or 25 °C unless otherwise specified).

The test samples are then saturated in SFW using vacuum saturation to assist the saturation process, then at pressures around 2000 psi for 3–5 days. Vacuum saturation should be limited to around 30 min to prevent water evaporation increasing the brine salinity. The sample pore volume is calculated from gravimetric measurements before and after saturation, and bulk volume is determined from Archimedes immersion in SFW. The re-saturation pore volume must agree with the helium pore volume within ±0.2 p.u. before continuing with the tests. If undersaturated, then the samples must be replaced in the pressure saturator. If the sample appears oversaturated, then grain loss should be checked and grain loss corrections applied accordingly.

On completion of the test sequence, and provided the sample is not scheduled for any other electrical property tests, the sample is cleaned in methanol (solvent flush or Soxhlet), dried and bulk volume finally determined, using mercury immersion methods. To check for any sample damage, helium grain volume and base permeability at nominal stress should also be remeasured.

8.2.2 Test Equipment

8.2.2.1 Ambient Conditions

Tests carried out at ambient (unconfined or nominal pressure) test conditions are used solely to ensure that resistivity (ionic) equilibrium has been achieved, prior to commencing the test under overburden conditions. Data from these ambient condition tests should not be used in petrophysical interpretation.

Measurements of rock resistance at ambient conditions can be made using either two- or four-electrode systems in a rock resistivity cell. Examples of two- and four-electrode systems are shown in Fig. 8.1. In two-electrode systems, platinised silver gauze electrodes are applied to the ends of the sample and these act as both voltage and current electrodes. Any air space between the electrodes, created by a poor contact between the electrodes and the end faces of the samples, will increase the measured resistance so that a contact is required which permits current to flow and conforms exactly to the end faces. This is achieved either by using saturated filter pads or by conductive rubber (rubber impregnated with silver-coated aluminium balls).

Four-electrode systems minimise contact resistance effects on the measurement by having a separate pair of voltage electrodes which are slightly offset from the current (terminal) electrodes along the length of the core. Although ample literature exists, the advantages and disadvantages of the two methods remain relatively unexplored (Bona et al., 2008). Four-contact systems are believed to be less affected by electrode polarisation and high contact resistances at low frequencies (<10 kHz) than two-contact systems. In general, two-contact measurements are preferable above 10 kHz, while four-contact systems seem more reliable at lower frequencies. Contact resistance artefacts in two-electrode systems are minimised by design of the core contacts, the application of a constant pneumatic pressure to the contacts and the ability to 'trim' out electrode and lead resistance at the contact electrodes.

A multi-frequency meter is used to determine resistance, frequency and phase angle. Data are acquired at minimum phase angle (normally less than 0.5°). Where the voltage and current sine waves are significantly out of phase—defined by the phase angle—then what is measured is not resistance but impedance: resistance plus capacitance. As the voltage and current sine waves get closer into phase, the phase angle reduces and the capacitance effect diminishes. This means that the impedance measurements made at low phase angles—0.5° or less—essentially represent resistance. Figure 8.2 plots impedance and phase angle versus AC frequency. The standard for SCAL used to be around 1 kHz but, in this case, the phase angle would be around −2.0° so there is capacitive effect. The current standard is between 20–50 kHz which would normally result in a low phase angle and minimal capacitance.

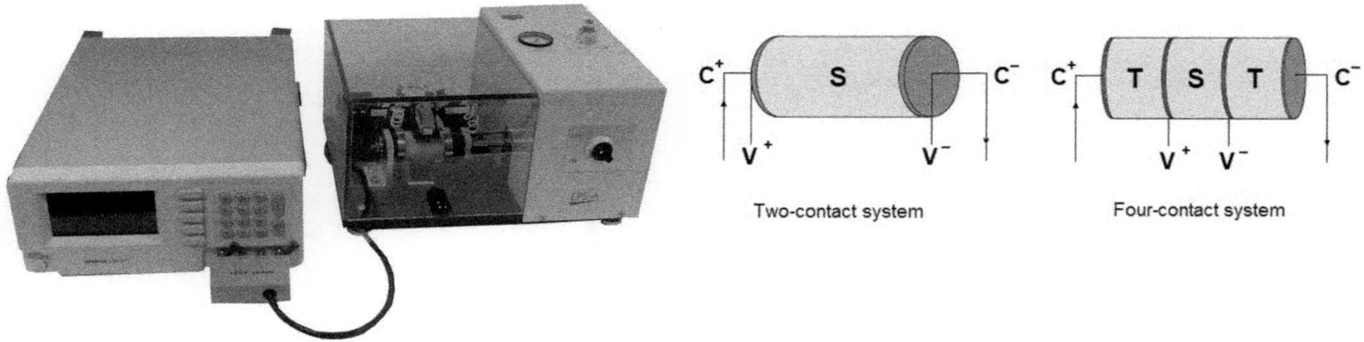

FIGURE 8.1 Example ambient condition rock resistivity cells: 2 electrode (left - *Courtesy of Vinci Technologies*) and 4 electrode (right - *After Bona et al. (2008)*).

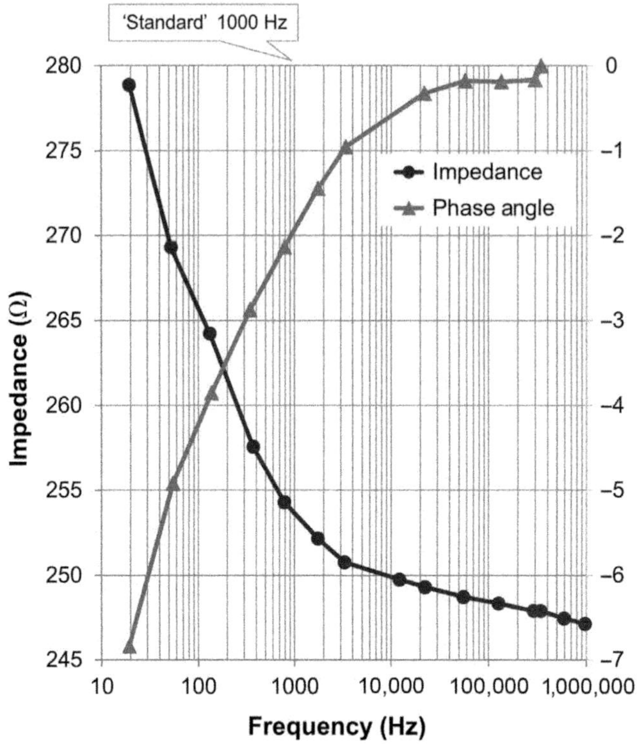

FIGURE 8.2 Effects of AC current phase angle on impedance. *After Al-Mjeni et al. (2002).*

8.2.2.2 Stress (Overburden) Conditions

The tests are performed normally using a two-electrode isostatic overburden cell capable of operating at the specified maximum test stress. Four-electrode systems are rarely used as they are difficult to manufacture in an overburden cell and the potential electrodes are not robust.

The cell (example shown in Fig. 8.3) is designed such that the stress acting on the core end faces is equal to the stress acting radially around the plug (isostatic loading). The end platens are electrically insulated from the coreholder body and are provided with fluid inlet and outlet pipes. Electrical connections are normally made to the stainless steel pipes. The outlet pipe is connected to a graduated micropipette to enable measurement of brine expulsion volumes, and hence pore volume compaction, on application of overburden stress. Brine evaporation from the pipette is prevented by a thin oil film on top of the water.

A platinum- or silver-impregnated membrane is placed between each end stem to ensure good electrical contact. In resistivity index tests, a porous plate or semi-permeable membrane is located at one end of the cell. These are saturated in SFW. They can be designed such that they allow a consistent

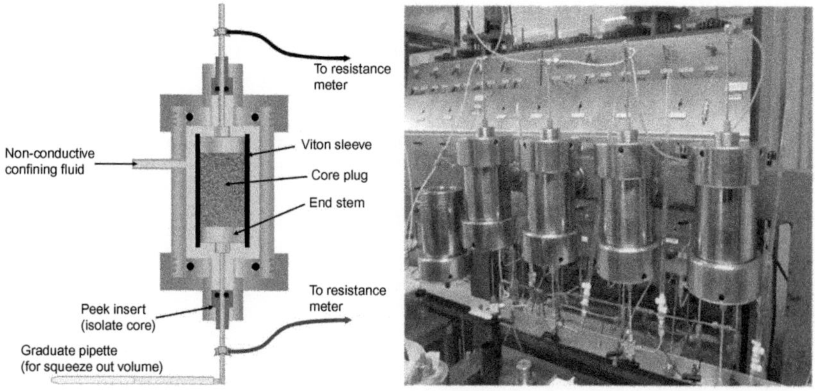

FIGURE 8.3 Example overburden condition rock resistivity cells: schematic (left) and typical lab set-up (right).

electrical flux to be applied to the end face of the plug sample. The pore volume reduction and resistivity increase of the porous plate as a function of confining stress may have to be determined to isolate the pore volume and resistivity response of the plug sample from the composite plug/membrane dataset.

A multi-frequency meter is used to determine resistance, frequency and phase angle. Data are acquired at minimum phase angle.

8.2.3 Test Procedures—Ambient (Base) Conditions

Following saturation, the plugs are stored in SFW for at least 3 days before commencing FRF measurements to allow ionic equilibrium to become established. Measurements of resistivity are made on a regular basis, over a period of at least 5 days, until temperature-corrected resistivity, Ro, becomes stable ($\pm 2\%$, or better) and such that successive measurements neither monotonically increase or decrease.

The Arps (1953) equation is used to correct both brine and rock resistivity data (R_l) measured at the prevalent lab temperature (T_l) to a standard or reference temperature (T_r) in degrees Celsius:

$$R_r = R_l\left(\frac{T_l + 21.5}{T_r + 21.5}\right)$$

As noted by McPhee (2012) for ambient measurements of fully saturated plugs, brine can cling to the surface of the plug after saturation. This must be removed since it is not part of the sample pore volume. Otherwise, a layer of brine on the sample surface will provide a conduit for current flow so that the measured resistance (and resistivity) will be too low. Worthington (1975) estimated that a 2.5-μm thick film of excess brine will produce an underestimation of F by nearly 30% in low-porosity samples. Thus, failure to remove excess brine results in a calculated ambient condition Ro that is too low. This results in an *m* that is too low, and an ambient condition saturation exponent, *n* that is too high.

Unfortunately, the methods used by different laboratories to remove excess brine can be substantially different and are subjective. Excess brine should be removed using a consistent method, either:

- by placing the plug in a porous plate cell and applying a very small capillary pressure sufficient to remove brine from the surface of the plugs. This applied pressure must be below the threshold pressure (capillary entry pressure) of the plug, or by;
- by rolling the sample (two circumferential rolls) on an unbleached paper towel.

Electrical measurements are made over several frequencies to ensure that the resistance is measured at minimum phase angle, and the test temperature data recorded for each measurement. The rock resistivity is determined from data measured at a frequency which produces the minimum phase angle ($<0.5°$) and is normalised to a standard temperature (20 or 25 °C or as otherwise specified by the lab) using the Arps equation.

Rock resistivity, Ro, is calculated from sample resistance normalised to the standard temperature (Ω m) and the length (L in m) and cross-sectional area (A in m^2):

$$\text{Ro} = z\left(\frac{A}{L}\right)$$

In four-electrode systems, the length, L, corresponds to the length between the current electrodes, which will be less than the plug length.

Regular measurements of saturated weight are required to assess the extent of grain loss during this stage.

8.2.4 Test Procedures—Stress (Overburden) Conditions

8.2.4.1 Sample Loading

Following the ambient condition tests, the saturated test plug is loaded into the overburden test cell. To ensure that full plug saturation and brine continuity in the test system are maintained, and to prevent air entering the system, the plugs are loaded under SFW and are initially confined at low pressure (e.g. 20 psig), then flushed with brine to remove any air from the annulus between the plug and the sleeve, the end stems and measurement system. The samples are then loaded to the desired test stresses by increasing the confining pressure on the rubber sleeve. The volume of brine produced during stress loading is used to determine stressed pore volume and stressed porosity.

8.2.4.2 Annulus Volume/Conformance Pressure Correction

There will be brine in the annulus between the surface of the core plug and the inside of the rubber confining sleeve at the beginning of the overburden tests. As stress is applied, the sleeve will progressively conform to the surface of the sample, and this annulus brine will be expelled and collected in the pipette. The accurate determination of the annulus volume expelled until the core sleeve exactly conforms to the plug surface, and the stress at which that is achieved—the sleeve conformance pressure (SCP)—are essential elements of these tests and must be satisfactorily established for each test plug. Once the annulus volume (Vann) at SCP has been determined, a volume correction is applied to the raw brine expulsion data (Vexp) to determine the corrected expelled volume at each overburden stress.

$$\text{Vexp}_{\text{corr}} = \text{Vexp} - \text{Vann}$$

Many laboratories assume that the SCP and annulus volume are the same for all samples. An example is shown in Fig. 8.4 where the normalised porosity compaction factor (PCF) is plotted versus confining stress. The lab has assumed that the SCP is about 320 psi such that the porosity at 320 psi is set equal to the ambient porosity (0 psi). In fact, SCP is probably much less than 320 psi so that the corrected volume Vexp will be too low and calculated stressed porosity too high. In reality, SCP and Vann depend on the thickness of the rubber sleeve and the surface topology and shape of the plug, which

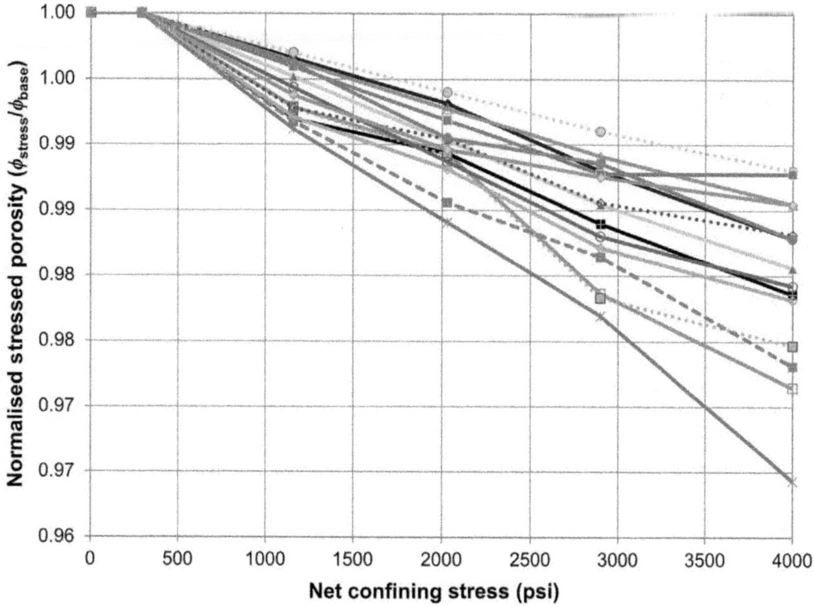

FIGURE 8.4 Normalised stressed porosity ($\phi_{stress}/\phi_{base}$) versus net confining stress.

will be different for each plug. Therefore, the SCP/annulus volume must be determined for each plug.

The procedure used to determine the annulus volume and SCP is to increase the confining pressure in small increments (e.g. 20–30 psi) and record both the brine volume expulsion values, and Ro, until the SCP/annulus volume has been defined. This is normally indicated by a change in the slopes of both resistance and/or expelled volume versus confining pressure curves. An example of expelled volume versus stress is provided in Fig. 8.5. There are two volume-confining pressure relationships: the first principally relates to production of brine from the sleeve/plug annulus (with some limited pore volume compaction) and the second (at higher stresses) relates only to pore volume compaction. The point of inflexion between the two curves defines the SCP (around 105 psi in this case). The annulus volume can be obtained by extrapolating the pore volume compaction curve at confining pressures greater than 105 psi to zero confining stress (1.75 ml in this case). The annulus volume must be subtracted from the cumulative volume expelled to estimate the true pore volume reduction.

At least four pressure increments below SCP and three increments above SCP are required to fully define the volume–pressure curve. A minimum period of 5 min should be allowed at each pressure, and resistance and volume expelled monitored as a function of time at each pressure.

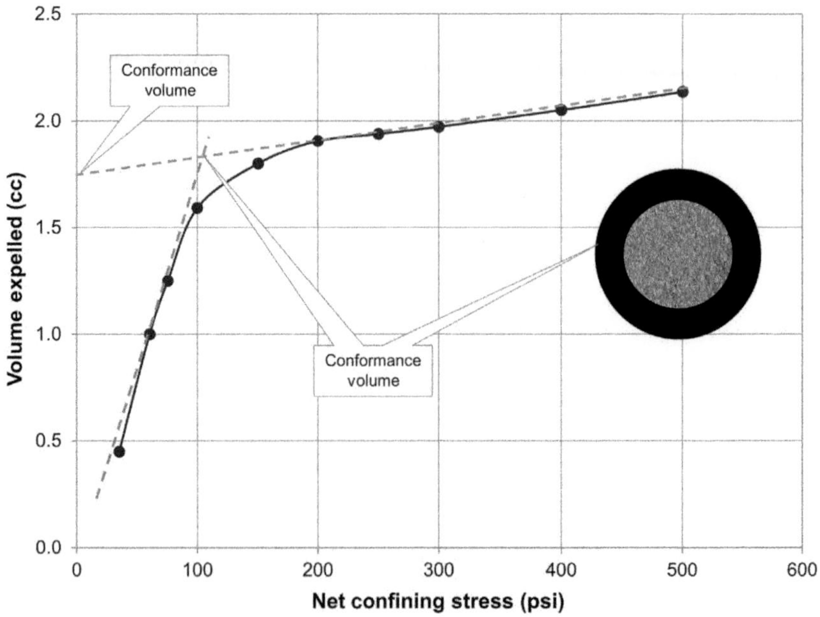

FIGURE 8.5 Example of sleeve conformance pressure determination.

Determination of the annulus volume can be extremely subjective and will vary depending on plug size, plug shape, plug surface topography and the thickness of the rubber sleeve. The expelled volume and resistance data, used to define the SCP, must be reported for each plug tested, so that alternative interpretations of the annulus volume can be made if required.

8.2.4.3 Stress Stations

Once SCP and annulus volume are established, the net confining stress is then increased. It is recommended that the measurements are made over a range of net confining stresses to define the full compaction/resistivity trend. These data are essential for test quality control and can be used to assess pore collapse. The data can also be used to interpolate (and possibly extrapolate) to intermediate (or higher) stresses if the equivalent effective isostatic reservoir stress is different from the stress measurement station. Measurements made at a single overburden test station are not recommended as they are impossible to quality control, and measurements might have been made under overstressed conditions where the rock might have failed on compaction.

Measurements are normally made over five or six test stress stations up to and including the estimated reservoir effective stress. An example is shown in Table 8.1. Stress stations should cover the range of uncertainty in the reservoir equivalent isostatic stress estimates. Note that the loading condition of increasing confining stress is not recommended for establishing petrophysical parameters (porosity, resistivity) on depletion as a completely different loading mechanism is involved. If these data are required, modified pore volume compressibility (K0) tests (see Chapter 12) are recommended.

TABLE 8.1 Example of Stress Stations (Reservoir Stress ~4000 psi NCS)

Test Station	Isostatic Confining Pressure (psi)	Formation Factor	Porosity	Brine Permeability
1	400	✓	✓	✓
2	800	✓	✓	✓
3	1200	✓	✓	✓
4	40% reservoir[a]	✓	✓	✓
5	60% reservoir[a]	✓	✓	✓
6	Reservoir	✓	✓	✓

[a]*If greater than 1200 psi.*

The loading rate should not exceed 1000 psi/h, and at least 24 h should be allowed to achieve stability at each stated overburden stress if the stress response is inelastic. In elastic materials, stress equilibration is achieved more rapidly. Fluid production and Ro are monitored at each specified stress station to ensure compaction stability before taking the final measurements.

Electrical measurements are made over a range of frequencies to identify the minimum phase angle, and the test temperature data are recorded for each measurement. All Ro data are normalised to a standard temperature (20 or 25 °C unless otherwise specified) using the Arps equation.

8.2.4.4 Stressed Porosity Determination

Porosity at stress (ϕ_{stress}) is determined from

$$\phi_{\text{stress}} = \frac{\text{Vp}_{\text{stress}}}{V_{\text{g}} + \text{Vp}_{\text{stress}}}$$

The pore volume at stress is determined from ambient condition pore volume (Vp_{amb}) and the annulus volume corrected volume expulsion data:

$$\text{Vp}_{\text{stress}} = \text{Vp}_{\text{amb}} - \text{Vexp}_{\text{corr}}$$

Reservoir rock grain compressibilities are very low, around $0.16\text{–}0.18 \times 10^{-6}\,\text{psi}^{-1}$ for sandstones, and so the grain volume, Vg, is assumed to be incompressible and constant in this analysis.

When using stressed core porosity data from FRF tests to convert RCA porosities to reservoir stress conditions, the SCAL test must be carried out a stress station identical to that used for RCA. Thus, if the RCA porosities rep resent unconfined measurements (e.g. $\phi_{0\text{psi}}$ from helium grain volume and mercury bulk volume), then the SCAL test should include unconfined measurements at 0 psi. If RCA porosities represent helium grain volume and helium pore volume at stress (e.g. 400 or 800 psi), then the SCAL tests must also include measurements at the same stress. The PCF can therefore be obtained from the multi-stress SCAL measurements. For example: $\text{PCF} = \frac{\phi_{\text{stress}}}{\phi_{0\text{ psi}}}$ or $\text{PCF} = \frac{\phi_{\text{stress}}}{\phi_{400\text{ psi}}}$ depending on the RCA confining stress.

8.2.4.5 Pore Collapse

If the increase in stress carried by the load-bearing grain framework of the sample is too high, it triggers micron-scale deformation mechanisms such as elastic grain-contact spreading, microcrack growth and closure, cement breakage, grain rotation and sliding and crystal plastic deformation in clay and mica grains (Schutjens et al., 2004). For example, Fig. 8.6 plots deviatoric stress (y-axis) versus effective mean isostatic stress (x-axis) for a typical quartz-rich sandstone. The solid line is termed the cap curve and represents the boundary between stability and instability.

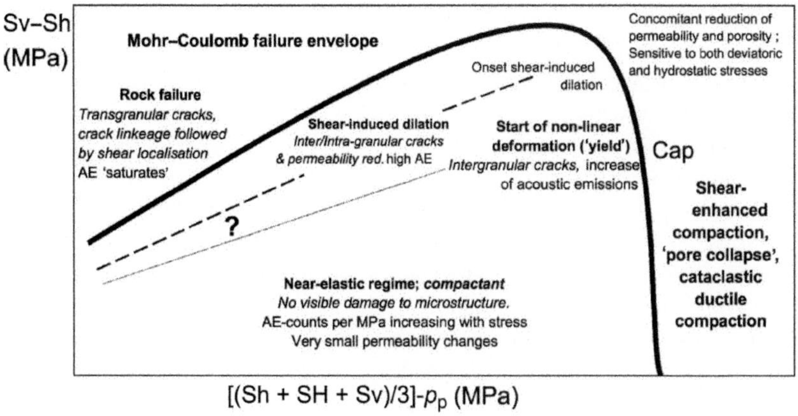

FIGURE 8.6 Conceptual picture of micromechanical and petrophysical response in 3D stress-space. *From Schutjens et al. (2004).*

In the stressed porosity/FRF test, the increase in confining stress produces an increase in effective isostatic stress (x-axis) as the pore pressure is held constant. The deviatoric stress (difference between axial and radial confining stress) is essentially zero under standard isostatic SCAL loading conditions. In an elastic domain, most deformation is recovered upon reloading along the same stress path, except perhaps for some coring-induced microcrack closure. However, if the isostatic compaction stress is high enough, then the rock can enter an inelastic domain where there is an increase in the activity of inelastic mechanisms: grain sliding and grain rotation triggered by intergranular microcracks, intragranular cracks (contact spalling) and cement breakage. This pore collapse causes grain rearrangement and repacking to conditions that are not representative of the intact (elastic) reservoir properties. This can occur under isostatic stress conditions if the mean effective stress is high enough to breach the cap curve.

Pore collapse is progressive and non-recoverable which means that not only are data obtained at stresses higher than the initiation stress invalid, but also the plugs cannot be used for other tests. As a rule of thumb, the initiation stress for pore collapse under isostatic loading is around 8–10 times the unconfined compressive strength value. However, Kaye et al. (1990) found that isostatic loading and stress cycling can cause pore collapse at stresses as low as 250 psi in unconsolidated sands.

It is essential that tests are performed at several stress stations so that the compaction behaviour can be assessed and any signs of pore collapse identified. Data acquired prior to pore collapse will not necessarily be invalid. If petrophysical damage as a result of stress cycling is considered a risk, then stressed tests should be performed on individual plug suites and no further tests should be scheduled on these samples. Repeated stress cycles can weaken the core which may therefore induce pore collapse on subsequent loading cycles.

8.2.4.6 Brine Permeability Measurements

If required, and provided the coreholder is not fitted with a semi-permeable membrane for resistivity index measurements which would block flow, liquid permeability can be determined following Ro and porosity determination. This can be performed at any stress station, but is usually carried out at the equivalent reservoir stress.

Detailed procedures for liquid permeability are provided in Chapter 5. The initial flow rate should be no more than 1 ml/min to prevent possible fines migration. When the flow rate and pressure drop are stable for at least two-pore volumes of brine injected, liquid permeability (kb) is determined from Darcy's law. If required multi-rate measurements might be performed to optimise the accuracy of the measurement but it is important trying to ensure that the flow velocity does not exceed the critical velocity for fines mobilisation.

8.2.4.7 Unloading

The sample is then de-stressed and unloaded. In some cases, measurements of resistivity and porosity are made during the unloading cycle to check for hysteresis. If the plug is not scheduled for other tests, it is cleaned by either Soxhlet extraction or solvent flush cleaning in methanol and dried. The silver nitrate test is used to determine the end point of cleaning.

The plugs are then dried in a convection or humidity oven to constant weight (±0.01 g or better) and base air or Klinkenberg permeability and helium grain volume are measured to check for any alteration. Final sample bulk volume can be determined from mercury immersion if the plug is suitable.

8.2.5 Data Utilisation

8.2.5.1 Porosity Compaction Factor

The SCAL porosity data, measured at a representative reservoir net isostatic confining stress, ϕ_{res}, can be plotted against the equivalent RCA base condition porosity, ϕ_{amb}, for the same plugs and a transform is determined. An example is provided in Fig. 8.7. Transforms are normally linear functions of the form:

$$\Phi_{res} = M_{\Phi\ amb} + C \quad \text{free regression}$$

$$\Phi_{res} = M_{\Phi\ amb} \quad \text{forced (constrained) regression}$$

A constrained (forced-fit) regression algorithm is applied to correct ambient porosity to reservoir conditions if the porosity of the core plugs in the RCA database is less than the lowest ambient porosity of the overburden porosity SCAL database. An unconstrained (free fit) regression better honours data in the SCAL sub-population intervals, but it can be overly levered by

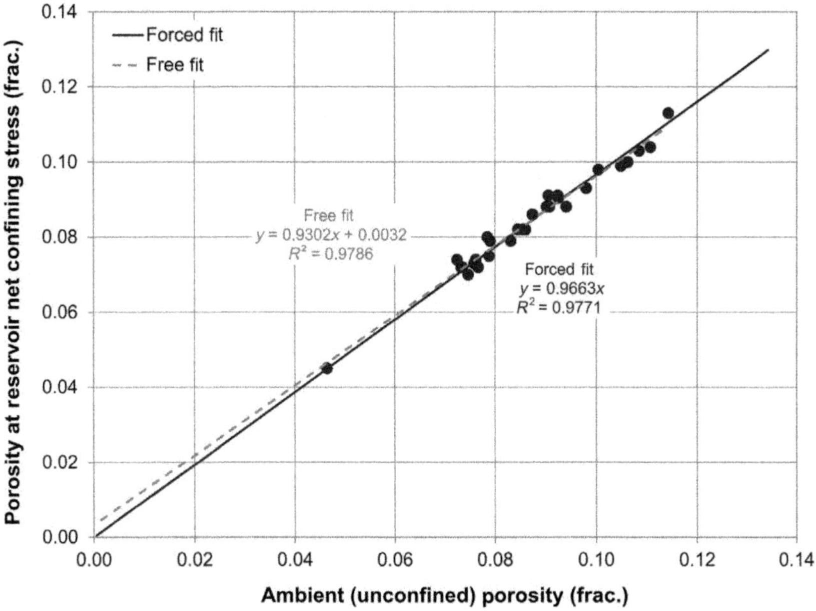

FIGURE 8.7 Example of stressed porosity versus ambient porosity.

inherent uncertainties remaining in the quality-controlled database. Consequently, the constrained regression algorithm is usually preferred, especially if the database is sparse or there is a limited range of porosities.

8.2.5.2 Cementation Exponent

The FRF, *F*, is determined at each confining stress station from:

$$F = \frac{\mathrm{Ro}}{\mathrm{Rw}}$$

where both Ro and Rw are normalised to standard temperature conditions.

For individual plugs, the cementation exponent, *m*, is found from FRF and stressed porosity (fractional) at each specified test station for each sample:

$$m = -\frac{\log(F)}{\log(\phi_{\mathrm{stress}})}$$

An example output is provided in Fig. 8.8.

Normally labs have little knowledge of the reservoir zonation scheme. It is the job of the petrophysicist to identify appropriate reservoir zones (e.g. in terms of facies, hydrocarbon zones or flow units) from log, core and test data and group the core FRF test data accordingly.

The composite cementation exponent *m* for a particular grouping of samples (for example by zone or lithofacies) is found from linear regression

Formation Resistivity Factor
as a Function of Confining Stress, Plus Permeability to Brine at Stress

Saturant, mg/l: 120,995
Resistivity of Saturant, Ω-m @ 25°C: 0.067

Sample ID	Depth (ft.)	Permeability (mD) to Air	to Brine	Porosity (Fraction)	Effective Confining Stress (psig)	Formation Resistivity Factor (F)	Cementation Exponent (m)
20	8728.20	–	–	0.148	0	32.76	1.82
		0.34	–	0.145	400	37.96	1.88
		–	–	0.143	800	40.09	1.90
		–	–	0.142	1500	41.83	1.91
		–	0.02	0.139	3600	44.78	1.93
44	8752.20	–	–	0.154	0	30.61	1.83
		1.4	–	0.148	400	33.92	1.85
		–	–	0.145	800	35.69	1.85
		–	–	0.143	1500	36.84	1.85
		–	0.21	0.140	3600	38.98	1.87
64	8776.25	–	–	0.230	0	15.86	1.88
		183	–	0.224	400	15.92	1.85
		–	–	0.222	800	16.26	1.85
		–	–	0.220	1500	16.50	1.85
		–	74	0.217	3600	17.10	1.86
79	8791.20	–	–	0.279	0	12.67	1.99
		2849	–	0.270	400	11.47	1.87
		–	–	0.268	800	12.70	1.93
		–	–	0.265	1500	13.15	1.94
		–	865	0.262	3600	12.72	1.90
101	8813.00	–	–	0.218	0	17.17	1.87
		294	–	0.211	400	16.20	1.79
		–	–	0.209	800	16.58	1.79
		–	–	0.206	1500	17.12	1.80
		–	77	0.204	3600	18.12	1.82

FIGURE 8.8 Example of suggested data output for porosity compaction, cementation exponent (*m*) and brine permeability tests.

analysis of bi-logarithmic F and ϕ_{stress} data pairs, at the selected *in situ* stress, using the Archie equation:

$$F = \frac{a}{\phi^m}, \text{ or}$$

$$\log(F) = \log(a) - m\log(\phi)$$

where m is the absolute value of the slope of the best-fit regression to the logarithmic data points, and log(a) is the intercept if logarithmic data are plotted on Cartesian coordinates.

If the constant a is set equal to 1, then m is found by 'forced-fit' regression techniques. If the data are plotted on Cartesian coordinates—log(F) versus

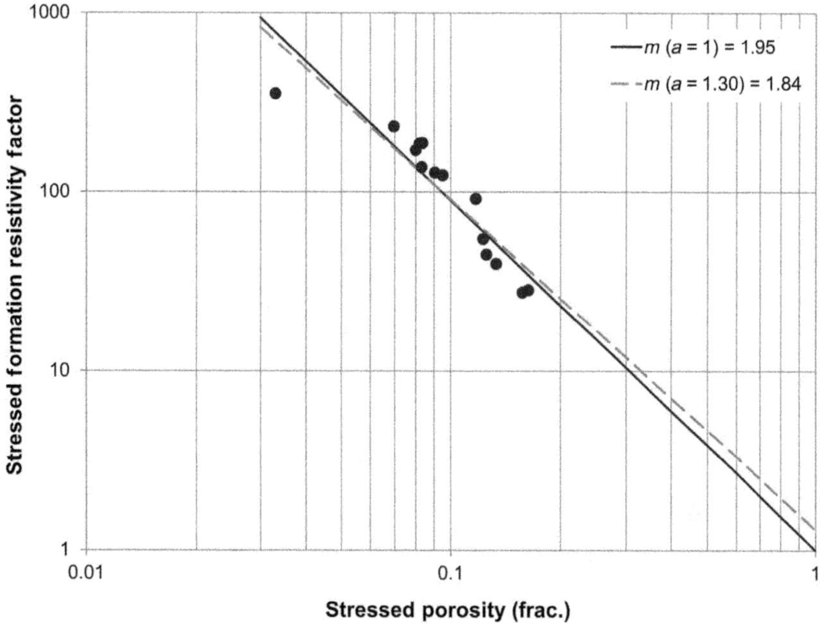

FIGURE 8.9 Cementation exponent plot presentation.

log(ϕ), then the regression is forced through $\phi=1$ and $F=1$ (i.e. log(ϕ)=0 and log(F)=0).

If the generalised form of the Archie equation is used, then a is found by 'free' regression. If a is not equal to 1, then this implies that Ro at 100% porosity is not equal to Rw, as defined by the Archie equation. Debate has raged in the industry over whether a must be unity or not, and whether the value of a has any physical significance. Maute et al. (1992) maintain that a is meaningless when viewed from the perspective of error in water saturation calculations and should always be set equal to unity.

An example of data plot, showing both forced ($a=1$) and free ($a\neq 1$) regressions to the same F–ϕ data set, is provided in Fig. 8.9.

8.2.6 Data Reporting Requirements

The data that need to be reported by the test laboratory to ensure adequate quality control of the FRF tests and test results are listed in Table 8.2.

If brine permeability measurements are also scheduled within the FRF tests, the data reporting requirements are provided in Chapter 5.

8.2.7 Advantages and Drawbacks/Issues

Summaries of the advantages and drawbacks/issues associated with FRF tests are provided in Table 8.3.

TABLE 8.2 Data Reporting Requirements for FRF Tests (for Each Sample)

Data	Comments
Brief description of procedures and experimental apparatus	Include plug preparation method and details of experimental equipment (electrode type, resistivity meter), etc.
Plug depth and orientation	Horizontal or vertical (or other)
Plug length and diameter	*A/L* ratio
Synthetic formation water (SFW) composition	If not already reported
SFW density and viscosity as function of temperature	3 or 4 measurements over lab range
SFW resistivity (Rw) at standard temperature	Ω m. Temperature in °F or °C
Measured initial sample dry weight	
Initial base permeability (Klinkenberg or air permeability) and helium porosity	Porosity determined from He Vg (or He Vp) and SFW immersion Vb
Measured saturated sample weight	
Calculated sample resaturated pore volume	
Calculated sample immersed bulk volume	
Calculated sample saturated porosity (fraction)	Based on saturated Vp and Vb
Calculated sample helium porosity (pre-test)	Within ±0.2 p.u., of saturated ϕ
Resistivity (Ro) at ambient conditions	Per day until stable
FRF at ambient conditions	Based on stable Ro
Loading rate (or total time) to each stress station	psi/min or minutes
Cumulative brine production per stress station	Raw data before applying SCP/annulus volume correction calculation
Sleeve conformance pressure (SCP) and annulus volume	Plot of volume versus stress required to define SCP and Vann. See Fig. 8.5
Calculated porosity (fraction), pore volume and bulk volume at each stress station, based on corrected expelled brine volume	Porosity calculated on basis of incompressible grain volume
Temperature, frequency (minimum phase angle), rock resistance, porous plate disc resistance (if used), phase angle, resistivity (Ro) on completion of the equilibration period, at each stress station	Ro at standard temperature

Continued

TABLE 8.2 Data Reporting Requirements for FRF Tests (for Each Sample)—Cont'd

Data	Comments
FRF at standard temperature per stress station	
Calculated cementation exponent for each plug (*m*) and for the composite plug dataset at each specified stress station	Plot *F* versus ϕ (log scale) for plug dataset at each stress station
Final saturated weight	Establish grain loss
Final dry weight, helium porosity and base permeability (Klinkenberg or air permeability)	Post-test. Compare to pre-test. Porosity from final mercury immersion Vb (provided sample is suitable)
Sample grain loss correction procedures (if applied)	Include typical worked example for one sample

8.2.8 FRF Test Quality Control Issues, Checks and Diagnostics

Analysis of the results should start with a thorough review of the laboratory report to check that acceptable procedures and test methods have been followed.

Ensure that the method used for cleaning and drying the test samples has not caused damage and depreservation of clay structures.

SCP annulus volume data and plots must be provided for each plug sample, and the corrected expelled volumes agreed between the end user and the test lab (see Fig. 8.5).

All experimental data listed in Table 8.2 must be supplied and lab calculations verified.

Phase angles for resistivity measurements must be less than 0.5°. High values could indicate poor contact between the sample and the electrodes, or that the sample has a low length to area ratio.

Porosity as a function of the logarithm of stress; PCF (ratio of stressed to base/ambient porosity) as a function of stress (linear); and PCF as a function of base porosity are the principal diagnostic tools used to audit overburden porosity data. Figure 8.10 provides an example of porosity versus the logarithm of stress. Well-behaved and consistent data should be able to be matched by either first- or second-order polynomial models. The polynomial function parameters can be used to interpolate (or possibly extrapolate) to any stress, for example, if the estimated reservoir stress is different from the test stress stations. Figure 8.11 plots stress-normalised porosity (ϕ_{stress}/ϕ_{amb})—the PCF factor—as a function of stress. Samples which are

TABLE 8.3 Advantages and Drawbacks/Issues Associated with FRF Tests

Advantages	Drawbacks and Issues
• Relatively simple test that normally provides reliable results. • Key test requirement for all SCAL programmes. • Provides two or three data sets (*m*, porosity and optional kw at stress) in one single loading operation. • Porosity compaction data on brine-saturated samples are more representative than stress tests on dry samples (helium pore volume). • Multi-stress station measurements define full compaction curve and can highlight potential experimental artefacts such as excess brine and pore collapse features. • Brine permeability measurements (optional) provide valuable data at little incremental cost.	• If the rock matrix is conductive (e.g. clays), FRF will be a function of brine resistivity. • Correct preparation of brine and its resistivity measurement are critical to the acquisition of reliable data. • Resistivity measurements are sensitive to the frequency used (phase angle should be less than 0.5°). • In PCF determination, it is essential that the ambient (or base) porosity of the overburden test SCAL plugs is measured under identical conditions to the ambient (or base) porosity in the RCA database. The appropriate base porosity should be measured and reported. • Excess brine (especially at FRF ambient conditions) produces artificially low resistivities and low FRF. • Correct establishment of electrical contact between the sample and the electrodes can be difficult, depending on the experimental set-up, the shape and competency of the sample and the skill of the operator. • Accurate determination of SCP/annulus volume is an essential element for porosity calculation and must be satisfactorily established for each test plug. This is especially important for vuggy carbonates. • Porosity data should not be used for pore volume compressibility estimates on depletion as the loading conditions for depletion experiments are different and compressibility will be influenced by micro-fracture closure on initial loading cycle. • Friable samples require to be protected against grain loss and sample failure. Stresses have to be carefully managed to prevent pore collapse in poorly consolidated samples.

outliers or demonstrate a progressive and continual decline in PCF with stress indicate either pore collapse or experimental problems such as loss of the effective seal between the rubber sleeve, and plug or system leaks. The data need to be scrutinised carefully to differentiate between valid rock-type behaviour and experimental artefacts. In the example in Fig. 8.11, stressed porosity measurements were made on samples from the same rock type. Lab A determined SCP and annulus volume for each sample as defined by

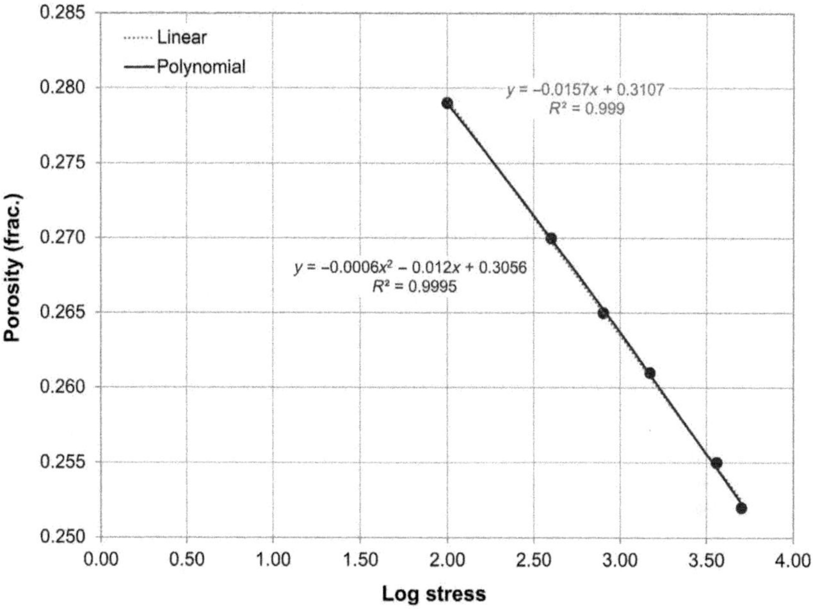

FIGURE 8.10 Porosity versus logarithm of stress for single sample.

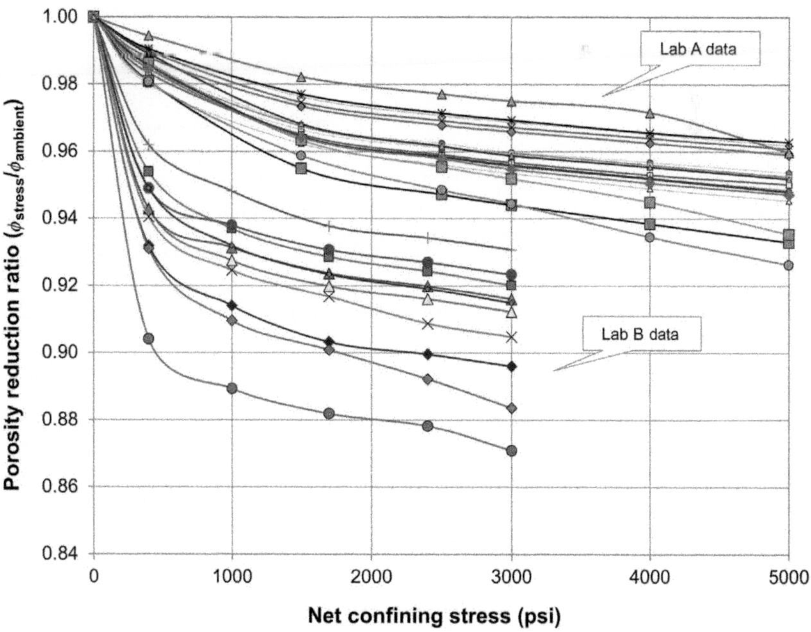

FIGURE 8.11 Porosity compaction factor versus stress.

Fig. 8.5. Lab B assumed an SCP (and corresponding annulus volume) that was too low, which resulted in an apparently lower porosity and more pessimistic PCF at stress. When the Lab B results were used to calibrate the density log interpretation in this gas reservoir (prior to the Lab A results), porosity was underestimated by 7% and gas initially in place by 4%. Some plugs also show signs of pore collapse. Data acquired at stresses higher than the pore collapse initiation stress must be disregarded.

There is an increasing tendency to carry out FRF and porosity tests at a single overburden stress, principally to save money in SCAL programmes. However, this makes it impossible to determine compaction trends which might indicate experimental issues, such as annulus volume correction and pore collapse, as it is only possible to fit the data with a straight line (Fig. 8.12). If measurements were made at stresses which exceed the pore collapse stress, then the data would not be representative and worse, pore collapse could not be identified. In addition, it is not possible to extrapolate to higher stresses nor to interpolate the data to lower stresses if there is uncertainty in the reservoir stress estimate.

Figure 8.13 plots PCF at a selected stress versus ambient porosity. This diagnostic plot is extremely useful in identifying anomalous data sets. In this case, the highlighted data were eliminated as they were subject to experimental error but they could have represented a different, more stress-sensitive rock type.

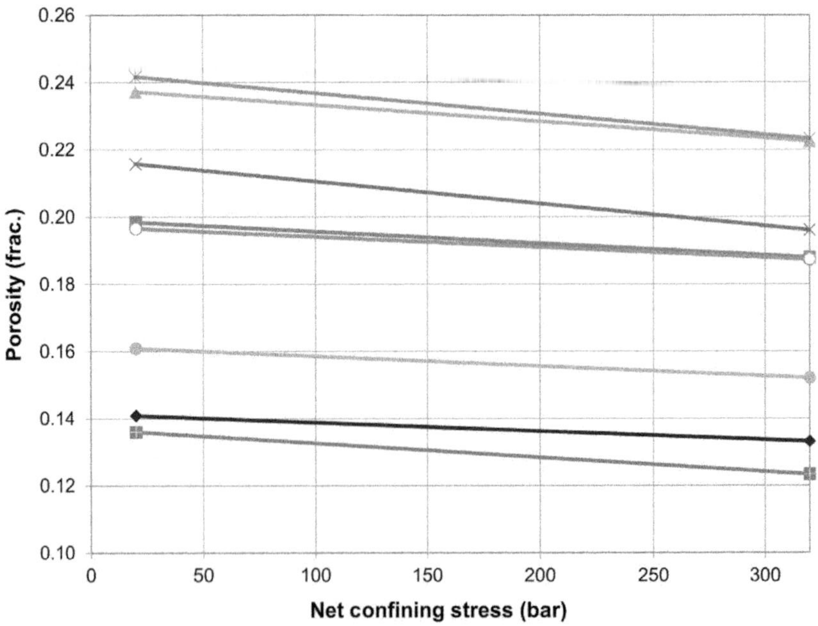

FIGURE 8.12 Porosity at stress—two-point measurements.

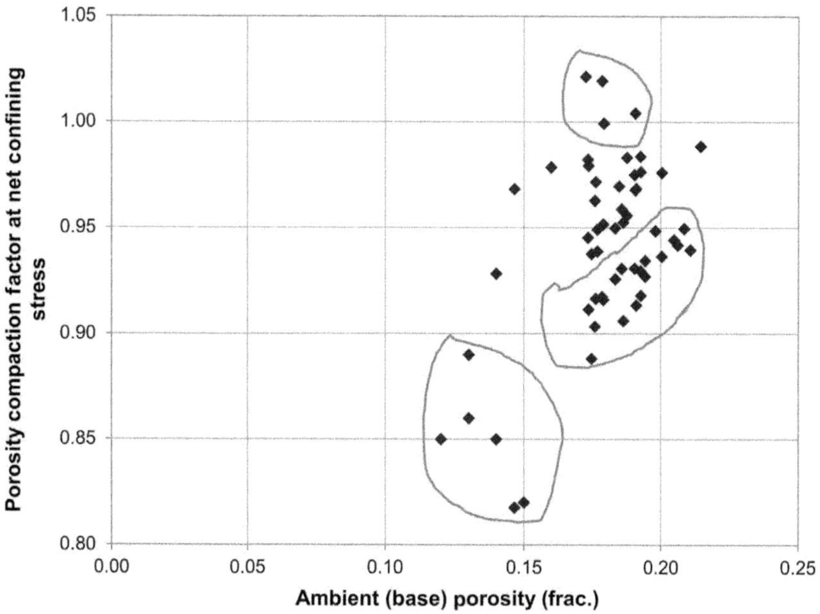

FIGURE 8.13 Porosity compaction factor (at specified stress) versus ambient porosity.

Samples omitted from the quality-controlled porosity compaction database must also be omitted from cementation exponent database in order not to degrade the linkage between porosity and resistivity in the determination of m.

Stress-normalised FRF (ratio of FRF at stress to base or ambient FRF) versus stress can also indicate experimental problems such as excess brine in ambient measurements, potential pore collapse, poor electrical contact and air in the test system. Generally, smooth curves are expected with normalised FRF becoming broadly asymptotic at higher stresses. Values less than 1 (e.g. Fig. 8.14) may be related to poor electrical contact due to sample shape at low confining stress. Samples should not be immediately rejected as electrode contact may have improved at higher stresses. Abrupt changes in slope could be due to measurement errors, air in the system at low stress or pore collapse at high stress.

Surface brine effects are clearly evident in Fig. 8.15, where stress-normalised formation factor is plotted versus stress. FRF apparently increases abruptly between 0 and 200 psi, but this is a direct result of ambient Ro being too low due to surface brine on the plugs. At 200 psi, the surface brine has been expelled by the core sleeve conforming to the plug surface. In this case, excess brine effects result in an average error of −30% error in ambient condition cementation exponent values. The ambient data set can be corrected by determining the 'true' Ro from extrapolation of the F-stress trend to 0 psi (McPhee, 2012).

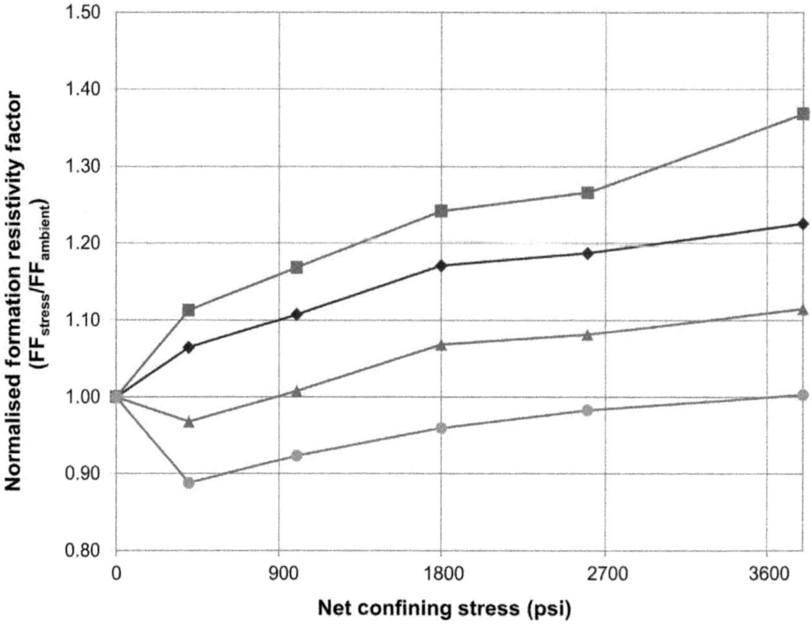

FIGURE 8.14 Normalised formation resistivity factor versus stress (Example 1).

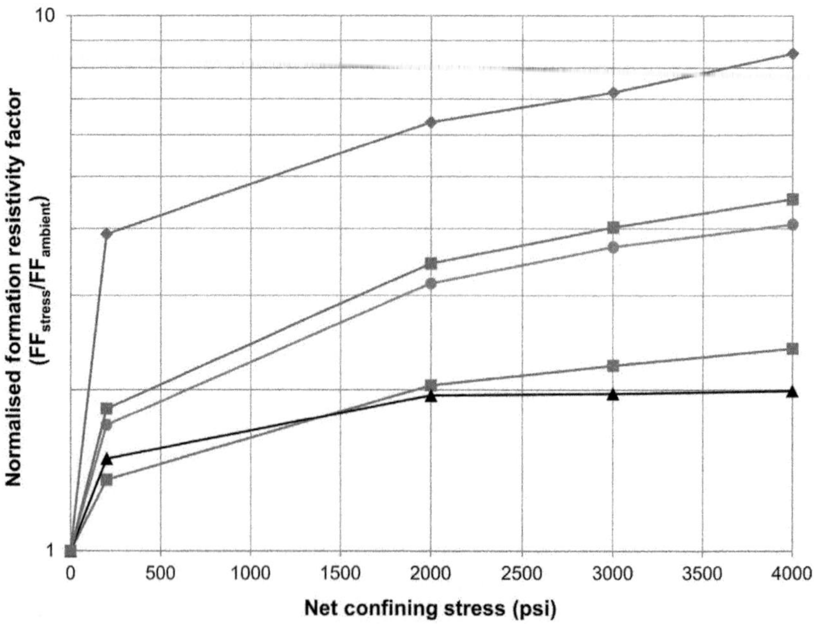

FIGURE 8.15 Normalised formation resistivity factor versus stress (Example 2).

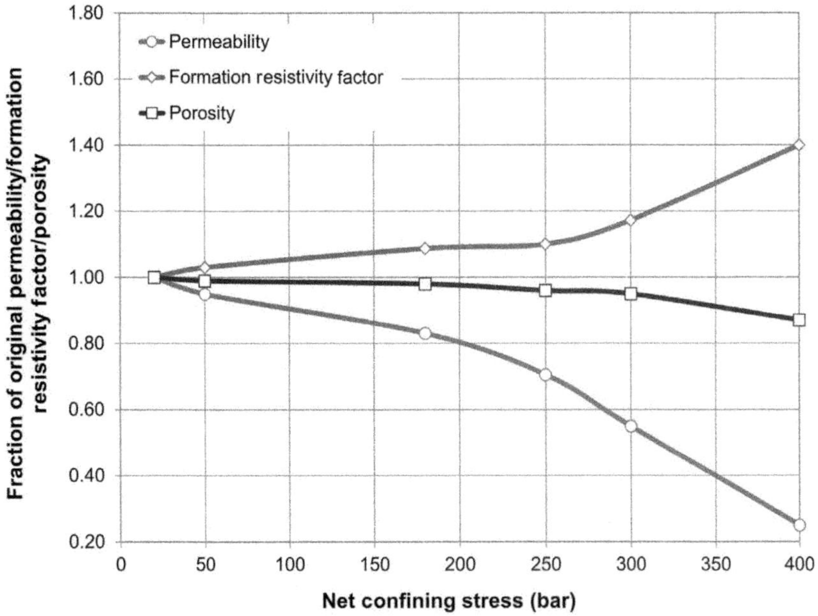

FIGURE 8.16 Example of normalised multi-parameter stress plot.

If multi-parameter measurements of FRF, porosity and brine permeability measurements are made, the normalised data should be plotted as a function of stress on the same chart. The impact of stress on the inter-related values of pore volume compaction, permeability and resistivity can then be clearly identified. In this laboratory's standard presentation shown in Fig. 8.16, porosity, formation factor and brine permeability data at stress are normalised to base values at 20 bar (~300 psi). The increase in resistivity and reduction in porosity and permeability above ~300 bar NCS are indicative of suspected pore collapse which causes an abnormal reduction in pore volume, an excessive reduction in permeability and an anomalous increase in resistivity. Nevertheless, the data acquired at NCS below 300 bar NCS may still be viable.

Outliers on the FRF versus porosity plot (see Fig. 8.17) further need to be scrutinised to differentiate between valid rock-type behaviour and experimental artefacts.

In general, in most siliciclastics as overburden stress increases, the rock pore volume compacts, porosity reduces and, in accordance with Archie's law, formation factor increases. However, the cementation exponent, like permeability, is related to pore tortuosity. A small change in pore volume reduction causes a large increase in tortuosity as smaller pore channels, which are open at lower stress conditions, may be closed off as pore volume compacts in response to overburden stress. Even though cementation exponent at overburden is calculated using overburden porosity, it will increase with

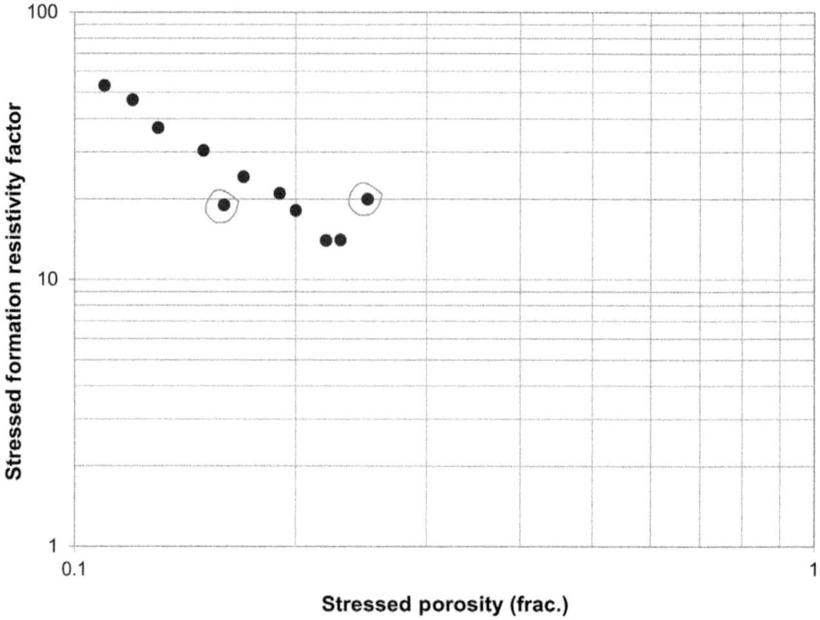

FIGURE 8.17 Formation resistivity factor versus porosity showing outliers.

overburden stress and often exhibits strong correlations with rock quality (via permeability and/or RQI). Cementation exponent in carbonates is often a function of the effective/total porosity ratio and in stiff carbonates can remain more or less constant with increasing stress.

In the literature, temperature appears to have little or no significant effect on m, so most experiments can be performed at ambient temperature.

8.3 RESISTIVITY INDEX TESTS

Resistivity index (I or RI) relationships are measured in the laboratory by progressively introducing a non-conductive phase into the test plug under capillary control and measuring the increase in resistivity. The tests are carried out, often on the same plugs, after FRF tests. However, if brine permeability measurements are scheduled within the FRF test programme, then either the plugs have to be unloaded to insert the porous plate, or the tests are conducted on separate plugs.

Normally, RI tests are carried out in conjunction with porous plate capillary pressure tests as capillary desaturation is required to ensure uniform water saturation in the sample.

Samples for RI may be processed on a batch basis (ambient air–brine system) or in individual coreholders at nominal or reservoir overburden stress using oil–brine or air–brine systems.

Tests can be carried out in both drainage (water saturation decreasing) and imbibition (water saturation increasing) modes.

8.3.1 Test Equipment

The test equipment is the same as used for FRF tests, but the test cell now incorporates a porous plate or semi-permeable membrane to allow controlled, capillary desaturation of the wetting phase. This means that RI tests can be run at the same time and in the same loading operation as air–brine or oil–brine capillary pressure tests. RI samples may be processed on a batch basis in a single-porous plate cell at ambient conditions, or in individual overburden coreholders which are similar to those used for formation factor measurements but are fitted with a porous plate on one end stem. The cells are attached to a pressure manifold which is used to regulate and monitor the pressure of the displacing phase. Where air is used it must be humidified to prevent water evaporation from the sample.

The semi-permeable membrane is selected such that it is only permeable to water under pressure differentials less than the pore entry pressure (Pce) of the membrane. The most common membrane is the porous ceramic plate (Fig. 8.18). This is a ceramic material with very small pore throats that is naturally water wet. Figure 8.19 provides an example of typical capillary pressure curves for the porous plate and a test sample. The porous plate entry pressure to air in an air–brine system is around 200 psi due to the very small pore throats. This means that the rock should have drained to, or close to, irreducible water saturation before the entry pressure is breached. This ensures that only brine and not air can drain from the plug through the membrane. For oil–water systems, the equivalent porous plate entry pressure is around 50% of the air entry pressure.

8.3.1.1 Ambient Condition Tests

The desaturation phase of ambient resistivity index tests is carried out in a batch porous plate cell (Fig. 8.20) under unconfined conditions. Sample

FIGURE 8.18 Porous plate for ambient condition (batch) cell.

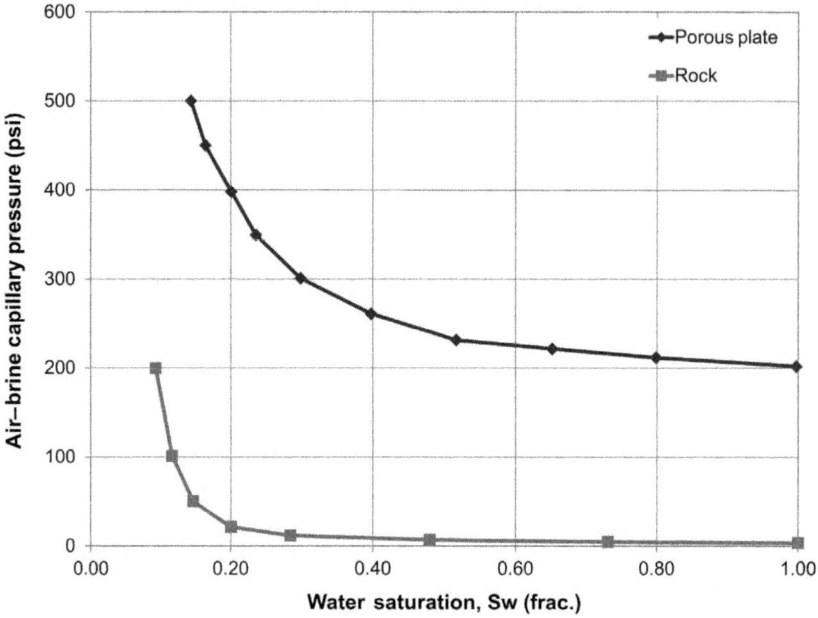

FIGURE 8.19 Relationship between rock (squares) and porous plate (diamonds) capillary pressure curves.

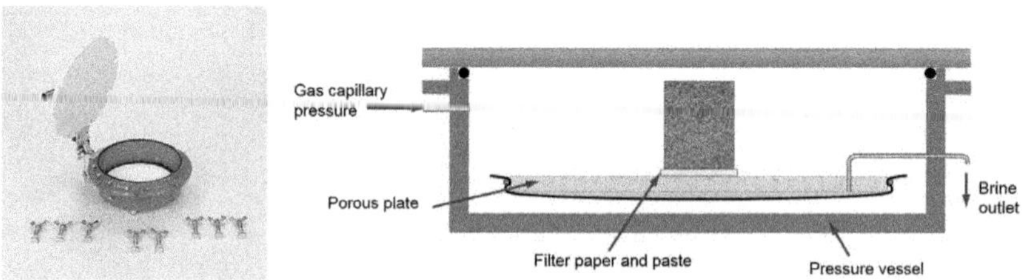

FIGURE 8.20 Batch porous plate equipment (ambient). (Left: *Courtesy of Eijkelkamp Soil & Water.*)

resistivity at each desaturation step is measured in the same ambient electrical system as used for ambient formation factor. This is normally done by removing the samples from the porous plate cell after each desaturation pressure step.

In the batch porous plate set-up, several samples can be run at the same time. For an air–brine ambient system, to ensure continuity between the brine in the rock and the brine saturating the plate, a capillary bridge is required between the sample and the plate. This ensures good capillary contact and typically consists of a thin layer of filter materials such as diatomaceous earth, kieselguhr or kaolin/celite mixture. The pore sizes of this filter aid (capillary bridge) must be smaller than the pore size in the rock that contains irreducible wetting phase saturation to ensure that the capillary bridge does not desaturate before the sample nears Swir. If the capillary contact between the plate and the sample is lost, the sample will not desaturate further at incremental pressures. Saturations are measured by gravimetric methods.

Sample resistivity at each desaturation step is measured using either a two-electrode or a four-electrode system. Two-electrode cells potentially suffer from contact resistance effects, but these can be minimised by good cell design. The problem with four-electrode cells is that the water saturation level and distribution in the sample between the voltage electrodes, and those in the bulk sample between the current electrodes may differ. Consequently, the resistivity that is measured may not correspond to the water saturation measured on the bulk sample. This can lead to some unusually shaped resistivity index versus Sw curves. The system must also be designed to ensure good electrical contact between the plug and the measurement electrodes, either using conductive rubber contacts or by the use of a brine-saturated filter aid.

8.3.1.2 Overburden Test Equipment

The overburden cell for resistivity index tests is the same as that used for formation factor (an example is shown in Fig. 8.21), but one end stem is fitted with a porous plate. The individual overburden cell is provided with the capability of monitoring water production and sample resistance continuously. Desaturation volumes are measured in a graduated pipette connected to the cell outlet, or using non-invasive or *in situ* saturation monitoring (ISSM), such as gamma ray or X-ray adsorption techniques. Air–brine and oil–brine tests can be run in the individual cell. A filter aid is often used to maintain capillary contact between the plate and the plug end face but, as the test is carried out at

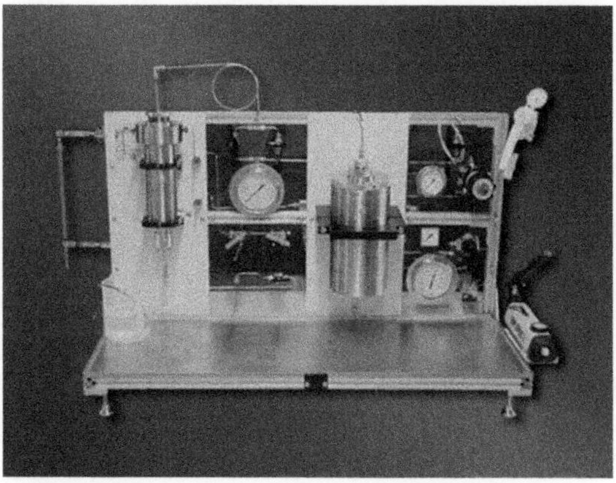

FIGURE 8.21 Example overburden porous plate resistivity index cell. *Courtesy of Coretest Systems, Inc.*

overburden stress, the axial pressure acting on the end stems is often sufficient to provide good capillary contact. The displacing phase (air or oil for a drainage test) can only enter the sample at the opposite plug face from the porous plate.

Since the presence of the capillary plate will increase the resistance of the plug/plate assembly between the current (end) electrodes, four-electrode cells are often used. In addition to potential saturation differences between the potential and current electrodes, four-electrode systems are difficult to manufacture and are often not robust. Two-electrode systems must therefore either account for the resistivity and pore volume compaction behaviour of the porous plate, or are designed such that they allow a consistent electrical flux to be applied to the end face of the plug sample.

A multi-frequency meter is used to determine resistance, frequency and phase angle for both ambient and overburden tests. Data are acquired at a minimum phase angle ($<0.5°$).

8.3.2 Test Procedures

Resistivity index tests are carried out either at ambient (unconfined conditions) in a batch cell or at stress conditions in individual overburden cells. As overburden pressure increases, the rock pore volume compacts, porosity reduces and, in accordance with Archie's law, formation factor increases. The pore tortuosity (and hence Ro and m) will also increase with overburden pressure. However, it is expected that Rt might be similarly affected by the change in tortuosity so that RI (RI=Rt/Ro) and hence n might not be expected to change significantly. In this case, the stress level may not be critical—the principal advantage is that tests at even nominal stress can eliminate excess brine effects and saturations are determined volumetrically rather than gravimetrically. Gravimetric measurements can suffer from grain loss effects.

The conventional test protocol is to carry out the experiments under primary drainage procedures—brine (wetting phase) reducing. It is also possible to conduct the tests under spontaneous imbibition conditions (brine saturation increasing) if that is considered appropriate for the petrophysical interpretation.

8.3.2.1 Ambient Condition Tests (Air–Brine)

Following electrical (ionic) equilibrium verification and after ambient FRF measurements, the SFW-saturated samples are placed in the porous plate cell.

The porous plate and the capillary bridge must be saturated in SFW to ensure no air in the pore systems, as air cannot or should not penetrate (break through) the pores in the filter aid or plate until the entry pressure is breached. Brine can then pass through the pores of the plate, to atmosphere, outside of the cell. In this way, a differential pressure is created across the plate which causes the brine in the samples to be displaced from the plug under capillary-controlled desaturation. The plates are selected on the basis of entry pressure—lower entry pressure plates may be used for lower capillary pressures and higher entry pressure plates for higher capillary pressures. This means that the plates may have to be changed at some point during the experiment.

Air is humidified to prevent brine in the core from evaporating. The humidified air pressure is increased over a number of controlled stages (capillary pressure steps) to the desired final pressure value, which should reduce the core saturation towards Swir.

In ambient measurements, carried out on a batch basis, it is impossible to monitor the expulsion rate of brine from individual samples, so that after total water production stability has been achieved at each pressure increment, the samples are removed, individual plug resistances measured and the saturated weight recorded. If the saturation and resistance are not stable, the sample is returned to the cell and the process continued. Consequently, the samples need to be removed repeatedly from the cells for weight measurement until the test is completed.

Electrical measurements are made over a range of frequencies to ensure measurement at minimum phase angle and all Rt data are normalised to a standard temperature (20 or 25 °C unless otherwise specified).

Typically six to eight pressure points are acquired depending on the sample permeability and test budget. Typical test times used by many laboratories are about 3–4 days per pressure point, for the first five or seven pressure increments and then 10–15 days at the final pressure increment. Longer periods are sometimes used for lower permeability samples. However, the criterion for saturation stability is cessation of brine production and resistivity equilibrium, not a fixed time period. This means that it can take many weeks to complete a full drainage curve ensuring both capillary and saturation equilibrium at each step.

It is recognised that there must be a compromise between achieving truly equilibrated data and excessive experimental time. However, the primary objective of this test is to determine the resistivity index–brine saturation relationship for each plug, not necessarily to determine an equilibrium air–brine capillary pressure relationship, since resistivity equilibrium is normally established more rapidly than capillary pressure equilibrium.

Calculation of plug water saturations from gravimetric measurements can suffer from the effects of grain loss. Therefore on completion of the tests, with the sample approaching, or at, Swir, the plug is removed and immediately extracted in a Dean-Stark apparatus as a check on calculated water saturations. The final (at maximum capillary pressure) average plug Sw should match the final Dean-Stark Sw within about ±3–4 saturation units. If on completion of the test samples are scheduled for another test (e.g. end-point effective permeability at Swir), Dean-Stark can still be performed on completion of these tests, taking into account water mobilised and collected during the tests.

8.3.2.2 Stress Condition Tests (Air–Brine and Oil–Brine)

Following electrical (ionic) equilibrium and after ambient FRF measurements, the SFW-saturated sample is placed into a pressure vessel (individual core-holder), along with the selected porous plate and filter pad which are both saturated in SFW.

As for the FRF test, the SCP and its associated annulus volume must be satisfactorily established for each test plug. Once the SCP has been determined, the overburden pressure is increased to the required test stress. Annulus volume corrections are applied to the raw data to determine the corrected pore volume (porosity) at the test stress. These corrections will have an effect on the pore volume and on subsequent water saturation calculations.

Normally, the tests are performed at a single-stress station which is usually equivalent to the reservoir effective isostatic stress. However, FRF measurements have been made at stress increments up to the reservoir stress value. For oil–brine systems, the tests can be performed using degassed stock tank oil and at reservoir temperature, if desired.

The humidified air or oil pressure is increased over a number of previously agreed capillary pressure increments to the desired final pressure value, which should reduce the core saturation towards Swir. The brine expulsion volumes (measured with a graduated pipette) and resistance are monitored continuously in the cell, to ensure that stability has been achieved at each pressure increment before increasing pressure to the next increment. Consequently, the samples need not be removed from the cells to estimate saturation volumes until the test is completed. This almost completely eliminates excess brine and grain loss issues. More sophisticated systems now use individual cell ISSM, such as gamma ray or X-ray adsorption techniques. Again, sufficient time must be allowed to achieve resistivity and (if Pc data are required) capillary equilibrium. This is discussed in detail in Chapter 9, but it is clear that a considerable time may be required to define a full RI versus Sw and Pc versus Sw curve. This can be several months for an oil–water primary drainage system, for example.

As for ambient condition tests, the final water saturation (at maximum applied capillary pressure) should be verified by Dean-Stark extraction. The final average plug Sw should match the final Dean-Stark Sw within about ±3–4 saturation units.

8.3.2.3 Continuous Injection Resistivity Index

The continuous injection resistivity index (CIRI) method (de Waal et al., 1989) is a variant of the capillary desaturation technique, but is designed to permit resistivity index data to be achieved more rapidly for an oil–water system, since capillary equilibrium is not necessarily a prerequisite for accurate resistivity index data. The method can provide the full I versus Sw curve, rather than the few data points that are possible using porous plate, in approximately 14 days.

The CIRI test utilises an overburden coreholder, fitted with a porous plate cell. A schematic of a CIRI set-up is shown in Fig. 8.22. The sample is saturated to 100% brine with formation water. Oil is injected by means of a very low flow rate pump so that the sample pore volume (around 10–15 ml) can be filled in 10–14 days. Although a dynamic test, the low flow rate means that desaturation is controlled more by capillary than by viscous forces. If the viscous force is too high, the displacement process may become unstable which will induce a non-uniform saturation in the sample, leading to erroneous results. An intelligent automatic control system is used to control oil injection and to ensure that the injection pressure never exceeds the breakthrough pressure of the plate. The resistivity of the sample is measured continuously during this injection period. The average brine saturation is determined simply from material balance of the produced brine. The water-wet porous plate/membrane prohibits production of hydrocarbons; therefore, the injected oil or the produced brine is a direct measure for the average saturation change in the plug.

The main advantages of CIRI tests are that the test is performed at overburden stress (minimising excess brine and grain loss issues) and that it is possible to acquire drainage *I*–Sw data in a few weeks as opposed to months with the porous plate resistivity index (PPRI). Unlike PPRI, however, the CIRI technique is not an equilibrium method. As the injection is not strictly under capillary control, viscous forces in the experiment can affect the desaturation process leading to non-uniform saturation and non-linear *I* versus Sw curves. Lyle and Mills (1989) demonstrate that correct measurements of *n* can only be made in uniformly saturated samples.

It is impossible to assess (and even correct for) non-uniform saturation without visualisation of the saturation profiles across the sample. This is why CIRI tests should only be carried out using ISSM scanning using gamma rays or X-rays. Figure 8.23 provides an example of ISSM scans for a CIRI

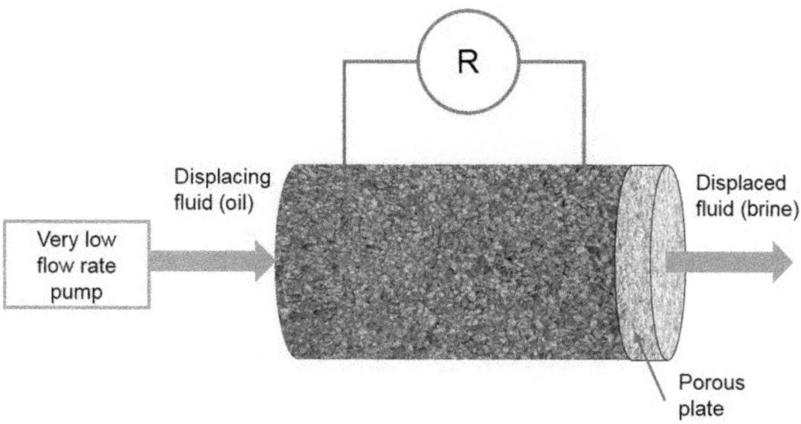

FIGURE 8.22 Schematic of CIRI test equipment.

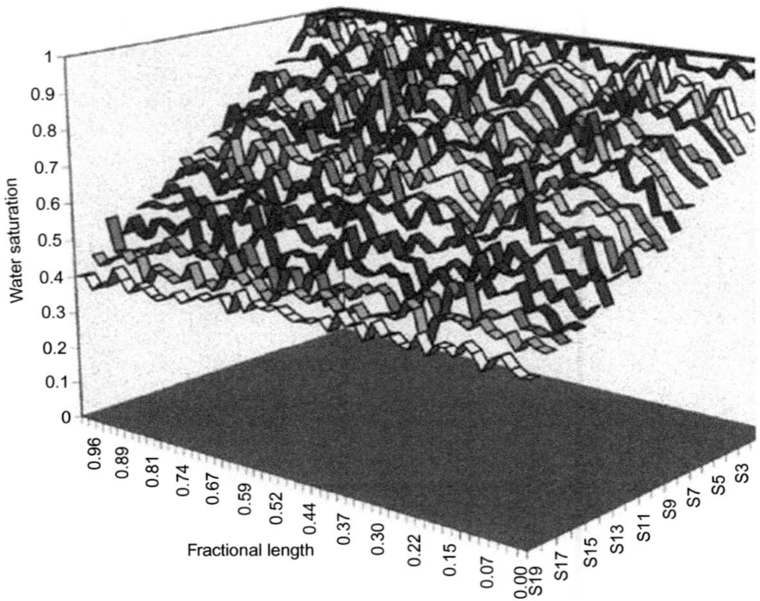

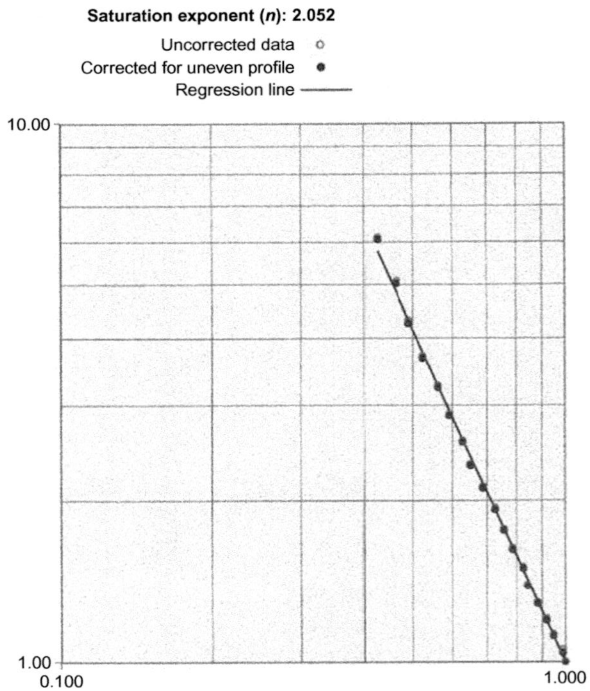

FIGURE 8.23 Example of ISSM scans for CIRI sample.

tests. The curves represent water saturation profiles along the (fractional) length of the plug as a function of time (each curve represents a different time step during the displacement process). In this case, the saturation profile is uniform which is represented by a well-behaved I versus Sw curve.

8.3.3 Data Assignment and Utilisation

8.3.3.1 Water Saturation Calculation

In circumstances (normally ambient condition RI tests) where gravimetric methods are used to determine water saturation, the water saturation, Sw, at any point during the desaturation sequence is determined from:

$$\mathrm{Sw} = \frac{\mathrm{Vw}}{\mathrm{Vp}} = \frac{\dfrac{W_{\mathrm{ps}} - W_{\mathrm{d}}}{\rho_{\mathrm{w}} - \rho_{\mathrm{g}}}}{\dfrac{W_{\mathrm{sat}} - W_{\mathrm{d_n}}}{\rho_{\mathrm{w}} - \rho_{\mathrm{g}}}}$$

for a water (w)–gas(g) test. The volume of water at any time is the partly saturated plug weight (W_{ps}) less the plug dry weight (W_{d}), divided by the density difference between water and gas ($\rho_{\mathrm{w}}-\rho_{\mathrm{g}}$). The pore volume (at ambient conditions) is the fully saturated plug weight (W_{sat}) minus the plug dry weight divided by the density difference.

The water saturation from volumetric measurements is:

$$\mathrm{Sw} = \frac{\mathrm{Vw}}{\mathrm{Vp}} = \frac{\mathrm{Vp} - V_{\mathrm{prod}}}{\mathrm{Vp}}$$

where Vp is the fully saturated pore volume (either at ambient or at stressed conditions) and V_{prod} is the cumulative volume of water expelled during the desaturation sequence, corrected for any temperature effects (density differences) and sleeve conformance annulus volume.

In both cases, and provided the test plug is not required for further testing, then Dean-Stark extraction measurements should be made after the final (maximum) pressure increment to verify volumetric and gravimetric saturation calculations.

8.3.3.2 Saturation Exponent

The resistivity index, at each water saturation, is calculated from:

$$I = \frac{\mathrm{Rt}}{\mathrm{Ro}}$$

The saturation exponent, n, is calculated for each plug at each brine saturation (Sw), using Archie's law:

$$I = \frac{1}{\mathrm{Sw}^n}$$

As with FRF tests, plug data should be grouped into representative zones, facies or rock types. The composite saturation exponent for a group of samples is found by linear regression analysis of logarithmic I vs. Sw data:

$$\log(I) = -n\log(\mathrm{Sw})$$

An example of a typical relationship between I and Sw for a composite sample group is shown in Fig. 8.24. This is a forced-fit regression as $I=1$ at $\mathrm{Sw}=1$. In Excel, the curve is forced through 1 by carrying out linear regression analysis of log(I) and log(Sw) data, forcing the intercept to 0 (log(1)). The data are then plotted on Cartesian coordinates and linear regression analysis performed on the log(I)/log(Sw) data pairs, as shown in Fig. 8.25.

Some laboratories simply calculate an n for each plug by averaging all the n data calculated at each individual water saturation. This is not recommended since any errors in the individual n calculations can have a large leverage on the composite value. For example, the calculation of n becomes very unstable at high water saturations and, if included in data averaging, will return an erroneous value for the composite n.

An example of a typical lab output is provided in Fig. 8.26. Note: In this example, base helium porosity and permeability are measured at 800 psi NCS.

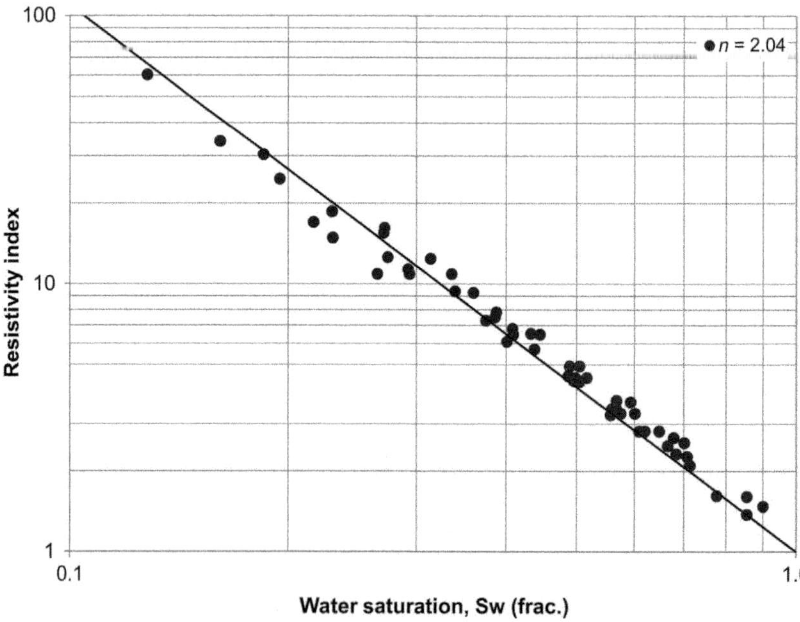

FIGURE 8.24 Example composite resistivity index versus Sw (logarithmic scale).

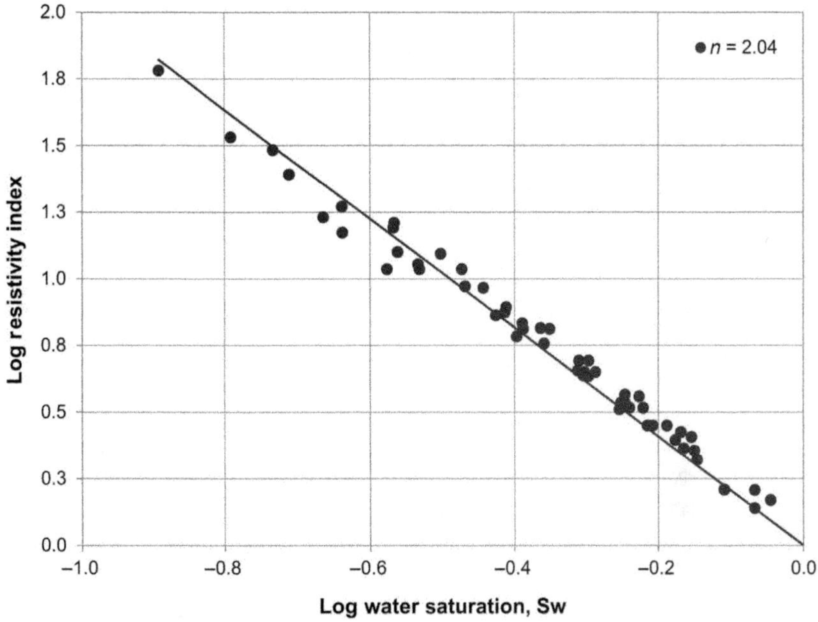

FIGURE 8.25 Example composite resistivity index versus Sw (Cartesian).

8.3.3.3 Inverse Tangent Method

In the conventional solution for Archie's saturation exponent:

$$I = \frac{1}{\mathrm{Sw}^n}$$

n is determined from linear regression of bi-logarithmic *I* (*Y*) versus Sw (*X*) values. Kennedy (2007) argues that *n* should be calculated using the inverse tangent average method (ITM). The conventional *Y*-on-*X* regression method finds an *n* value that minimises the error in resistivity (*I*) rather than Sw. The inverse tangent method seeks to find an *n* value that minimises the error in Sw. For example, the conventional method is initially used to solve for *n*, and this value is used to calculate Sw for each *I* datapoint (predicted $\mathrm{Sw_p}$).

$$\mathrm{Sw_p} = \left(\frac{1}{I}\right)^{(1/n)}$$

A solver function can then be used to determine an *n* value that minimises the sum of squares error between measured ($\mathrm{Sw_m}$) and predicted Sw ($\mathrm{Sw_p}$): i.e. $\sum(\mathrm{Sw_m} - \mathrm{Sw_p})^2$. Figure 8.27 provides an example of the ITM and conventional *n* solutions for the data in Fig. 8.24. Normally, close agreement between conventional and inverse tangent *n* exponents indicates well-behaved data with a consistent trend and high correlation coefficients. However, the

Saturant, mg/l:	21,930
Resistivity of Saturant, Ω-m @ 25°C:	0.268
Effective Confining Stress, psig:	3600

Sample ID	Depth (ft.)	Permeability to Air, (mD)	Stressed Porosity, Fraction	Formation Factor (*F*)	Cementation Exponent (*m*)	Brine Saturation, Fraction (Vp)	Resistivity Index (*I*)	Saturation Exponent (*n*)
2	8202.90	25.	0.264	8.48	1.61	1.000	1.00	
						0.798	1.28	1.11
						0.693	1.61	1.31
						0.640	1.80	1.32
						0.598	1.98	1.33
						0.413	2.95	1.22
						0.253	4.66	1.12
3	8228.00	135.	0.305	6.6	1.59	1.000	1.00	
						0.919	1.06	0.71
						0.715	1.69	1.56
						0.532	2.86	1.66
						0.421	4.23	1.67
						0.293	6.85	1.57
						0.172	12.57	1.44
4	8237.00	452.	0.315	5.69	1.51	1.000	1.00	
						0.899	1.07	0.66
						0.550	2.65	1.63
						0.420	4.33	1.69
						0.323	7.04	1.72
						0.227	12.99	1.73
						0.148	24.37	1.67
5	8256.00	498.	0.300	6.31	1.53	1.000	1.00	
						0.991	1.04	4.68
						0.855	1.05	0.30
						0.477	2.28	1.11
						0.337	3.73	1.21
						0.238	5.96	1.24
						0.130	10.76	1.16
						0.046	20.14	0.97
6	8262.00	286.	0.284	5.70	1.30	1.000	1.00	
						0.944	1.08	1.35
						0.793	1.22	0.85
						0.486	1.86	0.86
						0.222	5.75	1.16
						0.087	10.43	0.96

FIGURE 8.26 Example of typical tabular data output for resistivity index tests.

method should not be used where the *I*–Sw data relationships are real and are not just reflective of experimental uncertainty, for example, *I*–Sw curvature due to clays, multi-modal pore systems or wettability effects.

8.3.4 Saturation Exponent: Laboratory Artefacts

There are three main laboratory/experimental artefacts which influence resistivity index measurements:

- grain loss corrections
- excess brine
- non-uniform saturation

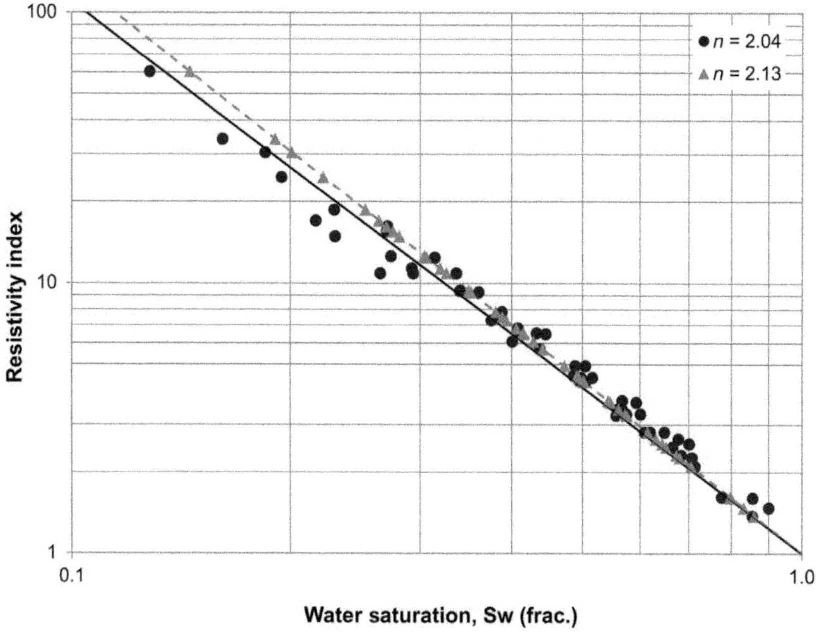

FIGURE 8.27 Example of conventional and ITM saturation exponent.

8.3.4.1 Grain Loss

The majority of modern resistivity index data have probably been acquired from experiments where the plug is held in an individual overburden cell and the water saturations calculated from volumetric measurements. However, a significant amount of legacy data, which may still be being used in petrophysical interpretation, will have been acquired from ambient condition measurements where plugs are desaturated in a batch cell. In this case, saturation calculations will have been performed on gravimetric measurements.

Grain loss is by far the biggest problem in resistivity index tests at ambient conditions where intermediate and final saturations are determined by weighing the samples. These also occur in most other SCAL tests where gravimetric saturation determination is used to determine saturations.

Since the test plugs are subjected to a considerable amount of handling during the test procedures, significant grain losses can occur. All plugs suffer from grain loss to different degrees, especially weak or friable samples. Even small amounts of grain loss can have a significant impact. Grain loss, if unrecognised and therefore unaccounted for, can result in considerable uncertainties in the calculation of fluid saturations from gravimetric measurements made on the plug samples, and so will affect the calculated value of n. For example, a loss in weight might be interpreted as a loss in water so that the calculated fluid saturation can be much less than the true value. The goal of grain loss correction is therefore to predict the liquid-filled pore volume at each stage of the handling and test procedures, and to validate these estimates, where possible, using measured data.

The total grain loss that has occurred is determined and then proportioned to each relevant stage in the experiment. For example, where data allow, the total grain loss, W_{gl}, is found from:

$$W_{gl} = W_{di} - W_{dcf}$$

where W_{di} is the initial plug dry weight before the test and W_{dcf} is the final cleaned and dry weight after the test.

The plugs may lose weight during the FRF measurements at ambient conditions. At this stage, the plugs remain fully saturated in brine so it is assumed that the total weight loss during the FRF stage results from a combination of grain and brine loss from the plugs. Consequently, both bulk volume and grain volume will reduce as the plug loses weight. It is assumed that the reduction in grain volume and bulk volume is in proportion to the sample porosity. That is, the porosity remains constant throughout the formation factor stage. This allows estimation of a new sample bulk volume and pore volume, and the prediction of a sample dry weight at the end of the FRF stage.

The total grain loss which is used to calculate intermediate water saturations is based on the estimated or predicted dry weight of the sample at 100% brine saturation (that is, after FRF measurement) and the measured final dry weight (after cleaning and drying). This grain loss is allocated to each step using:

$$\text{Grain loss/step} = \left(\frac{1}{x}\right)(W_{dn} - W_{dcf})$$

where x is the number of handling steps and W_{dn} is the estimated dry weight of plug at 100% saturation (after FRF). It is therefore assumed that $1/x$ of the total grain loss will occur during each handling step. A handling step may be defined as removing and replacing the plug after resistivity measurements after each capillary desaturation stage.

The 'dry weight' (W_d) in the gravimetric saturation mass balance equation

$$\text{Sw} = \frac{\text{Vw}}{\text{Vp}} = \frac{\dfrac{W_{ps} - W_d}{\rho_w - \rho_g}}{\dfrac{W_{sat} - W_{d_n}}{\rho_w - \rho_g}}$$

now represents the estimated dry weight at each capillary pressure increment.

The relatively small difference between oil and water densities, compared to air and water, for example, can lead to considerable uncertainties in calculating saturations when applying grain loss corrections.

Figure 8.28 shows an example of grain loss correction. The pre-test and post-test sample dry weights indicate a grain loss of 2.7 g which represents only 2% of the plug dry weight; yet the uncorrected final water saturation is around

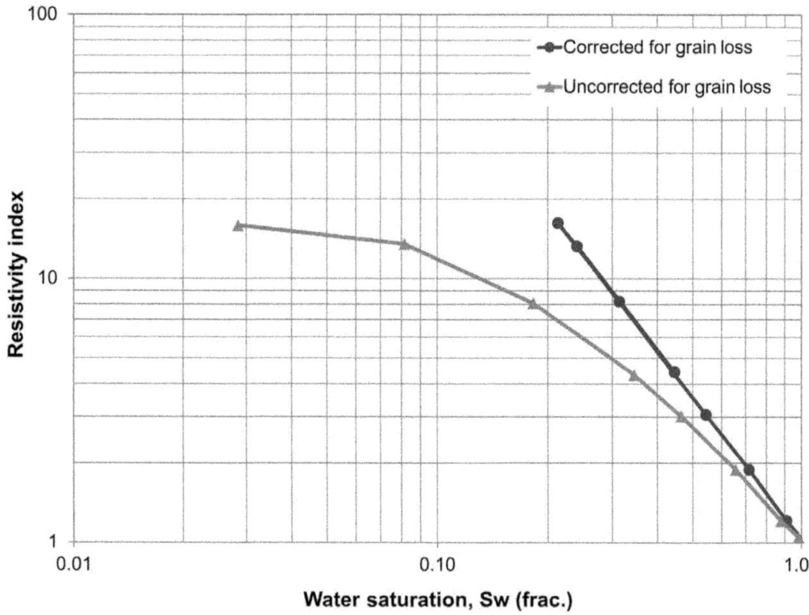

FIGURE 8.28 Example of grain loss correction on RI test data.

3% compared to a grain loss-corrected value of ~20%. The suspiciously perfect bi-logarithmic fit to the 'corrected' data is a result of creative data manipulation by the test laboratory who adjusted the grain loss data to give a faultless *I*–Sw relationship. This kind of unconstrained data makeover can mask real curvature in *I*–Sw curves such as clays and pore system modality.

It is essential therefore to remeasure helium grain volume and pore volume, bulk volume and dry weight data after final cleaning and drying, on completion of the test sequence. These data, together with the independent Dean-Stark water extraction data, can be used to check the grain loss-corrected water saturations.

8.3.4.2 Excess Brine

Excess brine not only has a large impact on ambient FRF, but it also affects saturation exponent where Ro is determined at ambient (unconfined) conditions.

Although excess brine effects do not impinge on Rt measurements as desaturation removes the brine clinging to the sample surface at low desaturation pressures, they transfer through to the calculation of RI data. If Ro is too low due to excess brine, then *I* and hence, *n*, will be too high.

Excess brine effects can be prevented either by performing the tests at overburden or, at ambient conditions, by applying very small desaturation

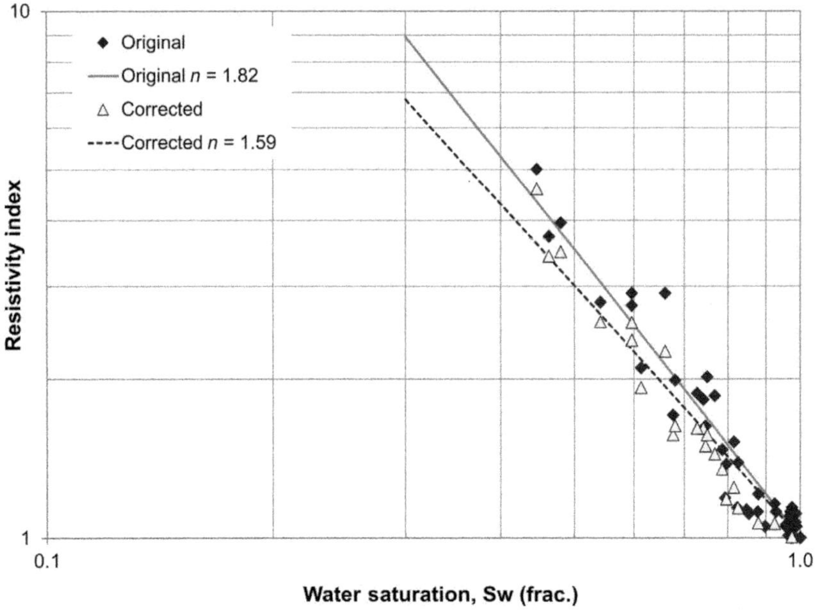

FIGURE 8.29 Example of excess brine correction on saturation exponent.

capillary pressures to drain the brine from the plug surface and surface pores. Unless experimental data are provided (which they rarely are), it is difficult to correct reported lab data. Where they are, it is possible to correct the data and recalculate *n*. In the example shown in Fig. 8.29, the resistivity index test was carried out at ambient conditions but previous measurements of Ro at stress allowed correction of the ambient condition Ro value which was affected by excess brine. In this case, the values of Rt were divided by the corrected Ro to correct *I* and hence *n*. In this example, excess brine produced an average error of −30% in *m* (Fig. 8.15) and +15% in *n*. According to McPhee (2012), these errors would result in uncertainties in hydrocarbon in place of 7–12% for a typical North Sea reservoir with 20% porosity, 20% water saturation and 100 MMstb OIIP.

8.3.4.3 Non-uniform Saturation

Reliable measurements of saturation exponent can only be made on samples where the water saturation is uniform.

Some laboratories carry out the tests by dynamic displacement. Normally only end-point data are available: Ro at 100% Sw prior to displacement with oil, and Rt at or near to irreducible water saturation, Swir, following oil flooding. Measurements of resistivity during displacement should not be made because there is no opportunity to establish a uniform or stable saturation during the displacement process. That is why only end-point drainage RI data are obtained. The method is much less controllable than porous plate and, more importantly, is likely to suffer from the adverse effects of capillary end effects. This can result in a saturation gradient developing across the core so that saturations are not uniform.

Although apparent non-linearity in *I* versus Sw curves can be emphasised by the number of data points generated in CIRI tests compared to typical porous plate measurements, the behaviour is real in many samples and can result from non-uniform saturations arising from multi-modal pore size distributions in the test samples or, in oil-displacing brine experiments, where some pores have an oil-wetting tendency (Gray et al., 1992). However, other reasons for non-linearity include differences in permeability between the core plug and the semi-permeable membrane, laboratory temperature fluctuation, oil–water viscosity ratio and viscous/capillary force balance where the viscous force is excessive during oil flooding. Temperature fluctuations not only cause changes in resistivity, which can be corrected to some extent, but can also result in pressure fluctuations. During the drainage process pressure oscillations can promote cyclic drainage–imbibition processes which result in non-uniform saturation, especially towards irreducible water saturations.

Figure 8.30 is an example of *I* versus Sw curves for a number of CIRI tests. The abrupt non-linearity evident in some samples is a result of flow rate changes during oil displacement and rate oscillations caused by temperature

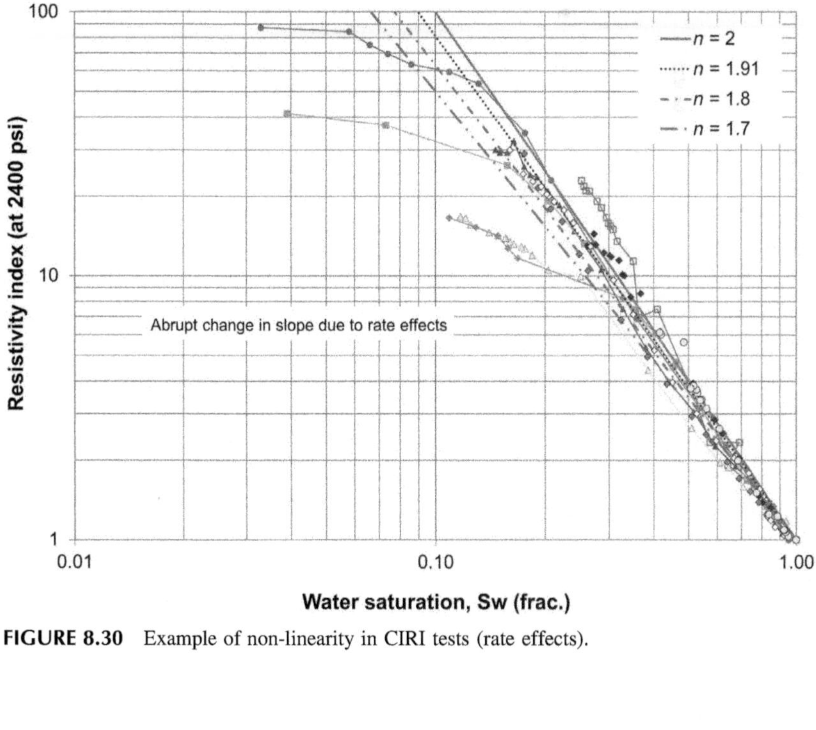

FIGURE 8.30 Example of non-linearity in CIRI tests (rate effects).

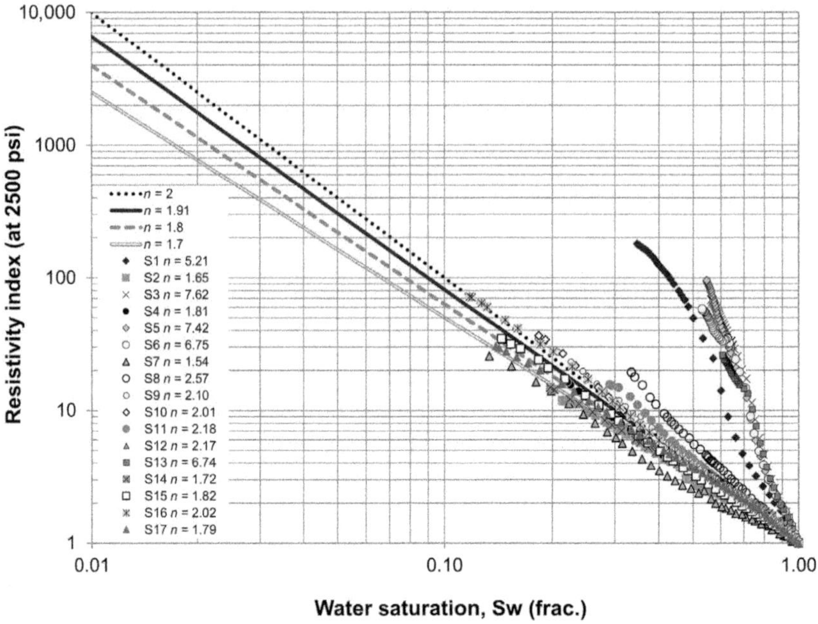

FIGURE 8.31 Example of non-linearity in CIRI tests.

fluctuations. The data in Fig. 8.31 indicate unfeasibly high values of n in some samples. Diagnosis of these anomalies is only possible with ISSM, but this was not used in these tests due to cost restrictions.

If ISSM scans are available, it is possible to distinguish non-uniform saturations and correct n accordingly. For example, Jing et al. (1993) present a finite element model (using measured resistivity and saturation distribution data) to calculate the resistivity index curves which would be measured if a uniform saturation distribution existed in the sample. Alternatively, data points corresponding to non-uniform saturations are ignored when regressing the Sw/RI data pairs. In a more rigorous approach, the underlying value of saturation exponent required to produce the measured resistivity index for each saturation profile is calculated. If the plug is viewed as being composed of a series of one-dimensional elements each obeying Archie's law, then the underlying value of saturation exponent required to give the measured bulk resistivity can be easily calculated. This assumes that saturation exponent is constant over the saturation range covered by a given saturation profile. This specific value is then used to calculate the average resistivity index for the specific saturation profile.

The RI for a particular profile (i) is determined by:

$$I = \sum_i \frac{1}{\mathrm{Sw}_i^n}$$

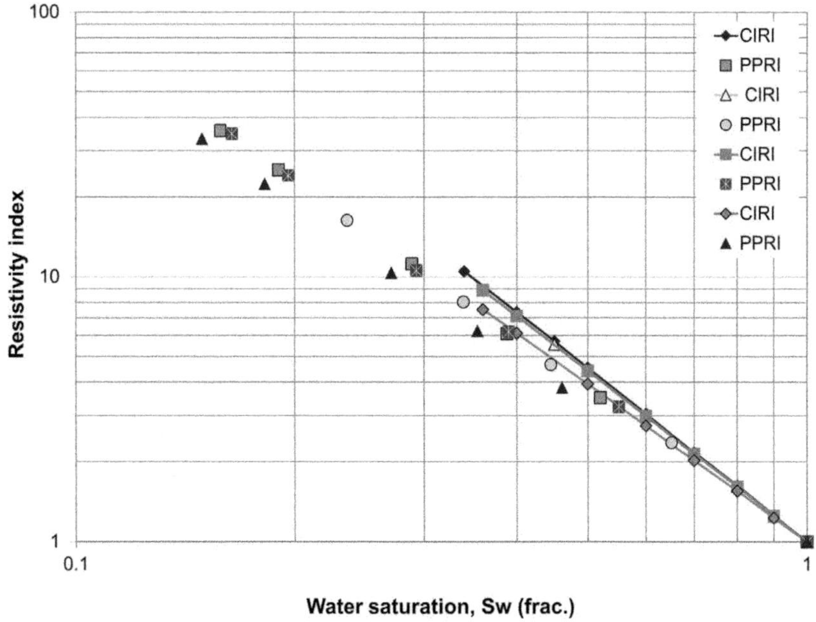

FIGURE 8.32 CIRI data corrected for non-uniform saturation.

This value amounts to a lower limiting value for RI as the saturation profile approaches uniformity for a given average saturation.

Figure 8.32 compares PPRI and CIRI data from parallel plugs in which the CIRI data have been corrected for the effects of non-uniform saturation. The CIRI and PPRI data agree well; yet the CIRI technique desaturation procedures fail to drain the samples to irreducible saturation. This is due to the higher viscous/capillary force ratio in CIRI compared to PPRI, and because the CIRI porous plate had a lower entry/breakthrough pressure rating than the PPRI membrane. However, provided the relationship between *I* and Sw is in accordance with Archie's law, the pore system is not multi-modal, and clay effects are negligible, then saturation exponent obtained from CIRI *I*–Sw data pairs at higher water saturations should be descriptive of *I*–Sw behaviour over the moveable wetting phase saturation range on drainage.

The non-uniform saturations associated with an excess viscous/capillary force ratio means that dynamic displacement methods should not be used in resistivity index experiments. CIRI can provide viable data but the need to identify and correct for non-uniform saturations using ISSM, and the requirement for strict temperature control makes these tests expensive to run. Centrifuge desaturation should never be used for resistivity index desaturation as a capillary pressure gradient is developed across the sample leading to highly non-uniform water saturations.

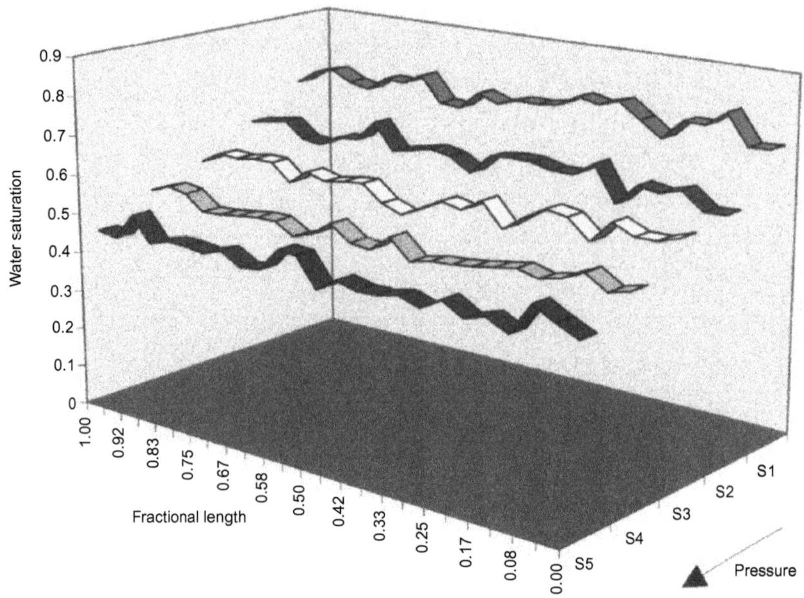

FIGURE 8.33 ISSM scans reveal uniform saturation profile in porous plate experiment.

The only way to virtually guarantee a uniform saturation is by using controlled porous plate desaturation ensuring sufficient time to attain capillary equilibrium. Figure 8.33 provides an ISSM scan example from a porous plate experiment. Each curve of water saturation versus fractional length along the plug represents an increasing capillary pressure.

8.3.5 Saturation Exponent: Rock Type and Pore/Fluid System Controls

Provided laboratory artefacts can be minimised or eliminated, then the saturation exponent n is a real measure of the influence of reservoir rock and fluid on resistivity.

The principal factors which control or influence saturation exponent are:

- rock texture/pore size modality;
- overburden stress;
- saturation history;
- wettability;
- conductive matrix.

8.3.5.1 Rock Texture/Pore Size Modality

The effects of bi- (or multi-) modal pore systems on saturation exponent have been amply demonstrated by many researchers but pivotally by Diederix

(1982), LaTorraca and Hall (1991) and Pallat et al. (1988). Diederix found that in certain formations containing filamentous illite which forms a 'rough' grain coating, the I versus Sw curve slope reduced at lower water saturations, and the calculated n values were lower than clean sandstones from the same facies. Diederix recreated the effect of illite grain coats using roughened and smooth glass beads in a resistivity index test. In a smooth bead pack, the conductive water pathways become increasingly isolated by non-wetting phase coming into contact with the smooth grain surface. In a bead pack with rough grains, for the same average water saturation, the 'valleys' in the grain surfaces retain a layer of water which can maintain a conductive pathway through the pore network, even at low water saturations. Consequently, for the same Sw level, the resistivity of the smooth bead pack (and hence n) will be higher than the rough bead pack (Fig. 8.34).

Inappropriate cleaning and drying can have similar effects and can markedly influence values of the Archie saturation exponent. Routine drying of

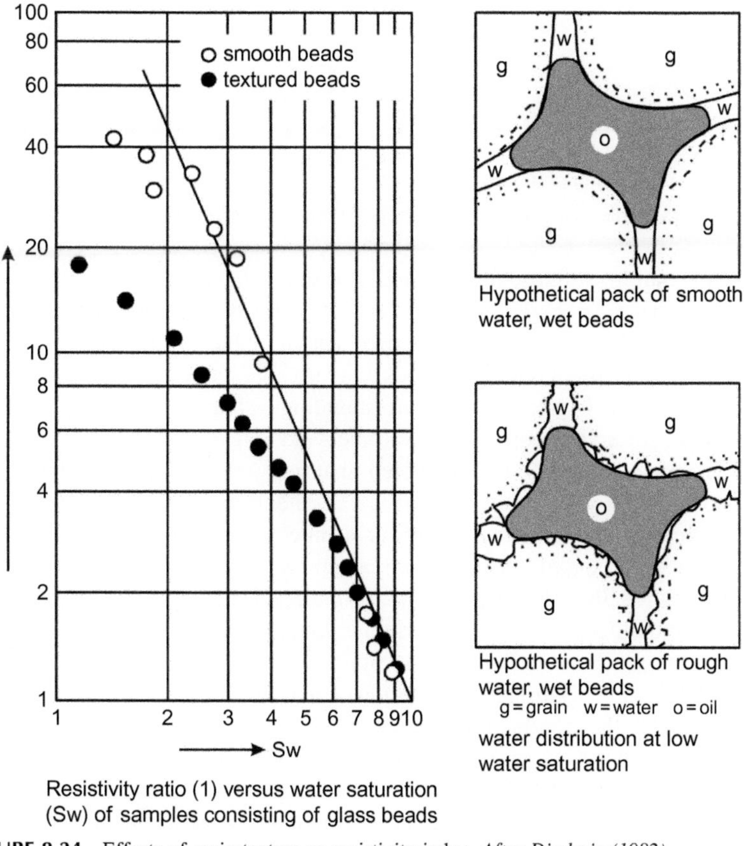

FIGURE 8.34 Effects of grain texture on resistivity index. *After Diederix (1982).*

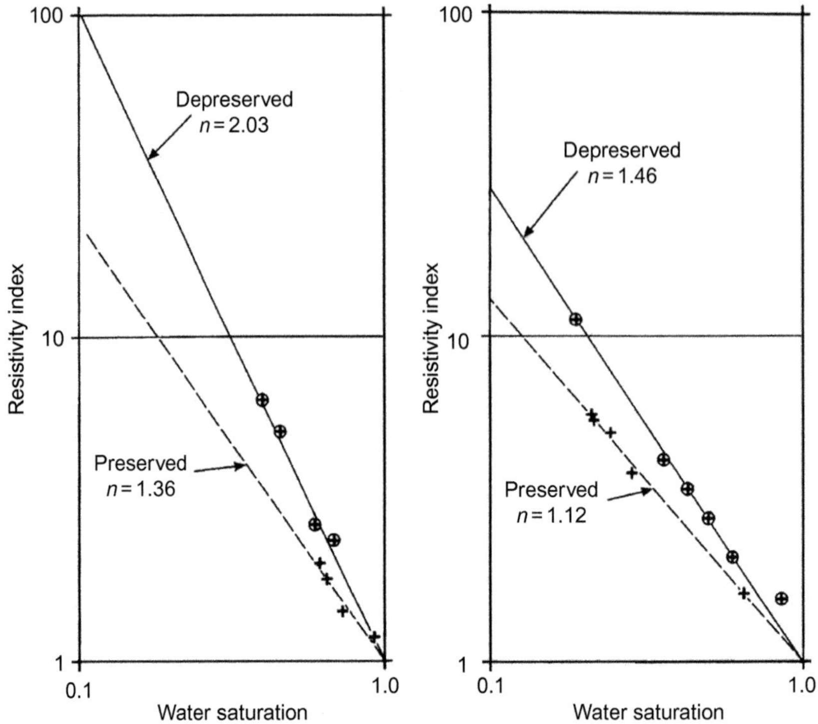

FIGURE 8.35 Effects of drying 'de-preserved' on resistivity index. *After Pallat et al. (1988).*

core can result in collapse of structurally delicate clay minerals such as 'hairy' illite thus changing 'rough' grains to 'smooth' grains, thereby increasing resistivity and n. Pallat et al. (1988) demonstrated that values of n obtained through routine preparation methods that involve the drying of core plugs can be up to 50% greater than corresponding values for miscibly cleaned, non-dried (preserved) core from the same field (Fig. 8.35).

LaTorraca and Hall (1991) noted that plug tests on finely laminated sandstones often yield relatively low n values and developed a simple model to explain this phenomenon. Consider a plug with well-defined coarse and fine-grained horizontal laminae, each occupying a certain volume fraction of the plug (Fig. 8.36). The fine- and coarse-grained laminae will have individual capillary pressure curves, but are assumed to have the same values of porosity and saturation exponent. For a given capillary pressure, the water saturation in the coarser grained laminae will be lower than in the finer grained laminae. Consequently, the conductivity of the finer grained lamina will be greater (and the resistivity lower) than the coarser grained lamina. The resistivity of the composite laminated plug sample can be modelled as two resistors in parallel. The bulk resistivity, R, of the laminated sample measured in the laboratory will be the parallel conductivity—the harmonic mean resistivity, weighted by the relative volumes of fine- (f) and coarse-grained lamina (c), thus:

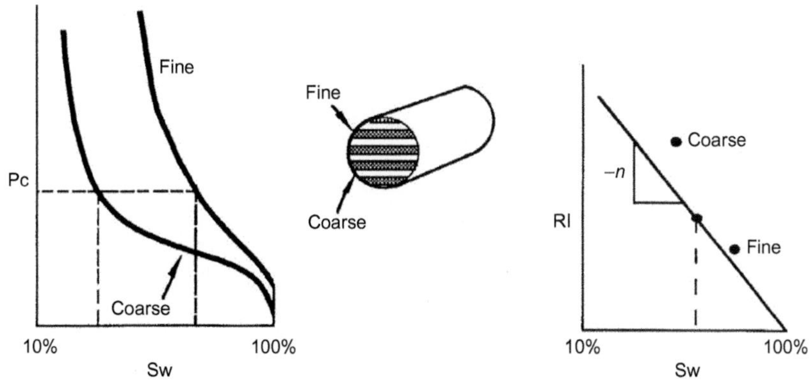

FIGURE 8.36 Laminated samples produce low *n*. *After LaTorraca and Hall (1991).*

$$\frac{1}{R}=\frac{f}{R_{\mathrm{f}}}+\frac{c}{R_{\mathrm{c}}}$$

The bulk plug brine saturation is the weighted arithmetic mean of the individual brine saturations in each lamina. The harmonic mean resistivity is dominated by the lower resistivity value in the finer grained lamina, so that the saturation exponent n for the composite plug (based on bulk I and Sw data) will be lower than the n value for the individual layers.

8.3.5.2 Overburden Stress

As overburden stress increases, the rock pore volume compacts, porosity reduces and, in accordance with Archie's law, formation factor increases. The pore tortuosity (and hence Ro and m) will also increase with overburden pressure. However, it is expected that Rt might be similarly affected by the change in tortuosity so that I and hence n might not be expected to change significantly. There is some lack of consistency reported in the literature but Sondena et al. (1992), Lewis et al. (1988) and the authors' experience shows that n at ambient conditions (in the absence of excess brine effects) is broadly similar to n measured at overburden stress. For example, Fig. 8.37 shows I–Sw relationships for parallel sample suites—one suite was tested at ambient condition, the other at 9000 psi overburden stress. The difference in n between ambient (1.66) and overburden (1.69) is negligible and within experimental error.

Temperature appears to have little or no effect on n, though in oil–water experiments where the sample is effectively aged in reservoir oil during the test, the reservoir temperature should be used.

8.3.5.3 Saturation History

Rock resistivity is a function of both brine saturation and its distribution within the pore network. Most resistivity index experiments are conducted in drainage mode in which the wetting phase saturation (usually presumed to be water) is decreasing. Drainage cycle experiments simulate hydrocarbon

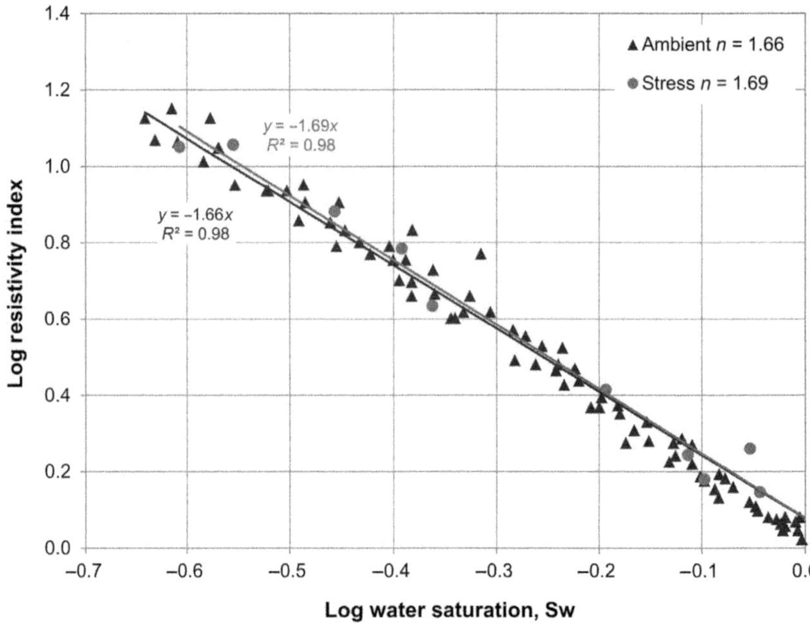

FIGURE 8.37 Example of effect of overburden stress on n.

accumulation and are used for water saturation interpretation in 'virgin', pre-production reservoirs. However, zones undergoing aquifer influx or waterflood are in an imbibition cycle (water saturation increasing), so that drainage data cannot be used to evaluate these zones because the water saturation distribution (for the same bulk water saturation level) will be different between drainage and imbibition. This is due to different trapping mechanisms between drainage and imbibition and contact angle hysteresis.

Imbibition experiments are difficult to perform. For example, forced imbibition systems require an oil-wet membrane (see Chapter 9), and spontaneous imbibition experiments (Fig. 8.38) require precise regulation of oil and water pressures to allow water to imbibe back through the water-wet membrane into the core plug under capillary control. Despite a considerable amount of research, the dependency of n on saturation history remains inconclusive and is probably rock type and wettability specific. Some results indicate that imbibition n for an oil–water system is considerably lower than the drainage value, whereas Al-Kaabi et al. (1997) identified a region of oil-displacing water (primary drainage) where the saturation exponent was found to be 1.8 compared to an imbibition n of 2.7. Imbibition and drainage n for gas–brine systems in water-wet rock appear broadly similar. Clearly, the only way to evaluate imbibition n is to carry out tests at representative reservoir and wettability conditions. However, this is time consuming and costly to do using

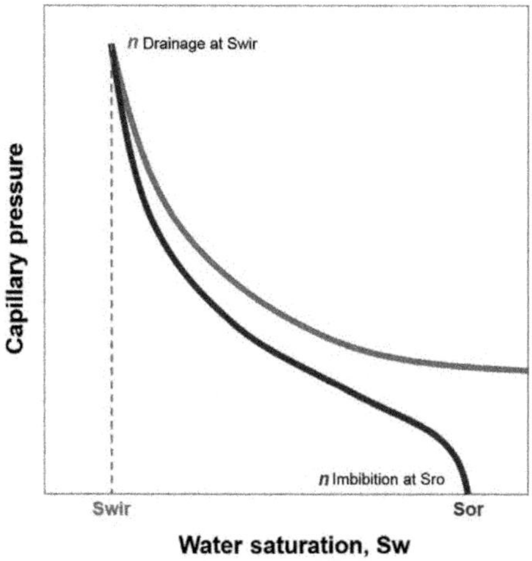

FIGURE 8.38 Primary drainage and spontaneous imbibition capillary pressure curves.

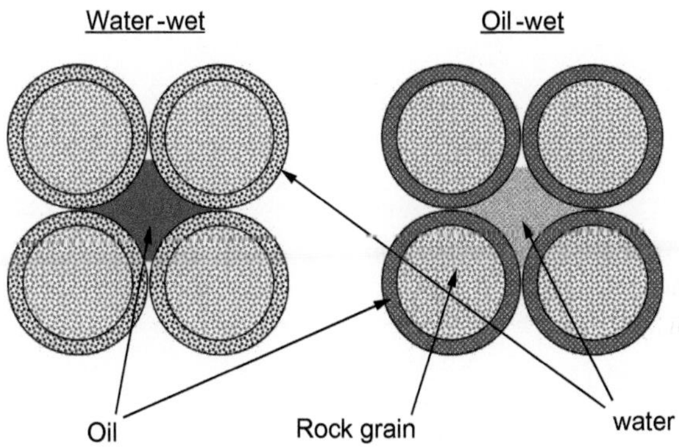

FIGURE 8.39 Wettability controls on distribution of oil and water in pore space.

porous plate methods and the forced imbibition curve is very restricted as it requires an oil-wet semi-permeable membrane.

8.3.5.4 Wettability

The distribution of water saturation in a porous medium with an oil phase present is also controlled by wettability. Simplistically, in a water-wet system, water coats the rock grains and fills the smaller pores, while the non-wetting phase occupies the centre of the larger pores (Fig. 8.39). In an oil-wet system, the reverse is true: oil coats the grains and occupies the smaller pores, whereas water fills the centre of the larger pores. In reality, wettability controls are much more subtle, especially in mixed-wet or neutrally wet formations.

As water is drained in a water-wet system, the strong wetting affinity of the grains maintains continuous water channels which allow current to pass, even though the effective conductive pathways become more tortuous. In an oil-wet rock, the water filling the larger pores becomes rapidly discontinuous as Sw reduces. The conductive flow paths are increasingly isolated and the ability of the pore system to conduct current is reduced. Consequently, for a unit reduction in Sw, the increase in resistivity of an oil-wet rock will be much greater than for a water-wet rock with similar pore system characteristics, and n will be greater. The increase in n with oil-wetting tendencies has been extensively reported (for example, Anderson, 1986; Lewis et al., 1988; Mungan and Moore, 1969; Sweeney and Jennings, 1960; Wang et al., 1991) and numerically modelled by Zhou et al. (1996) (Fig. 8.40).

The majority of SCAL tests have been historically performed by commercial laboratories using an air–brine system on cleaned cores, both of which are expected to force a water-wet condition. However, the majority of oil reservoirs are not water wet, so, ideally, the tests should be carried out using an appropriate fluid system, on restored wettability cores. Restoration occurs during the primary drainage process if reservoir oil is used and the test is

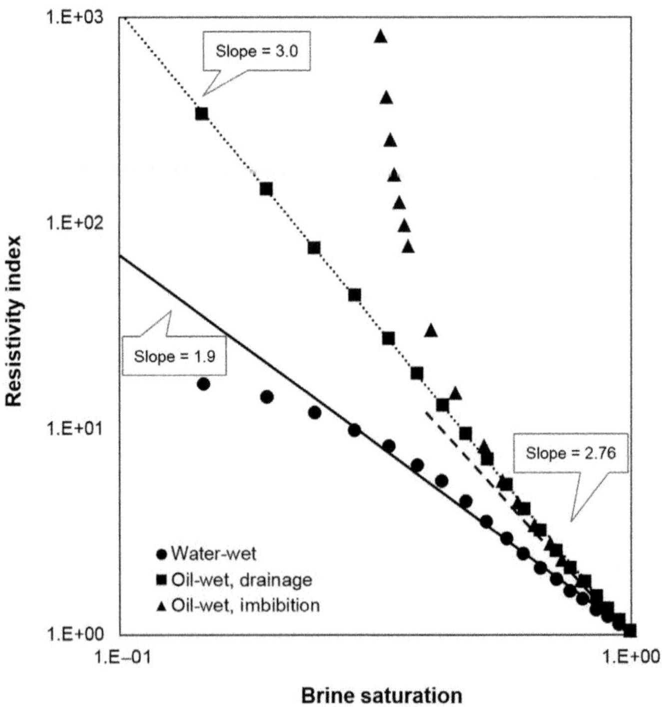

FIGURE 8.40 Modelled resistivity index for water-wet and oil-wet systems. *After Zhou et al. (1996).*

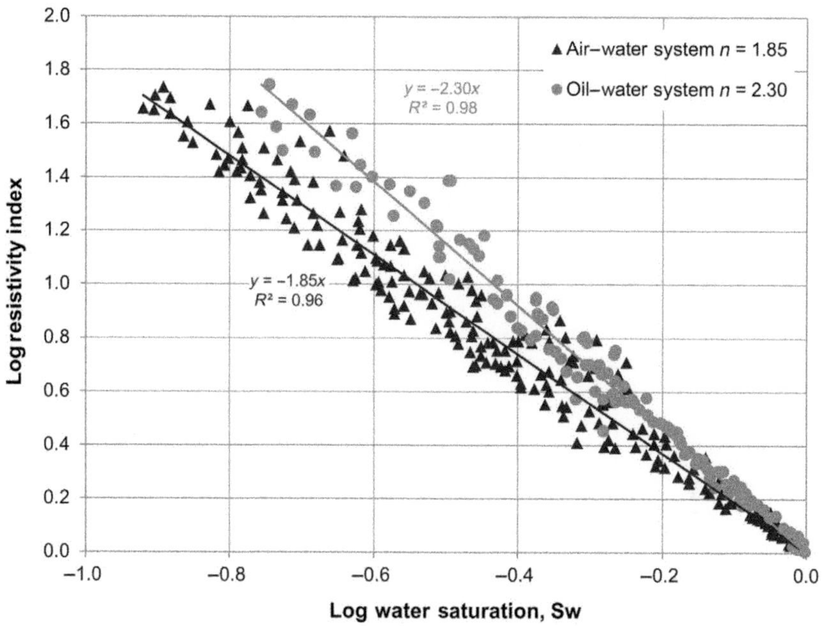

FIGURE 8.41 Comparison of air–water and oil–water composite saturation exponent.

performed at reservoir temperature. Unfortunately, oil–water reservoir condition tests using porous plate systems can take months to complete and are expensive. However, it is important to honour the reservoir wettability as this has a significant impact on n. Figure 8.41 shows composite I versus Sw relationships for parallel plug samples from the same reservoir formation. One set was tested using an air–water system (water-wet) at stress, while the other set was tested using a stock tank oil–water system at reservoir conditions. The air–water n is significantly lower and more optimistic as it would significantly overestimate oil saturation compared to the oil–water n.

8.3.5.5 Conductive Matrix

Archie's model implicitly assumes that the rock grains are non-conductive and ion transport only occurs through the pore system of the rock. However, this assumption is violated if the formation contains conductive minerals, such as pyrite and clays. The structure of clay minerals allows them to conduct electric current and so the matrix is not an insulator. However, clay effects are often suppressed by highly saline and highly conductive formation brine. The presence of clays can manifest itself on resistivity index relationships by a reduction in n, or curvature of the I versus Sw curve. For example, Waxman and Thomas (1974) demonstrated that as the brine becomes fresher, less saline and less conductive (more resistive) so the

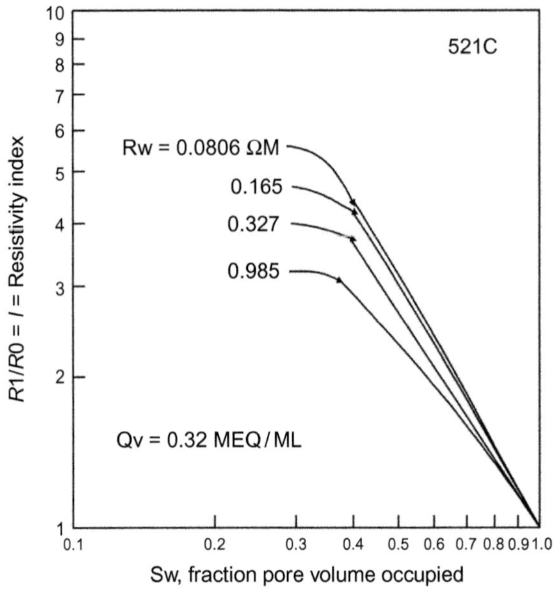

FIGURE 8.42 Effect of brine conductivity on *n*. *After Waxman and Thomas (1974).*

resistivity measured on a shaly sand sample, and hence the apparent *n*, reduces as the relative contribution to sample conductance from clays increases (Figs. 8.42 and 8.43). At high water saturations with a relatively saline brine, the clay conductance effects are suppressed. As the water saturation decreases, the relative amount of brine in the sample reduces and the clay conductance contributes increasingly to the measured sample conductance. The net effect is that the *I*/Sw curve slope reduces.

For relatively fresh formation water (low Cw) and shaly sands with high Qv, clay conductivity must be accounted for in petrophysical interpretation; otherwise, the application of a standard Archie model will underestimate hydrocarbon saturations.

The Waxman–Smits model, which accounts for clay effects on interpreted water saturation, and the core analysis tests used to acquire key model parameters, are discussed below.

8.3.6 Data Reporting Requirements

The data required to ensure adequate quality control of the resistivity index tests and test results are the same as those for FRF (Table 8.2). Additional data requirements related specifically to resistivity index tests are listed in Table 8.4.

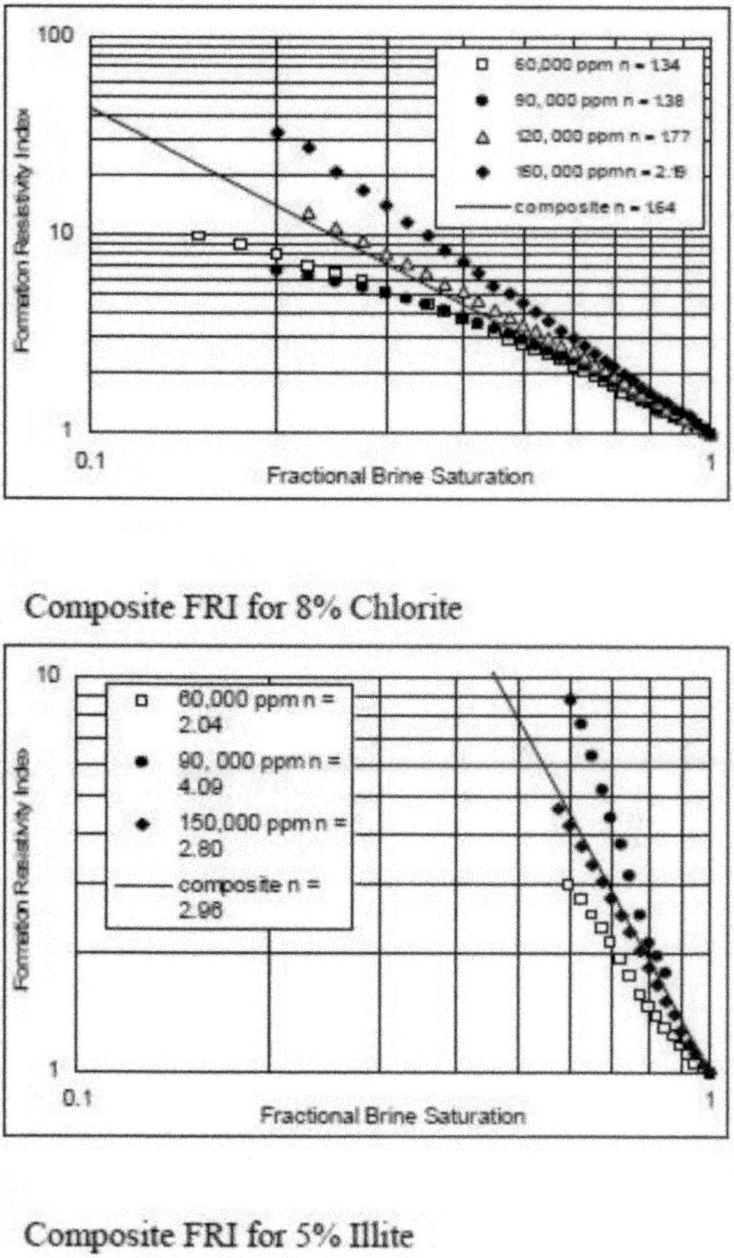

FIGURE 8.43 Composite resistivity index for shaly sands. *From Crain (2014).*

TABLE 8.4 Data Reporting Requirements for RI Tests (for Each Sample)

Data	Comments
Brief description of procedures and experimental apparatus	Ambient or overburden. Porous plate or CIRI. 2 or 4 electrode, etc.
Fluid system	Air–water or oil–water
Synthetic formation water (SFW) composition	If not already reported
Gas or oil properties	For example, stock tank oil composition
SFW resistivity (Rw) at standard temperature	Ω m. Temperature in °F or °C
Initial base permeability (Klinkenberg or air permeability) and helium porosity	Porosity determined from He Vg (or He Vp) and SFW immersion Vb
Saturated sample weights and sample pore and bulk volumes	To check lab calculations
Net confining stress	Overburden test
Calculated porosity (fraction), pore volume and bulk volume at net confining stress station, based on corrected expelled brine volume	Porosity calculated on basis of incompressible grain volume
Temperature, frequency (minimum phase angle), rock resistance, porous plate disc resistance (if used), phase angle, resistivity (Ro) on completion of the equilibration period, at test stress station	Ro at standard temperature
FRF at standard temperature	100% SFW saturation
Displaced (drainage) or imbibed (imbibition) brine volumes as a function of time during RI process. Stabilisation time allowed at each capillary pressure step	Per capillary pressure increment
Calculated plug resistivity and resistivity index at standard temperature at minimum phase angle	For each RI measurement
Calculated brine saturation for each RI measurement as a fraction of brine-saturated pore volume	Per pressure step
In situ saturation monitoring scans	If available
Resistivity index versus brine saturation data and plot for each sample. Tabulate versus capillary pressure	*n* calculated from linear regression analysis of bi-logarithmic *I* and Sw data
Final dry weight, helium porosity and base permeability (Klinkenberg or air permeability)	Post-test. Compare to pre-test

Continued

TABLE 8.4 Data Reporting Requirements for RI Tests (for Each Sample)—Cont'd

Data	Comments
Sample grain loss correction procedures (if applied)	Include typical worked example for one sample and all relevant sample weight data during entire test sequence
Final water saturation from Dean–Stark extraction	Corrected to test NCS

8.3.7 Advantages and Drawbacks/Issues

Summaries of the advantages and drawbacks/issues associated with resistivity index tests are provided in Tables 8.5 (porous plate ambient conditions) and 8.6 (stress conditions).

8.3.8 RI Test Quality Control Issues, Checks and Diagnostics

Analysis of the results should start with a thorough review of the laboratory report to check that acceptable procedures and test methods have been followed.

Ensure that the method used for cleaning and drying the test samples has not caused damage and depreservation of clay structures.

SCP annulus volume data and plots must be provided for each plug sample, and the corrected expelled volumes agreed between the end user and the test lab (see Fig. 8.5).

All experimental data listed in Tables 8.5 and 8.6 must be supplied and lab calculations verified.

Phase angles for resistivity measurements must be less than 0.5°. High values could indicate poor contact between the sample and the electrodes, or that the sample has a low length to area ratio.

If the RI test is performed immediately following FRF measurements at stress in the same loading sequence, then ensure FRF and porosity versus stress trends do not indicate pore collapse at the RI test stress.

Provided the plug is not scheduled for other tests, unload the sample on completion of the drainage or imbibition sequence and extract the water in a Dean-Stark apparatus. This will provide an independent check on volumetric or gravimetric water saturation calculations. The Dean-Stark Sw should match the final calculated Sw within about ±3–4 saturation units (s.u.). Figure 8.44 shows an example where samples S12 and S9 fall outside the minimum acceptable error between Dean-Stark Sw and the final RI test Sw which was calculated from volumetric measurements. Figure 8.45 compares the final Dean-Stark

TABLE 8.5 Advantages and Drawbacks/Issues Associated with RI Tests (Ambient Condition Tests)

Advantages	Drawbacks and Issues
Ambient conditions tests • Test performed under unconfined conditions either in batch cell or in individual cell. • Porous plate is the best method to achieve a uniform saturation profile. • Relatively simple. • Air-brine capillary pressure can also be measured simultaneously with RI. Time required to achieve capillary equilibrium is longer than that required to attain resistivity equilibrium. • Up to 12 plug samples can be run simultaneously in one standard batch cell.	Ambient conditions tests • Only air–brine can be run in batch cell. Oil–brine can be run in individual cell. • Resistivity measurements are sensitive to the frequency used (phase angle should be less than 0.5°). • There are limitations on the maximum attainable Pc values using standard porous plates (180 psi air–brine, 100 psi oil–brine). Higher pressure plates are available in some labs. • Benign cleaning and drying are required if the core plugs contain delicate clay structures. • In some cases, correct establishment of electrical contact between the sample and the electrodes can be difficult. This depends on the experimental set-up, the shape and competency of the sample and the skill of the operator. • Batch configuration does not allow for individual sample desaturation monitoring. Samples need to be removed continuously from the cell to measure water saturation gravimetrically and plug resistance. • Excess brine and grain loss (gravimetric saturation calculations) are two major issues associated with ambient condition, batch cell tests. Excess brine impacts on Ro, *m* and *n* at ambient conditions. Grain loss can result in significant uncertainties in calculated water saturations and grain loss corrections can be subjective and often indiscriminately applied. • Plugs run in batch mode at set capillary pressure increments should have similar permeabilities. A combination of high- and low-permeability samples in the same batch is not recommended. • Not suitable for poorly consolidated plugs. • Loss of capillary contact between plug and porous plate means that sample may not be drained to Swir (see Chapter 9).

Sw with the gravimetrically calculated final plug Sw after an imbibition RI test. Grain losses have a significant impact on the results and only 8 of 26 samples' Sw agree within acceptable limits. If the final saturation is in error, all intermediate saturations must equally be uncertain. Suspect data need to be carefully evaluated and, if necessary, eliminated from the quality-controlled data base.

For similar rock types, a log–log plot of water saturation versus permeability or rock quality index (RQI) should normally show a trend of water

TABLE 8.6 Advantages and Drawbacks/Issues Associated with PPRI and CIRI Tests (Stress Condition Tests)

Advantages	Drawbacks and Issues
Stress (overburden) conditions (PPRI) tests • Test can be run at equivalent isostatic reservoir stress but nominal stress is normally sufficient to prevent bypass of air/oil during the desaturation process, and to eliminate the effects of excess brine. • Recommended method for poorly consolidated samples. • The axial stress acting on the ends of the core end platens is usually sufficient to ensure good capillary contact; therefore, loss of capillary contact is normally not an issue. • Saturations and resistance can be monitored for each individual sample. • Use of reservoir oil (e.g. stock tank oil) at reservoir temperature can help restore wettability during the drainage process. Hence saturation exponent is more appropriate for the reservoir wettability and stress condition. • As sample manipulation is minimised, the method is less likely to require grain loss corrections. • Only method possible to determine resistivity index on spontaneous imbibition.	Stress (overburden) conditions (PPRI) tests • More expensive than the batch cell, as a dedicated coreholder is required for each sample, and experimental times can be prolonged. • SCP and annulus volume are essential for porosity and hence saturation at stress calculation and must be satisfactorily established for each test plug. • Friable samples require to be protected against grain loss and sample failure. Stress has to be carefully managed to prevent damage of poorly consolidated samples. • Grain loss corrections, although less than the batch cell, will also be required for friable samples. • Measurements using oil–water systems can take months to complete if capillary equilibrium is to be ensured. • Forced imbibition capillary pressure range is constrained and constricted by the need to ensure an oil-wet semi-permeable membrane. Spontaneous imbibition data may be sufficient for petrophysical analysis.
Continuous injection (CIRI) tests • Advantages are similar to overburden porous plate tests in terms of excess brine, grain loss and saturation determination. • The method is faster than the porous plate resistivity index (PPRI) method. Usually oil–water primary drainage tests take only a few weeks as opposed to months using PPRI.	Continuous injection (CIRI) tests • CIRI is a dynamic method and capillary equilibrium may not be attained at any saturation level. This leads to non-uniform saturation profiles that can result in non-linear *I*–Sw curves. • Although faster than PPRI, CIRI can be considerably more expensive as ISSM is required to establish and map water saturation profiles.

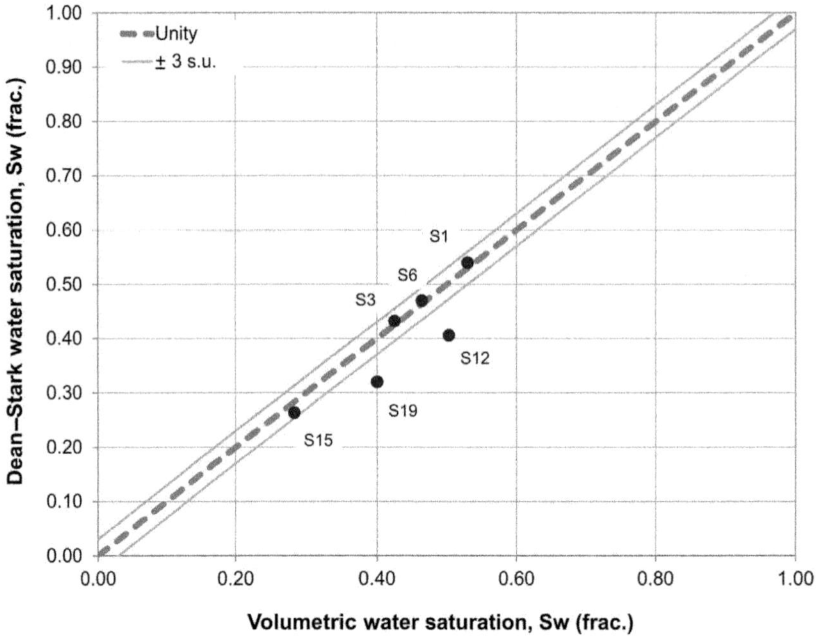

FIGURE 8.44 Water saturation check for end-point resistivity index test.

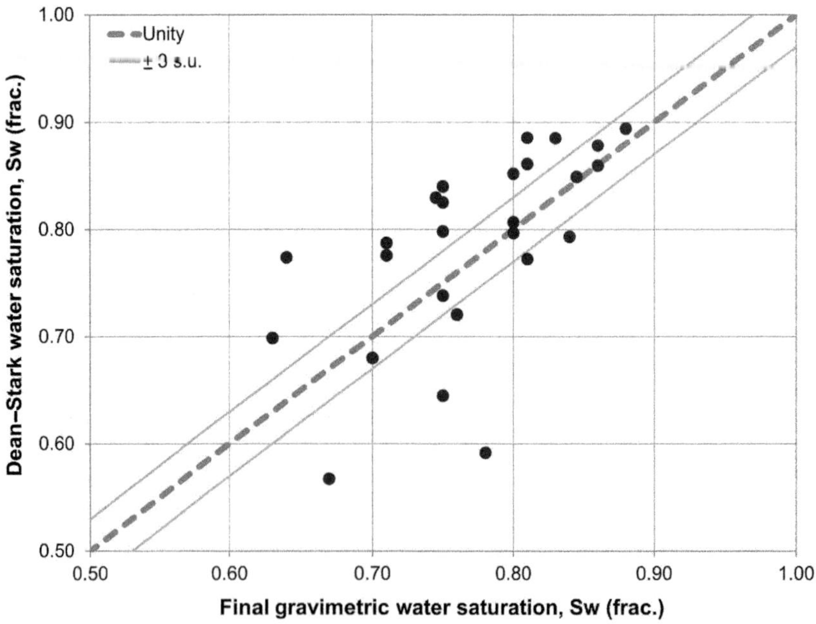

FIGURE 8.45 Example of water saturation check for end-point imbibition resistivity index test (gravimetric Sw).

saturation increasing with decreasing rock quality, as presented in Fig. 8.46. For sample S17, the Sw end point is lower than expected (20% Sw), compared to the general data trend. Data from this sample need to be carefully scrutinised to differentiate rock-type behaviour from experimental artefacts and saturation uncertainty. If available, RI end points (on drainage or on imbibition) are checked against data from other reliable sources such as capillary pressure tests and NMR tests. The data should form a consistent trend for similar rock types. Figure 8.47 provides an example of some resistivity index test data (highlighted in red; grey in the print version) where Swir is clearly not on trend with the other RI and capillary pressure data from the same reservoir interval. The *I*–Sw and saturation exponent data from these highlighted samples need not necessarily be in error, but the results must be carefully checked and verified.

When loss of capillary contact in ambient condition occurs in porous plate tests, it can be characteristically identified in the water saturation/capillary pressure curve, which often shows an abrupt change to become asymptotic (Fig. 8.48). Swir is much higher than the true value.

Excess brine in ambient condition experiments affects both resistivity index and water saturation calculations. Indications of excess brine are:

1. The saturated pore volume calculated from the brine density and the difference in weight between the dried and the brine-saturated sample exceeds the helium pore volume.

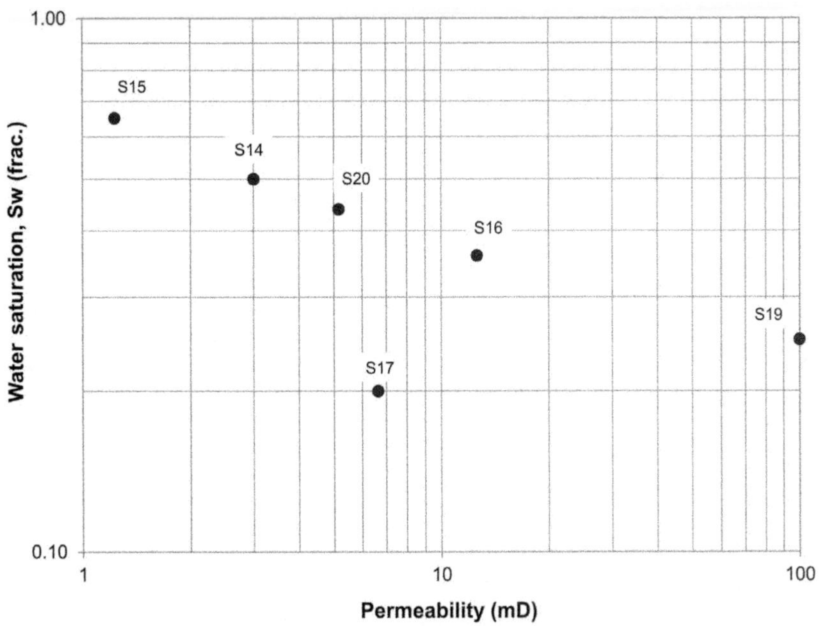

FIGURE 8.46 Example of permeability versus end-point Sw plot.

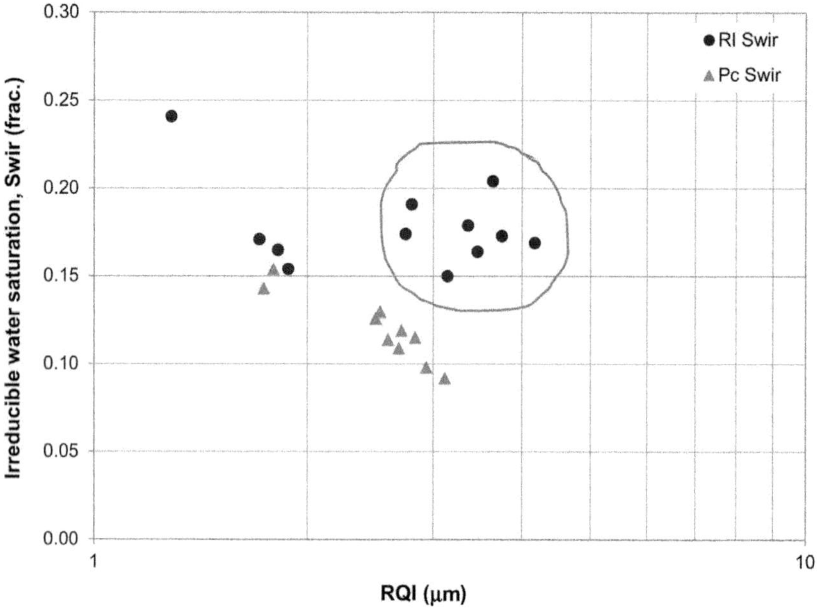

FIGURE 8.47 Comparison of Swir from RI and capillary pressure tests.

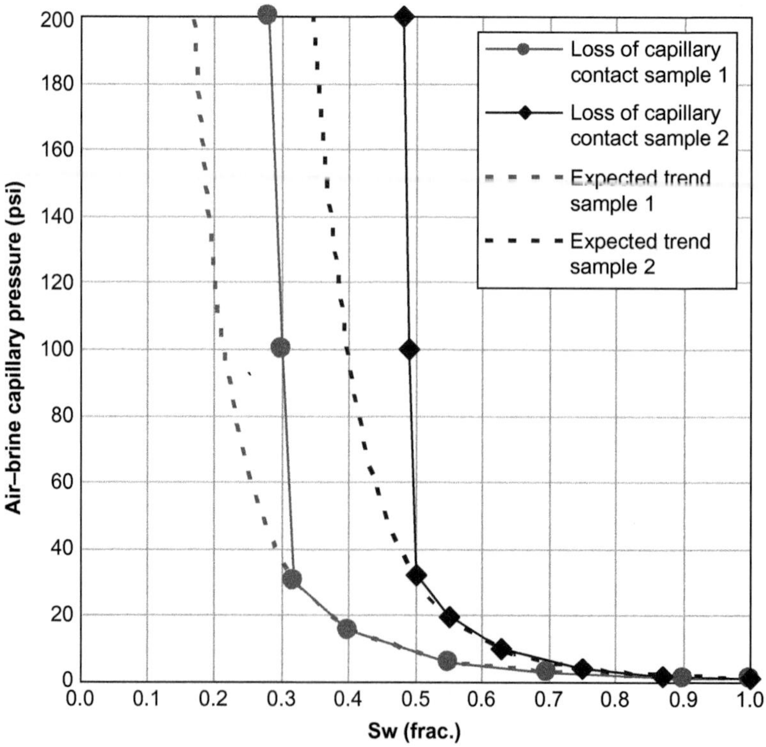

FIGURE 8.48 Hypothetical example of loss of capillary contact using ambient porous plate experiment.

2. A plot of sample mass (sample weight) versus capillary pressure shows an abnormally large mass decrease on desaturation at low capillary pressures. Once this excess brine is produced from the sample surface, the capillary pressure has to be raised above the entry pressure of the sample before further pore desaturation can occur. This produces the discontinuity evident in the mass–capillary pressure curve as exemplified by Fig. 8.49. The entry point, where $Sw=1$ and $I=1$, should be chosen as indicated and subsequent saturation and resistivity index data at higher pressures corrected based on the entry point values. If the entry pressure is low excess brine effects may be masked but the corrections for excess brine are much less significant in higher permeability than they are in lower permeability samples.

Grain loss can result in considerable uncertainties in the calculation of fluid saturation. Grain loss and gravimetric data are rarely (if ever) reported by commercial labs so the end user has to rely entirely on the laboratories' interpretation of the data, and have implicit faith in their grain loss corrections and saturation calculations. This is particularly true of legacy data. However in new SCAL programmes, if saturations are determined from gravimetric

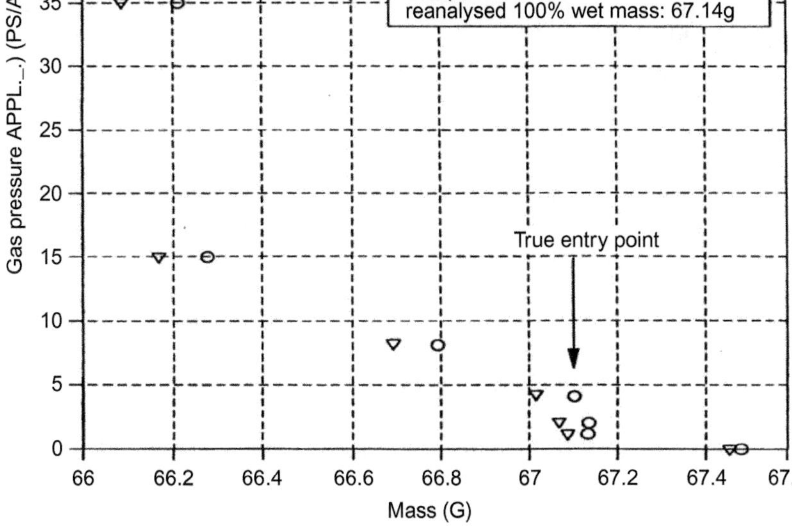

FIGURE 8.49 Excess brine effects on weight–capillary pressure curve. *After De Waal et al. (1989). https://www.onepetro.org/journal-paper/SPWLA-1991-v32n5a9.*

measurements, then all sample dry and saturated weight data, and Dean-Stark extraction measurements, must be reported to assess the magnitude of any grain loss, and the laboratory's grain loss corrections must be checked and verified.

8.4 WAXMAN–SMITS PARAMETERS

8.4.1 Shaly Sand Models

The presence of clay in reservoir rocks affects their effective porosity, resistivity and permeability. Most clastic reservoir rocks contain some clay. Depending on the type of clays, their distribution and volume, they can modify the following rock properties compared with those of clean, non-shaly sands:

- reduction in the effective porosity
- reduction in permeability
- reduction in true formation resistivity, Rt

Whether or not clays will significantly impact formation evaluation depends on a number of factors including the clays present and their volume, the distribution of the clays within the pore system and the formation water salinity.

The main objective of shaly sand interpretation is to correct water saturations calculated using the Archie equation which assumes the rock matrix is non-conductive. If the clay conductivity is sufficient and is not accounted for in the interpretation, the Archie model will underestimate the hydrocarbon saturation.

The Archie equations are based on the assumption that the rock matrix is non-conductive. In the case where clay minerals form part of the matrix they allow electrical current to flow through them reducing the apparent formation resistivity Fig. 8.50. Hence if the Archie equation is expressed in terms of the total conductivity, Ct:

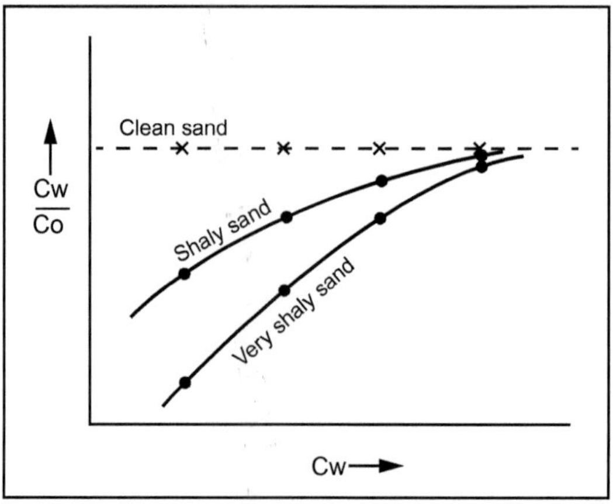

FIGURE 8.50 Effect of water resistivity on formation resistivity factor for a clean and shaly-sand. *After Worthington (1985).*

$$\mathrm{Ct}=\frac{\mathrm{Cw}}{F}*\mathrm{Sw}^{n}$$

then, for shaly sands, the extra conductivity introduced by the clays must be added:

$$\mathrm{Ct}=\frac{\mathrm{Cw}}{F}*\mathrm{Sw}^{n}+X$$

If the extra clay conductivity X is very small or Cw is very large, for example if the formation water is very saline, this equation reverts to the clean sand Archie equation.

The basis of the modification for hydrocarbon-bearing shaly sands is mathematically described by:

$$\frac{1}{\mathrm{Rt}}=\frac{1}{F\mathrm{Rw}}\mathrm{Sw}^{n}+X\mathrm{Sw}^{\mathrm{p}}$$

where n is the Archie saturation exponent and p is the exponent modified for clay effects. Two alternative approaches to the calculation of the shale conductivity have evolved (Thomas and Stieber, 1975). These are described in Section 8.1.3.

1. *Models based on the shale volume fraction Vsh*. These models are scientifically inexact but have the advantage that the required shale parameters can be determined relatively easily from log data alone. These models require effective porosity ϕ_e as input and calculate effective water saturation as a fraction of that porosity, Sw_e.
2. *Models based on an ionic double layer of bound water associated with the clays in shale*. These models provide a more scientific understanding of the mechanisms involved. However, for rigorous application, they require core data for calibration of the clay CEC (Qv) with respect to a log-derived parameter, usually porosity. In the absence of core data, the application of these methods can involve approximations which render them effectively equivalent to the Vsh-based equations. These models require total porosity ϕ_t as input and calculate water saturation as a fraction of that porosity Sw_t.

Vsh methods implicitly assume that the electrical properties of dispersed or laminated clays within the pore systems of sandstones are the same as the detrital clays in layers adjacent to the sandstones, and are exclusively log-based as Vsh cannot be determined from core.

In the double-layer models, the conductivity of the shale component is explained using the concept of an ionic double layer. Dispersed clay is distributed on the surface of sand grains. The structures of clays are such that their outside surfaces are negatively charged. Because of this charge, positive cations (such as Na^{+}) in the formation of water are attracted to the clay surfaces (Fig. 8.51). Water molecules are dipolar which means that

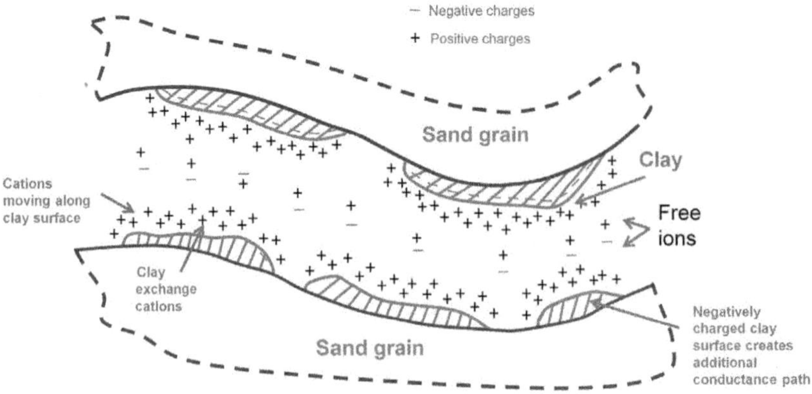

FIGURE 8.51 Sodium cations (+ve ions Na^+) attracted by negatively charged clay surfaces.

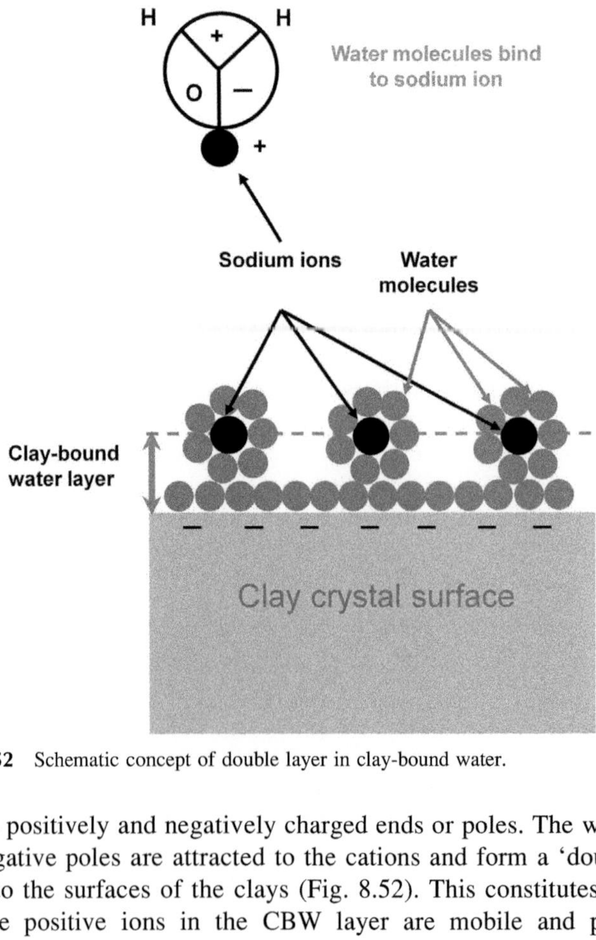

FIGURE 8.52 Schematic concept of double layer in clay-bound water.

they have positively and negatively charged ends or poles. The water molecules' negative poles are attracted to the cations and form a 'double layer' attached to the surfaces of the clays (Fig. 8.52). This constitutes the CBW layer. The positive ions in the CBW layer are mobile and provide an

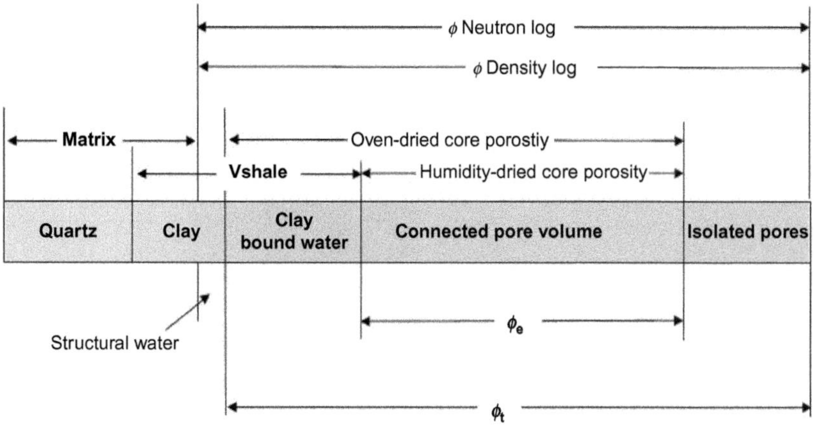

FIGURE 8.53 Clay-bound water is the difference between total and effective porosity (if no isolated pores). *After Eslinger and Pevear (1988).*

additional path for electrical current. The resistivity of the rock is thus reduced below that predicted by the Archie equation which would therefore predict too high a water saturation and too low a hydrocarbon saturation. The Waxman and Smits (1968) equation models the effect of the double layer on the formation conductivity to determine the correct water saturation.

Since the CBW is not free to move, and if there are no isolated pores in the pore system, the effective porosity ϕ_e is equal to the total porosity ϕ_t minus the CBW (Fig. 8.53). Because the CBW contributes to the total conductivity, the formation factor for the Waxman–Smits equation relates to the total porosity ϕ_t and the resulting water saturation Sw_t is relative to total porosity.

The thickness of the CBW layer relates to the salinity of the formation water; the more saline the water the thinner is the bound water layer. This phenomenon was investigated by Hill et al. (1979) who established the relationship between the ratio ϕ_e/ϕ_t, formation water salinity and Qv:

$$\frac{\phi_e}{\phi_t} = 1 - \left(0.22 + \frac{0.64}{\sqrt{S}}\right)\mathrm{Qv}$$

where S is the formation water salinity (g/l NaCl equivalent) and Qv is the normalised CEC, meq/ml.

This relationship is illustrated in Fig. 8.54. The fraction, and hence the thickness, of the CBW layer decreases with the salinity of the water. This dependence is explained by the increasing screening of water in the bulk of the porosity from the negatively charged clay surfaces by increasing numbers of positive ions available in the brine as the salinity increases.

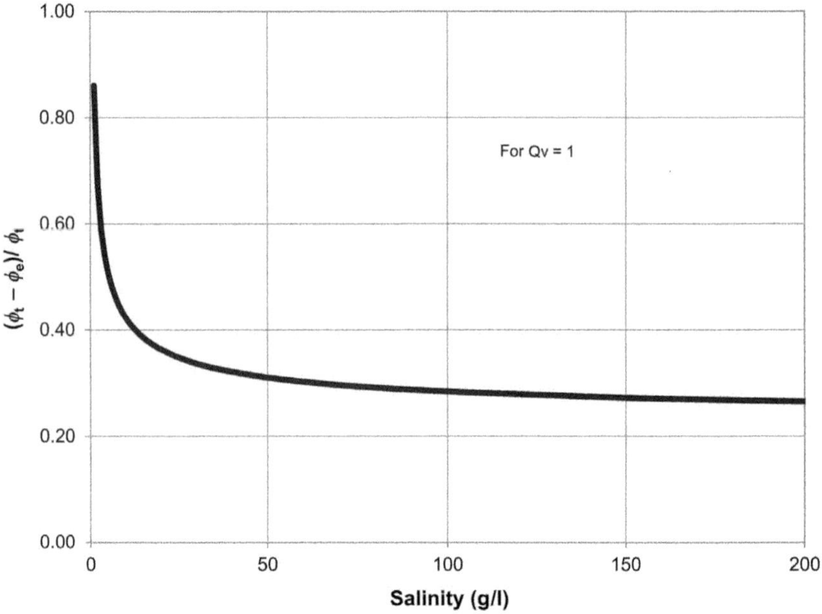

FIGURE 8.54 Salinity dependence of bound water fraction of pore volume (Qv = 1 meq/ml).

8.4.2 Waxman–Smits Equation

In order to add the effect of the conductivity of shales to derive the Waxman and Smits (1968) equation, it is easier to work with conductivities. Since $C = 1/R$ the Archie equation for water-bearing rock can be rewritten as:

$$\text{Co} = \frac{\phi^m}{a} \cdot \text{Cw}$$

where Co is the conductivity of brine-saturated rock (mho/m) and Cw is the conductivity of the formation brine (mho/m).

Waxman and Smits replaced Cw in the Archie equation for clean sands with the equivalent water conductivity of the formation pore water, plus additional clay conductivity (Ce, mho/m) resulting from the CBW layer in shaly sand (Cw + Ce).

8.4.2.1 Water-Bearing Shaly Sand

Waxman and Smits introduced a cementation exponent m^* and tortuosity constant a^*. The clay corrected or intrinsic cementation exponent m^* differs from the Archie constant m. The same formation factor $\phi_t^{m^*}/a$ is appropriate for both the clay conductivity and the pore brine conductivity because all the clay is assumed to be in the pores lining the pore walls. The product of CEC, Qv (meq/ml), and the equivalent conductance of sodium clay exchange cations, *B* ((mho/m)/(meq/ml)). is the additional clay conductivity, Ce.

Hence, the Waxman–Smits equation for water-saturated shaly sandstone becomes:

$$\mathrm{Co} = \frac{\phi_{\mathrm{t}}^{m*}}{a*} \cdot (\mathrm{Cw} + \mathrm{Ce}) = \frac{\phi_{\mathrm{t}}^{m*}}{a*} \cdot (\mathrm{Cw} + B \cdot \mathrm{Qv})$$

Returning to resistivity terms (R in Ω m):

$$\frac{\mathrm{Ro}}{\mathrm{Rw}} = \frac{a*}{\phi_{\mathrm{t}}^{m*}} \cdot \frac{1}{(1 + B \cdot \mathrm{Qv} \cdot \mathrm{Rw})}$$

If $F(=R\mathrm{o}/\mathrm{Rw})$ is plotted on a log–log plot versus ϕ, the data form a curve rather than a straight line because Qv decreases as porosity increases. To render the log–log plot linear and allow determination of $m*$, Waxman–Smits defined the intrinsic formation factor, $F*$:

$$F* = \frac{\mathrm{Ro}*}{\mathrm{Rw}} = \frac{a}{\phi^{m*}} = F \cdot (1 + B\mathrm{QvRw})$$

$F*$ is the Waxman–Smits (or intrinsic, clay-corrected) FRF. Hence, $m*$ can be determined from regression analysis of a log–log plot of $F*$ versus ϕ.

8.4.2.2 Hydrocarbon-Bearing Shaly Sand

The Archie equation for clean hydrocarbon-bearing rock is:

$$\mathrm{Rt} = a\phi^{-m}\mathrm{Sw}^{-n}\mathrm{Rw}$$

In terms of resistivity index I:

$$I = \frac{\mathrm{Rt}}{\mathrm{Ro}} = \mathrm{Sw}^{-n}$$

To derive the Waxman–Smits equation, the Archie equation is again converted to conductivity terms:

$$\mathrm{Ct} = a^{-1}\phi^{m}\mathrm{Sw}^{n}\mathrm{Cw}$$

where Ct is the conductivity of the hydrocarbon-bearing formation (mho/m).

Again Waxman and Smits replace Cw by an equivalent water conductivity (Cw+Ce/Sw). The additional Sw term was introduced to account for the increasing contribution to overall conductivity by the clays as the pore water saturation reduces. Hence the Waxman–Smits equation is:

$$\mathrm{Ct} = a*^{-1}\phi^{m*}\mathrm{Sw}^{n*}\left(\mathrm{Cw} + \frac{\mathrm{Ce}}{\mathrm{Sw}}\right)$$

Or, in terms of Qv:

$$\mathrm{Ct} = a^{*-1}\phi^{m^*}\mathrm{Sw}^{n^*}\left(\mathrm{Cw} + \frac{B\mathrm{Qv}}{\mathrm{Sw}}\right)$$

Expressed in resistivity terms this becomes:

$$\mathrm{Rt} = a^{*-1}\cdot\phi^{-m^*}\cdot\mathrm{Sw}^{-n^*}\cdot\left(\frac{\mathrm{Rw}}{1+\dfrac{B\mathrm{QvRw}}{\mathrm{Sw}}}\right)$$

For a water saturation of unity Rt becomes Ro and the equation simplifies to:

$$\mathrm{Ro} = a^{*-1}\cdot\phi^{-m^*}\cdot\left(\frac{\mathrm{Rw}}{1+\mathrm{Rw}B\mathrm{Qv}}\right)$$

Hence:

$$I = \frac{\mathrm{Rt}}{\mathrm{Ro}} = \mathrm{Sw}^{-n^*}\frac{(1+B\mathrm{QvRw})}{(1+B\mathrm{QvRw/Sw})}$$

A log–log plot of I versus Sw for a shaly sand will not be linear. Waxman and Smits introduced a modified (or intrinsic) resistivity index I^*:

$$I^* = \frac{\mathrm{Rt}^*}{\mathrm{Ro}^*} = \mathrm{Sw}^{-n^*} = I\frac{(1+B\mathrm{QvRw/Sw})}{(1+B\mathrm{QvRw})}$$

If I^* is plotted versus Sw in a log–log plot a linear plot with slope n^* results.

8.4.2.3 Utilisation of Waxman–Smits Equation

The Waxman–Smits equation for reservoir petrophysics applications can be written as:

$$\mathrm{Sw}_{\mathrm{t}}^{n^*} = \frac{a^*\cdot\mathrm{Rw}}{\mathrm{Rt}\phi_{\mathrm{t}}^{m^*}\cdot(1+B\mathrm{Qv}w/\mathrm{Sw})}$$

$\mathrm{Sw_t}$ is the fractional water saturation relative to total porosity, ϕ_t. Rw is the brine resistivity and B is the cation mobility both at the relevant formation temperature T, °C (Ω m), and m^* and n^* are the intrinsic cementation exponent and saturation exponent for shaly sand.

As water saturation decreases and hydrocarbon saturation (Sh) increases the larger the term BQv Rw/Sw becomes. If this term is <0.1 the effect of the shaliness can be neglected. Factors which render this term larger are:

- Fresh formation water—high Rw.
- High Qv—high clay content or a significant proportion of high CEC clay minerals (such as smectite).
- High B.

8.4.2.4 CEC (Qv in Waxman–Smits Model)

The number of positive ions (the cation concentration) attracted to the clay surfaces depends on both the amount and type of clays present. Two measures of the concentration of cations are available: the CEC, which is measured relative to mass, and Qv, which is measured relative to pore volume. CEC is expressed in milli-equivalent of exchangeable ions per 100 g of dry rock (meq/100 g). Qv is expressed in milli-equivalent per millilitre of pore volume (meq/ml).

The conversion from CEC to Qv is made by means of the following relationship:

$$\mathrm{Qv} = \frac{\mathrm{CEC}(1-\phi_t)\rho_g}{100\phi_t}$$

where is the ϕ total porosity (fraction) either at ambient or at stress conditions, depending on application, and ρ_g is grain density (g/cc).

8.4.2.5 Cation Mobility (B in Waxman–Smits Model)

The mobility of the cations is measured by the equivalent conductance per cation, B, which, for a given type of brine, is dependent only on the brine salinity and temperature. Waxman and Thomas (1974) made experimental measurements of B versus Rw for a series of temperatures and presented their results graphically in the figure reproduced as Fig. 8.55. Juhasz (1981) provided the following equation, which approximately fits the data:

$$B = \frac{-1.28 + 0.225T - 0.0004059T^2}{1 + \mathrm{Rw}^{1.23}(0.045T - 0.27)}$$

where T is the lab or formation temperature (°C), depending on application, and Rw is the formation brine resistivity (Ω m) at temperature, T. This equation strictly applies only for sodium cations. If the formation water has a significant amount of divalent cations, then B may be in error (typically ±10%).

Dacy and Martin (2008) proposed an improved and unifying equation for B:

$$B = \left[1 - 0.83\exp\left[-\frac{\left[-2.47 + 0.229\ln(T)^2 + \dfrac{1311}{T^2}\right]^{-1}}{\mathrm{Rw}}\right]\right]\left(-9.2431 + 2.6146T^{0.5}\right)$$

For a 0.303 Ω m, 19,667 mg/l brine at 25 °C, B from the Dacy and Martin model is 3.22 compared to 3.42 using Juhasz.

8.4.2.6 Waxman–Smits Application Workflow

To use the Waxman–Smits equation, the following workflow is generally used:

1. Calculate total porosity, ϕ_t

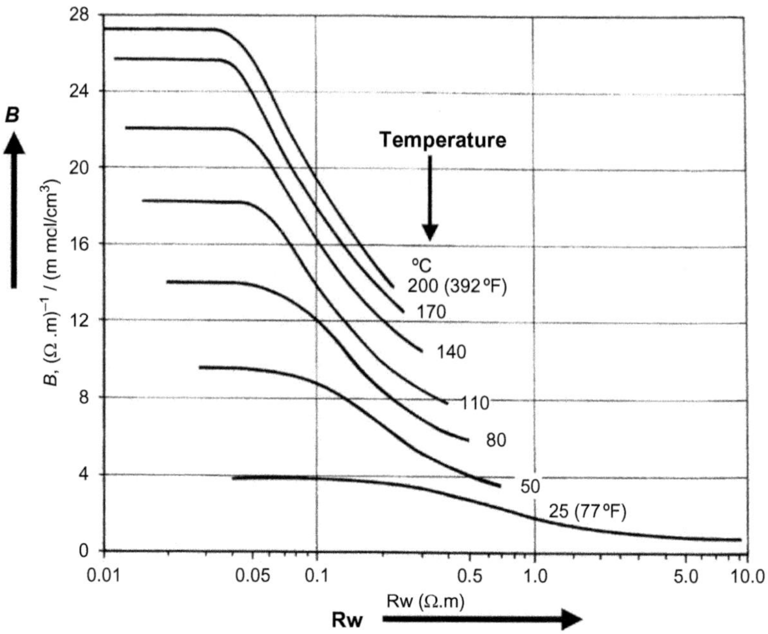

FIGURE 8.55 Dependence of cation mobility B on brine salinity and temperature. *After Waxman and Thomas (1974).*

2. Determine B at reservoir temperature, T, using the Waxman–Thomas/Juhasz relationship or the Dacy and Martin method
3. Determine Qv (ideally from core tests)
4. Establish a relationship between Qv and porosity
5. Determine m^* using core measurements of formation factor
6. Determine n^* using core measurements of resistivity index.

The following sections describe the input parameters Qv, m^* and n^* and how they are obtained in the laboratory.

8.4.3 CEC and Qv Tests

Qv is the CEC of the rock normalised by the pore volume. In commercial laboratories, it is normally determined using destructive (wet chemistry CEC) and non-destructive (multi-salinity Co–Cw) methods.

The non-destructive membrane potential method (Thomas, 1976) is based on determining the electromagnetic force (potential) generated across an ion-selective membrane when solutions on either side of the membrane have different salinities. Shales and clays are cationic membranes, since they allow the passage of cations but not anions. The membrane potential is reduced if the clay has a low cation-exchange capacity. Yuan and Diederix (1989) claim that the clay conductivities derived from membrane potential measurements provide more accurate estimates of the clay conductivity than those obtained from wet chemistry CEC conductometric titration. The reasons for this are that the membrane potential measurements are performed where the clays are maintained in the environment in which they contribute to the clay conductivity, and that the clay conductivity derived from membrane potential measurements does not depend upon empirical shaly sand parameters such as the shaly sand FRF, F^*. As a non-destructive method of measuring Qv, the technique is faster than the multiple salinity method, but care is needed in making the measurement and deriving the appropriate Qv. The test is difficult to perform and is not normally provided by commercial labs. Consequently, only wet chemistry and multiple salinity tests are described below.

8.4.3.1 Wet Chemistry CEC (Destructive Method)

This is the most widely used technique because it is cheap, convenient and rapid. Qv is derived from laboratory measurements of the CEC of the rock.

The ion-exchange capability of clay minerals, particularly smectites, influences their unique physical properties, such as the cation retention and diffusion processes of charged and uncharged molecules. These processes influence cation and molecule migration through clay-rich barriers. The numerical value of this property is described by the CEC. Methods for determining CEC involve the complete exchange of the naturally occurring cations by a cationic species, such as ammonium, potassium, sodium and barium, which are detected by conductometric titration or spectrometric methods.

A minimum 20 g of sand sample is required, normally taken from offcuts of plug samples whose porosity, permeability and Archie parameters m and n have been measured. Accuracy is decreased as sample size reduces, becoming more affected by test noise and by the sample becoming potentially unrepresentative.

Sample taken from whole core or plugs should be cleaved and not cut to reduce fines production, which will artificially increase the proportion of finer grains (clays). Ideally, cleaning and drying should be performed before sample disaggregation to minimise potential loss of the clay size fraction (clay). The samples are cleaned using hot Soxhlet apparatus (or immersion cleaning) until all residues from oil and brine are removed and then dried in a conventional extraction dry oven at 100 °C.

Once cleaned and dried, the samples are disaggregated using a mortar and pestle, or if the sample is from a friable or loosely consolidated sand, by sonication in an ultra-sonic bath. The use of a rotary motion of the mortar and pestle should be avoided, as this will crush the individual grains and artificially increase the number of exchange sites. A back and forth motion should be used, as this will separate the grains without crushing. An optical microscope should be used to ensure that there are no crushed grains. If the rock is well cemented, it might be necessary to use a rubber hammer to initiate disaggregation, but once the pieces are around 2 cm in size, it should be possible to continue with a mortar and pestle. The disaggregated sample is passed through a sieve (typically 100–150 μm) to isolate the fine size fraction.

Wet chemistry methods involve detection of exchangeable cations using chemical or conductometric titration equipment, a UV visible spectrophotometer or an atomic absorption spectrophotometer.

The cation exchange sites of the sample are saturated with specific cations by immersing the sample in a solution containing a known amount of cations for a period of about 48 h. This allows the natural exchange of cations present on the clays. The cations are then displaced with other cations. For example, the sample may be exposed to a solution like barium acetate, during which all the cation-exchange sites are replaced by barium (Ba^{2+}) ions. The solution is then titrated with another solution, such as $MgSO_4$, while observing the change in conductivity as the magnesium (Mg^{2+}) ions replace the Ba^{2+} ions. There are no 'standard' exchange solutions between laboratories—each lab tends to have their preferred method.

The titration data are used to determine CEC (meq/100 g of dry original rock sample), and the Qv (CEC per unit pore volume) is determined from:

$$\mathrm{Qv} = \frac{\mathrm{CEC}(1-\phi_t)\rho_g}{100\phi_t}$$

Note that the porosity, ϕ_t, and grain density, ρ_g, are better determined directly on the CEC sample, but can be measured on the parent plug if a small offcut is used. Since grain volume is assumed to be incompressible, stressed condition porosity can be used in the Qv equation. Ambient porosity can be converted to stressed porosity if the PCF is known for the suite of test samples, or porosity at stress is measured on the parent plug.

8.4.3.2 Multiple Salinity Co–Cw (Non-destructive Method)

The multiple salinity technique (Hill and Millburn, 1956) involves measuring the conductivity of samples saturated with brines of different salinities. Core conductivity, Co, is plotted against brine conductivity, Cw.

The Waxman and Thomas (1974) equation defines the relationship between Co and Cw:

$$\mathrm{Co} = \frac{\mathrm{Cw}}{F^*} + \frac{B\mathrm{Qv}}{F^*}$$

This is a linear equation of the form:

$$y = mx + c$$

Therefore, on a cross-plot of Co versus Cw (Fig. 8.56), $B\mathrm{Qv}/F^*$ is found directly from the intercept (c) of the regressed best-fit straight-line curve to

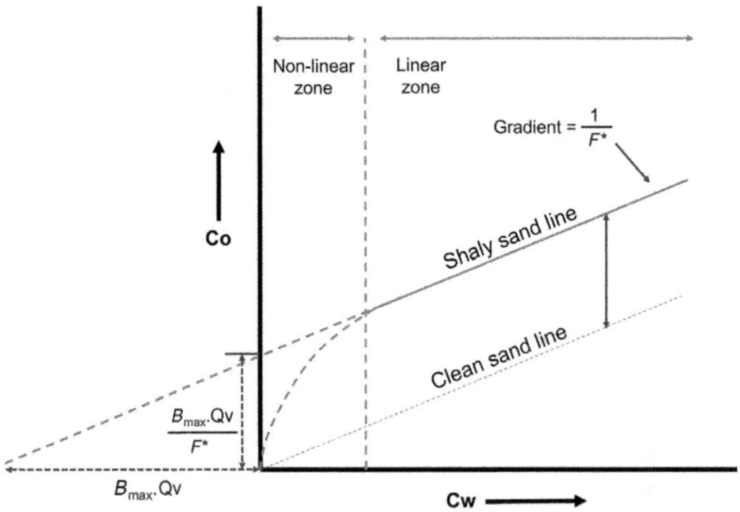

FIGURE 8.56 Determination of Qv from multiple salinity tests. *After Worthington (1985).*

the Co–Cw data pairs. The gradient (slope *m*) of the best-fit curve is 1/*F**. If *B* is known, or can be estimated from the Juhasz (1981) or Dacy and Martin (2008) equations as a function of lab temperature and formation water Rw, for example, then Qv can be determined from:

$$\text{Qv} = \frac{B\text{Qv}}{F^*} \cdot \frac{F^*}{B}$$

B is determined for the SFW brine Rw at the standard lab test temperature.

Figure 8.57 provides an example of a typical lab results presentation for a Co–Cw test. In this case, *F** (*FF** in this figure) is determined from the slope of the best-fit regression and *B*Qv from the intercept. In this case the regression analysis includes the Co–Cw pair associated corresponding to the Co and Cw measured with the formation water (SFW). This is obtained from standard Archie formation factor tests.

Using the data from this sample as an example, the *y*-on-*x* regression of Co–Cw datapairs (Fig. 8.58) yields a slope (1/*F**) of 0.1. Therefore *F** is 10.0. The intercept (*B*Qv/*F**) is 0.0257. *B* is calculated from the Juhasz equation using the temperature of the laboratory measurements. The resistivity of the simulated formation water (Rw = 1/Cw) for SFW is 1/1.4 = 0.714 Ω m. Hence, *B* = 4 ((mho/m)/(meq/ml)) and Qv = 0.0257/(0.1 × 4) = 0.064 (meq/ml).

The ratio of $F^*_{(\text{Co-Cw})}/F_{(\text{Archie})}$ is 1.1 and the Qv is relatively low. If such low values were characteristic of a formation, it would be regarded as a clean sand as *F**/*F* of 1.1 is within the experimental accuracy.

A cylindrical-shaped plug sample of 1″ or 1½″ diameter is required for the Co–Cw tests. The larger the sample, the larger the potential pore volume, and

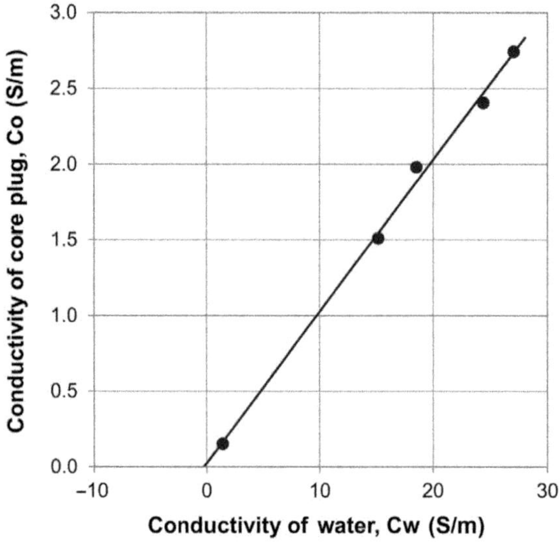

Salinity (ppm)	Cw (S/m)	Co (S/m)
250,000	27.030	2.744
200,000	24.390	2.407
150,000	18.520	1.982
100,000	15.150	1.511
SFW	1.400	0.155

FF: 9.03
FF*: 10
BQv: 0.257
Porosity: 0.29

FIGURE 8.57 Typical laboratory presentation of Co–Cw data.

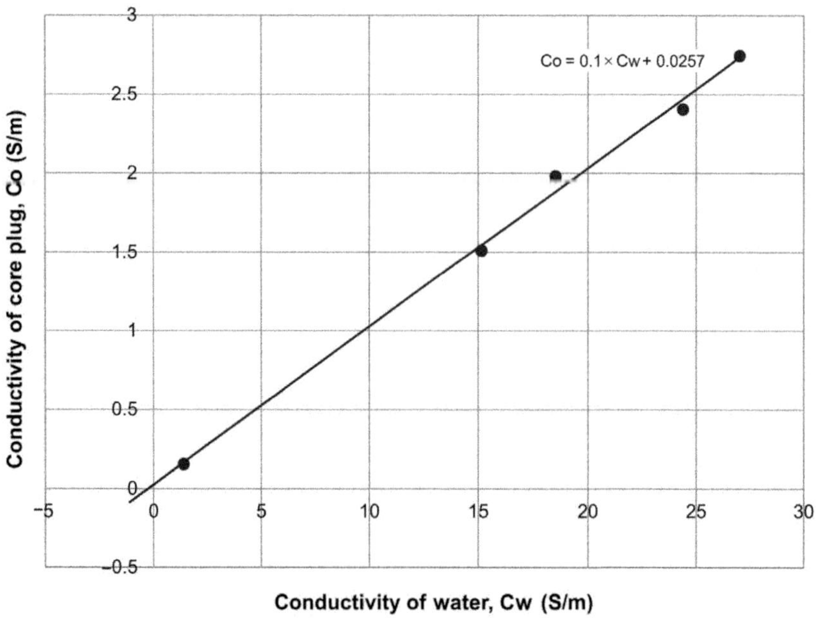

FIGURE 8.58 Determination of Qv and F^* from Co–Cw data for a single plug.

the less will be the impact of any volume measurement errors. SFW salinity and ionic composition should be as close as practically possible to the formation water in the reservoir, or should simulate the formation water resistivity at reservoir and lab temperature.

Sample selection, sample preparation and test procedures are the same as those used for FRF tests. Utmost care should be taken to avoid damage to the samples and preserve the clay structures during core cleaning and drying. Low-temperature injection (flush) cleaning, or total immersion cleaning and humidity oven drying is recommended. Base porosity and air/Klinkenberg permeability should then be measured for each plug, where possible. For clays that are sensitive to dehydration damage, such as 'hairy' illites, the plugs should not be dried but saturated in SFW immediately after flush cleaning with methanol. Permeability to brine can be measured and pore volume estimated from Fischer (1935) titration measurements.

In general, the test procedures are the same as those for FRF, but the conductivities of three, four or five high-salinity brines are measured. Multiple salinity tests are performed after stressed FRF tests where the test stress corresponds to the estimated isostatic reservoir *in situ* stress. SCP and annulus volume should also be determined, so stressed porosity can be calculated. It is recommended that Co for the SFW brine salinity is measured before the Co–Cw test. A typical test sequence might be:

TABLE 8.7 Typical Co–Cw Brine Salinity Test Sequence

Sequence	Brine (ppm NaCl)
1	SFW
2	200,000
3	150,000
4	125,000
5	75,000
Option	SFW

The key to the Co–Cw technique is to ensure that the salinity of the Co–Cw test brines at lab temperature is high enough that *B* for the brine Rw represents a maximum value and is constant for each of the multiple salinity brines. According to the Waxman–Thomas/Juhasz model in Fig. 8.55, *B* reaches a maximum and constant value at typical lab temperatures (25 °C) at brine resistivities of less than 0.1 Ω m. If Rw is too high, then *B* will be lower. The Waxman–Thomas Co–Cw regression analysis assumes *B* to be constant. If Cw is too low then *B* will not be constant for all Co–Cw data pairs. This can produce the downward curvature of the Co–Cw plot as indicated by the dotted line in Fig. 8.56.

The test plug is flooded with at least 10 pore volumes of the highest salinity brine at low rate to prevent fines movement (rates are typically less than 0.1 ml/min), and conductivity is monitored. Once conductivity stability is achieved, a measurement (Co) is taken. The normal criterion for conductivity measurement stability is that the change in *F* (defined as *F* = Cw/Co) between successive measurements should be less than ±2%. The next brine in sequence is then injected and the procedure is repeated for each, successively decreasing salinity, brine.

The option to test the Co of the SFW brine after the Co–Cw test sequence (Table 8.7) provides a check on the consistency of the pre-test SFW Co. It is often not advisable to determine Co for SFW directly after Co–Cw measurements if there is a high-salinity contrast between the final Co–Cw brine in the sequence and lower salinity SFW as this can initiate adverse clay/fluid interactions resulting in swelling and/or migration of clays.

On completion of the test, samples are unloaded and if not scheduled for further testing are placed in a Soxhlet apparatus and extracted in methanol to remove the water and salts. The silver nitrate test is used to determine the end point of cleaning. The base parameters of helium porosity and air or Klinkenberg permeability can be remeasured after drying. Comparison of pre- and post-test base poroperm properties can be useful in establishing any plug damage caused by clay interactions and/or harsh core cleaning and drying.

Core conductivity, Co, is plotted against brine conductivity, Cw, and the value of *B*Qv is determined by regression analysis of Co–Cw datapairs (e.g. Fig. 8.58).

8.4.4 Data Utilisation and Assignment

8.4.4.1 *Intrinsic Cementation Exponent,* m*

*F** is calculated from the following equation:

$$F^* = F(1 + B\text{QvRw})$$

F is the Archie formation factor and is determined from the equation:

$$F = \frac{\text{Cw}}{\text{Co}}$$

For multiple salinity tests, regression analysis of Co–Cw data pairs will yield *F** directly, from the inverse of the slope of the regression as shown in Fig. 8.58, and Qv from the intercept (BQv/*F**). The cation mobility *B* term is derived for the formation water Rw at lab temperature using the Juhasz or Dacy and Martin correlations.

For wet chemistry tests, Qv is determined from CEC measured on the FRF plug offcut. Both formation factor, *F*, and porosity ϕ will have (or should have) been measured on the parent plug. If the FRF test was carried out at stress, Qv is calculated based on the porosity measured for the same stress from:

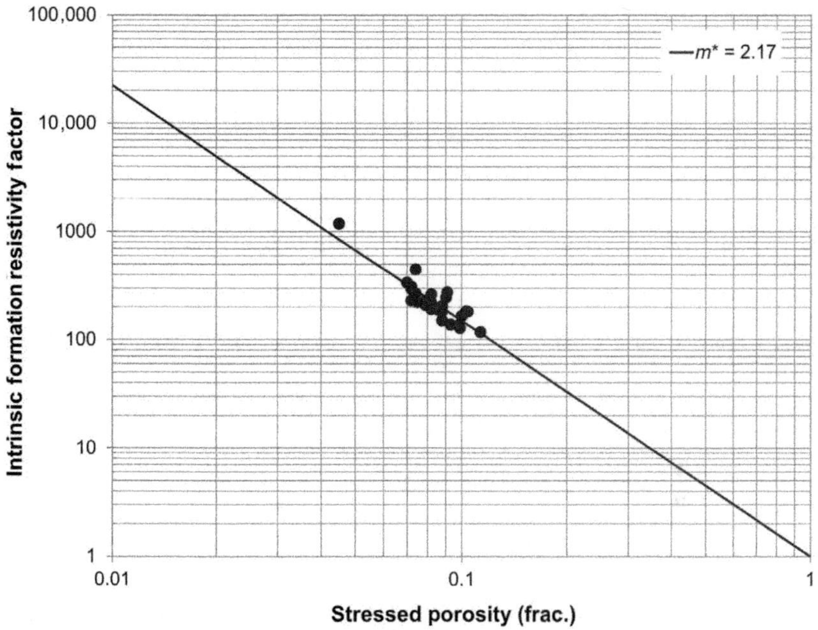

FIGURE 8.59 Example F^* versus porosity plot.

$$\mathrm{Qv_{res}} = \frac{\mathrm{CEC}\rho_g(1-\phi_{res})}{100\phi_{res}}$$

Alternatively, porosity at reservoir stress can be estimated if PCF data are available from similar samples.

The intrinsic cementation exponent, m^*, is determined from bi-logarithmic regression of intrinsic formation factor, F^*, and porosity, ϕ, in an identical fashion to that used to determine the Archie cementation exponent (Fig. 8.59).

8.4.4.2 Intrinsic Saturation Exponent, n*

The Keelan and McGinley (1979) method calculates a composite n^* directly from the equation:

$$n^* = -\frac{\log\left\{I\left[\dfrac{1+\dfrac{B\mathrm{QvRw}}{\mathrm{Sw}}}{1+B\mathrm{QvRw}}\right]\right\}}{\log \mathrm{Sw}}$$

In this case, average zonal values of B, Rw and Qv are input. Sw is evaluated at a fixed water saturation. This method implicitly assumes that n (and hence n^*) are constant throughout the saturation range. The n^* derived from this method may therefore be sensitive to the value of water saturation used to evaluate it. Further, variation in Rw and especially Qv within a zone can have a profound effect on n calculated using this method.

The more common (and recommended) method is to calculate to I^* for the resistivity index I versus Sw data set for each plug using the equation:

$$I^* = I\left\{\frac{\left(1+\frac{B\text{QvRw}}{\text{Sw}}\right)}{(1+B\text{QvRw})}\right\}$$

Qv can be determined from CEC measurements made on offcuts from, or on the same, test plug. The ambient condition Qv value must be corrected to the appropriate stressed porosity, if the resistivity index tests are carried out at stress. The B term can be derived for the SFW Rw at lab temperature from the Juhasz or Dacy and Martin correlations.

BQv can be obtained directly from regression analysis of Co–Cw datapairs in multiple salinity tests. However, resistivity index tests are often not performed on the same sample as multiple salinity tests, as in many cases the sample has to be unloaded to insert a porous plate in the coreholder, or there may be concerns over the effects of brine salinity changes on rock properties. In this case, Co–Cw and RI tests can be carried out on parallel plugs. The Qv data need to be corrected on the basis of any porosity and grain density differences between the RI test and Co–Cw plug pairs. For example, the Qv equation can be rearranged to predict an equivalent CEC for the Co–Cw test plug porosity and grain density:

$$\text{CEC} = \frac{100\phi_{\text{Co-Cw}}\text{Qv}_{\text{Co-Cw}}}{\rho_g(1-\phi_{\text{Co-Cw}})}$$

Then, the Qv of the resistivity index plug can be predicted knowing RI sample porosity and grain density:

$$\text{Qv}_{\text{RI}} = \frac{\text{CEC}\rho_g(1-\phi_{\text{RI}})}{100\phi_{\text{RI}}}$$

The intrinsic saturation exponent, n, is determined from bi-logarithmic regression of intrinsic resistivity index, I^*, and water saturation in an identical fashion to that used to determine the Archie saturation exponent.

The Rw and B and values used in this equation relate to the formation water resistivity and test temperature at which resistivity index was measured by the laboratory.

I^* is then cross-plotted against Sw to obtain n^* for each plug and, by grouping data from plugs from the same facies or rock type, composite n^* data can be obtained. An example is shown in Fig. 8.60.

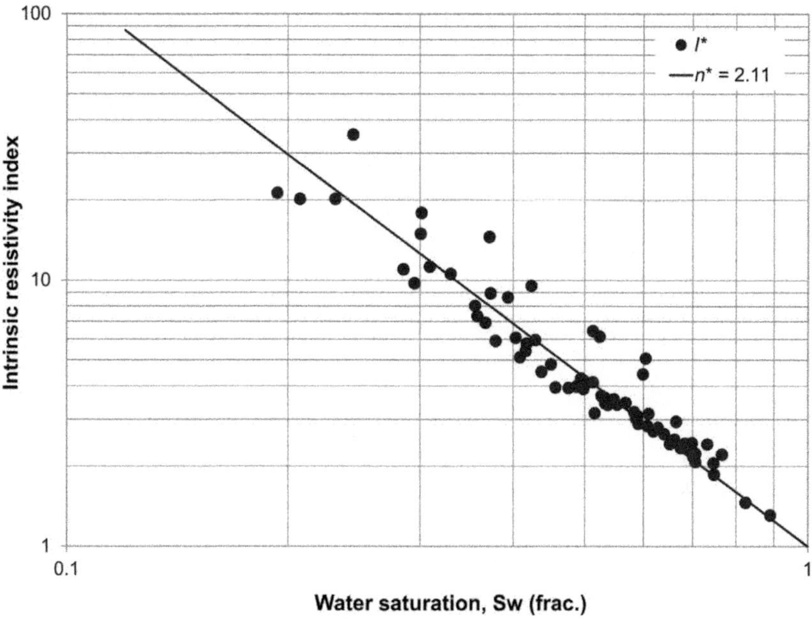

FIGURE 8.60 Example of typical I^* versus Sw plot ($n^*=2.11$).

8.4.5 Data Reporting Requirements

The data required to ensure adequate quality control of m^* and n^* measurements are the same as those for FRF (Table 8.2) and resistivity index (Table 8.4) tests. Additional data requirements related specifically to QC of wet chemistry and Co–Cw tests are listed in Table 8.8.

8.4.6 Advantages and Drawbacks/Issues

Summaries of the advantages and drawbacks/issues associated with CEC and Co–Cw measurements are provided in Tables 8.9 and 8.10.

8.4.7 Qv, m^* and n^* Test Quality Control Issues, Checks and Diagnostics

CEC tests are best performed on plugs as grain density, porosity and permeability are precisely known on the test sample. If offcuts are used, porosity and grain density should be measured on the offcut.

In direct comparison tests, Qv from wet chemistry tests is normally expected to be higher than from non-destructive multiple salinity tests. This is a particular problem where the sample is competent and consolidated and has to be excessively disaggregated to release the clay size fraction; or where the formation has a significant mica content. In its intact form, mica has a

TABLE 8.8 Data Reporting Requirements for Qv Tests

Data	Comments
Brief description of procedures and experimental apparatus	Co–Cw or CEC
Sample origin (end-trim, side-wall, plug); cation exchange solution types and concentrations	CEC
Sample cleaning and drying method	CEC and Co–Cw
Disaggregation equipment and time to fully disaggregate	CEC
Sample cation exchange capacity, meq/100 g dry rock	CEC
Test sample porosity measured or estimated at test stress	CEC and Co–Cw. CEC sample porosity to be reported if offcut from parent plug
Grain density, g/cc	CEC and Co–Cw
Parent (CEC) or test (Co–Cw) plug initial base permeability (Klinkenberg or air permeability) and helium porosity	CEC and Co–Cw
SFW Cw at standard temperature	Co–Cw
FRF (SFW) at standard temperature	Co–Cw and CEC
Cementation exponent (calculated from *F* and ϕ) for SFW	Co–Cw and CEC
Saturation exponent for SFW	Co–Cw
Co–Cw test brine salinities and corresponding Cw at standard temperature	Co–Cw
Co–Cw test net confining stress	Co–Cw
Calculated porosity (fraction), pore volume and bulk volume at net confining stress station, based on corrected expelled brine volume	Co–Cw
Brine injection sequence and brine flow rate	Co–Cw
Temperature, frequency (minimum phase angle), phase angle, core conductivity (Co) on completion of the equilibration period for each test brine	Co at standard temperature
BQv and F^* from regression analysis of Co–Cw data pairs	Co–Cw
F^*	CEC
Intrinsic cementation exponent, m^*, from regression analysis	Co–Cw and CEC
Intrinsic saturation exponent, n^*, from regression analysis	Co–Cw and CEC
Final dry weight, helium porosity and base permeability (Klinkenberg or air permeability)	Co–Cw Post-test. Compare to pre-test

TABLE 8.9 Advantages and Drawbacks/Issues Associated with CEC Tests

Advantages	Drawbacks and Issues
• This is the most widely used technique because it is cheap, convenient and rapid. • A large number of samples can be processed in a short time. • The method can be used as a first pass assessment of CEC and Qv and to identify intervals where non-destructive Qv tests may be performed. • Testing can be performed on end-trims or any available sample material, with no requirements for a specific shape.	• During sample preparation, incomplete disaggregation might not isolate clays and therefore would tend to reduce measured CEC. Excessive grinding and disaggregation increase the natural CEC by exposing more cation exchange sites than would be exposed in the intact reservoir formation. Huff (1987) correlated cation exchange capacity with the square root of the duration of the grinding period. CEC increased by a factor of over 3 when the duration of the grinding of unconsolidated samples was extended from 1 to 5 min. This underlines the inaccuracies and uncertainties that can be associated with the wet chemistry method. • If tests are performed on core plugs or offcuts from a parent plug, the samples must be dried under conventional, non-humidified conditions since any remaining adsorbed water will reduce the porosity and grain density. Data from humidity dried samples should not be used. The principal effect of a reduction in porosity and grain density on humidified drying would be to increase the apparent Qv values. • Qv cannot be determined at representative overburden pressure but can be converted to reservoir conditions if the porosity compaction factor of the parent plug (or for the rock type) is known. • Errors may also be introduced by improper leaching or washing, such that excess exchange solution remains in suspension and does not react with the clays. This will reduce the apparent CEC.

relatively low CEC, but if the mica platelets are broken or frayed by grinding they can release more CEC sites than would exist *in situ*. A typical example of Qv from destructive (wet chemistry) and non-destructive (Co–Cw) tests on the same samples is shown in Fig. 8.61. The effects of different test methods (titration or spectrometry) used by different labs are insignificant in comparison to the errors and uncertainties introduced by sample grinding and destruction.

Results from both CEC and Co–Cw tests should be considered in context with, and correlated against, petrographic test data—such as thin-section analysis, bulk rock and clay fraction XRD, and SEM. Given the problems associated with grinding the sample, the CEC test protocols often produce unrepresentative or inconsistent data, so the test data often show little or no

TABLE 8.10 Advantages and Drawbacks/Issues Associated with Co–Cw Tests

Advantages	Drawbacks and Issues
• Test can be run at equivalent isostatic reservoir stress but nominal stress is normally sufficient to eliminate the effects of excess brine and grain loss. • The method preserves the pore structure of the rock and the location of the clays within the 3D pore network. • Multiple salinity tests should yield results that are more appropriate to the reservoir formation than wet chemistry CEC tests. • Archie *F* and *F** can be determined on the same sample in the same loading operation.	• More expensive than CEC as tests can take several weeks to complete. Considerable time may be required to achieve ionic equilibrium in low-permeability/high-Qv samples. Long equilibration times for salinity changes in such samples then become a limiting factor. • In low-permeability samples, brine flow rates of only a few ml per hour or less may only be possible. This can make the procedure uneconomic for the test lab. In this case, tests might have to be carried out at lower stresses to limit permeability compaction effects or may have to be omitted altogether. Comparison of porosity between lower (test) stress and higher (net reservoir) stress measurements can be used to scale lower stress Qv to the appropriate stress condition: $$\left(\frac{Qv_{high}}{Qv_{low}}\right)=\left[\frac{\left(\frac{1}{PCF}\right)^{-\phi_{low}}}{(1-\phi_{low})}\right]$$ where PCF is the ratio of porosity at high stress to low stress. • With relatively low salinity brines, there is insufficient contrast between the pore water conductivity and the clay conductivity. This can reduce the clays' own conductivity which causes the Co–Cw curves to curve downwards, thereby under-predicting *B*Qv. This can occur at salinities less than 50,000 ppm (Austin and Ganley, 1991) or conductivities less than about 5 mho/m (Waxman and Smits, 1968). In addition, adverse clay reactions can occur if the sample is exposed to relatively fresh water (e.g. SFW) after higher salinity brine. • The Co–Cw test brine salinities should be greater than the limiting B_{max} Rw. • Imposition of a limiting brine salinity (Rw) can mean that a significant extrapolation of the Co–Cw regressed datapoints is required to define the intercept *B*Qv/*F**. Depending on the sample properties, small measurement errors can translate to large uncertainties and can adversely lever the *B*Qv/*F** intercept.

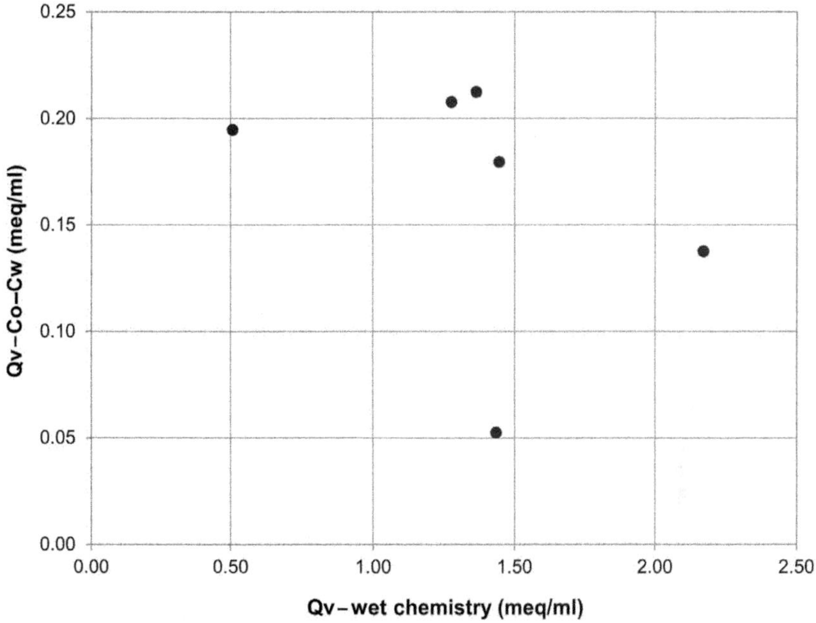

FIGURE 8.61 Comparison of Qv from Co–Cw (non-destructive) and wet chemistry (destructive) tests on same samples.

correlation between Qv and clay content or clay type in reservoir sands. An example is shown in Fig. 8.62. Samples containing significant proportions of higher CEC clays such as smectite should have higher measured CEC than samples comprising mainly lower CEC clays like kaolinite. Excess conductivity can also be introduced by low-resistivity minerals which do not have any CEC—such as disseminated pyrite and siderite.

CEC tests should not be used in petrophysical modelling without verification from non-destructive Qv tests.

The values of BQv and F^*, derived from the intercept and slope of the linear fit to the Co–Cw data, can be extremely sensitive to potential errors in Cw. The laboratory Cw data should be verified against standard industry correlations. For example, Worthington et al. (1990):

$$\text{Rw} = \frac{1686.212}{\text{NaCl(ppm)}^{0.8723}} \quad \text{for Rw at 25 °C}$$

Strictly, the SFW should not be included in the Co–Cw regression analysis if the brine salinity and Rw indicate that B is lower than B_{max} at the test temperature. This, coupled with the vagaries of Y-on-X regression analysis, can result in considerable uncertainty in Qv. For example, Fig. 8.63 compares the Qv derived from regression of only three high-salinity Co–Cw brines

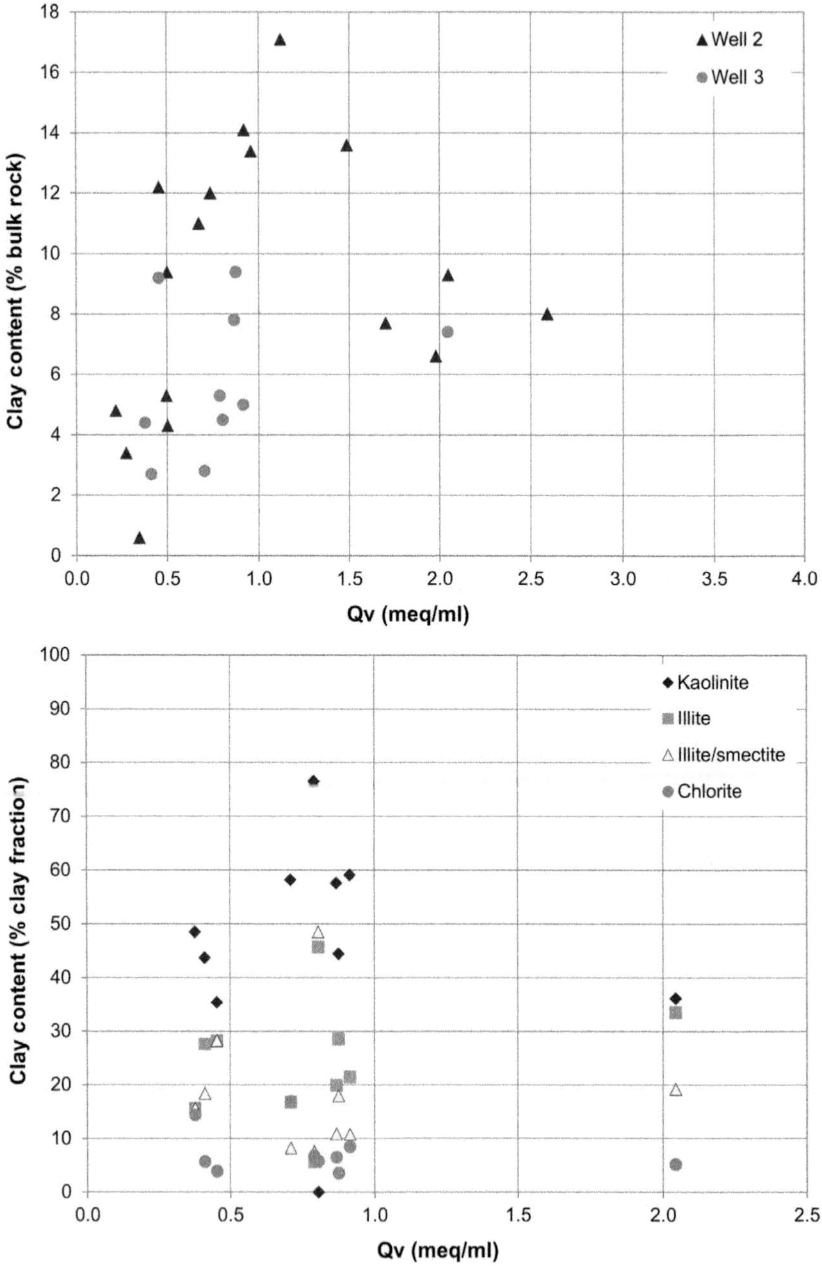

FIGURE 8.62 Poor relationship between clay content and type and Qv from wet chemistry CEC.

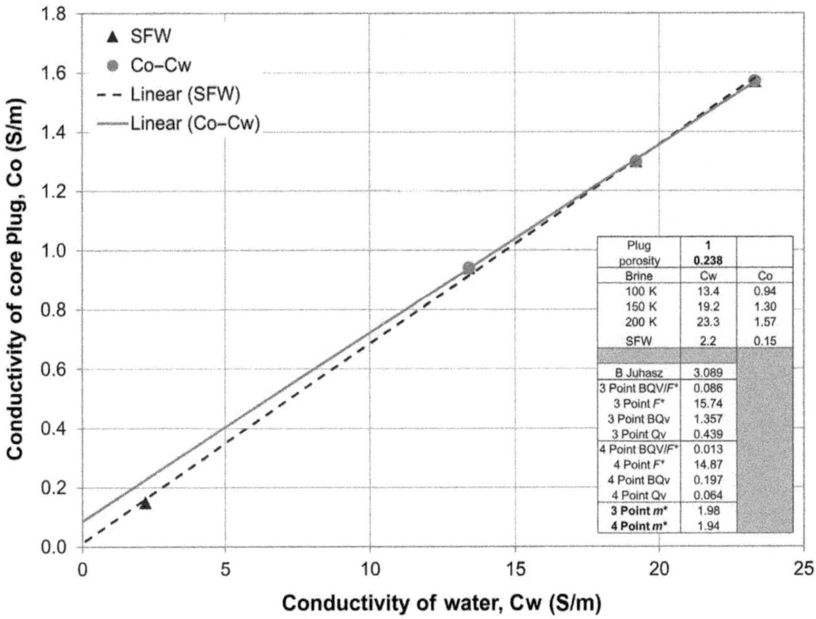

FIGURE 8.63 Uncertainty in Qv from 3 point (excluding SFW) and 4 point (including) Co–Cw *Y*-on-*X* regression analysis.

(100, 150 and 200 kppm) with a four-point regression which includes the low salinity, low Cw SFW. Based on the high-salinity brine data, the Qv is 0.44 meq/ml compared to a low 0.064 meq/ml from regression of the data from the three high-salinity brines and the SFW. Experience shows that the repeatability of the multiple salinity measurement is about 0.1 meq/ml so samples with apparent Qv less than this (or F^*/F less than 1.1) should be regarded as a clean sand.

Laboratory *B*Qv and F^* data derived from conventional *Y* on *X* or ordinary least squares (OLS) regression of Co–Cw data assumes no error in the dependent variable, Co. Reduced major axis (RMA) regression can be used (as opposed to conventional OLS) for this analysis since there can be equal uncertainty in both Co and Cw. The slope of an RMA regression is calculated from the slope of an OLS regression divided by the correlation coefficient. The example data set below (Table 8.11) compares the results of *B*Qv obtained from OLS and RMA regression. Clearly, the OLS technique can be extremely sensitive to data uncertainties, especially in sample #11.

8.5 ALTERNATIVE METHOD FOR SATURATION EXPONENT DETERMINATION

Resistivity index tests on clean sands or carbonates can be difficult to perform, are extremely time consuming, are sensitive to experimental artefacts

TABLE 8.11 Comparison of OLS (*Y*-on-*X*) and RMA Regression Results for *B*Qv

Sample	OLS Regression *B*Qv	RMA Regression *B*Qv
1	0.324	0.386
7	0.145	0.155
8	0.694	0.667
9	0.601	0.595
11	4.33	0.672
13	0.662	0.618

and, in oil–water reservoirs, are controlled by wettability which may be difficult to reproduce in SCAL test samples. There is an alternative that can provide reliable and representative reservoir condition saturation exponent data quickly, and at little extra cost: Dean–Stark water saturations.

This assumes that water saturations from Dean-Stark extraction are measured on a regular basis during the RCA programme, that the samples have not been allowed to dry out, that the core has been drilled with oil-based mud or that any invasion from the water filtrate phase of a water-based mud has been corrected using tracer analyses and that the Dean-Stark Sw data have been corrected for stress, water compressibility and formation water salinity. The best practice procedures for Dean-Stark Sw interpretation are discussed in Chapter 5. The true formation resistivity, Rt, from resistivity logs, total porosity (from density or sonic logs), the formation Rw and an estimate of cementation exponent (from SCAL or from Pickett plots) must be available from a robust petrophysical interpretation in the hydrocarbon leg. In addition, the core Dean-Stark data must be depth matched to the logs.

The basic of the technique is to solve for saturation exponent, n, over a particular cored interval using the Archie equation:

$$n = -\frac{\log(I)}{\log(\mathrm{Sw})}$$

where Sw represents the stress and salinity corrected Dean-Stark measurements.

The resistivity index, I, in the hydrocarbon leg is found from:

$$I = \frac{\mathrm{Rt}}{\mathrm{Ro}}$$

Rt represents the true resistivity from the resistivity logs (corrected for fluid invasion) which respond to, and are influenced by, the *in situ* stress and the rock native wettability: that is, Rt is a reservoir condition measurement.

Ro is the formation resistivity at *in situ* stress if the hydrocarbon leg was 100% saturated in formation water with resistivity Rw. Ro is determined from:

$$F = \frac{\mathrm{Ro}}{\mathrm{Rw}} = \frac{a}{\phi^m}$$

Porosity is derived from log measurements (with the formation at *in situ* stress conditions), a is set equal to 1 and m is the cementation exponent. In the absence of any SCAL data, m can be determined from Pickett plot analysis (Pickett, 1966) in a clean sand in the water leg and is assumed to be the same as in the hydrocarbon leg. Alternatively, m can be determined relatively rapidly from stressed FRF tests.

Figure 8.64 is an example Dean–Stark data set from an oil-bearing interval in a North Sea reservoir. The Dean-Stark data have been corrected for stress, salinity and water compressibility and have been depth matched against total porosity and true resistivity data from logs over the same interval. The core Sw and log Rt data are plotted versus depth through the interval in Fig. 8.65. The cementation exponent, m, is 1.81 (from Pickett plot), and the formation Rw at reservoir temperature is 0.0413 Ω m. The calculated resistivity index and saturation exponent are tabulated in Fig. 8.66 and the

Sample Number	Shifted Depth (ft.)	Log Total Porosity	Corrected Dean–Stark Sw (frac.)	Log Rt
1	9843.6	0.155	0.207	22.167
2	9844.8	0.145	0.237	21.719
3	9845.5	0.168	0.209	21.719
4	9846.6	0.151	0.216	24.318
5	9847.5	0.179	0.167	24.088
6	9848.5	0.165	0.178	25.144
7	9849.6	0.163	0.183	26.019
8	9852.3	0.163	0.177	28.453
9	9856.5	0.132	0.203	31.631
10	9863.5	0.149	0.258	17.306
11	9864.5	0.126	0.328	15.355
12	9865.5	0.169	0.220	17.333
13	9868.5	0.150	0.248	16.475
14	9869.5	0.159	0.219	17.195
15	9870.5	0.159	0.224	17.631
16	9871.5	0.152	0.215	19.573

FIGURE 8.64 Dean-Stark Sw dataset used for saturation exponent calculation.

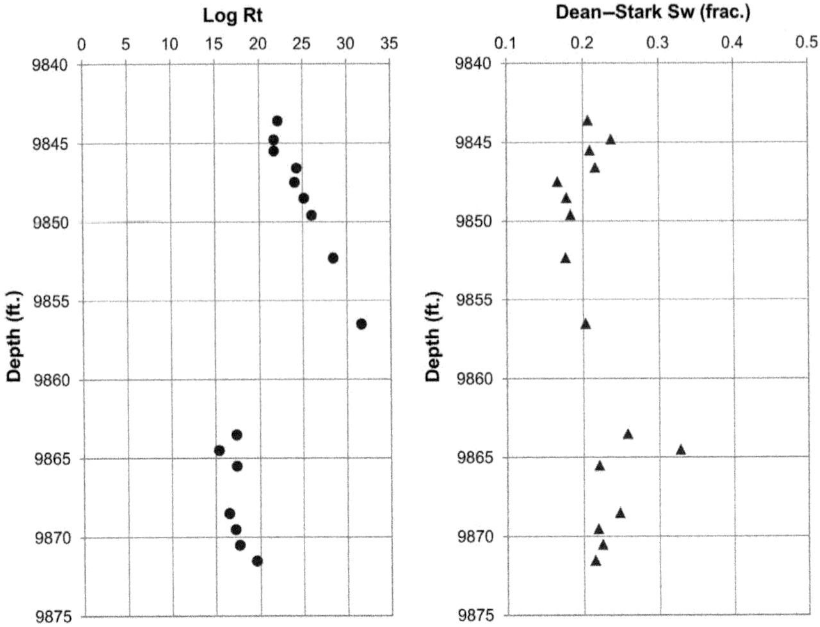

FIGURE 8.65 Dean-Stark Sw and depth-matched log Rt against depth.

Sample Number	Shifted Depth (ft.)	Log Total Porosity	Corrected Dean–Stark Sw (frac.)	Log Rt	Estimated Formation Resistivity Factor ($1/\phi^m$)	Estimated Ro (Ω m)	Resistivity Index	Saturation Exponent (*n*)
1	9843.6	0.155	0.207	22.167	29.36	1.21	18.26	1.84
2	9844.8	0.145	0.237	21.719	33.13	1.37	15.86	1.92
3	9845.5	0.168	0.209	21.719	25.35	1.05	20.73	1.94
4	9846.6	0.151	0.216	24.318	30.63	1.27	19.21	1.93
5	9847.5	0.179	0.167	24.088	22.62	0.94	25.76	1.81
6	9848.5	0.165	0.178	25.144	26.19	1.08	23.22	1.82
7	9849.6	0.163	0.183	26.019	26.73	1.11	23.55	1.86
8	9852.3	0.163	0.177	28.453	26.76	1.11	25.72	1.88
9	9856.5	0.132	0.203	31.631	39.12	1.62	19.56	1.87
10	9863.5	0.149	0.258	17.306	31.44	1.30	13.31	1.91
11	9864.5	0.126	0.328	15.355	42.43	1.75	8.75	1.95
12	9865.5	0.169	0.220	17.333	24.93	1.03	16.82	1.87
13	9868.5	0.150	0.248	16.475	31.09	1.29	12.82	1.83
14	9869.5	0.159	0.219	17.195	27.77	1.15	14.98	1.78
15	9870.5	0.159	0.224	17.631	28.00	1.16	15.23	1.82
16	9871.5	0.152	0.215	19.573	30.33	1.25	15.61	1.79

FIGURE 8.66 Calculated reservoir condition resistivity index and saturation exponent.

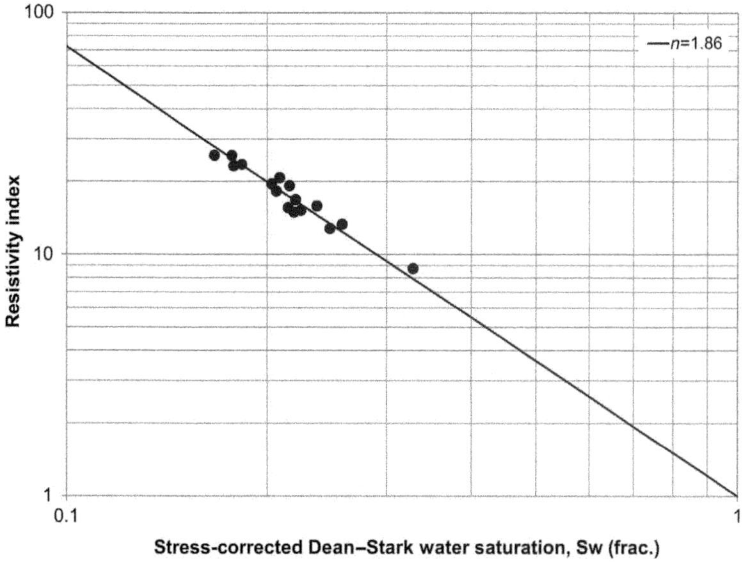

FIGURE 8.67 Composite saturation exponent for Dean-Stark core interval.

composite *I* versus Sw saturation exponent plot for the Dean-Stark interval is Fig. 8.67.

This method has the obvious advantages of simplicity, cost and speed—Dean-Stark data are normally available in the RCA programme at little additional cost. The principal disadvantages are the assumptions inherent in the analysis, that:

- the Dean-Stark data are reliable and representative across the cored interval.
- the petrophysical properties—*m* and Rw are the same across the cored interval.

The principal advantage is that the saturation exponent is essentially a reservoir condition estimate based on log data that represent the formation *in situ* stress and wettability conditions. To determine saturation exponent on individual SCAL plugs in the lab is expensive and can take months to complete—this method can provide data in around 3–4 weeks after core receipt—as soon as Dean-Stark data are available and have been quality checked.

In addition, individual saturation exponent data can be plotted as a function of depth, as shown in Fig. 8.68 for this example data set. This can be extremely useful in identifying saturation exponent trends. For example, in many reservoirs with relatively high structural relief, there is often a wettability gradient—trending from water wet just above the contact to mixed wet or slightly oil wet higher in the structure. If there is a wettability trend, then this method can be used to examine any corresponding saturation

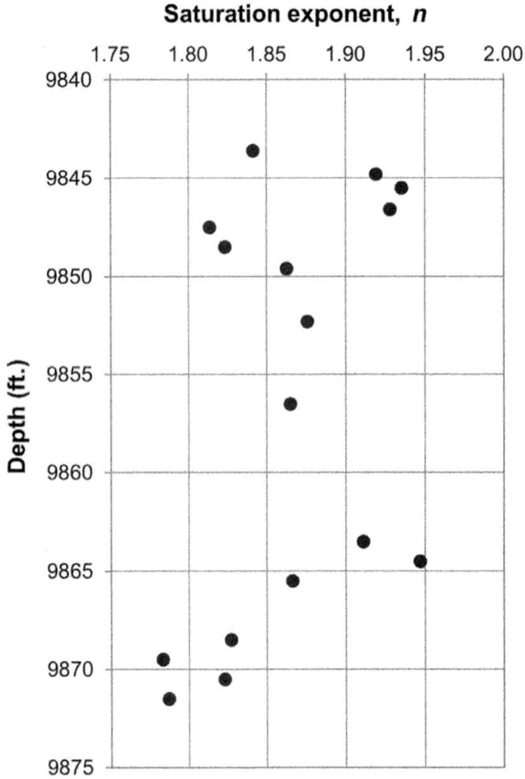

FIGURE 8.68 Saturation exponent versus depth for example dataset.

exponent trends—n is expected to increase with reduced water-wetting—and therefore define appropriate saturation exponents for different reservoir intervals.

REFERENCES

Al-Kaabi, A.U., Mimoune, K., Al-Yousef, H.Y., 1997. Effect of hysteresis on the Archie saturation exponent. In: Society of Petroleum Engineers Middle East Oil Show and Conference, 15–18 March, Bahrain, Paper SPE 37738.

Al-Mjeni, R., Gunzel, F., Jing, X., Gratonni, C., Zimmerman, R.W., 2002. The influence of clay fraction on the complex impedance of shale sands. Society of Core Analysts. Paper SCA 29.

Anderson, W.G., 1986. Wettability literature survey—part 3: the effects of wettability on the electrical properties of porous media. SPE J. Pet. Technol. 38 (12), 1371–1378.

Archie, G., 1942. The electrical resistivity log as an aid in determining some reservoir characteristics. Pet. Trans. AIME 146, 54–62.

Archie, G., 1947. Electrical resistivity as aid in core analysis interpretation. Am. Assoc. Pet. Geol. Bull. 31 (2), 350–366.

Arps, J.J., 1953. The effect of temperature on the density and electrical resistivity of sodium chloride solutions. J. Pet. Technol. 5 (10), 17–20, Technical Note 195.

Austin, S.P., Ganley, S.M., 1991. Measurement of the CEC of core plugs by a non-destructive wet chemical method. In: Advances in Core Evaluation II: Reservoir Appraisal: Reviewed Proceedings of the Second Society of Core Analysts European Core Analysis Symposium, London, UK, 20–22 May.

Bona, N., Rossi, E., Capaccioli, S., Lucchesi, M., 2008. Electrical measurements: considerations on the performance of 2- and 4-contact systems. In: International Symposium of the Society of Core Analysts, Abu Dhabi, UAE, 29 October–2 November, Paper SCA2008-07.

Clavier, C., Coates, G., Dumanoir, J., 1977. The theoretical and experimental bases for the "dual water" model for the interpretation of shaly sands. In: 52nd Annual Fall Technical Conference and Exhibition of the Society of Petroleum Engineers of The American Institute of Mining, Metallurgical, and Petroleum Engineers, Denver, Colorado, October 9–12.

Crain, R. Crain's Petrophysical Handbook. www.spec2000.net. (accessed December 2014).

Dacy, J., Martin, P., 2008. Practical advances in core-based water saturation analysis of shaly tight gas sands. Petrophysics 49 (1).

De Waal, J.A., Smits, R.M.M., de Graaf, J.D., Schipper, B.A., 1989. Measurement and evaluation of resistivity index curves. In: SPWLA 30th Annual Logging Symposium, 11–14 June, Denver, Colorado, Paper SPWLA-1989-II.

Diederix, K.M., 1982. Anomalous relationships between resistivity index and saturation in the Rotliegend Sandstone. In: SPWLA 23rd Annual Logging Symposium, June, Paper SPWLA-1982-X.

Eslinger, E., Pevear, D., 1988. Clay minerals for petroleum geologists and engineers. Society of Economic Paleontologists and Mineralogists Short Course Notes No. 22. Tulsa, Oklahoma.

Fischer, K., 1935. Neues verfahren zur maßanalytischen bestimmung des wassergehaltes von flüssigkeiten und festen körpern. Angew. Chem. 48 (26), 394–396.

Gray, R., Trewin, B., Pallat, N., Mitchell, P., 1992. Comparison of saturation exponent data by the porous plate and by the continuous injection technique with in-situ saturation monitoring. In: 3rd European Core Analysis Symposium of Society of Core Analysts, Paris, 14–16 September.

Hill, H.J., Millburn, J.D., 1956. Effect of clay and water salinity on electrical behaviour of reservoir rocks. In: Transactions of American Institute of Mining, Metallurgical, and Petroleum Engineers, 207, 65–72, Paper SPE 532.

Hill, H.J., Klein, G.E., Shirley, O.J., Thomas, E.C., Waxman, W.H., 1979. Bound water in shaly sands—its relation to Qv and other formation properties. Log Anal. 20 (3), 3–19.

Huff, G.F., 1987. A correction for the effect of comminution on the cation exchange capacity of clay-poor sandstones. SPE Form. Eval. 2 (3), 338–344.

Jing, X.-D., Gillespie, A., Trewin, B.M., 1993. Resistivity index from non-equilibrium measurements using detailed in-situ saturation monitoring. In: Offshore Europe Conference, Aberdeen, 7–10 September, Paper SPE 26798.

Juhasz, I., 1981. Normalised Qv—the key to shaly sand evaluation using the Waxman-Smits equation in the absence of core data. In: SPWLA 22nd Annual Logging Symposium, 23–26 June, Mexico City, Mexico, Paper SPWLA-1981-Z.

Kaye, L.N., Hacksma, J.D., Knudson, S., Smith, J.E., 1990. Special core analysis loading conditions for a friable, stress-sensitive sandstone. In: Worthington, P.F. (Ed.), Advances in Core Evaluation, Accuracy and Precision in Reserves Estimation. Gordon and Breach, New York.

Keelan, D.K., McGinley, D.C., 1979. Application of cation exchange capacity in a study of the Shannon Sand of Wyoming. In: SPWLA 20th Annual Logging Symposium, 3–6 June, Tulsa, Oklahoma, Paper SPWLA-1979-W.

Kennedy, D., 2007. The porosity-water saturation-conductivity relationship: an alternative to Archie's model. SPWLA Petrophys. 48 (5), 335–361.

LaTorraca, G.A., Hall, C.G., 1991. Observations of rock fabric controls on the electrical properties of sandstones. In: 3rd Society of Core Analysts Annual Technical Conference, SCA paper 9116.

Lewis, M.G., Sharma, M.M., Dunlap, H.F., 1988. Wettability and stress effects on saturation and cementation exponents. In: SPWLA 29th Annual Logging Symposium, 5–8 June, San Antonio, Texas, Paper SPWLA-1988-K.

Lyle, W.D., Mills, W.R., 1989. Effect of non-uniform core saturation on laboratory determination of the Archie saturation exponent. SPE Form. Eval. 4 (1), 49–52.

Maute, R.E., Lyle, W.D., Sprunt, E.S., 1992. Improved data analysis method determines Archie parameters from core data. J. Pet. Technol. 44 (1), 103–107.

McPhee, C.A., 2012. The core analysis elephant in the formation evaluation room. In: SPE Annual Technical Conference and Exhibition, 8–10 October, San Antonio, Texas, USA, Paper SPE 158087.

Mungan, N., Moore, E.J., 1969. Certain wettability effects on electrical resistivity in porous media. J. Can. Pet. Technol. 7 (1), 20–25.

Pallat, N., Toussaint-Jackson, J.E., Worthington, P.F., 1988. Effect of sample preparation on the saturation exponent in the Magnus field, North Sea. Log Anal. 29 (1), 48–53.

Pickett, G.R., 1966. Review of current techniques for determination of water saturation from logs. In: Rocky Mountain Regional Meeting, Denver, Colorado, USA, May 23–24, Paper SPE 1446.

Poupon, A., Leveaux, J., 1971. Evaluation of water saturation in shaly formations. In: Transactions SPWLA 12th Annual Logging Symposium, pp. 1–2.

Schutjens, P.M.T.M., Hansen, T.H., Hettema, M.H.H., Merour, J., de Bree, P., Coremans, J.W.A., Helliesen, G., 2004. Compaction-induced porosity/permeability reduction in sandstone reservoirs: data and model for elasticity-dominated deformation. SPE Reserv. Eval. Eng. 7 (03), 202–216.

Simandoux, P., 1963. Dielectric measurements in porous media and application to shaly formation. Rev. Inst. Fr. Pét. 18 (Supplementary Issue), 193–215.

Sondena, E., Bratteli, F., Normann, H.P., Killveit, K., 1992. A comparison between capillary pressure data and saturation exponent obtained at ambient conditions and at reservoir conditions. SPE Form. Eval. 7 (1), 34–40.

Spooner, P., 2014. Lifting the fog of confusion surrounding clay and shale in petrophysics. In: SPWLA 55th Annual Logging Symposium, 18–22 May, Abu Dhabi, United Arab Emirates, Paper SPWLA-2014-VV.

Sweeney, S.A., Jennings Jr., H.Y., 1960. Effect of wettability on the electrical resistivity of carbonate rock from a petroleum reservoir. J. Phys. Chem. 64 (1960), 551–553.

Thomas, E.C., 1976. The determination of Qv from membrane potential measurements on shaly sands. J. Pet. Technol. 28 (9), 1087–1096.

Thomas, E.C., Stieber, S.J., 1975. The distribution of shale in sandstones and its effect upon porosity. In: SPWLA 16th Annual Logging Symposium, 4–7 June, New Orleans, Louisiana, Paper SPWLA-1975-T.

Wang, Z., Hirsche, W.K., Sedgwick, G.E., 1991. Electrical and petrophysical properties of carbonate rocks. In: Proceedings of the 66th Society of Petroleum Engineers Annual Technical Conference and Exhibition, Dallas, Texas, 6–9 October, Paper SPE 22661.

Waxman, M.H., Smits, L.J.M., 1968. Electrical conductivities in oil-bearing shaly sands. Soc. Pet. Eng. J. 8 (02), 107–122.

Waxman, M.H., Thomas, E.C., 1974. Electrical conductivities in shaly sands—I: the relation between hydrocarbon saturation and resistivity index. II—the temperature coefficient of electrical conductivity. SPE J. Pet. Technol. 26 (2), 213–225.

Winsauer, W.O., Shearin, H.M., Masson, P.H., Williams, M., 1952. Resistivity of brine-saturated sands in relation to pore geometry. AAPG Am. Assoc. Pet. Geol. Bull. 36, 253–277.

Worthington, A.E., 1975. Errors in the laboratory measurement of formation resistivity factor. In: 16th Annual SPWLA Logging Symposium, 4–7 June, June, New Orleans, Louisiana, Paper SPWLA-1975-D.

Worthington, P.F., 1985. The evolution of shaly-sand concepts in reservoir evaluation. Log Anal. 26 (1), 23–40.

Worthington, A.E., Hedges, J.H., Pallatt, N., 1990. SCA guidelines for sample preparation and porosity measurement of electrical resistivity samples: part 1—guidelines for preparation of brine and determination of brine resistivity for use in electrical property measurements. Log Anal. 31 (1), 20–28.

Yuan, H.H., Diederix, K.M., 1989. The role of membrane potential measurements in shaly sand evaluation. Log Anal. 30 (6), 1989.

Zhou, D., Arbabai, S., Stenby, E.H., 1996. Effect of wettability on the electrical properties of reservoir rocks. In: Society of Core Analysts Conference, Paper SCA 9624.

RECOMMENDED READING

Amaefule, J.O., Keelan, D.K., 1989. Stochastic approach to computation on uncertainties in petrophysical properties. In: Society of Core Analysts Annual Technical Conference, Paper SCA 8907.

Baldwin, B.A., Chaves, A.A., LaTorraca, G.A., Maerefat, N.L., Swanson, B.F., 1990. SCA guidelines for saturating and desaturating core plugs during electrical resistivity measurements. Log Anal. 31 (2), 68–75.

Burck, L.J.S., 1983. A review of log and core methods for determining cation exchange capacity/Qv. In: 8th European Formation Evaluation Symposium, Paper F, 14–15 March.

Clerke, E.A., Martin, P.R., 1994. Absolute Qv: cost effective Juhasz correlations for shaly sand evaluation from integrated analysis of multiple lab measurements. In: SPWLA 35th Annual Logging Symposium, Mexico City, 19–22 June, Paper SPWLA-1994-P.

Dacy, J.M., Lerner, D.B., Raible, C.J., Rathmell, J.J., Swanson, G., Walls, J.D., 1990. SCA guidelines for sample preparation and porosity measurement of electrical resistivity samples. Part II—sample preparation and porosity measurement. Log Anal. 31 (2).

Dunlap, H.F., Bilhartz, H.L., Shuler, E., Bailey, C.R., 1949. The relation between electrical resistivity and brine saturation in reservoir rocks. J. Pet. Technol. 1 (10), 259–264.

Hamilton, J.M., Shafer, J.L., 1991. Measurement of pore compressibility characteristics in rock exhibiting 'pore collapse' and volumetric creep. In: Society of Core Analysts Annual Conference, San Antonio, Texas, 21–22 August, SCA Paper 9124.

Hoyer, W.A., Spann, M.M., 1975. Comments on obtaining accurate electrical properties of cores. In: SPWLA 16th Annual Logging Symposium, New Orleans, LA., June 4–7, Paper SPWLA-1975-B.

Juhasz, I., 1979. The central role of Qv and formation water salinity in the evaluation of shaly formations. Log Anal. 20 (4), 3–13.

Kurniawan, B., Bassiouni, Z., 2007. Use of CEC-dependent cementation and saturation exponents in shaly sand resistivity models. In: SPWLA 48th Annual Logging Symposium, 3–6 June, Austin, Texas, Paper SPWLA-2007-FFF.

Lasswell, P.M., 2006. Core analysis for electrical properties. Petrophysics 47 (3).

Martin, P., Dacy, J., 2004. Effective Qv by NMR core tests. In: SPWLA 45th Annual Logging Symposium, 6–9 June, Noordwijk, Netherlands, Paper SPWLA-2004-HH.

Maute, R.E., Rachers, C.L., Sprunt, E.S., 1990. An interpretation of the SCA resistivity survey. Log Anal. 31 (2), 76–88.

Mian, M.A., Hilchie, D.W., 1981. Comparison of results from three cation exchange capacity analysis techniques. Paper SPE 10550.

Nieto, J.A., Yale, D.P., Evans, R.J., 1994. Improved methods for correcting core porosity to reservoir conditions. Log Anal. 35 (3), 21–30.

Pallat, N., Palmer, T., 1991. The role of pore geometry in the interpretation of shaly sands. In: Worthington, P., Longeron, D. (Eds.), Advances in Core Evaluation. II: Reservoir Appraisal. Gordon and Breach Science Publishers, London.

Pratt, P.F., Holowaychuk, N., 1954. A comparison of ammonium acetate, barium acetate and buffered barium chloride methods of determining cation exchange capacity. Proc. Soil Sci. Soc. Am. 18, 365.

Ruhovets, N., Fertl, W.H., 1982. Digital shaly sand analysis based on Waxman-Smits model and log derived clay typing. Log Anal. 23 (3), 7–22.

Shaw, H.F., 1980. Clay minerals in sediments and sedimentary rocks. In: Hobson, G.D. (Ed.), Developments in Petroleum Geology. Allied Sciences Publication Ltd., London, pp. 53–85, (Chapter 2).

Worthington, P.F., 2011. The petrophysics of problematic reservoirs. J. Pet. Technol. 63 (12), 88–97.

Worthington, P.F., Daines, J.M., Bratli, R.K., Nicolaysen, R., 1997a. Comparative evaluation of the core compaction corrections for clastic reservoirs. Log Anal. 38 (5), 19–29.

Worthington, P.F., Daines, J.M., Moss, B.P., 1997b. Strengthening the estimation of Qv from porosity: the Achilles Heel of the total porosity model. DiaLog 5(2).

Worthington, P.F., Pallat, N., 1992. Effect of variable saturation exponent on the evaluation of hydrocarbon saturation. SPE Form. Eval. 7 (4), 331–336.

Chapter 9

Capillary Pressure

Chapter Outline

9.1 Introduction **449**
9.2 Primary Drainage Capillary Pressure **452**
9.3 High Speed Centrifuge **453**
9.3.1 Sample Preparation 454
9.3.2 Key Processes 455
9.3.3 Test Procedures (Primary Drainage) 458
9.3.4 Data Reporting Requirements 462
9.4 Porous Plate (Semi-Permeable Membrane): Primary Drainage **471**
9.4.1 Sample Preparation 472
9.4.2 Key Processes 472
9.4.3 Test Procedures 474
9.4.4 Alternatives to Porous Plate Membranes 478
9.4.5 Data Reporting Requirements 479
9.5 Mercury–Air (MICP) **484**
9.5.1 Sample Preparation 486
9.5.2 Test Equipment 486
9.5.3 Test Procedures 488
9.5.4 Pore Throat Size Distribution 490
9.5.5 Typical Output 493
9.5.6 Data Reporting Requirements 493
9.6 Capillary Pressure Methods: Equilibration Times **501**
9.7 Summary of Drainage Capillary Pressure Methods **501**
9.8 Data Corrections **502**
9.8.1 Closure Corrections 502
9.8.2 Stress Corrections 508
9.8.3 Clay-Bound Water Corrections 508
9.8.4 Fluid Corrections 510
9.8.5 J Function 513
References **516**
Recommended Reading **517**

9.1 INTRODUCTION

The ability of a formation to retain water is an important reservoir property and is a function of capillary sizes (i.e. the conduit sizes created between the matrix of the rock), rock texture, wettability, fluid properties (interfacial tension, IFT), density differences between fluid pairs and formation saturation history. Capillary pressures in a reservoir determine the saturation distribution, and hence the total in-situ volumes of fluids (oil/water/gas). Accurate knowledge of the capillary pressure distribution is one of the primary factors in the reliable estimation of hydrocarbon reserves.

Developments in Petroleum Science, Vol. 64. http://dx.doi.org/10.1016/B978-0-444-63533-4.00009-3

In carbonate reservoirs, there is often a high degree of large- and small-scale heterogeneity. This makes accurate appraisal of the capillary pressure distribution, in carbonate reservoirs, highly problematical. However, since heterogeneity has a strong impact on capillary pressure behaviour, reliable laboratory measurements of the small-scale heterogeneity (at the cm scale), as well as realistic up-scale modelling are an essential component to better understanding reservoir fluid distribution.

The saturation for capillary pressures of equal magnitude depends on saturation history, or the direction of the saturation change, i.e. whether the system is undergoing drainage (displacement of the wetting phase by the non-wetting phase) or imbibition (displacement of the non-wetting phase by the wetting phase). Wetting phase saturation is always found to be higher in drainage than in imbibition for the same capillary pressure. This is due to different contact angles between drainage and imbibition—receding and advancing—and different trapping mechanisms between the cycles; for example, snap off of the non-wetting phase during imbibition of the wetting phase in a strongly wetting complex pore system.

The key features of drainage and imbibition capillary pressure cycles are shown in Fig. 9.1 for a water-wet rock. For the water-wet case in the primary drainage cycle, pressure is required for the non-wetting hydrocarbon phase to

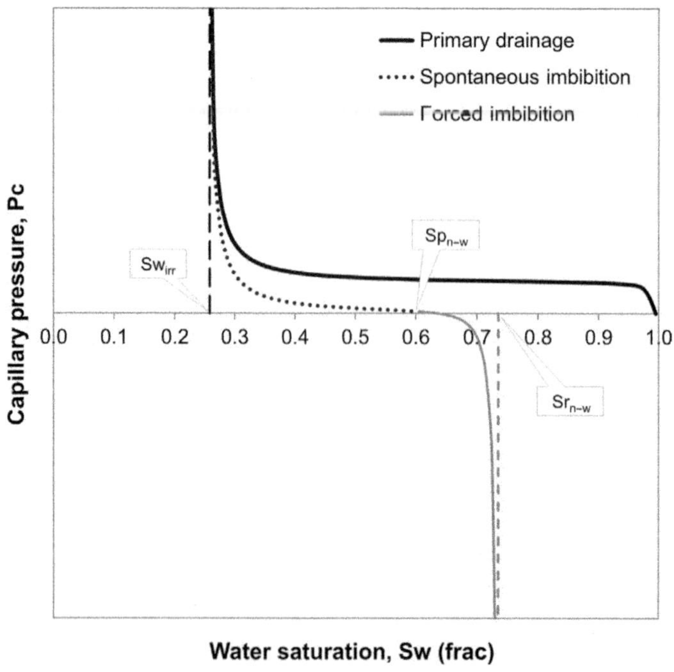

FIGURE 9.1 Primary drainage and imbibition capillary pressure curves.

displace the initially pore-filling water to irreducible water saturation, S_{wirr}. If the rock is then exposed to water at S_{wirr}, it will be spontaneously imbibed and achieve a certain water saturation at zero capillary pressure (and a non-wetting hydrocarbon phase saturation, $Sp_{n\text{-}w}$). This is referred to as spontaneous (positive or static) primary imbibition. Thereafter, pressure must be applied to the wetting phase to displace the non-wetting phase to residual values (e.g. $Sr_{n\text{-}w}$) (King et al., 1986). This is referred to as forced (negative or dynamic) imbibition.

Conventionally, water is referenced as the wetting phase since, for most reservoirs, the depositional environment is believed to have been water-filled. Thus, water-phase desaturation from 100% saturation to S_{wirr} is referred to as the primary drainage cycle, and water phase increasing from S_{wirr} to $Sr_{n\text{-}w}$ is referred to as the primary imbibition cycle, no matter the rock wettability. Strictly however, in an oil-wet rock, water phase decreasing is the imbibition cycle, so the definitions can be confusing. This is a throwback to the early days in the industry when all reservoir rocks were assumed to be water wet or made to be water wet by cleaning before carrying out the tests.

There are three main laboratory test methods for capillary pressure currently used in the industry:

1. *Centrifuge capillary pressure*: provides laboratory data which can be inverted to derive capillary pressure curves at laboratory conditions using an inversion procedure, which is uncertain and gives non-unique capillary pressure curves. The standard industry practice, however, has been to derive a single capillary pressure curve. Since its inception in 1945, the centrifuge technique has become quite popular in the oil industry because the technique is quite rapid. The drawback with the centrifuge technique, however, has been the interpretation of the results. Fully automated centrifuges, with video or photographic data collection systems may be used (and are recommended) to determine capillary pressure and wetting characteristics on 1″ or 1.5″ diameter core plugs. Gas–oil or gas–water fluid pairs can be used for liquid drainage cycles and oil–brine phases can be used in drainage and imbibition cycles (it is inadvisable to perform an imbibition cycle for a gas–liquid system in the centrifuge, due gas compressibility and gas dissolution into liquid as a function of pressure). Tests can be run at elevated stress and elevated temperature conditions. Ultra-centrifuges allow capillary pressure-saturation characterisation of low permeability rocks.
2. *Semi-permeable membrane* (*porous plate*): the method is used to determine capillary pressure curves and saturation relationships on cylindrical rock samples of either 1″ or 1.5″ diameter at ambient conditions or confining pressures up to 10,000 psi. Some laboratories can also use elevated temperature using a dead crude oil. The maximum capillary pressure that can be applied is dependent upon the properties of the membrane/plate

being used. The most commonly used/available have maximum pressures of approximately 200 psi (14 bar) (gas–water system) and about 80 psi (6 bar) (oil–water system). Usually sample permeability has to be above 1 mD for a successful drainage experiment. However, membranes are available with a maximum pressure of approximately 1000 psi (70 bar) for a gas–water system and about 400 psi (30 bar) for an oil–water system, which will allow use of core material with much lower permeability, though experimental time will be much longer. Resistivity index tests can be run simultaneously (Chapter 8).

3. *Mercury–air* (*mercury injection*): capillary pressure curves from 0 up to 60,000 psi air–mercury injection pressure can be obtained in automated equipment using core chips or, preferably, using the equipment maximum of a 1″ × 1″ core plug. This analysis gives capillary pressure data, pore throat distribution and pore level heterogeneity in a short time frame. Combined with petrography, the data can be used to investigate cap rock sealing capacity. Other parameters derived from mercury injection are theoretical permeability, reservoir quality and mercury entrapment. Certain laboratories also continue to maintain manual mercury injection equipment where 1.5″ diameter core plugs can be used, but with a more limited injection pressure (2000 psi or potentially 10,000 psi). Certain of these equipment may also enable measurements under confining stress.

Normally laboratory tests involve fluids that differ from those in the reservoir so that appropriate corrections have to be made to the data using interfacial (surface) tension and contact angle. It is recommended to measure these parameters.

Please note, that although the standard sample size for these analyses is either 1″ or 1.5″ diameter core plugs, a few laboratories have the capacity to perform these analyses on different sample sizes, e.g. 2″ diameter, or full diameter (whole core) samples of either 4″ or 5¾″ diameter (Lasswell, 2005). However, these tend to have more limited application of confining pressure, but can often be necessary for highly heterogeneous, vuggy carbonate and conglomerate systems.

9.2 PRIMARY DRAINAGE CAPILLARY PRESSURE

Primary drainage capillary pressure test provides essential data to:

- evaluate reservoir rock quality, pore throat size distributions and wetting characteristics;
- predict fluid saturations in the reservoir versus height; i.e.,
    - model saturation versus height for hydrocarbon in place determination;
    - calculate original saturations through (and identifying) depleted zones;
    - estimate saturations in thin bedded zones that resistivity logs cannot resolve well;
    - reference log derived saturations including checking irreducible water saturations derived from logs;
    - estimate the thickness of the transition zone;

  - derive the free water level (FWL) where formation pressure data is not available;
  - help explain differences in fluid contacts between wells;
- quantify sealing capacity of overlying rock.

Wettability is a critical parameter to consider when carrying out capillary pressure tests (Anderson, 1986, Literature survey part 1; Anderson, 1987, Literature survey part 4). Tests carried out using air/brine or air/mercury systems, as opposed to oil–brine, implicitly assume that wettability effects are not important. However, it is argued, due to the theory of the depositional environment of most reservoirs, that to define the primary drainage process in the laboratory the plug samples must be initiated in water-wet conditions as this is the natural wettability condition of the formation before hydrocarbon accumulation (excepting the presence of organic material occurring naturally in the matrix of the core). Consequently, the samples should be rendered water wet by appropriate cleaning techniques (Chapter 3). In many cases it would be more practical to use an air–brine system for the primary drainage cycle but not if resistivity index measurements are required for an oil reservoir (Chapter 8). Gas will always be strongly non-wetting relative to the liquid so any residual oil wetting remaining after cleaning should not affect the Pc–Sw relationships.

Primary imbibition tests, e.g. (water-displacing-oil) should be performed on plugs whose wettability is, or has been restored to be, representative of the reservoir wetting condition.

9.3 HIGH SPEED CENTRIFUGE

Capillary pressure can be determined iteratively from experiments performed in a high-speed centrifuge, in either drainage or imbibition mode. Measurements can be performed from 100% saturation or saturated with one fluid with an immobile second fluid present (e.g. oil saturated at Swi). The mobile saturant phase is displaced under centrifugal gravitational forces, i.e. denser fluids will be forced away from the centre of rotation, whilst lighter fluids move towards the centre of rotation. Centrifuge equipment allows the core plug to be placed either towards the centre of rotation (to allow production of a denser phase out of the core plug) or at the outer extremity of the rotor arm length (to allow production of the lighter phase out of the core plug). Consideration of fluid density is thus paramount in determining where the core plug should be positioned in the centrifuge.

For instance, for primary drainage, 100% water saturated core plugs are placed in rotor equipment towards the centre of rotation. Consecutively higher angular velocities are applied providing consecutively increasing gravitational forces. Water (as the denser fluid) is forced away from the centre of rotation, replaced by the less dense fluid (either oil or gas). However, for water imbibition, oil saturated core plugs (at Swir) are placed towards the outer extremity of rotor equipment surrounded by water. Under centrifugal gravity, water is forced away from the centre of rotation into the core plugs. Oil produced by this process, moves under gravity towards the centre of rotation.

Gravitational forces are significant and exponentially dependent upon angular rotation and distance from the centre of rotation; hence, pressures change significantly along the plug length. Liquid displacing gas experiments are therefore not recommended owing to gas compressibility and diffusion problems.

Purpose of test:	• To determine capillary pressure (Pc) as a function of saturation
Sample requirements:	• Cylindrical shaped core plug of 1″ or 1½″ diameter. Larger samples give larger potential pore volume, thus reducing saturation measurement errors. • Due to large gravitational forces, rock strength should be considered and maximum rotational velocity limited accordingly, especially for chalk samples. • *Ideally*—preserved core plug or taken from preserved core. • *Alternative*—core plug previously used in routine core analysis programme, provided the plug has not been pre-stressed nor preparation methods caused damage to delicate clay minerals.
Required input data for lab:	• Formation water (FW) salinity and ionic composition, as close as practically possible to the Formation Water in the reservoir or FW resistivity at reservoir and lab temperature. Lab will prepare synthetic FW (SFW) accordingly. • Number of pressure steps (normally 6–8); however, the Forbes interpretation method (industry recommended method) suggests at least 10 steps, because inappropriately selected steps may promote irregular changes in saturation versus changes in pressure, which could skew the results when differentiating these data. Note: pressure steps will be converted to rotational velocities. Requiring more pressure steps will cost more. • Estimated values of test pressure steps, which depend on the fluid system (air–brine, oil–brine, etc.). Data from mercury injection capillary pressure (MICP) tests can be used as an aid in test pressure design. Knowledge of the height above the free water level and fluid density difference can also be used as an aid to determine the highest Pc required, especially in the case of poorer quality samples (tight gas) (Newsham et al., 2004). If no other capillary pressure data is available, it is recommended to select the lowest to the highest centrifuge speed, with all other steps at even intervals between on a logarithmic scale. • Stabilisation criteria for each pressure step should be stipulated: the most appropriate is cessation of production. Many labs, however, will use set times (often 8 or 12 h).

9.3.1 Sample Preparation

The sample set must reflect the full range of rock types and poroperm properties in the reservoir. For more information refer to Chapter 6.3. The number of samples will depend on heterogeneity in the reservoir, and samples should be grouped by zone, facies, porosity class or hydraulic flow units. Normally, the range of porosity and permeability for core used in SCAL testing should reflect the porosity and permeability distribution of the RCA database. The derived capillary pressure curves could be affected if only a narrow range of porosity and/or permeability is covered, being unrepresentative. As most commercial centrifuge equipment can run three or four, 1½″ diameter or six, 1″ diameter plugs at a time, it is practical and economical to run this test with core plug multiples of three or four (many labs will charge for unused space in the centrifuge) ordered by permeability. Ten samples per facies/reservoir rock type is the minimum required for statistical validity.

Selected core plugs must be examined using a CT scanner (see examples in Section 6.3). This can identify any anomalous material or significant heterogeneity in the plugs. The plugs should also be photographed pre- and post-testing.

Primary drainage tests (e.g. oil-displacing-water) should be performed on water-wet samples (or using an air–brine system if residual wetting in still uncertain). Therefore, tests are performed on samples that have been cleaned and dried (Chapter 3) to a water-wet state. Care should be taken to avoid damage to the samples and preserve the clay structures during core cleaning and drying, as damaged to clay morphology may change its ability to hold water in microporosity, thus altering the low saturation capillary pressure correlation.

SFW is prepared and characterised (density, viscosity and resistivity)—refer to Section 6.1.

After cleaning and drying, air or Klinkenberg permeability at nominal confining pressure is measured using the methods described in Chapter 5. Helium grain volume is measured using Boyle's law procedure. Avoid mercury immersion bulk volume at this stage, especially if samples contain large pores or vugs: brine immersion bulk volume is preferred. The core plugs are saturated in SFW using the method employing vacuum saturation followed by pressurisation, to assist the saturation process. Saturated pore volume is calculated from gravimetric measurements before and after saturation, and bulk volume from Archimedes immersion in SFW. The re-saturation pore volume must agree with the helium pore volume (within ± 0.2 p.u.) before continuing with the tests. If under-saturated, then the samples must be replaced in the pressure saturator. If over-saturated, check for grain loss and apply grain loss corrections accordingly (McPhee, 2012). Porosity is calculated.

On completion of the capillary pressure tests, after Dean-Stark, porosity should be determined using mercury immersion bulk volume and helium grain volume. Air or Klinkenberg permeability at nominal stress should also be measured (Chapter 5).

9.3.2 Key Processes

1. Measure base porosity and base permeability and saturate samples in SFW (determine pore and bulk volume).

2. Spin samples in high speed centrifuge at each of the designed pressure steps
3. Dean-Stark samples (to check for end-point water saturation)
4. Determine capillary pressure curve using S (Forbes, 1991) or SPL (Spline) (Nordtvedt et al., 1990) to solve the centrifuge equation, including subsequent updates (Forbes, 1993, 1994, 1997; Nordtvedt and Kolltveit, 1991).

9.3.2.1 Test Equipment

Tests will usually be performed at ambient temperature using an ultra-centrifuge (Fig. 9.2). Elevated temperature may be applied, but will depend upon specific laboratories and equipment, but often up to 70–80 °C is possible. Centrifuges are capable of operating at high spin speeds, again dependent upon specific laboratory and equipment, but up to 20,000 rpm can be achieve in an ultra-centrifuge. Elevated temperature (and rotor wear from continual use) reduces the maximum speed limit (often to 10,000 rpm). Overburden pressure will be required in the case of friable/unconsolidated samples. Overburden centrifuges have lower spin speeds but longer rotor arm lengths.

The equipment comprises a core holder specially designed for operating in a centrifuge (Fig. 9.3) that includes a graduated tube where produced fluids are collected, a rotor where the core holders are placed (Fig. 9.4) and a non-intrusive optical or video monitoring apparatus (stroboscopic light or

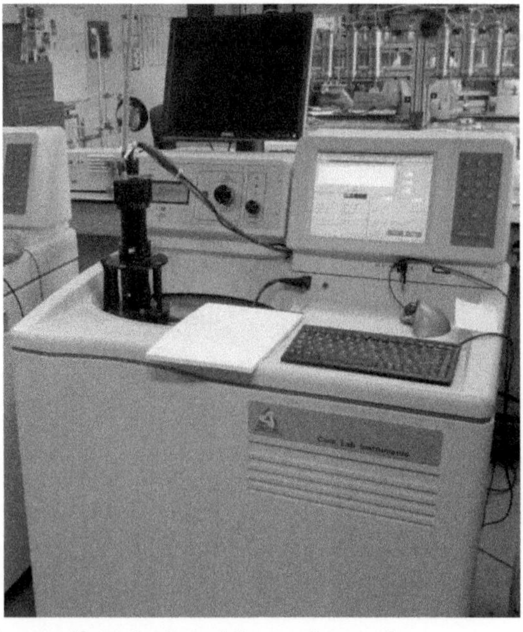

FIGURE 9.2 Ultra-centrifuge. *Courtesy of Coretest Systems, Inc.*

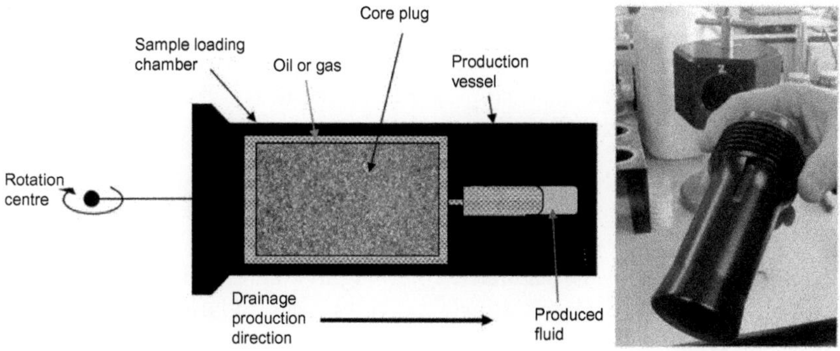

FIGURE 9.3 Centrifuge core holder.

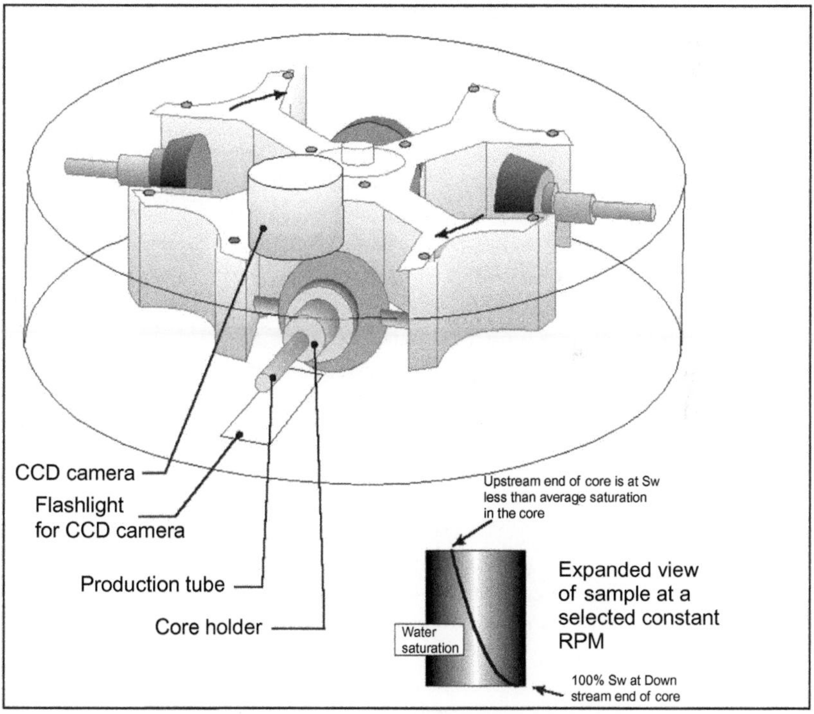

FIGURE 9.4 Illustration centrifuge rotor with CCTV. *Courtesy of Coretest Systems, Inc.*

camera) to read produced volumes while the centrifuge is still spinning, which eliminates the unacceptable practice of stopping the centrifuge to determine production volumes. This can lead to sample fracturing and hysteresis effects. If possible, a footbath should be used to ensure 100% saturation (zero capillary pressure) at the core outlet.

9.3.3 Test Procedures (Primary Drainage)

The lab should evaluate the selected plugs to ensure they can withstand the rigours of testing at elevated spin speeds. If necessary, the maximum spin speed (Pc) may be reduced.

The test procedures are:

1. Saturate the core plug in brine. Determine pore and bulk volume (porosity) and verify against helium porosity (within ±0.2 p.u.).
2. Place the 100% water saturated core plug in a core chamber surrounded by the less dense, non-wetting fluid (oil or gas). Place the core holder in the rotor, and the rotor in the centrifuge.
3. Initiate the first predetermined rotational speed. As a result of centrifugal force, the denser fluid (brine) is expelled from the outer radial end-face, r_e, and replaced by the less dense fluid (air or oil), entering the inner radial end-face, r_i, as shown in Fig. 9.5. Brine will be produced from the sample until equilibrium is achieved between the centrifugal and capillary forces. Collect the produced brine in the graduated tube. A stroboscopic light, set to the same frequency as the rotational velocity, produces the illusion of a static system and allows cumulative produced volumes to be read (either by eye or by camera). Automated systems with cameras allow for accuracy of produced fluid volumes of better than 0.05 cc. Production data should be plotted as a function of log time to ascertain production equilibrium. Fluid production curves can be used as input for core-flood simulators (such as Sendra, Scores, Cydarex), which can be used to improve estimates of capillary pressure through a history-matching approach.
4. Step 3 should be repeated for each subsequent, predetermined rotational speed.
5. The number and values of the pressure step increments can be decided on the basis of the mercury injection characterisation tests on end-trims, if available.
6. Bond number (Hagoort, 1980) should be calculated and plotted as a function of saturation to ascertain critical boundary conditions for singe-speed displacement experiments (excessive viscous forces may produce unrepresentative residual saturations—see Fig. 9.6). For single-speed analyses

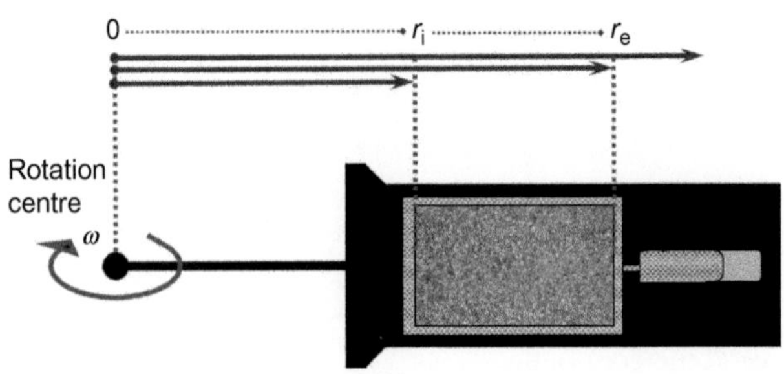

FIGURE 9.5 Diagram of centrifuge system.

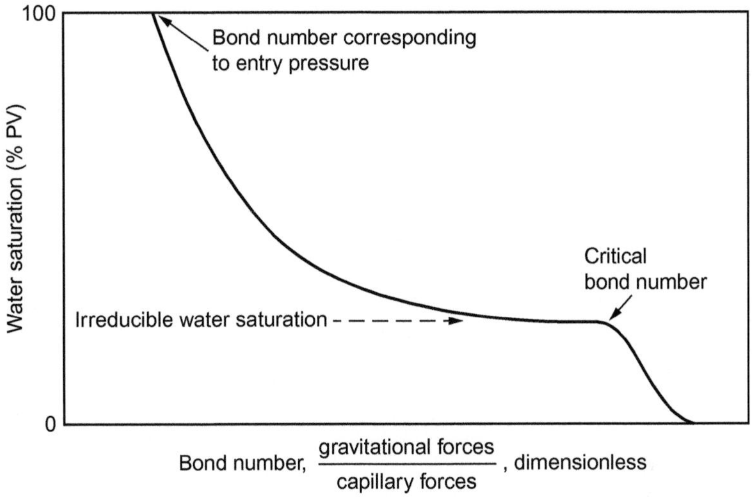

FIGURE 9.6 Critical bond number. *From Omoregle (1988).*

(centrifuge relative permeability), in strongly wetting systems, a bond number limit of 10^{-5}, is often used. However, bond numbers from centrifuge capillary pressure data can be used to directly determine this criterion. In some lithologies there is no noticeable change in curve shapes at high spin speeds. Bond number is defined as:

$$N_b = \frac{k \cdot \Delta\rho \cdot \omega^2 \cdot r}{\sigma}$$

where, k is permeability (m^2), $\Delta\rho$ is the density difference between the fluid phases (kg/m^3), ω is angular velocity (revolutions/sec), r is the distance from the centre of rotation to the centre of the plug (m) and σ is the IFT of the fluid phases (kg/m).

7. Test timing depends on absolute permeability, relative permeability and density difference of the two fluids. At least 24–48 h (depending on permeability) must be allowed to ensure capillary equilibrium at each pressure step. The initial speeds often require longer time. Typically 1–3 weeks are required to generate a curve (between 6 and 10 pressure points). The time allowed to achieve capillary equilibrium will govern the shape of the resultant Pc curve, hence the necessity to use produced volumes to decide when equilibrium has been achieved. The smaller the density difference between the test fluids the longer it takes to achieve equilibrium.
8. On completion of the primary drainage centrifuge test, following saturation stability at maximum rotational speed, the spin speed should be decelerated slowly to prevent sample failure, before being removed from the equipment.

TABLE 9.1 Example Data Output for Centrifuge Capillary Pressure Test (Single Sample)

Depth (m)				**2600.00**
Gas permeability (mD)				**1000**
Porosity (%)				**23.0**
RPM	**Pc (psig)**	**Sw (%)**	**Ave Sw**	**Bond No.**
-	0.00	100	100	-
750	0.74	80.5	96.5	3.86E-06
1150	1.74	48.2	81.8	9.07E-06
1625	3.48	36.2	52.8	1.81E-05
2300	6.97	30.2	43.5	3.63E-05
4600	27.9	25.7	33.1	1.45E-04
8000	84.4	24.7	28.7	4.39E-04

9. S (Forbes, 1991, preferred) or SPL (Nordtvedt et al., 1990) should be used as the preferred methods to calculate reservoir representative capillary pressure and saturations. Table 9.1 shows a typical table of centrifuge lab reported data.

10. The plug should be protected from fluid loss through evaporation.

11. Provided the plug is not required for further tests, it is strongly recommended that core plugs undergo Dean-Stark extraction on completion of the capillary pressure test, as a verification of water saturations (procedures are provided in Chapter 5). Plug final (max Pc) average Sw should match final Dean-Stark Sw within ±3 saturation units (experience-based). If on completion of the capillary pressure test samples are scheduled for other tests (Kg or Ko at Swir, etc.), Dean-Stark can also be performed at the end of those tests, taking into account water mobilised and collected during any subsequent test. The Dean-Stark leach method should be used: leaching with methanol in hot Soxhlet equipment and drying in a conventional oven at 105 °C until constant weights (±0.01 g or better) are achieved.

12. Following drying, re-measure core plug base porosity and permeability (Chapter 5).

9.3.3.1 Pc Curve Calculation

In the centrifuge method, the capillary pressure curve is not directly measured, but is calculated from fluid production measurements and the centrifuge rotational speed; various approximation methods can be used.

Using the sample pore volume, the liquid production can be converted to average saturation using the following equation. Note, for imbibition the negative sign becomes a positive:

$$S_{avg} = S_{avgi} - \frac{V_{displaced}}{V_p}$$

where S_{avg} is average saturation, S_{avgi} is initial average satuation (at start of test), V_p is pore volume (cc) and $V_{displaced}$ is volume of displaced fluid (cc).

Pressure difference across the sample is calculated from the rotational speed using:

$$\mathrm{Pc} = 1.578 \times 10^{-1}(\Delta\rho)\left(r_e^2 - r_i^2\right) \cdot \mathrm{RPM}^2$$

where, Pc is capillary pressure (psi), $\Delta\rho$ is the density difference between the fluid phases (g/cm^3), r_e is the distance from centre of rotation to outer face of the core plug (lab/equipment manufacturer dependant) (cm), r_i is the distance from centre of rotation to inner face of the core plug (lab/equipment manufacturer dependant) (cm) and RPM is revolutions per minute

NB: 1.578×10^{-7} is a unit conversion factor and will differ dependent upon the units used.

In the case of an oil/brine system, the density difference is lower than an air/brine system so that the capillary pressures will be lower for the same speed.

Capillary pressure and saturation, from the above equations are merely approximations, affected by the sample length, centrifuge geometry and uneven saturation distributions caused by boundary conditions. Reservoir representative capillary pressure and saturation must be iteratively derived for a distance along the core plug corresponding to a point of intercept for both capillary pressure and saturation. Several attempts have been made to obtain an exact method for calculating the inlet face saturations, but all are merely approximations since the equations are complex and have multiple roots. In 1993 the Society of Core Analysts, SCA, completed an inter-laboratory survey on drainage capillary pressure measurement by centrifuge. The study concluded that:

- the main source of inaccuracy in the drainage capillary pressure curve determination by centrifuge is related to the interpretation process and not to experimental procedures (assuming accepted procedures are in use) nor to the shape of the capillary pressure curve.
- the inaccuracy depends on the centrifuge geometry (and resulting contributions of centrifugal and radial effects) and on the method used for solving the centrifuge equation. The choice of the method, according to the centrifuge/core geometry, appears to be of prime importance, much more than improving experimental conditions. This is especially critical for short radius geometry, and includes the most common geometry of 1″ × 1″ sample run at 8.6 cm from the centrifuge axis.

- additional errors due to the inversion process (inlet pressure and saturation calculation) may be very large (±10 saturation units). They cannot be reduced significantly by increasing the number of rotation steps, or by reducing the experimental error below 2 saturation units, or by averaging the results from different interpretation processes. Because most of these processes introduce a negative bias, the common practice of using different interpretations and averaging the results is totally irrelevant. Errors being dominantly negative, no compensation is achieved by averaging.
- the only way to improve the capillary pressure curve determination is to use an appropriate solution of the centrifuge equation, consistent with the centrifuge and core geometry. Whatever the geometry in use, only a few solutions may insure a reasonable accuracy, i.e. below ±3 saturation units. These are S (Forbes, 1991) and SPL (Nordtvedt et al., 1990).
- these solutions require proper implementation and use of correction for radial effects and (or) the use of consistency constraints.

Full details of the S (Forbes, 1991) and SPL (Nordtvedt et al., 1990) methods can be consulted in the references provided. Core flood simulators provide methodology similar to Nordvedt for estimating capillary pressure and are a viable alternative.

9.3.4 Data Reporting Requirements

- description of procedures and experimental apparatus;
- SFW composition;
- fluid density (all pertinent fluids) at test temperature (g/cc);
- initial sample dry weight (g);
- initial sample saturated weight (g);
- calculated pore volume (cc);
- base permeability (mD) (pre-test);
- base porosity (fraction) (pre-test);
- pre-test permeability (mD) (if different from base permeability);
- centrifuge speeds used (RPM);
- calculated approximate capillary pressure steps (psi/bar/mbar/kPa);
- production volumes (cc) (preferably cumulative production as a function of test time);
- average saturations (fraction) (calculated from production volumes);
- Iterative saturations, per pressure step (fraction);
- Iterative capillary pressures (psi/bar/mbar/kPa);
- method used for capillary pressure and saturation iterative calculations;
- bond number (per pressure step);
- effective permeability (mD) (post-test);
- end-point average water saturation (fraction);
- Dean-Stark water saturation (fraction).

9.3.4.1 *Centrifuge Drainage Capillary Pressure Summary*

Advantages	Drawbacks and Issues
• Relatively simple and quick method. • Desaturation in a centrifuge is more rapid than with the porous plate. • Good method for most rock types. • Reservoir conditions tests possible: overburden pressure (friable/unconsolidated samples) and temperature. • Test is not destructive so samples can be used for any other tests (if not fractured after spinning). • Easily performed in both drainage and imbibition scenarios	• It is the most expensive method. • High speed centrifuge is needed to reproduce the forces that exist at the crests of reservoirs having very thick columns. • High spin-speed de-saturation may cause saturation regimes that do not occur under normal field conditions. Bond number as a function of saturation must be reviewed to ensure such regimes have not occurred. • The method creates an unusual stress pattern; where fracturing is possible, especially at higher spin velocities. Therefore, the method is not suitable for samples that cannot withstand the forces applied during rotation at high speeds. • Raw data needs to be corrected to account for the capillary pressure gradient across sample. The main source of inaccuracy is related to the data correction process and not to experimental procedures (assuming accepted procedures are in use) nor to shape of the capillary pressure curve. Only a few solutions may insure a reasonable accuracy, i.e. below ±3 saturation units (S (Forbes, 1991) and SPL (Nordtvedt et al., 1990)). • As with porous plate, representative drainage capillary pressure curves require the samples to be water wet for which the cleaning technique used is a critical aspect. For an air–brine system this should not be a problem as air will always be the non-wetting phase. Using oil–brine could be a potential problem as the rock can remain somewhat oil-wet if oil from oil-wet surfaces is not completely removed. • The time allowed to achieve capillary equilibrium can govern the shape of the resultant Pc curve. If inadequate time is allowed to reach equilibrium at each step, data could be compromised by indicating higher water content (thus lower hydrocarbon saturation and a pessimistic curve). Cessation of production is the key equilibrium criterion—production data should be monitored and plotted versus log time. • Stopping and restarting the centrifuge (when in-place production measurements are not available) can lead to sample fracturing and produce saturation hysteresis effects (continual drainage and imbibition cycles). Some labs follow this practice, which is unacceptable. We recommend that the optical measurement system is used, preferably with automated capture, since data is then traceable. • Samples with large or frequent surface vugs should be avoided if possible. Bimodal capillary pressure curves can be produced from these vugs desaturating at lower pressures than the remaining pore matrix.

QC Checks

- Analysis of the results should start with a thorough review of the laboratory report to check that acceptable procedures have been followed. This should be carefully evaluated.
- Check the method used for cleaning and drying and if care has been taken to avoid damage and preservation of the clay structures.
- Wettability tests (Chapter 7) should be performed to ensure core cleaning renders the samples water wet.
- Check that samples are fully saturated in brine before starting any test.
- Perform Dean-Stark extraction after completion of capillary pressure test. This will provide an independent check for water saturation. Final (max Pc) average Sw (not inlet-face saturation) should match final Dean-Stark Sw within about ±3 saturation units (experience-based). Figure 9.7 shows an example where samples S12 and S9 fall outside the minimum acceptable error between Dean-Stark and average Sw (end-point). Suspect data should be carefully evaluated and eliminated from the data base if need be. Additionally, inlet face Sw must be less than average Sw during a drainage process.
- A plot of logarithm of water saturation (inlet-face endpoint) versus logarithms of permeability (or rock quality index—RQI) should normally show a trend of water saturation increasing with decreasing permeability, as presented in

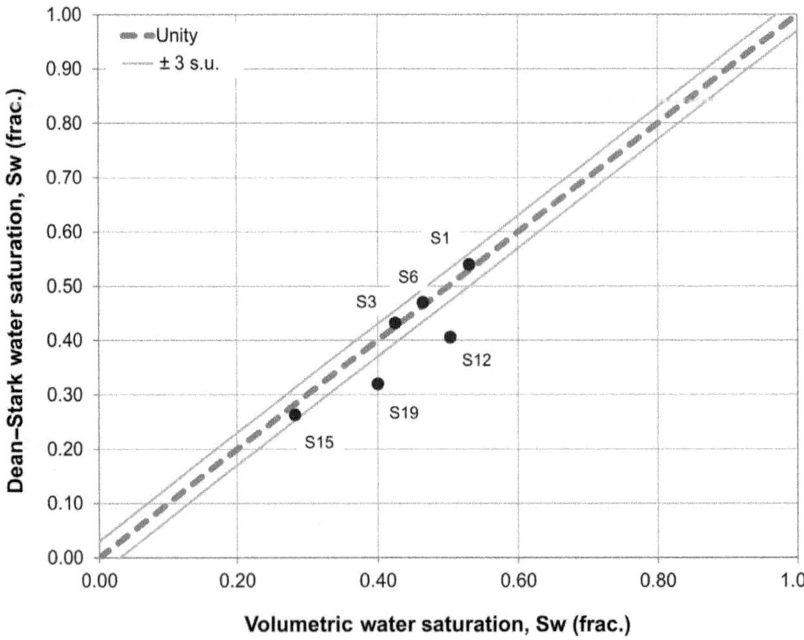

FIGURE 9.7 Example of water saturation check for end-point capillary pressure test.

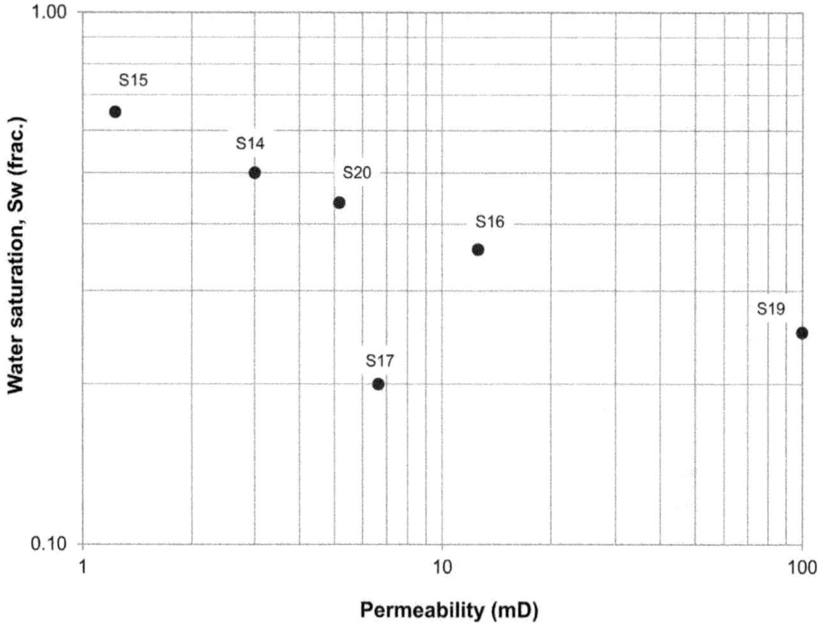

FIGURE 9.8 Example of permeability versus end-point Sw plot.

Fig. 9.8. Sample's S17 Sw end-point is lower than expected (20%), compared to the general trend. Data from this sample needs to be carefully scrutinised to differentiate rock type behaviour from experimental artefacts.

- If available, check end-points with data from other reliable sources (routine Dean-Stark, other Pc methods, Resistivity Index (at full capillary equilibrium), NMR, etc.).
- Lab should report both average and calculated end face (inlet face) saturations. Inlet-face saturations are used to generate the Pc curve—not average saturations. The average saturation data are used for quality control. For example, the final average saturation must match the Dean-Stark value and the average Sw versus Pc and calculated inlet face Sw versus Pc curves should not cross, as evidenced in Fig. 9.9.
- Poor design of pressure steps can produce curves that are missing part of the transition zone. This can also occur if batch-processed samples in the centrifuge (three or four samples per run) have large permeability variance but are de-saturated at the same speeds. A hypothetical example is shown in Fig. 9.10. This is normally evidenced by a straight line segment of the curve. Sample permeability for any batch run must be similar, or a greater number of speeds used to define the full curves of each different permeability value—possibly certain steps might not require full stability

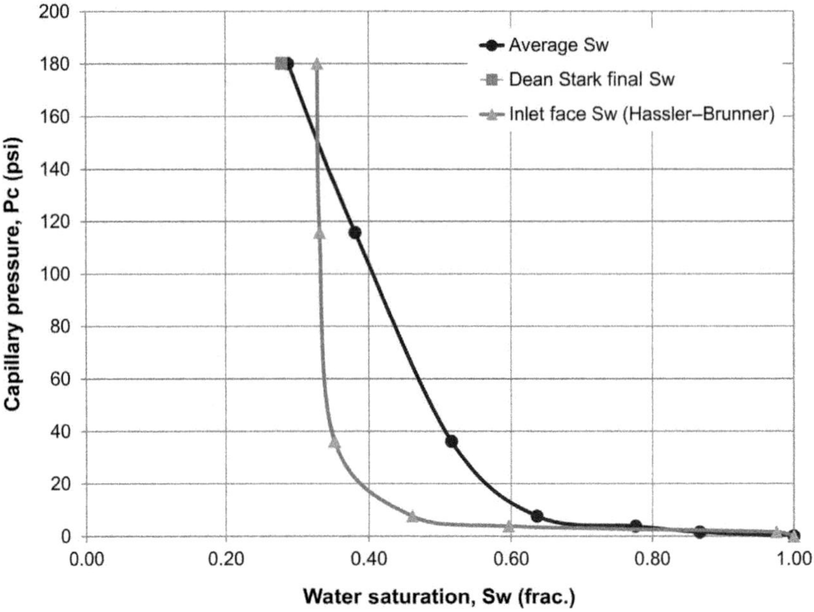

FIGURE 9.9 Average and inlet face Sw–Pc curves example.

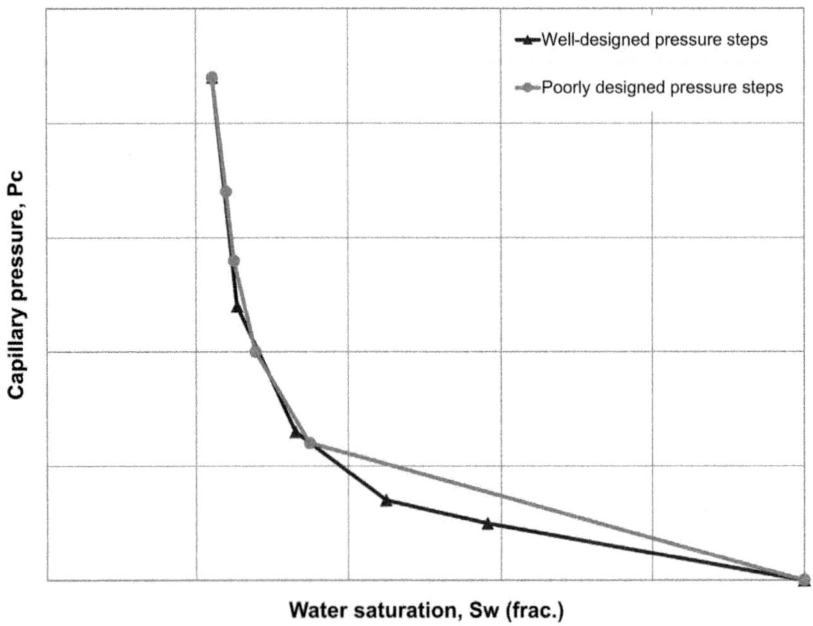

FIGURE 9.10 Hypothetical poorly designed Pc steps.

in certain samples, but unstable steps must not be included in the iterative solution of Pc and Sw.

- The method used by labs to process the data and calculate inlet face Pc and Sw should be clearly reported. As discussed, the Forbes method is the preferred analytical solution. For example, in a lab audit three labs were provided with production data and centrifuge geometry data which had been inverted from an invented Pc curve. Some minor but typical experimental measurement data were added. The true end face Pc curve, and true and "measured" average Pc curves are shown in Fig. 9.11. The labs processed the data using "standard" analytical solutions (Forbes, 1991; Hassler and Brunner, 1945; Rajan, 1986; Van Domselaar, 1984 and the results (Fig. 9.12) confirm the likely error bar introduced simply by different processing methodologies and protocols.
- Capillary pressure curves can be analysed using a method based on the model developed by Brooks and Corey (1964, 1966). The method can be used to identify well or bad behaved data and portions of the capillary pressure curves that are unreliable and need to be refined or rejected.
- In this method, primary drainage is defined using the equation:

$$\mathrm{Pc} = \mathrm{Pe} \cdot S_{\mathrm{wn}}^{-\lambda},$$

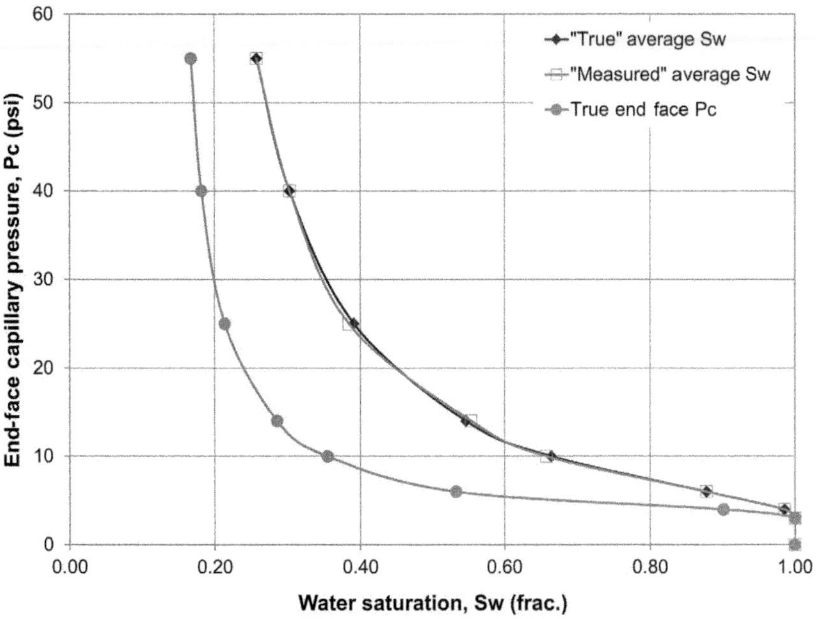

FIGURE 9.11 Input curves for audit dataset.

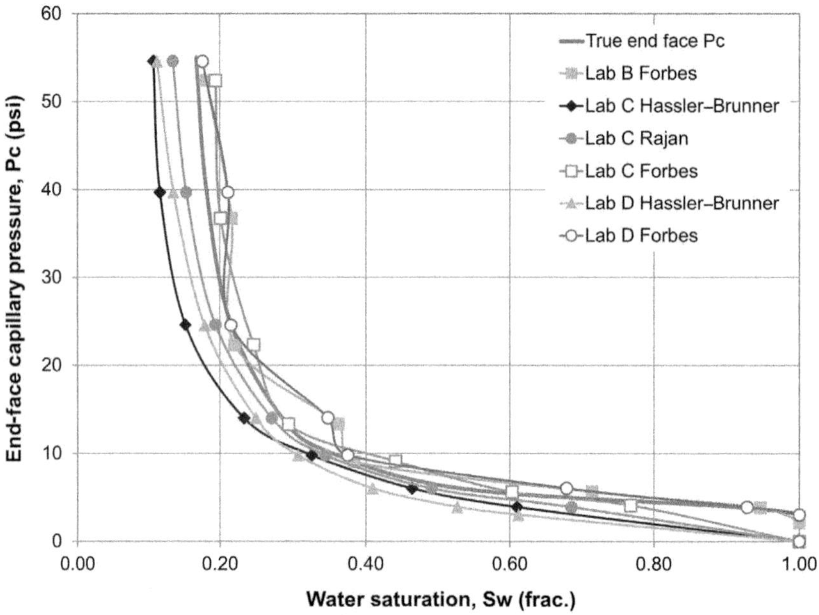

FIGURE 9.12 Lab calculated Pc curves.

where Pc is the capillary pressure, Pe is the capillary entry (threshold) pressure and S_{wn} is normalised water saturation and λ is the pore size distribution index. S_{wn} is given by:

$$S_{wn} = \frac{(S_w - S_{wirr})}{(1 - S_{wirr})}$$

where, S_{wirr} is irreducible water saturation, S_{wirr} is the saturation at which the water relative permeability (k_{rw}) approaches zero, and will inevitably be much lower than the value achieved at the end of any laboratory test, as very high pressures and long drainage times are required.

Analysis of the data should start by plotting capillary pressure versus wetting phase saturation, with capillary pressure plotted both on a linear and on a logarithmic scale. The resulting curves should be smooth and concave upward, in both instances. Examples are shown in Figs. 9.13–9.15. The curve is concave upwards; nevertheless there is some irregularity at about 80% water saturation and at the end of the test.

The next diagnostic plot consists of the logarithms of capillary pressure versus the logarithms of S_{wn}. The plot is constructed assuming different values of S_{wirr} in the S_{wn} calculation. The value of S_{wirr} that produces a line, which is approximately straight, will be the estimated

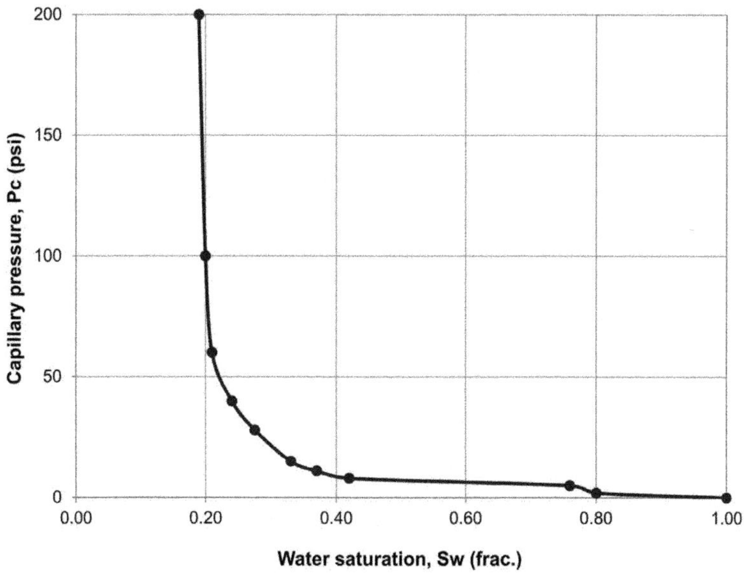

FIGURE 9.13 Linear plot of capillary pressure versus water saturation.

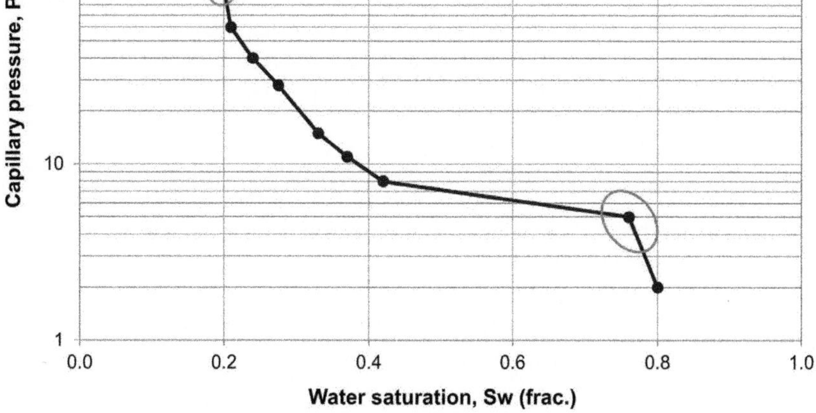

FIGURE 9.14 Semi-log plot of capillary pressure versus water saturation.

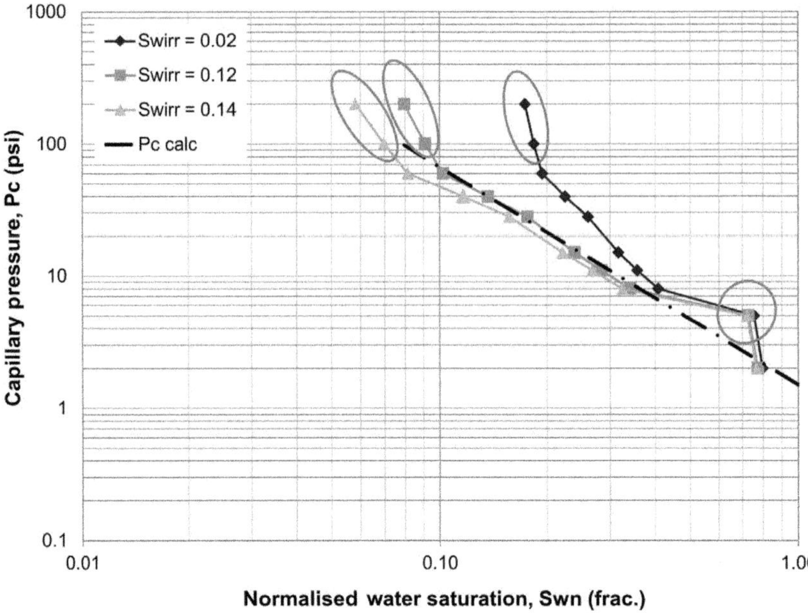

FIGURE 9.15 Log–log plot of capillary pressure versus normalised irreducible water saturation.

value for S_{wirr}. Difficulties in achieving a straight line with any value of S_{wirr} raises serious questions as to the reliability of the data and needs to be carefully evaluated. Figure 9.15 illustrates the methodology, showing that a value $S_{wirr}=0.12$ produces a straight line fitting values from the middle of the test and shows the irregularities at the start and the end (low and high Sw).

This procedure can also be used to extend data to higher Pc values if necessary, to define initial water saturations high in the oil column, when sufficient Pc cannot be applied in the lab due to experimental limitations.

Suitable Laboratories

- Most commercial SCAL labs. Beware of labs who do not have automated data acquisition equipment and are forced to stop the centrifuge to determine plug saturation.
- The lack of overburden centrifuge capability is a drawback in case unconsolidated or friable samples need to be tested.

9.4 POROUS PLATE (SEMI-PERMEABLE MEMBRANE): PRIMARY DRAINAGE

This is commonly referred to as the porous plate method but strictly it is the semi-permeable membrane method. In older datasets it is also known as the restored-state technique.

The samples may be processed on a batch basis (ambient) or in individual coreholders (ambient or overburden). The recommended approach is to use individual samples at reservoir net confining stress.

The method for de-saturating a saturated core plug involves placing one end in capillary contact with a saturated porous (usually ceramic) plate/semi-permeable membrane and applying step-wise increasing gas or oil under pressure to the open surface(s) not in contact with the plate. The porous plate is saturated with the same formation water as that used to saturate the plug sample. The porous plate is has a controlled, and fairly homogenous, unique pore (capillary) size, which when water-saturated will have a capillary entry pressure equivalent to the capillary radius. Different porous plates may be selected depending upon the maximum displacement pressure required, usually a 200 psi (15 bar) G–W threshold, equivalent to 75 psi (5 bar) O–W threshold, is used. If studying low permeability rock, porous plates are available up to 1000 psi (70 bar) G–W threshold, but permeability is reduced by approximately the same factor of threshold increase (hence why these are not in standard use).

Since water is the wetting phase in the sample and in the porous plate, and since the water in the sample is in contact with the water in the porous plate along water-wetting films covering all grain/mineral surfaces of the plug matrix, the formation water can be expelled from the plug capillaries through the porous plate capillaries, at increasing pressures up to the plate threshold pressure. At each applied oil or gas pressure, the oil or gas will fill all pore throats (and the larger pore spaces controlled by those throats) with equivalent radius (or larger) to the applied pressure. Desaturation at each pressure stage (step) thus continues until no more water production is observed, or weight loss if the batch porous plate method is used—though this method is not recommended. When production has ceased the sample is assumed to be at final (if not irreducible) water saturation at the maximum capillary pressure applied.

The porous plate is also used to condition samples towards Swir for relative permeability tests (Chapter 10), to provide capillary-equilibrated saturations for resistivity index (Chapter 8) and for nuclear magnetic resonance measurements (Chapter 11).

The porous plate pore sizes impose constraints on the maximum capillary pressure. Typically this is limited to around 75 psi for oil–brine systems and 180 psi for air–brine systems. Some labs have high pressure plates capable of running air–brine Pc tests to up to 1000 psi.

Purpose of test:	• To determine capillary pressure (Pc) versus water saturation (Sw). • Optional. To determine resistivity index simultaneously.
Sample requirements:	• Cylindrical shaped plug sample of 1″ or 1½″ diameter. The larger the sample, the larger the potential pore volume, and the less will be the impact of saturation measurement errors. • Ideally—sample taken from preserved core or preserved plug. Alternative—plug previously used in routine core analysis programme, provided that plug has not been pre-stressed and/or that plug preparation methods have not caused damage to delicate clay minerals.
Required input data for lab:	• FW salinity and ionic composition, as close as practically possible to the formation water in the reservoir or FW resistivity at reservoir and lab temperature. Lab will prepare SFW accordingly. • Number (normally 6–8) and estimated values of test pressure stations which will depend on the plug permeability and the fluid system (air–brine, oil–brine, etc.). Data from MICP tests can be used as an aid in test pressure design. Knowledge of the height above the free water level and fluid density difference can also be used as an aid to determine the maximum Pc required, especially for lower permeability samples. • Time required for each pressure step or criterion for cessation of production. • Test confining stress (for individual porous plate).

9.4.1 Sample Preparation

The same sample selection and preparation procedures described for the high speed centrifuge (Section 9.3), apply for the batch porous plate, with the exception that the number of plugs run in a single test will be limited to one sample per individual cell and to the size of the desaturation vessel for the batch porous plate (around 8–12 1.5″ diameter plugs.)

9.4.2 Key Processes

1. Saturate samples in SFW (determine pore and bulk volume (porosity))
2. De-saturate samples with air or oil at stepwise pressures
3. Plot capillary pressure versus water saturation
4. Dean-Stark extraction (to check for end-point water saturation)

9.4.2.1 Test Equipment

The samples may be processed on a batch basis (air–brine at ambient conditions) or in individual core-holders (air–brine or oil–brine ambient, overburden, or reservoir conditions). Capillary pressure cells for drainage experiments consist of a pressure vessel fitted with a strongly water-wet porous ceramic plate and are attached to a pressure manifold, which regulates and monitors the pressure of the displacing phase. Where air is used it must be humidified to prevent water evaporation from the sample.

The porous ceramic plates are normally provided in a range of entry pressures. The entry pressure reflects the largest pore throat size in the porous plate. Lower entry pressure plates may be used for lower applied pressures/higher permeability plugs and higher entry pressure plates for higher applied pressures/lower permeability plugs. This means that the plates may be changed at some point in the experiment, once samples have been offloaded for weight measurements after stabilising at a particular applied pressure. Lower pressure plates are used because they will allow faster flow through them at the low pressure.

The essence of the experiment is that the sample should have drained to final (irreducible) water saturation before the porous plate entry pressure is breached by the non-wetting phase.

9.4.2.2 Batch Porous Plate (Ambient)

In the batch porous plate setup (Fig. 9.16), several samples can be run at the same time but these systems are exclusively used for air–brine experiments. To ensure compliance between the plug and the porous plate and continuity between the brine in the rock and the brine saturating the plate, a capillary bridge (filter aid or contact medium) is required between the sample and the plate. This typically consists of a filter paper and a thin layer of clay powder (e.g. Keiselguhr or kaolin/celite mixture) which is compacted (rolled) to ensure that the pore sizes of the capillary bridge are smaller than most of the pore sizes in the rock. This ensures that the capillary bridge does not de-saturate before the rock. If the capillary contact between the plate and the sample is lost, the sample will not de-saturate further at incremental pressures. Desaturation is measured by removing the plugs and weighing them. The weight change represents the volume of brine drained at each pressure step.

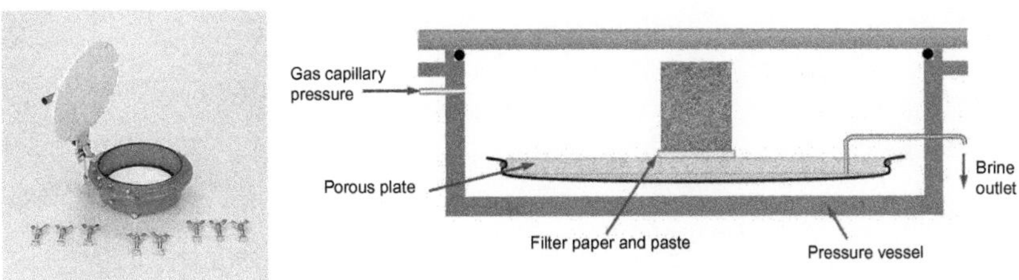

FIGURE 9.16 Batch porous plate equipment (ambient). (Left: *Courtesy of Eijkelkamp Soil & Water.*)

9.4.2.3 *Individual Cell Porous Plate (Ambient Or Overburden)*

In the individual cell (Fig. 9.17), only one sample can be accommodated per core-holder and it is possible to apply an overburden pressure and to carry out the test at reservoir temperature if required. The axial pressure acting on the ends of the core end platens is usually sufficient to ensure good capillary contact, and so no bridge is required between the sample and the porous plate. The individual cell method has the capability of monitoring water production and sample resistance continuously. Desaturation is measured volumetrically as a graduated pipette is connected to the cell effluent or using non-invasive saturation monitoring, such as gamma ray or X-ray adsorption techniques. Air–brine and oil–brine tests can be run in the individual cell porous plate.

9.4.3 Test Procedures

9.4.3.1 *Batch Porous Plate (Ambient)*

The batch method is not the recommended approach, but may be a viable method if the individual plate method is not possible. The SFW saturated samples are placed in the pressure vessel (soil moisture cell), onto a properly selected, porous plate (SFW saturated) with a capillary bridge. The porous plate cell is strongly water-wet and has pore throats that are smaller than most of the pores sizes in the test sample. Air and/or oil cannot penetrate (break through) the pores in the plate until the entry pressure is reached (which depends on the plate pore size); yet brine can pass through the pores of the plate since it is the saturant phase and fully wets every

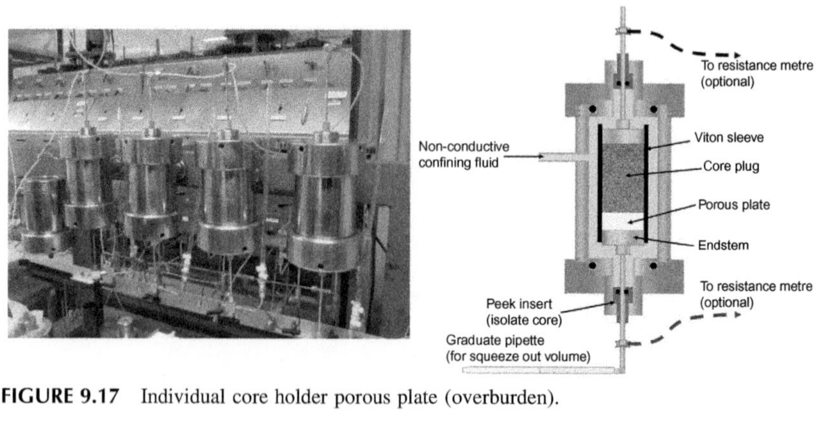

FIGURE 9.17 Individual core holder porous plate (overburden).

surface of the plate, to atmosphere, outside of the cell. In this way a differential pressure is created across the plate, which causes the brine in the samples to be displaced and allows controlled desaturation.

The humidified air pressure is increased over a number of previously designed stages to the desired final pressure value, which should reduce the core saturation towards Swir. In ambient measurements, carried out by this batch basis, it is impossible to monitor the expulsion rate of brine from individual samples, therefore after each pressure increment, the samples are removed and the weight recorded. Once weighed, the sample is returned to the cell and the process continued. Consequently the samples are repeatedly removed from and replaced into the cells throughout testing. At lower saturations there is the potential that the water film in the core plug may not connect again with the water in filter aid (e.g. celite and SFW mix paste). If capillary continuity fails to be achieved the plug remaining saturation will no longer drain at higher pressures. To overcome this the filter aid needs to be re-saturated each time so that, at the base of the plug, there will be a small imbibition of SFW into the plug to help capillary contact.

Permeability of the batch processed samples should be similar, otherwise important curve information may not be captured on certain samples.

Typically 6–8 pressure points are applied. Typical equilibration times used by most laboratories are about 3–4 days per pressure point, though longer periods are sometimes used for lower permeability samples. Recommended minimum stabilisation times per pressure step are 10 days for high permeability samples and 30–40 days for low permeability samples, although cessation of brine production from the batch cell is the normal criterion. It is not possible to determine whether individual plugs in the batch have achieved stability without removing and weighing the samples and samples of different permeability will have different stabilisation times. Excessive handling should be avoided to prevent grain losses.

It is highly recommended that on completion of the capillary pressure test, the samples undergo Dean-Stark as a check on calculated water saturations (procedures are provided in Chapter 5). Final (maximum Pc) plug Sw should match final Dean-Stark Sw within about ±3 saturation units (experience based). If, on completion of the capillary pressure test, the samples are scheduled for another test (e.g. Kg or Ko at Swi, etc.), Dean-Stark can also be performed at the end of these tests, taking into account water mobilised and collected during all of the test process.

Unless plugs are to be used for other tests, samples are leached in hot Soxhlet apparatus in methanol before being dried in a conventional oven at 105 °C to constant weight (±0.01 g or better). Following drying, the base porosity and permeability is determined (refer to Chapter 5).

9.4.3.2 *Individual Porous Plate (Overburden)*

The SFW saturated sample is placed in the pressure vessel (individual core holder), along with a properly selected individual porous plate (SFW saturated).

Similarly to the formation resistivity factor (Chapter 8), the accurate determination of sleeve conformance pressure (SCP) and its associated annulus volume are essential elements of this test and must be satisfactorily established for each test plug. Once the SCP has been determined, the overburden pressure is gradually raised to the required test stress. Annulus volume corrections are applied to the raw data to determine the correct pore volume (porosity) at the test stress. This will have an effect on the pore volume, and on subsequent water saturation calculations.

Normally, the test is performed at a single confining stress. This should be at, or close to, the equivalent reservoir stress (Chapter 6).

As with the batch porous plate, when the plate is saturated with brine, air or oil cannot penetrate (break through) the brine saturated pores in the plate until the plate threshold pressure is reached (which depends on the plate pore size), yet brine can pass through the pores of the plate, to atmosphere, outside of the cell.

The brine expulsion (measured volumetrically with a graduated pipette) and resistance (if RI tests are run in conjunction) are monitored continuously in the cell, to ensure that stability is reached. Consequently, the samples need not be removed from the cells until the test is completed. More sophisticated systems now use individual cell in-situ saturation monitoring (ISSM), such as gamma ray or X-ray adsorption techniques. Figure 9.18 is an ISSM scan of a porous plate experiment with saturation plotted as function of length along plug. Each line represents a saturation equilibrium at each capillary pressure and confirms uniform saturation profiles.

Typically 6–8 pressure points are acquired. Unlike the batch technique, individual cells allow production to be monitored as a function of time. The criterion for capillary equilibrium is cessation of brine production over an extended period (most appropriately determined by plotting production as a function of logarithmic time), but the same minimum stabilisation times apply. It is recommended that saturation is monitored either by ISSM or, if tests are run in conjunction with resistivity index, resistivity until an acceptable stability criterion has been achieved. Small changes in production volumes may be difficult to distinguish but, as core resistivity is an inverse logarithmic function of saturation, the resistivity response is better attenuated, hence cessation resistivity increase is a better determiner of saturation stabilisation. In the oil–brine porous plate example shown in Fig. 9.19 it takes around 180 days (half a year) to define the full drainage curve over 6 capillary pressures.

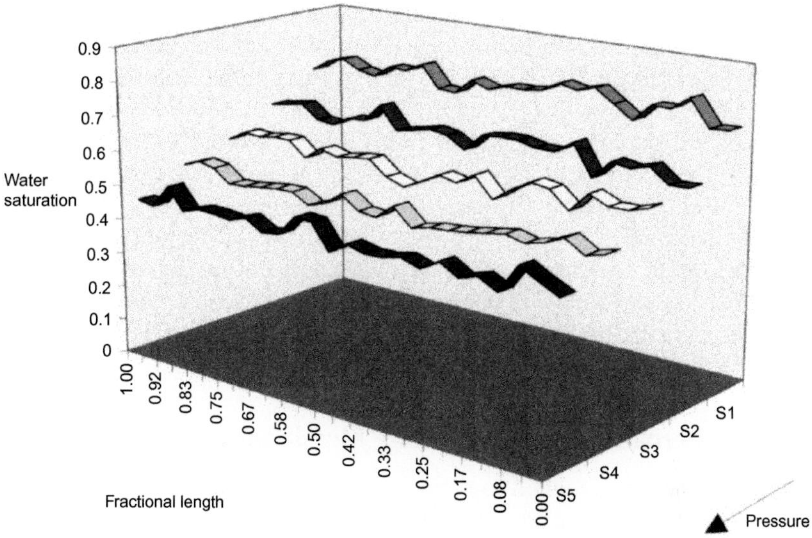

FIGURE 9.18 ISSM monitoring of porous plate test.

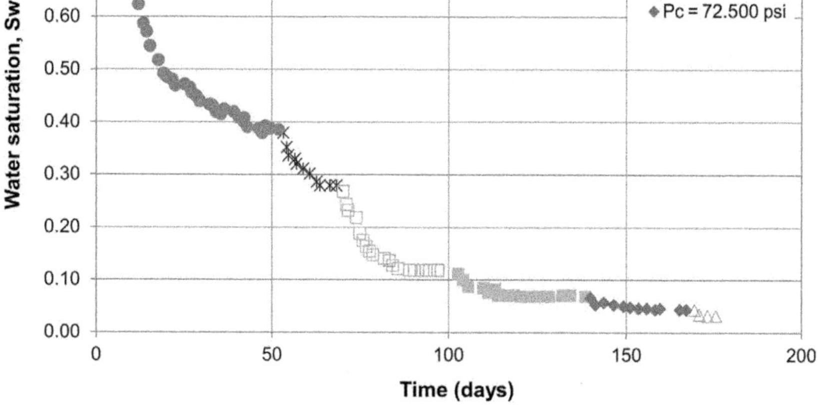

FIGURE 9.19 Porous plate saturation versus time.

It is highly recommended that on completion of the capillary pressure test, the samples undergo Dean-Stark extraction as a check on calculated water saturations (procedures are provided in Chapter 5). Final (max Pc) average Sw should match final Dean-Stark Sw within about ±3 saturation units (experience based). If on completion of the capillary pressure test, samples are scheduled for another test (Kg or Ko at Swi, etc,), Dean-Stark can also be performed at the end, taking into account water mobilised and collected during any previous test.

Unless used for other tests, samples are leached in hot Soxhlet apparatus with methanol and dried in a conventional oven at 105 °C to constant weight (±0.01 g or better). Following drying, the base porosity and permeability is determined (refer to Chapter 5).

9.4.4 Alternatives to Porous Plate Membranes

In view of the time required to drain core samples through ultra-low permeability ceramic porous plates, several attempts have been made to speed up the process.

A laminated ceramic porous plate called Keraflux was developed in 2001 by ResLab (Wilson and Tjetland, 2001; Wilson and Skjaeveland, 2002) (now Weatherford Labs). Initial results indicated that this semi-permeable membrane yielded a faster turnaround to obtain equilibrium at a fixed capillary pressure. Typical experimental times for drainage capillary pressure (7–12 capillary pressure steps) of relatively homogeneous samples was typically 5–9 weeks, compared to ~30–40 weeks for the conventional porous plate membrane. Recent Senergy experience uncovered experimental uncertainties related to the use of Keraflux membranes during porous plate desaturation for a friable gas reservoir in the Norwegian North Sea, including:

- extra liquid accounted for, due to gas production that could have drained the outlet end tube (including valve);
- Keraflux membrane weight loss, indicating that gas entered the membrane during the test leading to loss of capillary contact.
- the presence of gas rather than brine in the porous plate potentially has an influence on electrical measurements (higher n values).

Some issues related to ability to tolerate mechanical stress with time and wettability alteration issues during long term tests have also been raised. However, this technology is no longer offered by Weatherford due to quality control issues with the manufacturer.

Fleury and Longeron (1997), Shafer and Lasswell (2007) and Dernaika et al. (2010) have developed several mathematical models to speed up the collection of porous plate data by not waiting for capillary equilibrium at each pressure step. Equilibrium capillary pressure curves are calculated using extrapolated pseudo-equilibrium data. These methodologies seem promising but are not widely applied yet. They require the collection and reporting of a large amount of data used in the calculations that are not normally acquired by most commercial labs, therefore, increasing substantially the cost of the tests. Not many operators are willing to pay for additional data with no warranty that the results will be reliable.

9.4.5 Data Reporting Requirements

- brief description of procedures and experimental apparatus;
- sample length and diameter (cm);
- SFW composition;
- SFW density at test temperature (g/cc);
- initial sample dry weight (g);
- initial sample saturated weight (g);
- final sample dry weight (g);
- base permeability (mD) (pre and post test);
- base porosity (frac) (pre and post test);
- Dean-Stark water saturation (fraction).
- calculated pore volume (cc);
- capillary pressure (psi);
- volume of brine produced from batch cell versus time;
- calculated brine saturation as a fraction of brine saturated pore volume, per pressure step (fraction);
- sample grain loss correction procedures, including typical worked example for one sample from sample database. Raw data for calculation;

9.4.5.1 For Individual Porous Plate (in Addition to Batch Cell)

- overburden pressure (psi);
- pipette initial and final read per pressure step (raw data before applying SCP correction calculation) (cc);
- cumulative brine production per pressure step (raw data before applying SCP correction calculation) (cc);
- sleeve conformance pressure (SCP) (psi) and annulus volume (ml);
- volume of brine produced from individual cell versus time;
- capillary pressure (psi);
- calculated porosity (fraction), pore volume (cc) and bulk volume (cc) (calculated on basis of incompressible grain volume) at overburden pressure, based on SCP-corrected expelled brine volume.

9.4.5.2 *Porous Plate Drainage Capillary Pressure Summary*

Advantages (general):	Drawbacks and Issues (general):
• Provided sufficient time is allowed to achieve saturation stability, a uniform saturation can be guaranteed. • No corrections need be applied for non-uniform saturation (unlike centrifuge). • If the sample is relatively clay rich the method is preferable to others. • Test is not destructive so samples can be used for any other tests. • Run in conjunction with resistivity index (Chapter 8).	• The main disadvantage is time required to define a full Pc curve. Equilibration time is governed by the rate of brine drainage through the porous plate. This can be time consuming as this material has an extremely low permeability and will act to restrict the flow of brine from the samples, brine flux rates extremely slow, even for high permeability plugs. This is exacerbated as the brine saturation reduces—the relative permeability to brine in the sample reduces and the rate of brine drainage decreases. Consequently, if insufficient time is allowed at each pressure increment, the brine in the sample will not reach equilibrium saturations, especially at higher capillary pressures. This is accentuated in low permeability samples, where both absolute and effective brine permeabilities are very low. Therefore the method is not recommended for low permeability samples. Even in high permeability samples, insufficient desaturation times can lead to pessimistic Pc curves where Sw is too high. Figure 9.20 shows Pc curves for centrifuge (considered definitive) and porous plate tests run for 4 days per pressure and 10 days per pressure. The difference is significant. • There are limitations on the maximum attainable Pc values, depending on the system (air–brine or oil–brine). This may not define the full Pc curve in low permeability samples. • As with centrifuge, representative drainage capillary pressure curves require the samples to be water wet, for which the cleaning technique used, is a critical aspect. For an air–brine system this should not be a problem as air will always be the non-wetting phase. Using oil–brine could be a potential problem as the rock can remain somewhat oil–wet if oil from oil–wet surfaces is not completely removed. • Benign cleaning is required if the core plugs contain delicate clay structures.

Advantages (batch cell): • Several samples can be run simultaneously. Generally cheaper than individual cell porous plate.	**Drawbacks and Issues (batch cell):** • Batch configuration does not allow for individual sample desaturation monitoring. Samples need to be removed repeatedly from the cell to measure water saturation gravimetrically leading to sample grain loss. This will affect the saturation calculations, as mass loss will be wrongly interpreted as water saturation reduction, so that calculated fluid saturation can be much less than the true value. • Loss of capillary contact can occur due to poor contact between the sample and the porous plate. If the medium used in the capillary bridge has a pore size greater than the smallest pores in the sample, then this will desaturate before the rock has reached final saturation. The result is that air enters the bridge preventing brine from draining so that the saturation remains constant despite an increase in pressure. The saturation/pressure curve often shows an abrupt change to become asymptotic (as shown in Fig. 9.21 so that Sw may be much higher than the true value. Figure 9.22 shows data from a field-wide case where most sample data exhibited indication of this capillary discontinuity. • As all samples in the batch are exposed to the same set of capillary pressures, the samples should have similar permeability to describe the curves properly. Combination of high and low permeability samples in the same batch is not recommended as important Pc curve information may not be defined or captured. • Only air–brine tests can be run. • Not suited for unconsolidated samples.
Advantages (individual cell): • Tests can be run at overburden conditions and at elevated temperatures with crude oil or STO (for RI). • The axial pressure acting on the ends of the core end platens is usually sufficient to ensure good capillary contact. • Saturations (and resistance) can be monitored in line for each individual sample. • Suitable for air–brine and oil–brine systems • Testing can be run simultaneously with resistivity index. • This is the preferred method for unconsolidated samples as overburden pressure can be applied.	**Drawbacks and Issues (individual cell):** • Only 1 sample can be run in a core-holder at a time. • Generally more expensive than the batch cell, as a dedicated core holder is required for each sample. • Accurate determination of SCP is an essential element for porosity at stress calculation, hence saturation, and must be satisfactorily established for each test plug. • Oil–brine tests are susceptible to wettability alterations if crude oil or STO is used. Samples can "age" during the long duration tests. An initially water-wet rock (drainage) could become increasingly less water-wet during the desaturation process. Arguably, this is a more realistic representation of the reservoir processes and conditions if the test is run in conjunction with resistivity index.

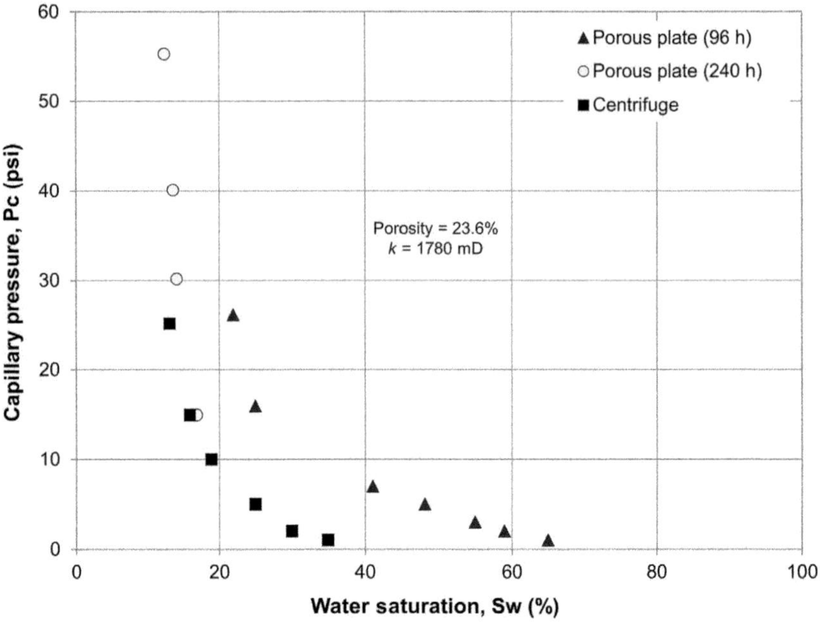

FIGURE 9.20 Example of effects of short equilibration times in porous plate test.

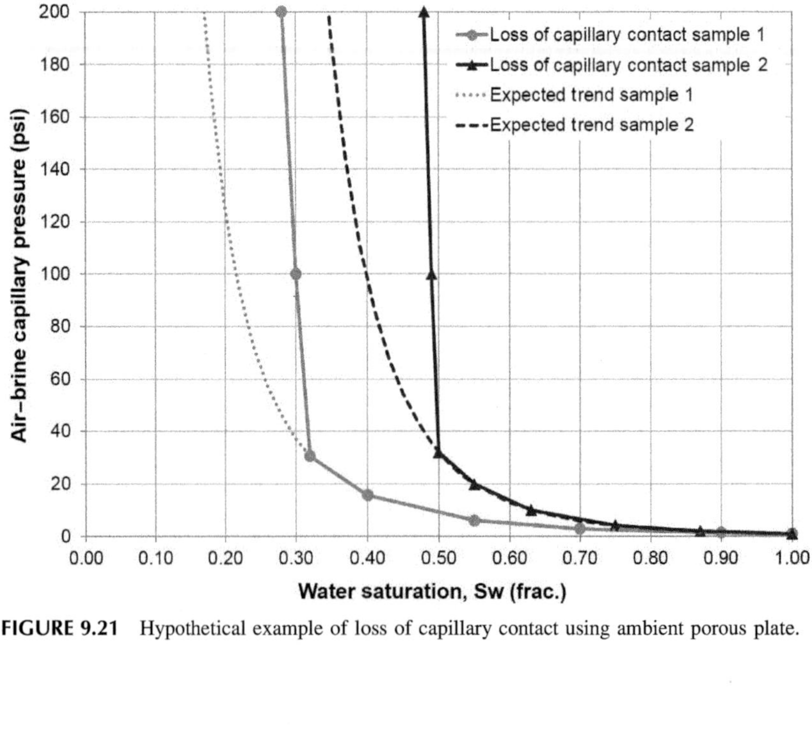

FIGURE 9.21 Hypothetical example of loss of capillary contact using ambient porous plate.

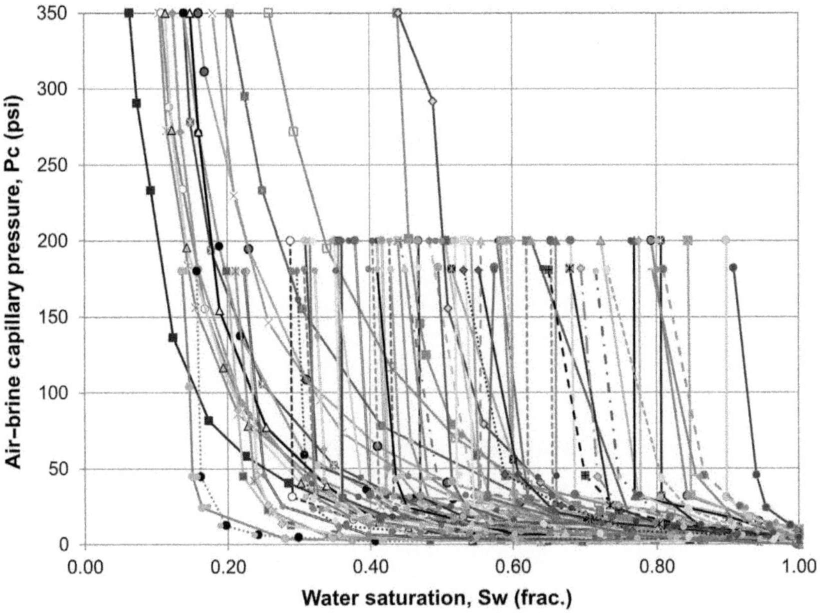

FIGURE 9.22 Actual field example of loss of capillary contact.

QC Checks (in addition to those for centrifuge):

- Laboratory raw data should be provided to enable and independent verification/check of the results (porosity, pore volume, saturations) and to ensure sufficient time has been allowed for equilibrium saturation.
- Perform Dean-Stark extraction after completion of capillary pressure test. This will provide an independent check for water saturation. Final Sw should match final Dean-Stark Sw within about ±3 saturation units (experience based).
- If available, check end-points with data from other reliable sources (routine Dean-Stark, other Pc methods, Resistivity Index (at full capillary equilibrium), NMR, etc.).
- Poor design of pressure steps or testing different permeability samples in the same batch may not fully define or capture key curve features.
- Loss of capillary contact and insufficient equilibration times can be assessed by comparing porous plate Pc curves with centrifuge and mercury injection tests. Data obtained using different fluid systems must be scaled to the same fluid system (see Section 9.8.4). The J function (where Pc is normalised by fluid system and rock quality) is a very useful tool in curve diagnostics. This is discussed in Section 9.8.5.
- If samples to test are friable or to suffer from grain loss, then a correction should be applied to account for it, otherwise it can result in considerable

uncertainties in the calculation of fluid saturation. A loss in weight might be interpreted as a loss in fluid so that the calculated fluid saturation can be much less than the true value. The goal of grain loss correction is therefore to predict the fluid-filled pore volume at each stage of the handling and test procedures. This is not normally reported by labs. The total grain loss that has occurred should be determined, and then proportioned to each relevant stage in the experiment. Details of grain loss corrections are the same as those for resistivity index (Chapter 8).

- Sleeve conformance pressure/volume (SCP) data and plots must be provided for each plug sample, and the corrected expelled volumes agreed between the client and test lab (see Chapter 6).

Suitable Laboratories:

- Most commercial SCAL labs offer ambient condition batch tests. Many commercial labs offer individual cell tests at stress but few offer tests at elevated temperatures with crude oil.

9.5 MERCURY–AIR (MICP)

The test involves injecting mercury into an evacuated and dry core sample under controlled pressures (capillary pressure). Mercury is the non-wetting phase fluid which forces air (mercury vapour) to be the "wetting" phase. The volume of mercury injected at each pressure step determines the non-wetting phase saturation (Shg). 1-Shg defines the wetting phase saturation.

Unlike the porous plate technique, this method is very fast, often requiring only hours rather than days or weeks. Injection pressures up to 60,000 psi are possible with modern equipment enabling all the connected pore volume to be filled, even in low permeability samples.

Although air–mercury capillary injection curves are used extensively to characterise the drainage case and are an excellent tool for quantifying pore sizes and distribution, this process has several disadvantages:

- high pressures and high IFT forces may destroy or "bulldoze" sensitive clays lining or filling pores;
- lack of standards for correcting the apparatus for system compressibility at high pressures resulting in water saturation/capillary pressure measurement errors;
- no "irreducible" wetting phase saturation as air will be compressed at high pressures and the pore volume will be completely filled with mercury;
- tests are destructive as mercury will be retained in the sample on completion of the tests.

The two principal mercury injection test methods are:

1. Low pressure manual (LPM) systems which can accommodate standard SCAL plugs but are limited to about 2000–3000 psi injection pressure and are manually operated. Some may enable the use of confinging stress (overburden), with total system pressure of up to 10,000 psi.
2. High pressure automated (HPMI) systems which are capable of injecting mercury into small samples at pressures up to 60,000 psi. These instruments are principally designed for measuring pore size distribution—not capillary pressure.

Few labs now offer the LPM system, preferring the HPMI system as it is automated, can run up to four samples at once, and provides very accurate injection volumes. Most legacy data (prior to 1994) has been measured using the LPM system.

Note that mercury is recognised as a highly toxic substance and precautions must be taken when using mercury in core analysis.

Purpose of test:	• To determine capillary pressure (Pc) versus wetting phase saturation (Shg). • To determine pore throat size distribution functions
Sample requirements:	• Equipment to determine MICP can accommodate small and irregularly shaped samples (e.g. core chips or cuttings). • Nevertheless, sample size and shape can affect capillary pressure data, with smaller samples having higher "surface area/sample bulk volume" ratios tending to cause slightly lower initial pore aperture entry pressures, and hence, a more optimistic capillary pressure curve. Additionally, the smaller the sample the bigger the saturation and porosity uncertainties arising from errors involved in measuring pore and bulk volumes. • Manual LPM equipment can accommodate plug samples of 1.5″ dia. × 2.5″ length. • The automated high pressure system has a maximum sample size limit of 1″ dia. × 1″ length (10–11 ml) but much smaller plug offcuts are typically used (around 5–6 ml bulk volume). • Samples must be cleaned and dried at high temperature to remove any water in the pore system as water will act as a wetting phase in preference to air. • Samples previously used in routine core analysis programme are commonly used, provided that it has not been pre-stressed and/or fracture.
Required input data for lab:	• Minimum time required for each pressure step. • Maximum Pc for LPM tests. For HPMI Pc should ensure connected pore volume is filled).

9.5.1 Sample Preparation

Since air is the wetting fluid in a mercury–air system, the rock to be analysed should be free of any water. Therefore the recommended cleaning and drying methodologies for MICP are hot Soxhlet extraction and convection oven dry at 105 °C (Chapter 4). This harsh preparation will most probably remove any clay-bound water (CBW) from the system. In this case, CBW corrections should be considered in the petrophysical interpretation, but the method is not recommended for shaly sands with sensitive or delicate clays.

9.5.2 Test Equipment

9.5.2.1 Low Pressure LPM

Most laboratories use variations of the same equipment. Figure 9.23 shows a typical setup which comprises:

- a mercury volume pump;
- a sample holder/pressure cell with a window system for observing the mercury level (Fig. 9.24);
- pressure gauges/manometers;
- a vacuum pump;
- Manifold system allowing vacuum evacuation of sample and pressurisation of mercury.

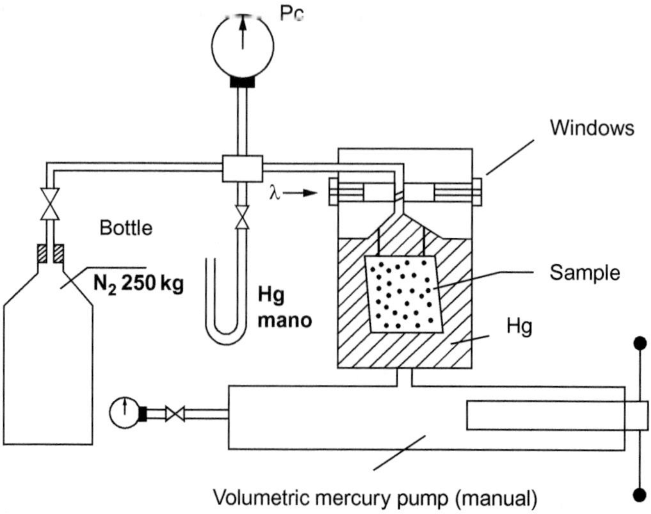

FIGURE 9.23 Example low pressure mercury injection apparatus. *Courtesy of Core Laboratories and its Petroleum Services and Core Lab Instruments divisions. All Other Rights Reserved.*

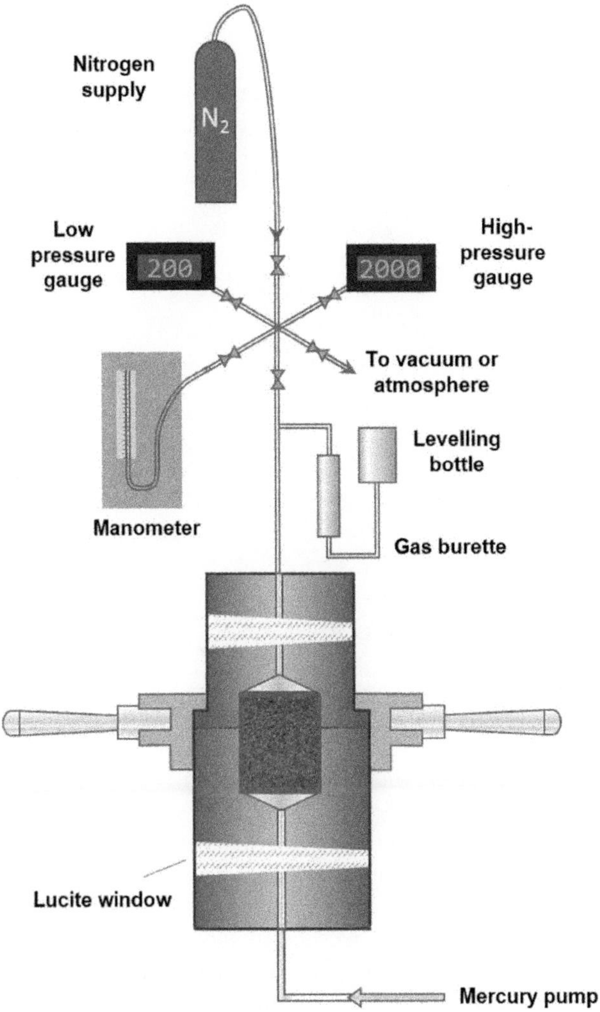

FIGURE 9.24 Example LPM mercury injection pressure cell.

Typically this equipment is operationally limited to a maximum pressure of 2000 psi. Historically, mercury injection has been performed upon full size core plugs, 1.5″ diameter by up to 2.5″ long.

9.5.2.2 Automated HPMI

Automated mercury porosimeters can accurately generate capillary pressure curves at high pressures and volume resolutions. Typical equipment is shown in Fig. 9.25. Typically this equipment has a maximum injection pressure of 60,000 psi.

FIGURE 9.25 Automated mercury injection capillary pressure apparatus. *http://www.micromeritics.com*

The sample is contained within a penetrometer (Fig. 9.26), which consists of a sample cup with an electrical contact cap, where the material to be analysed is placed. It is connected to a metal-clad, precision-bore, glass capillary stem. The position of the mercury meniscus in the penetrometer stem is monitored. As the pressure on the mercury in the glass capillary stem is increased, mercury enters the sample and recedes in the capillary tube. The volume injected is obtained from the change in capacitance between the metal sheath and the mercury in capillary.

Sample size in the penetrometer is limited to a maximum of 1″ dia. × 1″ length (10 ml) but much smaller samples (around 5 ml bulk volume) are normally tested and cut from a parent core plug whose porosity and permeability are known.

9.5.3 Test Procedures

The mercury should be clean and pure—if not the IFT and contact angle will be altered. Data conversion requires that these data for mercury are precisely known.

The measurements are relatively fast (1–2 days). Rapidity of the measurements is a function of the rapid capillary pressure equilibrium rates attained

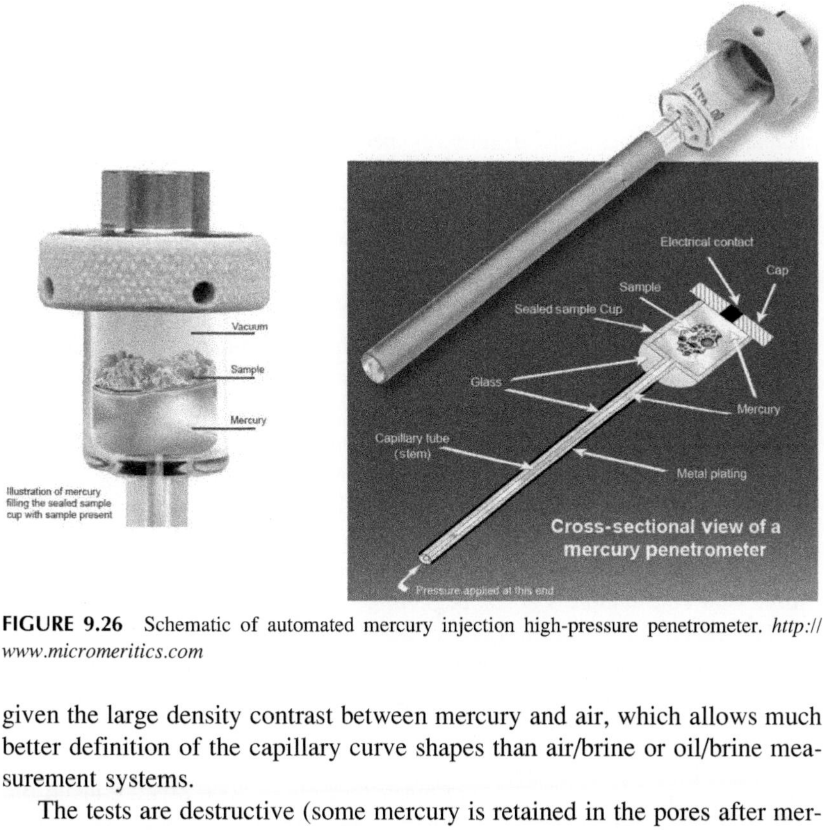

FIGURE 9.26 Schematic of automated mercury injection high-pressure penetrometer. *http://www.micromeritics.com*

given the large density contrast between mercury and air, which allows much better definition of the capillary curve shapes than air/brine or oil/brine measurement systems.

The tests are destructive (some mercury is retained in the pores after mercury withdrawal) so that the sample cannot be used for any other tests.

9.5.3.1 LPM Method

A cleaned and dried sample of known pore volume (normally measured in a separate helium porosity test) is introduced in the cell and subjected to a very high vacuum. The mercury level is brought to the mark in front of the window with the sample completely immersed. The pump is set to zero. The vacuum pressure is reduced in controlled stages and mercury is drawn into the sample and the volume monitored. Once atmospheric conditions are reached, the pump the constant level is re-established in the window cell by means of the mercury pump. The pressure acting on the top of the mercury in the cell is increased by applying nitrogen pressure via a control manifold with the pump locked. The mercury enters the sample and the level in the top window drops. The mercury is then injected from the pump to re-establish the level in the top window. The volume of mercury injected, read by the pump, is noted for each state of equilibrium and for each pressure plateau until the maximum pressure is reached (usually 2000 psi). Typically 20–30 pressure steps are used to define the drainage curve.

To account for the volume expansion of the test cell at high pressures, a blank run is normally carried out with no sample in place and the volume versus pressure curve obtained. These data are used to correct the sample injection volumes. The injected mercury volume is obtained for each pressure plateau by means of the differences between readings with and without samples.

Mercury saturation is calculated as a percentage of pore volume occupied for each pressure in order to establish capillary pressure curves.

9.5.3.2 Automated Method

A cleaned and dried sample is placed in the penetrometer, which holds a known amount of mercury at the start of the test. It is sealed and placed in a low pressure port, where the sample is evacuated to remove air and moisture—the user controls the speed of the evacuation and there's no need for a separate preparation unit. The penetrometer cup and capillary stem are then automatically backfilled with mercury. Excess mercury is automatically drained back into the internal reservoir; only a small amount remains in the penetrometer. As pressure on the filled penetrometer increases, mercury intrudes into the sample's pores, beginning with those pores of largest diameter. This requires that mercury moves from the stem into the cup, resulting in a decreased capacitance between the now shorter mercury column inside the stem and the metal cladding on the outer surface of the stem. The instrument automatically collects low pressure measurements over the range of pressures specified by the operator. Then, the penetrometer is moved to the high pressure chamber, where high pressure measurements (>2000 psi) are taken (up to 60,000 psi). Data are automatically reduced using the low and high pressure data points, along with values entered by the operator, such as the weight of the sample and the weight of the penetrometer loaded with mercury.

An example of MICP curves for a range of permeabilities and rock qualities is shown in Fig. 9.27.

9.5.4 Pore Throat Size Distribution

Pore throat size can be directly calculated from capillary pressure curves (Pickell et al., 1966; Swanson, 1977). Due to the large number of datapoints associated with high pressure mercury injection (HPMI) tests, mercury injection tests are normally preferred and are used routinely to define pore throat size distribution relationships. This is based on the assumption that the capillary pressure (Pc) of the non-wetting phase (mercury) required to penetrate a pore throat of radius, *r*, is:

$$r = \frac{C2\sigma\cos\theta}{\mathrm{Pc}}$$

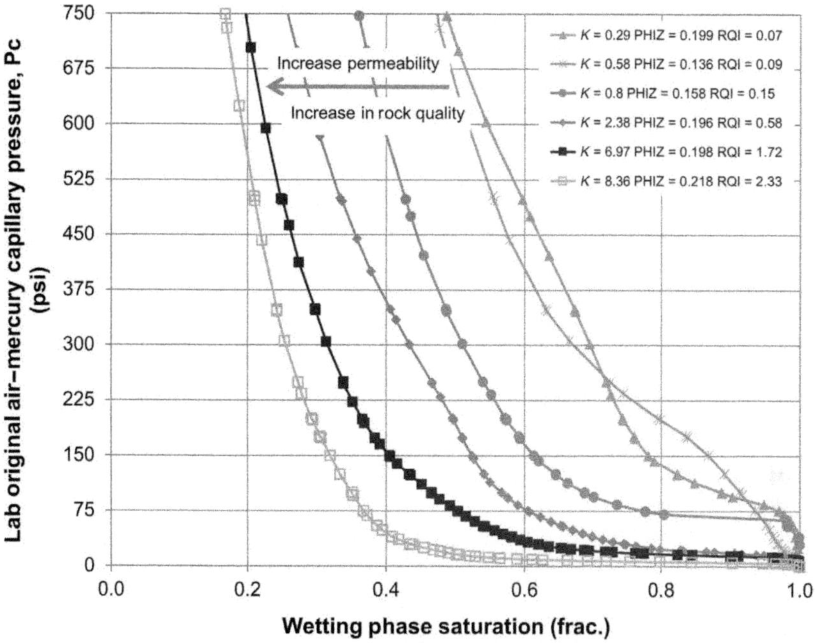

FIGURE 9.27 Example of MICP curves for samples of increasing permeability/rock quality.

where, C is a units conversion constant, Pc is capillary pressure (psi), σ is the IFT, mercury vapour/mercury (dyne/cm), θ is the contact angle (degrees) and r is pore throat radius (micron).

The equivalent pore throat radius penetrated by mercury at a given capillary pressure is plotted versus fractional pore volume injected (v). This is used to calculate a pore size distribution function (PSD) from the derivative:

$$\text{PSD} = \frac{\text{d}v}{\text{d}[\log(r)]}$$

The derivative function is then smoothed, e.g.:

$$\text{PSD}_i = \frac{[\text{PSD}_{i-1} + 2\text{PSD}_i + \text{PSD}_{i+1}]}{4}$$

and normalised by the maximum PSD:

$$\text{PSD}_{\text{normal}} = \frac{\text{PSD}_i}{\text{PSD}_{\text{max}}}$$

$\text{PSD}_{\text{normal}}$ is normally plotted against pore throat radius, r, (Fig. 9.28) and r is normally plotted as a function of mercury saturation (Fig. 9.29).

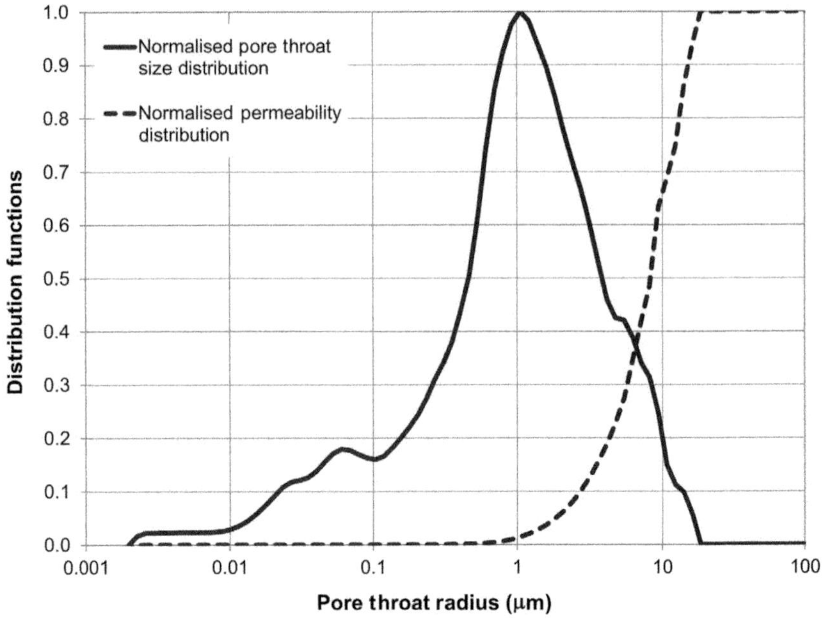

FIGURE 9.28 Example pore throat size frequency distribution from HPMI.

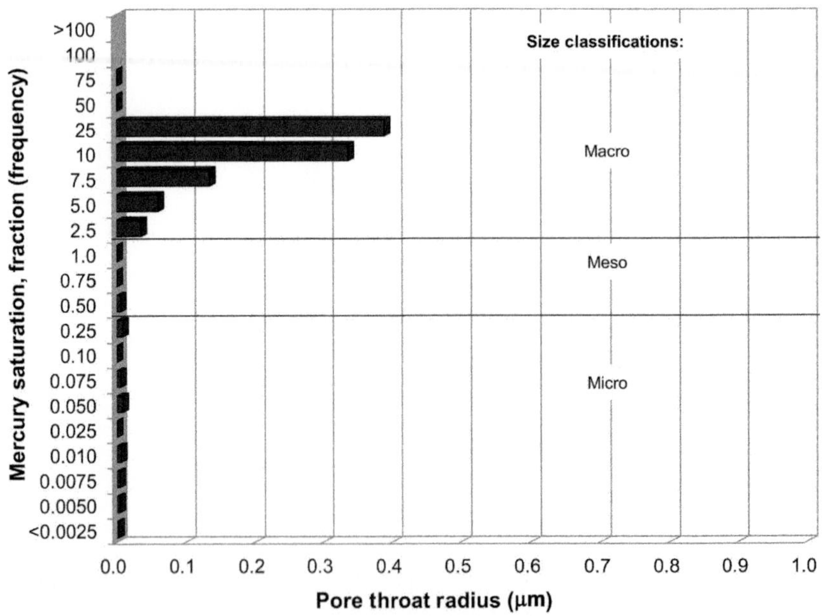

FIGURE 9.29 Example pore throat size volumetric distribution.

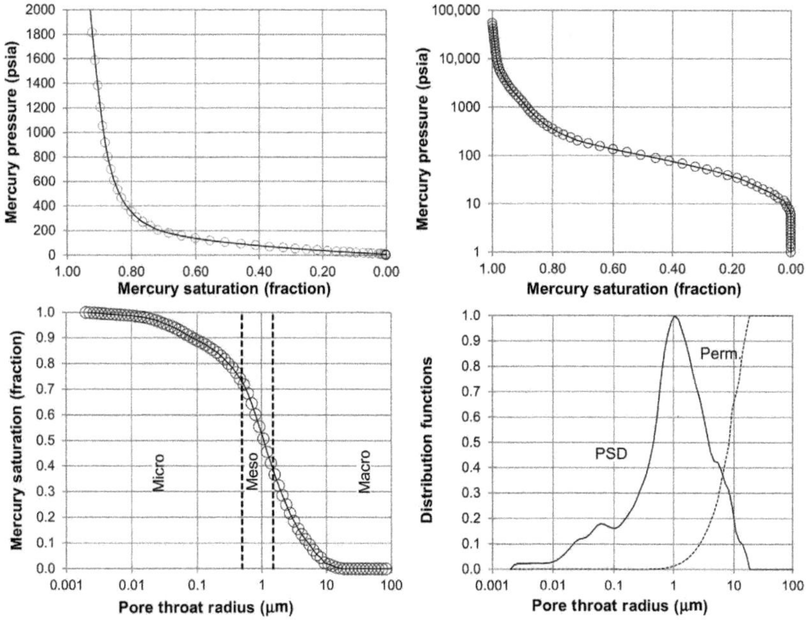

FIGURE 9.30 Example graphical output from an HPMI test.

9.5.5 Typical Output

Examples of typical laboratory data presentations are shown in Fig. 9.30 (graphical) and Fig. 9.31 (excerpt of tabular data). The graphical output includes: wetting phase or mercury saturation versus pressure on a linear scale (to 2000 psi) and on a logarithmic scale (to maximum injection pressure); and cumulative and incremental PSD functions.

The tabular data includes the porosity of the plug from which the injection sample was taken and the injection sample porosity (this is determined from the total volume of mercury injected at the maximum pressure). The two should agree within ±0.2 p.u. In this case the porosity difference (2.2 p.u.) signals heterogeneity in the sample, and or errors in the pore volume determination. The fluid properties (density, IFT and contact angle) are used to scale mercury Pc to other fluid systems (see Section 5.4.4.4). The injection pressure cell highlighted in yellow is the lab's interpretation of entry pressure. The "Raw" data is not normally supplied but is essential for QC.

9.5.6 Data Reporting Requirements

As determination of entry pressure is closely linked to saturation calculations and is subjective, it is imperative that the lab provide the necessary data to allow alternative interpretations/recalculations. This includes:

Mercury injection
High-pressure method

Company: XXXX
Well: XXXX
Field: XXXX
Formation: XXXX
Country: XXXX
Job ID: XXXX

Sample identification: 124
Sample depth, feet: XXXX
Permeability to air, mD: 26.40
Plug porosity, fraction: 0.187
Injection sample porosity, fraction: 0.162
Injection sample pore volume, cm³: 1.400
Injection sample bulk volume, cm³: 8.637
Brine density gradient, psig/foot:
Oil density gradient, psig/foot:
Mean hydraulic radius, μm: 2.647
Swanson's parameter: 8.87E−02
FZI: –

Grain density, gm/cc: 2.71
Conformance vol, cm³/g: 0.005
Sample wt, g: 19.600

IFT * Cosine contact angle	Air–brine	Air–oil	Oil–brine	Air–Hg
Lab --->	72	24	42	372
Res --->	50		26	

Injection Pressure, psia	Mercury Satn, Fraction Vp	Equiv. Water Satn, Fraction Vp	Pore Throat Radius, μm	Normalised Pore Size Distribution Function	Equivalent Injection Pressure, psia A/B (Lab)	O/B (Lab)	O/B (Res)	Height Above Free Water, ft.	Normalised Permeability Distribution Function	J Function	Raw data Injection Pressure, psia	Incremental Hg Injected, cm³/g	Pore Radii μm
0.5	0.000	1.000	218.442	0.000	0	0	0	–	1.000	–	0.500	0.0000000000	218.442
0.6	0.000	1.000	195.037	0.000	0	0	0	–	1.000	–	0.560	0.0000000000	195.037
0.6	0.000	1.000	173.367	0.000	0	0	0	–	1.000	–	0.630	0.0009393758	173.367
0.7	0.000	1.000	153.832	0.000	0	0	0	–	1.000	–	0.710	0.0008973909	153.832
0.8	0.000	1.000	136.526	0.000	0	0	0	–	1.000	–	0.800	0.0004864797	136.526
0.9	0.000	1.000	121.357	0.000	0	0	0	–	1.000	–	0.900	0.0003142364	121.357
1.0	0.000	1.000	108.140	0.000	0	0	0	–	1.000	–	1.010	0.0004682646	108.140
1.3	0.000	1.000	83.375	0.000	0	0	0	–	1.000	–	1.310	0.0005520266	83.375
1.5	0.000	1.000	72.332	0.000	0	0	0	–	1.000	–	1.510	0.0002027822	72.332
1.7	0.000	1.000	63.134	0.000	0	0	0	–	1.000	–	1.730	0.0001348860	63.134
2.0	0.000	1.000	55.162	0.000	0	0	0	–	1.000	–	1.980	0.0001438549	55.162
2.3	0.000	1.000	48.115	0.000	0	0	0	–	1.000	–	2.270	0.0001158575	48.115
2.6	0.000	1.000	42.008	0.000	1	0	0	–	1.000	–	2.600	0.0000990713	42.008
3.0	0.000	1.000	36.651	0.000	1	0	0	–	1.000	–	2.980	0.0001162654	36.651
3.4	0.000	1.000	32.030	0.000	1	0	0	–	1.000	–	3.410	0.0000936347	32.030
3.9	0.000	1.000	27.934	0.000	1	0	0	–	1.000	–	3.910	0.0001305533	27.934
4.5	0.000	1.000	24.325	0.000	1	1	0	–	1.000	–	4.490	0.0001616231	24.325
5.1	0.000	1.000	21.249	0.000	1	1	0	–	1.000	–	5.140	0.0001036944	21.249
5.9	0.000	1.000	18.543	0.000	1	1	0	–	1.000	–	5.890	0.0001081270	18.543
6.8	0.002	0.998	16.181	0.057	1	1	0	–	0.937	–	6.750	0.0001568608	16.181
7.7	0.006	0.994	14.111	0.098	2	1	1	–	0.859	–	7.740	0.0002570860	14.111
8.9	0.013	0.987	12.314	0.112	2	1	1	–	0.747	–	8.870	0.0004823278	12.314
10.2	0.017	0.983	10.708	0.150	2	1	1	–	0.687	–	10.200	0.0003374759	10.708
11.7	0.023	0.977	9.335	0.246	2	1	1	–	0.633	–	11.700	0.0004022471	9.335
13.4	0.043	0.957	8.151	0.313	3	2	1	–	0.486	–	13.400	0.0014499156	8.151
15.3	0.055	0.945	7.139	0.340	3	2	1	–	0.420	–	15.300	0.0008453624	7.139
17.6	0.073	0.927	6.206	0.390	3	2	1	–	0.346	–	17.600	0.0012521651	6.206
20.1	0.095	0.905	5.434	0.421	4	2	1	–	0.275	–	20.100	0.0015758965	5.434
23.1	0.114	0.886	4.728	0.426	4	3	2	–	0.227	–	23.100	0.0013878215	4.728

FIGURE 9.31 Example tabular output from an HPMI test.

- brief description of procedures and experimental apparatus;
- sample size and shape;
- plug or parent plug (if end-trim) porosity (fraction) and permeability (mD);
- sample weight (g);
- injection pressure (psi);
- incremental injected mercury volume (cc);
- calculated cumulative injected mercury volume (cc);
- pore diameter (μm);
- any corrections applied when changing from low pressure to high pressure penetrometers;
- closure (conformance) pressure (psi) and closure conformance volume;
- closure corrected calculated cumulative injected mercury volume (cc);

- closure corrected calculated non-wetting (mercury) saturation (fraction);
- closure corrected calculated wetting (water) saturation (fraction);
- closure corrected calculated pore volume (cc);
- closure corrected calculated bulk volume (cc);
- closure corrected calculated porosity (cc);
- calculated mean hydraulic radius (μm);
- calculated pore throat radius (micron);
- calculated normalised pore size distribution function;
- "skeletal" density (g/cc). This is a standard output from the HPMI manufacturer's software and is equivalent to grain density.

9.5.6.1 *Mercury MICP Summary*

Advantages: LPM and HPMI	Drawbacks and Issues: LPM and HPMI
• It is a relatively low-cost technique. • Capillary equilibrium is attained after a short period (minutes) not hours (centrifuge) or weeks (porous plate), hence tests are very quick. • The method enables large amount of data to be acquired quickly at relatively low cost. These data should capture the key Pc curve features. • Provides useful input data to properly design pressure steps (entry pressures) for centrifuge and porous plate capillary pressure tests, which last longer and are more expensive but utilise reservoir appropriate fluids. It can also provide information quickly about the presence (or not) of a unimodal system. • High Pc values are attainable (38,000 ft of oil column), much higher than those for porous plate and centrifuge. • The mercury injection technique is well suited for low permeability samples (clastics and carbonates) and samples from high relief reservoirs because of the high injection pressures that can be obtained. • Mercury injection provides essential information on pore network throat sizes and their distribution. • Dual porosity/multi-modal pore systems can be identified given the high resolution of the test (Fig. 9.32). • LPM can accommodate much larger samples (up to 1.5″ dia. × 2.5″ length) than the HPMI system. This gives the advantage of reduced uncertainty in volumes, hence, saturations and lessens the impact of volume errors on calculated saturations. • HPMI equipment can accommodate samples of any shape (up to a maximum of about 10 ml bulk volume) and does not require a calibration run. • High resolution data can be obtained easily in HPMI and provides an excellent good definition of key curve features and pore throat sizes. • Operator intervention is minimal in HPMI	• Air–mercury tests are not strictly capillary pressure tests since there is no strong wetting phase involved (air is assumed to be the wetting phase since mercury is totally non-wetting), and there is no irreducible saturation since all the air will be compressed by the mercury if the pressure is raised high enough. Consequently, in most samples, the "capillary pressure curve" (up to 2000 psi, for example) does not reach an irreducible wetting phase saturation, but will trend to zero wetting phase saturations at high injection pressures. Air–mercury capillary pressure curves therefore tend to be displaced to slightly lower wetting phase saturations at higher pressures when compared to centrifuge or porous plate tests (for example) where an irreducible saturation is effectively attained. This is a particular problem in low permeability rocks and must be addressed in data interpretation. • The method is not representative of the reservoir fluids. • Tests are sensitive to sample size hence any small errors in pore volume or bulk volume measurements can translate into significant errors in saturation and porosity estimation. This is especially important for HPMI systems which utilise much smaller (4 ml bulk volume in the example shown in Fig. 9.33). • If very small samples are used (end-trims) and the reservoir is heterogeneous, the analysis may not be representative. • In shaly formations (especially in formations with pore lining clays), MICP tests will destroy any clay present due to the excessive IFTs between air and mercury (485 dyne/cm), especially at high pressures. This effectively increases pore throat diameters and pore volume, and results in displacement of the Pc curves to lower wetting phase saturations compared to tests in which the clay morphology has been preserved (centrifuge and porous plate).

Mercury withdrawal tests can be carried out on LPM to define the spontaneous imbibition curve by progressively reducing the Pc back to zero. The mercury saturation at 0 Pc is presumed to be the "residual" non-wetting phase (hydrocarbon) saturation (Shr). However this is only relevant for a strongly water wet system and the Shr depends on the initial wetting phase saturation (Swi) which is conventionally determined at 2000 psi capillary pressure.

- Harsh sample preparation to remove any water can also cause clay dehydration and potential collapse.
- The test is destructive (mercury is retained at residual saturation when mercury pressure is reduced back to zero) so that the sample cannot be used for any other tests..
- The method is not valid for unconsolidated samples as high pressures and high IFT forces can disrupt grain/grain contacts.
- A closure (or conformance) correction is required to account for the packing of mercury around the rock sample prior to mercury entering the pore volume. This is usually done by the laboratory so data is normally provided "corrected". The determination of the closure correction is extremely subjective.
- A pressure-volume correction has to be applied for the volume change of the LPM measurement cell caused by increasing the injection pressure. A calibration run, with no sample in place, is usually done before each test (or every two tests).
- LPM operation requires experienced operators. HPMI equipment is automated with minimal operator intervention.
- In HPMI tests on small samples, pore volume determined from helium grain volume and mercury immersion bulk volume is subject to error which can have a large impact on calculated saturations. This means that pore volume is determined by injection at high pressure (in excess of 25,000 psi usually). An example is shown in Fig. 9.33. For the "original" curve (on the right) saturations were calculated based on helium/mercury immersion pore volume. For the "repeat" curve (on the left) saturations were determined from the high pressure mercury-filled pore volume. This is a substantial and wholly unacceptable error.
- The HPMI test cannot be used for spontaneous imbibition Pc curves as the test will initiate at a wholly unrepresentative Swi of zero as the pore volume is filled with mercury at maximum drainage pressure.

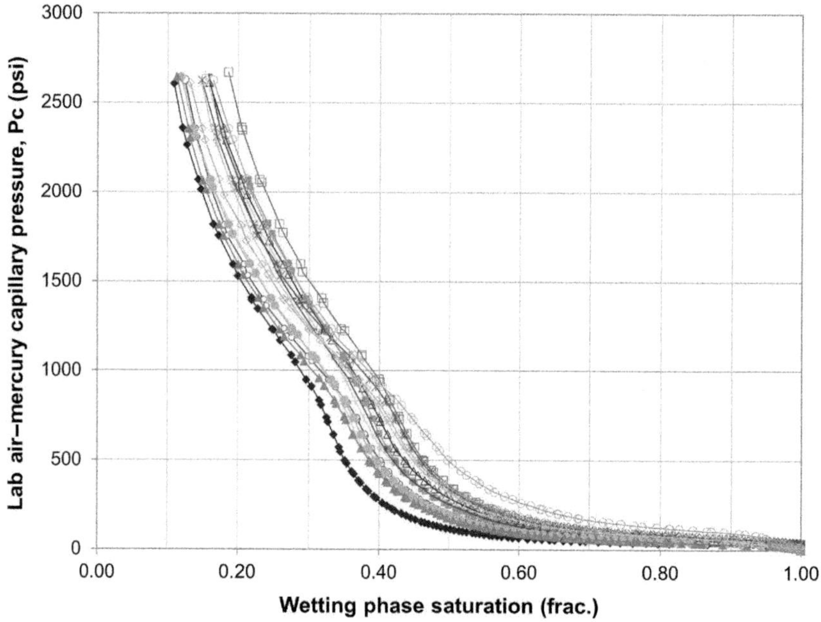

FIGURE 9.32 Example of high-pressure MICP curves with dual porosity systems.

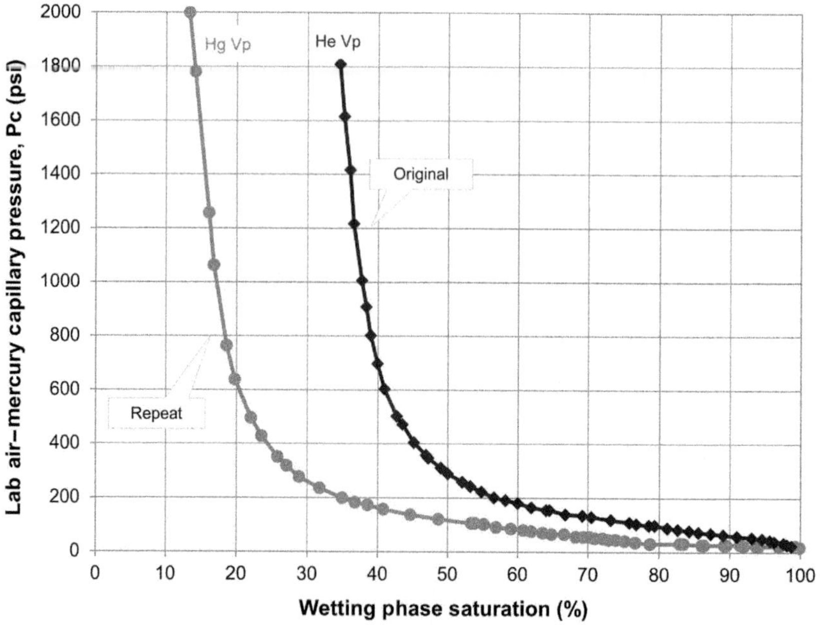

FIGURE 9.33 Impact of pore volume errors in HPMI.

QC Checks (In addition to porous plate and centrifuge)

- Analysis of the results should start with a thorough review of the laboratory report to check that acceptable procedures have been followed. This should be carefully evaluated.
- If the surface of the test sample is rough and, if the largest pore throats are smaller than the voids created by surface irregularities, the latter will be filled with mercury before it enters the true pore system. In a semi-log plot of cumulative injected mercury volume versus injection pressure (Fig. 9.34) (raw data—not closure corrected) this appears as an entry of mercury into the sample observed in the lower values of volume injected. In fact the mercury is closing around, or conforming to the sample. Closure effects are more severe for samples with large external surface areas to volume ratios (e.g. offcut samples) than for larger (plug) samples. The data are corrected by subtracting the volume of mercury associated to closure correction (0.04 ml at 90 psi entry pressure in the example) from the total volumetric injection data at each point. Determining the point at which mercury enters the pore system (entry pressure) is found from the discontinuity in the volume-pressure data. In many cases, the closure correction can be subjective and most labs do not describe the techniques they have used. Therefore it is essential to check the closure (or conformance) correction applied by the lab (appropriate raw data should be provided as

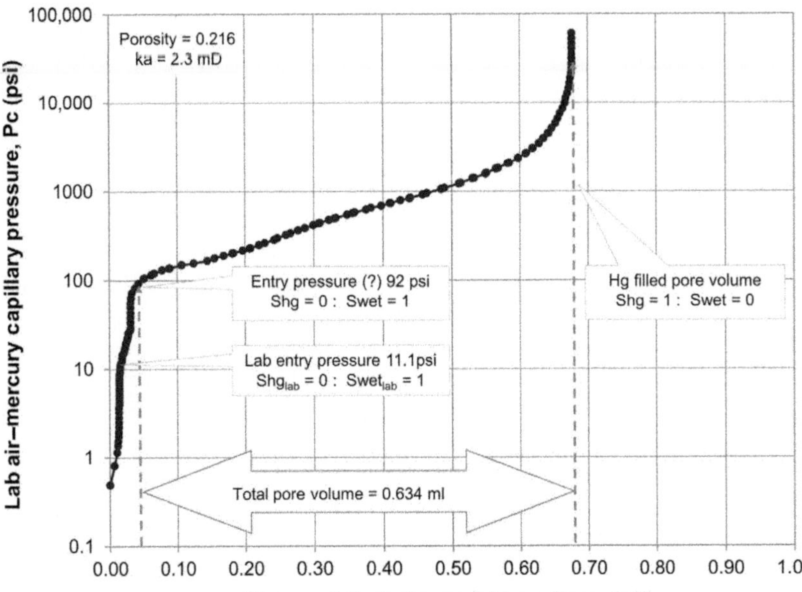

FIGURE 9.34 Example of a volume injected versus injection pressure curve to determine closure correction.

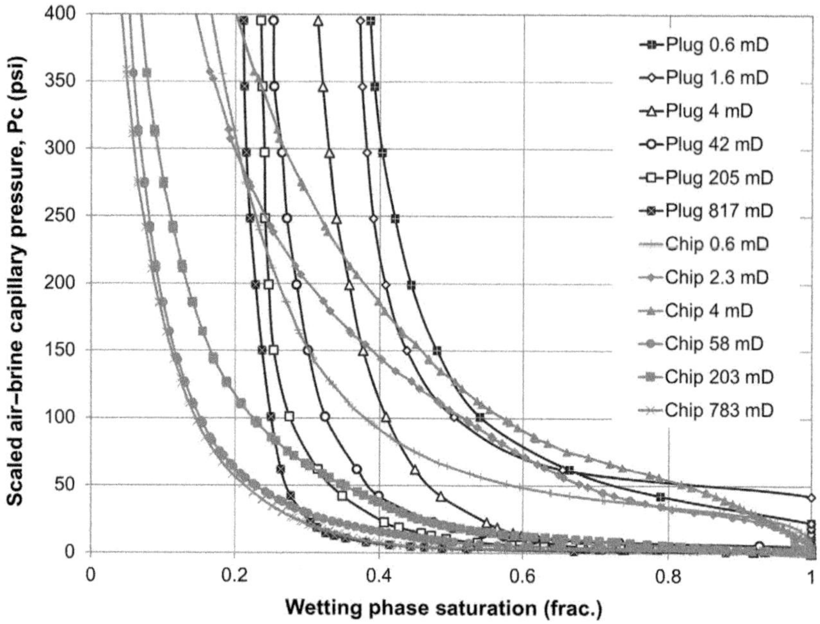

FIGURE 9.35 Capillary pressure curve distortion from high-pressure mercury injection in small chip samples.

shown in Fig. 9.32. Service laboratories rarely provide "raw" data and they will have already made the closure correction prior to submitting the pressure and saturation data to the end-user. In the example the lab has selected a closure correction pressure of 11 psi, whereas a pressure of 90 psi has been selected after re-interpreting the closure correction point. This makes a difference of about 0.02 ml in the volume associated with the closure corrections, which impacts on calculated saturation data.

- If available, check end-points with data from other reliable sources (routine Dean-Stark, other Pc methods, Resistivity Index (at full capillary equilibrium), NMR, etc.). Pc curves are best compared using the J function.
- In certain formations HPMI appears to cause distortion of the capillary pressure versus water saturation curves. In the example shown on Fig. 9.35 mercury injection tests were run using conventional manual equipment on 1.5″ plugs whereas the tests on the "chip" samples were run on 4–6 ml size specimens. In both cases the test plugs were unconfined (3D injection). The distortion phenomena involved are not as yet clearly understood. They may be related to sample size percolation dependencies but they appear worse on samples containing clay-filled micropores which are not normally accessed by mercury at 2000 psi. Continued higher pressure injection appears to progressively and permanently damage the pore system effectively increasing the pore volume during the test. Obviously this has a significant impact when the end user

generates saturation-height curves from HPMI data. If unquestioningly applied in isolation, HPMI Pc curves can result in misleadingly optimistic hydrocarbon saturations.

- In view of the potential for volumetric errors and pore volume alteration, Pc curves from HPMI tests can be erroneous. As a rule of thumb, if the injection sample porosity differs from the parent plug porosity by more than $\pm$ 0.2 p.u. and the plug is not heterogeneous, then the Pc curve generated from HPMI should only be used with caution after comparison with other data.
- For this reason, the LPM method is the preferred method to generate drainage Pc curves. HPMI is preferred for pore size distribution tests.

Suitable Laboratories

- Most commercial SCAL labs now only offer HPMI tests.
- Very few labs continue to provide the LPM service but these labs should be encouraged as it is the preferred method for mercury injection drainage capillary pressure curve acquisition. If overburden is available too, it negates the need to perform empirical corrections for the effect of stress.

9.6 CAPILLARY PRESSURE METHODS: EQUILIBRATION TIMES

The time allowed to achieve capillary equilibrium can completely govern the shape of the resultant capillary pressure curve. The smaller the density difference between the test fluids the longer it takes to achieve equilibrium. Therefore, equilibrium is achieved much more rapidly in mercury injection tests than in air–water or oil–water porous plate or centrifuge tests. It takes about twice as long to achieve Pc equilibrium for an oil–water system than it does for an air–brine system.

In porous plate tests, equilibrium is governed by the rate of brine drainage through the porous plate. This material has extremely low permeability and brine flux rates are extremely low. Figure 9.20 is a good example of the effects of short equilibration times on porous plate data. Clearly, the 4 day per pressure increment is insufficient to drain the sample to equilibrium saturations even for this high permeability sample.

The typical, "industry standard" desaturation times that have been historically used by commercial laboratories are listed in Table 9.2 where they are compared to the recommended minimum stabilisation times. The cessation of production is the preferred criterion, as indicated by the example in Fig. 9.19.

9.7 SUMMARY OF DRAINAGE CAPILLARY PRESSURE METHODS

A summary of the advantages, drawbacks and issues of the various capillary pressure methods is presented in Table 9.3.

TABLE 9.2 Summary of Equilibration Times for Capillary Pressure Tests

Method	Industry Standard Duration	Recommended Duration (Minimum)
Centrifuge	1–2 h (per pressure)	24 h (high permeability—per pressure) 48 h (low permeability—per pressure)
Porous plate	3–4 days (per pressure)	10 days (high permeability—per pressure) 30–40 days (low permeability—per pressure)
Mercury injection	3–4 h (complete curve)	8 h (complete curve)

9.8 DATA CORRECTIONS

Before capillary pressure data can be applied to the reservoir situation, several corrections have to be applied to the data. These are:

- closure correction;
- stress correction;
- clay-bound water correction;
- wettability and IFT correction.

The order in which the corrections are made can make a difference to the calculations. The closure, stress and clay-bound water corrections should be applied before smoothing the data. Corrections for wettability (fluids) and IFTs can be applied after smoothing. When capillary pressure curves measured using different fluids and methods are used, wettability and IFT corrections have to be applied before any data reduction takes place.

In addition, and prior to reservoir saturation-height modelling, it is essential to compare data measured on different samples with different fluid systems using different techniques, often at different times and by different laboratories. In clastics and most carbonate formations the J function can prove to be an excellent comparison tool and key quality-control diagnostic.

9.8.1 Closure Corrections

The majority of the air–mercury data exhibit discontinuities in the volume-pressure curves at high wetting phase saturations, as previously described. An example is shown in Fig. 9.36 where the apparent volume of mercury injected is plotted against injection pressure.

The surface of the sample is rough and, if the largest pore throats are smaller than the voids created by surface irregularities, the latter will be filled with

TABLE 9.3 Summary of Capillary Pressure Tests

Method	Advantages	Drawbacks and Issues
Centrifuge	• Relatively simple and quick • Uses reservoir appropriate (synthetic) fluids • Desaturation is more rapid than for the porous plate • Good for most rock types • Tests at reservoir overburden pressure and elevated temperature are possible • Non-destructive test, re-usable samples.	• It is the most expensive method. • Raw production data is corrected by the test lab to account for the capillary pressure gradient across sample. This is the main source of inaccuracy of the method. Only a few solutions may insure a reasonable accuracy, i.e. below ±3 saturation units (S (Forbes, 1991) and SPL (Nordtvedt et al., 1990)). • High speed centrifuge needed to reproduce pressures at crests of thick column reservoirs • At high spin speeds desaturation affects due to excessive viscous forces can cause changes in the capillary pressure that do not occur under normal field conditions (not capillary dominated flow at the pore level). • The method creates an unusual stress regime; where fracturing is possible, especially at higher spin velocities (more important for friable samples). • Primary drainage curves require the samples to be water wet, for which the cleaning technique used is a critical aspect. • The time allowed to achieve capillary equilibrium can completely govern the shape of the resultant Pc curve. • Stopping and restarting the centrifuge is not accepted as can lead to sample fracturing and produce saturation hysteresis effects. • Benign cleaning is required to preserve delicate clay structures.
Porous plate	**Advantages (general):** • It is the best method/most representative to achieve a uniform saturation profile during the drainage phase. • Uses reservoir appropriate fluids	**Drawbacks and Issues (general):** • The main disadvantage lies in its length of time (average 30–50 weeks) for the description of a complete curve. There are limitations on the maximum attainable Pc values, depending on the system (air–brine or oil–brine).

Continued

TABLE 9.3 Summary of Capillary Pressure Tests—Cont'd

Method	Advantages	Drawbacks and Issues
Porous plate—cont'd	• If the sample is relatively clay rich the method is preferable to others. • Not destructive test, reusable samples. • Test can be run in conjunction with resistivity index. • In general cheaper than centrifuge.	• Curves require the samples to be water wet, for which the cleaning technique used is a critical aspect. • The time allowed to achieve capillary equilibrium can completely govern the shape of the resultant Pc curve. • Benign cleaning is required if the core plugs contain delicate clay structures.
	Advantages (batch cell): • Various samples can be run simultaneously. • Generally cheaper than individual cell porous plate. • Test can be run in conjunction with resistivity index.	**Drawbacks and Issues (batch cell):** • Batch processing does not allow for individual sample desaturation monitoring. • Sample grain loss, due to repeated removal of the samples from the cell to measure water saturation gravimetrically. • Loss of capillary contact, which occurs due to poor contact between the sample and the porous plate via the capillary bridge • Pressure steps applied will be the same for all samples. Processed samples should have similar permeabilities. If not the full Pc curve and key curve features may not be defined/captured. • Only air–brine tests can be run. • Not suited for unconsolidated samples die to grain loss.
	Advantages (individual cell): • Tests can be run at overburden conditions and at elevated temperatures. • Loss of capillary contact is normally not an issue. • Saturations (and resistance) can be monitored in line for each individual sample. • Suitable for air–brine and oil–brine systems • Test can be run in conjunction with resistivity index, and resistivity monitoring of individual samples can help assess saturation equilibrium. • This is the preferred method for both consolidated and unconsolidated samples as overburden pressure can be applied and the test protocols minimise handling and optimise saturation determination and stability.	**Drawbacks and Issues (individual cell):** • Only 1 sample can be run in a core-holder at a time. • Generally more expensive than the batch cell, as a dedicated core holder is required for each sample. • Accurate determination of SCP is an essential element for porosity at stress calculation. • Oil-brine tests using crude or STO are susceptible to wettability alterations where water wet samples could become increasingly less water wet during the desaturation process, effectively affecting the test results and influencing resistivity response.

Mercury injection	**Advantages (general):** • It is a relatively low-cost technique. • Capillary equilibrium is attained after a short period hence tests are very rapid. • The method enables large amount of data to be acquired quickly at relatively low cost. • Provides data on pore throat size distribution • Provides useful input data to properly design pressure steps (entry pressures) for centrifuge and porous plate tests. • The mercury injection technique is well suited for low permeability samples (clean clastics and carbonates) and samples from high relief gas reservoirs. • Mercury injection gives a good understanding of the pore networks which can be interpreted into distributions of pore access radius. • Dual porosity systems can be identified given the high resolution of the test	**Drawbacks and Issues (general):** • Air–mercury tests are not strictly capillary pressure tests since there is no strong wetting phase involved and "irreducible saturation" is never attained—the wetting phase saturation will be zero if the pressure is raised high enough. • Not suited for samples with sensitive and/or reactive clays: CBW is eliminated and clays can collapse during high temperature drying; and the high IFT forces (485 dyne/cm) during testing can bulldoze and destroy pore lining and filling clay structures. • The method is not representative of reservoir fluids. • Tests are sensitive to sample size hence any small errors in pore volume or bulk volume measurements can translate into significant errors in saturation and porosity estimation. • The method assumes a strong water-wetness. • The test is destructive so that the sample cannot be used for any other tests. Mercury is retained in the sample. • The method is not recommended for unconsolidated samples. • A closure (or conformance) correction is often required. This is subjective and lab raw data are required to confirm or re-interpret the closure corrections and saturations.
	Advantages: LPM • Can accommodate standard SCAL plugs so minimising impact of volume measurement errors. • As pressures are lower than the automated system (2000 vs. 60,000 psi) there is less risk of pore system damage and distortion of Pc curves. • Better system than HPMI for spontaneous imbibition curves. • It is the recommended method for acquiring mercury injection capillary pressure curves.	**Drawbacks and Issues: LPM** • A pressure-volume correction has to be applied for the volume change of the measurements apparatus caused by increasing the injection pressure. • Rarely run at overburden stress. • Results can depend on experience and expertise of the operator.

Continued

TABLE 9.3 Summary of Capillary Pressure Tests—Cont'd

Method	Advantages	Drawbacks and Issues
Mercury injection—cont'd	**Advantages: HPMI** • Ideally suited for small (maximum 10 ml) and irregularly shaped samples. • Does not require a calibration run for volume/pressure corrections. • High resolution data can be obtained which helps definition of entry pressures and key curve features • The tests are run automatically and so operator intervention is minimal. • Pressure can be raised up to 60,000 psi enabling mercury to fill the connected pore volume.	**Drawbacks and Issues: HPMI** • Sample size is much smaller than with LPM systems—for the same porosity pore volumes are typically 10–14 times lower than for standard (1.5″ × 2.5″) SCAL plugs. Any small errors in pore volume or bulk volume measurements can translate into significant errors in saturation and porosity estimation. • For this reason, tests are run at high pressure to define the mercury-filled pore volume. This can result in damage to the pore system in certain lithologies. • If very small samples are used (end-trims) and the reservoir core is heterogeneous, the analysis may not be representative.

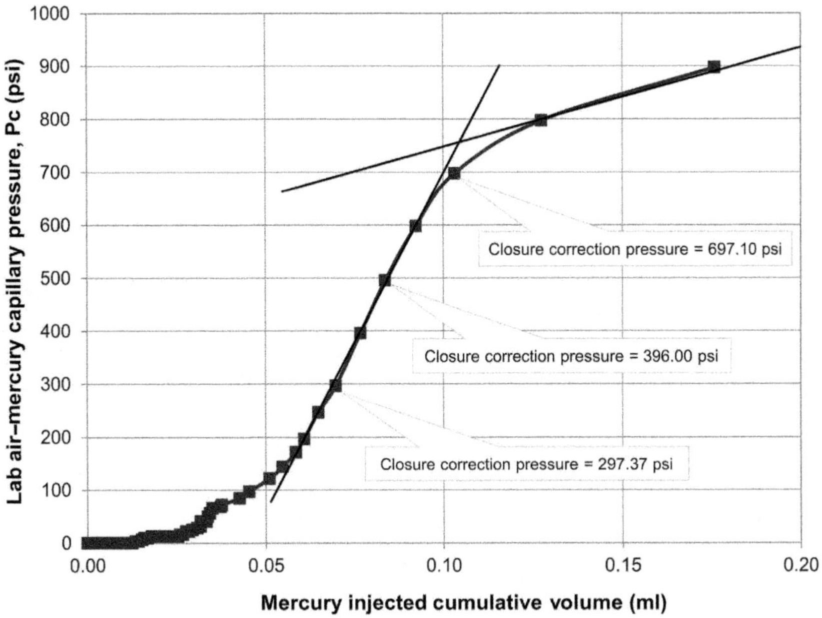

FIGURE 9.36 Closure correction uncertainty from cumulative mercury injection raw data.

mercury before it enters the true pore system. On the capillary pressure curve, this appears as an entry of mercury into the sample. In fact the mercury is closing around, or conforming to, the sample. Closure effects are more severe for samples with large external surface areas to volume ratios, so that tests run on small samples have larger closure effects than large samples, and for samples with significant surface voids or vugs.

The data are corrected by subtracting the apparent volume of mercury injected at the defined entry pressure from the total volumetric injection data. Determining the point at which mercury enters the pore system is found from the discontinuity in the volume-pressure data. As discussed above, service laboratories rarely provide the "raw" data shown on Tabular data provided in Fig. 9.31 and they have already made the closure correction prior to submitting the pressure and saturation data to the end-user. Figure 9.36 is a good example of how subjective the closure correction can be: three different entry pressures have been interpreted by different analysts. The subjectivity can be reduced by plotting incremental as well as cumulative mercury volume versus pressure, and by using a semi-log scale.

It must be emphasised again, that uncorrected data must be reported by the laboratory contractor—their corrections may be subjective and possibly invalid.

9.8.2 Stress Corrections

A rock at reservoir conditions will be subjected to a different stress system then a rock at laboratory conditions. Stress relief at atmospheric conditions results in an increase in both porosity and permeability, and affects capillary curve-derived saturations. The pore entry pressure is effectively reduced. As permeability decreases, the effects of stress on capillary curves become more pronounced. For very low permeability samples, it is preferable to use stressed capillary curves.

Centrifuge and porous plate equipment has been developed that allows capillary curves to be measured on stressed samples. This is not normally possible for LPM mercury injection equipment and impossible for HPMI equipment.

Ideally, capillary pressure should be determined at representative stress and needs to be investigated for each reservoir. If not, ambient data can be corrected on the assumption that the porosity-permeability relationship and trends are similar at stressed and unstressed conditions. The Juhasz method uses the ratio of stressed to ambient porosity to correct the data:

$$\mathrm{Pc}^* = \mathrm{Pc}\left[\frac{\phi_{\mathrm{res}}}{\phi_{\mathrm{lab}}}\right]^{-0.5} \rightarrow S_{\mathrm{nw}}^* = S_{\mathrm{nw}}\left[\frac{\phi_{\mathrm{res}}}{\phi_{\mathrm{lab}}}\right]$$

where, Pc is ambient capillary pressure, Pc* is stress-corrected capillary pressure, S_{nw} is ambient non-wetting phase saturation, S_{nw}^* is stress-corrected non-wetting phase saturation, ϕ_{lab} is total porosity as measured in the laboratory (under ambient conditions) and ϕ_{res} is stress-corrected porosity (e.g. from zonal porosity compaction factor).

9.8.3 Clay-Bound Water Corrections

The presence of clays will also have an effect on capillary curve behaviour. The delicate structure of some clays can be easily damaged by cleaning and drying which influences the micro-porosity and therefore the measured capillary curve. These clays can be damaged (collapsed) on sample preparation for mercury injection tests. The presence of clays also reduces the pore throat radius and hence has the effect of increasing the pore entry pressure. Clays can be preserved by suitable cleaning methods and critical point drying, but this may be redundant in capillary pressure measurements as the liquid/liquid or liquid/vapour IFT in capillary pressure experiments can destroy fine clay structures anyway.

At reservoir conditions, water that is associated with the clay minerals as bound water exists in the pore spaces. This Clay-Bound Water (CBW) is removed during the cleaning and drying of the sample prior to the mercury MICP measurement.

The effect that this CBW has on the capillary curve is largely dependent on the wetting fluid used to obtain the capillary curve. If the wetting phase does not interact with the clay, then CBW is not accounted for, and the capillary curve generates water saturations that are too low. This is the case for mercury/air capillary curves. Hence, a particular point on these curves has to be corrected to both a higher pressure and higher wetting phase saturation air–brine and oil–brine capillary curves do incorporate the effect of CBW, at least when the correct brine salinity is used, and the samples have been prepared to preserve the clay structures and CBW.

A correction for mercury–air capillary curves on sands with an appreciable clay content may be determined if the air/brine capillary curves have been measured as well. The difference in saturations at a given pressure between these two capillary curves is considered a measure of the amount of CBW. A height dependent CBW correction can be derived. This is valid as long as the only difference between the mercury/air and air/brine curves is the CBW correction. There may also be a rock quality dependency.

The correction is normally applied to mercury injection curves where the sample has been dried and the CBW removed. This is described by Hill et al. (1979) and assumes a uniform distribution of CBW over all the pores. This method is based on an empirical relationship between the amount of CBW and Qv, derived from a limited set of samples. The cation exchange capacity (CEC) and the water salinity are required for the correction.

$$\mathrm{Pc}^* = \mathrm{Pc}\left[\frac{\phi_\mathrm{e}}{\phi_\mathrm{t}}\right]^{-0.5} \rightarrow S_{\mathrm{Hg}}^* = S_{\mathrm{Hg}}\left[\frac{\phi_\mathrm{e}}{\phi_\mathrm{t}}\right]$$

where, Pc is measured capillary pressure, Pc* is clay-corrected capillary pressure, Shg is measured mercury saturation, Shg* is clay-corrected mercury saturation, ϕ_t is total porosity as measured in the laboratory, ϕ_e is effective porosity, usually estimated using:

$$\frac{\phi_\mathrm{e}}{\phi_\mathrm{t}} = 1 - \left(0.6425 \times \mathrm{Sal}^{-0.5} + 0.22\right) \times \mathrm{Qv}$$

where, Sal is formation water salinity in g/l NaCl equivalent, and Qv is cation exchange capacity per total pore volume, in meq/ml. A suitable relation may be derived from laboratory data.

The equation shows that the CBW correction decreases with the water salinity. Arguably, mercury injection tests should not be carried out on shaly sands in any case as clay structure damage (which is not accounted for in the CBW correction) will have a significant impact on the pore network.

Figure 9.37 provides an example of the generic stress and CBW correction on a mercury injection dataset.

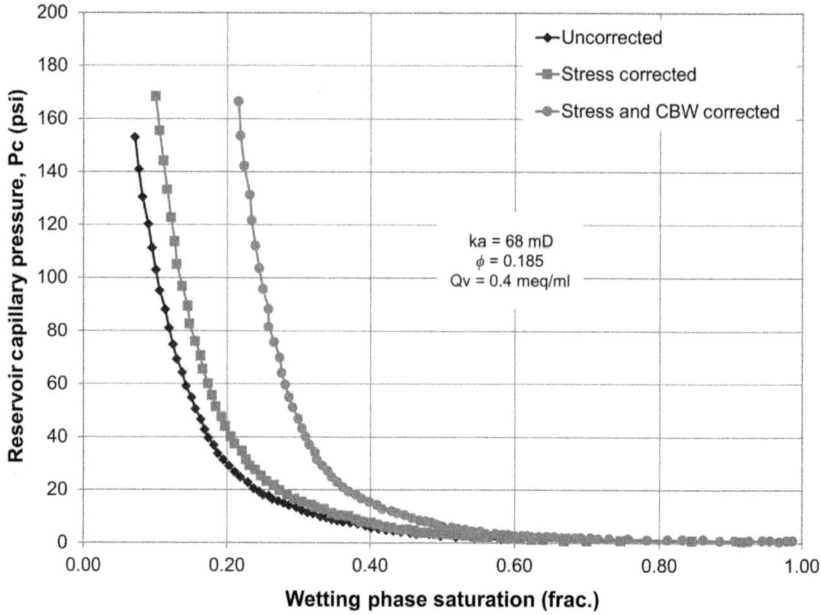

FIGURE 9.37 Effects of stress and CBW correction on an ambient condition Pc curve.

9.8.4 Fluid Corrections

Different fluids to the actual reservoir fluids are used in the capillary pressure measurements. Therefore, when converting measured laboratory capillary curves to the reservoir conditions, differences in rock-fluid interaction, IFT and contact angle have to be considered. The corrections are based on the assumption that capillary pressure in real rocks can be modelled as a bundle of capillary tubes. Capillary pressure is taken to be proportional to the IFT and the cosine of the contact angle. Consequently, lab data measured using an air–mercury (A–Hg) system for example can be converted to an oil–water (O–W) system at lab conditions using the following conversion:

$$\mathrm{Pc}_{(\mathrm{o-w})} = \frac{(\sigma\cos\theta)_{(\mathrm{o-w})}}{(\sigma\cos\theta)_{(\mathrm{A-Hg})}} \mathrm{Pc}_{(\mathrm{o-w})}.$$

Similarly, lab data can be converted to reservoir conditions:

$$P_{\mathrm{c,res}} = \frac{(\sigma\cos\theta)_{\mathrm{res}}}{(\sigma\cos\theta)_{\mathrm{lab}}} P_{\mathrm{c,lab}}$$

Typical, and widely used, values for σ and cos θ for the various lab and reservoir systems are given in Table 9.4. As it is assumed that primary drainage curves are representative of an initial water-wet formation condition contact angles of 0° and 30° are assumed for gas–water and oil–water systems, respectively.

TABLE 9.4 Typical Values for σ and Cos θ for the Various Systems

	System	Cos θ	σ (dyne/cm)	σ Cos θ (dyne/cm)
Laboratory	Air/brine	1.0	72	72
	Oil/brine	0.866	48	42
	Mercury/air	0.765	480	367
Reservoir*	Oil/brine	0.866	30	26
	Gas/brine	1.0	50	50

Whereas the lab "values" may be reasonably well defined and do not vary that much if nitrogen and synthetic oil and brines are used, the reservoir values can vary considerably depending on fluid type, composition, pressure and temperature. However, IFT in the lab may be affected by impurities in the fluids. Labs attempt to control these, but the addition of multivalent cations in lab brines may affect lab IFT values. It is recommended to measure lab fluid pairs under lab test conditions as well as reservoir fluid pairs, under reservoir conditions (though not many labs offer this service). This can lead to considerable error of *in place* estimates based on core-derived saturation-height functions. Figure 9.38 provides an indication of the variation in oil–water IFT from a literature review. This does not take into account differences in fluid properties and pressure and temperature. Nevertheless the data show a range from 4 to 34 dyne/cm with a P50 of 24 dyne/cm, compared to the "standard" value of 30 dyne/cm. Figure 9.39 provides gas–water IFT measurements on a variety of gas compositions (including N_2 and CO_2 content) as a function of pressure and temperature. Again the measured IFTs are all lower than the standard value.

Consequently, it is strongly recommended to determine IFT at reservoir pressure and reservoir pressure using the actual (or recombined) reservoir fluids. There are three main methods used to determine IFT in the PVT laboratory:

- Pendant drop
- Du Noüy ring (Noüy and Lecomte, 1925)
- Wilhelmy plate

These are discussed in Chapter 6. The method that is considered to be the best is the pendant drop method.

Care should be taken when using/reviewing IFT data. IFT changes with time from first contact of the fluids until a stable/equilibrated fluid system is obtained—even if fluids have been brought into equilibrium outside of the test equipment. This may be due to differences in dynamic and static IFT. As an example, an equilibrated oil–water system may produce IFT in the order 35–40 dyne/cm within the first few seconds of test contact, but

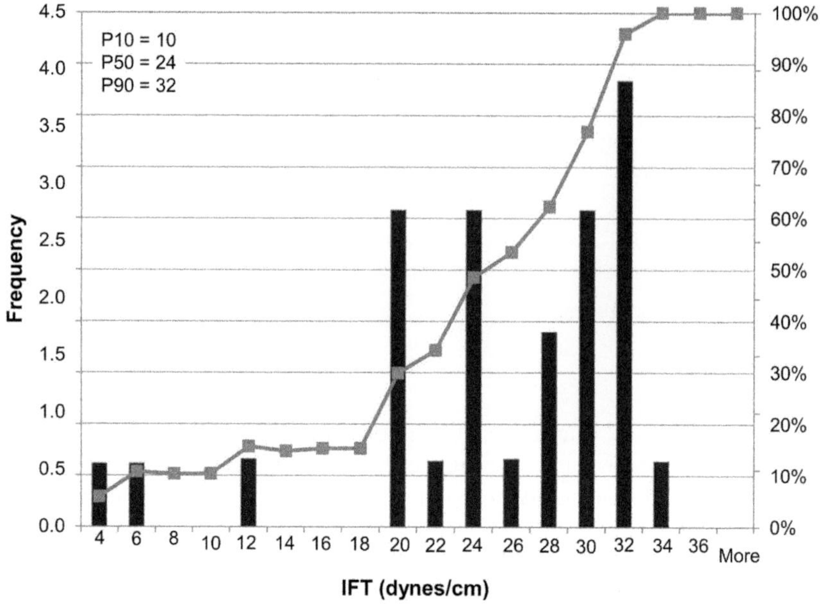

FIGURE 9.38 Oil–water IFT variation.

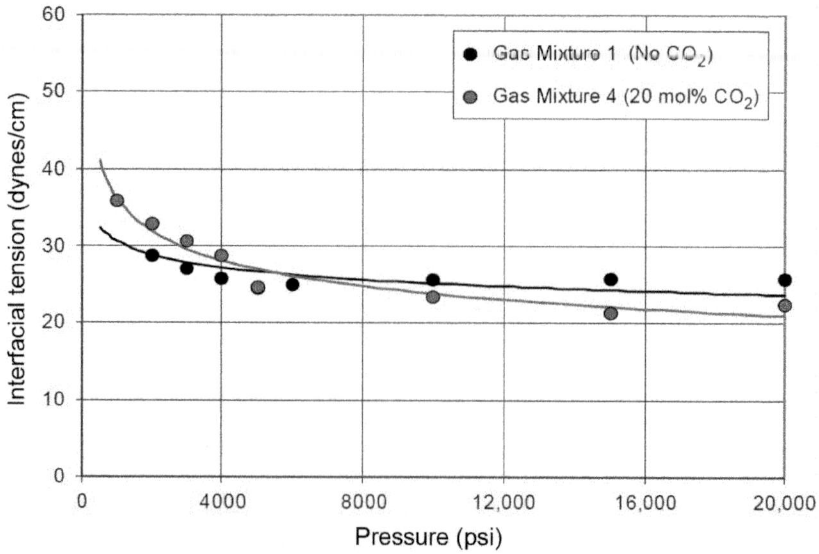

FIGURE 9.39 Gas–water IFT for various gas mixtures (CO_2 and N_2 content). *From Rushing et al. (2008).*

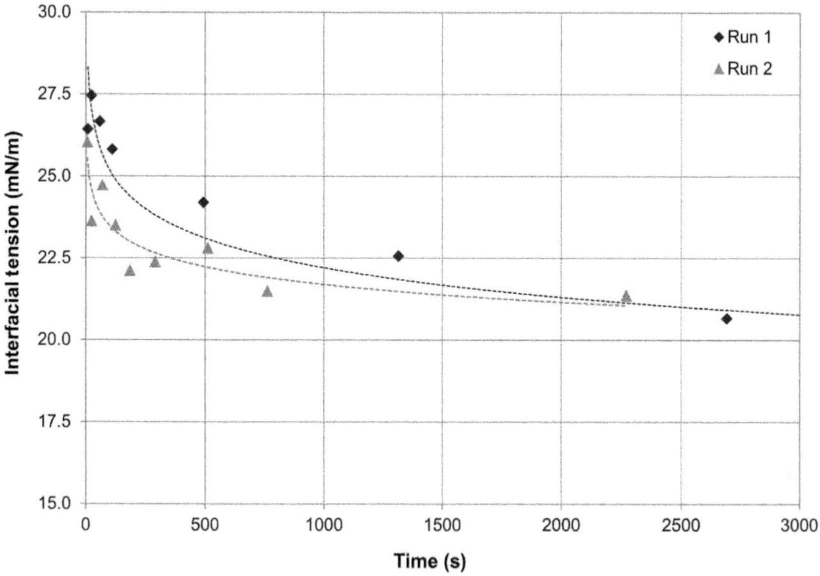

FIGURE 9.40 IFT as a function of time during measurements. *From Rushing et al (2008).*

decreases (as a decay curve) to 35–30 dyne/cm over the course of a few hours. An example is shown in Fig. 9.40. Laboratories should be asked to provide information about their procedures.

IFT can also be estimated from fluid compositional data and PVT properties, such as Firoozabadi and Ramey (1988):

$$\sigma = \left[6.95 \frac{\Delta\rho}{\mathrm{Tr}^{0.3125}}\right]^4$$

where, Tr is reduced temperature, $\frac{T}{\mathrm{Tc}}$, T is reservoir temperature (°R), Tc is the mixture critical temperature (°R), and $\Delta\rho$ is water–oil density difference at reservoir temperature (lb/ft^3) and the parachor method proposed by Schecter and Guo (1998).

9.8.5 J Function

The Leverett J method allows capillary pressures to be normalised to account for differences in fluid systems and sample permeability and porosity.

$$J_{\mathrm{Sw}} = \frac{0.2166\mathrm{Pc}}{\sigma\cos\theta}\sqrt{\frac{\mathrm{ka}}{\phi}}$$

where, Pc is measured capillary pressure (psi), σ is interfacial (surface) tension of the fluid pair (dyne/cm), θ is contact angle (degrees), ka is plug/sample air or Klinkenberg permeability (mD), and ϕ is core porosity (fractional).

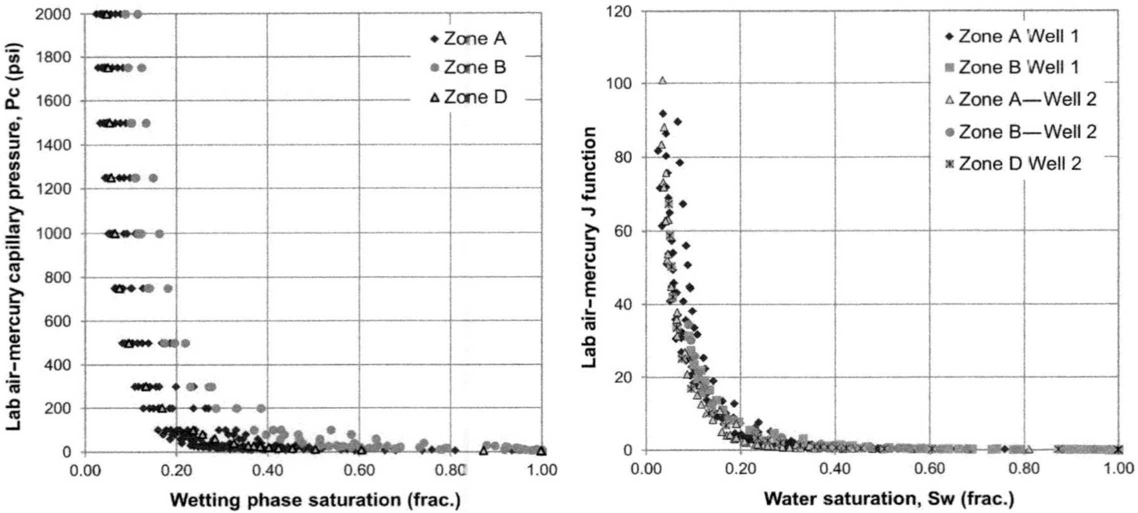

FIGURE 9.41 Capillary pressure and J function curves—well-behaved data.

The normalisation assumes that porosity and permeability are closely related. The J function is then plotted against wetting phase saturation (Sw). If porosity and permeability are indeed related, a consistent curve results.

Capillary pressure measurements are performed on each core plug and are then converted to J values for each sample and plotted against water saturation. For a set of samples with similar pore size distributions the J–Sw curves should fall on the same trend defined by:

$$\mathrm{Sw} = \frac{a}{J^b}$$

which is linearised as follows:

$$\log(\mathrm{Sw}) = \log a - b\log(J)$$

The Leverett J function assumes a parallel capillary tube model and has been widely used as a correlating group for all capillary pressure measurements using different fluid systems. It is a powerful diagnostic tool. For example, Fig. 9.41 plots the original Pc curves and the J curves for a set of lab data. The Pc curves reflect the variation in sample porosity and permeability but the J curves fall on the same trend. This indicates that the samples have similar pore size distributions or pore geometry and that the lab data are well behaved.

In contrast Fig. 9.42 plots the J function for a set of LPM mercury injection and porous plate tests. The discrepancy between the datasets is systematic

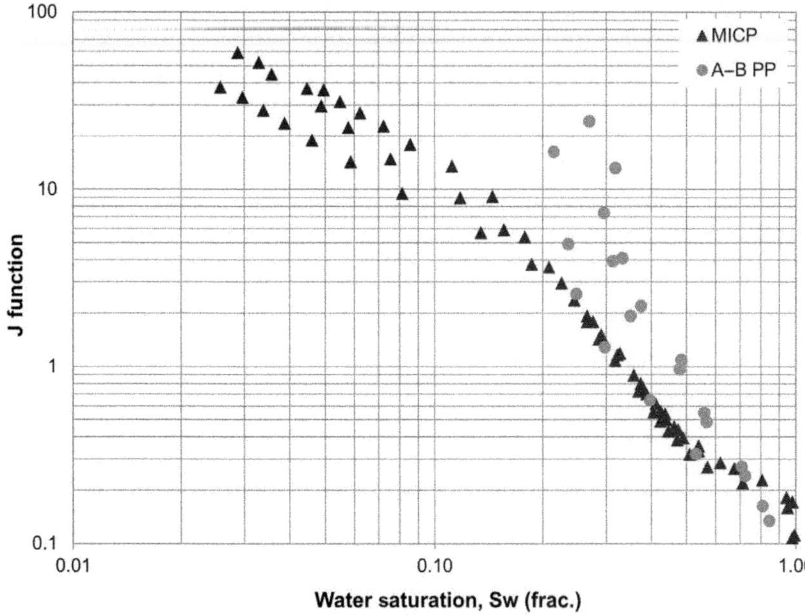

FIGURE 9.42 J function trends highlight un-equilibrated porous plate Pc data.

and could be indicative of a different rock type, but in this case the samples were from the same rock type so the J function indicates lack of equilibrium saturation for the porous plate tests.

REFERENCES

Anderson, W.G., 1986. Wettability literature survey part 1: rock/oil/brine interactions and the effects of core handling on wettability. JPT 38 (11), 1125–1144.

Anderson, W.G., 1987. Wettability literature survey—part 4: the effects of wettability on capillary pressure. SPE JPT 39, 1283–1300.

Brooks, R.H., Corey, A.T., 1964. Hydraulic properties of porous media, Hydraulic Paper Number 3, Colorado State University.

Brooks, R.H., Corey, A.T., 1966. Properties of porous media affecting fluid flow. J. Irrig. Drain. Div. 92 (2), 61–88. Proc. of ASCE.

Dernaika, M., et al., 2010. Drainage capillary pressure and resistivity index from short-wait porous plate experiments. In: SCA, Halifax, Nova Scotia, Canada, 4–7 October, 2010.

Firoozabadi, A., Ramey, H.J., 1988. Surface tension of water-hydrocarbon systems at reservoir conditions. J. Can. Pet. Technol. 27 (3).

Fleury, M., Longeron, D.G., 1997. Full imbibition capillary pressure measurements on preserved samples using the membrane technique. In: SCA 9716—Calgary, Canada.

Forbes, P.L., 1991. Simple and accurate methods for converting centrifuge data into drainage and imbibition capillary pressure curves. In: Society of core analysts annual technical conference August 20-22, San Antonio, paper 9107.

Forbes, P.L., 1993. Centrifuge determination of Capillary Pressure - Review of data analysis procedures. Workshop. SCA - 7th Annual Technical Conference of the Society of Core Analysts, Houston, Aug. 9–11, 1993, Proceedings.

Forbes, P.L., 1994. Simple and accurate methods for converting centrifuge data into drainage and imbibition capillary pressure curves. Log. Anal. 35 (4), 31–53.

Forbes, P.L., 1997. Centrifuge data analysis techniques: an SCA survey on the calculation of drainage capillary pressure curves from centrifuge measurements. In: Society of Core Analysts, SCA-9714.

Hagoort, J., 1980. Oil recovery by gravity drainage. Soc. Petr. Eng. J. (June), 139–150.

Hassler, G.L., Brunner, E., 1945. Measurements of capillary pressures in small core samples. Trans. AIME 192, 10.

Hill, H.J., Shirley, O.J., Klein, G.E., 1979. Bound water in shaly sands—its relation to Qv and other formation properties. Log. Anal. 20 (3), 3–19.

King, M.J., Flazone, A.J., Cook, W.R., Jennings Jr., J.W., Mills, W.H., 1986. Simultaneous determination of residual saturation and capillary pressure curves utilizing the ultracentrifuge. In: Paper SPE 15595 Presented at the 1986 SPE. Annual Technical Conference and Exhibition, New Orleans, Louisiana, October 5-8.

Lasswell, P.M., 2005. Laboratory analysis of electrical rock properties and capillary pressure in full diameter core samples. In: International Symposium of the Society of Core Analysts, Toronto, Ontario, Canada, Paper 2005-75.

McPhee, 2012. The core analysis elephant in the formation evaluation room. In: Paper SPE 158087 Presented at the SPE Annual Technical Conference and Exhibition held in San Antonio, Texas, USA, 8-10 October 2012.

Newsham, K.E., et al., 2004. A comparative study of laboratory techniques for measuring capillary pressures in tight gas sands. In: Paper SPE 89866, SPE Annual Technical Conference and Exhibition, Houston, TX, September 2004.

Nordtvedt, J.E., Watson, A.T., Mejia, G., Yang, P., 1990. Estimation of capillary pressure and relative permeability functions From centrifuge experiments. SPE unsolicited paper 20805.

Nordtvedt, J.E., Kolltveit, K., 1991. Capillary pressure curves from centrifuge data by use of spline functions. In: SPE Reservoir Engineering Paper, November 1991, pp. 497–501.

Noüy, D., Lecomte, P., 1925. An interfacial tensiometer for universal use. J. Gen. Physiol. 7, 625–633.

Omoregle, Z.S., 1988. Factors Affecting the Equivalency of Different Capillary Pressure Measurement Techniques. Soc. Petr. Eng. J. http://dx.doi.org/10.2118/15384-PA.

Pickell, J.J., Swanson, B.F., Hickman, W.B., 1966. Application of air-mercury and Oil-Air capillary pressure data in the study of pore structure and fluid distribution, Soc. Petroleum Eng. Paper No. 1227.

Rajan, R.R., 1986. Theoretically correct analytic solution for calculating capillary pressures-saturation from centrifuge experiments. In: SPWLA Trans 27th Annual Logging Symposium, June, 1986, Paper O.

Rushing, J.A., et al., 2008. Laboratory measurements of gas-water interfacial tension at HP/HT conditions. In: Paper SPE 114516.

SCA, 1993. Capillary pressure determination using the centrifuge. In: Workshop. SCA—7th Annual Technical Conference of the Society of Core Analysts, Houston, Aug. 9-11, 1993, Proceedings.

Schecter, D.S., Guo, B., 1998. Parachors based on modern physics and their uses in IFT prediction of reservoir fluids. SPE Reserv. Eval. Eng. (June).

Shafer, J., Lasswell, P., 2007. Modeling porous plate capillary pressure production data: shortening test duration and quality controlling data. In: Paper presented at the SPWLA 48th Annual Logging Symposium, Held in Austin, Texas, June 3–6, 2007.

Swanson, B.F., 1977. Visualizing pores and non-wetting phase in porous rocks. In: Society of Petroleum Engineers, Annual Fall Technical Conference. SPE Paper 6857.

Van Domselaar, H.R., 1984. An exact equation to calculate actual saturations from centrifuge experiments. Rev. Tech. INTEVEP. 4 (1), 55–62, Caracas, Venezuela.

Wilson, O.B., Skjaeveland, S.M., 2002. Porous plate influence on effective imbibition rates in capillary pressure experiments. In: International Symposium of the Society of Core Analysts, Monterey, California, September 22-25, Paper 2002-16.

Wilson, O.B., Tjetland, B.G., 2001. Porous plates influence on effective drainage rates in capillary pressure experiments. In: International Symposium of the Society of Core Analysts, SCA 2001-30.

RECOMMENDED READING

Hagoort, J., 1980. Oil recovery by gravity drainage. Soc. Petr. Eng. J. (June), 139–150.

Hirsch, L.M., Thompson, A.H., 1994. Size-dependent scaling of capillary invasion including buoyancy and pore size distribution effects. Phys. Rev. 50, 2069–2086.

Holmberg, K. (Ed.), 2002. Handbook of Applied Surface and Colloid Chemistry". Wiley and Sons, New York.

Lasswell, P.M., 2006. Core analysis for electrical properties. Petrophysics 47 (3), 191–213.

Pittman, E.D., 1992. Relationship of porosity and permeability to various parameters derived from mercury injection-capillary pressure curves for sandstone. AAPG Bull. (Issue 2-February).

Subbey, S., Christie, M., Sambridge, M., 2004. The impact if uncertain centrifuge capillary pressure on reservoir simulation. J. Soc. Ind. Appl. Math. 26 (2), 537–557.

Chapter 10

Relative Permeability

Chapter Outline

10.1 Introduction 520
10.1.1 Section Outline 520
10.1.2 Definitions 521
10.1.3 Irreducible Water Saturation 523
10.1.4 Residual Oil Saturations: Water–Oil System 525
10.1.5 Residual Oil Saturations: Gas–Oil System 527
10.1.6 Residual Gas Saturation: Water–Gas System 528
10.1.7 Critical Gas Saturation: Gas–Oil System 530
10.1.8 Critical Gas Saturation: Water–Gas System 530
10.1.9 Critical Water Saturation 532
10.1.10 Critical Oil Saturation: Gas Condensate Systems 533
10.1.11 Wettability 535
10.1.12 Corey Exponents 541
10.2 Measurement Techniques 543
10.2.1 USS Tests 543
10.2.2 SS Tests 545
10.2.3 Centrifuge Tests 547
10.2.4 Curve Definition as a Function of Test Method 547
10.2.5 Test Equipment 548
10.2.6 *In Situ* Saturation Monitoring (ISSM) 551
10.3 Test Data Interpretation Methodology 554
10.3.1 Unsteady State 554
10.3.2 Steady State 557
10.3.3 Centrifuge 557
10.3.4 Flow Rate and Capillary End Effects 558
10.3.5 Coreflood Simulation 562
10.4 Sample Selection, Test State and Test Conditions 565
10.4.1 Sample Selection 565
10.4.2 Test State 568
10.4.3 Overburden Stress 568
10.4.4 Initial Water Saturation 569
10.5 Measurement Descriptions 572
10.5.1 Unsteady State Water-Oil Relative Permeability (USSWO) 574
10.5.2 Steady State Water–Oil Relative Permeability (SSWO) 584
10.5.3 Centrifuge Water–Oil Relative Permeability (CentKrWO) 589
10.5.4 Unsteady State Gas–Oil Relative Permeability (USSGO) 592

Developments in Petroleum Science, Vol. 64. http://dx.doi.org/10.1016/B978-0-444-63533-4.00010-X

10.5.5 Steady State Gas–Oil Relative Permeability (SSGO) 598
10.5.6 Centrifuge Gas–Oil Relative Permeability (CentKrGO) 603
10.5.7 Gas–Oil Relative Permeability Curve Generation 606
10.5.8 Unsteady State Gas–Water Relative Permeability (USSGW) 607
10.5.9 Steady State Gas–Water Relative Permeability (SSGW) 610
10.5.10 Unsteady State Water–Gas Relative Permeability (USSWG) 613
10.5.11 Steady State Water–Gas Relative Permeability (SSWG) 618
10.5.12 Centrifuge Water–Gas Relative Permeability (CentKrWG) 622
10.5.13 Residual Gas Saturation: Imbibition Centrifuge Capillary Pressure (Sgr-CentPc) 626
10.5.14 Counter-current Imbibition Residual (Trapped) Gas Saturation (CCI-Sgr) 629
10.6 Critical Gas Saturation (Depressurisation) 634
10.6.1 Sample Preparation 635
10.6.2 Test Equipment 635
10.6.3 Test Procedures 636
10.7 Experimental Data Provision 639
10.7.1 Samples and Fluids (for All Tests) 639
10.7.2 Drainage Capillary Pressure (Desaturation to Swir) 640
10.7.3 Centrifuge Relative Permeability (Water–Oil, Gas–Oil, Decane–Water) 641
10.7.4 Counter-Current Imbibition 642
10.7.5 USS and SS Flood Tests (Water–Oil, Gas–Oil and Water–Gas) 642
10.8 Summary and Recommendations 643
10.8.1 Oil Reservoirs 644
10.8.2 Gas and Gas Condensate Reservoirs 647
References 650
Recommended Reading 651

10.1 INTRODUCTION

10.1.1 Section Outline

Although a wide range of relative permeability tests can be used to obtain engineering parameters for oil and gas field design purposes, the aim of this section is to highlight and explain only those tests which are undertaken on a relatively routine basis across the industry. This document does not address the wider range of more advanced tests and testing conditions which might be used in research programmes, such as three-phase flow measurements, secondary drainage, WAG, SWAG, low salinity flooding, gas-condensate relative permeability, CO_2 flooding and EOR/IOR corefloods.

This section concentrates on those tests most commonly used to determine relative permeability used in flow calculations to predict or match reservoir production and pressure response, that is, the fundamental rock relative permeability parameters and associated residual saturations which might be needed to model:

- water injection;
- gas injection, solution gas drive and gas cap drive;
- gas storage;
- aquifer influx in gas reservoirs;
- mobilisation of trapped gas from below the gas–water contact by depressurisation from an initial reservoir pressure.

The tests described include:

- Water–oil relative permeability, including end points; initial water saturation (Swi) and residual oil saturation (Srow);
- Gas–oil relative permeability, including end points; critical gas saturation (Sgc) and end-point residual oil saturation (Srog);
- Gas–water relative permeability, including end-point residual water saturation after gasflood (Srwg);
- Water–gas relative permeability, including end points; initial water saturation (Swir) and end-point residual or trapped gas saturation (Sgr);
- Critical gas saturation and mobilisation of trapped gas.

10.1.2 Definitions

Absolute permeability (K) is defined as the permeability of the porous medium to a fluid at 100% pore space saturation of that fluid, typically air (Ka) or water/brine (Kw/Kb). Klinkenberg permeability (Kl or K_L) is also an absolute permeability to gas at infinite pressure (a theoretical non-reactive, totally inert liquid). Refer to Chapter 5 for a full definition and explanation of Klinkenberg permeability.

Note: Throughout the chapter, reference is made to water and/or brine permeability. These terms are used interchangeably. The term, "water" refers to formation water (FW), which will be a saline solution of varying salinity dependent upon the particular reservoir system in question. "Brine" is a saline solution, usually of 3–5% salinity or higher. Many FWs are some form of brine (though some may be less saline, i.e. brackish water). In the oil and gas industry, the terms have become virtually interchangeable, though there is division of terminology, i.e., some will invariably use the term "water", while others use "brine".

Effective permeability (Keff) is a measure of the resistance of the porous medium to one fluid phase when more than one fluid phase is present. The effective permeability is a function of the saturation and distribution of the fluid phases.

- effective permeability to oil—Ko′
- effective permeability to water—Kw′—or brine—Kb′
- effective permeability to gas—Kg′

End-point effective permeability refers to the effective permeability of one phase at the residual or irreducible saturation of a second phase. For example:

- end-point effective permeability to oil—Ko′ at Swir (water–oil);
- end-point effective permeability to water—Kw′ at Srow (water–oil) or Kw′ at Sgr (water–gas);
- end-point effective permeability to gas—Kg′ at Swir (gas–water) or Kg′ at Srog (gas–oil).

Relative permeability (kr) is one permeability value normalised to another permeability value, thus scaling relative permeability from zero (0) to a maximum of one (1). It is normal convention, especially in literature, to normalise values to an absolute permeability. The value of absolute permeability used to normalise effective permeability is referred to as the base permeability, and the application of use will determine which absolute permeability value will be used. However, it must be noted absolute permeability is not always used as the base permeability; in fact, many laboratories will use an effective permeability so that the resultant graph passes through kr = 1. Thus, care must be taken when using relative permeability data to determine what value has been used as base permeability and to re-scale all relative permeability values, if necessary. Thus, base permeability may be:

- air or Klinkenberg permeability $kr = \frac{Keff}{Ka}$ or $kr = \frac{Keff}{Kl}$
- water (or brine) permeability $kr = \frac{Keff}{Kw}$ or $kr = \frac{Keff}{Kb}$
- end-point oil permeability at Swir $kr = \frac{Keff}{Ko'}$ or;
- end-point gas permeability at Swir $kr = \frac{Keff}{Kg'}$

Normally, effective permeability is less than the absolute permeability, although there are numerous documented cases, especially in strongly wetting systems with significant clays present, where effective permeability of the non-wetting phase at irreducible saturation of the wetting phase can be higher than the absolute wetting phase permeability (Baptist and Sweeney, 1954; Ramstad and Øren, 2010; Tanikawa and Shimamoto, 2006).

Normalisation to the laboratory measured end-point effective permeability (kr_{lab}) is typically used for SCAL reporting purposes (and in various literature). Re-normalisation to absolute permeability is often required before implementation in reservoir models (kr_{model}). To avoid mistakes, it is essential to define the absolute permeability base reference (Kabs) in the model (it may be Klinkenberg permeability at nominal stress, for example) and to establish the relationship between Kabs and the end-point permeability for each lab test prior to re-normalisation. For example, for an oil–water system where the lab has used effective oil permeability at Swi as the base permeability:

$$kr_{model} = kr_{lab}\left(\frac{Ko'}{Kabs}\right)$$

For a gas reservoir, for example, using effective permeability as the base, the end-point gas and water and relative permeabilities are:

$$krg' = \frac{Kg}{Kg'} = 1 \quad \text{and} \quad krw' = \frac{Kw}{Kg'}$$

where Kg′ is the effective gas permeability at Swir.

Using absolute air permeability as the base, the end-point gas and water and relative permeability are:

$$krg' = \frac{Kg}{Ka} < 1 \quad \text{and} \quad krw' = \frac{Kw}{Ka}$$

Relative permeability can be converted using:

$$krg_{(abs)} = \frac{Kg'}{Ka} krg_{(eff)} \quad \text{and} \quad krw_{(abs)} = \frac{Kg'}{Ka} krw_{(eff)}$$

Because fluids interact with one another within a porous media, the sum of relative permeability of the various phases is less than 1. This is due to mutual interference of the fluids when flowing simultaneously. There are a number of reasons for this interference—part of the pore channels available for flow to one fluid is reduced in size by the presence of the other fluid(s); immobilised droplets of one fluid may completely occlude some pores; and adverse capillary forces may obstruct pore constrictions if the pressure gradient is too low to force an interface through the restriction.

Relative permeability decreases from a maximum value to zero as the saturation of the particular fluid phase decreases and reaches residual or irreducible values. This is due to decreasing cross-sectional area available for flow as saturation decreases and to the development of more tortuous flow paths. Relative permeability is, however, a non-linear function of fluid saturation for immiscible fluid flow through porous media. The non-linear relation of relative permeability and saturation arises from competing gravity, viscous and capillary forces within the porous media.

10.1.3 Irreducible Water Saturation

In a water-wet core, initially saturated with water, water is surrounded by oil or gas (non-wetting phase). If the pressure in both water (wetting—Pw) and hydrocarbon (non-wetting) phases (Pnw) is equal, then capillary pressure (Pc) is zero and the non-wetting phase will not enter the pore spaces filled with water (Sw = 1.0). Capillary pressure is defined as Pc = Pnw − Pw and for water saturation decreasing, Pnw outside the pore throat is greater than Pw inside the pore throat, resulting in a positive pressure. If the pressure in the hydrocarbon (HC) phase is increased, the interfaces on all inlet pores will bend inward until a critical curvature is achieved at a critical pressure, where the applied pressure is equal to the capillary forces of the largest pore throats. Upon increasing applied pressure in the HC phase, HC will then move into the largest pore throats (least capillary pressure) forcing water from pores; water saturation decreases. The point of initial entry of the HC phase is called entry (or threshold) pressure. Further increases in HC pressure will result in the invasion of progressively smaller pores. Capillary pressure of a porous media containing a range of differently sized pore throats is exponentially related to saturation; thus at low saturations, enormous increases in pressure are necessary for small changes in saturation. These pressure changes become physically impractical for the gravity drainage process of reservoir systems (based on water–HC density difference and the height of the HC column). At this stage, water saturation comprises of water trapped in the smallest pore throats and a film of water coating the rock surfaces. This is known as "irreducible water saturation", Swir or Swirr. The entire process is defined as "drainage", i.e., decreasing the wetting phase saturation. An example drainage capillary pressure curve (for different rock qualities) is shown in Fig. 10.1.

This specific process from a system initially fully water saturated is called primary drainage, where HCs first migrated into the water-filled (and hence water-wet) reservoir system.

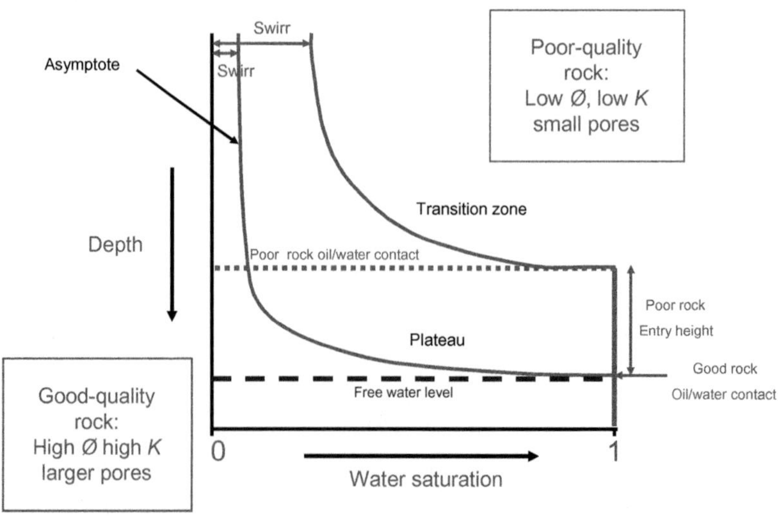

FIGURE 10.1 Example drainage capillary pressure curve.

10.1.4 Residual Oil Saturations: Water–Oil System

Once reservoir HCs have entered the pore system, it is possible, at elevated temperatures, that certain HC components may begin to wet certain surfaces of the rock. Wettability is a complex mechanism which depends on the composition of the rock, composition of HCs, and composition and pH of the FW. The important components of the reservoir HCs involved in wetting are considered to be the natural surfactants (polar components): resins (or NSOs—nitrogen, sulphur and oxygen containing HCs) and asphaltenes.

Alterations to wettability will alter the relationship of capillary pressure to saturation. Subsequent saturation changes (water saturation increasing [imbibition] followed by subsequent water decreasing [secondary drainage]) will not follow the same saturation path as primary drainage. This is known as saturation hysteresis, since water may be no longer the only wetting phase. Although water may not be now the only wetting phase, all language describing subsequent processes assumes water as the wetting phase; hence, imbibition always discusses water imbibition (water saturation increasing) and secondary drainage always describes the subsequent water drainage (water saturation decreasing) of previously imbibed water.

Water saturation has a major controlling impact on wettability; for instance, a 100% water-saturated sample has no contact with HCs and thus cannot be anything other than water-wet. Progressively lower water saturations allow progressively greater access of HCs to the rock surface and thus progressively greater potential for wetting change. The greater the wetting change, the larger the hysteresis effect (i.e. larger difference to primary drainage curve). Wettability is a strong controlling factor on capillary pressure and relative permeability, and it is therefore imperative that reservoir representative wettability is established prior to initiating laboratory flooding experiments. Since water saturation is a controlling factor in establishing wettability, it is vital that reservoir representative water saturation is used in establishing the reservoir representative wettability: this may not be Swir or may not be the lowest water saturation attainable by laboratory applied pressures.

Following primary drainage and after ageing to establish representative wettability, if the pressure in the HC phase (Pnw) reduces, water may flow back under capillary suction or "imbibition". This process is illustrated for an oil–water system in Fig. 10.2 where the drainage curve (positive Pc) is also shown for reference. When the wetting (Pw) and non-wetting phase (Pnw) pressures are equal, $Pc = 0$ and the oil saturation at this point is defined as the oil saturation remaining after spontaneous imbibition (Spo). Unless the water phase pressure is increased ($Pw > Pnw$), no more water enters the core. If Pw is greater than Pnw, then Pc will be negative and water is forcibly imbibed (injected). The section of the Pc curve defining the saturation change as a result of negative capillary pressure is referred to as

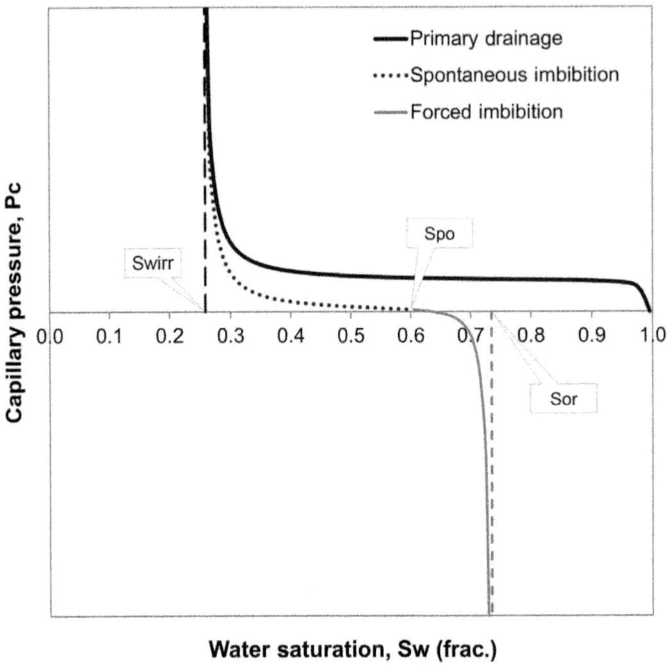

FIGURE 10.2 Residual oil saturations defined by water–oil imbibition capillary pressure.

"forced imbibition", and depending on wettability and displacement mechanism, oil saturation may reduce further—eventually reaching residual oil saturation (Sor in this figure—also referred to as Sro in some lab reports) which is defined by the asymptote of the forced imbibition (negative Pc) curve. Sro is defined by the entire spontaneous and forced imbibition and also depends on Swi (the starting point for imbibition) and the displacement process. In this example, only capillary forces are taken into consideration. The end-point oil saturation as obtained from displacement floods (viscous force dominates) will differ somewhat from those obtained from capillary pressure measurements.

Residual oil saturation depends on a number of factors and has different definitions. It is contingent on the rock type and properties, wettability, initial oil saturation (1 – Swi), the type of test and the test conditions.

There are three principal definitions of residual oil saturation:

1. True residual oil saturation (Srot) is a function only of the rock/fluid system. It is the final oil saturation at the microscopic level that can be achieved under the influence of viscous, capillary and gravitational forces. It is the saturation at which the imbibition capillary curve becomes asymptotic and will provide the microscopic sweep efficiency.

2. Remaining oil saturation (ROS) is the oil saturation that is actually achieved at the end of a field life within the pores contacted and swept by water. The value is dependent on the true residual oil saturation and microscopic sweep but also dependent upon areal sweep efficiency and the number of pore volumes flooded through the pores. Different areas of the reservoir, even in the same rock type, can therefore drain to different remaining saturations, e.g. ROS close to injector wells will be lower than close to producing wells.
3. Residual oil saturation (Srow) is the final value of oil saturation from laboratory waterflood displacement experiments and depends primarily on the flow rate (or differential pressure) applied, permeability and capillary end effects, test methodology and (in many cases) the test laboratory.

Inevitably, residual oil saturations obtained from correctly designed centrifuge imbibition tests are lower, and closer to true residual oil saturation, than from waterflood tests. This results from the relatively higher displacement forces that can be imposed on the sample, which overcome capillary forces and allow the rock to drain to lower oil saturations. The difference between the waterflood Sro and true Srot in intermediate or oil-wet conditions can be substantial, especially in high-permeability rock where viscous forces (differential pressure) are small and oil continues to drain under capillary action. In low-permeability rock, the difference between waterflood Sro and Srot may be smaller due to the imposition of relatively higher differential pressures (viscous forces).

Sro and Srot are also dependent on rock type (pore geometry). However, discrimination of the influence of rock properties is often obscured by experimental influences, such as capillary effects on low-rate waterfloods. Normally, high-permeability rocks are expected to drain to lower Sro than low-permeability rocks.

Wettability exerts a fundamental influence on Sro. For similar rock qualities, lower Sro values are associated with neutral-wet (intermediate) rocks compared to strongly water-wet rocks. Thus, recovery efficiency is at a maximum in intermediate-wet rocks. This is consistent with the concept of maintenance of a continuous oil phase to low oil saturations.

10.1.5 Residual Oil Saturations: Gas–Oil System

Gas-displacing-oil unsteady-state (USS) experiments on short core plugs tend to be dominated by viscous instabilities (too high effective gas flow rate) and capillary end effects (too short core plugs). In addition, many tests use an immiscible gas (nitrogen) as the displacing phase. As a result, the residual oil saturation to gas (Srog) from most laboratory tests is influenced by these lab artefacts.

Generally, in immiscible gasflooding operations, high oil recovery (and low Srog) cannot be attained, especially in waterflood reservoirs. However, in miscible or partially/dynamically miscible flooding, if the relationship of interfacial tensions (IFTs) between the three phases (oil, gas and water) is optimum, thin oil film drainage mechanisms can occur. Oil film drainage phenomena involve the formation of a thin film by residual oil droplets in water-wet media, which becomes continuous between water and gas under certain conditions and consequently allows greater oil recovery.

In the absence of end effects and in the presence of thin film and miscibility effects, gravity drainage occurs and nearly all the oil may drain from the rock with time. Consequently, "true" Srog could be zero or close to zero.

Centrifuge gas–oil displacement is an effective method to study Srog and obtain results approaching Srot by gas. However, an important consideration is to ensure the gravity forces applied do not exceed the forces used to establish initial water saturation, else water will also be mobilised, negating the results. If water is produced during an initial run, the samples should be re-saturated with oil (undersaturated with the test gas), by flooding using pore pressure (back pressure)—at least 75 psi (5 bar). This will dissolve gas into the injected oil allowing it to be produced and replaced with fresh undersaturated oil. The centrifuge experiment can be then easily repeated, though from a different initial water saturation.

10.1.6 Residual Gas Saturation: Water–Gas System

The process of gas displacement by water will lead to forced imbibition in areas of high drawdown and spontaneous imbibition in areas of low drawdown. It is further assumed that in the bulk of the reservoir, spontaneous imbibition will prevail and the reservoir will be water-wet.

Residual gas saturation (Sgr; also known as trapped gas saturation, Sgt) is known to be dependent upon initial gas saturation, Sgi, and pore network characteristics. Correlations between Sgr and Sgi (1 − Swir) have been developed, such as Agarwal (1967): $\mathrm{Sgr} = 0.1813\mathrm{Sgi} + 0.096071$ and Land (1968): $\mathrm{Sgr} = \frac{1-\mathrm{Swir}}{1+C(1-\mathrm{Swir})}$, where C is a constant describing the curvature.

Suzanne et al. (2003) describe how Sgr versus Sgi relationships can be decomposed into two distinct parts: a linear Sgr–Sgi region and a constant Sgr region. The gas saturation corresponding to the breakpoint between the regions is termed Sgo:

$$\text{If Sgi} < \text{Sgo}: \text{Sgr} = \frac{\text{Sgrm}}{\text{Sgo}}\text{Sgi}, \quad \text{else Sgr} = \text{Sgrm}$$

Sgrm is the maximum residual gas saturation obtained from an initial Sgi = 1 (Swir = 0). In effect, this means that up to a certain value of Sg (Sgo), the Sgi–Sgr relationship is linear. Above Sgo, Sgr will be constant.

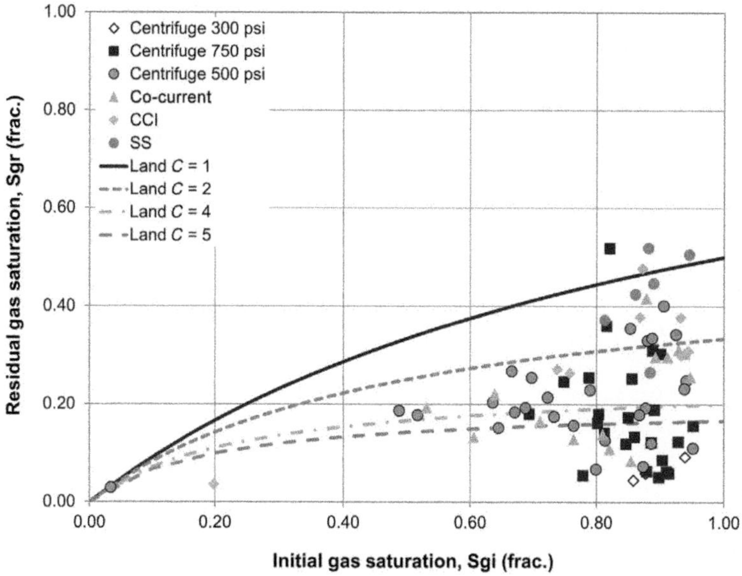

FIGURE 10.3 Typical Sgr versus Sgi relationships.

Unfortunately, attempts to correlate Sgr with rock properties have often proved futile and Sgi-Sgr correlations are often not directly applicable or reproducible in all reservoirs. This is partly a consequence of the different experimental methods used and differences in sample wettability. For example, Fig. 10.3 provides Sgi versus Sgr relationships for different test methods (counter-current imbibition (CCI), co-current imbibition, centrifuge and steady-state (SS) flooding). The Agarwal and Land models (with a range of *C* values) are shown as solid lines.

Unfortunately, it is difficult to give clear and cogent recommendations on the "best" test method to determine Sgr though, arguably, static tests which are dominated by capillary forces (CCI and low-rate co-current imbibition) should provide potentially more representative data than dynamic tests which are dominated by viscous forces (water–gas SS- and USS tests) and may be liable to disequilibria effects. However, these static tests often lead to compression of gas into the middle of the core plug, producing artificially high Sgr. Further details of static and dynamic test procedures, including advantages and disadvantages, are described in Sections 10.5.11–10.5.14.

More recent dynamic tests use elevated pore pressures, usually in the order of 300 psi (20 bar), to reduce gas compressibility. Injection rates should be controlled to maintain low differential pressures (DPs). Initial flow rates reproduce reservoir advancement rates (1–2 ft./day) but if DP begins to surpass 2% of the pore pressure, injection rate is dropped to limit DP. Pressure limits are used to reduce disequilibria effects: dissolution of gas into the water at the higher inlet pressure and subsequent gas dropout at the lower outlet pressure, which ultimately may produce all gas volume but usually will result in Sgr values that are too low. Another aspect of such testing is to limit injection volumes. Disequilibria effects are only minimised in this method, not eradicated; limited injection volumes further limit the extent of disequilibria effects. Since the process is strongly controlled by spontaneous water imbibition, it is expected that production should be all but complete at water breakthrough (BT). Thus, the test should be limited to a maximum of 1–2 pore volumes injected (PVI).

10.1.7 Critical Gas Saturation: Gas–Oil System

The critical gas saturation, Sgc, for a gas–oil system is the gas saturation at which gas bubbles coalesce and gas begins to flow. A number of experimental depletion studies have been undertaken to study depressurisation phenomena, most focussing on evaluating the critical gas saturation at which gas becomes mobile under virgin conditions. These have shown that the critical gas saturation is a power law function of the depletion rate. The critical gas saturation also depends on the gas–oil IFT, which controls the nucleation process. To extrapolate laboratory results to field conditions, it is extremely important that the experiments are conducted with reservoir fluids and reservoir cores under reservoir conditions (temperature and pressure) and with representative wettability conditions.

In the absence of such data which require tests at reservoir conditions using live fluids, Sgc can often be estimated from a logarithmic plot of the krg/kro ratio versus gas saturation. Figure 10.4 plots example krg/kro data from the SCAL tests. Ideally, in semi-log space, the krg/kro curves should become asymptotic at low gas saturations, approaching Sgc. In some cases, Sgc cannot be reliably extrapolated due to the lack of curvature of the krg/kro curves at low gas saturations, but most data indicate that Sgc will be low, approximately approaching Sgc = 0.03 which is a common value normally assumed for gas–oil tests and in simulation.

10.1.8 Critical Gas Saturation: Water–Gas System

Invasion by water above the original gas/water contact leaves potentially irrecoverable reserves in the form of trapped or residual gas saturation (Sgr or Sgt) below the new gas/water contact. This trapped gas saturation could represent significant incremental resources on depressurisation. Overall recovery could depend on the extent to which this trapped gas mobilises during depletion. Knowledge of critical gas saturation and subsequent relative permeability has a significant impact on anticipated gas recovery.

In this context, the accumulation of gas represents a primary drainage case. Gas trapping behind an advancing waterfront (either over geological

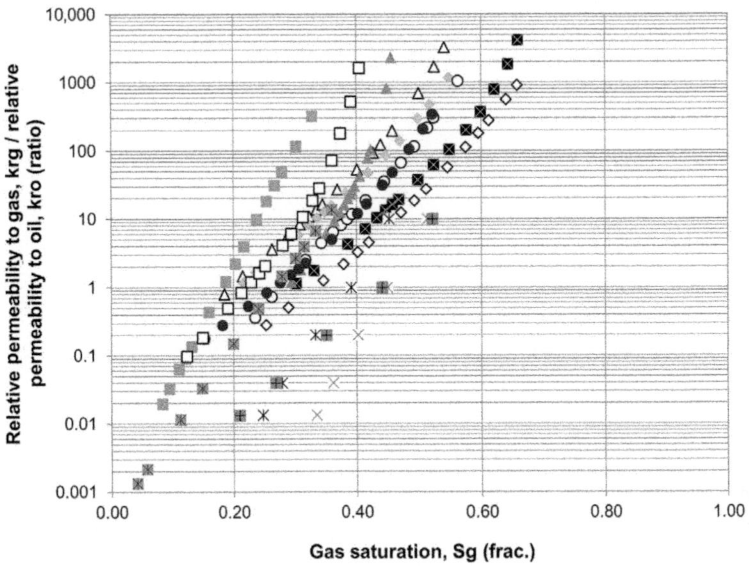

FIGURE 10.4 Example krg/kro versus Sg curves used to estimate Sgc.

time or as a result of reservoir production) represents a primary imbibition cycle. Consequently, mobilisation of trapped gas on blowdown would characterise a secondary drainage case. Typical imbibition and drainage curves for a water-wet system are illustrated in Fig. 10.5. There is hysteresis between the drainage and imbibition gas (non-wetting phase) relative permeability curves but often not in the water curves (wetting phase).

Critical gas saturation is defined as the threshold saturation where the gas mobilises during secondary drainage initiated by depressurisation. Based on the conventionally determined (dynamic flood test) curves exemplified by Fig. 10.5 (where gas is injected into core plugs at Sgr or Sgt to initiate the secondary drainage cycle), it is assumed that gas would become mobile immediately because of gas expansion, and a small increase in gas saturation from the residual or trapped value would produce a high fractional flow of gas. However, blowdown experiments, in which the secondary drainage curve is initiated by direct depressurisation of a core at Sgt/Sgr, have found that trapped gas does not become mobile immediately as it expands but an appreciable increase is required before gas flow can be detected. These critical gas saturations were found to be around 3–15 saturation units higher than the initial trapped gas saturation, and were essentially independent of the rate of depletion. This suggests that water production on depressurisation will be higher and gas production lower than is predicted from conventional dynamic flood methods.

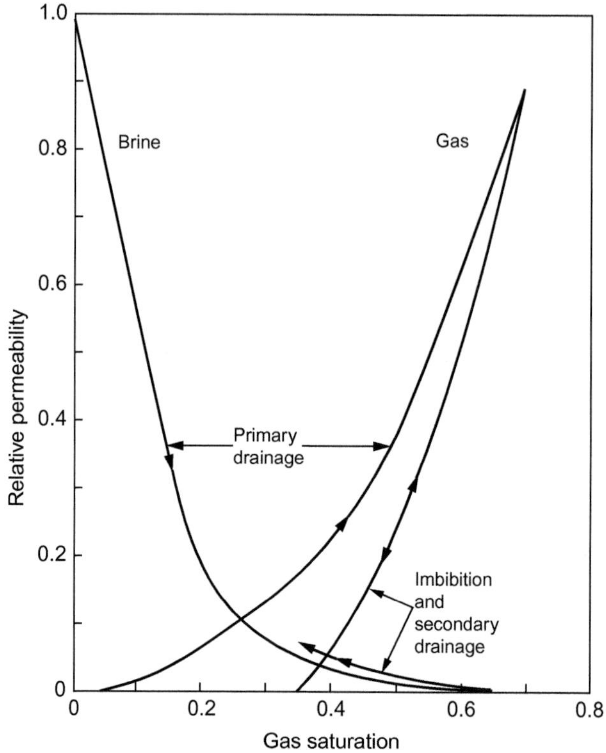

FIGURE 10.5 Example primary drainage and imbibition, and secondary drainage curves for gas–water systems. *After Fishlock et al. (1988).*

10.1.9 Critical Water Saturation

Critical water saturation (Swc) should not be considered in a similar manner to critical gas saturation (Sgc); since, unlike gas saturation (where Sg will increase without flow, due to nucleation and expansion through depressurisation), water flow is never zero, it is merely so low it is not measurable. It is most often used as a "fitting function" in reservoir models, but has no basis on real physical properties. It is therefore not a recommended parameter to use. It also represents the region of a saturation axis where HCs are most mobile. Critical water saturation is the upper limit for FW saturation above which a specific sample will produce a significant percentage of water with HCs. Below Swc, water flow is relatively negligible, on the timescale of the importance for the evaluation of flow properties.

In general, critical water saturation increases linearly with irreducible water saturation. In high-permeability (low Swir) reservoirs, Swc and Swir

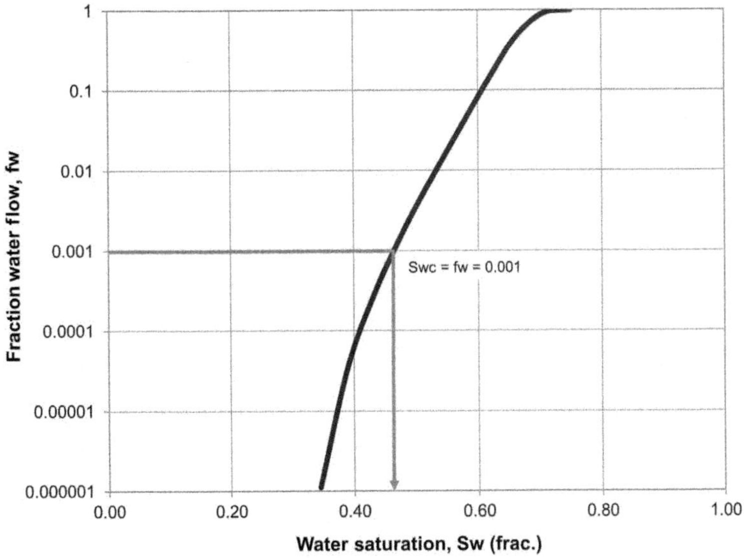

FIGURE 10.6 Water–gas fractional flow curve for Swc determination.

are similar but, in low-permeability reservoirs, Swc can be significantly higher than Swir (this partly explains the observations that well tests in "high" water saturation intervals often produce dry gas).

Swc has been experimentally defined as the saturation at which the ratio of the non-wetting phase (gas or oil) flow to water flow exceeds a given value. For gas reservoirs undergoing imbibition, this critical fractional flow is assumed to be 1000, i.e., the fractional water flow (fw) is less than 0.001. Fractional flow is estimated from the Welge equation for water–gas or water-oil systems:

$$\mathrm{fw} = \frac{1}{1 + \left(\frac{\mathrm{krg}}{\mathrm{krw}}\right)\left(\frac{\mu_w}{\mu_g}\right)} \qquad \mathrm{fw} = \frac{1}{1 + \left(\frac{\mathrm{kro}}{\mathrm{krw}}\right)\left(\frac{\mu_w}{\mu_o}\right)}$$

where μ_g, μ_w and μ_o are gas, water and oil viscosities, respectively.

Figure 10.6 provides an example of the use of water–gas fractional flow curves to determine Swc.

10.1.10 Critical Oil Saturation: Gas Condensate Systems

Pressure depletion in gas condensate reservoirs can lead to the deposition of liquid HCs in the porous media in the near-wellbore region, reducing the effective permeability to gas and the recoverable reserves. The precipitated liquid HC will be drawn into the smallest pore spaces under capillary forces. Critical condensate (oil) saturation (where the condensate becomes mobile and blocks the flow of gas) is a key parameter in the evaluation of the impact of liquid dropout.

Gas-condensate relative permeability is sensitive to rock type, IFT (close to the critical point), capillary forces, gravity forces, wettability, depletion rates, flow velocity, connate water and the method of condensate generation or condensate injection. Tests are therefore usually carried out at reservoir conditions, using reservoir fluids, and require sophisticated experimental equipment and controls. Both pressure reduction and temperature reduction techniques have been used in laboratory experiments to cause condensation within core plugs. The pressure reduction technique most closely resembles the actual reservoir process.

A quicker and lower cost estimate of the critical condensate saturation can be obtained from oil-displacing-gas imbibition tests with the core at Swir. A SS oil-displacing-gas relative permeability test is offered by many commercial labs (without the facilities for reservoir condition controlled pressure depletion tests). Initially, a mixture with a very low proportion of oil and high proportion of gas is used. Figure 10.7 provides an example of an *in situ*

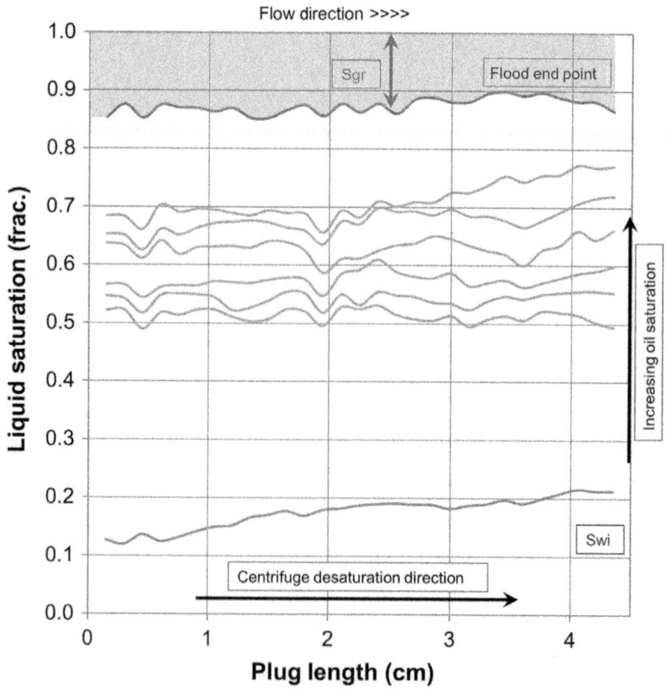

FIGURE 10.7 ISSM saturation profile for steady-state oil–gas test.

saturation monitoring (ISSM) scan for an oil–gas test at lab conditions. The oil ratio in the SS process is increased in stages until 100% oil flow. However, since the condensate-gas relative permeability is dependent upon numerous other factors, these data should be used with care, allowing for some unknown uncertainty.

Unfortunately, the analogue test is compromised by two factors:

1. There is no clear definition of critical oil saturation. Rigorously, critical saturation is defined as "the largest saturation for which relative permeability (kro in this case) is zero", which is neither useful nor practical for taking measurements. Some define the critical saturation as the saturation at which the imbibing phase will "just" begin to flow. The issue here is the definition of "just". Soc can be taken as the first point on the kro imbibition curve but this is partly a consequence of constraining oil and gas flow in the SS technique rather than representing a critical value. It can also be defined as the saturation at which oil is effectively immobile, e.g. the oil saturation where the fractional oil flow is 0.001 (as for critical water saturation).
2. It should be noted that the process of condensation is very different from the dynamic displacement process in this analogue dynamic test protocol, since the condensate is created in a distributed fashion within the pore network as the gas pressure depletes.

A further approach is to utilise a combination of iterative modelling and laboratory experimentation (after Saevereid et al., SCA1999-30); the laboratory data being used to validate the veracity of the model. In this method, a model fluid system is designed to represent the same condensate-gas ratios as the live reservoir system. Laboratory analysis is performed as a constant injection experiment using the model fluid system (at the designated temperature and pressure) until SS (constant rate and pressure) is achieved. The results are then feed-looped back into the model to ascertain the next model fluid system required to represent a different condensate-gas ratio. The next laboratory analysis is performed to validate the understanding of the EOS of the condensate system. These laboratory analyses are more controllable than performing at full reservoir conditions with live fluids, yet remain fairly complex and require application of elevated temperature and pressure. The analysis provides confirmation of condensate-gas relative permeability for a given condensate saturation, which can be directly related to the reservoir.

10.1.11 Wettability

10.1.11.1 Introduction

Wettability is a fundamental rock property that controls fluid distributions within porous media. It is the result of a complex interaction of forces which are related to both the fluid system type and the rock type, and has a major impact on capillary behaviour, core electrical measurements, relative permeability tests and residual saturations.

It is important to maintain, or attain, the appropriate reservoir wettability conditions if core analysis tests are to provide reliable results. Despite many studies on the effects of wettability on rock properties measured in core analysis experiments, appropriate wettability conditioning is frequently still ignored or misunderstood. Laboratories often seem quite happy to offer to carry out tests on "fresh-state" or cleaned-state cores knowing that the results of the tests may not be wholly appropriate. It is essential therefore to ensure that the wettability of the test samples is representative of the reservoir wetting state.

In a rock/oil–water system, many wetting states are thought to occur. Some of these are:

Water-wet	• Where water coats the grains and fully occupies the smaller pores. Oil would occupy only the centre of the larger pores
Oil-wet	• Where the reverse of the above is the case and oil coats the grains and occupies the smaller pore spaces
Neutrally-wet	• Where the system has a uniform non-preference for oil or water, i.e., no spontaneous imbibition of either fluid
Fractional wettability	• Where certain areas of the rock are water-wet, while other areas are oil-wet
Mixed wettability	• Mixed wettability falls into two categories: **Intermediate-Wet** • Where certain rock surfaces (often clay minerals) are oil-wet, while the smallest pores remain water-wet (or vice versa, with oil filling the smallest pores and water wetting certain surfaces of the larger pores) **Neutral-Wet** • Where there is no preference for either water or oil. This results in no spontaneous imbibition of either water or oil

In a rock/gas/water system, the native formation is assumed to be water-wet as it has never been exposed to oil. The exception is where there has been a previous oil charge which has been displaced by gas or water or has matured into the current gas phase. In some fields, the residual oil saturation can be high in asphaltenes which are known to reduce water wetness.

10.1.11.2 Native Wettability

The original wettability of reservoir rock is controlled by a number of factors including:

Oil type	• Certain components of reservoir oil can wet certain rock surfaces. These are: • Polar compounds (often referred to as resins or NSOs: where NSO stands for nitrogen, sulphur and oxygen, forming functional groups within the organic compound structure) which are natural surfactant components of reservoir oil. The polar end will chemically bond with the opposite pole of polar rock surfaces. These are usually relatively strong bonds and require time/energy to create the bond. The bond is usually broken by the abundance of stronger polar chemicals • Solid components (most often asphaltenes and bitumens, sometimes kerogens) can precipitate and physically deposit on surfaces. These are usually relatively weak bonds, wetting surfaces quickly and easily broken by dissolution of the solid particles. For instance, asphaltenes are readily dissolved in toluene
Water chemistry	• Changes in pH can cause changes in wettability or high concentrations of divalent cations in the FW can promote surfactant adsorption on the rock surfaces, resulting in less water-wetting tendencies
Grain type	• Carbonates tend to be more oil-wet than silicate grain surfaces. This is related to the surface charge of the rock and the nature of the oil and water surrounding them. Generally, under similar fluid conditions, adsorption of asphaltenes on silica surfaces (which can lead to oil wetting) is an order of magnitude less than adsorption of asphaltenes on carbonates. For silicates, evidence points to minerals being the major rock surfaces involved in oil wetting (e.g. kaolinite, smectite and carbonate content)

Unfortunately, there are currently no tools available which can be used to definitively establish native formation wettability (especially since the processes involved in drilling/coring may alter native wettability). Although inferences can be drawn from core, tracer and log resistivity measurements, NMR currently offers the best prospect for a measurement that is conducted downhole, and RFT data may prove useful. At present, there is no technology available to definitively establish *in situ* wettability.

10.1.11.3 Native Wettability Alteration

Native reservoir wettability estimates from reservoir core samples can be further complicated as a result of interactions between the rock and fluids during coring, core recovery, core handling and testing, including:

- contact with drilling mud
- pressure and temperature losses on core recovery
- certain core stabilisation materials used (e.g. epoxy resin)
- oxidation of the oil through lack of appropriate core preservation

- oxidation of the oil through lack of appropriate stock tank oil (STO) preservation
- inappropriate core cleaning

All can alter the native wettability to different degrees.

The literature on wettability, wettability alteration and its effects on rock properties is extensive and interested users of this guide should refer to the Society of Petroleum Engineers' search engine (http://www.onepetro.org/mslib/app/newSearch.do).

For example, Anderson (1986, 1987) provides an exhaustive and definitive discussion of the effects of wettability and waterflooding, and how core wettability can be altered during its journey between the reservoir and the lab.

The key points are:

1. *Drilling Mud*: Wettability alteration is normally more severe for oil-based and invert emulsion mud fluids (especially those using diesel as the base oil) because of the oil-wetting surfactants commonly used. OBM contact can induce an extremely strong oil-wetting tendency that can be difficult to remove, even by harsh cleaning. Synthetic oil-based muds and water-based muds, in general, appear to be less damaging.
2. *Core Recovery*: Water-wet rock can be made more oil-wet during core retrieval as the core is brought to the surface. Pressure reduction results in the liberation of gas as the core pressure is reduced below the bubble point of the crude oil. This increases the relative concentration of heavier end components (increases surfactant concentration) in the oil which can be adsorbed on the rock surfaces. Temperature reduction reduces the solubility of the surfactants of the oil and increases the potential for adsorption.
3. *Oxidation of Oil*: Through exposure of the core to air during handling at wellsite and in the laboratory. This causes a potential increase in asphaltene content (relative to the content in the live reservoir oil) and provides an opportunity to increase the oil-wettability. Correct core preservation, designed to eliminate or minimise core exposure to air, will prevent this form of wettability damage. The use of liners for core retrieval helps minimise core exposure, provided they are filled with liquid and sealed. On removal from the liners, whole core sections (around 10–15 cm long), or core plugs, can be preserved. Core preservation methods are either "wet" or "dry". Dry methods encapsulate the whole core section or core plug in a material intended to keep formation fluids from evaporating and which should prevent exposure to air during handling and storage. The most widely used method (Seal Peel) is to enclose the core section in plastic film wrap, then in aluminium foil, then dip the core in paraffin wax or plastic to form the seal (Fig. 10.8). With the right sealant, water loss can

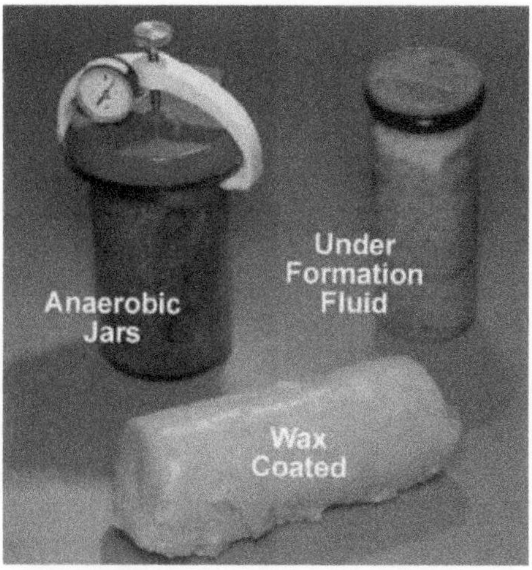

FIGURE 10.8 Wet and dry methods for core preservation. *Courtesy of Corex.*

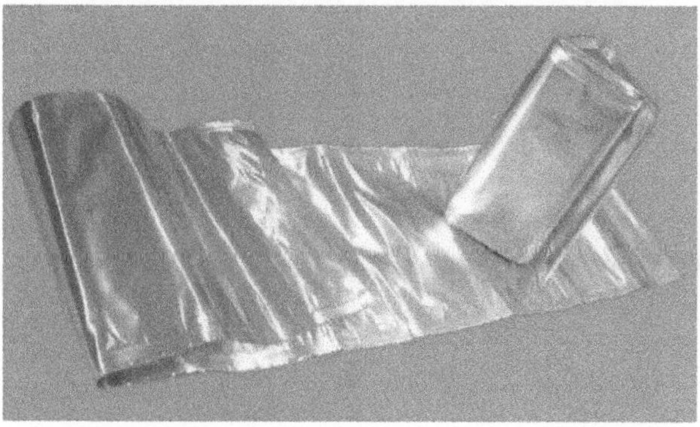

FIGURE 10.9 Plastic-aluminium laminate ProtecCore. *Courtesy of Core Laboratories and its Petroleum Services and Core Lab Instruments divisions. All Other Rights Reserved.*

be reduced to practically zero with this method. Another method is to use plastic-aluminium laminate to seal the core sections (Fig. 10.9). The core section is wrapped firstly in plastic film then placed in the laminate, which is then evacuated (in some applications) then sealed with a heat gun. The laminate is less permeable to gases and fluids than the combination of film/foil/wax. The main problem with this method is the fragility of the laminate—they can be subject to pinholes and cracks, and some operators have found them to be too delicate for field use. Wet methods involve submerging the core section or plugs in deoxygenated FW (Fig. 10.8) or depolarised kerosene or other mineral oil. Anaerobic jars (Fig. 10.8) can be evacuated or purged with helium or nitrogen to remove air, before the containers are sealed. PVC containers are not optimal because they permit diffusion of water and oxygen. Glass containers are inert, but experience has shown that they can be easily damaged in transport such that the preservation fluid is lost.

4. *Core Cleaning*: Cleaning methods are designed to remove oil and water (plus other drilling mud contaminants) from a plug sample. Invariably therefore, core cleaning tends to increase the water wetness. In fact, core cleaning as part of wettability restoration is designed to induce a strong wetting state in the test samples, and the cleaning methods and solvents are selected primarily on this basis. Anderson (1986) provides an excellent description of the methods and techniques used. Depending upon the oil and rock types, extremely harsh cleaning methods and solvents may be required. Recent (mostly unpublished) evidence shows that hot soxhlet extraction in toluene and methanol often fails to remove oil-wetting contaminants so that the cleaned core is not wholly water-wet. Hot soxhlet extraction in toluene evaporates water contacting the grains and can permit the heavier (polar) oil fraction (which contain surfactants and asphaltenes) to come in contact directly with mineral surfaces and alter wettability. The phenomenon may be exacerbated by low initial water saturation.
5. *Test Conditions*: Tests run with reservoir crude oil at reservoir pressure and temperature can provide a closer representation of reservoir wettability. An increase in temperature and pressure will increase the solubility of wettability altering compounds. The IFT of the fluid phases and the oil/water contact angle on the solid grain surfaces will decrease as temperature increases.

10.1.11.4 Wettability Tests

Amott (1959) and/or USBM Donaldson et al. (1969) tests should be carried out on parallel plug samples to determine the wettability of the samples. The tests are best carried out on cleaned-state, fresh-state and restored-state samples, and the data compared. These analyses are performed to investigate: the efficiency of the cleaning process to render core water-wet, the current wettability (whether native or altered), and the wettability after restoration, respectively. It should also be realised that most reservoirs exhibit a wettability gradient with depth. Consequently, wettability tests should establish these depth trends. Wettability tests are explained in detail in Chapter 7.

There is no definitive test to determine wettability at wellsite. Even if there were, alteration of native wettability due to contact with oil-based mud (for example) or a change in wettability due to pressure and temperature reduction on core recovery might invalidate the results.

10.1.12 Corey Exponents

It has been recognised that relative permeability curves between the residual end points of each phase can often be described by a simple exponential function, based on normalised phase saturation. The most widely used functional form is that described by Corey et al. (1956) who initially provided the empirical models for imbibition oil and water relative permeability curves.

10.1.12.1 Water–Oil Systems

In the Corey expression, the normalised oil relative permeability, over the moveable HC saturation range (from $Sw = Swir$ to $Sw = 1 - Sro$) is defined by: $kron = Son^{No}$ where the normalised oil saturation, Son, is defined by: $Son = \frac{1-Sw-Sro}{1-Swir-Sro} = 1 - Sw$, and $kron = \frac{kro}{kro'}$.

For laboratory data where the end-point oil relative permeability, kro′, is referenced to end-point Ko′ at Swir, then kron = kro.

The normalised water relative permeability is defined by: $krwn = Swn^{Nw}$, where the normalised water saturation, Swn, is given by: $Swn = \frac{Sw-Swir}{1-Swir-Sro} = 1 - Son$, and $krwn = \frac{krw}{krw'}$, where Son is normalised oil saturation, No is Corey exponent for oil, Swn is normalised water saturation, Nw is the Corey exponent for water, Sw is water saturation at a point during waterflood and Swir is irreducible water saturation.

The Corey equations are valuable in interpolating and extrapolating relative permeability curves, generating curves if only end-point data are available, and in assessing the validity of laboratory data.

In the absence of experimental artefacts, the Corey exponents, No and Nw, broadly correlate with the rock wettability and are typically expected to be in the range (Table 10.1):

Curve shapes, other than those described by the Corey expressions, are frequently observed. For example, Ibrahim and Koederitz (2000) and Honarpour et al. (1986) provide empirical correlations for water–oil and gas–oil systems.

TABLE 10.1 Typical Oil and Water Corey Exponents

Wettability	No (kro)	Nw (krw)
Water-wet	2–4	5–8
Intermediate-wet	4–6	3–5
Oil-wet	6–8	2–3

10.1.12.2 Gas–Oil Systems

A Corey-type approach can also be used to characterise gas–oil curves. Relative permeability to oil can be expressed as: $\frac{kro}{kro'} = kron = Son^{No}$ and gas relative permeability by: $krg = Sgn^{Ng}$, where krg is referenced to Ko′ at Swir. Son is the normalised oil saturation given by: $Son = \frac{1-Sg-Swir-Srog}{1-Swir-Srog}$ and the normalised gas saturation, Sgn, is given by: $Sgn = \frac{Sg-Sgc}{1-Swir-Srog-Sgc}$, where Sg is gas saturation, Swir is initial water saturation, Srog is residual oil saturation after gas injection and Sgc is the critical gas saturation.

The relationship between the oil (No) and gas (Ng) Corey exponents is normally quite well defined (compared to water–oil Corey exponents where changes in wettability play a more influential role), since the oil is always the wetting phase, so No is much larger than Ng. Typical ranges for both are:

Corey Exponent	Values
No	4–7
Ng	1.3–3.0

However, Jerauld (1997) showed that relative permeability curves (particularly gas) may be better described by a sigmoid model. Lomeland et al. (2005) provided a model that allows sigmoid description of relative permeability curves, called the LET model.

10.1.12.3 Water–Gas Systems

The Corey approach can also be adapted to evaluate imbibition water–gas systems. In this approach, krg and krw are related to the base effective gas permeability, Kg′ at Swir, and are normalised as follows: $krgn = krg = (1 - Swn)^{Ng}$ and $krwn = \frac{krw}{krw'} = Swn^{Nw}$, where krgn is normalised gas relative permeability, krwn is normalised water relative permeability, Ng is the Corey exponent for gas (water–gas system) and Nw is the Corey exponent for water (water–gas). Normalised water saturation is given by: $Swn = \frac{Sw-Swir}{1-Swir-Sgr}$.

The intermediate krw–krg curves can also be generated using the MAK correlation (Mulyadi et al., 2000). This calculates krw and krg referenced to Kg′ base (effective gas permeability at Swir) as a function of end-point residual saturations, test fluid viscosities, and rock properties (porosity and permeability).

$$krw = \frac{2Sw(S^*)^a}{\alpha} \quad \text{and} \quad krg = \left(1 - \frac{Sw - Swir}{1 - Swir - Sgr}\right)^b \left(1 - (S^*)^c\right)$$

where $S^* = \frac{Sw-Swir}{1-Swir-Sgr}$. The parameters α, a, b and c are defined by the following equations:

$$\alpha = -0.0466\text{Ln}(\text{Ka}) - 10.544\phi^2 + 10.263\rho_g{}^{-0.0326}\phi$$

$$a = -9 \times 10^{-5}\,\text{Ka} - 9.034\phi^2 + 1.6758\phi + 3.0779$$

$$b = -72.092\text{Swir}^2 + (38.0845 - 0.0343\text{Ln}(\text{Ka}))\text{Swir} + 0.0098\sigma - 2.6332$$

$$c = 1.20287b^{-0.8117}$$

where Ka is air permeability (mD) and ϕ is porosity (v/v).

10.2 MEASUREMENT TECHNIQUES

Relative permeability on core samples can be determined using three principal test methods:

1. Unsteady state
2. Steady state
3. Centrifuge

10.2.1 USS Tests

The USS technique is a dynamic displacement test. The principal objectives are to define end-point effective permeability and, if incremental production data are acquired, a portion of the relative permeability curves. In this test, a single phase is injected at the inlet of the core. Initially, at the outlet, only the displaced phase is produced, at the same rate as the injected phase (assuming incompressible fluids) thus observing a linear production stage. This continues until the injected (displacing) phase reaches the outlet. The point at which the injected phase is first observed exiting the outlet face is called BT. After BT, both displaced and displacing phases are produced and the displaced phase production becomes a curve.

The test sample is located in a coreholder under some predetermined net confining stress (net overburden pressure) and should be pre-conditioned to achieve representative initial water saturation and wettability. The initial "end-point" effective permeability (to oil or gas at Swi) is measured. The displacing phase (e.g. water or gas) is injected, either at constant pressure or constant rate until some predefined cessation criteria (e.g. many laboratories use 99.9% produced water cut—although different labs observe different cessation criteria) and the final end-point effective permeability to the displacing phase is measured at the residual saturation of the displaced phase. If end points only are required, only total production volumes are recorded. For relative permeability curves, incremental production data (vs. injection volumes) must be acquired.

The tests can be performed on cleaned, fresh-state and restored-state cores and at elevated reservoir pressure and temperature, and at reservoir

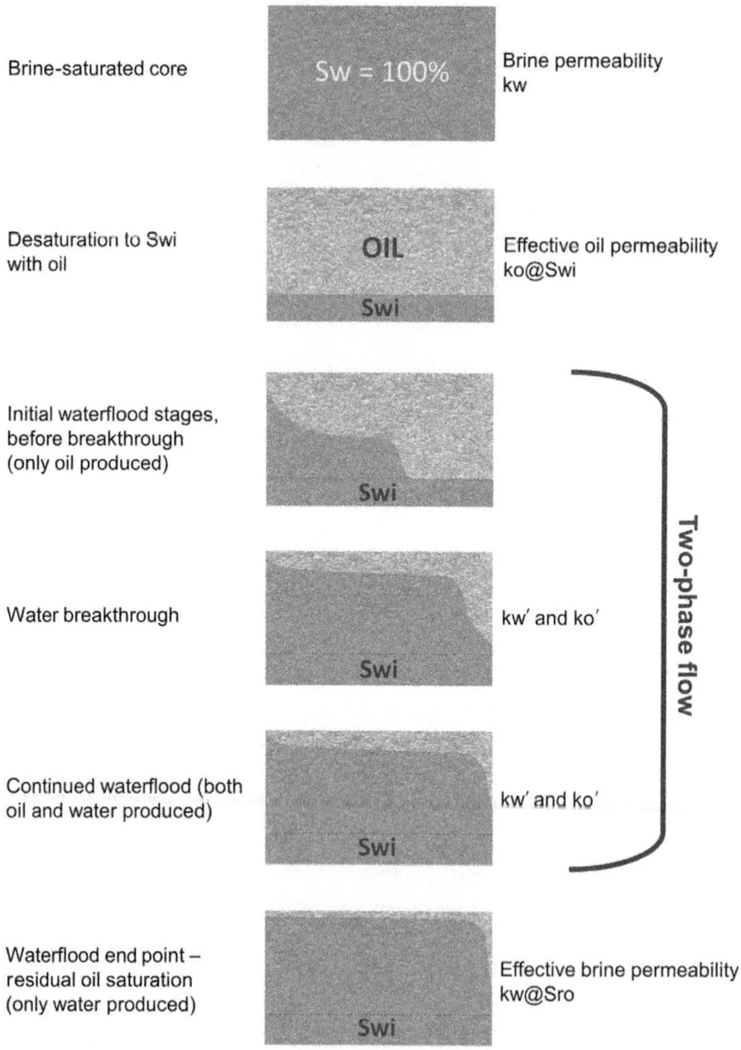

FIGURE 10.10 Schematic of water–oil unsteady-state test. *Courtesy of Paul Glover.*

conditions. Figure 10.10 provides a schematic of a water–oil USS test. In this figure, Keo and Kew represent effective oil and water permeability.

End-point permeability is determined using Darcy's law. The intermediate relative permeability data are determined either analytically (using JBN or Jones and Roszelle methods (Jones and Roszelle, 1978)) or numerically (via coreflood simulation).

DPs and flow rates are recorded throughout the test—saturations are most often determined by a combination of volumetric and gravimetric measurements, though non-invasive ISSM can also be used to visualise and measure saturations at the end and/or during the flood.

In situ monitoring is recommended in addition to volumetric or gravimetric measurement methods, since it provides indication of the extent of capillary end effects and provides additional input to numerical simulation processes to improve the match between how capillary pressure and relative permeability controls the production history (production volume vs. injected volume and DP vs. injected volume).

Detailed descriptions of the USS tests are provided in Sections 10.5.1, 10.5.4, 10.5.8 and 10.5.10.

10.2.2 SS Tests

The SS technique is also a dynamic displacement test. The principal objectives are to define the intermediate relative permeability curves, especially at low saturation values of the displacing phase (since the USS method can only provide relative permeability once BT has occurred, it cannot provide relative permeability at early saturation data). It is argued that end-point permeability after SS displacement may not be representative since the flow mechanisms operating in the core are different to those used in USS tests and in the reservoir. A water "shock front" is not developed. In the SS test, a sequence of fixed ratios of fluids is injected through the core sample until saturation and pressure equilibrium is established at each ratio.

A schematic of the SS method for a water–oil relative permeability test is shown below. In this case, the initial end-point effective (or base) permeability is determined at initial water saturation, as for the USS method. The waterflood is commenced by injecting a fixed ratio of water and oil, commencing at a low water:oil ratio. The fraction of water is increased in fixed ratios, successively increasing the water fraction and eventually terminating at 100% water flow. Produced fluids are separated, rates measured, and are re-circulated prior to re-injection.

Like USS tests, SS tests can be performed on cleaned, fresh-state and restored-state cores and at elevated reservoir pressure and temperature. However, elevated pressure and reservoir conditions tests are constrained by the requirement to measure core saturations accurately, using ISSM. It is strongly recommended to perform SS tests with ISSM (gravimetric and/or volumetric measurements incorporate too many potential errors). It is also preferable to use a re-circulating fluid system, where production is gathered in a separator system, separating the two immiscible fluid phases, and the pumps draw fluids from the separator for re-injection to the core plug. The system thus becomes a closed volume system and fluids are more likely to maintain equilibrium. Figure 10.11 provides a schematic of a water–oil SS test.

Both end points and intermediate relative permeability to both phases are calculated using Darcy's law when the flow of fluids within the porous media

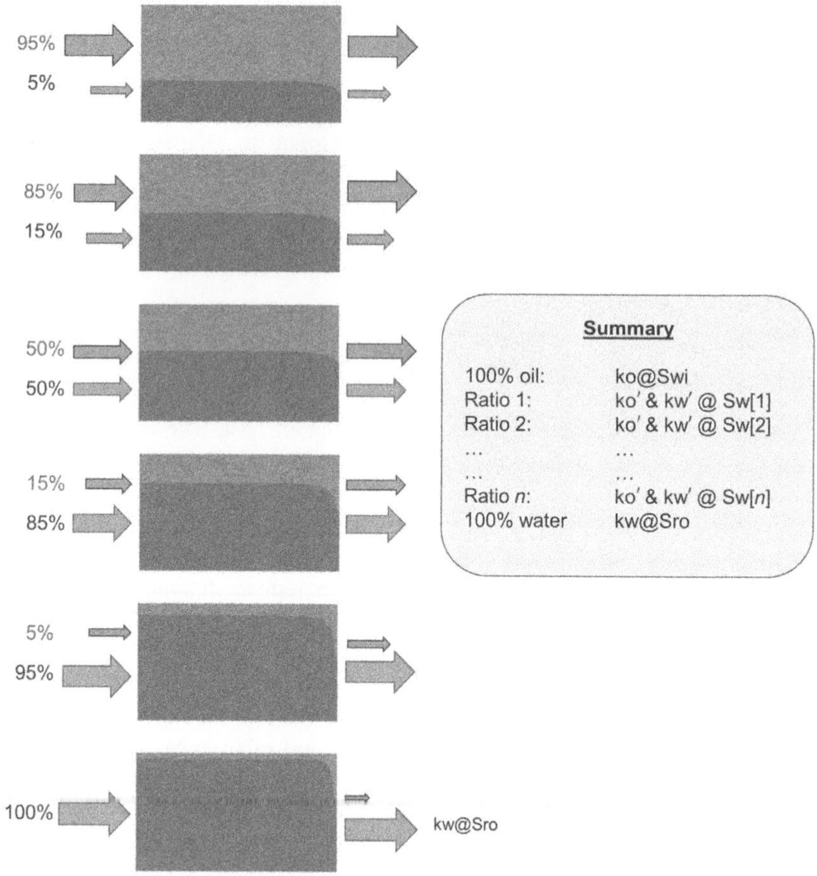

FIGURE 10.11 Schematic of steady-state test method (water–oil).

reaches SS (i.e. stable DP and stable saturation). This technique assures that flow and saturation equilibrium has been achieved before making the final readings of flow rate and pressure differential.

DPs, injection rates and/or production rates are recorded throughout the test. ISSM by use of provides the best solution for saturation determination but is more expensive. Saturations can be determined by volumetric and/or gravimetric measurements, though gravimetric measurements will compromise the data due to the requirement to remove and replace the sample at each stable injection ratio. Removal of the sample leads to a high probability of fluid redistribution, pressure hysteresis, grain loss and other undesirable artefacts. Volumetric measurements are often viable at the test extremities (i.e. low injection fractions and high injection fractions), but the assumptions made in the intermediate stages of volumetric distribution within the rig test lines can often skew saturation data for these intermediate fractions.

Detailed descriptions of the SS test are described in Sections 10.5.2, 10.5.5, 10.5.9 and 10.5.11.

10.2.3 Centrifuge Tests

The centrifuge can be used to generate relative permeability. The principal advantages of the centrifuge technique are faster and more efficient drainage, and the ability to measure very low relative permeability values close to residual saturations.

However, only relative permeability of the displaced phase is obtained, because only the displaced phase production can be measured. Thus, water-displacing-oil is used to generate the imbibition relative permeability to oil, kro, and oil-displacing-water is used to generate a drainage krw (though in a different saturation history than kro).

Liquid imbibition tests displacing gas (e.g. water–gas) are not recommended due to gas compressibility and gas diffusion effects. If the imbibition curve is required, then decane (or similar very light saturate oil) typically replaces the gas (non-wetting) phase, while water then represents the wetting phase.

As the sample is spun in a centrifuge, a gravity field is created (low to high gravity force as an exponential function of distance from the centre of rotation). Fluids migrate under gravity based on density differences; high-density fluids will move away from the centre of rotation. Fluid production versus time data is collected, either by visual observation or cameras (automated camera data collection is strongly recommended, providing a more accurate and robust dataset, as well as providing traceability). A calculation method such as that developed by Hagoort (1980) is most often used to transform effluent production data to relative permeability, but the preferred technique is to use simulation of the production data, together with a relevant capillary pressure, which many commercial laboratories do not have access to.

Detailed descriptions of the centrifuge test are described in Sections 10.5.3, 10.5.6, 10.5.12 and 10.5.13.

10.2.4 Curve Definition as a Function of Test Method

Each of the different methods provides description of different parts of the relative permeability curves. As discussed above, the USS method can only provide relative permeability definition after BT and thus only provides late saturation data. It will not usually achieve Srot, but will most often achieve lower Sro values than the SS method.

The SS method is the only method that can describe relative permeability over a wide saturation range, since fractional rates are controlled. However, during imbibition of a strongly wetting phase, the initial saturation change is mainly controlled by strong spontaneous imbibition capillary forces. In such cases, there may be a large change in saturation at the first (low)

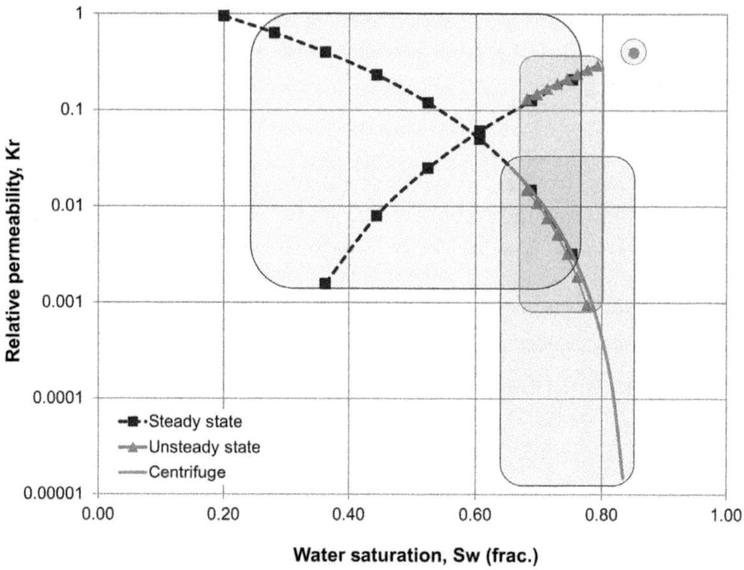

FIGURE 10.12 Relative permeability curves as a function of methodology.

fractional rate of the wetting phase. This will reduce the range of saturation that will be described and is typically problematic in liquid–gas SS floods. The SS method will not achieve Srot and, of the three methods, is most likely to exhibit higher Sro than the other methods.

The centrifuge method only provides relative permeability data of the displaced phase at late stage saturation—high saturation values. However, if the centrifuge test is correctly controlled and test parameters allow, it is the method most likely to achieve Srot.

Figure 10.12 presents a plot indicating the different range of relative permeability and saturation typically provided by the different methods.

10.2.5 Test Equipment

10.2.5.1 Corefloods (USS and SS)

Relative permeability test equipment can and does vary significantly in terms of sophistication and instrumentation, ranging from simple bench-top systems at ambient conditions with synthetic fluids to full reservoir conditions systems (for reference, see Fig. 10.13). This equipment consists of a coreholder with two or three inlet ports (one to two for fluid injection and one pressure port) and two outlet ports (one fluid production port and one pressure port). This design ensures that pressure is measured directly at the core end faces. This equipment is usually provided with three DP transducers to be able to log low, medium and high pressure drop. From analysis of the pressure response

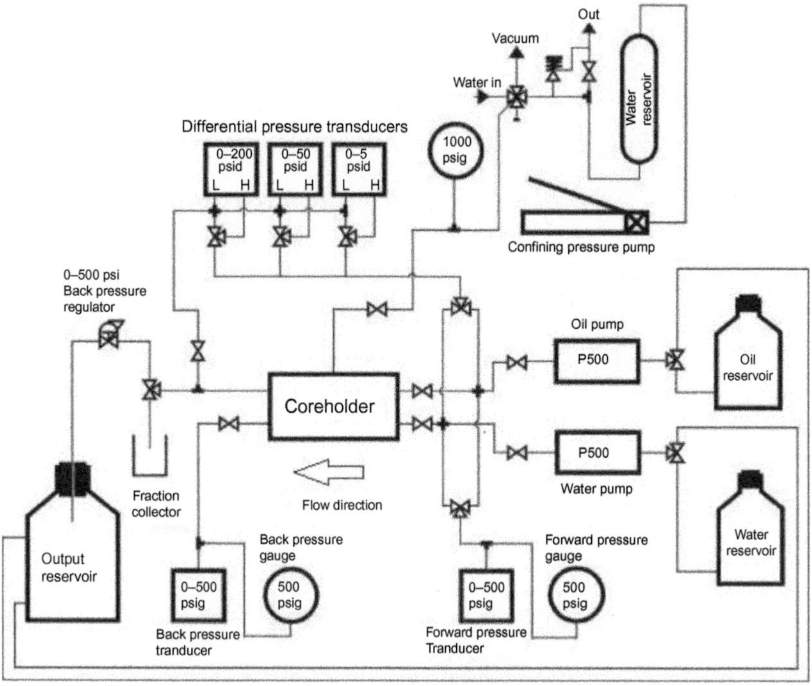

FIGURE 10.13 Schematic diagram of relative permeability equipment. *Courtesy of Paul Glover.*

of the core sample in relation to injected fluid volume and BT recovery oil volumes, the system dead volume can be identified and used to correct produced fluid volume data.

The displacement equipment has a set of pumps to inject the fluids into the sample and a hydraulic pump to set the overburden pressure. A backpressure regulator is located at the output of the system to maintain the fluid pressure above the bubble point of oil or to minimise gas expansion from core inlet to outlet. Generally, a pressurised endoscope, pressure-monitored separator or acoustic separator is used to separate and measure the produced fluids. A recirculation loop is also set up to recycle the fluids for SS tests. Figure 10.14 shows typical commercial equipment for relative permeability experiments.

The equipment required to measure relative permeability in SS is more complex than for USS measurements. The SS technique requires the capability to inject and measure fixed fractional flow of oil and water (or water and gas, or gas and oil) at low rates, and the resultant saturation changes. Pressure response is also determined at each fractional flow.

The equipment should incorporate inlet and outlet mixing (porous media) sections to promote fluid mixing and minimise capillary end effects as shown

FIGURE 10.14 Typical commercial equipment for liquid–liquid and liquid–gas relative permeability.

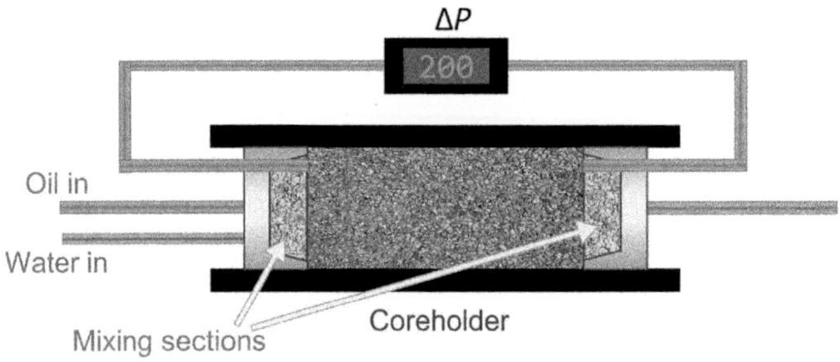

FIGURE 10.15 Steady-state optimised coreholder to promote fluid mixing.

in Fig. 10.15. These should be designed such that core pressure drop can be measured directly at the test core end faces. The coreholder is also provided with three inlet and two outlet ports—two for fluid injection at the inlet, one for fluid production at the outlet, and the remaining inlet and outlet pair dedicated to core pressure measurement. Constant rate pumps are required to precisely control the injection ratios.

10.2.5.2 Centrifuge

Centrifuge relative permeability tests are carried out on the same equipment used to generate drainage capillary pressure (Chapter 9). Figure 10.16 provides an example. In addition to the limitations on relative permeability curve acquisition imposed by the centrifuge displacement, test conditions tend to be limited to ambient (or moderate overburden) conditions using synthetic and not reservoir fluids (or possibly dead crude oil).

Fluid production versus time data is collected by visual observation or, and better, via cameras linked to automated data capture devices.

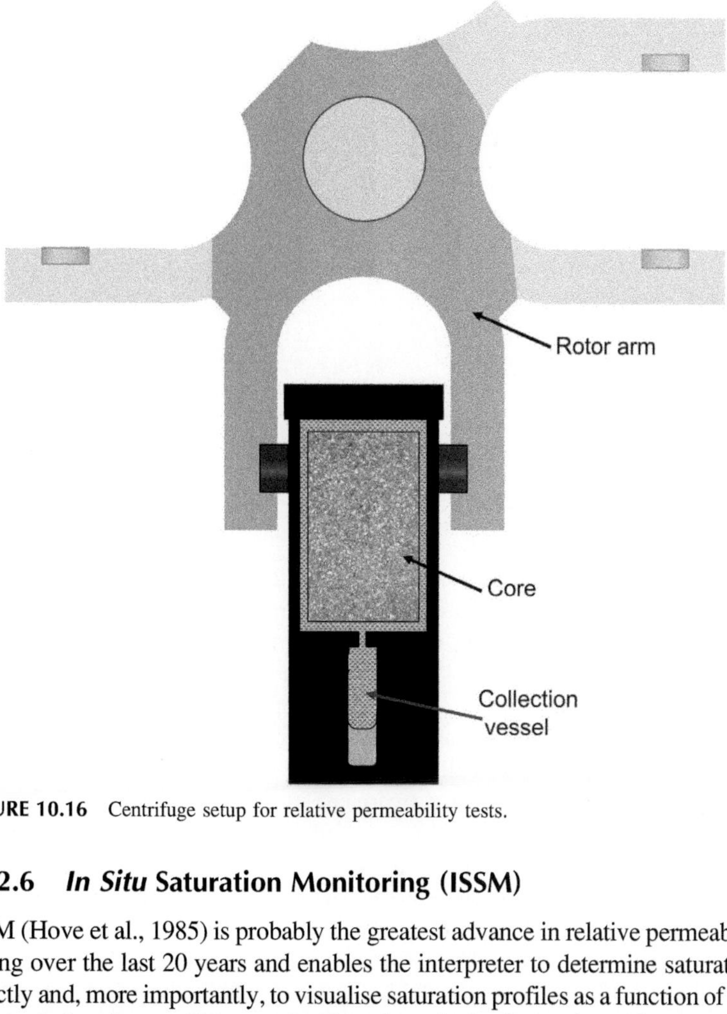

FIGURE 10.16 Centrifuge setup for relative permeability tests.

10.2.6 *In Situ* Saturation Monitoring (ISSM)

ISSM (Hove et al., 1985) is probably the greatest advance in relative permeability testing over the last 20 years and enables the interpreter to determine saturations directly and, more importantly, to visualise saturation profiles as a function of core length, during the test. This can significantly assist in diagnosis and interpretation of core floods where flood instability and capillary end effects may dominate.

To measure *in situ* water and oil saturation, ISSM uses a collimated sealed energy source to generate electromagnetic radiation (usually either gamma rays or x-rays) that penetrate the core holder and assembly. A collimated sodium iodide detector positioned directly opposite the coreholder, and connected to a multi-channel analyser, is used to detect and measure the intensity of the emerging rays. Intensity is measured as "counts" or "counts per unit time". The source and detector are connected to a platform where movement can be controlled by a computer operating a stepper motor drive system. The platform can be moved along a track parallel to the length of a core sample. This moves the source and detector to specific predetermined locations

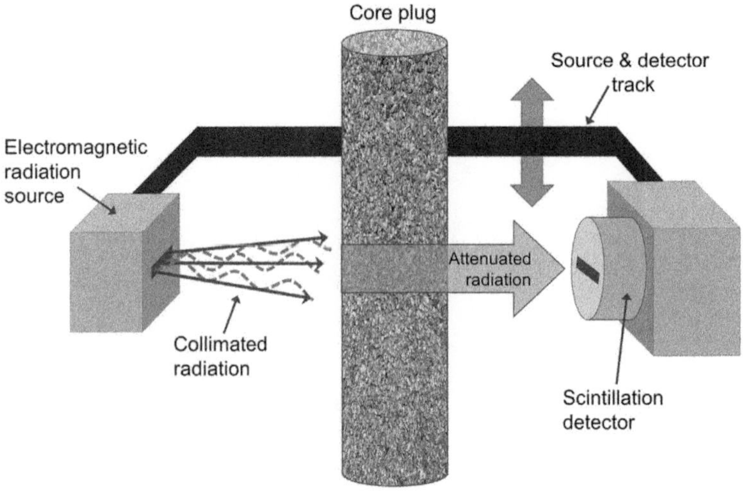

FIGURE 10.17 ISSM schematic.

along the length of the core, and the collimation of the electromagnetic radiation allows spatial resolution of around 2–4 mm. The process of moving the source and detector, opening the source and recording radiation intensity, is called "scanning". A typical ISSM schematic is illustrated in Fig. 10.17.

To improve the saturation accuracy, a dopant (sodium iodide or caesium chloride) is usually added to the water phase, and/or iododecane to the oil phase, to increase the difference in attenuation between the oil and water phases.

To quantify saturation (in a two-phase fluid system), the core plug must undergo a calibration process. This entails the core plug being saturated at 100% of each of the fluid pair, e.g. for a gas-water system, calibration scans must be performed when 100% gas saturated and 100% formation water saturated. Electromagnetic radiation intensity, received at the detector, is linearly proportional to the volume of doped fluid (e.g. NaI water) in the core sample (Fig. 10.18).

Figure 10.19 is an example of a lab system—in this case, Core Laboratories' SMAX (Saturation Monitoring by the Attenuation of X-rays) equipment. The core is located inside a test coreholder that is composed of a material (usually either aluminium or some composite fibre) that does not completely attenuate X-rays and which is mounted on a scanning track. An X-ray transmitter–detector pair moves along the track, transmitting X-rays at specific locations relevant to a certain length along a core plug. X-Rays that are not attenuated (blocked) by the coreholder, core plug and/or saturant fluids pass through and scintillate the detector. Detector scintillation levels are proportional to the level of non-attenuated ionising radiation, which in turn are proportional to changes in fluid saturation levels, allowing calculation of fluid saturation using the Lambert–Beer law.

Following measurement of Ko′ at Swi, ISSM scans are performed to determine the initial water saturation profile and average Swir. ISSM scans are continued at short time intervals, once the coreflood test is begun, to allow

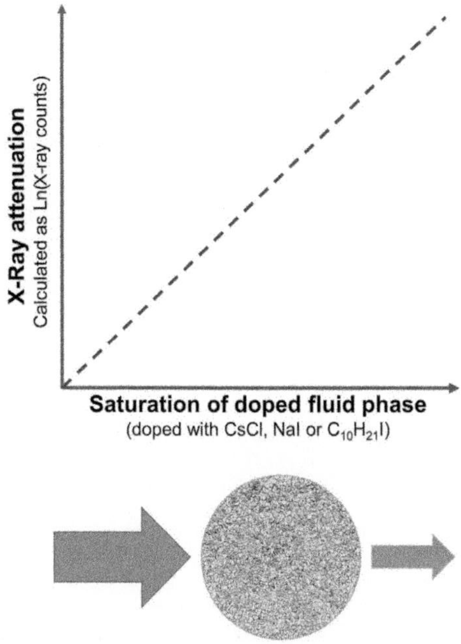

FIGURE 10.18 ISSM calibration principles.

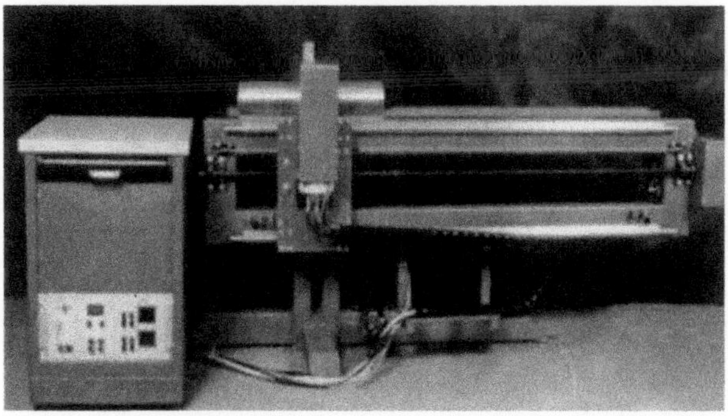

FIGURE 10.19 Core Laboratories' SMAX system. *Courtesy of Coretest Systems, Inc.*

tracking of the flood front during the waterflood before and up to BT and the location and magnitude of Sro on completion of the waterflood.

The ISSM example shown in Fig. 10.20 indicates a fairly uniform distribution of water at Swi, within ±3 saturation units. Scans were taken at the end of low-rate and high-rate (bump) waterfloods. An end effect (increase in So, and hence decrease in Sw, close to the core outlet) is clearly evident after both injection rates, but the increased viscous forces during the high rate (bump) flood has reduced the end effect. This suggests a non-water-wet core. The krw increased by a factor of 3 after the high-rate flood.

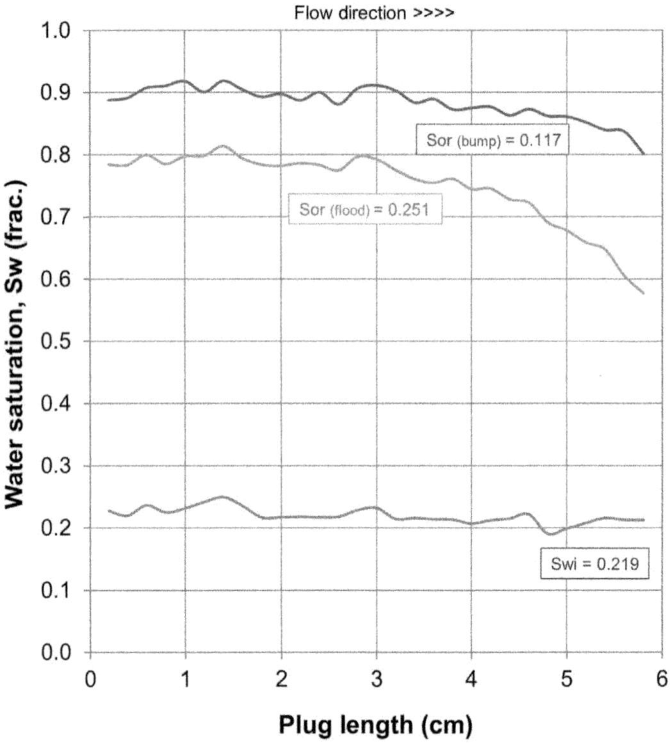

FIGURE 10.20 ISSM example—neutral-wet sample.

More sophisticated ISSM technology includes dual energy systems, CT and NMR/MRI which enables monitoring of three-phase fluid systems and/or visualisation of flow processes in 2D or 3D, but this technology tends to be restricted to research laboratories.

10.3 TEST DATA INTERPRETATION METHODOLOGY

10.3.1 Unsteady State

Darcy's law is used to determine end-point effective permeability data from USS tests.

USS relative permeability data are based on the theory of immiscible displacement behaviour of oil by water proposed by Buckley and Leverett (1942) and extended later by Welge (1952). This theory is based on a material balance equation which relates the rate of change of water saturation, S, at a given distance, x, along the core, at a given time, t; thus, $\frac{\mathrm{d}(S)}{\mathrm{d}(t)x} = q_\mathrm{t}\frac{\mathrm{d(fw)}}{\mathrm{d}(x)t}$.

The rate of advance of a constant saturation front can then be obtained by rearranging the equation and substitution: $\frac{\mathrm{d}(x)}{\mathrm{d}(t)}_{\mathrm{Sw}} = \frac{q_\mathrm{t}}{A\phi}\frac{\mathrm{d(fw)}}{\mathrm{d(Sw)}}$, where A is cross-section area of the porous medium, ϕ is porosity of porous medium, fw is fractional flow of water at total rate, q_t.

$$fw = \frac{1}{1 + \frac{kro}{krw}\frac{\mu_o}{\mu_w}} \quad \text{Welge equation}$$

where μ_o is oil viscosity, μ_w is water viscosity, kro is oil relative permeability and krw is water relative permeability.

The Welge equation can be expressed graphically as a function of Sw (Fig. 10.21). The height of a flood front is given by the intersecting saturation point of a tangential line from critical (irreducible) water saturation (Swc, equal to Swir in this case) to the fractional flow curve. The average water saturation at water BT (behind the flood front) is the intercept tangential line with fw = 1. The cumulative water injected, pore volumes, is given by the reciprocal of the derivative of fw with respect to Sw.

By measuring the fractional volumes and oil and water viscosity at the core outlet, it is possible to calculate fw and the relative permeability ratio, kro/krw. In USS tests, until BT, the fractional flow of water is zero, so only post-BT pressure and production data can be used in the analysis to generate relative permeability data. The saturation range over which saturation and relative permeability data are available might therefore be restricted, as at the core outlet fw could be close to 1 immediately after water BT. However, as pre-BT production is associated with the development of a Buckley–Leverett "shock front" (fractional flow theory shows that the shock-front saturation is the saturation with the fastest velocity, incipient shock fronts never develop because they would travel more slowly), relative permeability data at lower

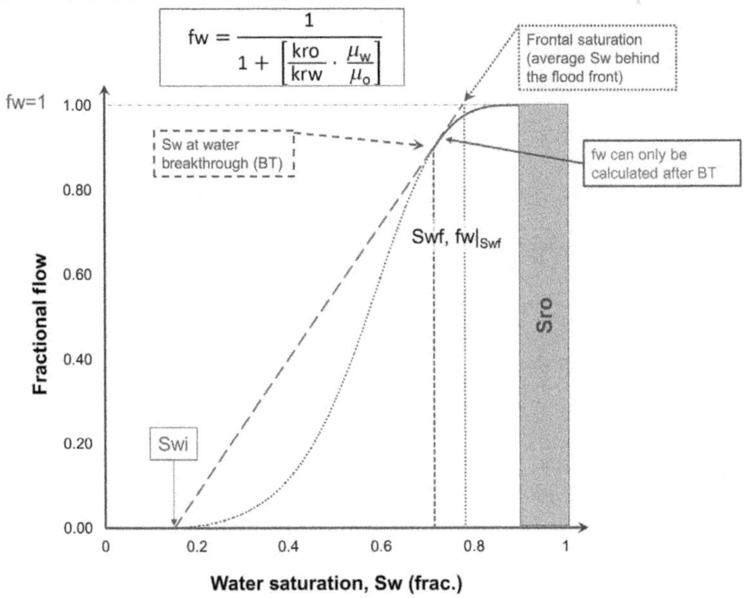

FIGURE 10.21 Water–oil fractional flow curve (Welge).

water saturations are not required to describe the displacement process, provided the displacement is dominated by viscous forces.

The Welge solution gives the relative permeability ratios as a function of Sw. The discrete oil and water relative permeabilities are decoupled from the kro/krw ratio using a method described by Johnson et al. (1959)—JBN method or by Jones–Roszelle. The following relationship can be applied to determine relative permeability:

$$\frac{d\left(\frac{1}{WI_r}\right)}{d\left(\frac{1}{W}\right)} = \frac{fo_2}{kro_2}$$

where $I_r = \frac{\Delta p \text{ at start of injection}}{\Delta p \text{ at time } t}$, W is the volume of water injected (pore volumes) at time t, fo_2 is fractional flow of oil at core outlet at time t and kro_2 is oil relative permeability at time t (at outlet).

The outlet saturation and fractional flow are calculated as above and kro_2 can then be calculated as the only unknown. The corresponding water relative permeability, krw_2, can then be determined using the kro/krw ratio from the Welge equation.

The Buckley–Leverett, Welge and JBN theories all involve the following assumptions:

- flow velocity must be large enough so that stabilised flow is achieved;
- viscous force is high compared with the capillary pressures between the flowing phases, thus capillary effects are negligible;
- gravity is ignored;
- the DP across the plug is small compared to pore pressure so that compressibility effects are insignificant;
- flow is linear (one-dimensional) and unidirectional;
- the plug is homogeneous;
- displacement force and fluid properties are constant during the test.

Where there is a significant phase density difference, the test should be carried out under gravity-stabilised conditions—the coreholder is held vertical and the denser phase injected from the bottom of the coreholder and the lighter phase from the top down. However, many corefloods invalidate the inherent assumptions in waterflood theory.

Many laboratories perform (and the majority of legacy data are based on) USS relative permeability tests with an adverse oil–water mobility ratio and performed at high rates to accelerate BT time and force extended two-phase flow production, and in order to overcome capillary end effects. These conditions enhance viscous fingering that invalidates the assumption that flow is one-dimensional. The use of matched viscosity ratios and appropriately scaled flow rates may better represent the reservoir situation, but capillary end effects in low rate/low mobility ratio corefloods also undermine

Buckley–Leverett theory. Unfortunately, the effects of capillarity and viscous fingering cannot be suppressed simultaneously. This makes the interpretation of imbibition floods extremely difficult, especially if *in situ* saturation profiles are unavailable. In view of the violations of the underlying assumptions upon which Buckley–Leverett/Welge and JBN/Jones and Roszelle methods are based, relative permeability data obtained from USS displacement floods must be viewed with caution.

For gas–water and gas–oil systems, the same equations are used but corrections are made to account for fluid compressibility. However, the use of gas specifically invalidates the Buckley–Leverett assumptions of incompressible fluids.

10.3.2 Steady State

Darcy's law is used to determine all permeability data (end points and intermediate relative permeability) from SS tests. Darcy's law for single-phase fluid flow in porous media is extended to two-phase flow. The volumetric flow rates of each phase under SS conditions of constant fractional flow rate and pressure are:

$$q_w = \frac{\mathrm{Kw}A\Delta P_w}{\mu_w L} \quad \text{and} \quad q_o = \frac{\mathrm{Ko}A\Delta P_o}{\mu_o L}$$

where q_o is volumetric oil flow rate (cm^3/s), q_w is volumetric water flow rate (cm^3/s), Kw is effective permeability of water (D), Ko is effective permeability of oil (D), A is cross-section area of the plug (cm^2), ΔP_w is DP of water (atm), ΔP_o is DP of oil (atm), L is sample length (cm), μ_w is water viscosity (atm/s), μ_o is oil viscosity (atm/s).

Unless the test equipment can measure the discrete phase pressures, the pressure differential to oil and water is assumed to be equal to the measured total pressure drop.

Although artefacts from sample heterogeneity are much reduced (dissipation of 1D flood front), SS relative permeability data are also prone to similar artefacts as USS tests, specifically due to capillary end effects. Without additional validation and correct by simulation/history matching, the analytical data remain uncertain.

10.3.3 Centrifuge

The relative permeability obtained is only the displaced phase since only the displaced phase production can be measured during centrifugation. The most widely used method for determining relative permeability is that developed by Hagoort (1980).

The example equations below assume oil to be the displaced phase. Relative permeability for oil (for example) is calculated as $\mathrm{kro} = \frac{1109 q_o \mu_o}{\mathrm{Ka}A\Delta\rho g'}$, where kro is oil relative permeability (fraction), Ka is absolute air permeability (mD),

q_o is oil drainage rate (cm^3/s), μ_o is oil viscosity (cp), A is cross-section area of core plug (cm^2), $\Delta\rho$ is density difference between oil and water (gm/cc) and g' is the centrifugal acceleration.

The term g' is defined as $g' = 0.00001117\omega^2 r$, where ω is centrifuge speed (rpm) and r is centrifugal radius, defined as the distance from centre of the core plug to centre of rotation (cm).

Uncertainties arise if the centrifuge technique is used to derive relative permeability from systems exhibiting "trapping" of residual oil. These uncertainties arise because of the difficulty associated with defining saturation distributions in centrifuged samples (i.e. saturation varies as a function of distance from the production face due to differences in applied gravitational displacement forces and the inherent capillary forces of the core plug). Conclusive resolution of these doubts by theoretical analysis of explicit volumetric and quantitative saturation data, via a simulation model that incorporates the physics directly, is required before centrifuge relative permeability can be better accepted. Many commercial laboratories do not have access to simulation technology and thus merely provide the basic analytical solution, incorporating all assumptions.

More basic concerns include (not limited): uncontrolled imbibition at Swi, uncontrolled hysteresis, production uncertainty during the acceleration period, time delay due to acceleration, mobilisation of trapped oil (if too high displacement forces are employed), difficulty in performing reservoir condition tests and potential for sample fracturing/damage.

10.3.4 Flow Rate and Capillary End Effects

The Buckley and Leverett frontal advance equation applies only to homogenous porous media under conditions where capillary pressure is negligible compared to the applied viscous pressure gradient. This implies that $\frac{d(Sw)}{d(x)} = 0$ and $\frac{d(Pc)}{d(Sw)}\frac{d(Sw)}{d(x)} = 0$.

The term $\frac{d(Sw)}{d(x)} = 0$ means that for a given saturation, water flows at the same rate through all locations in the core, and a stabilised zone is achieved. Capillary pressure effects can be negligible when the flowing system is large, such as in a reservoir, or where the flow rates are large. High flow rates are often used in most laboratory tests as an attempt to reduce capillary effects, with the assumption that they are made negligible. However, high rates are not representative of reservoir advancement rates and may begin to cause instability of the flood front.

A dimensionless group called the "capillary number" has been defined as the ratio of viscous to capillary forces. Various expressions have been proposed but generally have the following form (for a waterflood), $Nc = \frac{v\mu_w}{\sigma}$ or $Nc = \frac{K}{\sigma}\left|\frac{\Delta P}{\Delta x}\right|$ where μ_w is water viscosity, v is flood front velocity, σ is IFT, K is absolute permeability, ΔP is DP and Δx is sample length.

Most reservoir waterfloods are performed with Nc approximating 10^{-7} to 10^{-6}. Ideally, lab tests should honour the reservoir capillary number as relative permeability and residual saturations are a function of Nc. For example, Amaefule and Handy (1982) presented an empirical correlation relating two-phase water/oil relative permeability data and residual saturation with capillary number. For standard core plugs of 5–7 cm long, maintenance of reservoir Nc means flooding at very low rates (~4 ml/h).

However, in short core plug waterfloods at low rates, capillary pressure is no longer negligible when compared to viscous forces, and can cause both inlet and outlet end effects. Leverett and Lewis (1941) pointed out that upon exiting the core; the wetting phase has a tendency to be held back due to a discontinuity in capillary forces. This causes a buildup in the wetting phase saturation and a subsequent decrease in the non-wetting phase permeability near the outlet. The entire process results in an additional pressure drop that would not be present in a reservoir where this effect should be negligibly small.

In a waterflood, water BT occurs once oil saturation at the outlet of the core is near to Spo (oil saturation after spontaneous water imbibition), since the displacement pressure at the outlet face is zero and capillary forces dominate entirely. Towards the core plug inlet, DP increases, and capillary pressure effects are comparatively reduced. This results in the formation of a saturation gradient between the inlet and outlet. Although the flood front may be dispersed within the core, the retention of water by the end effect gives the production data the appearance of a piston-like process. The amount of oil produced after BT is largely associated with the outlet end effect, with greater production indicating a greater end effect. Since relative permeability is a function of saturation and since saturation is potentially changing as a function of distance from the plug outlet, flow at different distances along the core plug is defined by different portions of the relative permeability curve.

The influence of capillarity in low-rate floods can also result in an inlet end effect, in which the spontaneous imbibition of water in a water-wet rock can result in counter flow of oil out of the inlet end of the sample. This oil is then partially forced back into the sample as the water displacement proceeds. This results in a rapid rise in DP as the oil is forced back into the water-wet matrix. The DP then reduces as the oil re-injection terminates and the DP is then characterised by the imbibition process. However, small droplets of oil can remain "embedded" in the inlet surface and will result in a constant increased DP.

The principal effect of both the inlet and outlet end effects is to suppress relative permeability to water, calculated by the Buckley–Leverett–Welge–JBN methods, and in some cases, kro. In many low-rate floods, which are affected by capillary end effects, relative permeability calculated by the JBN method provides very low krw values after BT.

Rapaport and Leas (1953) defined a scaling coefficient to attempt to define more appropriate flow rates to use during flooding experiments, $\mathrm{Sc} = LV\mu_w$, where L is core length (cm), v is waterflood velocity (cm/min), μ_w is water viscosity (cp). This equation can be used to predict the presence (if not the magnitude) of the end effect, in water-wet conditions. They reported that $LV\mu_w > 1$ (cm^2/min/cp) resulted in a minimal end effect. This means that the use of high flow rates (increasing V) can help to minimise end effects. The magnitude of the end effect can be assessed using a dimensionless flow parameter, Nc_{end} (Sigmund and McCaffery, 1991) which is the ratio of a characteristic capillary pressure to the viscous pressure drop across the core, and can be approximated by:

$$Nc_{end} \approx \frac{\sigma\sqrt{k\phi}}{\mu_o L}$$

where k is absolute permeability of the core, ϕ is porosity, μ_o is oil viscosity and L is core length.

For $Nc_{end} > 0.1$, both oil and water permeability decrease as Nc_{end} increases (Mohanty and Miller, 1991). However, for $Nc_{end} < 0.1$, the end effect is negligible and may not affect relative permeability calculations.

Laboratories have historically carried out floods on short cores at high rates (400 ml/h) to increase the flood velocity (and increase the Sc scaling coefficient greater than 1). However, high flow rates in the lab are much higher than in the reservoir so the flood Nc at the pore scale (which controls residual saturations, for example) is very different from the reservoir Nc. Although laboratories might ensure that flow rates are high enough to suppress capillary end effects, the flow regime may bear little relation to that in the reservoir.

Heaviside et al. (1983, 1987), among others, investigated the effects of flow rate. USS displacement tests were conducted at typical laboratory rates of 400 ml/h. The tests were then repeated at a reservoir appropriate rate of 4 ml/h (Nc about 10^{-7}) until no more oil was produced after BT. Flow rate was then increased ("bump" flood) to 400 ml/h and any additional fluid production and pressure monitored to determine any changes in saturation and relative permeability. They found that, for water-wet rock, low-rate floods (honouring reservoir Nc values) will often obtain the appropriate value of Sro, but krw will be suppressed by end effects. The krw measured after the bump flood may be more appropriate (but not the associated post-bump Sro). For intermediate-wet rocks, Sro and end-point relative permeability to water (krw′) are expected to be dependent upon the ratio of capillary to viscous forces imposed during the flood: i.e., relative permeability data are rate dependent.

Striking a balance between attempting to honour reservoir Nc and trying to avoid end effects in short core plugs is almost impossible. However, increasing the length of the sample will have the effect of increasing Sc and reducing Nc_{end}, for any given flood rate. Capillary end effects are smaller relative to the viscous pressure drop. The use of longer core plugs is impractical in most reservoir cases—the core geometry often only allows long plugs to be taken in a vertical (perpendicular to bedding) direction, which may not be appropriate for reservoir flow. The use of composite core, in which a number of horizontal (parallel to bedding) plugs are butted together, effectively increases the length (Fig. 10.22). This requires special coreholders and saturation monitoring techniques to ensure that there are no capillary discontinuities between each core section (internal capillary artefacts, thus magnifying the effect rather than reducing it). Composite cores require strict criteria for core plug selection, to reduce possible heterogeneous effects—which in turn lead to capillary discontinuity. Criteria determining core selection include: CT image analysis, XRD/SEM/thin section analysis, MICP analysis (trim sample to determine similar Pc and PSD), K_L, porosity, GD, Kw, Swir and Ko′ at Swir. All these measurements must evidence each plug has basically the same properties (within measurement error of each measurement).

Capillary end effects are a particular problem in gasfloods (due to the adverse mobility ratio), and the potential presence, influence and impact of end effects should be carefully considered when designing a gas-displacement relative permeability test programme.

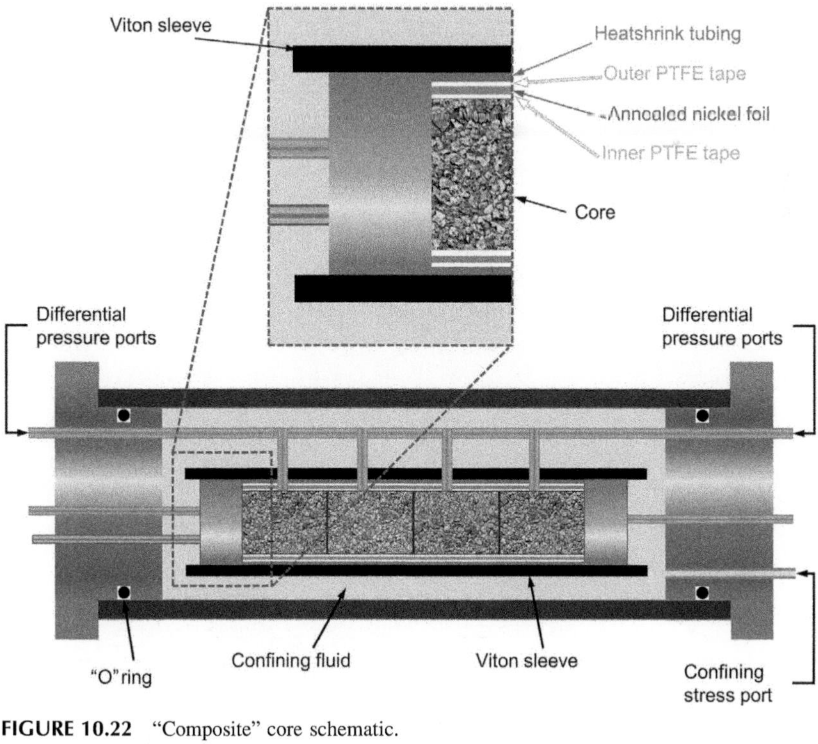

FIGURE 10.22 "Composite" core schematic.

10.3.5 Coreflood Simulation

The inability of classical waterflood theory (e.g. JBN) to account for boundary capillarity in lab experiments can invalidate the results from low-rate floods which attempt to honour the reservoir capillary number, Nc. Coreflood numerical simulation provides an opportunity to analyse data and interpret relative permeability results that account for capillarity. The approach is similar to reservoir simulation models, but is designed to operate at the core plug scale. Essentially, the simulator starts with estimate relative permeability values and adjusts the curves to history match production and pressure data (see example in Fig. 10.23) and ISSM profiles (Fig. 10.24).

Coreflood simulation has been restricted historically, since commercial labs rarely had access to the technology. However, open access/non-proprietary coreflood simulation models are now available. Examples are SCORES (http://www.jgmaas.com/scores), Sendra (http://www.sendra.no/) and CYDAR™ (http://www.cydarex.fr/), PORLAB (druth@cc.umanitoba.ca).

Sendra is a commercial 1D, two-phase immiscible core flood simulator specifically designed to simulate and verify SCAL experiments. It covers all common experimental approaches including USS and SS flow experiments and single- and multi-speed centrifuge experiments. It can be used for all standard fluid systems water–oil, gas–oil or gas–water experiments, and both imbibition and drainage processes. An example is shown in Fig. 10.25, where Sendra simulation results are compared to conventional JBN interpretation. The capillary effects, which are ignored in the JBN method, act to suppress

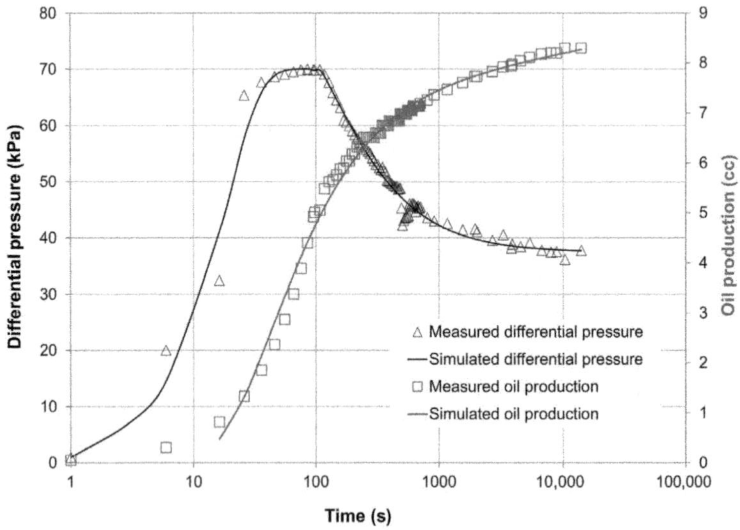

FIGURE 10.23 Coreflood simulation—history match to measured pressures and oil production.

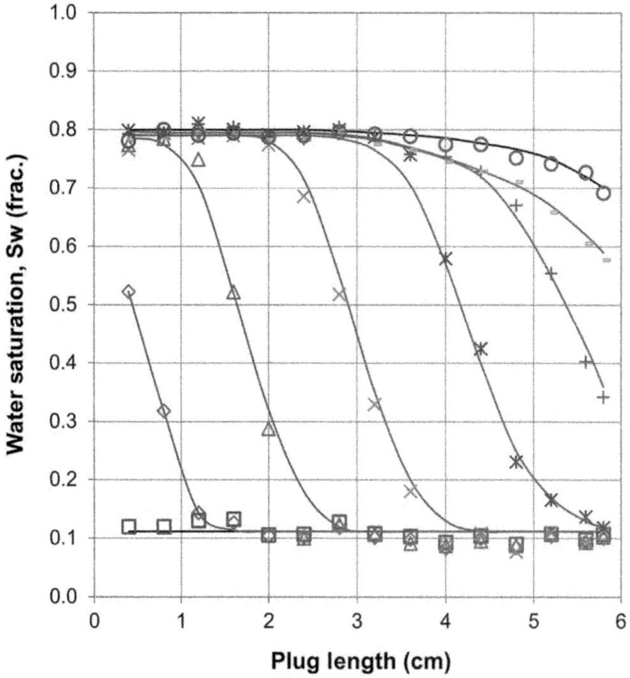

FIGURE 10.24 Coreflood simulation—history match (lines) to measured (symbols) saturation profiles.

water and oil relative permeability. Sendra can also account fluid compressibility, but care must be taken when employing compressible fluids in the simulator. Sendra incorporates a parameter estimation function (least squares regression), which will find a best estimate solution for any selected parameters, to match the production history data. This can be made very flexible allowing Sendra to estimate all relative permeability parameters and all capillary pressure parameters using a wide variety of different models. The user must be very careful, and knowledgeable, about which parameters to allow the simulator to estimate upon and which to constrain. It is best to constrain as much as possible, i.e., obtain capillary pressure from additional, sister samples, obtain end-point permeability from other testing where residual saturations were close to Srot. There are potentially multiple solutions, and the more constraints imposed on the simulation, the more unique the ultimate result.

CYDAR™ is also a commercial 1D, two-phase immiscible coreflood simulator, also incorporating a certain level of parameter estimation. Raw measurement data from corefloods cannot normally be used directly to calculate relative permeability nor input directly to most simulators, because of

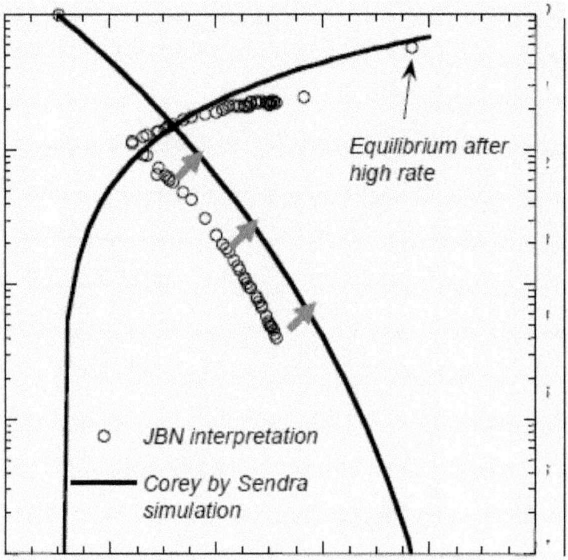

FIGURE 10.25 Sendra simulation of unsteady-state water and oil relative permeability test data. *Courtesy of Sendra (https://www.prores.no/sendra/).*

equipment properties (e.g. dead volumes of inlet and outlet lines, delay between actual outlet of plug and collection of produced volumes in a separator unit). However, Cydar™ can also perform standard correction for these additional equipment properties. It gives a reasonable clear schematic image of the test setup to allow the user to visualise what equipment was in use and the orientation and parameters surrounding the test. However, there are potential limitations in the models available for parameter estimation, which can be both a strength and a weakness.

SCORES is a similar 1D, simulator of two-phase immiscible, incompressible flow, able to simulate the same test scenarios but, unlike Sendra, is free access. Simulation results are provided by e-mail, in the form of a spreadsheet, tabulating data (such as oil and water production rates and cumulative volumes, pressure drop, average saturation) as a function of time. Saturation profiles are also generated in a separate spreadsheet. There is no parameter estimation feature on the freeware product. The user must alter model parameters as input to the forward simulation of production history profiles. It is therefore more important to have constraints on capillary pressure and endpoint data from additional testing, to leave only two or three parameters to alter—thus making a more robust, unique result.

Pre-simulations can also be used to design experiments. Using some initial estimate of relative permeability and capillary pressure, simulation will provide expected DP transients (to assess whether these are measurable in the lab), saturation data (to assess capillary end effects), production data and time to achieve end-point or SS saturation. Once initial relative permeability measurements have been made (e.g. on the first sample), the tool becomes a valuable aid for optimising further tests.

Simulation input data are not onerous, but represent significantly more information than is normally reported (or even acquired) by most commercial labs. Typical data input includes core details (absolute and end-point permeability data, porosity and pore volume, core dimensions), fluid properties (IFT, viscosity and density) and transient flood data (injection rates and/or volumes, production rates, DP as a function of experimental time, saturation profiles as a function of length).

Production data (DP and production volumes) are dependent on both relative permeability and capillary pressure. There are numerous (and non-unique) solutions/roots of the complex equations governing these relationships. It is therefore essential to define as many of the input parameters as possible (reduced input parameters increase the number and non-unique nature of solutions). Therefore, full Pc curve (relevant to the displacement scenario—e.g. water–oil imbibition Pc for water–oil imbibition relative permeability) and ISSM scans are strongly recommended to improve the results of the history-matching process. If coreflood simulation is to be used, test procedures must clearly specify the data requirements for input to the simulation.

Despite the caveat that several non-unique solutions to the same input data are possible, coreflood simulation represents the most appropriate method to analyse laboratory relative permeability data, and is preferred above the analytical approaches (e.g. JBN).

10.4 SAMPLE SELECTION, TEST STATE AND TEST CONDITIONS

10.4.1 Sample Selection

The sample set must reflect the full range of rock types and poroperm properties in the reservoir and should cover all the rock types comprising the significant reservoir flow units. The number of samples will depend on heterogeneity in the reservoir, and samples should be grouped by zone, facies, porosity class or hydraulic flow units. Normally, the range of core porosity and permeability used in SCAL tests should reflect the porosity and permeability distribution of the RCA database.

Mohammed and Corbett (2003) studied the number of relative permeability curves required for a specific reservoir, based on geological analysis, rock typing, heterogeneity analysis and flow simulation, which they considered most appropriately, represents a framework for the selection of the appropriate material for relative permeability tests. The selection of relative permeability curves from the more important rock types, defined as those making up the dominant transmissive and storage units in the reservoir, may be significant.

A simpler approach is to plot normalised cumulative RQI data as a function of depth. This could also include probe permeameter measurements if available. Both RQI and cumulative RQI are calculated for each core point and normalised by dividing the RQI at each point by the total cumulative RQI for the section of interest, i.e., RQI/Σ(RQI). This is effectively a pseudo-PLT, as indicated in Fig. 10.26 where the green (light grey in the print version) arrows pick out the units which will provide the majority of the contribution to flow.

All samples (both core sections and plugs) should be visually inspected and CT scanned prior to selection to attempt to ensure homogeneity. A grading system should be adopted, for example, grading samples from one (1) to five (5), where:

1 = good homogeneous core,
2 = some heterogeneity, but still usable,

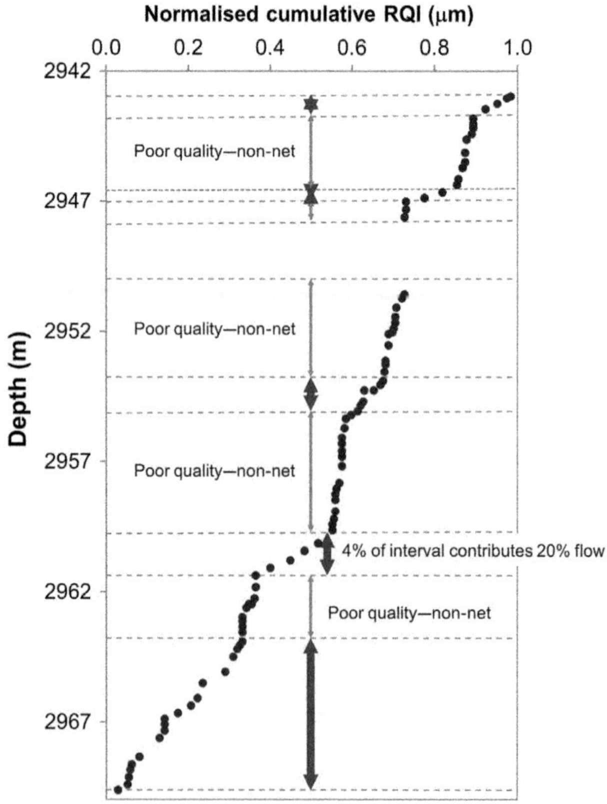

FIGURE 10.26 Example of normalised cumulative RQI plot.

3 = highly questionable, do not use except from necessity and in the understanding that using the sample entails more risk of failure than usual,
4 = do not use (under any circumstance) and
5 = damaged/fractured sample that physically cannot be used.

In addition, the samples should be characterised for pore size distribution (mercury injection test) and for petrography/mineralogy (thin section, SEM and XRD).

Despite adopting a rigorous approach to sample selection, bias can be introduced by core availability and core viability (from CT assessment). For example, if tests are to be carried out on "fresh-state" core, or where it can be shown that de-preservation may alter key rock properties, samples are selected from preserved core. Core preservation is often done on samples selected at regular intervals or to capture the best rock. Selective preservation can often mean that 80% of the samples are from the top 20% of rock quality. It is essential to ensure that whole core sections are representative and not biased. It is advised to preserve parallel SCAL plugs at every two or three RCA plug locations as this provides a better chance of acquiring more representative samples than whole core preservation alone. It also reduces the amount of "missing" material for sedimentological interpretation.

In addition, core recovery is often poor in high-permeability zones (especially in unconsolidated sands). Samples displaying heterogeneity should be avoided, since data analysis is ultimately dependent on a 1D flow regime and upscaling of results may not be particularly reliable. A strict grading regime is advisable, where only top grade samples are used for dynamic tests. If it becomes necessary to use lower-graded samples, one must be aware that there is a risk of not obtaining reliable data (and willing to pay for that risk).

In cases where reservoir condition mobility ratios are significantly greater than 1, composite cores may be required. In short cores, the velocity may be too high for the appropriate imbibition process to take place between fingers of invading water. Experimental evidence (Rapaport and Leas, 1953) suggests that core lengths of at least 25 cm are necessary to obtain consistent results. To obtain composite core sample of lengths greater than 25 cm parallel to the bedding plane, it is necessary to butt together small cores and construct a composite core (Section 10.3). Composite cores should have component sections of similar petrophysical character. Criteria for assessing similar petrophysical character are: porosity, absolute permeability (air and water), capillary pressure, pore size distribution, initial saturation and effective permeability at initial saturation. Note: The displacement method and force applied to acquire the initial saturation must be the same.

Individual core plug order within the composite stack is very important. It is recommended to use "retrogressive ordering", i.e., highest permeability at the injection face with descending permeability towards the production face.

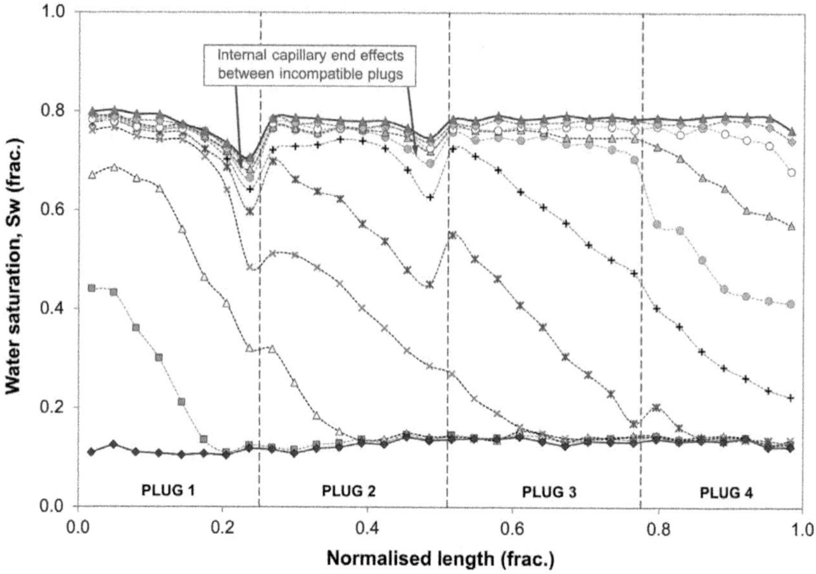

FIGURE 10.27 Example of capillary end effects in composite core (marked by discontinuities in saturation profiles at plug boundaries).

It is essential to ensure the samples are butted together (for example, under biaxial stress) to guarantee capillary continuity. If not, end effects may exist between butts and can adversely affect the flood front dynamic through the composite material (for example, Fig. 10.27).

10.4.2 Test State

Water–oil and gas–oil tests can be carried out in three principal core states:

- Fresh state (tested "as is" without cleaning or drying),
- Clean state (tested after cleaning and drying and assumed to be water-wet),
- Restored state (tested after attempting to restore wettability by ageing in crude oil).

A summary with the advantages and disadvantages of the different test states are presented in state summary tables (Tables 10.2–10.4).

10.4.3 Overburden Stress

Ideally, the tests should be carried out at effective confining stresses which are representative of the reservoir stresses (Refer to Chapter 6).

It may be argued that, although absolute permeability will be reduced by overburden pressure, relative permeability may be comparatively unaffected. In this case, tests could be carried out at more nominal overburden conditions,

TABLE 10.2 Fresh State: Initial Sample State Summary

Fresh State	
Description • Fresh state refers to core tested "as received" in the laboratory. Fresh state does not usually differentiate between oil- and water-based muds. This core could be preserved • Native state is a special case of fresh state. Proponents believe in minimal interference with the core prior to testing, and that changes induced during core cleaning are not reversible. One technique for obtaining a "native-state" core is by drilling with a crude oil followed by a special packing to prevent the core from drying out and exposure to oxygen	
Advantages	**Disadvantages**
• No previous treatment is required on the sample. The sample can be used directly for testing • Allows for a short testing period • Minimal interference with the core prior to testing. Good preservation of the core is required • Representative initial saturation conditions are established in the case of native state • Lab test can be performed at reservoir conditions using live oil reservoir and reservoir pressure and temperature	• Wettability may be altered as on drilling and core recovery, and if the core is exposed to air, can be rendered oil-wet. Cores should be cut using a low-invasion core bit and bland, water-based mud • The initial properties of the test sample, particularly pore volume, are unknown at the start. This can compromise lab quality control of saturation calculations • Lab testing needs to start as soon as the samples are received in case the preservation deteriorates • Not recommended in cores drilled with oil-based mud, or in heavy oil reservoirs, the core recovered at surface will almost certainly be more oil-wet than in the reservoir • Cores drilled with WBM may have suffered filtrate invasion, increasing Sw. The samples have to be returned to Swir by oil flooding so the fresh-state waterflood may represent a secondary imbibition cycle. Effects of saturation hysteresis may not be well defined • High cost if tested at reservoir conditions

avoiding the expense associated with overburden tests. Again, initial tests might be carried out at both ambient and overburden pressures. If the resulting relative permeability is the same, then both approaches are valid and either could be used in further testing programmes. Representative overburden pressures should always be used for tests on unconsolidated media.

10.4.4 Initial Water Saturation

It is essential that the initial water saturation is representative of water saturation appropriate to the rock quality of the test core plugs and height above the FWL for the zone of interest. It must be noted that Swi may not be at Swir and

TABLE 10.3 Clean State: Initial Sample State Summary

Clean State	
Description • Clean state refers to the core material where attempts are made to remove all the fluids and adsorbed organic material by flushing or refluxing in solvents. The objective of cleaning is to return the cores to a water-wet state—the condition in the reservoir prior to hydrocarbon accumulation	
Advantages	**Disadvantages**
• Test data can be acquired more rapidly that on restored-state cores • If the reservoir wettability is confirmed to be water-wet by appropriate wettability tests, cleaned-state tests will be representative. This is the normal condition for gas–water or water–gas tests. Care needs to be taken to ensure that the cleaning method removes wettability contaminants and does not induce an oil wetting tendency—as discussed in Chapter 4	• Tests that are performed using clean-state samples may not be representative of a hydrocarbon reservoir leading to unrepresentative data • Hot Soxhlet cleaning of oil cores (or cores drilled with some oil-based muds) can induce a non-water-wetting tendency

the reader is advised against applying the maximum available laboratory force to establish Swi for testing which has the objective of emulating the reservoir conditions. However, if Swi is too high (for instance, in cases where the laboratory forces are insufficient to achieve Swir), the waterflood may become unstable (due to a mobile water phase at the outset of testing) leading to early BT and the residual HC saturation phase may not be representative. Also, water saturation is one of the controlling factors in establishing reservoir wettability and incorrect Swi could lead to incorrect wettability and, hence, incorrect relative permeability. When commencing displacement floods, one should consider the forces employed to establish initial conditions; one should pre-simulate (either by coreflood simulator or simple calculations) expected experimental forces; one should mitigate adverse forces (i.e. forces higher than those used for establishment) by reducing injection forces (rate and/or pressure).

For fresh-state core, the initial water saturation is essentially as received. If the core has been cut with OBM, then the sample is initially flooded with (or centrifuged under) test oil to ensure Sw is at Swir. If the core is drilled with WBM, then filtrate invasion is most likely to have increased Sw above irreducible values (primary imbibition). The water saturation has to be returned towards Swir by oil flooding or centrifugation (secondary drainage) such that

TABLE 10.4 Restored State: Initial Sample State Summary

Restored State	
Description • Restored state assumes that core wettability changes will have occurred during coring and core recovery/handling. Oil-wetting contamination must be removed by thorough solvent cleaning, followed by restoration by ageing at representative initial water saturation in crude oil at reservoir conditions for several days or weeks. This is assumed to re-establish reservoir wetting character (but can never be definitively proved). Nevertheless, restored-state analysis is the most appropriate test state for oil reservoirs	
Advantages	**Disadvantages**
• Tests are performed on samples potentially restored to original wettability conditions • The displacement tests are ideally performed with real reservoir fluids at reservoir conditions but, provided wettability has been restored, could be carried out with synthetic fluids • Refined oils have been extensively used for measuring relative permeability for the reason of cleanliness and their wide range of viscosities to ensure good subordinate reduction following breakthrough. Care should be taken to ensure that they do not contain polar compounds which could alter the wettability of the rock surface. Flushing with decalin following ageing in crude oil is recommended if mineral oil is used • No specific restrictions on mud type but wettability contaminants must be removed by cleaning prior to restoration • If GOR is low, dead crude oil (stock tank oil) is adequate for ageing	• Testing takes some considerable time to complete as the sample needs to be cleaned then aged • Restoring the wettability can take at least 1200 h to establish equilibrium, although this can vary from reservoir to reservoir and can be much shorter (200–400 h) for lighter oils • Restored-state core wettability might be too water-wet due to irreversible cleaning effects, insufficient crude ageing time or a change in the crude oil composition (between accumulation and production) • Restored cores can be too oil-wet if any oil-wetting components have not been removed by cleaning and the core is not rendered wholly water-wet prior to restoration. If high GOR dead crude is used for ageing, the relative concentration of wettability-altering surfactants is increased • If GOR is high, then samples should be aged at reservoir conditions with recombined reservoir fluids • Using live crude for ageing and testing is expensive • Tests at reservoir conditions may preclude the use of some ISSM equipment (pressure and temperature limitations on system)

the subsequent waterflood represents a secondary imbibition cycle. In cleaned-state and restored-state tests, the core sample is conditioned to Swir and needs to be consistent with those estimated for the reservoir either from log analysis, core saturation measurements or from capillary pressures.

The initial water saturation can be established using three different techniques:

1. Dynamic displacement (oil flooding)
2. Centrifuge
3. Porous plate

Table 10.5 summarises these methods and gives the advantages and disadvantages of each.

The porous plate method is the only method that allows the establishment of a uniform fluid distribution at a specific capillary pressure. It is also possible to avoid water-trapping errors. However, McPhee and Arthur (1994), in a review of the comparative performance of four laboratories (Fig. 10.28) carrying out simple waterflood experiments, found that two laboratories using the porous plate method produced values of Swir that were almost three times higher than the appropriate value for the test rock (Fig. 10.28). They also found that dynamic displacement techniques failed to establish a representative Swir, the values again being three times higher than they should have been. Thus, the establishment of Swi should be a control point for all tests where this is achieved, to ensure that correct saturation values are attained, prior to continuing with ageing and subsequent analyses.

Nevertheless, time (and cost) constraints may favour the use of the centrifuge. Refer to Chapter 9 for further details and recommended practice. An ultra-centrifuge with automated data acquisition is recommended. Following desaturation at a specific capillary pressure (usually equivalent to the maximum height above the FWL), the plug is offloaded and stored under oil (for oil–water systems). Some time must be allowed for the non-uniform saturation profile to equilibrate and dissipate under capillary forces. For both oil–water and gas–water systems, the core is flooded with oil or gas for several pore volumes, and pressure drop and water production monitored, to ensure stable flow conditions prior to testing. Conventional analysis calculates the core inlet face saturation for centrifuge Pc curves. However, for relative permeability analysis, it is the average plug saturation that is used for Swir.

10.5 MEASUREMENT DESCRIPTIONS

This section provides full details of the principal relative permeability test methods for water–oil, gas–oil, gas–water and water–gas systems. The advantages and disadvantages of each test are discussed, and quality control procedures and diagnostics are provided. Each test is presented in a similar format.

Refer to Sections 10.1.2, 10.2 and 10.3 for definitions, test overviews and test interpretation methodologies.

TABLE 10.5 Methods to Obtain Initial Water Saturation

	Method Description	Advantages	Disadvantage
Dynamic Displacement	• Oil is injected at constant pressure or rate through a sample that has been initially saturated with water. The sample is located in a pressurised coreholder and is pre-conditioned to achieve representative wettability and initial water saturation. Usually, the sample is pre-flushed with a low-viscosity oil followed by a viscous oil to assure more rapid desaturation to Swir. After reaching Swir, the viscous oil is displaced with the test oil	• Rapid • Can also be used to provide primary drainage relative permeability for a gas–water or oil–water system	• A non-uniform saturation profile is developed in the sample • A linear or quasi-linear pressure drop will result from the oil flow • Swir can be higher than the true initial conditions • Sro distribution on subsequent waterflood is affected because of changeable initial saturation conditions and perturbation of the shock front • In oil-wet systems, water trapping may occur leading to a high initial water saturation, premature water breakthrough and suppressed oil effective permeability
Centrifuge	• The core sample is saturated in water and placed in a ultra-speed centrifuge (Chapter 9). The sample is then spun down at a rotational speed equivalent to a predefined level of drainage capillary pressure that should yield Swir. Incremental speeds can also be used to define a full Pc curve, if required. Air (gas–water) or oil can be used as the displacing fluid	• This technique is faster than porous plate method • Capillary pressures can be controlled to set a specific water saturation • The pressure gradient across the core is non-linear so that boundary effects are less than those experienced under dynamic displacement	• Swir can be non-uniform due to mild end effects. In this case, the core is flooded with oil or gas at Swir in a test coreholder to allow the end effects to dissipate • Friable samples can be damaged and fracturing can occur during centrifuging. The presence of microfractures can dominate the flood behaviour
Porous Plate	• The sample is desaturated in a porous plate cell under a single or incremental capillary pressure to define Swir (refer to Chapter 9). Air (gas–water) or oil can be used as the displacing fluid	• Swir is uniform as the pressure applied in the sample is equal • Capillary pressures can be controlled to set a specific water saturation • Good method for weak samples, provided desaturation is performed at nominal confinements	• Even for a single capillary pressure, it can take between 3–6 weeks before capillary and saturation equilibrium are achieved, as a consequence of the low permeability of the porous plate. Longer desaturation times may be required for low-permeability samples • Sample grain loss and poor capillary contact can be a problem at ambient conditions • Swir may be too high if sufficient desaturation time is not allowed

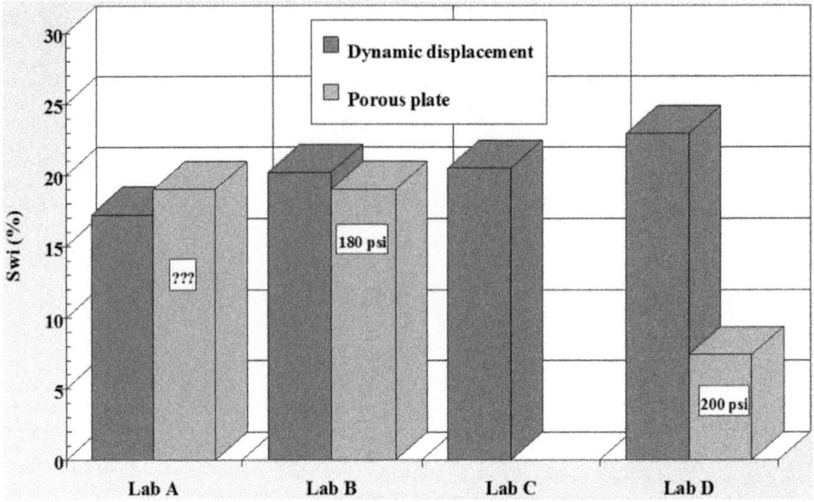

FIGURE 10.28 Lab variation in Swir. *McPhee and Arthur (1994).*

10.5.1 Unsteady State Water-Oil Relative Permeability (USSWO)

Purpose of Test:	• To determine water–oil imbibition relative permeability, kro and krw. A "waterflood" test is designed to yield only the end points—Ko′ at Swir and Kw′ at Sro—and total oil recovery. A "relative permeability" test in addition is designed to provide the intermediate relative permeability curves between the residual saturations and the incremental oil production
Sample Requirements:	• Cylindrical-shaped plug sample of 1″ or 1½″ diameter. The larger the sample, the larger the potential pore volume, and the less will be the impact of saturation measurement errors • *Ideally*—sample taken from preserved core or preserved plug. *Alternative*—plug previously used in routine core analysis programme, provided that plug has not been pre-stressed and/or that plug preparation methods have not caused damage to delicate clay minerals • Composite core (3–6 core plugs butted together) will improve the precision of saturation determination and may help minimise end effects
Required Input Data for Lab:	• Swir appropriate to maximum height above FWL for the sample porosity and permeability • FW salinity and ionic composition, as close as practically possible to the FW in the reservoir or FW resistivity at reservoir and lab temperature. Lab will prepare synthetic FW (SFW) accordingly. SFW (or oil) may be doped with NaI (or iododecane) to facilitate ISSM

- For non-reservoir condition tests: reservoir STO can be used for ageing and/or testing. *Alternatively*, the lab may use mineral oil of appropriate viscosity. For reservoir condition tests, reservoir oil and gas samples and PVT data are required to allow recombination
- Physical properties of the oil (i.e. viscosity, API)
- Overburden stress
- Ideally—Reservoir conditions (pressure and temperature) where possible
- Flow rate (or reservoir capillary number)

10.5.1.1 USSWO Sample Preparation

The initial water saturation of the sample set should be established as described in Section 10.4.1.

10.5.1.2 USSWO Key Processes

1. Prepare core
2. Condition sample to appropriate Swir and restore wettability (if required)
3. Overburden stress application
4. Determine end-point oil permeability, Ko′ at Swir
5. Displacement (waterflood)
6. Determine end-point water permeability, Kw′ at Sro
7. Possible bump flood—re-determine Kw′ at Sro (post-bump)
8. Dean–Stark extraction to check calculated end-point water saturation (if ISSM scans are not required)
9. Processing and refining lab-calculated relative permeability (JBN or coreflood simulation)
10. Plot relative permeabilities versus water saturation.

10.5.1.3 USSWO Test Procedures

The sample is located in a pressurised coreholder and is pre-conditioned to achieve the conditions at which the test will be performed (e.g. fresh state, cleaned state or restored state). The waterflood tests are normally gravity-stabilised: the coreholder is held vertical, and the denser phase (water) is injected into the bottom of the coreholder, displacing the lighter phase (oil) from the top of the cell. Gravity-stabilised floods prevent gravity override, where there is a substantial density difference between phases, or where injection rates are so low that the ratio of gravitational forces to viscous forces could become significant.

In fresh-state or restored-state tests where mineral oil is used as the test fluid and the core has been exposed to or aged in STO, it is advisable to flow 2–3 pore volumes of decalin through the sample before oil is injected to prevent possible asphaltene deposition arising from an adverse reaction between mineral oil and STO.

Fluid can be injected either at constant rate (using a pump) or at constant pressure. *Constant rate is preferred* (for liquid systems) since precise control on flow rate is possible and relative permeability changes during the test are reflected by pressure and fractional flow changes which are easier to measure than changes in total flow and fractional flow (constant pressure).

Fluid injection rate, pressure drop and fluid temperature are monitored and recorded against time. Normally, produced fluids are collected in a series of specially designed volumetric flasks of increasing volumes. The produced oil and water phases are allowed to separate under gravity and the relevant volumes read off. In reservoir condition tests, the fluid pressure is maintained above the bubble point of oil using a backpressure regulator. Accurate separation and measurement of produced fluids are best carried out at backpressure using a pressurised endoscope, pressure-monitored separator, or acoustic separator. Increasingly, ISSM is used to determine the saturation level during these tests, but the techniques are expensive and elevated pressure or reservoir condition systems may not be widely available.

Flooding at an excessively high rate in low-permeability samples can produce very high DP which requires high sleeve confining pressure to maintain a seal. This can create a significant net overburden pressure gradient across the core sample, which will produce differential rock compaction effects that, at best, may lead to permeability data inconsistency and, at worse, could cause plug failure.

Waterflooding should be continued until pressure drop stabilises and, where experimental constraints allowed, water cut exceeds 99.95%. End-point effective water permeability, Kw′, is then determined at Srow. If the flood is conducted at low rate (typically 4 ml/h), capillary end effects may dominate. Consequently, the flow rate is often ramped up (bump flood) in magnitude increases, i.e., to 40 ml/h, then 400 ml/h, if significant additional production was observed during the initial bump rate (e.g. +5 saturation units). Any incremental oil production and permeability increase (DP decrease) should be monitored throughout. In some cases, the sample may be removed, placed in a centrifuge under SFW and spun at rotational speeds sufficient to define the ultimate Sro (see Fig. 10.5). The core is then replaced in the test coreholder and the final effective water permeability is measured.

Plug saturations, at the beginning and at the end of the waterflood, are calculated on the basis of the final dry weights and porosities, measured on completion of the total test cycle, the saturated plug weights at each test stage, and/or volumetric production data measured during the test sequences. The change in the saturated weight in the plug between the beginning and end of the waterflood is used to determine and reconcile end-point saturations, and is assumed to be a direct result of the decrease in oil during the test. Otherwise, ISSM can be used to verify and visualise saturations at the end points and/or during the flood. Unless a final ISSM scan is required, final water saturations should be checked using Dean–Stark extraction.

Water–Oil Relative Permeability USS Summary

Advantages:	Drawbacks and Issues:
• Less time consuming and expensive than SS • Most labs can perform the test at ambient conditions • Appropriate "shock-front" development • Ability to flow at appropriate reservoir flow rates • Relatively low throughput (less prone to fines migration) except at high rates	• Long core plugs are preferred to optimise measurement accuracy • Data optimisation dictates that test should ideally be carried out using a high-pressure capacity rig and using ISSM. This demands a high level of sophistication in lab equipment and resources • Test costs would be high due to requirements for ISSM and/or reservoir conditions • Relatively complex interpretation (JBN) • Capillary instability invalidates Buckley–Leverett theory and affects residual saturations, and krw curve • Coreflood simulation may be required to adequately model intermediate kro and krw • Friable samples require to be protected against grain loss and sample failure • kro curve can be distorted by imbibition Pc if viscous force is not large • Difficult to measure kro values at low oil saturations • Careful evaluation and selection of test conditions is essential for representative data. For example, select from: fresh-, cleaned- or restored-state cores; ambient condition experiments with synthetic fluids, or reservoir condition tests with live fluids; matched or enhanced oil/water viscosity ratios; reservoir-equivalent or artificially high flood rates • Residual saturation is dependent of the flood rate used. Unfortunately, even when test conditions have been correctly selected, the lack of standardisation of commercial laboratory test techniques provides further opportunity for significant data variation (McPhee and Arthur, 1994)

QC Checks

- Quality control of relative permeability data is essential due to the variability of raw data even on tests performed on the same core samples using identical conditions by different laboratories. Nevertheless, laboratory audits, the use of check samples, detailed quality control and quality assurance procedures, and fully documented waterflood data records can increase the confidence in contractors' data.
- Analysis of the results should start with a thorough review of the laboratory report to check that acceptable procedures have been followed. This should be carefully evaluated both in terms of lab procedures and the representativeness of the test to the reservoir wettability condition.
- Check what type of drilling mud was used to obtain the core material. Use of OBM and/or inefficient cleaning could lead to a non-water-wet condition
- Check timing and method of core preservation.
- Check how the laboratory handles the core, i.e., plugging, cleaning and drying procedures. Ensure care has been taken to avoid damage or de-preservation of clay structures.
- Check if the test state, fluids and test conditions are representative of the reservoir.
- If composite core plugs are used, ensure that they are correctly arranged in permeability order (permeability decreasing towards inlet) and evaluate any end effects between individual plug butts using ISSM if available.
- The end points should be checked by using the data obtained from different methods. Swir should be verified against primary drainage capillary pressure, and Sro from imbibition Pc tests.
- USS tests on short core samples with intermediate wettability and high permeability (>500 mD) may not always generate sufficient viscous force (DP) across the core to minimise capillary force instability.
- The Rapaport and Leas scaling criterion (see Section 10.3) can be used to give an initial assessment of the likelihood of end effects.
- Lab results need to be refined before use in reservoir simulation. This can be done analytically (see below) or by coreflood simulation (preferred).

Suitable Laboratories

- Most commercial laboratories can provide basic waterflood tests. Typically, these may be done at ambient or elevated confining stress with synthetic fluids. Normally, tests are done at high rates and, for relative permeability tests, adverse viscosity ratios to promote two-phase flow and calculation of kro/krw using Welge and JBN.
- Reservoir condition test facilities are available in the major laboratories (e.g. Core Lab, Houston and UK, and Weatherford, Norway, UK and USA). These labs also have access to ISSM. Both Weatherford and Core Lab can provide coreflood simulation on request.

10.5.1.4 Refining krw and kro – USSWO

Even in cases when tests are conducted with great care, inaccuracies associated with flooding short plugs and problems associated with capillary end effects and equilibrium must be considered. It is very rare to find laboratory results that do not have to be refined in some way. This refinement technique makes several simplifying assumptions but represents a reasonably viable approach to correct and refine legacy laboratory data in the absence of the experimental data required for full coreflood numerical simulation. The following procedures can be used to enhance and improve an USS water–oil dataset.

10.5.1.5 USSWO Curve Shape

- Check if the USS data show that krw curves increase too rapidly at low water saturations. A characteristic concave upward curvature of krw–Sw, when plotted in a semi-logarithmic scale, is generated due to lack of capillary equilibrium at the beginning of the flood (as shown in Fig. 10.29).
- A characteristic concave downward shape, or flattening, of the krw at higher water saturations may occur if capillary end effects are manifest at higher water saturations. These create an additional pressure differential at the core boundaries (e.g. "skin") that is not properly allowed for in the conventional calculation of relative permeability. It results in values of krw that are too low but can also easily be confused with fines migration

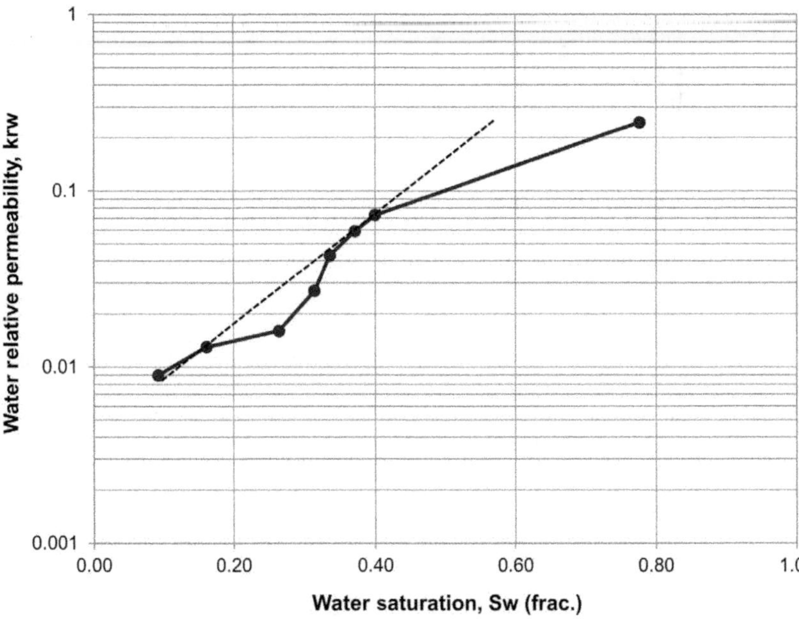

FIGURE 10.29 Example of poor relative permeability curve.

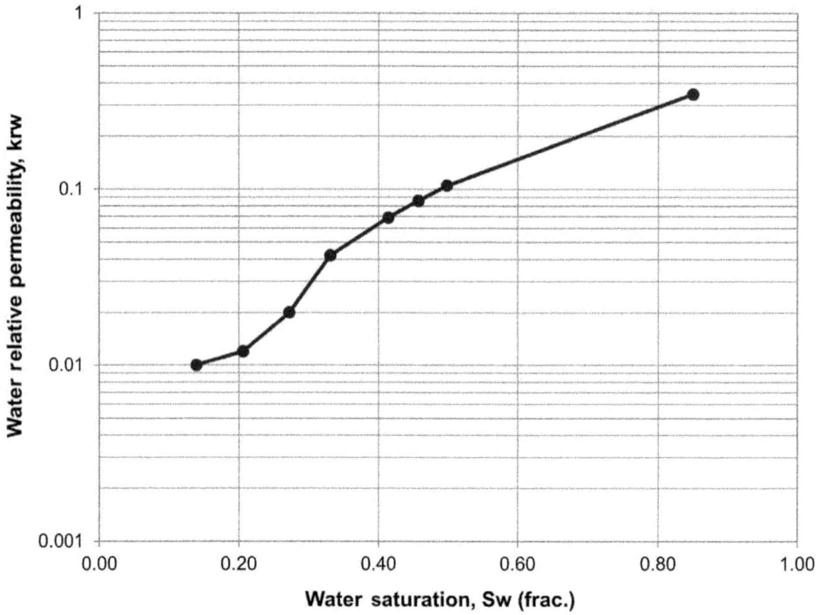

FIGURE 10.30 Example of a well-behaved krw curve.

where krw reduces at higher rates. Figure 10.30 provides an example of a reasonably well-behaved krw curve.

- Laboratory data are analysed and refined using the Corey approach. Refer to Section 10.1.12. This assumes that relative permeability curves can be described by a Corey-type function.

10.5.1.6 USSWO – Analysing, Refining and Extending, krw

- Estimate a value for residual oil saturation. Capillary effects (Section 10.3.4) can cause premature trapping of oil in some cases and viscous instabilities can cause bypassing of oil, so in many cases Sro values reported at the end of most waterfloods are often significantly higher than the true Sro of the rock (Srot). To correct by refining, firstly, the oil saturation is normalised by assuming different values for Sro (based on the lab results and other data) and plotting Son versus kro for each Sro estimate on a log–log plot (Fig. 10.31). The value of Sro that results in a straight line (slope represents the Corey exponent, No) is used as the best estimate of residual saturation. As shown in Fig. 10.31, Sro 2 = 0.320 gives a straight line.
- Refine the krw end point. The water saturation is normalised (Swn) using the best estimate of Sro, above:

$$\text{Swn} = \frac{(\text{Sw} - \text{Swir})}{(1 - \text{Swir} - \text{Sro})}$$

- Plot krw versus Swn on a log–log scale as shown in Fig. 10.32.

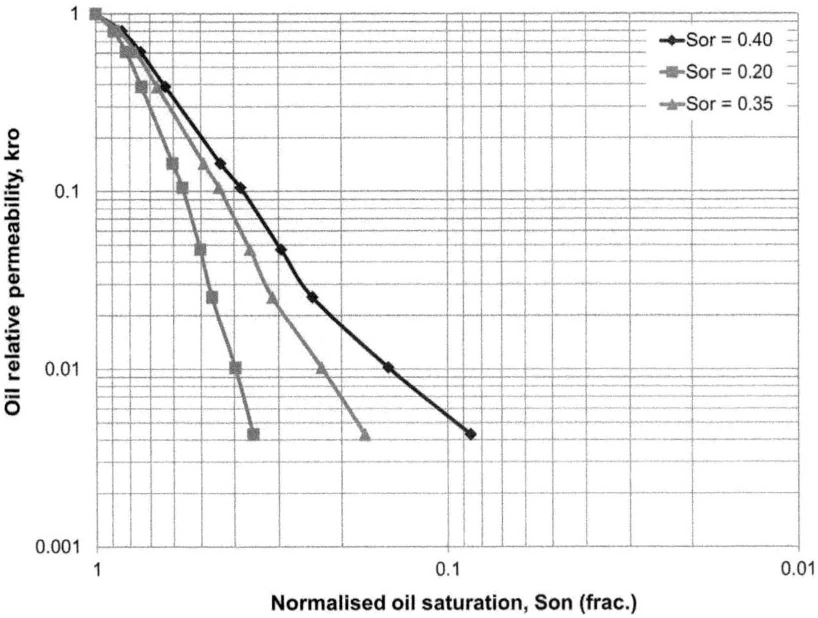

FIGURE 10.31 Normalised oil saturation (Son) as a function of oil relative permeability (kro) on a log–log scale.

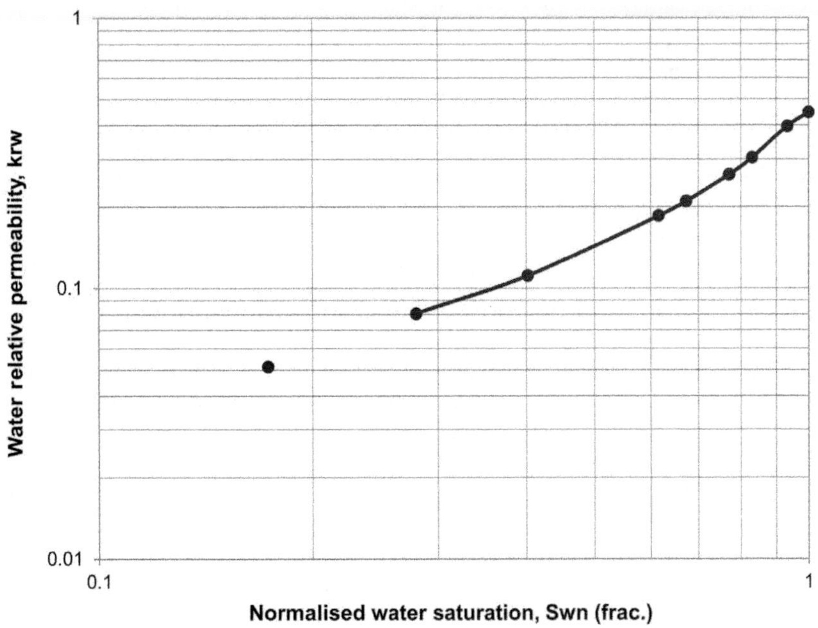

FIGURE 10.32 Normalised water saturation as a function of krw in a log–log scale.

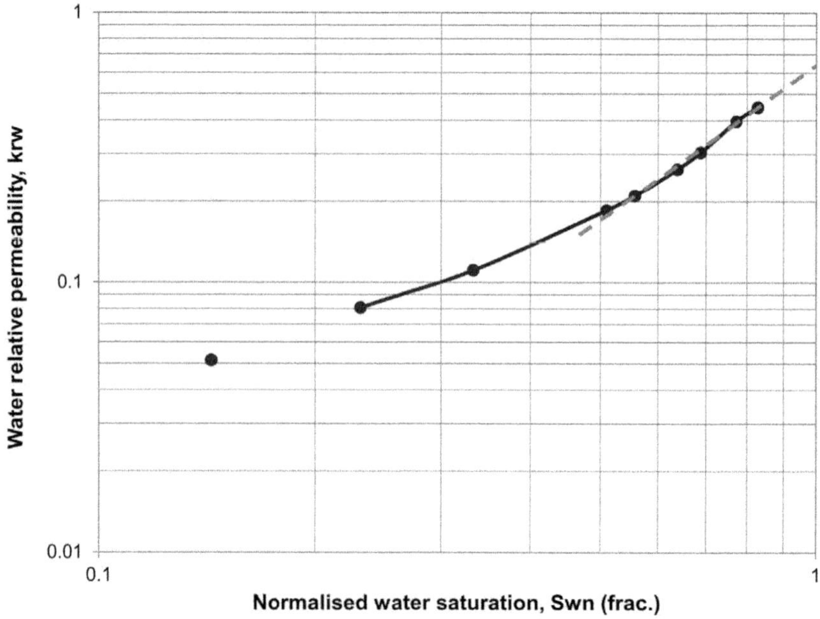

FIGURE 10.33 krw versus Swn.

The plot of krw versus Swn will be a straight line if the data follow a Corey-type curve. Deviation from this relationship could signal the influence of capillary end effects and/or flood instability. In the example above, the data at lower Swn seem to be affected by instability problems, but the remainder of the data (at higher Swn and higher viscous force) may still be valid.

- The instantaneous Corey exponent for the krw curve is determined for each data point using:

$$Nw' = \frac{\log(krw') - \log(krw)}{\log(1.0) - \log(Swn)}$$

This equation is equivalent to building a straight line from the end-point krw′ through each individual data point, starting each time from krw′ on a log–log plot. An example is shown in Fig. 10.33, where the krw curve for the last three data points is extended to define $krw'_{refined}$ at Swn = 1, where Swn is based on the best estimate (refined) Sro.

- The Nw Corey exponent is determined graphically by plotting Nw′ values against water saturation. Stiles (2007) suggests that Nw can be determined

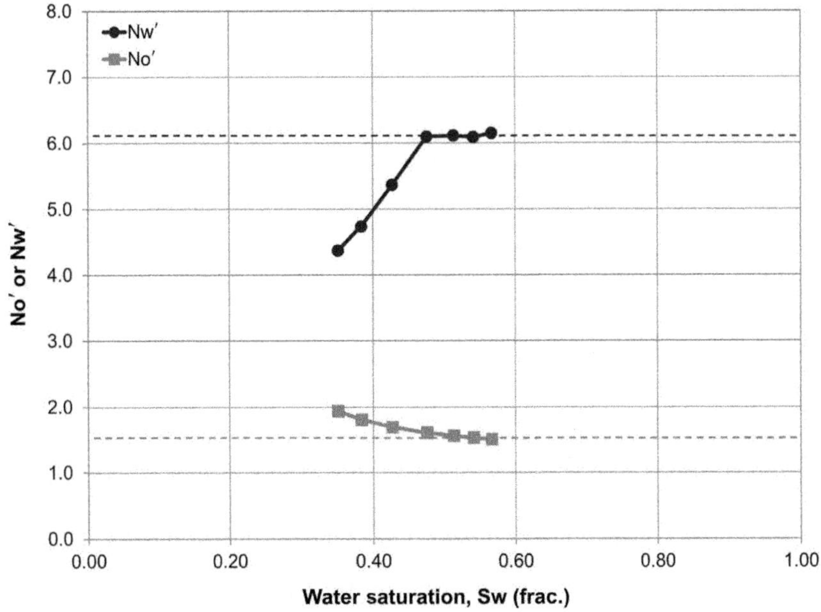

FIGURE 10.34 Water saturation versus No′ and Nw′.

by plotting Nw′ as a function of Sw. Figure 10.34 presents and example of this plot, which indicates that Nw′ is low at low saturations, but increases and stabilises. In this case, the Nw′ and $krw'_{refined}$ at Swn=1 might be used to improve the krw curve:

$$krw = krw'_{refined}(Swn)^{Nw}$$

Another example is shown in Fig. 10.35 where Nw′ stabilises over the last two data points. Nw can be estimated from these flat-line sections.

10.5.1.7 USSWO – Analysing, Refining and Extending kro

- kro is also refined and extended beyond the range measured in the laboratory.
- As kron=kro for kro based on end-point Ko′ at Swir, the instantaneous No′ Corey exponent can be determined for each lab data point using $No' = \frac{\log(kro)}{\log(Son)}$, where $Son = \frac{(1-Sw-Sro)}{(1-Swir-Sro)} = 1 - Swn$, based on the refined (best estimate) Sro.
- The Corey exponent No can be estimated graphically from the stable section of the No′ versus Sw plots (e.g. Figs. 10.34 and 10.35).

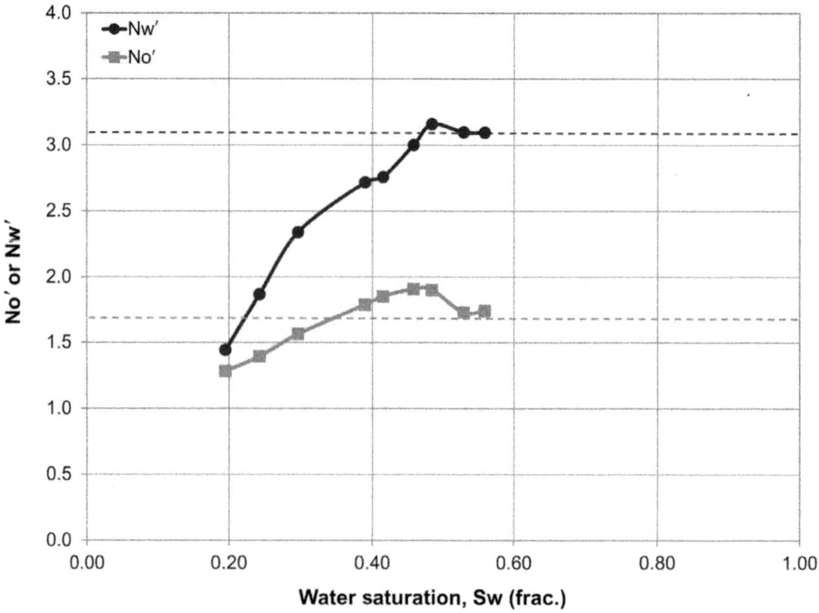

FIGURE 10.35 No′ and Nw′ stabilisation.

kro and krw refined curves

- The extended and refined krw and kro curves are generated based on the best estimate Corey exponents and the refined krw′ and Sro defined above. These are:

$$\text{kro}_{(\text{refined})} = \text{Son}^{\text{No}} \quad \text{and} \quad \text{krw}_{(\text{refined})} = \text{krw}'_{\text{refined}}\text{Swn}^{\text{Nw}}$$

- An example of the refined kro and krw curves is given in Fig. 10.36. In this case, the effects of capillary end effects and flood instability on the lab-calculated JBN data are evident. The refined curves extend to lower residual oil saturations and higher krw′ than the lab data.

10.5.2 Steady State Water–Oil Relative Permeability (SSWO)

In the steady-state (SS) method (Section 10.2.2), a fixed ratio of fluids is injected through the core sample until saturation and pressure equilibria are established. Water–oil relative permeability can be calculated using Darcy's law when the flow of fluids within the porous media reaches SS (Section 10.3.2).

SS techniques assume that capillary equilibrium has been achieved before making the final readings of flow rate and DP.

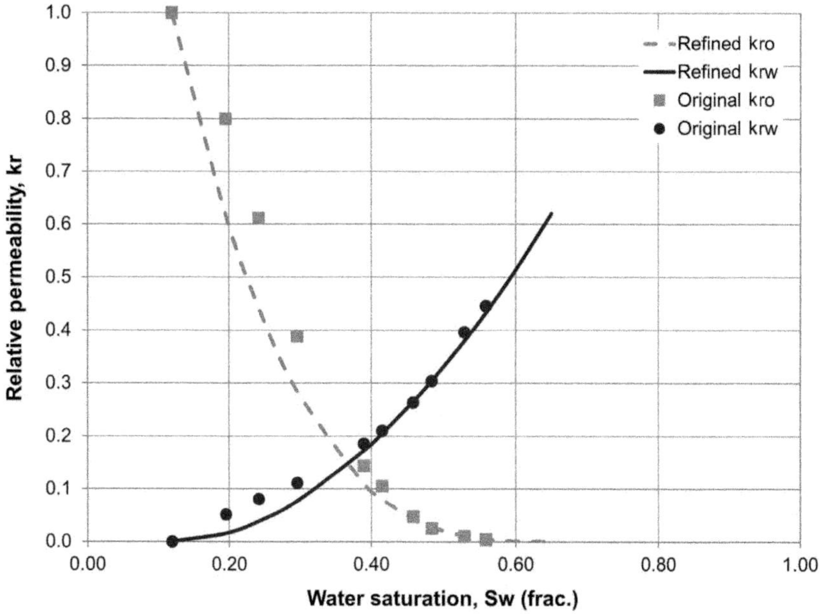

FIGURE 10.36 Comparison of original and refined krw/kro as a function of water saturation.

Purpose of Test:	• To determine intermediate imbibition water–oil relative permeability krw/kro. The end-point krw′ may differ from that established by USS or centrifuge techniques
Sample Requirements:	• Cylindrical-shaped plug sample of 1″ or 1½″ diameter. The larger the sample, the larger the potential pore volume, and the less will be the impact of saturation measurement errors • *Ideally*—sample taken from preserved core or preserved plug. *Alternative*—plug previously used in routine core analysis programme, provided that plug has not been pre-stressed and/or that plug preparation methods have not caused damage to delicate clay minerals • Composite core (3–6 core plugs butted together) will improve the precision of saturation determination and help minimise end effects
Required Input Data for Lab:	• Swir appropriate to maximum height above FWL for the sample porosity and permeability • FW salinity and ionic composition, as close as practically possible to the FW in the reservoir or FW resistivity at reservoir and lab temperature. Lab will prepare SFW accordingly. SFW (or oil) may be doped with NaI (or iododecane) to facilitate ISSM

- For non-reservoir condition tests: reservoir STO for ageing and/or testing. *Alternatively*, lab may use mineral oil of appropriate viscosity. For reservoir condition tests, reservoir oil and gas samples and PVT data to allow recombination
- Physical properties of the oil (i.e. viscosity, API, PVT)
- Overburden stress
- Ideally—Reservoir conditions (pressure and temperature) where possible
- Flow rate and water:oil injection ratios

10.5.2.1 SSWO Sample Preparation

The initial water saturation of the sample set should be established as described in Section 10.4.4.

10.5.2.2 SSWO Key Processes

1. Prepare core
2. Condition sample to appropriate Swir and restore wettability if required
3. Overburden stress application (as for USS tests)
4. Determine end-point oil permeability, Ko′ at Swir
5. Intermediate relative permeability curves:
   a. Inject oil and water simultaneously in steps
   b. Determine So and Sw at SS conditions
   c. Kw′ at Srow (Darcy analysis)
   d. Relative permeability (Darcy analysis)
6. Dean–Stark extraction for final water saturation check (if ISSM scans are not required)
7. Plot relative permeabilities versus water saturation.

10.5.2.3 SSWO Test Procedures

The test equipment is described in Section 10.2.5 and is similar to that used for USS. The key difference is that the test coreholder should incorporate inlet and outlet mixing (porous media) sections to promote fluid mixing and minimise capillary end effects as shown in Fig. 10.15. These should be designed such that core pressure drop can be measured directly at the test core end faces. Constant rate pumps are required to precisely control the injection ratios. Fluids are best separated and re-circulated.

Test procedures are similar to those for USS tests (Section 10.5.1) with the exception that the waterflood is carried out using incremental water:oil injection ratios (fractional water rates).

A decalin pre-flush is used when exchanging to and from laboratory oil and STO, to prevent asphaltene deposition from adverse reactions between the paraffinic mineral oil and STO.

The initial permeability to oil at Swi (most often used as the reference for relative permeability calculations by laboratories) is initially measured as for USS tests.

The imbibition cycle (waterflood) is commenced by flowing oil and water together through the sample, initially at a low water fraction (say 2% water:98% oil) to effect the first change in saturation. To avoid saturation "shocks", the water rate should be increased slowly to the desired flow rate. SS flow is assumed to have been achieved when no further change in saturation or DP has been observed. The oil and water rates, DP, and saturation are then used to calculate relative permeability versus saturation using Darcy's law.

The fractional flow of water (fw) is then increased and when the system measurements are again stable, relative permeability and saturation at this fractional flow level are calculated. The procedure is repeated until fw=1 and residual oil saturation is achieved.

The following incremental injection ratios might typically be used:

Ratio	Oil Fraction (%)	Water Fraction (%)
1	98	2
2	95	5
3	80	20
4	50	50
5	20	80
6	5	95
7	1	99
8	0	100

A high total flow rate is most often used, since equipment limitations make it difficult to inject at very low fractional injection rates, and since the flow regime does not represent the expected regime in the reservoir, there is little point in attempting to utilise reservoir-equivalent flow rates. At low rates, it would also be difficult to establish continuous flow (tendency towards slug flow) and equilibration at each fractional flow could take weeks to achieve, making the test prohibitively expensive and increasing the opportunity for technical failure.

Plug saturations, at the beginning and at the end of the waterflood, are calculated on the basis of the final dry weights and porosities, measured on completion of the total test cycle, the saturated plug weights at each test stage, and/or volumetric production data measured during the test sequences. However, intermediate saturations can be much more problematic. Essentially, the water saturation between injection ratios may change by very small amounts and the difficulty is detecting such a small increase when circulating many pore volumes (litres) of fluids. Consequently, ISSM is the only viable method to determine intermediate saturations in SS tests. The alternative is to unload the sample after stability at each injection ratio, weigh it and determine saturations gravimetrically. Grain loss and fluid expansion on unloading will lead to large uncertainties in saturation, and *gravimetric methods are not recommended.* Unless a final ISSM scan is required, final water saturations should be checked using Dean–Stark extraction or miscible solvent displacement and Karl Fischer titration of the solvent effluent for water content.

Water–Oil Relative Permeability SS Summary

Advantages:

- Relative permeability is calculated over a wide saturation range
- krw better defined at low Sw values
- Avoids krw instability problems
- Simple calculation (Darcy)
- Recommended method for adverse oil: water viscosity ratio tests as USS tests suffer from viscous instability

Drawbacks and Issues:

- Long core plugs are preferred to optimise measurement accuracy
- Data optimisation dictates that tests must be carried out using a high-pressure capacity rig and using ISSM. This demands a high level of sophistication in lab equipment and resources
- Test costs would be high due to requirements for ISSM and/or reservoir conditions
- Time consuming and expensive
- Flow mechanisms not representative of reservoir waterflood as little or no shock front are developed. This has an impact on end-point Kw′ and Sro
- Must re-start if test is disrupted
- kro definition limited near residual oil saturations
- Tests can only be performed at high rate—this too has an impact on Sro
- The test is prone to fines migration since large volumes of fluid are injected during the test

QC Checks

- The same QC checks used for water–oil USS (Section 10.5.1) are also valid for the steady date method.
- Lab results do not usually need to be refined as capillary end effects are minimised by the test system.

Suitable Laboratories

- Most commercial laboratories can provide basic SS tests. However, it is essential that all SS tests are carried out using ISSM to minimise saturation uncertainties. This will limit the test to the major laboratories (e.g. Core Lab, Houston and UK, and Weatherford, Norway, UK and USA) who have ISSM facilities.

10.5.3 Centrifuge Water–Oil Relative Permeability (CentKrWO)

As discussed in Sections 10.2.3 and 10.3.3, water–oil centrifuge displacements *only provide imbibition kro curves*. However, the end-point krw′ can be determined by transferring the sample into a coreholder and flooding with SFW until a stable pressure drop is achieved. The test equipment is described in Section 10.2.3 and must utilise automated data capture for production volumes.

The tests are carried out at a single rotational speed. This should be set such that the critical bond number (Nb—the ratio between gravitational and capillary forces in the experiment) should not exceed a widely accepted value of 10^{-5} (Section 10.3.3). However, if centrifuge capillary pressure tests were performed, a plot of Sro versus Nb will determine whether a critical Nb exists or will exhibit at which value of Nb, Sro begins to asymptote. This should then be the selected centrifuge speed.

The test may not be suitable for weak or friable samples, especially if end-point permeability is measured in a coreholder (repeated stress cycling can cause samples to fail).

Purpose of Test:	• To determine imbibition oil relative permeability, kro, at low oil saturations, approaching Sro
Sample Requirements:	• Cylindrical-shaped plug sample of 1″ or 1½″ diameter. The larger the sample, the larger the potential pore volume, and the less will be the impact of saturation measurement errors • *Ideally*—sample taken from preserved core or preserved plug. *Alternative*—plug previously used in routine core analysis programme, provided that plug has not been pre-stressed and/or that plug preparation methods have not caused damage to delicate clay minerals
Required Input Data for Lab:	• Swir appropriate to maximum height above FWL for the sample porosity and permeability • FW salinity and ionic composition, as close as practically possible to the FW in the reservoir or FW resistivity at reservoir and lab temperature. Lab will prepare SFW accordingly • As tests cannot currently be performed at reservoir conditions, tests tend to be restricted to synthetic fluids at ambient temperature, and in some equipment, at moderate overburden pressure. Lab may use mineral oil of appropriate viscosity • Overburden pressure if required and if possible • Spin speed (or critical bond number) and time intervals to acquire production data

10.5.3.1 CentKrWO Sample Preparation

The initial water saturation of the sample set should be established as described in Section 10.4.4.

10.5.3.2 CentKrWO Key Processes

1. Condition core to Swir (oil–water centrifuge primary drainage)
2. Determine end-point effective oil permeability (as for SS and USS) using same test fluids and overburden stress as used for centrifuge tests
3. Spin samples at the set speed
4. Measure the effluent product at specified time intervals time until residual conditions are established
5. Calculate the oil relative permeability curve using the equations derived from Hagoort (1980) or other models, or via coreflood simulation
6. Determine end-point krw′ at ultimate Sro in a coreholder
7. Dean–Stark extraction to check calculated end-point water saturation
8. Plot oil relative permeability versus water saturation.

10.5.3.3 CentKrWO Test Procedures

The centrifugal forces create a capillary pressure which causes the fluid in the sample to be displaced. Fluid production versus time data are collected by automated camera systems. The effluent produced at the single set speed is measured at specified time intervals until residual conditions are established.

On completion, the sample is loaded into a coreholder, confined at the predetermined stress, and flushed with SFW until pressure differential and rate are stable, to determine end-point krw′ at Srot.

Methods such as those developed by Hagoort are used to analyse the effluent production data, but the preferred technique is from computer simulation of flow behaviour, which many commercial laboratories do not have access to.

Relative Permeability Centrifuge Method Summary

Advantages:

- Relatively simple and quick method for kro curve close to Sro
- Good method for most rock types
- More efficient drainage than dynamic methods and the ability to measured very low kro values
- The pressure gradient is non-linear so that boundary effects are less than those experienced under dynamic displacement
- Although some uncertainty will be involved with these tests, the data should complement the USS and SS tests and will help define the kro curve towards ultimate residual oil saturation

Drawbacks and Issues:

- Use de-gassed oil or mineral oil only
- Although end effects are less, they still occur and result in a non-uniform distribution of saturation along the core sample length. Supporting the sample on a porous semi-permeable membrane can reduce these problems, but reduces sample length
- Friable samples can be damaged and fracturing can occur during centrifuging and end-point permeability determination. Friable samples must be tested at overburden pressure which can limit the spin speed in some equipment. The presence of microfractures can dominate the flood behaviour
- More basic concerns include uncontrolled imbibition at Swir, uncontrolled hysteresis, mobilisation of trapped oil, the difficulty in performing reservoir condition tests and the uncertain validity of accelerated testing
- Uncertainties arise if the centrifuge technique is used to derive relative permeability from systems exhibiting "trapping" of residual oil. These uncertainties arise because of the difficulty associated with defining saturation distributions in centrifuged samples
- Maintaining Nb less than the critical value is likely to be difficult in higher-permeability samples, so applied forces will be low, increasing the capillary end effects

QC Checks

- The same QC checks used for water–oil USS and SS analysis (Sections 10.5.1 and 10.5.2) are also valid for the centrifuge method.
- Experimental data are required (core dimensions, distance to inlet and outlet of sample, rotational speeds, production volumes and rates) to independently validate the results.
- The high spin speeds may invalidate the critical bond number (Nb) criterion normally imposed on centrifuge imbibition experiments. Associated desaturation effects can cause changes in capillary pressure and relative permeability which (usually) do not occur under normal field conditions.
- As the capillary-dominated displacement may be more efficient than dynamic displacement (especially in higher-viscosity oils), the end-point Sro may be lower and (from subsequent coreflood) the end-point krw′ may be higher than from dynamic displacement tests.

Suitable Laboratories

- Most commercial laboratories can provide basic centrifuge tests where data are analysed by conventional analytical solutions. Typically, these are done at ambient or elevated confining stress with synthetic fluids. However, it is recommended that the tests are carried out in a centrifuge fitted with automated data acquisition, so only laboratories with this equipment should be considered.

10.5.4 Unsteady State Gas–Oil Relative Permeability (USSGO)

Gas–oil tests are conducted with the sample at irreducible water saturation (Swi); however, this plays no significant part in displacement (as long as displacement forces are controlled to be less than the equivalent Swi establishment forces). Gas is a non-wetting phase and it initially follows the path of least resistance through the largest pores (Fig. 10.37). Compared to waterflooding a water-wet system, gas relative permeability will be higher than water relative permeability, and there is a more rapid decrease of oil relative permeability (Fig. 10.38) than when water displaces oil.

In gas–oil systems, gas permeability remains zero until gas saturation reaches a "critical value". The critical gas saturation, Sgc, is the point at which the gas bubbles become large enough to break away from the rock surfaces and move. In a reservoir, the critical gas saturation is reached soon after gas is evolved as the fluid pressure is reduced below the bubble point of oil. In conventional core displacement tests (at most laboratories), increasing the gas saturation by pressure depletion is rarely performed, and Sgc is determined by extrapolation of the gas relative permeability curve to a zero value at the saturation axis, or from analysis of the gas/oil relative permeability ratios, as discussed in Section 10.1.7. Typically, Sgc for most rocks is between 2% and 5% of pore space.

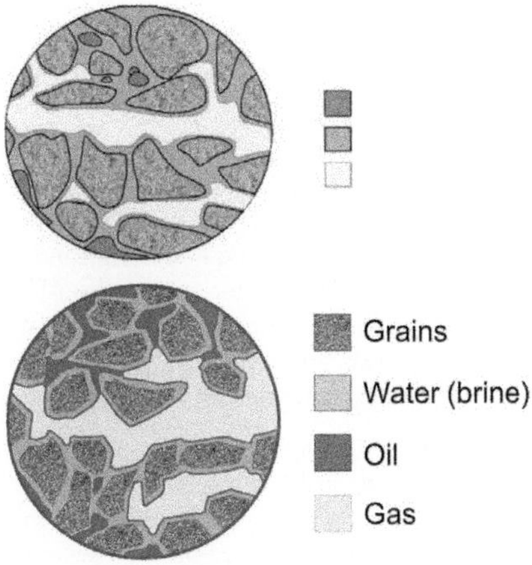

FIGURE 10.37 Gas flow in porous media.

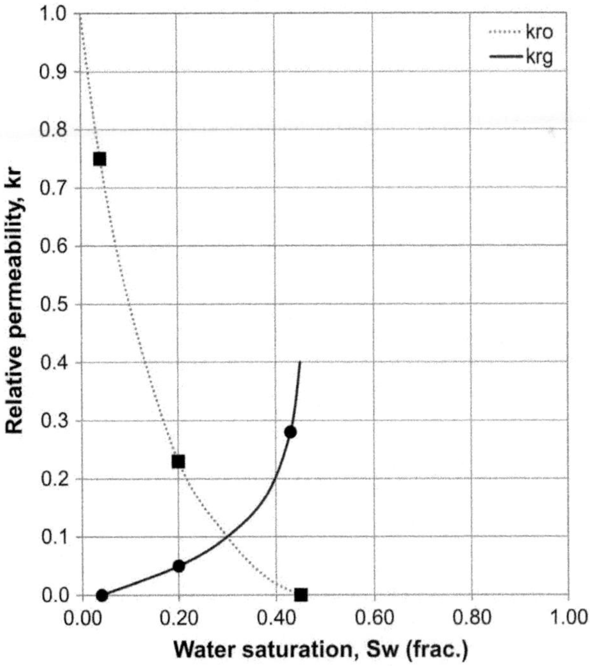

FIGURE 10.38 Example gas–oil relative permeability curve (linear).

Purpose of Test:	• Measure gas–oil relative permeability curves (kro and krg) as a function of gas or liquid saturation
Sample Requirements:	• Cylindrical-shaped plug sample of 1″ or 1½″ diameter. The larger the sample, the larger the potential pore volume, and the less will be the impact of saturation measurement errors • Composite core is preferred • *Ideally*—sample taken from preserved core or preserved plug. *Alternative*—plug previously used in routine core analysis programme, provided that plug has not been pre-stressed and/or that plug preparation methods have not caused damage to delicate clay minerals
Required Input Data for Lab:	• Swir appropriate to maximum height above FWL for the sample porosity and permeability • FW salinity and ionic composition, as close as practically possible to the FW in the reservoir or FW resistivity at reservoir and lab temperature. Lab will prepare SFW accordingly. Oil may be doped (with iododecane) to facilitate ISSM • Reservoir oil • Composition of the gas (if test at reservoir conditions) • Physical properties of the oil (i.e. viscosity, API, PVT) • Overburden stress • Reservoir conditions (pressure and temperature) • Flow rate (or reservoir capillary number)

10.5.4.1 USSGO Sample Preparation

As gas–oil tests are often conducted prior to water–oil tests on the same sample, the preparation requirements are the same as those described for water oil systems (e.g. Section 10.5.1).

10.5.4.2 USSGO Key Processes

1. Prepare core
2. Condition sample to appropriate Swir and restore wettability (if required)
3. Overburden stress application
4. Determine end-point Ko′ at Swir
5. Displacement (gasflood)
6. End-point Kg′ at Srog
7. Calculate gas and oil relative permeability (JBN or coreflood simulation)
8. Plot relative gas and oil relative permeability versus gas, oil or liquid (So + Swir) saturation.

10.5.4.3 USSGO Test Equipment

Unless tests are carried out under reservoir conditions, gasfloods are carried out using nitrogen and STO or mineral oil. The equipment required to

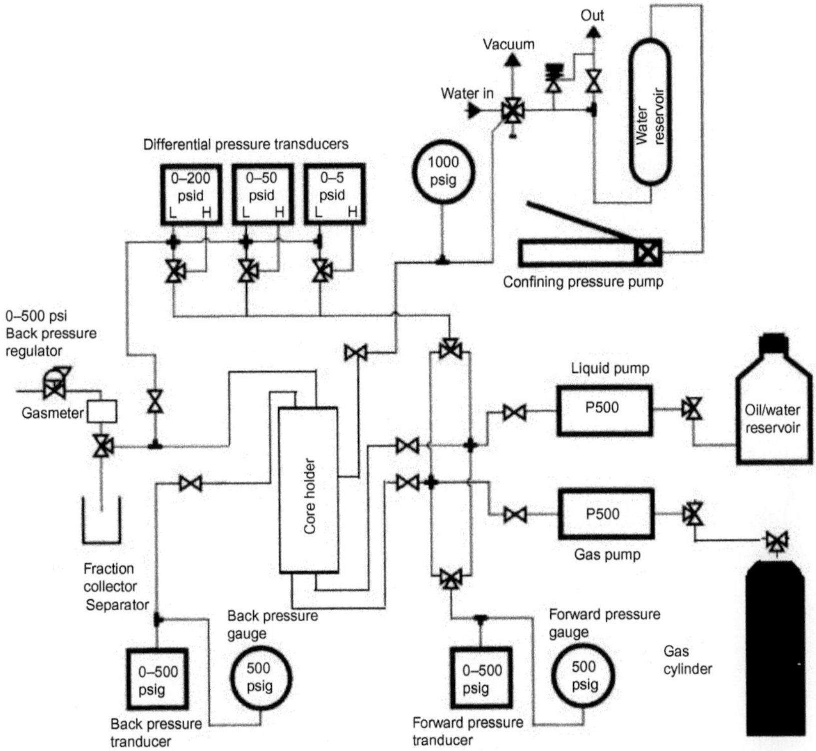

FIGURE 10.39 Typical gas–oil relative permeability setup. *Courtesy of Paul Glover.*

measure gas–oil relative permeability by the USS method is similar to the one described for water–oil relative permeability (e.g. Section 10.5.1). Tests should be carried out under backpressure to reduce gas compressibility artefacts and prevent gas saturation gradients developing across the test core.

For gas–oil measurements, the flood should be gravity-stabilised—the coreholder is vertically positioned as shown in Fig. 10.39, and gas injected from the top down. The coreholder confining sleeve should be constructed from material that prevents gas permeation; alternatively, annealed nickel foil may be used to encase the circumferential surface of the core plug (or composite stack), providing a viable gas block.

10.5.4.4 USSGO Test Procedures

The USS procedure for gasflooding is similar to that for waterflooding. For gas–oil tests, the core is initially conditioned to Swir and then saturated in oil. Oil permeability at Swir is measured at the test confining stress and is normally used (by most laboratories) as the reference/base for relative permeability calculations.

Typically, gas–oil relative permeability is conducted on samples prior to water–oil tests. Following gas injection, where it is assumed that water in the sample is retained at Swir, the sample is re-saturated in oil (often by flooding undersaturated oil, under backpressure to dissolve gas), and the waterflood commenced from Swir.

The gas should be humidified by passing through oil to prevent mass transfer between phases.

The injection pressure is normally held constant throughout the test and selected to provide optimum flow rate measurement. Produced oil or water is separated and measured. The gas flow is most often measured at ambient conditions (where volumes can be corrected to core pressure conditions, if required, using the ideal gas laws). Most mass flowmeters provide gas flow rate as normal volume per time—i.e., as an ambient pressure mass flow rate.

Gas injection should be continued until gas production is observed to be in excess of 99.99%. Periods of around 12–24 h should be allowed to ensure equilibrium. Unfortunately, many laboratories only allow a few hours, at most.

Gas relative permeability data can be calculated using the same JBN method as used for the water–oil tests, but modified to account for gas compressibility and constant pressure injection. It should be noted that the JBN method calculates end-face gas saturations. If there is a significant absolute pressure difference between the sample inlet and outlet, then a gas saturation gradient will exist in the core (in accordance with Boyle's law). This can create problems in the calculation of relative injections. Due to unfavourable mobility ratios, the gasflood front is inevitably unstable and viscous fingering occurs, weakening the viscous drive force. Capillary forces often dominate gasfloods and simulation becomes a necessity, since the analytical calculations assuming these forces to be negligible are incorrect.

Since gasfloods are always inadequate to achieve Srot, it is recommended that centrifugation is used after flooding has completed, in order to reduce to Srot, then reload to a coreholder and measure effective gas permeability at Srot.

Gas–Oil Relative Permeability (USS) Method Summary

Advantages:

- In theory provides kro–krg drainage curves at ambient or reservoir conditions
- Adverse mobility ratio ($\mu_g \ll \mu_o$) promotes prolonged two-phase flow data after BT

Drawbacks and Issues:

- Long core plugs are preferred to optimise measurement accuracy
- Data optimisation dictates that test should ideally be carried out using a high-pressure capacity rig and using ISSM. This demands a high level of sophistication in lab equipment and resources

	• Test costs would be high due to requirements for ISSM and/or reservoir conditions • Data are significantly constrained if test is done using synthetic fluids (e.g. N2 is immiscible with oil where methane is at least partly miscible). This can significantly affect relative permeabilities and oil recovery • krg is often too low and Sro to gas (Srog) is too high in most gasflood tests • End effects dominate flood behaviour in short cores • This test is only recommended if carried out at reservoir conditions with reservoir fluids on long cores

QC Checks

- The same generic QC checks used for water–oil USS and SS analysis (Sections 10.5.1 and 10.5.2) are also valid for gas–oil tests.
- Compared to a water–oil system, gas relative permeability is higher than krw and oil relative permeability to gas is lower than kro.
- In short core floods, the residual saturation value (Srog) is often higher than in a water–oil system (Srow) primarily due to the strong influence of capillary end effects. In the reservoir (in the absence of these end effects), thin film and gravity drainage occurs and nearly all the oil will drain from the rock with time. "True" Srog will therefore be zero or very close to zero. Since Srog approaches zero, the end-point gas permeability (referenced to Ko′ at Swir) will approach unity.
- The removal of concerns over wettability and knowledge of true residual oil saturation and krg end points are useful in analysing lab data and making refinements.
- Floods of individual core plugs often encounter often problems arising from early gas BT due to flood instability and from capillary end effects. Gasfloods conducted on long composite cores created by butting together 3–6 or more plugs from the same rock type overcome many of these problems and can provide more reliable data. However, thin film drainage and gas miscibility influences on residual oil saturation are only possible using reservoir fluids at reservoir conditions.

Suitable Laboratories

- Most commercial laboratories can provide basic ambient condition tests on short core plugs. However, it is recommended that all USS gasflood tests should be carried out on composite cores, using ISSM to minimise saturation uncertainties and to evaluate end effects, and ideally using reservoir fluids. This will limit the test to the major laboratories (e.g. Core Lab, Houston and UK, and Weatherford, Norway, UK and USA) who have reservoir condition and/or ISSM facilities.

10.5.4.5 USSGO Refining and Extending krg and kro

Gas–oil USS experiments on short core plugs are dominated by viscous instabilities and capillary end effects. Reliable data are only possible on long composite cores where the Pc end effects (which scale with the inverse of core length) are relatively insignificant, or from centrifuge tests.

Usually, it is necessary to extend the gas–oil curves to lower oil saturations than obtained from lab tests as oil is expected to drain to very low values through thin film drainage and gravitational forces. The recommended method of extending the curves is through use of the appropriate Corey exponents. However, simulation is a more appropriate interpretation tool to allow extension of the curves by defining the production history using a single relative permeability set together with capillary pressure (drainage gas–oil equivalent).

10.5.4.6 USSGO Corey Exponents

In view of the uncertainties introduced by the potential for capillary end effects in the dynamic core flood tests and the lack of thin film and gravity drainage, the recommended curves should principally be based on Corey exponents (Section 10.1.12).

For example, the Corey exponents should be constant (in terms of Son and Sgn), but lab data could indicate exponents that increase or decrease during the gasflood, as indicated in Figs. 10.40 and 10.41. This occurs since the entire test is affected by viscous fingering and capillary effects.

10.5.5 Steady State Gas–Oil Relative Permeability (SSGO)

In the steady-state method, a fixed ratio of fluids is injected through the core sample until saturation and DP equilibria are established at each injection ratio. Both gas and oil relative permeability can be calculated using Darcy's law when the flow of fluids within the porous media reaches SS, as described in Section 10.5.2 for water–oil SS tests. However, as the tests need to be conducted under backpressure to minimise gas compressibility and gas expansion effects, the tests must be carried out using ISSM to monitor and determine oil and gas saturations. The requirement to use composite core and the potential need to conduct the test at reservoir conditions with reservoir fluids, coupled with the requirement for ISSM, means that this test is rarely performed, other than by major commercial laboratories or research institutions.

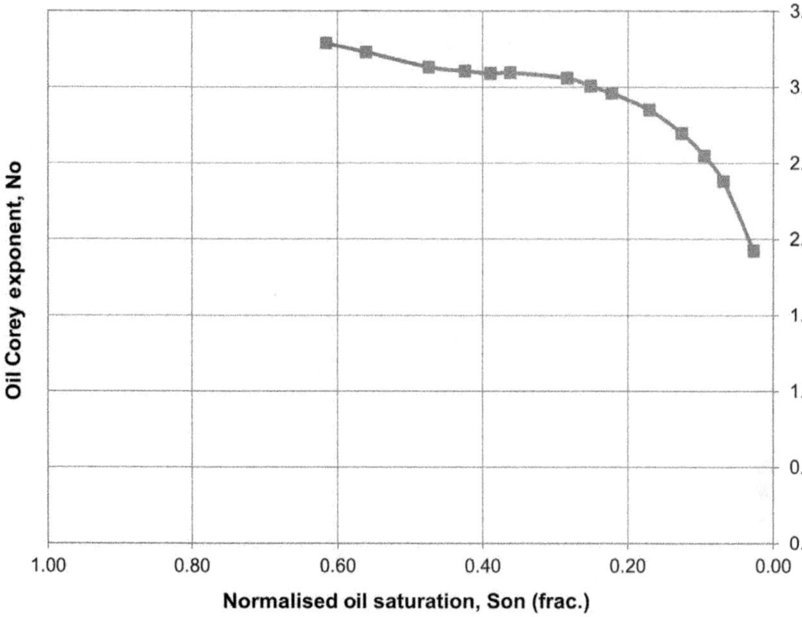

FIGURE 10.40 Normalised oil saturation as a function of No.

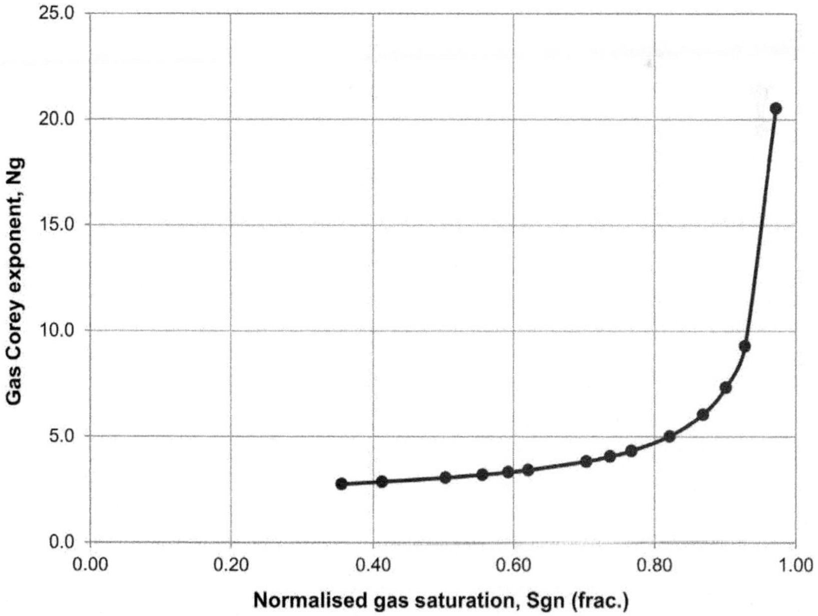

FIGURE 10.41 Normalised gas saturation as a function of Ng.

Purpose of Test:	• Measure gas–oil relative permeability curves (kro and krg) as a function of gas or liquid saturation
Sample Requirements:	• Cylindrical-shaped plug sample of 1″ or 1½″ diameter. The larger the sample, the larger the potential pore volume, and the less will be the impact of saturation measurement errors • Composite core is preferred • *Ideally*—sample taken from preserved core or preserved plug. *Alternative*—plug previously used in routine core analysis programme, provided that plug has not been pre-stressed and/or that plug preparation methods have not caused damage to delicate clay minerals
Required Input Data for Lab:	• Swir appropriate to maximum height above FWL for the sample porosity and permeability • FW salinity and ionic composition, as close as practically possible to the FW in the reservoir or FW resistivity at reservoir and lab temperature. Lab will prepare SFW accordingly. Oil may be doped (with iododecane) to facilitate ISSM • Reservoir oil • Composition of the gas (if test at reservoir conditions) • Physical properties of the oil (i.e. viscosity, API, PVT) • Overburden stress • Reservoir conditions (pressure and temperature) • Gas:oil injection ratios

10.5.5.1 SSGO Sample Preparation

The test samples are prepared in the same manner as USS test samples (Section 10.5.4).

10.5.5.2 SSGO Key Processes

1. Prepare core
2. Condition sample to appropriate Swir and restore wettability (if required)
3. Overburden stress application
4. Determine end-point oil permeability, Ko′ at Swir
5. SS displacement (gasflood at incremental gas:oil ratios)
6. Calculate gas and oil relative permeability (Darcy or coreflood simulation)
7. Determine end-point gas permeability, Kg′ at Srog
8. Plot relative gas and oil relative permeability versus gas, oil or liquid (So+Swir) saturation.

10.5.5.3 SSGO Test Equipment

The equipment required to measure relative permeability by the SS method is similar to that described in Sections 10.2.5 and 10.5.2 for water–oil SS tests. Fluids may be injected either at constant rate or constant pressure and it is essential that the tests are conducted at elevated pore pressure (backpressure) to minimise gas expansion artefacts. ISSM and use of long composite cores are often required for reliable saturation data and measurable DP.

10.5.5.4 SSGO Test Procedures

The cores are pre-conditioned to an appropriate initial water saturation and wettability. The initial effective permeability to oil at Swir (used as the reference/base for relative permeability calculations) is measured. The gas and oil should be brought into equilibrium prior to flooding, at the test system backpressure, by injecting through the equipment flowlines while bypassing the sample.

The imbibition cycle is commenced by flowing oil and gas together through the sample, initially at a low gas fraction to effect the first change in saturation and to capture the reduction in kro for a small change in gas saturation. SS flow is assumed to have been achieved when no further change in saturation or DP has been observed. Moderate backpressure must be used in this experiment to minimise saturation gradients across the test sample. It will be necessary to attempt to regulate (minimise) the DP. While DP must be measurable, the nature of a pressure gradient across the sample initiates a disequilibria between the gas and oil phases. Equilibrium gas:oil ratios at the injection face (backpressure+DP) are different from gas:oil ratios at the outlet face (backpressure). This may create saturation gradients across the core. Reducing DP will reduce the disequilibria effects. It is recommended to limit DP to +2% of backpressure.

Typical injection ratios are:

Ratio	Gas:Oil Ratio
1	1:75
2	1:25
3	1:5
4	1:1
5	10:1
Kg′ at Srog	100% gas

For ISSM, the oil is doped with suitable X-ray or gamma ray adsorber (usually iododecane). Pre- and post-flood calibration measurements are used to quantify core saturations during the floods from radiation adsorption measurements. The scan data acquired during the tests are also used to determine the existence and possible influence of capillary end effects which may occur during the floods.

The oil and gas rates, DP and saturation are then used to calculate relative permeability versus saturation using Darcy's law (Section 10.3.2), based on the SS flow rates, pressure drops and viscosities of each fluid phase.

Gas–Oil Relative Permeability (SS) Method Summary	
Advantages: • Relative permeability kro and krg is calculated over a wide saturation range • Define krw at low Sw values • Capillary end effects often minimised, especially in long composite cores	**Drawbacks and Issues:** • Long core plugs are preferred to optimise measurement accuracy • Data optimisation dictates that test should be carried out using a high-pressure capacity rig and using ISSM. This demands a high level of sophistication in lab equipment and resources • Test costs would be high due to requirements for ISSM and/or reservoir conditions • Data are significantly constrained if test done using synthetic fluids (e.g. N2 is immiscible with oil where methane is at least partly miscible). This can significantly affect relative permeabilities and oil recovery • End effects can dominate flood behaviour in short cores • SS method has similar problems because of viscous instability at high rate and low DP at low rates. In addition, gravity segregation is always a question if the core is horizontal even when rotated during flow • Much slower than USS tests • Difficult to gain value from such expensive tests

QC Checks

- The same generic QC checks used for water–oil USS and SS analysis (Sections 10.5.1 and 10.5.2) and USS gas–oil tests are also valid for SS tests.
- Floods of individual core plugs often encounter often problems arising from early gas BT due to flood instability and from capillary end effects. Gasfloods conducted on long composite cores created by butting together 6 or more plugs from the same rock type overcome many of these problems and can provide more reliable data. However, thin film drainage and gas miscibility influences on residual oil saturation are only possible using reservoir fluids at reservoir conditions.
- Viscous instability is still an issue at high rate and low DPs.

Suitable Laboratories

- It is recommended that if SS gasflood tests are considered at all, they should be carried out on composite cores, using ISSM to minimise saturation uncertainties and to evaluate end effects, and ideally using reservoir fluids. This will limit the test to the major laboratories (e.g. Core Lab, Houston and UK, and Weatherford, Norway, UK and USA) who have reservoir condition and/or ISSM facilities.

10.5.6 Centrifuge Gas–Oil Relative Permeability (CentKrGO)

Gas–oil USS experiments on short core plugs are dominated by viscous instabilities and capillary end effects. One advantage of the centrifuge method is that viscous forces are replaced by centrifugal gravitational forces, which act against capillary pressure, so kro curves from centrifuge tests are less likely to be influenced by end effects. In addition, very low kro can be measured (although no data are gained on krg).

Purpose of Test:	• Measure gas-displacement oil relative permeability curves (kro), especially at low oil saturations
Sample Requirements:	• Cylindrical-shaped plug sample of 1″ or 1½″ diameter. The larger the sample, the larger the potential pore volume, and the less will be the impact of saturation measurement errors • *Ideally*—sample taken from preserved core or preserved plug. *Alternative*—plug previously used in routine core analysis programme, provided that plug has not been pre-stressed and/or that plug preparation methods have not caused damage to delicate clay minerals
Required Input Data for Lab:	• Swir appropriate to maximum height above FWL for the sample porosity and permeability • FW salinity and ionic composition, as close as practically possible to the FW in the reservoir or FW resistivity at reservoir and lab temperature • As tests cannot currently be performed at reservoir conditions, tests tend to be restricted to synthetic fluids at ambient temperature, and in some equipment, at moderate overburden pressure. Lab may use mineral oil of appropriate viscosity • Overburden pressure if required and if possible • Spin speed (or critical bond number) and time intervals to acquire production data

10.5.6.1 CentKrGO Sample Preparation

The test samples are prepared in a similar manner as USS test samples (Section 10.5.4) and water–oil centrifuge tests (Section 10.5.3) and initially reduced to Swir by centrifuge (oil-displacing-water).

10.5.6.2 CentKrGO Key Processes

1. Condition core to Swir (oil–water centrifuge primary drainage)
2. Determine end-point effective oil permeability (as for SS and USS) using same test fluids and overburden stress as used for centrifuge tests
3. Load into centrifuge and spin samples at the set speed under gas displacement
4. Measure the effluent produced at specified time intervals until residual conditions are established
5. Calculate the oil relative permeability curve using the equations derived from Hagoort (1980) or other models, or via coreflood simulation
6. Determine end-point krg′ at Sro in a coreholder at confining stress
7. Plot oil relative permeability versus gas, oil or liquid (So+Swir) saturation.

10.5.6.3 CentKrGO Test Equipment

The equipment required to measure oil relative permeability by the centrifuge method is the same as that described in Section 10.2.5. It is essential that production data are acquired using automated data capture.

10.5.6.4 CentKrGO Test Procedures

The sample is reduced to Swir by centrifuge desaturation with oil (mineral oil) either at a single maximum speed or over a series of incremental pressures to define a full drainage curve. The sample is loaded into a core holder and Ko′ at Swir determined at the appropriate overburden stress.

Each sample is then loaded into the centrifuge core holder and spun at a set single speed. The effluent produced is measured as a function of time until residual conditions are established.

On completion, the sample is loaded into a coreholder, confined at a nominated stress, and flushed with gas until pressure differential and rate are stable. End-point krg′ at ultimate Sro is determined.

Analytical solutions are used to determine kro, but the preferred technique is from computer simulation of flow behaviour, which many commercial laboratories do not have access to.

Gas–Oil Relative Permeability Centrifuge Method Summary	
Advantages: • Relatively simple and quick method for drainage kro curve at low oil saturations • Good method for most rock types • More efficient drainage than dynamic methods and the ability to measured very low kro values • Viscous forces are replaced by centrifugal forces, which act against capillary pressure, so kro curves from centrifuge tests are less likely to be influenced by end effects than USS or SS dynamic tests	**Drawbacks and Issues:** • Only drainage kro curve is determined • Use de-gassed mineral oil only • Although end effects are less, they still occur and result in a non-uniform distribution of saturation along the core sample length. Supporting the sample on a porous semi-permeable membrane can reduce these problems • Friable samples can be damaged and fracturing can occur during centrifugation. Friable samples must be tested at overburden pressure, which can limit the spin speed in some equipment. The presence of microfractures can dominate the flood behaviour • Maintaining Nb less than the critical value of 10^{-5} is likely to be difficult in higher-permeability samples, so applied capillary pressure may be low

QC Checks

- The same QC checks used for gas–oil SS and USS, and water–oil centrifuge are also valid for the centrifuge method.
- Experimental data are required (core dimensions, distance to inlet and outlet of sample, rotational speeds, production volumes and rates) to independently validate the results.
- The high spin speeds may invalidate the critical bond number (Nb) criterion normally imposed on centrifuge imbibition experiments. Associated desaturation effects can cause changes in capillary pressure and relative permeability, which (usually) do not occur under normal field conditions.
- As the capillary-dominated displacement may be more efficient than dynamic displacement (especially in higher-viscosity oils), the end-point Srog may be lower and (from subsequent coreflood) the end-point krg′ may be higher than from dynamic displacement tests.

Suitable Laboratories

- Most commercial laboratories can provide basic centrifuge tests where data are analysed by conventional analytical solutions. Typically, these

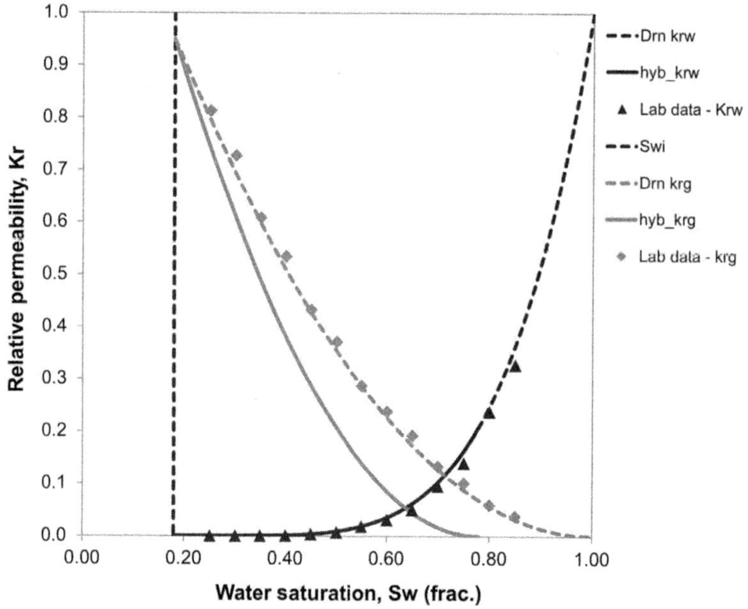

FIGURE 10.42 Example drainage and imbibition krw and krg curves.

are done at ambient or elevated confining stress with synthetic fluids. However, it is recommended that the tests are carried out in a centrifuge fitted with automated data acquisition, so only laboratories with this equipment should be considered.

10.5.7 Gas–Oil Relative Permeability Curve Generation

In view of the experimental difficulties and artefacts associated with both SS and USS gas–oil tests, and the expense involved in carrying out tests at reservoir conditions on composite plugs with ISSM, it is often better to invent gas–oil relative permeability curves rather than attempt to measure them using lower costs and typically non-optimised procedures and equipment. This is based primarily on Corey exponents (see Section 10.1.12) and that for miscible or part-miscible displacements, nearly all the oil will drain from the rock with time. Consequently, "true" Srog will be zero or close to zero and the end-point gas permeability will approach unity.

The following parameters might be considered when generating krg–kro data. Sensitivities (e.g. base, high and low cases) should be performed as indicated by the parameter ranges.

Parameter	Value/Source
Irreducible water saturation, Swir	From Pc model
Critical gas saturation, Sgc	0.02–0.05
True residual oil saturation, Srog	0.10–0.00
End-point gas relative permeability, krg′	1.0
Oil relative permeability, kro	$kro = Son^{No}$
Gas relative permeability, krg	$krg = Sgn^{Ng}$
Oil Corey, No	4–7
Gas Corey, Ng	1.3–3.0

10.5.8 Unsteady State Gas–Water Relative Permeability (USSGW)

Gas–water relative permeability tests provide drainage krg and krw curves and are generally performed for gas storage modelling. However, drainage tests can also be used to generate or constrain imbibition krw curves. This is predicated on the assumption that there is no hysteresis in the wetting phase (water) curve. Drainage tests can then be used to provide part of the imbibition krw curve (from Swir to Sgr), as indicated in Fig. 10.42.

In these tests, water is the strong wetting phase, so samples need not be wettability restored. Tests on cleaned samples are usually acceptable.

Due to the high gas:water viscosity ratio, early gas BT and fractional flow of gas occur over an extended saturation range so that full relative permeability curves can be generated (though dominated by capillary artefacts).

Purpose of Test:	• Measure drainage gas–water relative permeability curves (krg and krw)
Sample Requirements:	• Cylindrical-shaped plug sample of 1″ or 1½″ diameter. The larger the sample, the larger the potential pore volume, and the less will be the impact of saturation measurement errors • Composite core is preferred • *Ideally*—sample taken from preserved core or preserved plug. *Alternative*—plug previously used in routine core analysis programme, provided that plug has not been pre-stressed and/or that plug preparation methods have not caused damage to delicate clay minerals

	• Tests are carried out on cleaned-state, water-wet core, appropriately cleaned to prevent clay damage and to remove any OBM contamination
Required Input Data for Lab:	• Swir appropriate to maximum height above FWL for the sample porosity and permeability • FW salinity and ionic composition should be as close as practically possible to the FW in the reservoir. SFW may be doped (with NaI) to facilitate ISSM • Overburden stress • Flow rate (or reservoir capillary number)

10.5.8.1 USSGW Sample Preparation

A composite core might be preferred for gas–water relative permeability measurements (to attempt to minimise end effects and viscous instability).

10.5.8.2 USSGW Key Processes

1. Saturate in FW (SFW)
2. Determine brine permeability, Kb
3. Water drainage by gasflood
4. Monitor production volumes
5. Determination of krg and krw relative permeabilities using the modified JBN method or via coreflood simulation
6. Dean–Stark extraction to verify calculated Swir (if imbibition tests or extended drainage capillary desaturation is not required)
7. Plot relative permeability as a function of the gas or water saturations.

10.5.8.3 USSGW Test Equipment

The equipment required to measure relative permeability by the USS method is similar to that described in Section 10.5.4 for gas–oil tests. Ideally, the tests should be performed on composite cores and equipment should include ISSM at reservoir conditions, but reasonably reliable results might be obtained from tests at ambient conditions with synthetic fluids.

10.5.8.4 USSGW Test Procedures

Following saturation in SFW, the plugs are loaded into an USS coreholder and the water permeability, Kw, established at net overburden stress. Samples are flooded with humidified gas. Water and gas production are monitored until no more water is produced and the sample is assumed to be at Swrg (some remaining gas saturation after gasflooding). Drainage krw and krg are calculated from the production data. In laboratory reports, the brine permeability is normally used as the reference for relative permeability calculations but this should be verified with the lab. For water–gas tests, the convention is to use the end-point gas permeability Kg' at Swir as the reference permeability.

Gasflood tests should be gravity-stabilised: the coreholder held vertical and gas injected into the top of the coreholder, displacing the denser phase (water) from the bottom of the cell.

The injection gas is passed through a water humidifier to prevent mass transfer between phases and evaporation of water. The injection pressure is normally held constant throughout the test and should be selected to provide optimum flow rate measurement. Produced water is separated and measured.

The gas flow is measured at ambient conditions (where volumes can be corrected to core pressure conditions, if required). Gas injection should be continued until water production virtually ceases. Periods of around 12–24 h should be allowed to ensure equilibrium flood-out. Unfortunately, many laboratories only allow a few hours, at most.

The gas–water relative permeability is calculated using JBN method as used for the water–oil tests, but as explained for gas–oil systems the method is modified to account for gas compressibility and constant pressure injection. It should be noted that the JBN method calculates end-face gas saturations.

Gas–Water Relative Permeability (USS) Method Summary	
Advantages: • Determination of drainage krw and krg curves • Reasonably reliable data can be obtained at ambient conditions with synthetic fluids	**Drawbacks and Issues:** • Viscous instability can be an issue at high rates and low DPs, especially in short core plugs • Capillary end effects can predominate and affect end-point Swir (and end-point Kg′) due to non-uniform saturation distribution • Desaturation with gas flow does not adequately drive plugs to uniform or representative irreducible water saturation; therefore, capillary pressure desaturation may be necessary to complete this process • If there is a significant absolute pressure difference between the sample inlet and outlet, then a gas saturation gradient will exist in the core (in accordance with Boyle's law). This can create problems in the calculation of relative injectivities

QC Checks

- The same generic QC checks used for water–oil and gas–oil USS and SS analysis (Sections 10.5.1 and 10.5.2) are also valid for SS tests.
- Ensure the method used for cleaning and drying does not cause damage or de-preservation of clay structures.

- Ensure that Swir after gasflooding is consistent with capillary pressure data for similar rock qualities.
- Samples scheduled for imbibition tests following drainage experiments may require capillary pressure desaturation to establish a uniform and representative Swir.

Suitable Laboratories

- Most commercial laboratories can provide basic ambient condition tests using synthetic fluids on short core plugs. However, if USS gasflood tests are to be carried out on composite cores, using ISSM to minimise saturation uncertainties and to evaluate end effects, then only major laboratories (e.g. Core Lab, Houston and UK, and Weatherford, Norway, UK and USA) are likely to have the necessary capabilities and resources

10.5.9 Steady State Gas–Water Relative Permeability (SSGW)

Although rarely performed, the SS procedure for gas–water relative permeability can provide reliable results for high-permeability core (viscous instability is reduced compared to USS). For low-permeability cores, however, SS methods are difficult to apply because the measurement of the extremely small flow rate is difficult and time consuming. Equipment and procedures are similar to those for SS gas–oil relative permeability (see Section 10.5.5) but tests can be performed on cleaned cores as water is the strong wetting phase.

Purpose of Test:	• Measure drainage gas–water relative permeability curves in high-permeability cores
Sample Requirements:	• Cylindrical-shaped plug sample of 1″ or 1½″ diameter. The larger the sample, the larger the potential pore volume, and the less will be the impact of saturation measurement errors • Composite core is preferred • *Ideally*—sample taken from preserved core or preserved plug. *Alternative*—plug previously used in routine core analysis programme, provided that plug has not been pre-stressed and/or that plug preparation methods have not caused damage to delicate clay minerals • Tests are carried out on cleaned-state, water-wet core, appropriately cleaned to prevent clay damage and to remove any OBM contamination
Required Input Data for Lab:	• Swir appropriate to maximum height above FWL for the sample porosity and permeability • FW salinity and ionic composition should be as close as practically possible to the FW in the reservoir. SFW may be doped (with NaI) to facilitate ISSM • Overburden or confining stress • Gas:water injection ratios

10.5.9.1 SSGW Sample Preparation

Sample preparation for SS gas–water relative permeability measurements is the same as that for the USS method (see Section 10.5.8). A composite core is preferred for gas–water relative permeability measurements (to minimise end effects and viscous instability).

10.5.9.2 SSGW Key Processes

1. Saturate in FW (SFW)
2. Determine brine permeability, Kb
3. Gasflood at incremental, fixed gas:water injection ratios
4. Monitor production volumes (ISSM)
5. Determination of krg and krw relative permeabilities using Darcy's law
6. Dean–Stark extraction to verify calculated Swir (if imbibition tests or extended drainage capillary desaturation is not required)
7. Plot relative permeability as a function of the gas or water saturations.

10.5.9.3 SSGW Test Equipment

The equipment required to measure relative permeability by the SS method is similar to that described in Section 10.5.5. Fluids may be injected either at constant rate or constant pressure, and it is essential that the tests are conducted at elevated pore pressure (backpressure) to minimise gas expansion artefacts. Limiting flow through the backpressure regulator to single-phase gas flow provides more stable pressures than those that would have resulted with two-phase flow across a regulator. Flowing only gas through the regulator also minimises corrosion and plugging within the backpressure regulator. ISSM and use of long composite cores are recommended for increased reliability of saturation data and to allow improved data input to coreflood simulations.

10.5.9.4 SSGW Test Procedures

Following saturation in SFW, the plugs are loaded into an SS coreholder and the water permeability, Kw, established at net overburden stress. The imbibition cycle is commenced by flowing water and gas together through the sample, initially at a low gas fraction. DP, production volume and flow rates are monitored throughout until stable readings are achieved. SS flow is measured to have been achieved when no further change in saturation or DP has been observed. Moderate backpressure must be used in this experiment to minimise saturation gradients across the test sample. For ISSM, the water is doped with a suitable X-ray or gamma ray adsorber (usually sodium iodide).

Typical injection ratios are:

Ratio	Gas:Water Ratio
1	1:75
2	1:25
3	1:5
4	1:1
5	10:1
Kg′ at Swir	100% gas

The water and gas rates, DP and saturation are used to calculate relative permeability versus saturation using Darcy's law, based on the SS flow rates and viscosities of each fluid phase. Desaturation with gas flow towards Swir does not adequately drive plugs to irreducible water saturation; therefore, capillary pressure desaturation may be necessary to complete this process prior to imbibition tests, for example.

Unless a final ISSM scan is required or subsequent centrifuge desaturation is continued, final water saturations should be checked using Dean–Stark extraction or miscible solvent displacement with Karl Fischer titration of the solvent effluent for water content.

Gas–Water Relative Permeability (SS) Method Summary

Advantages:

- Determination of drainage krw and krg curves in high-permeability cores
- Reasonably reliable data can be obtained at ambient conditions with synthetic fluids

Drawbacks and Issues:

- Long core plugs are preferred to optimise measurement accuracy
- Data optimisation dictates that test should ideally be carried out using a high-pressure capacity rig and by using ISSM. This demands a high level of sophistication in lab equipment and resources
- Test costs would be high due to requirements for ISSM
- For high-permeability cores, stabilisation times are reasonable and the SS method is acceptable. For low-permeability cores, however, the use of SS methods is not feasible. To establish gas and water injection rates at reasonable pressure differentials across these tight cores, the injection

	rates are extremely small. Difficulty arises when attempting to accurately measure low flow rates. Also, stabilisation times can be prohibitively long

QC Checks

- The same generic QC checks used for water–oil, gas–oil and gas–water USS and SS analysis (Sections 10.5.1, 10.5.2, 10.5.4, 10.5.5 and 10.5.8) are also valid for SS tests.
- Ensure the method used for cleaning and drying does not cause damage or de-preservation of clay structures.
- Ensure that Swir after gasflooding is consistent with capillary pressure data for similar rock qualities.
- Samples scheduled for imbibition tests following drainage experiments will probably require capillary pressure desaturation to establish a uniform and representative Swir.

Suitable Laboratories

- It is recommended that if SS gasflood tests are considered at all, they should be carried out on composite cores, using ISSM to minimise saturation uncertainties and to evaluate end effects. This will limit the test to the major laboratories (e.g. Core Lab, Houston and UK, and Weatherford, Norway, UK and USA) who have the necessary equipment and ISSM facilities.

10.5.10 Unsteady State Water–Gas Relative Permeability (USSWG)

Imbibition tests can be set up to give the full relative permeability curves (krw & krg), but are extremely difficult and costly to perform. Dynamic tests using USS water-displacing gas methods on short core plugs can often yield a residual gas saturation that is too low due to the overly favourable mobility of the flood and owing to disequilibria effects, if DP and injection volumes are not effectively controlled. In these tests, water as the strongly wetting phase is often drawn into the sample under capillary forces (spontaneous imbibition). If performing these tests at equivalent reservoir advancement rates (approx. conversion to 4 ml/h core plug flow rate), it is often observed that water is drawn in ahead of the flood front. This leads to film thickening of the water film coating grains as water encroaches on the system, and eventually to "snap-off" of the gas column by the encroaching water at the smallest pore throats, where the thickening film eventually attract across the throat radius.

Co-current imbibition tests are a variation of the USS tests but are performed at low rates using four or five plug composite cores. In these tests, water is the strong wetting phase, so samples need not be wettability restored. Tests on cleaned samples, using synthetic fluids, are usually acceptable.

Purpose of Test:	• Measure imbibition water–gas end points and potentially relative permeability curves
Sample Requirements:	• Cylindrical-shaped plug sample of 1″ or 1½″ diameter. The larger the sample, the larger the potential pore volume, and the less will be the impact of saturation measurement errors • Composite core is preferred and used for co-current tests • *Ideally*—sample taken from preserved core or preserved plug. *Alternative*—plug previously used in routine core analysis programme, provided that plug has not been pre-stressed and/or that plug preparation methods have not caused damage to delicate clay minerals
Required Input Data for Lab:	• Swir appropriate to maximum height above FWL for the sample porosity and permeability • FW salinity and ionic composition should be as close as practically possible to the FW in the reservoir • Overburden or confining stress • Flow rate (or reservoir capillary number)

10.5.10.1 USSWG Sample Preparation

Co-current USS gasflood tests using composite cores are highly recommended to overcome the issues associated with flood instability and capillary effects. The composite core is prepared as described in Section 10.4.1.

For water-displacing-gas, the core sample (and individual plug samples for composite cores) is initially desaturated to Swir in a porous plate cell or centrifuge (faster), and the gas permeability at Swir end point is measured (on the individual plug or composite). This value is normally used by laboratories to calculate relative permeability, while many Corey-type curve relationships are based on the Kg. Care must be taken to ensure which value is used as the base reference.

For tests where both drainage and imbibition gas/water curves are scheduled, extending the drainage process via capillary pressure desaturation (porous plate or centrifuge) may be necessary to ensure a representative Swir prior to imbibition tests.

10.5.10.2 USSWG Key Processes

1. Condition the sample to Swir in a porous plate cell or centrifuge
2. Overburden stress application
3. Determine end-point relative permeability to gas, Kg′ at Swir

4. Waterflood
5. Measured gas production volumes at backpressure or via ISSM
6. Determination of the relative permeability using the modified JBN method or coreflood simulation
7. Determine end-point relative permeability, Kw′ at Sgr at confining stress
8. Dean–Stark extraction to verify calculated final water saturation (if ISSM scans not required)
9. Plot krw and krg as a function of water saturation.

10.5.10.3 USSWG Test Equipment

Imbibition waterflood (water-displacing-gas) tests are carried out using similar equipment to the gasflood test described in Section 10.2.5. The use of ISSM and composite cores is preferred.

10.5.10.4 USSWG Test Procedures

The test plugs are desaturated to Swir at an appropriate capillary pressure, sufficient to attain a representative Swir, and loaded into the test coreholder. Humidified nitrogen gas is flowed through each sample and the effective end-point permeability to gas (Kg′) at irreducible water saturation (Swir) is determined at overburden stress. The water and gas are brought into equilibrium at the system backpressure, prior to initiating the flood, by flowing both fluids in a closed-loop sample bypass around the equipment.

Water injection is gravity-stabilised—the coreholder is held vertical and water injected from the base to the top of the plug. The floods are carried out under moderate backpressure to prevent gas saturation gradients developing across the sample. It is essential to also moderate DP as a function of the backpressure, to minimise disequilibria effects, which may cause dissolution of gas at the injection face, thus unrepresentatively reducing Sgr.

Since imbibition in a gas reservoir is controlled more by capillary forces than by viscous forces, the tests are best carried out at low rates—typically about 4 cc/h (co-current imbibition) and for limited injection volumes (maximum 1.5–2 PVI).

Gas production volumes are measured at backpressure using a pressurised sight glass or separator, designed such that produced gas is collected and water allowed to flow via a backpressure regulator to atmosphere.

As the initially dominant gas phase is replaced by more viscous water, the pressure differential rapidly increases to a maximum. It then falls slowly as gas dissolves in the flowing water. This dissolution is unavoidable to some extent, but can be reduced by equilibrating the injected water with the gas at backpressure prior to injection. It should be noted that the injected water will not completely displace the gas, and a trapped gas saturation will always remain.

On completion, the final permeability to water at residual gas saturation, Kw′ at Sgr, is measured.

Dean–Stark extraction (if ISSM scans are not required) is used to verify/check final water saturations.

Water–Gas Relative Permeability (USS) Method Summary	
Advantages: • Imbibition krg and krw end points can be acquired using co-current technique at reasonable cost in a relatively short time frame	**Drawbacks and Issues:** • Not recommended for short core plugs • The mobility ratio of the water-displacing gasflood front is so favourable in USS tests that very little two-phase flow data are available after water BT, so end points are often only attainable • Flood instability at high rates and capillary end effects at low rates • End-point Sgr and krw′ from dynamic tests on short cores can be significantly different from co-current, counter-current or centrifuge tests • Preferential flow can occur through higher-permeability pathways, and gas pockets in more isolated pores can be bypassed without the stabilising influence of capillarity • Potential gas–water mass transfer effects if gas is not humidified • The reservoir situation is slightly, but significantly, different from the laboratory technique. In the reservoir, water migrates into the gas zone as pressure declines, but unlike the core test, the gas saturation does not necessarily decline. It tends to remain high or increase slightly, since the trapped gas expands as pressure falls. This maintenance or even increase in Sg tends to keep krw low or reduce it even further. This scenario operates in many reservoirs even if some of the gas migrates onwards and upwards

QC Checks

- The same generic QC checks used for water–oil, gas–oil and gas–water USS and SS analysis are also valid for water–gas tests.
- Ensure the method used for cleaning and drying does not cause damage or de-preservation of clay structures.
- Ensure that Swir is consistent with capillary pressure data for similar rock qualities. For example, Fig. 10.43 plots the Swir–rock quality relationships from co-current (COC), counter-current (CCI), SS tests and drainage centrifuge tests. The composite COC samples from some COC tests appear to have slightly lower Swir values than other samples of similar rock quality. This arises from differences in air permeability between the COC and centrifuge samples—the COC samples have much lower apparent Ka, possibly as a result of permeability barriers between butted plugs.
- Sgr from dynamic tests should not be used in simulation in isolation from data from other tests (centrifuge and CCI).

Suitable Laboratories

- Most commercial laboratories can provide basic ambient condition tests using synthetic fluids on short core plugs. However, if co-current tests are to be carried out on composite cores, using ISSM to minimise saturation uncertainties and to evaluate end effects, then only major laboratories (e.g. Core Lab, Houston and UK, and Weatherford,

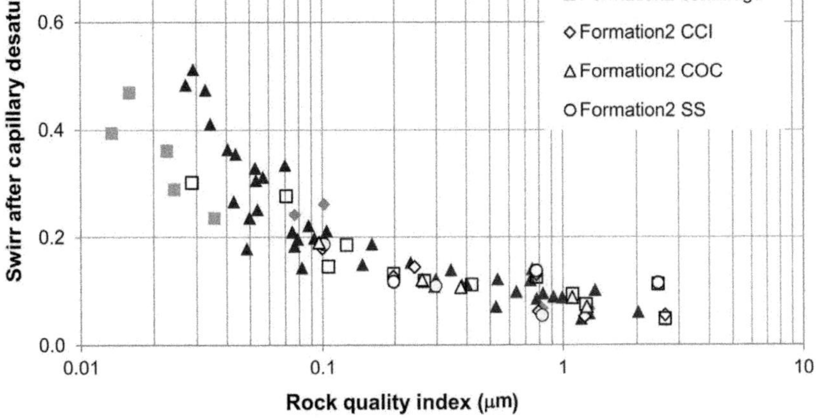

FIGURE 10.43 Example Swir versus RQI for Kw–Kg test samples.

Norway, UK and USA) are likely to have the necessary capabilities and resources.

10.5.10.5 USSWG Intermediate Relative Permeability Data

The intermediate krw–krg curves may not be determined experimentally, but generated using the MAK correlation (Mulyadi et al., 2000) or Corey exponents (Section 10.1.12).

In the Corey approach, krg and krw are related to the base effective gas permeability, Kg′ at Swir, and are normalised as follows:

$$\mathrm{krgn} = \mathrm{krg} = (1 - \mathrm{Swn})^{\mathrm{Ng}} \quad \text{and} \quad \mathrm{krwn} = \frac{\mathrm{krw}}{\mathrm{krw}'} = \mathrm{Swn}^{\mathrm{Nw}}$$

$$\mathrm{Swn} = \frac{\mathrm{Sw} - \mathrm{Swir}}{1 - \mathrm{Swir} - \mathrm{Sgr}}$$

where krgn is normalised gas relative permeability, krwn is normalised water relative permeability, Ng is Corex exponent for gas (water–gas), Nw is Corey exponent for water (water–gas). The Corey Ng has a narrow range from 2 to 3 as gas is non-wetting. The potential range for Nw is wider: typically 4–8.

10.5.11 Steady State Water–Gas Relative Permeability (SSWG)

SS tests in which water and gas are injected simultaneously can provide reliable intermediate krw and krg imbibition curves but the tests are expensive as they do require non-invasive saturation monitoring to determine accurate saturations.

Purpose of Test:	• To determine intermediate krw and krg imbibition curves
Sample Requirements:	• Cylindrical-shaped plug sample of 1″ or 1½″ diameter. The larger the sample, the larger the potential pore volume, and the less will be the impact of saturation measurement errors • Composite core is preferred • *Ideally*—sample taken from preserved core or preserved plug. *Alternative*—plug previously used in routine core analysis programme, provided that plug has not been pre-stressed and/or that plug preparation methods have not caused damage to delicate clay minerals • Tests are carried out on cleaned-state, water-wet core, appropriately cleaned to prevent clay damage and to remove any OBM contamination
Required Input Data for Lab:	• Swir appropriate to maximum height above FWL for the sample porosity and permeability • FW salinity and ionic composition should be as close as practically possible to the FW in the reservoir. SFW may be doped (with NaI) to facilitate ISSM • Overburden or confining stress • Water:gas injection ratios

10.5.11.1 SSWG Sample Preparation

The samples for water–gas SS relative permeability measurements are prepared using the same procedures as for USS tests (described in Section 10.5.10).

10.5.11.2 SSWG Key Processes

1. Desaturate the sample to Swir in a porous plate cell or centrifuge
2. Apply confining stress
3. Determine end-point relative permeability to gas, Kg′ at Swir
4. Waterflood displacement by SS methods at incremental water:gas ratios
5. Monitor and determine saturations using ISSM
6. Determine krw and krg after flow and pressure stability after each injection ratio (Darcy)
7. Determine end-point relative permeability, Kw′ at Sgr
8. Dean–Stark extraction to verify calculated final water saturation (if ISSM scans are not required)
9. Plot krw and krg as a function of water saturation.

10.5.11.3 SSWG Test Equipment

The equipment required is similar to that described in Section 10.2.5. Fluids may be injected either at constant rate or constant pressure, and it is essential that the tests are conducted at elevated pore pressure (backpressure) to minimise gas expansion artefacts using ISSM. The use of long composite cores is recommended for reliable saturation data.

10.5.11.4 SSWG Test Procedures

The test procedures are similar to those used for USS water–gas tests. The water is doped with sodium iodide. Reference Kg at Swir is determined at overburden stress. Humidified nitrogen gas and SFW, which has been pre-equilibrated with gas, are injected simultaneously at several increasing water–gas injection ratios to allow the water saturation within the samples to increase. Typical injection ratios are:

Ratio	Water:Gas Ratio
1	1:75
2	1:25
3	1:5
4	1:1
5	10:1
Kw′ at Sgr	100% water

Saturation changes are monitored by ISSM, and injection is continued at each ratio until an equilibrated, SS condition within the core plugs is established, based on the consistency of the ISSM saturation profile and DP (see Fig. 10.44).

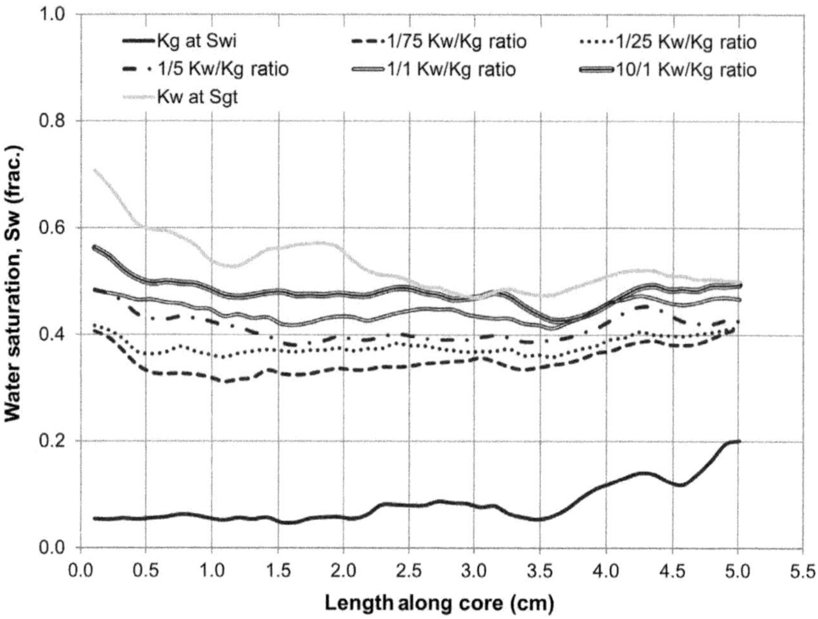

FIGURE 10.44 Example ISSM saturation results from SS water–gas experiment.

Finally, SFW alone is injected and Kw′ at Sgr is determined while scanning the samples. If a final ISSM scan is not required, samples should be extracted in a Dean–Stark apparatus as a final check on water saturation.

Water–Gas Relative Permeability (SS) Method Summary

Advantages:	Drawbacks and Issues:
• Intermediate imbibition krg and krw curves can be acquired	• Long core plugs are preferred to optimise measurement accuracy • Data optimisation dictates that testing should ideally be carried out using a high-pressure capacity rig and using ISSM. This demands a high level of sophistication in lab equipment and resources • Test costs would be high due to requirements for ISSM • The key data from SS tests are the curve shapes, and the Corey exponent values that can be generated from them. The tests are not normally run to ultimate residual gas saturation • Preferential flow can occur through higher-permeability pathways, and gas pockets in more isolated pores can be bypassed without the stabilising influence of capillarity • Potential gas–water mass transfer effects if the gas is not humidified

- The reservoir situation is slightly, but significantly, different from the laboratory technique. In the reservoir, water migrates into the gas zone as pressure declines, but unlike the core test, the gas saturation does not necessarily decline. It tends to remain high or increase slightly, since the trapped gas expands as pressure falls. This maintenance or even increase in Sg tends to keep krw low or reduce it even further. This scenario operates in many reservoirs even if some of the gas migrates onwards and upwards

QC Checks

- The same generic QC checks used for water–oil, gas–oil and gas–water USS and SS analysis are also valid for water–gas tests.
- Ensure the method used for cleaning and drying does not cause damage or de-preservation of clay structures.
- Ensure that Swir is consistent with capillary pressure data for similar rock qualities.
- Sgr from dynamic tests should not be used in simulation in isolation from data from other tests (centrifuge and CCI).
- Verify/check krw and krg curves against Corey exponents (e.g. Figs. 10.45 and 10.46).

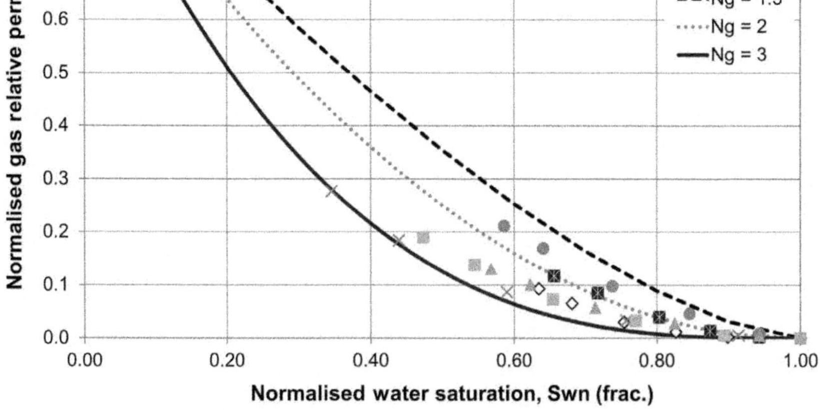

FIGURE 10.45 Comparison of experimental krg data and Corey exponents.

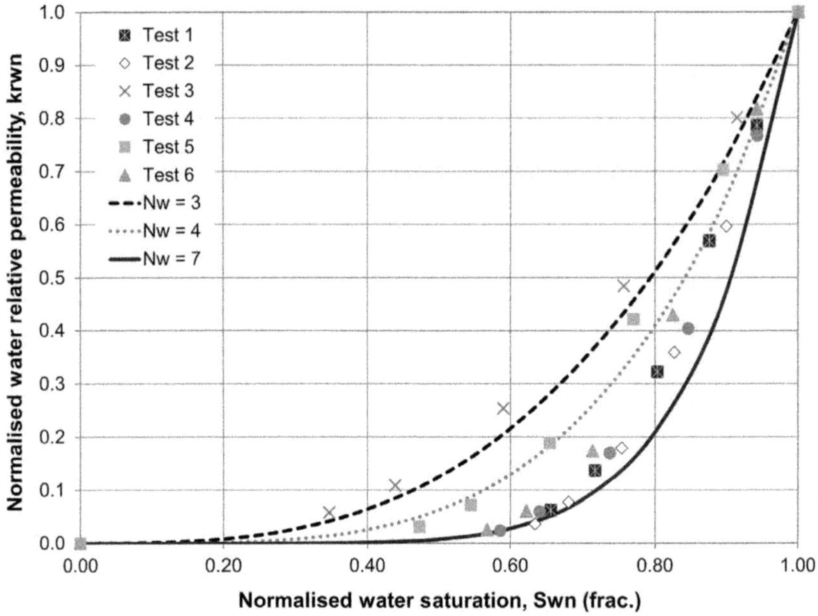

FIGURE 10.46 Comparison of experimental krw data and Corey exponents.

Suitable Laboratories

- It is recommended that SS gasflood tests should be carried out on composite cores, using ISSM to minimise saturation uncertainties and to evaluate end effects. This will limit the test to the major laboratories (e.g. Core Lab, Houston and UK, and Weatherford, Norway, UK and USA) who have the necessary equipment and ISSM facilities.

10.5.12 Centrifuge Water–Gas Relative Permeability (CentKrWG)

Centrifuge tests can be used to determine gas relative permeability. However, the *relative permeability obtained* is that of the *phase that is being displaced (gas)*—not the displacing phase (water). If it is assumed that krw drainage is the same as krw imbibition, then *a combination of drainage USS gasflood* and *centrifuge imbibition* tests can be used to define both *imbibition krg and krw* curves.

It is argued that water displacement of gas in centrifuge tests cannot be achieved under the backpressure necessary to maintain equilibrium between the test fluids. Gas dissolution into water and gas diffusion from water could compromise test results. Additionally, gas compressibility under centrifugal forces could artificially reduce gas volume during testing and expand gas volume again once the centrifuge is stopped. Consequently, for imbibition centrifuge tests the gas phase can be replaced by a light, non-wetting saturate oil, such as decane. Although a water–decane system may not accurately reflect a water-displacing-gas system, it does have the advantage that gas compressibility and gas dissolution/diffusion issues are avoided. However, theoretically, relative permeability is independent of viscosity and as long as the two fluids have similar wettability, the exchange may be viable. The difference to the system is the interfacial effects on capillary forces; however, these may be simulated.

Gas-equivalent *imbibition krg* relative permeability can be obtained from water–decane centrifuge experiments using a single pre-selected speed, within the limitations of the bond number ($Nb < 10^{-5}$), from the speed, time and production data using the Hagoort (1980) method or coreflood simulation. End-point permeability to water at residual decane saturation (essentially Kw′ at Sgr) can be measured in a coreholder.

In a gas–water system, there is little or no hysteresis in the wetting phase (water) relative permeability curve so *the imbibition water relative permeability, krw*, can be generated on the same samples from a drainage gasflood (as described in Section 10.5.8) or from a single-speed centrifuge test (using a decane–water system). Figure 10.47 provides an example. The drainage and imbibition krw curves overlie one another so the drainage USS

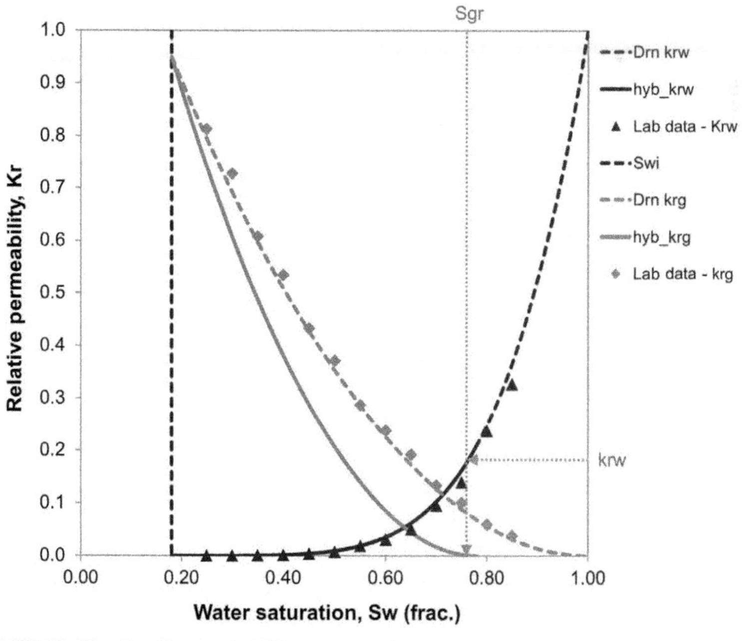

FIGURE 10.47 Combination imbibition and drainage water–gas relative permeability.

gasflood can be used to mimic the imbibition krw curve which is truncated at krw′ at Sgr.

Purpose of Test:	• Acquire drainage krw from USS gasfloods or centrifuge displacement (assuming no hysteresis occurs between drainage and imbibition cycles) and krg from imbibition centrifuge displacement tests
Sample Requirements:	• Cylindrical-shaped plug sample of 1″ or 1½″ diameter. The larger the sample, the larger the potential pore volume, and the less will be the impact of saturation measurement errors • Single core plug • *Ideally*—sample taken from preserved core or preserved plug. *Alternative*—plug previously used in routine core analysis programme, provided that plug has not been pre-stressed and/or that plug preparation methods have not caused damage to delicate clay minerals • Tests are carried out on cleaned-state, water-wet core, appropriately cleaned to prevent clay damage and to remove any OBM contamination
Required Input Data for Lab:	• FW salinity and ionic composition should be as close as practically possible to the FW in the reservoir • Overburden pressure if required and if possible • Spin speed (or critical bond number) and time intervals to acquire production data for decane–water displacement

10.5.12.1 CentKrWG Sample Preparation

Samples are prepared and saturated in SFW as discussed in Section 10.5.8.

10.5.12.2 CentKrWG Key Processes

1. Saturate in SFW
2. Determine brine permeability, Kb at overburden stress
3. Flood with gas until Swir attained (alternative use single-speed centrifuge desaturation)
4. Determine drainage krg and krw from modified JBN analysis or core-flood simulation
5. Determine end-point Kg′ at Swir at overburden stress
6. Desaturate in centrifuge (if USS drainage test performed) then flood with decane and re-establish Kg′ (equivalent) at Swir
7. Load into centrifuge under SFW and spin samples at the set single speed. Measure the effluent product until residual conditions (Sgr) are achieved
8. Calculate the imbibition krg (decane) relative permeability curve using the equations derived from Hagoort or via computer simulation
9. Load into coreholder and measure end-point Kw′ and Sgr at stress
10. Dean–Stark samples (to check for end-point water saturation).

10.5.12.3 CentKrWG Test Equipment

The equipment required to measure relative permeability by the centrifuge method is described in Section 10.2.3. Tests should only be performed on equipment with automated data capture systems.

10.5.12.4 CentKrWG Test Procedures

Following saturation in SFW, the plugs are loaded into an USS coreholder and brine permeability, Kb, established at net overburden stress. The samples are then flooded with humidified gas, and water and gas production are monitored until no more water is produced and the sample is assumed to be at Swir. Drainage krw and krg are calculated from the production data using modified JBN or coreflood simulation.

As an alternative, the drainage process (decane–water) can be carried out in a centrifuge at a single pre-selected speed, within the limitations of the bond number ($Nb < 10^{-5}$). Relative permeability is determined using the Hagoort method or coreflood simulation.

The sample is loaded into a coreholder and end-point Kg' is measured before flooding with decane under backpressure to remove the remaining gas.

Gas-equivalent *imbibition* krg relative permeability is then acquired from water–decane centrifuge experiments at a single pre-selected speed (at $Nb < 10^{-5}$) from the speed, time and production data using the Hagoort method or via coreflood simulation.

The samples are removed and end-point permeability to SFW at residual decane saturation (essentially Kw' at Sgr) is measured in a coreholder at net overburden stress.

Dean–Stark extraction is used to verify final water saturation calculations.

On the assumption of no hysteresis between the drainage and imbibition krw, the imbibition krw curve can be generated by truncating the drainage krw curve at Sgr (rather than $Sw = 1$).

Relative Permeability Centrifuge Method Summary	
Advantages (Imbibition krg): • Relatively simple and quick method • Good method for most rock types • More efficient drainage and the ability to measured very low imbibition krg values • The pressure gradient is non-linear, so boundary effects are less than those experienced under dynamic displacement	**Drawbacks and Issues:** • Although end effects are less, they still occur and result in a non-uniform distribution of saturation along the core sample length. Supporting the sample on a porous semi-permeable membrane can reduce these problems • Friable samples can be damaged and fracturing can occur during centrifuging • The tests are only possible at ambient conditions and at moderate overburden stress

	• A decane–water system may not accurately reflect a water-displacing-gas system. It does have the advantage that gas compressibility and gas dissolution/diffusion issues are avoided • Centrifuge Sgr data maybe too low if DP (spin speed) is too high

QC Checks

- The same generic QC checks used for water–oil, gas–oil and gas–water USS and SS analysis are also valid for water–gas tests.
- Ensure the method used for cleaning and drying does not cause damage or de-preservation of clay structures.
- Ensure that Swir is consistent with capillary pressure data for similar rock qualities.
- The high spin speeds may invalidate the critical bond number (Nb) criterion normally imposed on centrifuge imbibition experiments. The critical bond number should not exceed a widely accepted value of 10^{-5}.
- Sgr from centrifuge relative permeability tests may be obtained at a much higher DP than might occur in the reservoir and should not be used in simulation in isolation from data from other tests (centrifuge imbibition Pc and CCI).

Suitable Laboratories

- Most commercial laboratories can provide basic centrifuge tests where data are analysed by conventional analytical solutions. It is recommended that the tests are carried out in a centrifuge fitted with automated data acquisition, so only laboratories with this equipment should be considered.

10.5.13 Residual Gas Saturation: Imbibition Centrifuge Capillary Pressure (Sgr-CentPc)

These tests are designed to establish residual gas saturation under capillary dominated, as opposed to viscous dominated, flow. This arguably provides a closer representation of the reservoir processes. The tests are carried out using a *decane* and SFW system to minimise compressibility issues associated with gas–water systems in imbibition experiments. The centrifuge displacements can be carried out at ambient stress conditions (the imposition of overburden stress may be difficult to achieve in some labs and is not strictly warranted).

Purpose of Test:	• Determine residual gas saturation under capillary-dominated displacement
Sample Requirements:	• Cylindrical-shaped plug sample of 1″ or 1½″ diameter. The larger the sample, the larger the potential pore volume, and the less will be the impact of saturation measurement errors • Single core plug • *Ideally*—sample taken from preserved core or preserved plug. *Alternative*—plug previously used in routine core analysis programme, provided that plug has not been pre-stressed and/or that plug preparation methods have not caused damage to delicate clay minerals • Tests are carried out on cleaned-state, water-wet core, appropriately cleaned to prevent clay damage and to remove any OBM contamination
Required Input Data for Lab:	• FW salinity and ionic composition should be as close as practically possible to the FW in the reservoir • Overburden pressure if required and if possible • Spin speeds (and critical bond number)

10.5.13.1 Sgr-CentPc Sample Preparation

Samples are prepared and saturated in SFW as discussed in Section 10.5.8.

10.5.13.2 Sgr-CentPc Key Processes

1. Saturate in SFW
2. Determine brine permeability, Kb at overburden stress (if required)
3. Drainage capillary pressure curve using decane (or air) to Swir
4. Determine end-point Kg′ at Swir (decane or air) at overburden stress (if required)
5. Imbibition Pc curve using decane–water system at incremental pressures
6. Load into coreholder and measure Kw′ and final Sgr at stress (if required)
7. Dean–Stark samples (to check end-point water saturation)
8. Plot Sgr as a function of capillary pressure/bond number.

10.5.13.3 Sgr-CentPc Test Equipment

The equipment required is described in Section 10.2.3. Tests should only be performed on equipment with automated data capture systems. The primary drainage curve is obtained by decane displacement or air displacement to Swir using similar pressure increments as the legacy drainage air–water tests, allowing at least 24 h per drainage pressure. These data can be compared with the legacy capillary pressure data to ensure Swir is representative. Forbes (or similar) method is used to interpret the drainage Pc tests.

If end-point relative permeability is required, then the sample is loaded into a coreholder and the equivalent Kg′ at Swir (decane or air) is determined at net overburden stress. If air is used for the drainage Pc curve, the core must be flushed with decane under backpressure to remove any air.

10.5.13.4 Sgr-CentPc Test Procedures

The samples are removed and loaded into the centrifuge imbibition test cell or an Amott cell under SFW. Spontaneous imbibition of SFW is monitored until complete. Samples should be allowed to spontaneously imbibe water in the centrifuge cell and imbibition monitored before the imposition of the first capillary pressure. Otherwise, spontaneous imbibition will be masked as the centrifuge builds up to the test speed.

The imbibition cycle data are then acquired over specific pressures, focussing on the low-pressure range. Gas saturation (and Sgr) is determined as a function of imbibition pressure and bond number. A typical test pressure sequence might be:

- 1 psi
- 2 psi
- 5 psi
- 10 psi
- 25 psi
- 50 psi
- 100 psi
- or maximum speed as defined by a critical bond number of 10^{-5}.

Alternatively, one should plot the centrifuge speeds, initially plotting the selected minimum and maximum speeds (determined by either maximum centrifuge speed or maximum acceptable bond number); the remaining speeds should be distributed between the extremities such that they are spaced evenly on a logarithmic scale.

Sg is evaluated as a function of Pc and Nb. Sgr should be established at a Pc (or corresponding Nb) appropriate to the pressure drop across the advancing aquifer water front in the reservoir for different permeability zones. This pressure drop is considered to be low—less than 10 psi typically.

On completion, the plugs are removed and the final (and maximum) equivalent water (SFW) permeability, Kw′, is determined at final Sgr at overburden stress (if required). The DP for this flood must be less than the maximum imposed by the centrifuge.

A Dean–Stark extraction is used to verify final water saturation.

Residual Gas Saturation Centrifuge Method Summary	
Advantages (Imbibition krg): • Relatively simple and quick method • Good method for most rock types • Provides both drainage and imbibition Pc curves and end-point krg′ and krw′ • Sgr can be obtained as a function of DP and bond number. Sgr is established at a Pc appropriate to the pressure drop across the advancing aquifer water front in the reservoir for different permeability zones	**Drawbacks and Issues:** • The test may not be suitable for weak or friable samples • The tests are only possible at ambient conditions and at moderate overburden stress • A decane–water system may not accurately reflect a water-displacing-gas system. It does have the advantage that gas compressibility and gas dissolution/diffusion issues are avoided

QC Checks

- The same generic QC checks used for water–gas relative permeability and centrifuge tests are also valid for imbibition Pc tests.
- Ensure the method used for cleaning and drying does not cause damage or de-preservation of clay structures.
- Ensure that Swir is consistent with capillary pressure data for similar rock qualities.
- Residual gas saturation centrifuge data obtained at possibly unrepresentative imbibition pressures or bond numbers may be too low.
- There is often a robust relationship between initial gas saturation (Sgi = 1 − Swir) and residual gas saturation (Sgr). Refer to Section 10.1.6. Unfortunately, Sgr is sensitive to the method used to acquire data (e.g. dynamic flood, CCI, centrifuge displacement) and Sgi–Sgr correlations are often not directly applicable or reproducible in all reservoirs. For example, see Fig. 10.3.

Suitable Laboratories

- Most commercial laboratories can provide basic centrifuge tests where data are analysed by conventional analytical solutions. It is recommended that the tests are carried out in a centrifuge fitted with automated data acquisition, so only laboratories with this equipment should be considered.

10.5.14 Counter-current Imbibition Residual (Trapped) Gas Saturation (CCI-Sgr)

Residual gas saturation can also be determined using the CCI technique, which can provide remarkably consistent and reliable data—despite its simplicity. These provide a relatively cheap and rapid way to obtain information

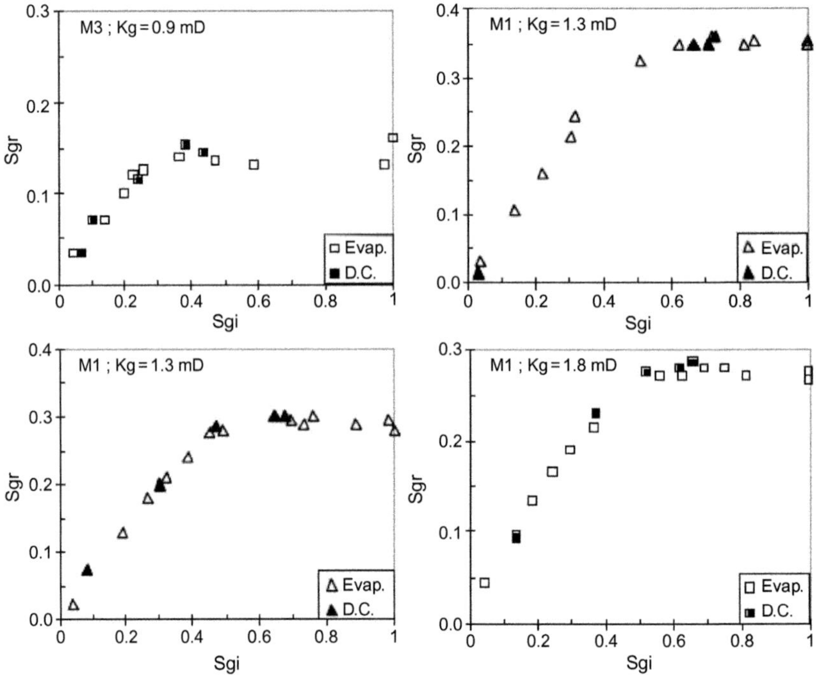

FIGURE 10.48 Examples of Sg–Sgr relationships from CCI tests. *Suzanne et al. (2003).*

on the relationship between initial (Sgi) and residual (Sgr) gas saturations, as indicated in Fig. 10.48.

The test mimics an imbibition process dominated by capillary forces and involves repeated imbibition of liquid from various controlled initial gas saturation values, covering the range in the productive reservoir zones. An air–toluene system is used because the initial gas saturation in the test cores can be controlled by evaporation rather than by the more time-consuming capillary desaturation. The data from a range of plugs can be used to establish a composite Sgi versus Sgr relationship, often as a function of rock quality.

Purpose of Test:	• To determine residual gas saturation, Sgr, under capillary-dominated processes
Sample Requirements:	• Cylindrical-shaped plug sample of 1″ or 1½″ diameter. The larger the sample, the larger the potential pore volume, and the less will be the impact of saturation measurement errors • Single core plug

	• *Ideally*—sample taken from preserved core or preserved plug. *Alternative*—plug previously used in routine core analysis programme, provided that plug has not been pre-stressed and/or that plug preparation methods have not caused damage to delicate clay minerals
Required Input Data for Lab:	• Target initial gas saturations

10.5.14.1 CCI-Sgr Sample Preparation

The sample is initially cleaned and dried and the base poroperm properties measured. The sample is saturated in toluene. Weight difference is used to determine toluene saturations.

10.5.14.2 CCI-Sgr Key Processes

1. Saturate in toluene
2. Desaturated (evaporate toluene) to target initial gas saturation, Sgi
3. Spontaneous imbibition of toluene
4. Residual gas saturation (analysis of weight change)
5. Composite Sgi versus Sgr plot for each sample.

10.5.14.3 CCI-Sgr Test Equipment

The CCI test setup is extremely simple. The test sample (at the target Sgi) is suspended on a wire which is attached to an electronic balance as shown in Fig. 10.49. The output from the digital balance is fed to a computer with data acquisition software to record weight changes as a function of time. The change in weight represents the weight of toluene imbibed.

10.5.14.4 CCI-Sgr Test Procedures

The plug is initially saturated in toluene under vacuum and/or pressure. The toluene is allowed to evaporate under controlled conditions to ensure uniform toluene saturation in the plug. This may be best achieved by slow rate (cool temperature) evaporation, with the rate and amount of evaporation continuously monitored on a weigh balance.

Evaporation is continued until the plug toluene saturation is about 5% PV or so above the first target initial gas saturation (e.g. Sgi = 80%). The plugs are wrapped in film and foil (to prevent further evaporation) before being refrigerated for 24 h to allow the toluene to redistribute under capillary forces. The wrapped samples are removed from the fridge and allowed to equilibrate at ambient conditions for a minimum period of 4 h before commencing the tests.

The plug is unwrapped and loaded onto a cradle, which is connected to an electronic balance, and is immersed under toluene. Since imbibition is

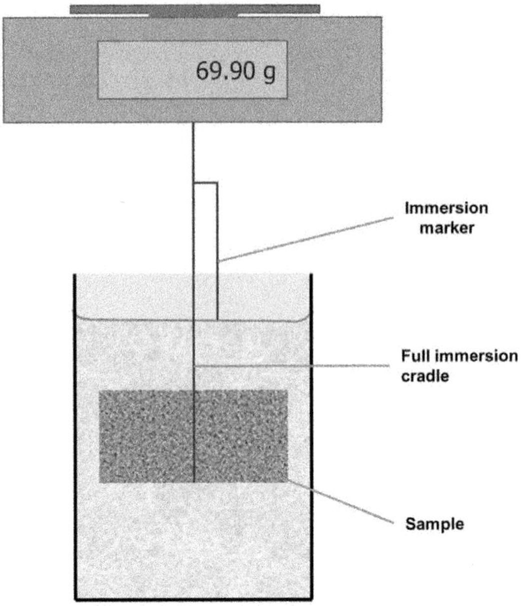

FIGURE 10.49 Counter-current imbibition (CCI) setup.

expected to occur during the period that the plug is being lowered to the reference level under toluene, the initial immersion should be carried out as quickly as possible, and data recording commenced immediately.

Immersed weight is monitored against time until sufficient data is collected to differentiate between toluene imbibition and air diffusion curves, so that the residual gas saturation, Sgr, can be established. This is best achieved by plotting gas saturation versus either the square root of time or logarithmic time (see example in Fig. 10.50).

The CCI test is repeated to establish a total of 4 or 5 Sgi conditions covering the range in the reservoir, for example:

Target Sgi	Sgi (v/v)
1	0.80
2	0.60
3	0.40
4	0.20

The data from each test are used to establish a composite Sgi versus Sgr relationship for each plug. Consequently, for any value of Sgi (1 – Sw) in the

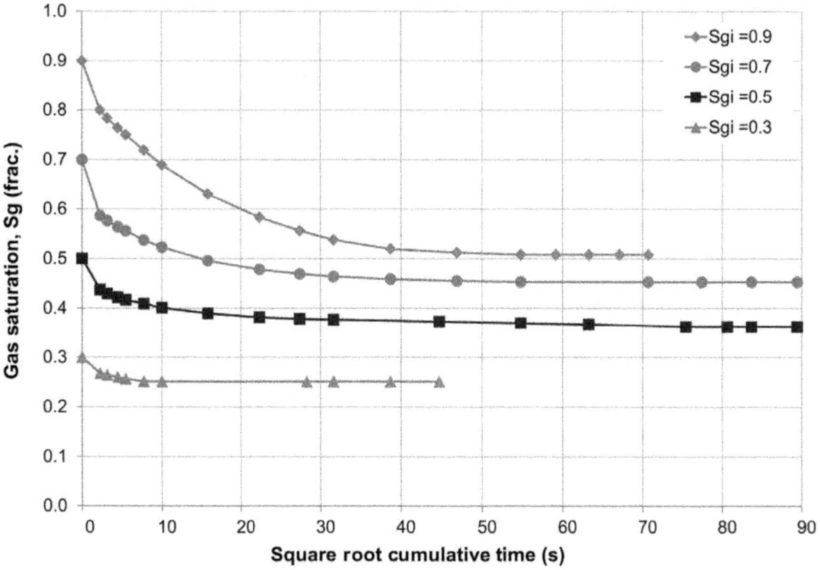

FIGURE 10.50 CCI experimental data.

reservoir, the residual gas saturation (e.g. on aquifer influx) can be predicted from this relationship.

On completion of the tests, the plugs should be dried to a constant weight and the final dry weight measured to check for grain loss.

Residual Gas Saturation (CCI) Method Summary

Advantages:	Drawbacks and Issues:
• Simple, fast, reliable and repeatable test • Upper bound estimate of Sgr as a function of Sgi	• CCI tests can often provide a pessimistic value of residual gas saturation

QC Checks

- Experimental data from the imbibition phase are essential to differentiate between imbibition and diffusion processes.
- Sgr is sensitive to the method used to acquire data and Sgi–Sgr correlations are often not directly applicable nor reproducible in all reservoirs.
- CCI tests often provide a pessimistic value of residual gas saturation, compared to centrifuge or dynamic flood data, as a result of rapid imbibition from the outside of the sample preventing gas from escaping from the centre, and as the test method replicates spontaneous imbibition as

opposed to forced imbibition in the centrifuge tests. However, the test does allow capillary forces to dominate and diffusion flow in microporous spaces.

Suitable Laboratories

- Most commercial laboratories are equipped to carry out this simple test. The imbibition stage (weight change) should be monitored using digital data acquisition software.

10.6 CRITICAL GAS SATURATION (DEPRESSURISATION)

In this context, critical gas saturation is defined as the critical saturation for gas mobilisation under secondary drainage processes, for example, blow-down of trapped gas on reservoir depletion. The test procedures proposed are designed to mimic the reservoir processes as closely as possible.

Purpose of Test:	• Trapped gas saturation (Sgt/Sgr) and end-point water permeability—Kw′ at Sgt/Sgr • Trapped gas mobility and critical gas saturation (Sgc) under simulated depressurisation conditions • End-point (final) gas permeability (Kg′) and end-point (final) water permeability (Kw′) at final gas saturation at final reservoir pressure • (Option) intermediate gas (Kg) and water permeability (Kw) and relative permeability (krg and krw) at test pressures below initial pressure
Sample Requirements:	• Cylindrical-shaped 1½″ diameter plug sample. To avoid influence of capillary end effects on short core plug flood performance, tests are better performed on 10- to 15-cm-long composite core • *Ideally*—samples taken from preserved core. *Alternative*—plug previously used in routine core analysis programme, provided that plug has not been pre-stressed and/or that plug preparation methods have not caused damage to delicate clay minerals • Tests are carried out on cleaned-state, water-wet core, appropriately cleaned to prevent clay damage and to remove any OBM contamination
Required Input Data for Lab:	• Swir appropriate to maximum height above FWL for the sample porosity and permeability • FW salinity and ionic composition should be as close as practically possible to the FW in the reservoir. SFW may be doped (with NaI) to facilitate ISSM • Overburden or confining stress • Height/length of trapped gas column • Field depressurisation rate (psi/year)

10.6.1 Sample Preparation

Core analysis samples selected for measurement should be representative of the lithologies in the reservoir. As there is often a dependency between Sgc and permeability (Sgc increases with permeability), it may be worthwhile conducting experiments on a range of formation permeabilities (e.g. 1, 10, 100 mD).

By using longer plugs, it is possible to minimise the influence of capillary end effects and maximise pore volume and saturation accuracy.

10.6.2 Test Equipment

Experimental pressure, temperature and flow rates require to be precisely controlled, and gas and water saturations carefully determined. In isolation, gravimetric and/or volumetric fluid production measurements are not sufficiently accurate to detect quite subtle changes in gas saturation on depletion. Consequently, best practice dictates that this test should only be carried out using a high-pressure capacity rig and using ISSM.

A typical experimental setup for depressurisation studies is shown in Fig. 10.51.

A range of facilities should be available capable of test temperatures of up to reservoir temperature and to the maximum net effective isostatic reservoir stress. The facility must include the necessary pressure controllers, absolute pressure transducers and DP transducers to cover the expected pressure range during flooding.

The rig should be equipped with positive displacement pumps for laboratory flow rates in the range from 4 to 400 ml/h. Samples are mounted vertically and

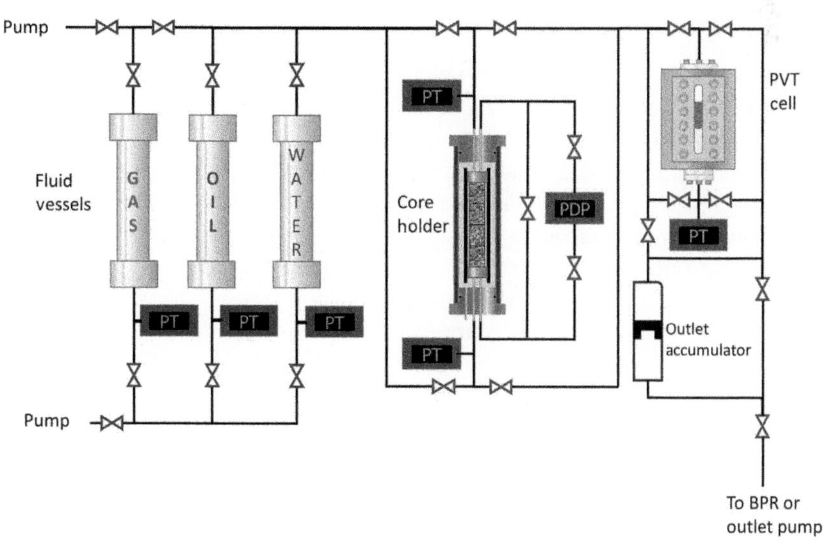

FIGURE 10.51 Typical critical gas saturation experimental setup.

the flood direction is usually bottom to top (i.e. oil gravity stable) or top to bottom (gas gravity stable). A visual PVT cell or high-pressure (acoustic) separator is used to observe or record gas production from the core (in addition to ISSM).

10.6.3 Test Procedures

10.6.3.1 Swir and Sgr

As Sgc is largely a function of residual or trapped gas saturation (Sgr/Sgt), and as Sgr/Sgt often depends on Sgi, establishing a reliable and representative irreducible water saturation is essential. Either a porous plate or centrifuge system can be used.

The test samples are replaced back into the rig and end-point effective gas permeability (Kg′ at Swir) is determined at ambient conditions.

The waterflood is carried out using equilibrated (and doped) water at a reservoir advance rate of approximately 1–2 ft/day. During the flood, gas production is determined using both ISSM and from volumetric production data acquired using either a high-pressure separator or from visual inspection or a high-pressure cell positioned at the core outlet. Flooding is continued either until no more gas is produced (core at Sgr) or the desired level of trapped gas saturation (consistent with log analysis, for example) is attained.

ISSM is essential to establish Swir and Sgr and to ensure uniform desaturation and imbibition.

At the trapped gas saturation, end-point effective water permeability—Kw′ at Sgt/Sgr—is measured.

10.6.3.2 Depressurisation

The pressure decline rate should be low enough to prevent the gas saturation from increasing above the re-mobilisation saturation in order to allow the mobile gas to migrate to the outlet. As an example, to experimentally represent a 10-m column with residual gas in a reservoir in which the pressure declines to two-third of its initial pressure in 25 years, the gas pressure in the experiment should decline to two-third of the initial value in about 6 weeks (42 days). It is possible that a faster decline rate (say over 14 days) may result in premature re-mobilisation of trapped gas saturation, although there is evidence that critical gas saturations (at least in higher-permeability core) are independent of pressure decline rate.

Depressurisation from the trapped gas saturation is commenced by isolating the water injection pump and allowing either:

- a constant rate extraction from the outlet separator using a second pump or
- a constant pressure decline rate (by extraction of fluid from an outlet accumulator).

The core is depressurised continuously over a 14- or 42-day period. The redistribution and production of gas is measured using ISSM and by visual or acoustic measurements of produced volumes. Gas production data are used to define the critical gas saturation. Absolute pressures are measured at the core inlet and core outlet, together with direct measurement of DP.

10.6.3.3 Secondary Drainage End-Point Kg′ and Kw′

The effective "end-point" gas permeability—Kg′—is determined at a final gas saturation attained at the end of the depressurisation stage at the final reservoir pressure and final gas saturation. The system is allowed to equilibrate before gas is flowed at low flow rate and Kg′ determined. Gas production is monitored and final saturations observed using ISSM.

The "end-point" water permeability at final reservoir pressure—Kw′—is determined by flooding water at very low flow rate (at decreasing rates if possible) to measure the pressure gradient and water permeability (Darcy's law) at the final gas saturation Sg. However, it is possible that water imbibition may redistribute and/or mobilise gas, which may introduce uncertainty in the depressurisation data (observation and ISSM will should detect this).

10.6.3.4 Secondary Drainage Intermediate Relative Permeabilities

Although end-point relative permeability measurements are relatively straightforward, acquisition of intermediate USS secondary drainage gas and water relative permeability measurements are very difficult to measure during depressurisation and will require coreflood simulation to model accurately. This essentially leaves two options:

1. SS tests
2. Quasi-steady-state (qSS) tests

In the *SS* method, the core is depressurised at a constant extraction rate until critical gas saturation is observed. Fractional flow tests are then undertaken at 4 or 5 incremental gas:water fractional flows. The total flow rate must be kept as low as is practicable to minimise viscous forces, since capillary forces are dominant for the depressurisation process. Very low fractional flow is desirable, but is often found to be impractical. At each SS condition, the fractional flow, total flow rate, measured DP and *in situ* gas saturation are used to calculate relative permeability from Darcy analysis.

In the *qSS* method, the core is depressurised at a constant extraction rate until critical gas saturation is observed. The depressurisation process is stopped at intermediate pressure steps, and the system is allowed to equilibrate. Gas permeability at the given gas saturation is determined at ultra-low flow rates using the same method as for end-point Kg′. Saturations are carefully monitored during each measurement. Water permeability can be determined following Kg′ measurements but water imbibition hysteresis effects can have an adverse influence on the data. In this case, either:

1. the potential uncertainty is accepted or
2. the depressurisation sequence is completed without water permeability measurements, then the core is re-pressurised, re-flooded with water to

re-establish a similar trapped gas saturation, and the depressurisation process repeated with the water permeability measurements undertaken at the intermediate pressures.

Given these uncertainties (and additional costs), end-point (final reservoir pressure) gas and water permeabilities may prove to be sufficient for most simulation applications.

Critical Gas Saturation (Depressurisation) Summary

Advantages:	Drawbacks and Issues:
• Only method that can reliably provide critical gas saturation and trapped gas mobilisation data on reservoir depressurisation. Conventional secondary drainage dynamic flood tests assume that gas becomes immediately mobile (i.e. Sgc = Sgt/Sgr) • Provides end-point gas permeability (Kg′ at Swir) and end-point water permeability (Kw′ at Sgt/Sgr) under imbibition conditions and at representative reservoir pressure conditions • Provides Sgc and secondary drainage end-point gas and water permeability (Kg′ and Kw′ at final gas saturation) at final reservoir pressure under simulated reservoir depletion conditions • Intermediate gas and water permeability relative permeability measurements under depressurisation conditions (optional) from SS or qSS conditions	• Long core plugs are preferred to optimise measurement accuracy • Data optimisation dictates that test should only be carried out using a high-pressure capacity rig and using ISSM. This demands a high level of sophistication in lab equipment and resources • Test costs are high due to requirements for ISSM and depressurisation periods • Intermediate secondary drainage relative permeabilities are difficult (and expensive) to perform and can be affected by saturation hysteresis. Coreflood simulation may be required to adequately model intermediate krw and krg • Requirement for calibration scans for ISSM normally precludes the use of Dean–Stark for final saturation verification • Friable samples require being protected against grain loss and sample failure

QC Checks

- Analysis of the results should start with a thorough review of the laboratory report to check that acceptable procedures have been followed. This should be carefully evaluated.
- Check the method used for cleaning and drying and if care has been taken to avoid damage and preservation of the clay structures.
- As Sgc is largely a function of Sgr/Sgt, and as Sgr/Sgt often depends on Sgi, establishing a reliable and representative irreducible water saturation is essential. Ensure Swir for plug or composite is representative for given rock quality.

- The pressure decline rate should be low enough to prevent the gas saturation from increasing above the re-mobilisation saturation in order to allow the mobile gas to migrate to the outlet. Depressurisation periods range from 12 to 42 days.
- Ensure Sgr/Sgt is representative by comparison with log data and/or other residual gas saturation estimates (e.g. CCI or co-current tests).
- Evaluate Kw′ at Sgr against other data (imbibition centrifuge and SS tests).
- All end-point and intermediate relative data should be normalised to both air (or Klinkenberg) and end-point gas permeability (Kg′ at Swir). This can help identify anomalous permeabilities.
- Critical gas saturation is expected to increase with permeability and be between 0.03 and 0.15 saturation units greater than Sgc/Sgt. Values out with this range should be more carefully evaluated.

Suitable Laboratories

- The requirement for sophisticated test equipment, ISSM and potential coreflood simulation will limit the test to the major laboratories (TBC) who have the necessary equipment and ISSM facilities.

10.7 EXPERIMENTAL DATA PROVISION

Laboratory reports need to be improved by ensuring that complete records of the test methods and procedures together with laboratory experimental data are available, so that there is a complete audit trail. Experimental data are not normally provided by service laboratories but are essential to a detailed understanding of the results of the core tests, enable alternative interpretations and may prove invaluable in future reservoir evaluation or unitisation.

The following represent the key data reporting requirements for water–oil, gas–oil and gas–water dynamic and centrifuge experiments.

10.7.1 Samples and Fluids (for All Tests)

Data Required	Comments
• sample length and diameter (cm) and plug orientation (horizontal/vertical)	
• test oil, water and gas density (g/cc) and viscosity (cP) at test temperature and initial and final pressures	Recommended to perform these parameters at a range of temperatures and calculate equations of curve fits
• calculated sample porosity (fraction) and grain density (g/cc) on dry samples (pre-test)	Data should be cross-checked by the various calculation methods, and against dimension data above

Data Required	Comments
• calculated sample saturated porosity (fraction) for each test sample	QC control within ±0.2 p.u.
• calculated sample saturated air permeability Ka (mD) for each test sample and permeability ordering if composite used	
• calculated porosity, pore volume and bulk volume from water saturation (individual plug or total composite)	QC checks should agree with dry plug properties
• dopant type and concentration in SFW and/or oil (if ISSM used)	
• test fluid IFT at test temperature and pressure	IFT of reservoir fluids also recommended for converting lab to reservoir data
• post-test final sample saturated weight, dry weight (after final cleaning and drying), helium porosity and permeability	Data must be assessed for potential change and comments provided
• sample grain loss correction procedures, including typical worked example for one sample from sample database	

10.7.2 Drainage Capillary Pressure (Desaturation to Swir)

Data Required	Comments
• **description of procedures and experimental apparatus	
• pressure and time allowed to achieve saturation stability (porous plate)	Plot of production versus log time or square root time is recommended
• distance from centre of rotation to inlet and outlet of core and other relevant data sufficient to allow alternative Pc interpretation (centrifuge)	
• RPM, volume of water produced at each capillary pressure increment, and time to stabilise (centrifuge)	Also recommended final production volume (from collection cup) and gravimetric measurements to QC data.

Data Required	Comments
• method used to smooth data and to convert average to end-face saturations (e.g. Forbes, coreflood simulation)	
• bond number for each capillary pressure step with worked example (centrifuge)	
• test temperature	
• test overburden pressure	

10.7.3 Centrifuge Relative Permeability (Water–Oil, Gas–Oil, Decane–Water)

Data Required	Comments
• description of procedures and experimental apparatus	
• end-point oil or gas or decane permeability at Swir and confining pressure	QC correlation to absolute permeability measurements
• spin speed and equivalent capillary pressure for relative permeability test	Criteria for speed selection should also be reported
• test temperature	
• test overburden pressure	
• calculated bond number at spin speed	
• fluid production data as a function of time	
• calculated relative permeability as a function of water, oil or gas saturation, and method used to analyse data (e.g. Hagoort, simulation)	
• end-point water or gas permeability at residual saturation	Including correlation to absolute permeability and/or base permeability
• final air permeability, helium porosity and dry weight	QC against pre-test measurements, discussion of observations
• final water saturation from Dean–Stark analysis (water–oil and water–decane)	Correlation against expected saturation

10.7.4 Counter-Current Imbibition

Data Required	Comments
• description of procedures and experimental apparatus	
• sample weight at target saturation	
• plot/data of sample immersed weight (and interpreted gas saturation) versus square root and/or logarithmic time	
• initial gas saturation	Actual Sgi calculated from sample weight (not merely stipulating the target saturation)
• residual gas saturation	
• composite Sgi versus Sgr plot for each sample	

10.7.5 USS and SS Flood Tests (Water–Oil, Gas–Oil and Water–Gas)

Data Required	Comments
• description of procedures and experimental apparatus	
• initial individual plug porosity and air permeability	
• method used to decide order of butted plugs in composite core and method used to ensure capillary continuity	
• test temperature	
• test overburden pressure	
• initial end-point gas or oil permeability at Swir and flow rate	QC correlation against absolute permeability
• flood rate for USS waterflood (both low rate and bump)	
• flood rate for gas–oil USS	Or differential pressure used for constant pressure displacements
• flood rate for water–gas USS	
• produced volumes, DP and pore pressure as function of injection volumes and time	

Data Required	Comments
• end-point water or gas permeability at Srow, Srog or Sro (low rate)	QC correlation to absolute permeability and/or base permeability
• flood rate for high-rate bump flood	Or increased DP for constant pressure displacements
• produced volumes and DP as function of injection volume for bump flood	
• method used to calculate intermediate relative permeability curves (with example)	
• water:oil, gas:oil, gas:water and water: gas injection ratios and total flow rates for SS tests	
• time allowed/required to achieve SS conditions	Preferable to provide production history (volume and pressures versus time) to indicate steady state was achieved at each fraction flow rate
• DP versus time at and between each injection ratio	
• ISSM scans at each key experimental stage (e.g. at end points and intermediate points)	In addition to saturations, the measured radiation counts should be provided for each stage and should include the calibration scans performed
• final saturated weights for individual plugs	
• final Dean–Stark water saturation (water–oil, water–gas) for individual plugs (if no ISSM scan)	It may be possible to acquire effluents for measuring water content by Karl Fischer (KF) titration, thus KF saturation should be provided and cross-checked.
• final air permeability, helium porosity and dry weight of individual plugs	QC checks against pre-test properties should be performed and any observations discussed.

10.8 SUMMARY AND RECOMMENDATIONS

The following section provides a summary and recommendations of which relative permeability measurements are important in the context of any oil and gas field developments, and provides recommendations on the optimum strategies to obtain these measurements.

The key question that should be asked when designing a SCAL programme for relative permeability tests is: What is the impact on HCs in place and ultimate recovery? For example, simulation modelling may show that recovery factors may be largely insensitive to the intermediate relative permeability curves—only end points might be important. The intermediate relative permeability curves are normally the first things that are adjusted in history matching the simulation model (whether this ought to be the case or not). In this case, is there any need to acquire intermediate curves, especially since the curves may only be viable if acquired on wettability-restored cores, at reservoir conditions and at significant cost and time? Especially so, if the curves are ultimately going to be ignored. In this case, perhaps only end points might be required, so the SCAL programme should concentrate on defining these and their relationships with rock type, initial saturations and wettability. The intermediate curves between the end points could be invented, for example, using Corey exponents for a specific wetting state or rock type, or using analogue information. For FDP planning and production forecasting, with no production data available, the intermediate relative permeability curves might assume greater importance and need to be established experimentally.

The SCAL data should cover the full range of reservoir quality—avoid the typical 80/20 situation where 80% of the data are from the best 20% of the reservoir. Refer to Section 10.4.1 for guidance.

However, expect ambiguous results across the range. Experience has shown that non-optimised SCAL experiments can reproduce poorly, particularly at low permeability Byrnes (2008), and that different experimental methods and different labs both produce different results.

The results require sense checking. For example, if water relative permeability approaching Sro is unusually high for a water–oil test in a water-wet system, then the results need to be carefully evaluated in the context of the test state, core condition, petrophysical properties and lab artefacts. Are there consistent and intuitive trends in the data—for example with permeability or wettability? Does the new data agree with analogue or legacy data—if not, assess reasons why they are, or might be, different. Is Swir uniform and representative for the height above the FWL and rock type and rock quality?

Lastly, and perhaps most importantly, it is essential not to sacrifice representative and useable results at the altar of expediency and cost reduction. Reliable relative permeability data take time and are expensive to acquire.

10.8.1 Oil Reservoirs

10.8.1.1 Model Initialisation

1. Drainage capillary pressure SCAL tests for STOIIP initialisation in simulation models must provide full coverage of the HC column height and full coverage of the net reservoir quality range.
2. Although porous plate data are acceptable (and in some cases preferable), centrifuge experiments (oil–water or air–water) provide data in a

reasonable time frame with significantly less artefacts. Tests are carried out on cleaned core as drainage is assumed to occur under water-wet conditions (unless there has been a previous oil fill). It is important to:

- ensure bond numbers are less than 10^{-5} in centrifuge tests if at all possible (Section 10.3.3)
- use a robust analytical model (e.g. Forbes) or simulation models to calculate the end-face saturations
- to verify the core Pc data against log-derived saturation-height models.

3. The use of overburden stress can limit rotational speeds so in some low-permeability cases, it may not be possible to define the full Pc curve.
4. If imbibition Pc is required, centrifuge provides the only viable method to define the forced imbibition curves. These must be performed on wettability-restored core.

10.8.1.2 Relative Permeability

The requirements depend on the development concept and expected recovery mechanism.

Waterflood

1. Waterflood relative permeability tests (Sections 10.5.1 and 10.5.2) should only be carried out on wettability-restored samples. Prior to Amott/USBM wettability tests, ensure that core cleaning removes any wettability contamination, and a restored-state wettability condition should be established.
2. Ensure Swir is representative for the particular rock type and rock quality. Even for transition zone applications, Swir should be equivalent to the maximum height above FWL for the rock type, and relative permeabilities for higher Sw in the transition zone can be determined from scanning curves.
3. If the reservoir is judged to be water-wet, tests carried out using synthetic fluids (SFW and mineral oil of the appropriate viscosity) should provide reliable data. For end points, it is important to ensure that the lab oil:water viscosity ratio is the same as the reservoir ratio.
4. If the reservoir is intermediate or slightly oil-wet, tests should be carried out at reservoir conditions using reservoir fluids.
5. Where possible, waterflood tests (SS or USS) should be carried out on long composite cores using ISSM to monitor and resolve saturations.
6. USS methods (Section 10.5.1) are preferred to define the end points—kro′ and krw′—and should use a matched oil:water viscosity ratio. The tests are conducted at reservoir appropriate capillary number and on completion, a bump flood (increasing the rate and/or centrifuge displacement with water) might be performed to establish the presence and impact of end effects.

7. Intermediate relative permeability curves are better defined by SS tests (Section 10.5.2) especially if the oil:water viscosity ratio is adverse. SS tests can only provide reliable data if saturations are monitored with ISSM.
8. As JBN methods are invalid for most corefloods with significant capillary effects, coreflood simulation (Section 10.3.5) should be used to analyse and determine relative permeability for all USS tests. Simulation should also be considered for SS and centrifuge tests, and can also be used to design the test flood sequence.
9. Centrifuge displacement tests (Section 10.2.3) can help resolve low kro close to Sro and should be used in conjunction with data from SS and USS.
10. Lab data should be compared with Corey models (Section 10.1.12). The match parameters should be plotted versus absolute permeability (or flow zone indicator or RQI) to establish whether there are any clear trends or whether averaging is required. In addition, results should be correlated with wettability.
11. End-point scaling can be used to combine relative permeability data from different tests which allows modelling of spatial differences in relative permeability and capillary pressure end points from a single set of generic saturation functions. This requires merging USS end points, SS intermediate kro and krw, and centrifuge kro and Sro. Normalised functions are generated versus reduced saturations, end-point values are assigned and the corresponding functions are derived. The rock curves can be generated by denormalising the curves using the appropriate end points (kro′ and absolute air permeability).
12. Composite relative permeability curves used in reservoir simulation should be based on what is considered to be the "best" data. Any uncertainties must be reduced by additional testing and evaluation of end points and intermediate curves.

Depletion (solution gas drive)

1. The key parameters are critical gas saturation, Sgc (Section 10.1.7), and residual oil saturation to gas, Srog (Section 10.1.5).
2. Sgc is notoriously difficult to measure or extrapolate from krg/kro ratio trends from gas–oil displacements. Default values of 0.02–0.05 should be used.
3. Srog is generally approached away from the well in secondary gas caps. The residual oil saturation to gas from most laboratory gas–oil tests is influenced by lab artefacts (Section 10.1.5). In the absence of end effects and in the presence of thin film and miscibility effects, gravity drainage occurs and nearly all the oil will drain from the rock with time. Consequently, "true" Srog will be zero or close to zero.

Gas-cap drive

1. Gas–oil displacement is highly unstable. Away from the well, the displacement is generally segregated and gas–oil displacement efficiency depends mainly on end-point relative permeabilities. Near the well, the gas–oil saturation distribution is very complex. The fractional flow of gas and oil is dictated by displacement efficiency away from the well. Modelling displacements require the end-point relative permeabilities krg′ (Srog) and kro (Swir).
2. Gas–oil USS experiments (Section 10.5.4) on short core plugs are dominated by viscous instabilities and capillary end effects. Reliable flood data are only possible on long composite cores using reservoir fluids where the Pc end effects (which scale with the inverse of core length) are relatively insignificant or from centrifuge tests (Section 10.5.6). One advantage of the centrifuge method is that very low kro can be measured (although no data are gained on krg). Viscous forces are replaced by centrifugal forces which act against capillary pressure, so kro curves from centrifuge tests are less likely to be influenced by end effects.
3. The recommended relative permeability curves for gas–oil systems should principally be based on centrifuge tests (kro) and Corey exponents (krg). The range of Corey exponents for gas relative permeability is relatively narrow (1.3–3.0) as gas is non-wetting.
4. In the absence of reliable relative permeability data, gas–oil curves can be invented using a Corey approach and sensitivities to Sgc, Srog, krg′ evaluated (Section 10.5.7).

10.8.2 Gas and Gas Condensate Reservoirs

10.8.2.1 Model Initialisation

1. Requirements are similar to those for oil reservoirs (Section 10.8.1) but it is particularly important to ensure full coverage of all net intervals, as the largest GIIP uncertainty is generally in poor-quality rock.
2. Air–water centrifuge Pc drainage curves on cleaned, water-wet rock are preferred. The pressure range should be sufficient to define the full column height, even if this involves experiments exceeding the critical bond number. Pc curves should be checked and verified against analogues or field models to ensure they are consistent.

10.8.2.2 Relative Permeability

1. For reservoir simulation and modelling purposes, lab gas–water and water–gas relative permeabilities must be re-normalised to absolute permeability—normally air or Klinkenberg permeability.
2. Gas is always the non-wetting phase, so there is no need for restored-state or preserved core.

3. End-point relative permeability for gas (i.e. drainage curve) is required for differential depletion in poor-quality rock, and sealing capacity of faults. In lower permeability formations, Kg at Swir can be significantly less than the absolute air permeability due to water occupying the critical pore space.
4. End-point krg′ can be easily and cheaply determined on completion of the air–water drainage Pc tests described above. The samples, at Swir, are loaded in a coreholder and flushed with air or decane to determine krg′.
5. End-point krg′ should be sense-checked by plotting as a function of permeability and comparing with analogue data. Even with a large database, trends may be difficult to distinguish. For example, Fig. 10.52 shows a surprisingly abrupt reduction in krg′ below 1 mD air permeability: at 0.1 mD, krg′ is less than 0.1. This may be consistent with the presence of clays or water occupying critical pore space. Clay dehydration and destruction during cleaning and drying will enhance subsequent Ka measurements. Saturating the samples in SFW will result in clay rehydration and clay flocculation such that the clay distribution may no longer be representative with the result that subsequent krg measurements may be suppressed. The effects on subsequent krg and krw measurements in a physically disturbed clay/pore system are unknown.
6. End-point relative permeability to water is required for aquifer influx. The pressure drop over the encroached zone is a significant choke on aquifer influx in large reservoirs.

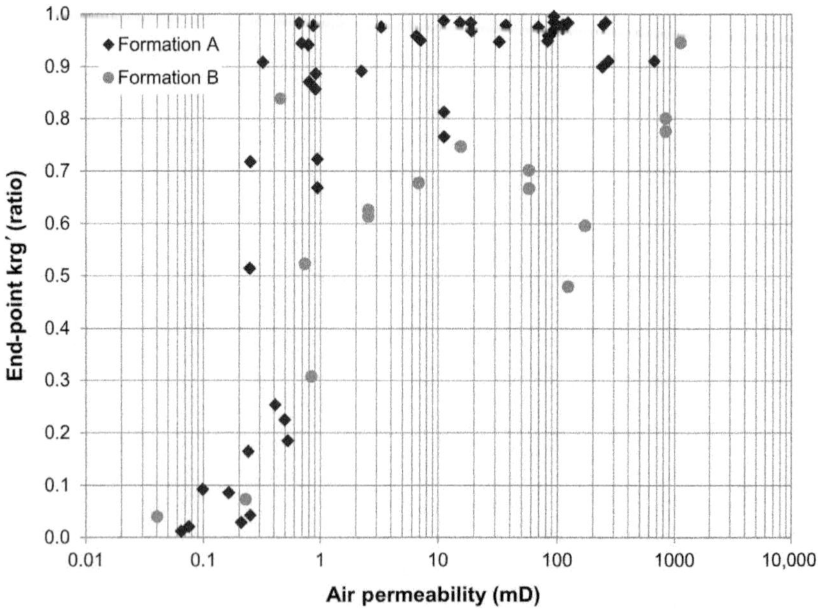

FIGURE 10.52 End-point krg′ as a function of air permeability.

7. The preferred technique is to spin samples (at Swir) to residual HC saturation (Sgr equivalent) in a decane–water imbibition centrifuge experiment (Section 10.5.12). The spin speed is selected such that the capillary pressure is similar to the pressure drop across encroached zone. End-point krg′ (at Swr) and krw′ (at Sgr equivalent) can be determined at both end points of this imbibition experiment.
8. Lab data should be sense-checked against analogue data, correlations and expectations. The movable saturation range (1–Sgr–Swc) becomes very small in low-permeability formations so both krg′ and krw′ reduce with reducing permeability. End-point krw′ is much less important than krg′ at low permeabilities as this rock will not be waterflooded.
9. In reasonable quality rock, residual gas saturation, Sgr, is typically in the range from 0.20 to 0.35, but can vary significantly, depending on the method used to acquire it. Imbibition decane–water centrifuge tests (see Section 10.5.3) can be useful in evaluating Sgr as a function of pressure differential across an encroachment zone, and CCI tests (Section 10.5.14) often provide a very cheap and useful estimate of Sgi as a function of Sgr and Sgr correlations (if any) with rock quality.
10. For dry gas reservoirs undergoing aquifer influx, the water–gas movement represents a stable displacement and largely independent of shape of the relative permeability curves. Consequently, only end-point data are normally required. Drainage gas–water relative permeability curves (Sections 10.5.8 and 10.5.9) might be required for highly unstable gas–water displacements, such as underground gas storage developments or CO_2 sequestration.
11. For gas-condensate reservoirs, pressure depletion can lead to the deposition of liquid HCs in the porous media in the near-wellbore region reducing the effective permeability to gas and the recoverable reserves (condensate banking). Both krw–krg and krg–kro curves are often combined to predict condensate relative permeability behaviour. However, the process of condensation is very different from the dynamic displacement of synthetic analogue fluids at lab conditions since the condensate is created in a distributed fashion within the pore network as the gas pressure depletes.
12. Gas-condensate relative permeability is sensitive to rock type, IFT (close to the critical point), capillary forces, gravity forces, wettability, depletion rates, flow velocity, connate water and the method of condensate generation or condensate injection. Tests should therefore be carried out at reservoir conditions, using reservoir fluids, and require sophisticated experimental equipment and controls. Velocity-dependent relative permeability curves capture the effects of coupling (Kr increase by velocity increase and/or IFT decrease) and inertia (Kr reduction by velocity increase).
13. For carbonate reservoirs, any interbedded vugs in a dense carbonate mix can lead to anomalous relative permeability curve shapes, which are usually associated with early water BT.

14. Carbonate rock often exhibits a wide pore size distribution. This is often caused by a mixture of reasonably homogeneous pore matrix and vugs. The pore matrix and vugs are interconnected in a complex manner, which leads to an equally complex linear displacement behaviour during fluid flow. Even small-scale heterogeneities can have a marked effect on fluid flow through the sample. The use of ISSM is highly recommended in any sample where there is any sign of heterogeneity. ISSM can reveal the water front deformation ahead of the vugs and therefore can help interpret the anomalous curve shapes produced (JBN), which are usually associated with water BT ahead of the main flood front.

REFERENCES

Agarwal, R.G., 1967. Unsteady-state performance of water-drive gas reservoirs. Ph.D. Thesis, Texas A&M University.

Amaefule, J.O., Handy, L.L., 1982. Relative permeability of consolidated porous media. JPT 22, 371–381.

Amott, E., 1959. Observations relating to the wettability of porous rocks. Trans. AIME 216, 156–162.

Anderson, W.G., 1986. Wettability literature survey—part 1: rock/oil/brine interactions and the effects of core handling on wettability. JPT 38, 1125–1144.

Anderson, W.G., 1987. Wettability literature survey—part 6: the effects of wettability on waterflooding. JPT 39, 1605–1622.

Baptist, O.C., Sweeney, S.A., 1954. The effect of clays on the permeability of reservoir sands to waters of different saline contents. In: USBM 1954, Coast Regional Conference on Clays and Clay, pp. 505–515.

Buckley, S.E., Leverett, M.C., 1942. Mechanism of fluid displacement in sands. Trans. AIME 146, 107–116.

Byrnes, A.P., 2008. Issues with gas relative permeability in low-permeability sandstones. In: Cumella, S.P., Shanley, K.W., Camp, W.K. (Eds.), Understanding, Exploring, and Developing Tight-Gas Sands—2005 Vail Hedberg Conference: AAPG Hedberg Series, no. 3, pp. 63–76.

Corey, A., et al., 1956. Three-phase relative permeability. Trans. AIME 207, 349–351.

Donaldson, E.C., et al., 1969. Wettability determination and its effect on recovery efficiency. SPE J. 9, 13–20.

Fishlock, T.P., Smith, R.A., Soper, B.M., Wood, R.W., (1988, May 1). Experimental Studies on the Waterflood Residual Gas Saturation and Its Production by Blowdown. Society of Petroleum Engineers. http://dx.doi.org/10.2118/15455-PA.

Hagoort, J., 1980. Oil recovery by gravity drainage. SPE J. 20, 139–149.

Heaviside, J., et al., 1983. Fundamentals of relative permeability: experimental and theoretical considerations. In: SPE 12173, October.

Heaviside, J., Brown, C.E., Gamble, I.J., 1987. Relative permeability for intermediate wettability reservoirs. In: SPE 16968.

Honarpour, M., et al., 1986. Relative Permeability of Petroleum Reservoirs. CRC Press, Boca Raton, FL.

Hove, A.O., et al., 1985. Visualisation of laboratory corefloods with the aid of computerised tomography of X-rays. In: SPE 13654.
Ibrahim, M.N.M., Koederitz, L.F., 2000. Two-phase relative permeability prediction using a linear regression model. In: SPE 65631, Presented at the 2000 SPE Eastern Regional Meeting, Virginia, October.
Jerauld, G.R., 1997. Prudhoe bay gas/oil relative permeability. SPE Res. Eng. 12 (1), 66–73. SPE-35718-PA.
Johnson, E.F., Bossler, D.P., Naumann, V.O., 1959. Calculation of relative permeability from displacement experiments. Trans. AIME V216, 370.
Jones, S.C., Roszelle, W.O., 1978. Graphical techniques for determining relative permeability from displacement experiments. JPT 30, 2905–2908.
Land, C.S., 1968. Calculation of imbibition relative permeability for two- and three-phase flow from rock properties. SPE J. 8, 149–156.
Leverett, M.C., Lewis, W.B., 1941. Steady flow of gas-oil mixture through unconsolidated sands. Trans. AIME 142, 107.
Lomeland, F., Ebeltoft, E., Thomas, W.H., 2005. A new versatile relative permeability correlation. Reviewed paper at the 2005 International Symposium of the SCA, Toronto, 21–25 Aug (SCA2005-32).
McPhee, C.A., Arthur, K.G., 1994. Relative permeability measurements: an inter-laboratory comparison. In: SPE 28826.
Mohammed, K., Corbett, P., 2003. How many relative permeability measurements do you need? A case study from a North African reservoir. Petrophysics 44 (4), 262–270.
Mohanty, K.K., Miller, A.E., 1991. Factors influencing relative permeability of a mixed wet reservoir rock. SPE Form. Eval. 6, 349–358.
Mulyadi, H., et al., 2000. Measurement of residual gas saturation in water-driven gas reservoirs: comparison of various core analysis techniques. In: SPE 64710, November.
Ramstad, T., Øren, P.E., Bakke, S., 2010. Simulation of two-phase flow in reservoir rocks using a Lattice-Boltzmann method. In: SPE 124617.
Rapaport, L.A., Leas, W.J., 1953. Properties of linear waterfloods. Trans. AIME 198, 139–148.
Sigmund, P.M., McCaffery, F.G., 1991. An improved unsteady-state procedure for determining relative permeability characteristics of heterogeneous porous media. SPE J. 19, 15–28.
Stiles, J., 2007. Using special core analysis in reservoir engineering. Imperial College course notes.
Suzanne, K., Hamon, G., Billiotte, J., Vincent, T., 2003. Residual gas saturation of sample originally at residual water saturation in heterogeneous sandstone reservoirs. International Symposium of the SCA, Pau, France, pp. 21–24, Sept. (SCA2003-14).
Tanikawa, W., Shimamoto, T., 2006. Klinkenberg effect for gas permeability and its comparison to water permeability for porous sedimentary rocks. Hydrol. Earth Syst. Sci. 3, 1315–1338.
Welge, H.J., 1952. A simplified method for computing oil recovery by gas or water method. Trans. AIME 195, 91–98.

RECOMMENDED READING

Amyx, J.W., Whiting, D.M., Bass, R.L., 1960. Petroleum Reservoir Engineering. McGraw-Hill, New York.
Baker, P.E., 1960. Discussion of effect of viscosity ratio on relative permeability. JPT, Technical Note.

Bardon, C., Longeron, D.G., 1980. The influence of very low interfacial tension on relative permeability. SPE J. 20, 391–401.

Batycky, J.P., et al., 1981. Interpreting relative permeability and wettability from unsteady-state displacement measurements. SPE J. 21, 296–308.

Cable, A., Mogford, D., Wannell, M., 2004. Mobilisation of gas from below the gas-water contact. In: Society of Core Analysts, Paper SCA2004-29.

Craig, F.F., 1971. The reservoir engineering aspects of waterflooding. In: SPE-AIME Monograph Series, vol. 3. Society of Petroleum Engineers

Cuiec, L.E., 1977. Study of problems related to the restoration of the natural state of core samples. J. Can. Petrol. Technol. 29, 68–80.

Delclaud, J.P., 1972. New results on the displacement of a fluid by another in a porous medium. In: SPE 4103.

Donnez, P., 2007. Essentials of reservoir engineering. Editions Technip.

Downie, J., Crane, F.E., 1961. Effect of Viscosity on Relative Permeability. Society of Petroleum Engineers. http://dx.doi.org/10.2118/1629-G.

Foster, W.R., 1973. A low-tension waterflooding process. JPT 25, 205–210.

Goode, P.A., 1991. Momentum transfer across fluid-fluid interfaces in porous media. Ph.D. Thesis, Heriot Watt University.

Grist, D.M., et al., 1975. The dependence of water permeability on core cleaning methods. J. Can. Petrol. Technol 14, 48–52.

Hirsch, D.E., et al., 1972. Separation of high boiling point distillates using gradient elution. Anal. Chem. 44, 915–919.

Kantzas, A., Ding, M., Lee, J., 2001. Residual gas saturation revisited. SPE Reserv. Eval. Eng. 4, 467–477.

Kyle, J.R., Rapaport, L.A., 1958. Linear waterflood behaviour and end effects in water-wet porous media. Trans. AIME 213, 423.

Lefebvre du Prey, E.J., 1973. Factors Affecting Liquid-Liquid Relative Permeabilities of a Consolidated Porous Medium. Society of Petroleum Engineers. http://dx.doi.org/10.2118/3039-PA.

McCaffery, F.G., Bennion, D.W., 1974. The effect of wettability on two-phase relative permeabilities. J. Can. Petrol. Technol. 13, 42–53.

McPhee, C., 1994. Effects of strong wetting on end point relative permeability. Dialog (Lond. Petrophys. Soc.) 2, 6–7.

Mikkelson, M., et al., 1991. Abnormal permeability behaviour of a North Sea sandstone. In: Paper SPE 22600.

Miller, M.A., Ramey, H.J., 1985. Effect of temperature on oil/water relative permeabilities of unconsolidated and consolidated sands. SPE J. 945–953. http://dx.doi.org/10.2118/12116-PA.

Moore, T.F., Slobod, R.L., 1956. The effect of viscosity and capillarity on the displacement of oil by water. Prod. Monthly 20, 20–30.

Morgan, J.T., Gordon, D.T., 1970. Influence of pore geometry on water-oil relative permeability. JPT 22, 1199–1208.

Mulyadi, H., Amin, R., Kennaird, A.F., 2001. Practical approach to determine residual gas saturation and gas-water relative permeability. In: SPE 71523.

Mungan, N., 1966. Certain wettability effects in laboratory waterfloods. JPT 18, 247–252.

Odeh, A.S., 1959. Effect of viscosity ratio on relative permeability. Trans. AIME 216, 346–353.

Owens, W.W., Archer, D.L., 1971. The effect of rock wettability on oil-water relative permeability relationships. JPT 23, 873–878.

Owens, W.W., et al., 1956. An evaluation of a gas drive method for determining relative permeability relationships. Trans. AIME 207, 275–280.

Rathmell, J.J., et al., 1973. Reservoir waterflood residual oil saturation from laboratory tests. JPT 25, 175–185.

Ritter, H.L., Drake, L.C., 1945. Pore size distributions of porous materials. Ind. Eng. Chem. 17, 782–791.

Rose, W., 1972. Some problems connected with the use of classical description of fluid/fluid displacement processes. In: Fundamentals of Transport Phenomena in Porous Media. Elsevier, New York, pp. 229–240.

Sævareid, A., Whitson, C.H., Fevang, Ø, 1999. An engineering approach to measuring and modeling gas condensate relative permeabilities. SCA 1999-30.

Sandberg, C.R., et al., 1958. The effect of fluid flow rate and viscosity on laboratory determination of oil-water relative permeability. Trans. AIME 213, 36–43.

Sarem, A.M., 1966. Three-phase relative permeabilities by unsteady-state methods. SPE J. 6, 199–205.

Yuster, S.T., 1951. Theoretical considerations of multi-phase flow in idealised capillary systems. In: Proceedings of the 3rd World Petroleum Conference, The Hague.

Chapter 11

Nuclear Magnetic Resonance (NMR)

Chapter Outline

11.1 Introduction	**655**	11.7.3 Test Equipment	664
11.2 Nuclear Spin Relaxation in Rocks	**657**	11.7.4 Recommended NMR Measurement Parameters	665
11.3 NMR Relaxation and Pore Size	**659**	11.7.5 Test Procedures	666
11.4 Porosity from NMR	**661**	11.7.6 Data Reporting Requirements	667
11.5 Clay-Bound Water	**662**	**11.8 NMR Core Measurement Summary**	**668**
11.6 Permeability Estimation	**662**	**References**	**668**
11.7 NMR Tests on Core	**663**	**Recommended Reading**	**669**
11.7.1 Sample Preparation	663		
11.7.2 Key Processes	664		

11.1 INTRODUCTION

Nuclear magnetic resonance (NMR) describes the response of nuclei to an applied magnetic field. Protons within the nucleus exhibit behaviour like spinning magnets, which interact with the magnetic field to produce a measurable response. The NMR tools used in the oil and gas industry, both downhole and desk-top laboratory equipment, are tuned to measure the response of the protons within the hydrogen nucleus, as hydrogen is the dominant atom found in reservoir fluids.

In typical NMR measurements, a substantial magnetic field is applied. The protons within the hydrogen atoms align parallel to the magnetic field orientation (Fig. 11.1). Once the protons align, an oscillating magnetic field (B_1) is applied, which oscillates at a frequency of ~2.3 MHz (optimum frequency for hydrogen atoms). The application of this secondary field causes the protons to change orientation to 90°. The amplitude and decay are measured during oscillations. The signal amplitude is proportional to the number of hydrogen atoms present which, when calibrated, gives porosity. The decay of the NMR signal during measurements is referred to as the relaxation time.

Developments in Petroleum Science, Vol. 64. http://dx.doi.org/10.1016/B978-0-444-63533-4.00011-1

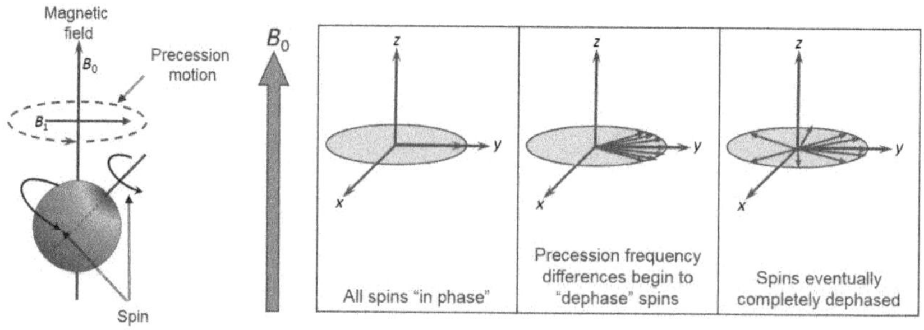

FIGURE 11.1 Precession of the hydrogen nuclei after application of the B_1 magnetic field (left). T_2 relaxation process (right).

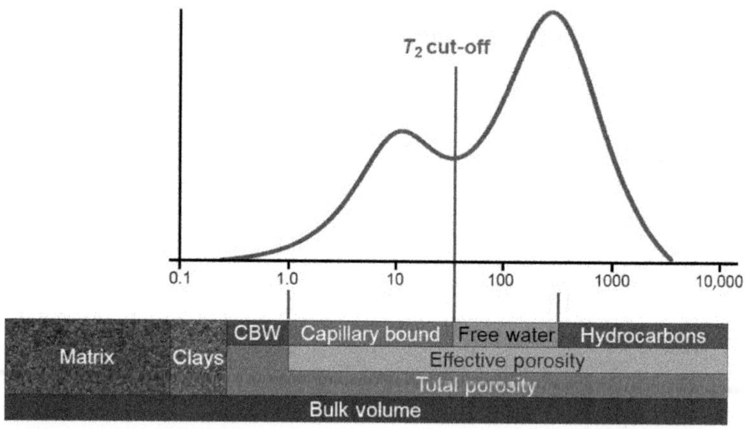

FIGURE 11.2 A typical NMR T_2 distribution trace and how this corresponds to the volume of fluids in place.

This relaxation time is dependent on the pore sizes and so generates data on the distribution of pore sizes. This distribution can be used to extrapolate the amount of clay- and capillary-bound water as well as free water and hydrocarbons (Fig. 11.2), and permeability.

NMR measurements on core are used to improve the characterisation of the fluid content in core material and provide a unique look at the interaction of pore fluids within reservoir rock fabric. It is possible to determine porosity, permeability, bound- and free-fluid volumes, which are used to establish a reasonable estimate of residual oil volumes, from a single measurement of signal amplitude and transverse relaxation time (T_2). Smaller pores have a short relaxation time (corresponding to the smaller fluid volumes); conversely, the larger pores have a longer relaxation time (corresponding to the larger fluid volumes). Using these basic concepts, the trace can then be used to start to describe the fluid, as well as the pore size, distribution within the sample. These are important characteristics as they are used to estimate permeability from NMR logs.

11.2 NUCLEAR SPIN RELAXATION IN ROCKS

NMR relaxation studies can tell us about long distance atomic and molecular motions in condensed matters. NMR relaxation times are also sensitive to localised movement such as motion of atoms bound to defects or surfaces. To understand why large internal magnetic field gradients can cause complications, it is necessary to consider the three T_2 relaxation mechanisms in rocks. These are due to (1) molecular motion in fluids, (2) surface relaxivity at the pore wall and (3) molecular diffusion in the internal gradients. The first mechanism, due to local motions such as molecular tumbling, is equally effective whether or not the fluid is in a rock. This is called bulk liquid relaxation.

For the bulk liquid, the spin–spin relaxation time, T_2, decreases as fluid viscosity, μ, increases:

$$T_{2B} = \frac{T}{\mu}$$

where T is absolute temperature.

The second mechanism is the relaxation of hydrogen nuclei when they closely approach paramagnetic ions such as iron or manganese, which reside on grain surfaces. Fluid molecules that diffuse to the pore surface have a probability of relaxing there and diffusing away. For water in an individual pore, the relaxation time constant is:

$$T_{2S} = \frac{1}{\rho_2\left(\frac{S}{V}\right)}$$

where (S/V) is the surface-to-volume ratio of the pore system, ρ_2 is the transverse surface relaxivity.

Figure 11.3 illustrates how excited protons move about pore space, colliding with other protons and grain surfaces. Every time a proton collides, there is a possibility of relaxation. The probability of colliding with pore walls is high.

The third mechanism is due to the diffusion of molecules in magnetic field gradients. Using the Carr–Purcell–Meiboom–Gill (CPMG) pulse sequence, the relaxation due to diffusion (T_{2D}) is:

$$T_{2D} = \frac{3}{\gamma^2 G^2 D \tau^2}$$

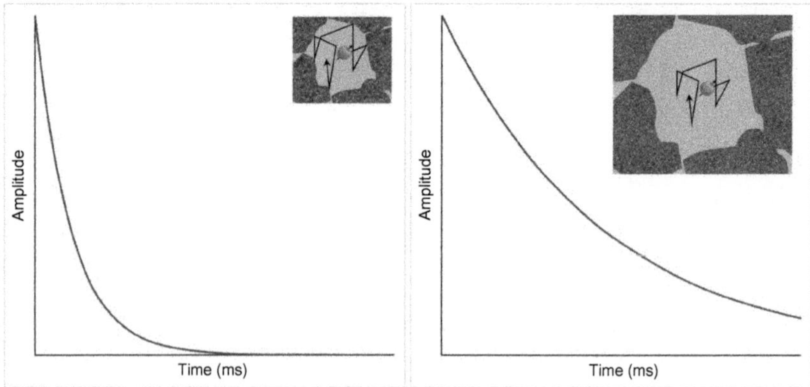

FIGURE 11.3 Illustrations of hydrogen spin relaxing at the grain surface.

where the constant γ is the hydrogen gyromagnetic ratio, D is the diffusion coefficient and τ is the CPMG pulse spacing. The time dependence of the signal, $A(t)$, from all three relaxation processes and measured at the time t is:

$$A(t) = A(0)e^{-(-t/T_{2B})}e^{-(-t/T_{2S})}e^{-\left(1/3\gamma^2 G^2 D\tau^2 t\right)}$$

where $A(0)$, the signal at time $t=0$, is proportional to net magnetisation and therefore fluid-filled porosity.

For quantitative analysis $A(0)$, T_{2B} and T_{2S} are of interest, but the gradient diffusional relaxation is considered a nuisance. In rocks with potentially high internal gradients, e.g., very shaly sands (or measurements made at high fields), the extrapolation to find $A(0)$ can be inaccurate and the relaxation curves for the bulk fluid or surface relaxation can be completely obscured. Fortunately, the relaxation due to the gradients can be minimised by shortening τ. However, because of spectrometer hardware constraints at high fields, it is not always possible to achieve the required inter-echo spacing and for many rocks it is better to work at low magnetic field strengths (where G is also smaller).

NMR transverse (spin–spin) relaxation data are expressed as a sum of exponentials in the following expression:

$$A(t) = \sum_i A(0)_i e^{-(t/T_{2i})}$$

The previous equation can be inverted into a T_2 relaxation time distribution, where all T_{2i} are a pre-selected set of relaxation time constants and all $A(0)_i$ are signal amplitudes associated with each time constant by fitting. Single-component fluids are characterised by a single value of T_2. For example, bulk water has a T_2 of approximately 3 s at room temperature.

Multi-component fluids such as crude oils often exhibit a complex distribution of relaxation times.

11.3 NMR RELAXATION AND PORE SIZE

Brownstein and Tarr (1979) first discussed the relationship between NMR relaxation and pore size in saturated porous media. Nuclei of the saturating medium in a pre-prepared, excited, high-energy state are allowed to freely diffuse across an isolated, saturated pore. As the nuclei diffuse, they interact with other nuclei forcing the saturating medium to exchange or liberate energy through mechanisms of relaxation. NMR spin relaxation (T_1 and T_2) in porous media is enhanced by very strong interactions at the pore–grain surface. This enhancement occurs because the excited hydrogen spins diffuse and eventually reaches the grain surface where they have a finite probability of being relaxed. This relaxation may be limited by relaxation (interaction) at the surface (grain properties) or by transport of the unrelaxed spins to the surface (fluid properties). As shown in Fig. 11.3, hydrogen nuclei (protons) move about pore space colliding with other protons and grain surfaces. When protons collide, there is a possibility of relaxation. It is obvious that the probability of colliding with a grain surface is high in small pores but low in larger pores. The relaxation is rapid when the probability of colliding with pore walls is high. Thus, the NMR spin relaxation times are shorter in small pores. Where the relaxation is limited only by the interaction between protons and the surface walls (the so-called fast diffusion limit), the case is that of one fluid-filled rock core. In this case, NMR relaxation (T_1 and T_2) depends upon the surface-to-volume ratio (*S/V*) and consequently pore size such that:

$$\frac{1}{T_{1,2}} = \rho_{1,2}\left(\frac{S}{V}\right)$$

where $T_{1,2}$ represents the T_1 and T_2 relaxation time, $\rho_{1,2}$ is the corresponding pore surface relaxation strength (proton relaxivity) and S/V is the pore surface-to-volume ratio. Note that ρ may be different for T_1 and T_2 relaxation processes. Since an isolated pore has a single T_1 and T_2 value in this limit, a sample with a pore size distribution would have a corresponding distribution of T_1 and T_2 values. The conversion factor between relaxation and pore size is ρ, the proton relaxivity. For many rocks, the T_1/T_2 ratio is roughly constant. Although T_1 measurements are more common in the literature, they are more time consuming to measure than T_2 measurements. T_2 carries the same pore size distribution information as T_1 and is easier and quicker to measure. In water-saturated rocks, the bulk water relaxation is often negligible, and the distribution of T_2 arises from the distribution of surface-to-volume ratio in the pores. Because T_2 depends linearly upon pore size, the T_2 distribution corresponds to a pore size distribution with the large pores having the longest relaxation times. When oil and water coexist in the pore space, the general principle is that the relaxation of fluid in contact with the grain surface is enhanced by those surfaces. Conversely, fluid that does not wet the surface relaxes at its bulk rate, as relaxation does not occur at the oil–water interface. Because oil and water have similar proton density, the T_2 distribution is a linear summation of the individual contributions from the oil and water.

Different techniques have been used for generating the pore size distributions necessary for comparison with the relaxation distributions in order to determine ρ. Such techniques include scanning electron microscopy, mercury injection and surface adsorption. These techniques interrogate different pore scales and therefore produce different estimates of ρ for a fixed NMR relaxation distribution.

For other diffusion limits (intermediate through to slow), nuclei relaxing at the pore wall will be distinguished from those relaxing in the bulk fluid, and there will be an increasing amount of higher order term relaxation processes being observed. This means that the relaxation processes occurring in a single pore will no longer be described by a single relaxation time for T_1 and T_2, but by a relaxation distribution related to the square of the pore size (Fig. 11.4). Such effect overlaps with the normal relaxation distribution of nature porous

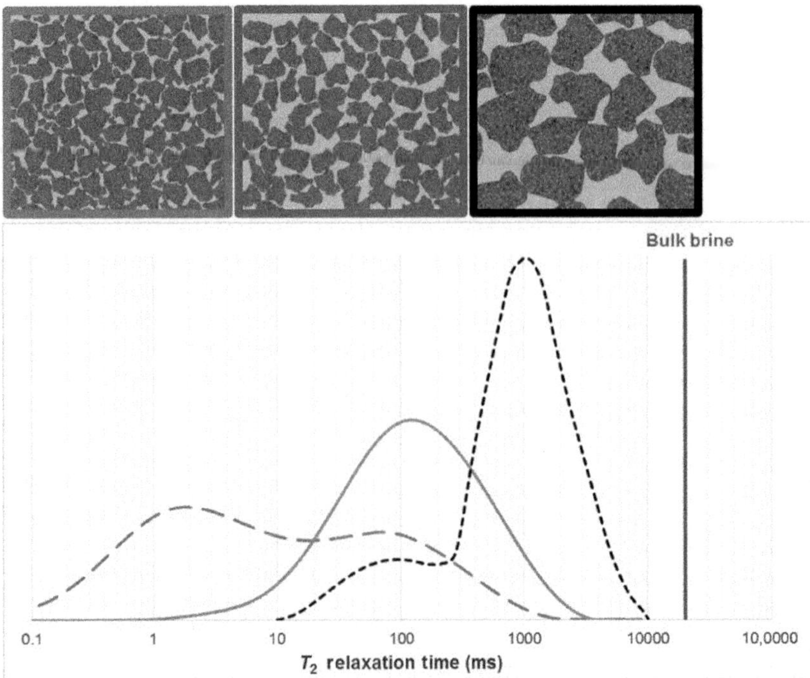

FIGURE 11.4 NMR T_2 relaxation and pore size.

media, which means that the conversion of relaxation to useful physical or chemical parameters becomes complicated.

11.4 POROSITY FROM NMR

All the pore space within a volume of rock is the total porosity (ϕ_t), which includes water in clays and shales. Effective porosity (ϕ_e) is defined as the total porosity excluding the clay-bound water volume (S_{wb}). Effective porosity is further subdivided into pore space containing capillary-bound water and pore space containing free fluid (FFI). Capillary-bound water is also referred to as the bulk volume irreducible fluid (BVI).

BVI, FFI and porosity (MPHI) are defined as:

$$\mathrm{BVI} = (S\mathrm{wir} - S\mathrm{wb})\phi\mathrm{t} \le \mathrm{BVW}$$

$$\mathrm{FFI} = (1 - S\mathrm{wir})\phi\mathrm{t}$$

$$\mathrm{MPHI} = (1 - S_{\mathrm{wb}})\phi\mathrm{t}$$

where S_{wir} is the irreducible fluid saturation (water), S_{wb} is clay-bound water saturation and BVW is the bulk volume of water $\left(\phi_t{}^* S_w\right)$.

In theory, NMR is able to see all the fluid porosity, including clay-bound water, and it is probable that NMR bench-top experiments capture the total porosity (ϕ_t). However, NMR borehole measurements contain instrument "dead time" at the beginning of the T_2 measurement and consequently cannot detect the part of the signal originating from clay-bound water. The instrument dead time is the time lag between switching the RF pulse and receiving a signal in the RF coil. Low-field bench-top NMR rock core analysers are able to capture the clay-bound water in the NMR signal. The shortest times in the NMR relaxation distribution (smallest pores) contain the BVI and the longest times (largest pores) contain the FFI which can contain hydrocarbon, water or both.

The division between BVI and FFI is referred to as the T_2 cut-off that is expressed in time units (ms) (see Fig. 11.5). At lower values than the T_2 cut-off, fluids are considered to be immobile, either held by capillary forces in small pores or hydrostatically bound in the clay. It has been found that the average values of T_2 cut-off are 33 and 92 ms for sandstones and carbonates, respectively. However, the T_2 cut-off is known to vary and in sandstones can range from 1.5 to 75 ms. Clearly, variable T_2 cut-off values have significant implications in the calculation of FFI and bound water volumes.

An alternative to using a default value is to derive T_2 cut-offs from desaturation centrifuge experiments. The value of the T_2 cut-off is determined by the best agreement between the FFI porosity and the volume of fluid that is centrifuged from the sample at 100 psi air–brine equivalent capillary pressure.

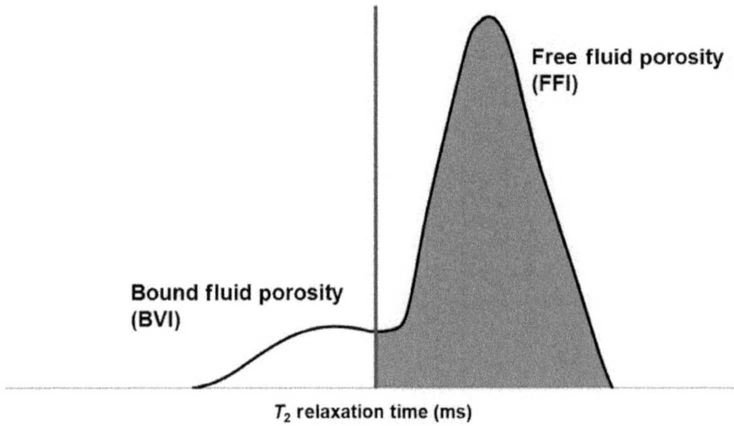

FIGURE 11.5 Vertical cut-off on a T_2 distribution separating the distribution into the long T_2 porosity (shaded) and the short T_2 porosity.

11.5 CLAY-BOUND WATER

Estimation of clay-bound water volume from the T_2 distribution is based on an assumption that clay-bound water has the shortest T_2 values in the distribution because clays have high surface area and small pores. Hence, the pore volume with T_2 values shorter than a certain cut-off can be correlated with clay-bound water estimates from electrical or other measurements.

11.6 PERMEABILITY ESTIMATION

There are two primary approaches for predicting permeability from NMR data, one is based on the estimation of an average pore size (SDR) and the other on irreducible water saturation (Machado et al., 2008).

The SDR approach (for example) relies on the T_2 distribution as being a pseudo-pore size distribution of the formation. The approach is analogous to empirical studies of capillary pressure data by Leverett (1941), Purcell (1949), Thomeer (1960) and Swanson (1981) who showed permeability in sandstones, and indeed many carbonates can be predicted from the shape of the capillary pressure curve which is controlled by pore size. The SDR model is:

$$k = aT_{2\text{LM}}^{b}\phi^{c}$$

where $T_{2\text{LM}}$ is the logarithmic mean T_2 distribution, K is permeability permeability (mD) and ϕ is porosity (%).

The second approach is based on the observation that permeability is related to irreducible water saturation. The Timur–Coates equation (Coates et al., 1991) uses the free-fluid to bound-fluid ratio, determined from the T_2 distribution, and the porosity to predict permeability:

$$k = \left(\frac{\text{FFI}}{\text{BVI}}\right)^2 \left(\frac{\phi}{c}\right)^b$$

The parameters a, b and c in both models are adjustable and can be determined from NMR measurements on samples of known permeability and porosity. However, selection of FFI/BVI ratios and T_2 cut-offs can cause inaccuracies in the model predictions. Where optimised NMR permeability equations differ significantly from the standard equations, then NMR, MICP and petrographic analyses should be conducted on core samples to establish depositional and diagenetic controls on pore geometry.

NMR permeability equations incorporate the entire pore size distribution, so the resultant permeability relates to the average permeability for the sample. It follows, therefore, that where permeability is accurately predicted from NMR logs, the pore system must be homogeneous over the volume range sampled by the log and the permeability measurement to which it is compared.

11.7 NMR TESTS ON CORE

Purpose	• Determine NMR response of core samples for calibration of NMR logs.
Sample selection	• Cylindrical-shaped plug sample of 1″ or 1½″ diameter. The larger the sample, the larger the potential pore volume, and the less will be the impact of saturation measurement errors. • *Ideally*—sample taken from preserved core or preserved plug. *Alternative*—plug previously used in routine core analysis programme provided that plug has not been pre-stressed and/or that plug preparation methods have not caused damage to delicate clay minerals.
Sample orientation	• Horizontal or vertical plugs (depending upon application)
Required input data for lab	• Formation water (FW) salinity and ionic composition should be as close as practically possible to the FW in the reservoir or FW resistivity at reservoir and lab temperature. Lab will prepare synthetic FW (SFW) accordingly. • Test confining stress (most tests are done at ambient—unconfined conditions—though a few major labs do offer stressed condition measurements).

11.7.1 Sample Preparation

Core analysis samples selected for measurement should be representative of the lithologies in the reservoir. The range of core porosities and permeabilities used in SCAL tests should reflect the porosity and permeability distribution of the RCA database. Ten samples per facies/reservoir rock type is the minimum required for statistical validity.

Selected plugs are examined using a CT scanner. This can identify any anomalous material or significant heterogeneity in the plugs (see Chapter 6). The plugs should also be photographed pre- and post-testing.

Tests are performed on samples that have been cleaned and dried (refer to Chapter 4) and base porosity and permeability measured. Utmost care should be taken to avoid damage to the samples and preserve the clay structures during core cleaning and drying. Low-temperature flush cleaning or immersion cleaning and humidity oven drying is recommended, although for very sensitive clays the plugs should not be dried but saturated in SFW immediately after flush cleaning with methanol.

SFW is prepared and characterised (density, viscosity and resistivity) as described in Chapter 6.

After cleaning and drying, base porosity and air/Klinkenberg permeability are measured at base conditions, as described in Chapter 5. The sample pore volume is calculated from helium grain volume and SFW-immersion bulk volume and validated by gravimetric measurements before and after saturation. The re-saturation pore volume must agree with the helium pore volume within ± 0.2 pu before continuing with the tests. If undersaturated, then the samples must be replaced in the pressure saturator. If oversaturated, check for grain loss and apply grain loss corrections accordingly.

On completion of the tests, and provided no other tests are scheduled, porosity should be determined using mercury immersion bulk volume and helium grain volume, and the air/Klinkenberg permeability at base confining stress should also be measured (Chapter 5).

11.7.2 Key Processes

1. Saturate sample in SFW and determine pore and bulk volume (porosity).
2. Determine T_2 distribution on 100% brine saturated plugs.
3. Determine brine permeability (optional) at nominal confining stress.
4. Desaturate to irreducible water saturation in porous plate capillary cell.
5. Determine T_2 distribution at S_{wir}.
6. Calculate T_2 cut-offs and tune permeability parameters.

The processes are broadly similar to those for resistivity index (RI) tests (Chapter 8) and NMR measurements can often be conveniently scheduled into the RI test workflow, with minimal additional time.

11.7.3 Test Equipment

The NMR analyses are to be carried out using a desk-top instrument with the following specifications:

- 2 MHz instrument or similar;
- uniform gradient spectrometer.

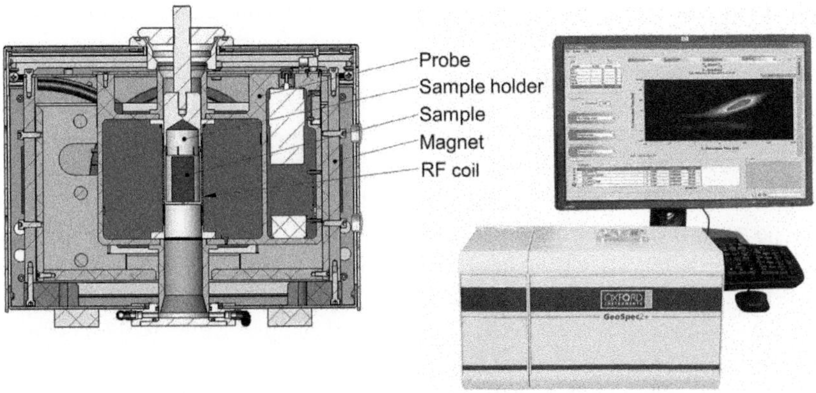

FIGURE 11.6 Typical NMR desk-top instrument. *Courtesy of Oxford Instruments and Green Imaging Technologies Inc.*

An example of ambient condition equipment is shown in Fig. 11.6.

The maximum length of the samples for MR analysis is dictated by the height of the "sweet spot" (the sensitive volume in which the magnetic field is homogeneous) of the particular measurement equipment—these vary depending upon the equipment manufacturer different laboratories have chosen to purchase their NMR equipment from. Also, it is recommended to run a series of quick NMR experiments with the plug at different heights in the instrument, in order to optimise the signal from the core. It is important to note that if the plug sample is longer than the equipment limit, the BVW will not agree with plug pore volume and some length correction factor will be required, which assumes that the sample contains homogeneous porosity throughout its length.

In ambient condition experiments, cores are initially weighed, wrapped in regular Clingfilm, wrapped in Saran and sealed with PTFE tape to attempt to eliminate any evaporation during the analysis cycle. Evaporation is never fully eliminated since the films and tape are gas permeable and allow some mass transfer—hence measurement times should be limited.

To preserve sample integrity, unconsolidated samples must be encapsulated with specific (and non-metallic) packaging materials such as Teflon, that lack hydrogen, or have no MR detectable hydrogen and that will not perturb the linearity of the spectrometer's magnetic field. It should be noted that some plastic materials can give a significant MR signal. NMR signals are temperature dependent; therefore, it is advised to maintain a constant temperature throughout the test. Running NMR tests at temperature increases the risk of brine evaporation and salt precipitation, causing MR porosity to be too low. Procedures should be optimised to minimise the risk of brine evaporation and precipitation.

11.7.4 Recommended NMR Measurement Parameters

1. *Echospacing range* (TE). It is advisable to use significantly short *TE* to capture all the visible NMR porosity (microporosity): 0.2–4.8 ms.

2. *Polarisation time* (TW)—*delay time*. The signal amplitude is proportional to the number of hydrogen nuclei associated with the pore fluids in the measurement volume. The number of hydrogen nuclei polarised by the NMR experiment depends on the delay or wait time (TW). Typically the TW is $5 \times T_1$. Furthermore, a rule of thumb is, for Sw = 1.0, TW = 10 s; however, for Sw <1.0, then TW <10 s, and this parameter is usually left at 10 s, unless specified. If TW is long enough, all the hydrogen nuclei will be polarised and total amplitude–porosity. If TW is too short, there will be incomplete polarisation of the hydrogen nuclei, and the porosity will be underestimated. Changing TW will affect the amplitude of the NMR signal and, for a series of samples which require different TW, must be set to the longest required value for the sample set. Recommended optimum TW is 10 s to cover the full T_2 spectrum.
3. *Number of echoes* (*NECH*). The number of echoes has to be such that the echo train completely decays to zero. Initially for the fully saturated samples (good geometrical pore shapes) with average pore size, 16,000 echoes could be a good starting point. The larger the pore size, the slower the decay rate and more echoes—up to 32,000—may be needed to complete the full relaxation decay cycle.
4. *Number of scans for targeted signal-to-noise ratio* (*NS*). In general, the less the liquid volume and/or closer the components, the higher the SNR required to resolve them, and the higher the number of scans required. Targeting SNR is normally 200:1 or more, for the fully saturated samples; and at least 100:1 or more for partly saturated sample (subject to excessive running times, that can compromise the accuracy of provided data). SNR doubles for every quadrupling of the number of scans (NS), i.e. 16 scans have twice the SNR of 4 scans, but of course the time to perform the experiment increases by a factor of 4; therefore, NS is always set to a multiple of 4 (e.g. 4, 8, 16, 60, …, 120, etc.). Setting NS to 4 cancels receiver artefacts in the NMR data.
5. *Time inversion* (*TI*). Inversion of the CPMG pulse sequence to T_2 requires the use of a variable number of bins and a variable bin range. The bin range should cover 0.1 ms to 10 s and should be distributed logarithmically. Typically 30–50 bins are used to define the frequency of distribution. The use of more bins (120–200 logarithmically distributed) is good for particularly high-permeability samples (>1000 mD), although its use is rare and often not necessary.

11.7.5 Test Procedures

Measure base porosity and permeability (Chapter 5).

Saturate in SFW (Chapter 6).

Measure T_2 distribution on 100% brine saturated plugs using the parameters and test conditions specified in Section 11.7.4: TE = 0.2 ms; TW = 10 s; NECH = up to 32,000; TI—invert CPMG using 30–50 bins.

Drain to S_{wir} in porous plate cell under nominal confining stress (~1000 psi) using matched hydrogen index mineral oil at a single (maximum) capillary pressure sufficient to drain sample to S_{wir}. Capillary pressure data (see Chapter 8) can be used to define the maximum Pc. Allow at least 2–4 weeks (depending on permeability) to achieve saturation equilibrium but monitor brine production versus time to ensure full equilibrium.

Measure T_2 distribution on plug at S_{wir} using the parameters and test conditions specified in Section 7.4 in Chapter 7: TE = 0.2 ms; TW = 10 s; NECH = up to 32,000; TI—invert CPMG using 30–50 bins. Multiple scanning is required for maximum signal-to-noise ratio. NS must always be set to a multiple of 4.

On completion of the test, samples are unloaded and, if not scheduled for further testing, are placed in a Dean-Stark apparatus and extracted in toluene to provide a final water saturation check. The plug is then extracted in methanol to remove any salts. The silver nitrate test is used to determine the end point of cleaning (Chapter 5).

The plugs are removed and dried in a convection or vacuum oven (conventional oven) at 105 °C to constant weight (±0.01 g or better).

Following drying, air/Klinkenberg permeability and helium grain volume are measured. Final sample bulk volume is determined from mercury immersion (refer to Chapter 5). Comparison of pre- and post-test base poroperm properties can be useful in establishing any plug damage caused by clay interactions and/or harsh core cleaning and drying.

11.7.6 Data Reporting Requirements

- brief description of procedures and experimental apparatus;
- sample length and diameter (cm);
- SFW density at test temperature (g/cc);
- base helium porosity and permeability at base confining stress;
- measured initial sample dry weight (g);
- measured saturated sample weight (g);
- measured immersed sample weight in brine (g);
- calculated sample resaturated pore volume (cc);
- calculated sample immersed bulk volume (cc);
- calculated sample saturated porosity (fraction) (pre-test);
- T_2 distributions and estimated BVI/FFI cut-off at 100% brine saturation;
- net confining stress for capillary desaturation;
- maximum capillary pressure and brine production data as a function of time at maximum Pc;
- capillary pressure net confinement stress (psi);
- T_2 distributions at S_{wir};
- final sample saturated weight (g);
- final Dean-Stark water volume (corrected for salinity);
- final sample dry weight (g), helium porosity (fraction) and permeability (mD);
- sample grain loss correction procedures, including typical worked example for one sample from sample database.

11.8 NMR CORE MEASUREMENT SUMMARY

Advantages	Drawbacks and issues
• Generates total porosity, clay-bound water (CBW), irreducible water (capillary-bound water, BVI) and mobile fluids (hydrocarbon plus water, BVM). • Can be used to generate pseudo-capillary pressure curves and permeability data. • Helps to verify permeability estimation and determine T_2 cut-off times for NMR log calibration.	• The T_2 distribution is often referred to as pore size distribution but actually reflects a distribution of S/V ratios. MICP tests (for example) provide pore throat size distribution data and the two should not be confused. • NMR response is dependent on temperature and pressure. Core NMR measurements are rarely performed at reservoir conditions. As a consequence interpretation parameters derived from core analysis, such as the T_2 cut-off for bound fluid, often require shifting for application to logging data. • T_2 distribution is influenced by hydrocarbons and so T_2 must be measured at 100% brine saturation and at S_{wir} (at least).

QC Checks

- Analysis of the results should start with a thorough review of the laboratory report to check that acceptable procedures have been followed. These should be carefully evaluated.
- Check that samples are fully saturated in brine before starting any test.
- Carry out mercury injection pore size distribution test on sample post-test or sample offcut. Although data are not strictly comparable, mercury injection data can be used to assess the pore size distributions from T_2 analysis.

Suitable Laboratories

- Only a limited number of commercial labs have NMR equipment.

REFERENCES

Brownstein, K.R., Tarr, C.E., 1979. Importance of classical diffusion in NMR studies of water in biological cells. Phys. Rev. A 19 (6), 2446.

Coates, G.R., Miler, M., Gillen, M., Henderson, G., 1991. The MRIL in Conoco 33-1: an investigation of a new magnetic resonance imaging Log. In: Annual Logging Symposium of the SPWLA. vol. 32. , pp. 1–24.

Leverett, M.C., 1941. Capillary behaviour in porous solids. Trans. Am. Inst. Min. Metall. Pet. Eng. 142, 152–169.

Machado, V.F., Azeredo, R.B.V., Ramos, P.F.O., Coutinho, M.R., Reveriego Carneiro, S.R., 2008. A NMR permeability model based on partial least square (PLS) regression for a low permeability gas sand in Santos basin. In: Paper SCA2008–43 Presented at SCA Annual Symposium, Abu Dhabi, November.

Purcell, W.R., 1949. Capillary pressures—their measurement using mercury and the calculation of permeability therefrom. Pet. Trans. AIME 186, 39–48.

Swanson, B.F., 1981. A simple correlation between permeabilities and mercury capillary pressures. J. Pet. Technol. 33, 2498–2504.

Thomeer, J.H.M., 1960. Introduction of a pore geometrical factor defined by the capillary pressure curve. J Pet Technol 12 (3), 73–77. SPE-1324-G.

RECOMMENDED READING

Coates, G.R., Dumanoir, J.L., 1973. A New approach to improved log-derived permeability. In: Transactions of the SPWLA 14th Annual Logging Symposium, Lafayette, USA, Paper R.

Coates, G.R., Xiao, L., Prammer, G., 1999. NMR Logging, Principles & Application. Halliburton Energy Services, Houston.

Kenyon, W.E., Day, P.I., Straley, C., Willemsen, J.F., 1986. Compact and consistent representation of rock NMR data from permeability estimation, SPE-15643: society of petroleum engineers, 61st annual meeting preprint, 22 p. later published in 1988 as. A three-part study of NMR longitudinal relaxation properties of water-saturated sandstones: SPE Formation Evaluation 3 (3), 622–636.

Kenyon, W., Kleinberg, R., Straley, C., Gubelin, G., Morriss, C., 1995. Nuclear Magnetic Resonance Imaging Technology for the 21st Century. Oilfield Review, Autumn, pp. 19–33.

Kleinberg, R.L., 1996. Utility of NMR T2 distributions, connection with capillary pressure, clay effect, and determination of the surface relaxivity parameter rho2. Magn. Reson. Imaging 14, 761–767.

Lowden, B., 2000. Some Simple Methods for Refining Permeability Estimates from NMR Logs and Generating Capillary Pressure Curves. Dialog, London Petrophysical Society, Piccadilly, London.

Marschall, D., Gardner, J.S., Mardon, D., Coates, G.R., 1995. Method for correlating NMR relaxometrry and mercury injection data. In: Transactions International Symposium of the SCA, San Fransisco, California. September 12–15.

The NMR Carbonate Rock Catalogue, 1999. Applied Reservoir Technology Ltd.

The NMR Sandstone Rock Catalogue, 1997. Applied Reservoir Technology Ltd.

Timur, A., 1968. Effective porosity and permeability of sandstones investigated through nuclear magnetic resonance principles. In: Transactions of the SPWLA 9th Annual Logging Symposium, New Orleans, USA, Paper K.

Volokitin, Y., Looyestijn, W., Slijkerman, W., Hofman, J., 1999. Constructing capillary pressure curves from NMR Log data in the presence of hydrocarbons. In: Trans 40th Annual Logging Symposium, SPWLA, Oslo, Norway, Paper KKK.

Chapter 12

Geomechanics Tests

Chapter Outline

12.1 Introduction **672**

12.2 Sample Selection and Preparation **673**

12.2.1 Test Sites 674

12.2.2 Sample Orientation 678

12.2.3 Plugging Fluids 679

12.2.4 Sample Dimensions 679

12.2.5 Sample Saturation 680

12.3 Unconfined Compressive Strength Tests **681**

12.3.1 Purpose and Sample Requirements 681

12.3.2 Test Equipment 681

12.3.3 Test Procedures 682

12.3.4 Data Utilisation 682

12.3.5 Data Reporting Requirements 683

12.3.6 Advantages and Drawbacks/Issues 683

12.3.7 UCS Quality Control Issues, Checks and Diagnostics 684

12.4 Triaxial Compression Strength Tests **687**

12.4.1 Purpose and Sample Requirements 687

12.4.2 Test Equipment 688

12.4.3 Test Procedures 688

12.4.4 Data Utilisation 695

12.4.5 Data Reporting Requirements 700

12.4.6 Advantages and Drawbacks/Issues 700

12.4.7 Triaxial Test Quality Control Issues, Checks and Diagnostics 700

12.5 Triaxial Testing of Shales **706**

12.5.1 Purpose and Sample Requirements 706

12.5.2 Sample Preparation 706

12.5.3 Test Equipment 709

12.5.4 Test Procedures 709

12.5.5 Data Utilisation 712

12.5.6 Data Reporting Requirements 713

12.5.7 Advantages and Drawbacks/Issues 714

12.5.8 Shale Triaxial Test Quality Control Issues, Checks and Diagnostics 714

12.6 Thick-Wall Cylinder Tests **717**

12.6.1 Purpose and Sample Requirements 717

12.6.2 Test Equipment 719

12.6.3 Test Procedures (Standard TWC) 722

12.6.4 Data Utilisation 722

12.6.5 Data Reporting Requirements 724

12.6.6 Advantages and Drawbacks/Issues 724

12.6.7 TWC Test Quality Control Issues, Checks and Diagnostics 725

Developments in Petroleum Science, Vol. 64. http://dx.doi.org/10.1016/B978-0-444-63533-4.00012-3

12.7 Tensile Strength Tests 728
12.7.1 Purpose and Sample Requirements 728
12.7.2 Test Equipment and Procedures 729
12.7.3 Data Utilisation 730
12.7.4 Data Reporting Requirements 731
12.7.5 Advantages and Drawbacks/Issues 731
12.7.6 Tensile Strength Test Quality Control Issues, Checks and Diagnostics 732
12.8 Acoustic Velocity (Travel Time) Tests 732
12.8.1 Purpose and Sample Requirements 732
12.8.2 Sample Preparation 733
12.8.3 Test Equipment 733
12.8.4 Test Procedures 735
12.8.5 Data Utilisation 735
12.8.6 Data Reporting Requirements 736
12.8.7 Advantages and Drawbacks/Issues 736
12.8.8 ATT Test Quality Control Issues, Checks and Diagnostics 736
12.9 DSCA Tests 739
12.9.1 Purpose and Sample Requirements 739
12.9.2 Test Equipment 739
12.9.3 Test Procedures 740
12.9.4 Data Utilisation 742
12.9.5 Data Reporting Requirements 744
12.9.6 Advantages and Drawbacks/Issues 744
12.9.7 DSCA Test Quality Control Issues, Checks and Diagnostics 744
12.10 Pore Volume Compressibility Tests 746
12.10.1 Purpose and Compressibility Definitions 746
12.10.2 Compressibility Test Loading Conditions 747
12.10.3 Sample Preparation 751
12.10.4 Uniaxial K_0 Test Equipment 751
12.10.5 Uniaxial K_0 Test Procedures 751
12.10.6 Data Utilisation 754
12.10.7 Data Reporting Requirements 757
12.10.8 Advantages and Drawbacks/Issues 757
12.10.9 Uniaxial K_0 Compressibility Test Quality Control Issues, Checks and Diagnostics 757
12.11 Particle Size Analysis Tests 761
12.11.1 Purpose 761
12.11.2 Mechanical Particle Size Analysis 762
12.11.3 Laser Particle Size Analysis 769
References 776
Recommended Reading 778

12.1 INTRODUCTION

Although a wide range of rock mechanics tests can be used to obtain engineering parameters for oilfield design purposes, the aim of this section is to highlight and explain only those tests which are undertaken on a routine basis across the industry. This chapter does not address the wider range of more advanced tests and loading conditions which might be used in research institutions, such as stress- and strain-path studies, true triaxial loading, cyclic loading, hollow cylinder or poly-axial loading.

This chapter contains suggestions and recommendations based on the authors' experience of those tests most commonly used to determine fundamental rock mechanics parameters such as strength, elastic properties, hardness, ultrasonic behaviour, index characteristics and *in situ* stresses. In addition, this chapter includes information on particle size analysis/distribution tests which are commonly run in conjunction with geomechanics tests for sand failure valuation and sand control system design. In addition, modification of the reservoir effective stress due to production causes volumetric changes in pore space in a reservoir. The engineering parameters quantifying these volumetric variations are compressibilities. Reliable compressibility values are essential for resource estimation, reservoir maintenance and drive assessments, as well as subsidence evaluations.

The tests described in this chapter include:

- unconfined compressive strength (UCS), with and without measurement of sample deformation;
- thick-walled cylinder (TWC) collapse tests;
- single-stage (SST) and multi-stage triaxial (MST) tests;
- tensile strength tests (TT);
- ultrasonic (acoustic travel time) velocity (ATT) measurements;
- *in situ* stress determination by differential strain curve analysis (DSCA);
- compressibility (compaction) tests; and
- particle size analysis (PSA) tests by sieve analysis and laser analysis.

and are used for:

- drillability assessment;
- wellbore stability analyses and open-hole stability prediction;
- solids and sand failure prediction;
- completions and sand control completion design and selection;
- log strength and stress correlations;
- estimating *in situ* stress;
- formation characterisation; and
- strength characterisation.

The test descriptions should be viewed as guide only as test equipment and specific test procedures may differ between laboratories.

12.2 SAMPLE SELECTION AND PREPARATION

Full details of the requirements for RCA, SCAL and rock mechanics test plug preparation, including sample inspection, are presented in Chapters 3 and 6. The key requirements for rock mechanics tests samples are summarised below.

12.2.1 Test Sites

The selection of test sites for rock mechanics tests should be based on both visual and/or CT core examination of core and a qualitative strength indicator derived from logs run over the cored intervals. The objective for wellbore stability and sand failure analysis, in particular, is to concentrate on the weaker intervals yet include stronger sections to optimise the dynamic range of rock strength for log–core calibration. This is particularly important as most geomechanical models rely on log-derived strength models: few wells are cored and core is discontinuous across the reservoir and can often be lost in weaker formations.

The application of non-destructive hardness testing can greatly assist in defining both the location and number of test samples. For example, the Equotip portable hardness tester (Daniels et al., 2012), as shown in Fig. 12.1, works on the principle that the height of rebound (Ls) of a small steel ball after its collision with a rock surface depends on the elasticity of the surface, which in turn reflects the mechanical strength of the rock (Fig. 12.2). The data are processed and plotted against depth on the core and logs, as indicated in Fig. 12.3.

The scratch tester (Suárez-Rivera et al., 2002) is an instrument which cuts a groove of fixed depth and width on the surface of a rock sample (normally a core slab) with a diamond cutting tool at constant velocity. An example is shown in Fig. 12.4. The normal and tangential components of the force acting

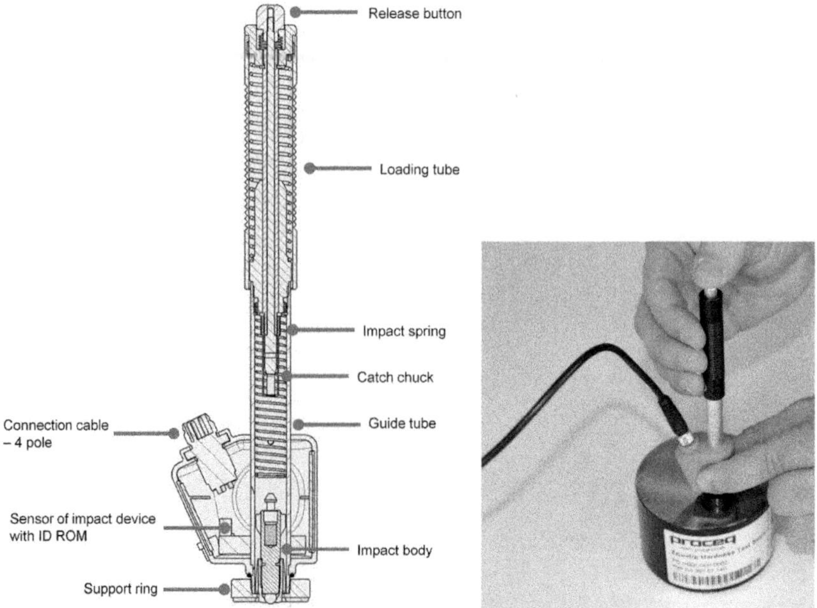

FIGURE 12.1 Equotip strength index tester principle of operation. *Copyright Proceq SA.*

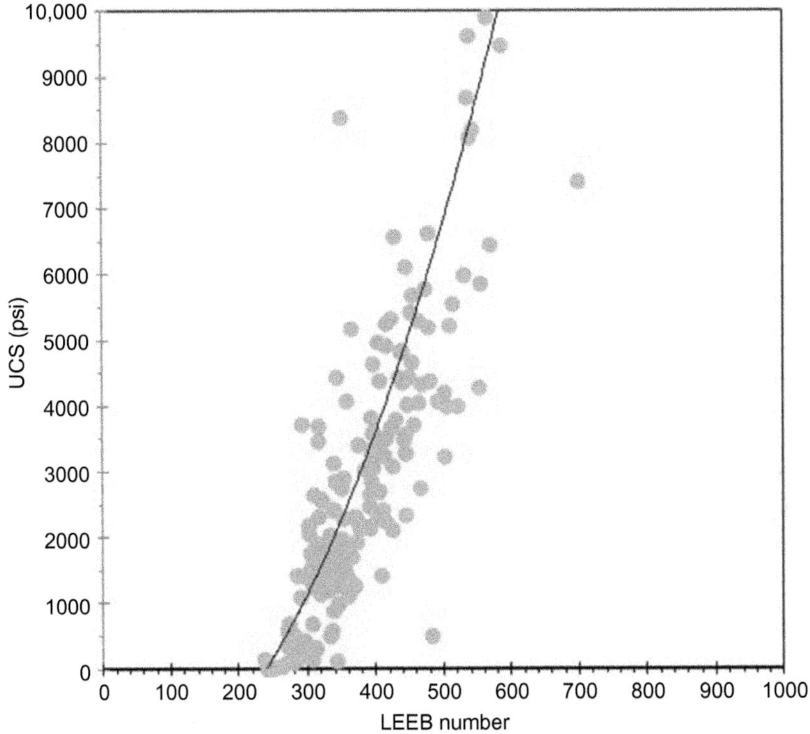

FIGURE 12.2 Equotip rebound response (Leeb number) with UCS strength. *From Daniels et al. (2012).*

on the cutter are recorded and the specific energy required to cut the groove is correlated against UCS strength (Fig. 12.5). Therefore, it is possible to obtain a semi-quantitative estimate of UCS strength continuously over the cored interval (Fig. 12.6).

The high-resolution Equotip and scratch test data are used for core plug site selection and can be incorporated in, or used to guide the development of, log-based strength models.

Further details of core viewing and test site selection for geomechanics tests are provided by McCurdy (2013).

For shales, the use of radial acoustic (P wave) velocity measurements can be used to assess anisotropy, and core samples must be inspected for micro-cracks, fractures and other stress-release features. Shale cores should be preserved as soon as possible after coring (often at wellsite) and never frozen. Cores drilled with water-based mud must not be frozen if sand strength measurements are required.

The number of samples selected depends not only on rock strength and strength variation but also on core availability, core geometry and core

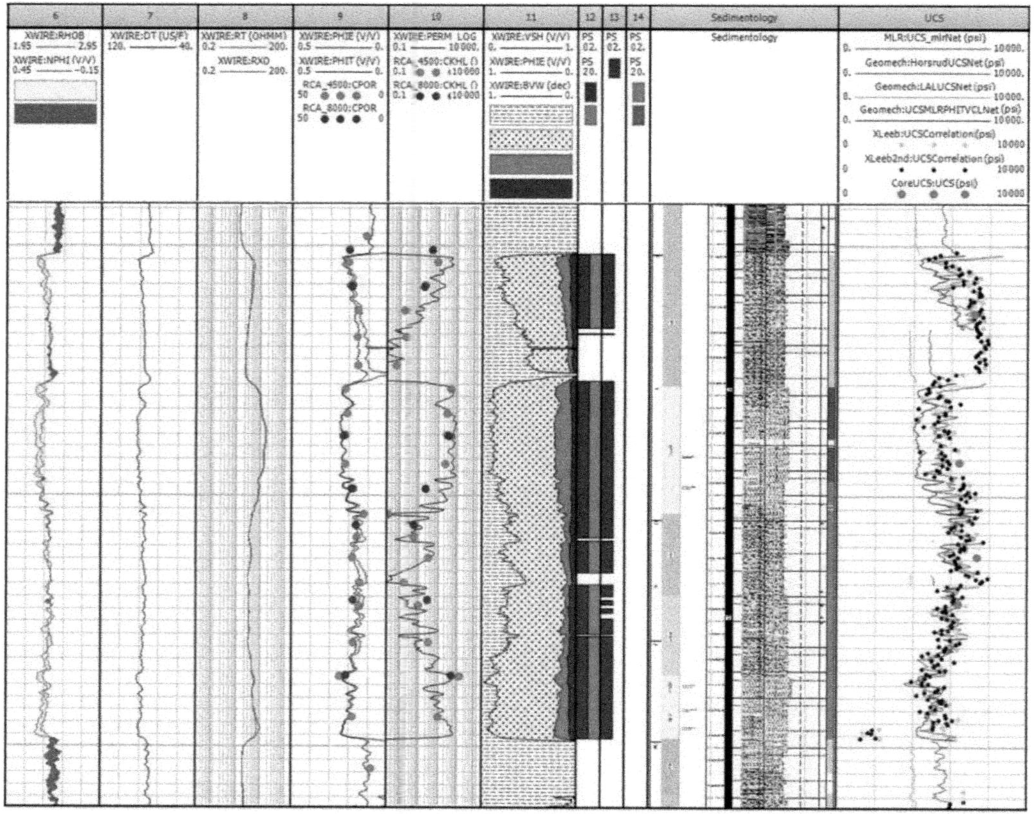

FIGURE 12.3 Equotip (Xleeb) UCS strength profile comparison with logs and core UCS.

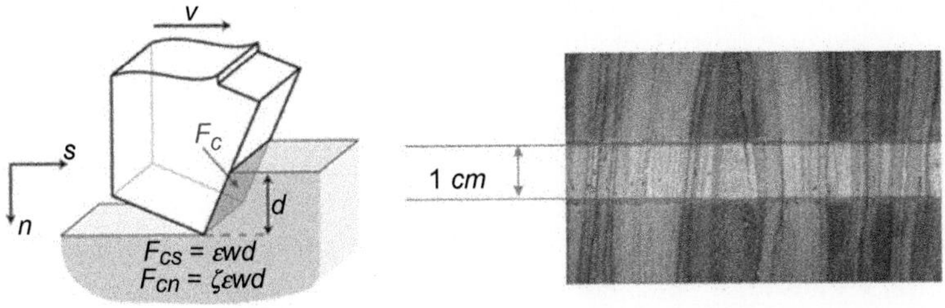

FIGURE 12.4 Core scratch test cutting tool and groove cut in slab surface. *Courtesy of Epslog.*

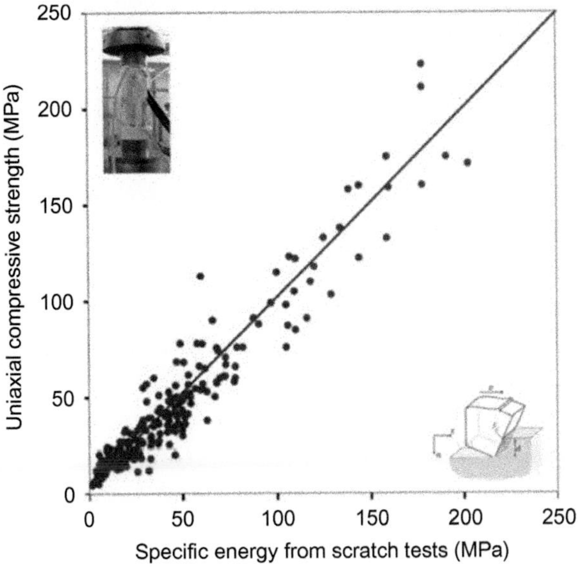

FIGURE 12.5 Correlation between specific cutter energy and UCS. Crossplot shows results of scratch tests and conventional UCS tests performed in different sedimentary rocks by several independent laboratories. *Courtesy of Epslog.*

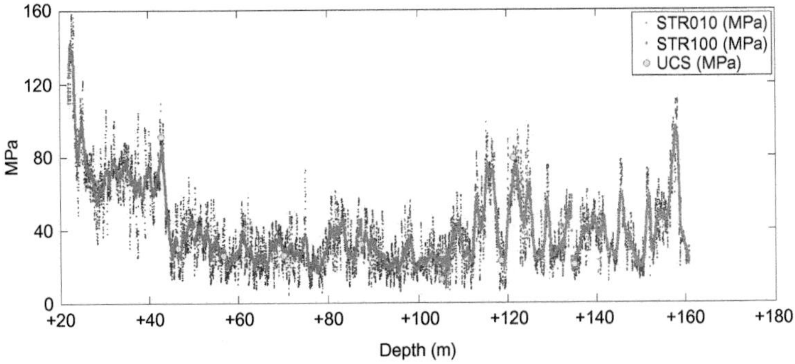

FIGURE 12.6 Example of UCS strength profile from scratch test. *Courtesy of Repsol Technology Centre and Epslog.*

condition. Ideally, plugs should be taken from preserved core sections (this is an essential prerequisite for shale cores) but this is often not possible and they often have to be taken from slabbed core or core which has been exposed to the atmosphere for some time. Visual inspection and SEM analysis are recommended to evaluate any core deterioration during storage. If the core has been slabbed, then potential plugging sites may be constrained. For example, it would not be possible to obtain a 1.5″ diameter plug of the correct dimensions from 2/3 slabbed, 4″ diameter core. It may also be difficult to take vertical plugs (for UCS and SST/MST, for example). It is advisable to take plugs for rock mechanics tests at the same time as the core is plugged for RCA.

Rock mechanics tests should not be performed on plugs which have been previously used for RCA and SCAL as stress cycling can impact on the results, and the plugs may be petrophysically damaged. Figure 12.7 is a plot of Young's modulus (from sonic logs) plotted against UCS for samples from four wells. The sample highlighted (circled) were previously used for SCAL where confining stresses of around 5000 psi are used. The plugs are clearly weakened and UCS data suspect as a result of stress cycling.

12.2.2 Sample Orientation

Unless otherwise stated, UCS, tensile strength, triaxial tests and compressibility tests are normally performed on 90° vertical (bedding plane perpendicular)

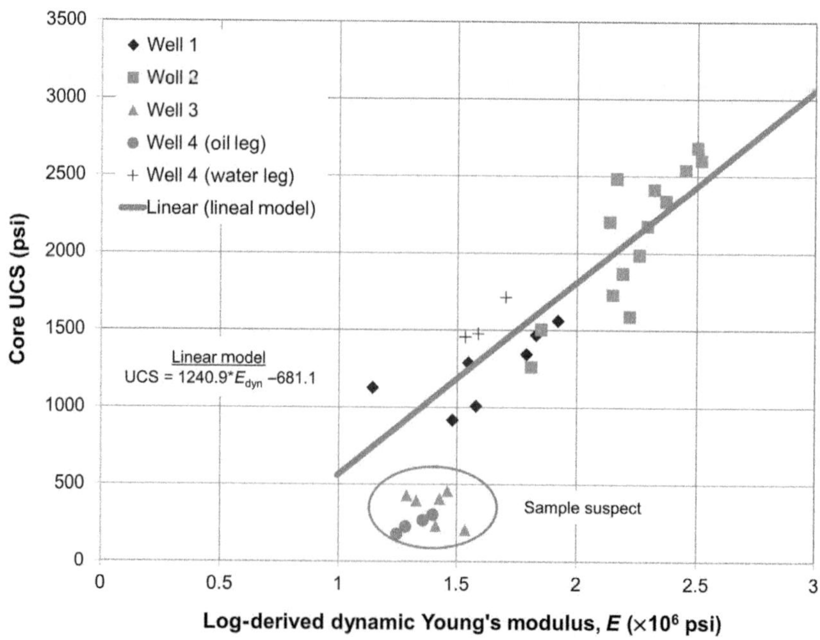

FIGURE 12.7 Plugs weakened due to stress cycling.

samples; and ATT and TWC on 0° horizontal (bedding plane parallel) samples. There are exceptions: some shales can exhibit significant mechanical strength anisotropy across the bedding planes so plug orientation with respect to bedding planes can be important. DSCA tests are performed on 25 or 40 mm core cubes cut orthogonal to the core axes and normally oriented to a datum line (for example orientation scribe) on the core.

Where UCS or SST/MST and TWC plugs are required at the same (or similar) depths, a horizontal plug and vertical plugs are ideally cut at closely adjacent locations. Where three plugs are required for an SST test set, the plugs are ideally cut as parallel plug 'triplets' immediately adjacent to one another.

12.2.3 Plugging Fluids

Mineral oil or kerosene should be used as the plugging fluid to prevent potential rock–fluid incompatibility problems often associated with brine. For weaker formations, plugs may have to be cut with chilled air or liquid nitrogen.

12.2.4 Sample Dimensions

Plugs cut for geomechanics tests require much stricter control on plug geometry and end face allowance than either RCA or SCAL plugs.

1.5″ diameter plugs are preferred, especially in coarse-grained material where the grain diameter/plug diameter ratio must respect the International Society for Rock Mechanics (ISRM) recommendations. However, it is recognised that the core condition, geometry and heterogeneity may prevent this, so tests on 1″ diameter plugs are acceptable. Shale plugs are often cut as 1″ diameter to allow for faster permeation or drainage of pore fluid. SST test samples should be cut parallel to one another from closely adjacent locations. The samples are end lathed to produce a sample that meets ISRM (1979) standards. The tolerances required are:

- end surfaces parallel and flat within 0.1 mm;
- end surfaces perpendicular to sample axis within 0.5°; and
- length/diameter (*L*:*D*) ratio of at least 2:1.

Failure to ensure flat end faces can lead to tensile splitting, as indicated by the example in Fig. 12.8. Loading a sample with uneven end faces will produce a point load rather than a load which is evenly distributed across the end face, so that the test loading conditions more closely resemble a Brazilian disk test (tensile strength test) than a shear test.

In view of these tolerances on plug geometry, it is normally recommended that plugs are cut from the whole-core sections by the rock mechanics test lab and not by conventional core analysis labs. In general, rock mechanics test laboratories have much stricter criteria on plug dimensions that routine core

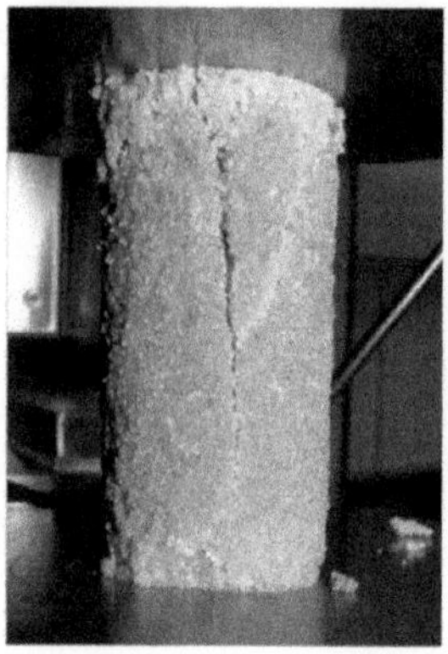

FIGURE 12.8 Tensile splitting of UCS sample due to uneven end faces.

labs. Where the logistics of, and regulatory controls on, sample shipping are difficult, then it might be possible for the local core analysis lab to cut plug 'sticks' then ship the plugs to the rock mechanics lab for final preparation.

Photographs of all rock mechanics test plugs must be taken before testing commences. In most cases, the plugs should also be CT-scanned, especially in heterogeneous lithologies.

12.2.5 Sample Saturation

Strength tests performed on cleaned and dried plugs are not representative of *in situ* reservoir conditions and can cause petrophysical damage in sensitive formations. Plugs should always be tested saturated in liquid (oil or water). Cleaning and drying then resaturating in formation water can cause issues associated with clay rehydration in sensitive lithologies. For oil well applications, tests should be ideally performed on saturated samples so that the sample contains a liquid phase that effectively lubricates the grains (as in the reservoir). Oil is used rather than brine to prevent any deleterious chemical reactions between brine and rock cements and clays. In special circumstances, tests can be performed on samples fully or partly saturated in SFW or synthetic sea water to evaluate effects of rock weakening on contact with produced or injected water.

Although tests can be performed on resaturated clean and dried plugs (after porosity and permeability measurements, for example), plugs are normally tested in 'fresh state' (as received) condition without cleaning or drying. They are simply saturated under bland mineral oil or kerosene by evacuation, and then immersion under moderate pressure. On completion, the samples are removed and the saturated weight is recorded (accounting for any sleeve material). Initial dry and saturated weight measurements are used to estimate the saturated bulk density and porosity. If there is any evidence of oil or oil-based mud contamination in the core, the samples might be processed by soaking in a bath of toluene at room temperature. The samples are then left to dry by evaporation at room temperature prior to saturation with bland mineral oil or kerosene.

For gas well applications, the plugs can be tested in fresh state (as received) without saturating in oil although the core should not have been allowed to dry out so that connate water is still present. For shales, saturation is better achieved during the initial stages of the testing process. The saturant must be compatible with the shale. Sample preparation and saturation requirements for ATT and compressibility tests are the same as for other SCAL tests, as discussed in Chapters 4 and 6.

12.3 UNCONFINED COMPRESSIVE STRENGTH TESTS

12.3.1 Purpose and Sample Requirements

The principal application of UCS tests is as a qualitative strength indicator or index. The data are used, in conjunction with friction angle, in wellbore stability and some sand failure calculations. Unless otherwise specified, UCS tests are normally performed on vertical (bedding plane perpendicular) samples with a length:diameter ratio of at least 2:1. If this is not possible, corrections to account for boundary effects on short samples are normally required.

In some applications, the 'unconfined' tests are performed at nominal confining stress (150 or 300 psi). Often referred to as quasi-UCS (q-UCS) test, effectively this is a triaxial test but strength test data measured at nominal confinement are much less influenced by plug irregularities and discontinuities.

12.3.2 Test Equipment

The unconfined compressive strength (UCS) tests are normally performed using a stiff compression machine with servo-controlled actuators and intensifiers. If sample deformation characteristics are required under uniaxial compression, then the equipment is instrumented to allow elastic moduli determination. Sample deformations are measured using linear variable differential transformers (LVDTs) and cantilever devices mounted onto stand-offs at mid-height around the circumference of the samples. LVDT systems are used for extremely brittle material where catastrophic failure is anticipated due to spalling (regardless of the fact that these are high-stiffness machines). The maximum radial strain, using cantilevers, is 10% and the minimum is 0.001%. The maximum axial strain, using cantilevers, is 20% and the minimum is 0.005%. There are no limits if LVDTs are used. Calibrations of all the measurement devices employed must be traceable to current national standards of measurement. Axial and radial strain measurements are used to calculate static Young's modulus and Poisson's ratio for each sample. However, as these will not be representative of the confined or stressed rock system instrumentation is optional.

FIGURE 12.9 UCS sample loading.

The samples are placed between hardened steel platens so that load can be transferred evenly over the sample end faces (Fig. 12.9).

12.3.3 Test Procedures

If the sample is instrumented with gauges the axial load is applied by means of a servo-hydraulic load machine at a constant rate of axial displacement, which ranges from 10^{-5}/s for sands to 10^{-6}/s or 10^{-7}/s for shales, until failure. If not instrumented, the loading rate is selected such that failure occurs within 15–20 min of the onset of loading and therefore is largely based on lab experience and operator expertise.

12.3.4 Data Utilisation

UCS strength is determined from the ratio of load, L, to sample cross section area, A

$$\text{UCS} = \frac{L}{A}$$

Correction factors are applied to account for short boundary effects if the sample has an $L{:}D$ ratio of less than the ISRM (1979) standard of 2:1. For example,

$$\text{UCS}_{\text{corrected}} = \frac{\text{UCS}_{\text{measured}}}{0.25\frac{D}{L_p} + 0.875}$$

where L_p is the plug length and D is the plug diameter, both in mm.

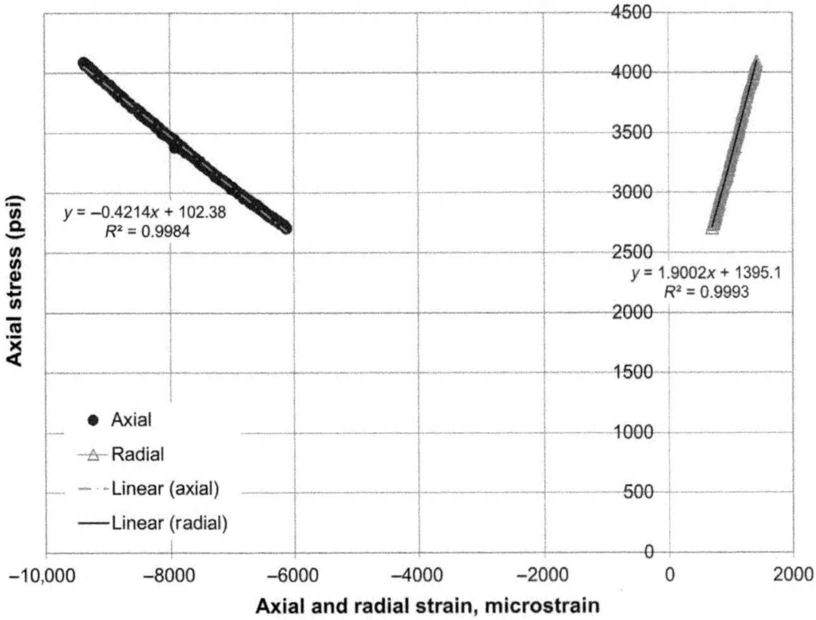

FIGURE 12.10 Typical stress/strain relationships used for Young's modulus.

The average Young's modulus, E, is determined from the slope of the axial stress (σ_a) versus axial/strain (ε_a) curve (Fig. 12.10):

$$E = \frac{\sigma_a}{\varepsilon_a}$$

and the average Poisson's ratio is determined from the radial strain (ε_r) versus axial strain curves (Fig. 12.11):

$$\nu = -\frac{\varepsilon_r}{\varepsilon_a}$$

both over an axial stress range between 40% and 60% of peak (failure) strength.

12.3.5 Data Reporting Requirements

The data that need to be reported by the test laboratory to ensure adequate quality control of the UCS tests and test results are listed in Table 12.1.

12.3.6 Advantages and Drawbacks/Issues

Summaries of the advantages and drawbacks/issues associated with UCS tests are provided in Table 12.2.

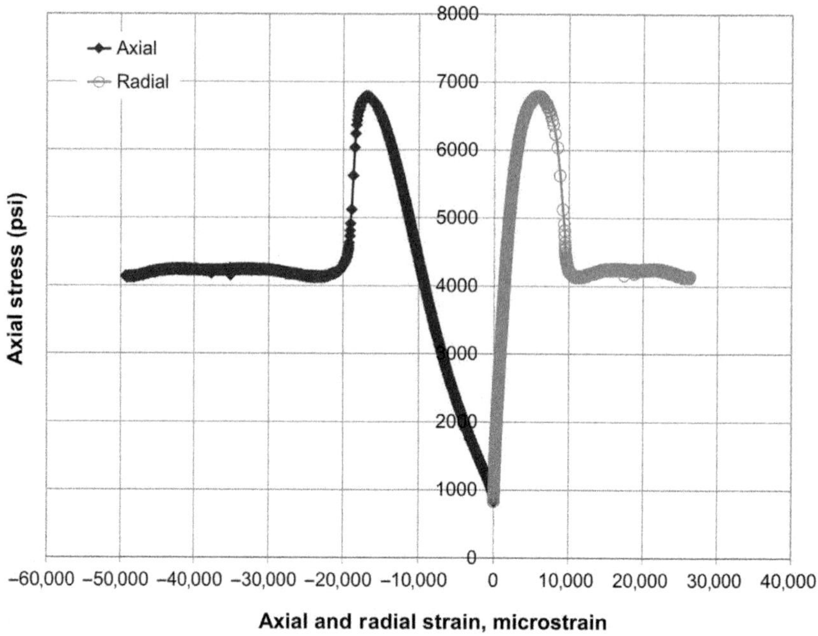

FIGURE 12.11 Typical stress/strain relationships used for Poisson's ratio.

12.3.7 UCS Quality Control Issues, Checks and Diagnostics

Cementing axial strain gauges to the sample surface is not recommended as pore fluid can cause the cement to degrade which will result in anomalous strain (and hence Poisson's ratio and Young's Modulus) data.

The UCS of a sample loaded parallel to any laminations may differ to that if loading is perpendicular to the laminations. The UCS of a dry sample will exceed that when saturated. Water-saturated samples are often weaker than oil-saturated samples, especially if they contain sensitive clays or reactive cements.

Low-permeability materials (i.e. shales) may exhibit undrained or partially drained behaviour if tested at the recommended loading rates for permeable formations.

Where UCS and triaxial tests are performed on adjacent samples, the UCS derived from extrapolation to zero confining stress should be higher than direct UCS measurements under zero confinement conditions. Lower measured UCS at unconfined conditions can be a consequence of the generally inelastic behaviour of the material, and sample surface irregularities and microcracks which do not impact on confined core strength measurements to the same degree.

UCS data should not be used in isolation for rock strength and failure calculations. Data should be verified against TWC and triaxial test data.

TABLE 12.1 Data Reporting Requirements for UCS Tests

Data	Comments
Brief description of procedures and experimental apparatus	Including how plugs were cut and prepared
Plug depth and orientation	Normally vertical
Plug length and diameter	Establish *L*:*D* ratio
Pore saturant	Oil or water
Sample bulk density and pore saturant density	Porosity estimate
Loading rate versus time under compression	If not instrumented
Data and plots of axial stress, radial strains versus mean axial strain for elastic moduli determination	If instrumented
Calculated values of Poisson's ratio and Young's modulus at 40–60% peak stress (and method used to calculate moduli)	If instrumented
Failure load	Peak strength
Uncorrected unconfined compressive strength	For shorter plugs (L:D<2)
Any corrections that have been applied should the test sample be shorter than the ISRM standard	Supply correction algorithm
Corrected unconfined compressive strength (if applicable)	
Description of failure mode (e.g. shear, planar shear, conjugate shear, etc.)	
Photographs of test plugs before and after testing. These should include a scale marker (e.g. ruler). Post-test plug photographs of UCS test samples should orient the core vertically and should clearly show the failure plane	To ensure shear failure and not tensile failure has occurred. An example is provided in Fig. 12.12

The measurement of Poisson's ratio is inherently more error prone than the measurement of the Young's modulus and depends on several factors including sample heterogeneity and anisotropy, sample size, measurement techniques and data processing. Poisson's ratio can be influenced by pre-existing microfractures and hairline cracks in the sample. The stresses are more

FIGURE 12.12 Post-test photograph of test plug showing clear shear failure surfaces.

TABLE 12.2 Advantages and Drawbacks/Issues Associated with UCS Tests

Advantages	Drawbacks and Issues
• Allows one of the most important rock index properties to be determined simply, directly and reliably. • Widely regarded as a rock strength estimator. Many log-derived strength models are based on large databases of UCS tests. • A good indicator of the general competence of a rock material.	• Provides only the most basic strength data. • Tests under unconfined conditions can produce misleadingly low-strength values, depending on the surface condition and surface topology of the test samples. Quasi-UCS tests, in which the core is confined at nominal stress (e.g. 150 or 300 psi), are routinely carried out by some companies. The loading conditions are similar to those for triaxial tests and can be measured with or without strain gauges. • Elastic moduli are not representative of the behaviour of a confined sample. • UCS for sand failure and wellbore instability prediction is better obtained from analysis of sets of single-stage triaxial tests. • Tests are destructive and cannot be repeated.

easily measured with the required accuracy, and the error is normally low. The measured strains may be quite small and the measuring errors can be more significant. As a result, the calculation of Young's modulus ($\delta_{\text{axial stress}}/\delta_{\text{axial strain}}$) has a lower error than the calculation of Poisson's ratio ($\delta_{\text{radial strain}}/\delta_{\text{axial strain}}$). It is common to reproduce the Young's modulus measurements reasonably well, to within 5–10%. However, the uncertainty range on Poisson's ratio measurements is always wider—around 25% or higher.

Test labs should be audited to ensure they have the capabilities, resources and experience to carry out rock mechanics tests on reservoir rock samples. Some laboratories offering a commercial service are more suitable for soil mechanics testing or rock mechanics testing for civil engineering applications, not for reservoir geomechanics applications.

12.4 TRIAXIAL COMPRESSION STRENGTH TESTS

12.4.1 Purpose and Sample Requirements

The primary purpose of triaxial tests is to determine:

- rock failure parameters, principally Mohr–Coulomb cohesive strength and friction angle, under representative reservoir loading/stress conditions,
- unconfined compressive strength (UCS) from extrapolation of confined failure strengths, and
- static elastic moduli—Young's modulus and Poisson's ratio—under representative reservoir loading/stress conditions.

Unless otherwise specified, triaxial tests are normally performed on 90° vertical (bedding plane perpendicular) samples with a length:diameter ratio of at least 2:1. If this is not possible, corrections to account for boundary effects on short samples may have to be made. SST tests are made on a set of at least three plugs (each tested at different confinements). MST tests are carried out on a single sample at multiple, incremental confinements. The plug must be regular. Figure 12.13 shows a photograph of a plug which exhibits variation in diameter along the length, probably caused by plugging bit wear or changes of weight on the bit when plugging this hard rock. The axial stress versus radial stress data for this sample exhibited significant non-linearity.

As rock peak strength increases with increasing confining stress, selecting the appropriate radial (confining) stresses is important. These will depend on data application. For elastic moduli determination for deformability characteristics—for example for reservoir compaction estimates—the radial stress is normally set equivalent to the average reservoir isostatic effective stress or average effective horizontal stress at virgin reservoir conditions. For wellbore/perforation failure applications, effective confining stresses are normally lower and will depend upon the rock strength. For example, in weak rocks, the sample may fail on initial confinement if the confining stress is high enough, prior to initiating the triaxial loading sequence.

FIGURE 12.13 Example of plug with irregular sides.

12.4.2 Test Equipment

The 'triaxial' test is a misnomer as, in almost all commercial labs where tests are conducted on cylindrical plug samples, the test cell is biaxial: that is stress is applied to the sample on two axes: the radial or confining stress around the sample (σ_3) and the axial stress along the sample (σ_1). For Mohr–Coulomb analysis, this is sufficient as the intermediate stress (σ_2) is ignored. Triaxial tests, in which $\sigma_1 \neq \sigma_2 \neq \sigma_3$, are only possible on cubic samples, which are extremely difficult to cut to the precise dimensions required, or using a 'true triaxial cell' (Smart et al., 1999) where the radial stress can be varied around the cylindrical plug. Almost all test cells in commercial labs are Hoek-type 'biaxial' cells.

Typically, triaxial tests are performed using a stiff compression machine with a closed-loop servo-control system incorporating an instrumented Hoek-type triaxial cell (Fig. 12.14) mounted on a load frame (Fig. 12.15). Confining stress on the samples is generated and maintained using the triaxial cell in conjunction with servo-controlled actuators and intensifiers.

12.4.3 Test Procedures

12.4.3.1 SST Tests

Each sample is initially confined at a known isostatic stress, by applying a cell pressure to the surrounding sleeve in the Hoek cell and by applying an equivalent axial stress along its axis. The samples are allowed to stabilise for at least 30 min before increasing the axial stress. Axial and radial strain measurements are required during isostatic loading to assess possible compactive failure and to ensure stabilisation (no creep) prior to deviatoric loading. Figure 12.16 shows volumetric strain data from a sample during loading to constant confining stress (1450 psi) in a relatively weak sand, and strain

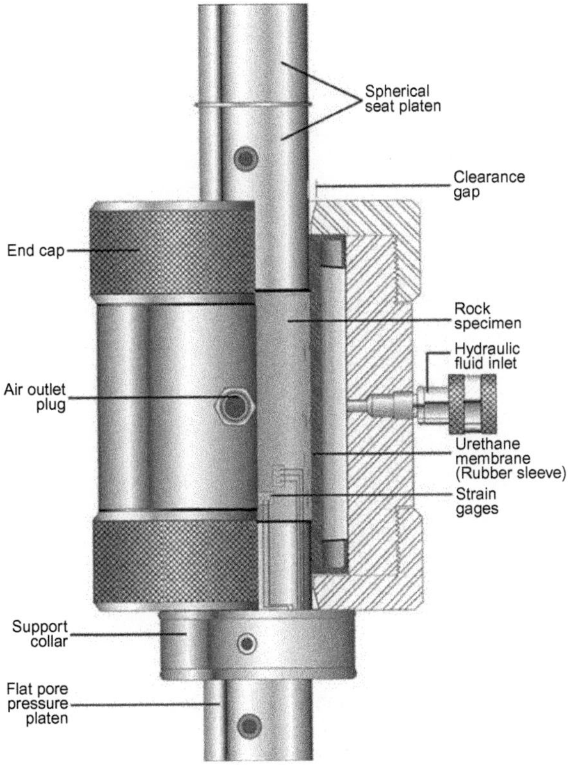

FIGURE 12.14 Hoek cell schematic. *Courtesy of Roctest.*

data during the stabilisation period. The sample should have been allowed to stabilise at constant isostatic loading but clearly volumetric strain is continuing to increase (creep) when the deviatoric loading phase was initiated. This can affect subsequent elastic moduli, and sometimes yield and peak strength.

In most test cases, pore pressure is maintained at ambient conditions (or at a nominal value) during loading. In these tests, it is imperative that samples are drained directly to atmosphere, and that any outlet flow pipework restrictions (e.g. valves) are minimised, to ensure fully drained conditions. Excess pore pressures generated during loading must be allowed to drain via fluid ports in the cell loading platens. When testing low-permeability samples, wire mesh side drains may be employed to improve pore pressure dissipation.

Deviatoric loading is then initiated by increasing the axial stress while keeping the radial stress constant until failure occurs. To ensure drained behaviour, loading is conducted under axial strain control at a nominal axial strain rate of around 10^{-5}/s for permeable samples and around 10^{-6}/s for low-permeability samples (e.g. siltstones) to 10^{-7}/s for shales.

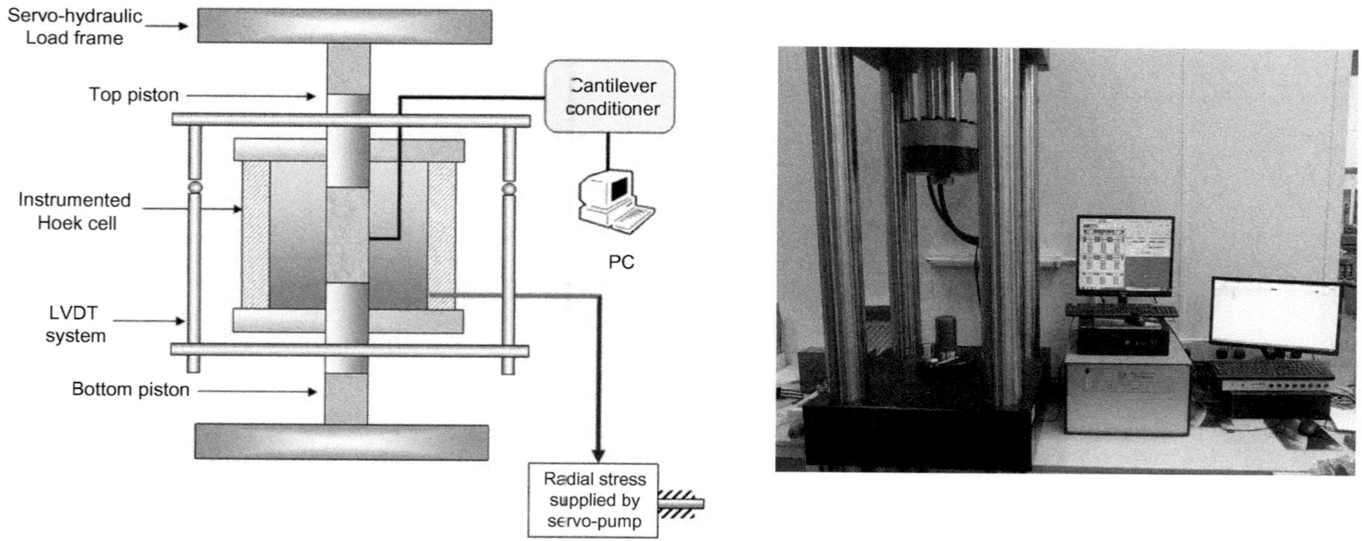

FIGURE 12.15 Triaxial test apparatus schematic and load frame. *Courtesy of FracTech Laboratories.*

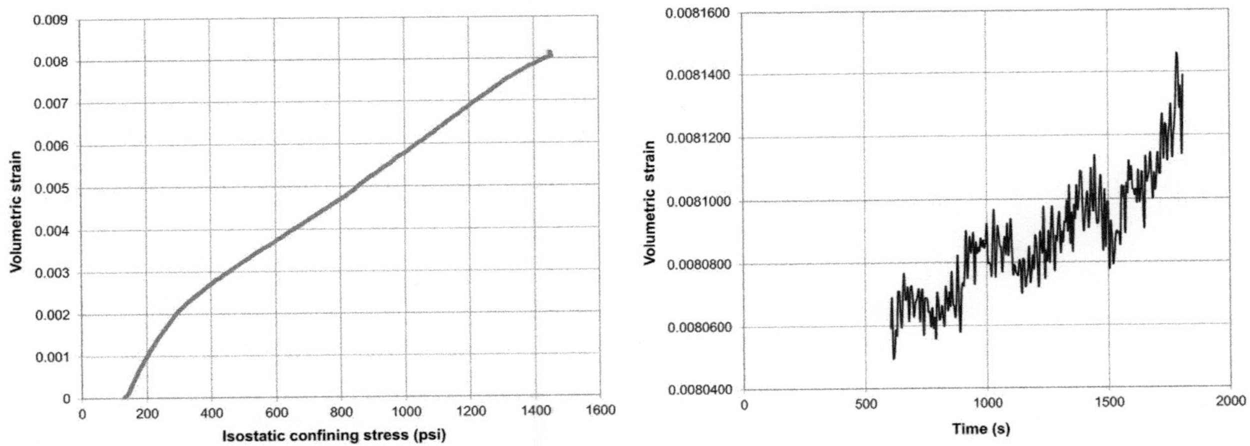

FIGURE 12.16 Example of volumetric strain behaviour during isostatic loading.

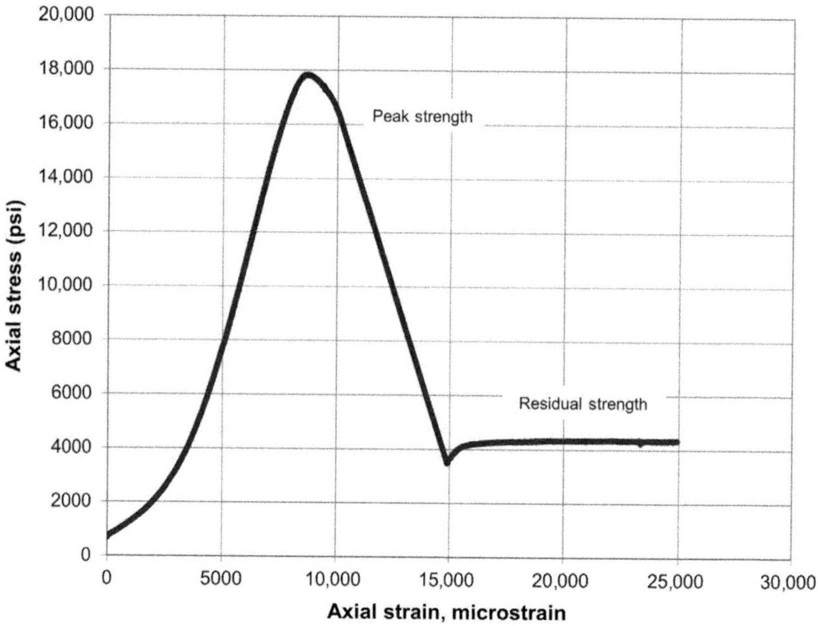

FIGURE 12.17 Peak and residual strength from axial stress vs axial strain relationship.

Post-yield stress–strain behaviour needs to be clearly defined. In brittle rocks, loading should be continued beyond peak strength (σ_P) until residual strength (σ_R) is achieved, as indicated in Fig. 12.17. In more plastic rocks, axial strain must be monitored post-yield to clearly define any strain-softening or strain hardening behaviour as these data will be required for finite element analysis model calibration.

The confining (radial) stresses for each sample of the SST set (samples A through C for a 3-sample set) will depend on the application of the data and the formation failure characteristics. Strength and elastic properties at low confinements are of greatest significance in hole and perforation stability. For wellbore stability and sand failure applications (for tests carried out under drained conditions), the effective radial stresses are low. An example for a relatively weak rock is provided in Table 12.3. In very weak samples, the lower radial stresses may be used. It is possible to carry out the tests using elevated pore pressures, similar to those at reservoir conditions. However, there are concerns that the outlet valve used to control pore pressure presents a tight restriction that could become blocked by sand or other debris. Excess pore water pressures generated due to applied loads must be allowed to dissipate rapidly (drained conditions).

Where triaxial tests are used to establish deformation characteristics and to calibrate log-derived dynamic elastic moduli with core-derived static elastic moduli, the confining stress should approximate the effective reservoir stress

TABLE 12.3 Example Confining Stresses for set of Three Triaxial Tests on Weak Rock

Sample	Total Confining (Radial) Stress (psi)
A	145 or 290
B	290 or 725
C	725 or 1450

at virgin reservoir conditions. In this case, the applied effective radial stress, σ'_{rad}, is often based on the average of the two *in situ* reservoir horizontal stresses, σ_H and σ_h, minus the reservoir pore pressure (p_p), i.e.:

$$\sigma'_{rad} = \left(\frac{\sigma_H + \sigma_h}{2}\right) - p_p$$

The maximum axial stress imposed on deviatoric loading should be broadly equivalent to the effective vertical stress, although higher axial stresses may be warranted to ensure sufficient data to determine elastic moduli. In most cases, a cap is set on the axial stress so the test may not be carried out to failure, or beyond the yield point.

12.4.3.2 MST Tests

The MST test is carried out where formation heterogeneity or sample availability precludes SST tests. There may not be sufficient core material to cut parallel SST triplets, for example. In most respects, the equipment and procedures are similar to the SST test but with one important exception: deviatoric loading is achieved by increasing the axial stress until imminent failure is detected as indicated by a rapid reduction in the rate of load increase, or by acoustic emissions picked up by detectors in the loading platens of the Hoek cell. If the test was permitted to proceed the sample would eventually rupture, however, by increasing the confining pressure to the next selected value, it is possible to return the sample to stable conditions and prevent macroscopic failure of the rock. The confining stress increase acts to reinforce the sample, thus preventing macroscopic failure. Consequently, the radial confining stress is increased to a higher value and axial loading proceeds until the new 'peak' or incipient failure strength is achieved. The radial stress is again increased to another known value, and the procedure is repeated to obtain a total of three or more peak strength values. An illustration of the multi-stage loading process (five stress stations) is shown in Fig. 12.18.

Usually, three to four stages are adequate to obtain the representative Mohr circles to allow a plot of the rock failure envelope in order to estimate

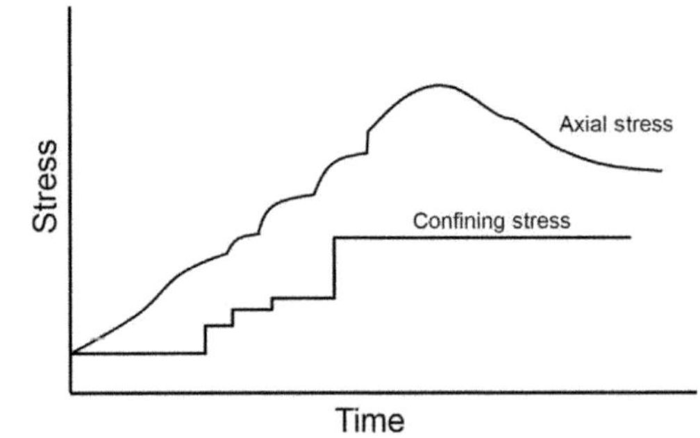

FIGURE 12.18 Schematic illustration of multi-stage triaxial test.

cohesion and friction angle of the sample. While Young's modulus and Poisson's ratio can still be determined from the resulting stress–strain curves, the failure envelope now must be estimated from one 'failure' Mohr's circle obtained from the last loading stage and several 'non-failure' Mohr circles obtained from the previous stages. The definition of the failure parameters from 'non-failure' Mohr circles and the single failure (final confinement) Mohr circle can be subjective. The conventional criterion is to stop the test at the point before the sample exhibits signs of approaching failure on the stress–strain curves. This can be difficult in non-linear material (especially in the absence of acoustic emission detectors) and there is a danger that sample deformation or incipient sample failure at lower confining stresses can affect the failure stresses (and elastic moduli) at higher confining stresses.

Tran et al. (2010) argue that if a 'critical' loading is exceeded then it is possible that this can result in irreversible deformation of the samples which could affect subsequent elastic moduli and possibly 'failure' stresses. For example, the volumetric strain inflection point represents the end of the stable (controlled) fracture propagation region. Above this point, any additional load would create new cracks and acoustic energy (emission rate) increases exponentially. The critical loading is defined by the volumetric strain inflection point. An example is shown in Fig. 12.19. The inflection point appears to have been exceeded at around 7600 psi axial stress.

By terminating loading at the inflection point and assuming a constant difference between the stresses at failure and at the inflection point ($\sigma_{fail} - \sigma_{inf}$) from conventional single-stage tests, the inflection point stress can be used to estimate a failure stress and hence Mohr circles for the MST non-failure loading stages. In tests on Berea sandstone, Tran et al. showed that the difference between stresses at failure and at the inflection point is relatively constant at lower confining stresses, whereas for tests at higher confining pressures, the stress difference was much lower.

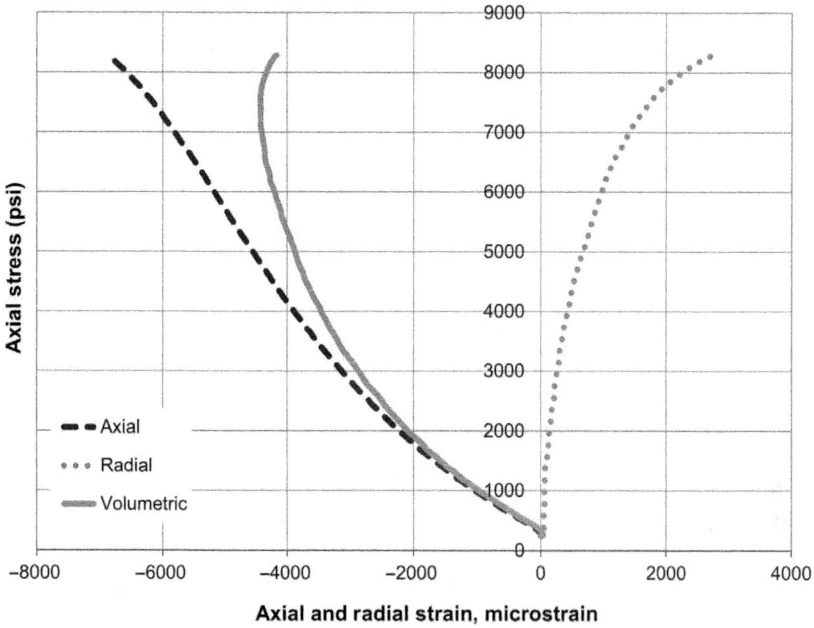

FIGURE 12.19 Example of volumetric strain plot for MST test (radial stress = 300 psi).

12.4.4 Data Utilisation

12.4.4.1 Elastic Moduli

In SST and MST tests, Young's modulus, E, and Poisson's ratio, υ, are normally determined for each confining stress station from data acquired between 40% and 60% of peak stress. The stress–strain behaviour must be within a linear elastic regime.

If the confining stresses are low, for example, for wellbore stability test applications, then the data may not be representative of the *in situ* rock properties. If elastic moduli data are required for determination of compaction or subsidence, or for calibration of dynamic elastic moduli from logs, the deviatoric loading tests may be carried out with the cell stresses selected to simulate the effective vertical and mean effective horizontal *in situ* reservoir stresses, at initial reservoir conditions. In these tests, deviatoric loading might be restricted to ensure data are acquired in a linear elastic regime such that samples might not be taken to peak failure. Non-linearity in the stress–strain response can result from a number of causes such as lithology (shales are more plastic than sandstones, for example), plug irregularities and the presence of microfractures. Distortion on the stress–strain curves is sometimes a consequence of microfractures opening during deviatoric loading.

At high confining stress, materials are less likely to shear fracture and lose load-bearing capacity because the pressure of the surroundings tends to hinder the formation of shear fractures. At lower confining stress, material behaviour therefore will tend to be more brittle. At higher confining stress, the stress–strain relationship appears to show less brittle, more plastic behaviour.

It is expected that Young's modulus, which is a measure of rock stiffness, will increase with confining stress, as indicated in Fig. 12.20. However, Poisson's ratio trends can be very variable and often show no systematic correlation with sample porosity, rock strength or confining stress. Figure 12.21 shows Poisson's ratio measured on the same samples as the Young's moduli. The data are variable and suggest two distinctly different rock types which is not the case. As discussed for UCS tests, Poisson's ratio is a rather uncertain measurement. Poisson's ratio is strongly influenced by mineral composition, and the presence of dry or wet cracks within the formation. The influence of cracks depends on the saturation state: cracks with low aspect ratio will lower (dry case) or increase (saturated case) the Poisson's ratio of the rock. The magnitude of the change depends on the volume and aspect ratio of the cracks. Effective stress will also affect the measured Poisson's ratio. At low effective stresses open cracks will lower the Poisson's ratio significantly, while at high effective stress the majority of cracks are closed and Poisson's ratio is primarily influenced by the

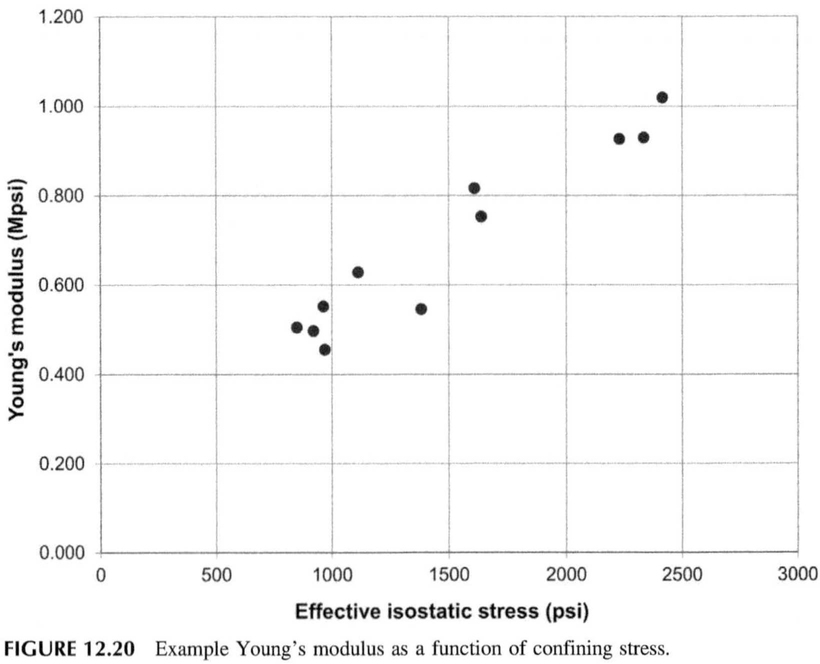

FIGURE 12.20 Example Young's modulus as a function of confining stress.

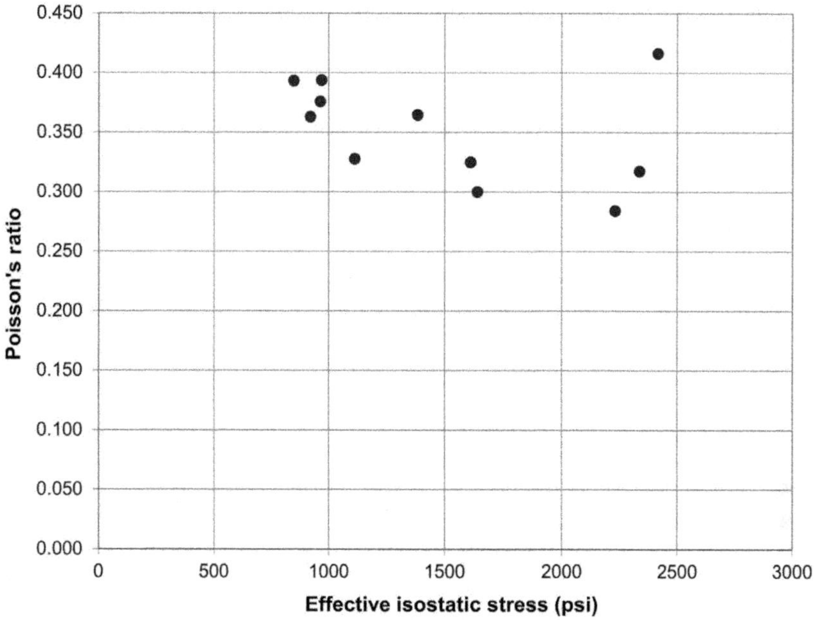

FIGURE 12.21 Example Poisson's ratio as a function of confining stress.

rock matrix and will tend to increase. Thus, static Poisson's ratio from core should be used with caution in geomechanics applications.

12.4.4.2 Mohr–Coulomb Parameters

Mohr–Coulomb theory mathematically describes the response of brittle materials to shear stress, τ, as a function of normal stress, σ. Generally, the theory applies to materials for which the compressive strength far exceeds the tensile strength. Mohr circles are a two-dimensional graphical representation of the state of shear stress and normal stress at any point, and the circumference of the circle is the locus of points that represent the state of shear and normal stress on individual planes. The Mohr–Coulomb failure model only considers the maximum (σ_1) and minimum principal (σ_3) stresses—the intermediate principal stress (σ_2) plays no part. The Mohr failure envelope is a line tangent to the maximum possible circles obtained at different stresses (Fig. 12.22) and no circle could have part of it above that tangent curved line. Combining the Mohr failure criterion with the Coulomb equation gave a straight line tangent to most of the Mohr circles, and the Mohr–Coulomb failure envelope is defined by the relationship:

$$\tau = S_0 + \sigma \tan \theta$$

where S_0 is the cohesive strength and θ is the angle of internal friction. Cohesive strength is defined as the inherent shear strength or cohesion of the material.

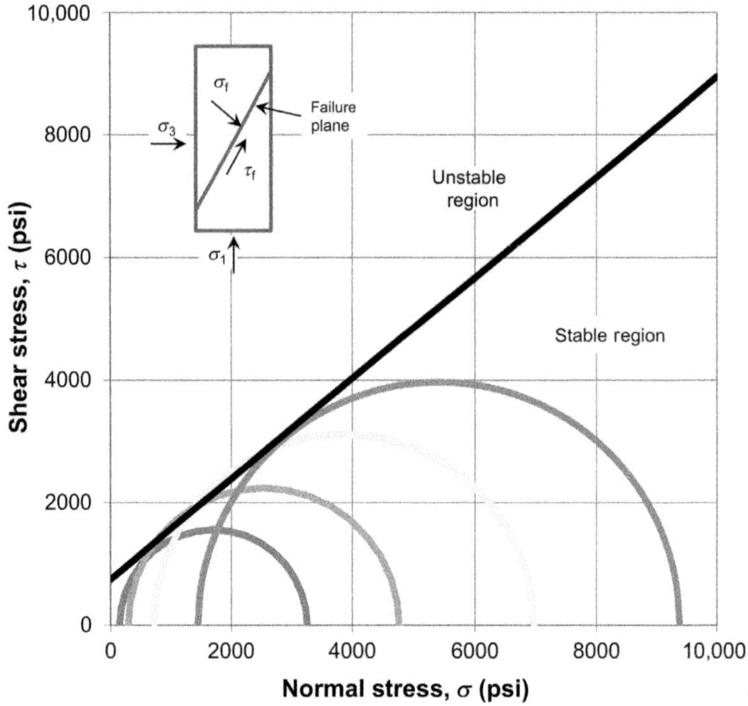

FIGURE 12.22 Example Mohr–Coulomb failure envelope.

For any given failure assessment, provided the deviatoric stress acting on the rock and represented by the Mohr circle with diameter $\sigma_1 - \sigma_3$, remains below the failure envelope, the material is stable. If, however, the deviatoric stress is high enough such that circle crosses the failure envelope, the rock will fail. Beyond the failure envelope is a region of plastic deformation. The linear failure envelope is just an approximation to simplify calculations. It is stress dependent and will exhibit curvature if shear strength tests on the same rock are carried out at different confining stresses.

The failure envelope is determined from analysis of triaxial tests on core. Normally, the Mohr circles are defined in terms of effective stress, that is:

$$\sigma' = \sigma - p_{\mathrm{p}}$$

So if pore pressure is at ambient conditions, the total and effective stress will be the same. Consider the example shown in Fig. 12.23 which includes data from four sets of SST samples. The effective confining stress, σ_3, and effective peak failure stress, σ_1, are collated for each sample. The Mohr circle plot is then constructed by drawing a circle for each sample with diameter $\sigma_1 - \sigma_3$. A tangent is then drawn to touch each circle and extended to the intercept on the shear stress axis. This defines the cohesive strength. The slope of the tangent line is $\tan \theta$.

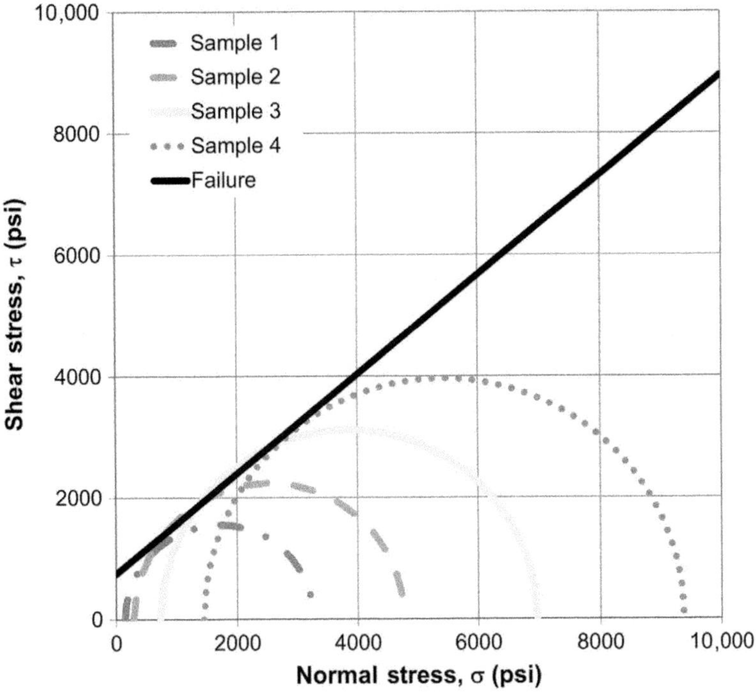

FIGURE 12.23 Example Mohr circle plot.

The Mohr–Coulomb failure criterion is a linear function which can be expressed in terms of principal stresses (σ_1 and σ_3):

$$\sigma_1 = \sigma_0 + \sigma_3 k$$

where σ_1 is the maximum principal stress—the axial stress at failure in the test configuration—and σ_3 is the confining or radial stress. The σ_1/σ_3 pairs from each sample are plotted on what is termed the principal stress axis plot. An example is shown in Fig. 12.24. The intercept on the σ_1 axis is equivalent to the unconfined (uniaxial) compressive strength, σ_0, which should correspond, or be close, to the value determined at zero confinement (UCS test).

The triaxial stress factor, k, is the slope of the linear best fit to the σ_1 versus σ_3 data where:

$$k = \frac{1 + \sin\theta}{1 - \sin\theta}$$

and θ is the angle of internal friction, degrees.

Cohesive strength and friction angle can also be found from analysis of the principal stresses:

$$S_0 = \frac{\sigma_0}{2\sqrt{k}}$$

and:

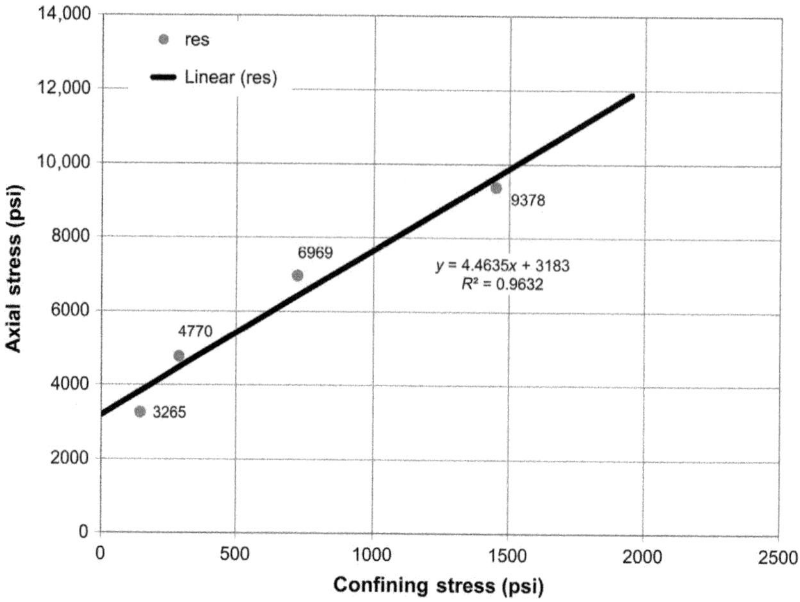

FIGURE 12.24 Example principal stress axis plot.

$$\tan\theta = \frac{k-1}{2\sqrt{k}}$$

The same analysis is performed for MST data where σ_1 represents the incipient failure stresses.

12.4.5 Data Reporting Requirements

The data that need to be reported by the test laboratory to ensure adequate quality control of the triaxial tests and test results are listed in Table 12.4.

12.4.6 Advantages and Drawbacks/Issues

Summaries of the advantages and drawbacks/issues associated with both MST and SST tests are provided in Table 12.5.

12.4.7 Triaxial Test Quality Control Issues, Checks and Diagnostics

Cementing axial strain gauges to the sample surface is not recommended as pore fluid can cause the cement to degrade which will result in anomalous strain (and hence Poisson's ratio and Young's Modulus) data.

TABLE 12.4 Data Reporting Requirements for Triaxial Tests

Data	Comments
Type of test	SST or MST
Brief description of procedures and experimental apparatus	Including how plugs were cut and prepared.
Plug depth(s) and orientation	Normally vertical
Plug length and diameter	Establish *L*:*D* ratio
Pore saturant	Oil or water/brine
Sample bulk density and pore saturant density	Porosity estimate
Pore pressure, p_p	For each sample in SST and each stress station in MST
Total and effective confining stress, σ_3	For each sample in SST and each stress station in MST
Data and plot of volumetric strain versus confining stress and versus time during isostatic loading	Establish strain stability
Axial loading rate during deviatoric loading	For each sample in SST and each stress station in MST
Peak (failure) stress	For each sample in SST and each stress station in MST. Define 'failure' criterion for MST tests
Any corrections that have been applied should the test sample be shorter than the ISRM standard	
Pore pressure as function of axial stress during deviatoric loading	Ensure drained conditions
Data and plots of axial stress versus axial strain, radial strain and volumetric strain during deviatoric loading. Data must clearly define post-failure behaviour and residual strength (in brittle rocks)	For each sample in SST and each stress station in MST
Calculated values of Poisson's ratio and Young's modulus over defined stress range (e.g. 40–60% peak stress)	Method used to calculate moduli
Mohr–Coulomb plot (normal stress vs. shear stress) with peak failure stress Mohr circle	Mohr circle for each sample in SST and each stress station in MST
Principal stress axis plot (peak axial stress vs. radial stress)	Analysis for each sample in SST and each stress station in MST
Mohr–Coulomb failure parameters for each plug set/MST plug test (including estimated UCS strength at zero confinement	Cohesive strength, friction angle and estimated UCS

Continued

TABLE 12.4 Data Reporting Requirements for Triaxial Tests—Cont'd

Data	Comments
Description of failure mode (e.g. shear, planar shear, conjugate shear, etc.)	For each sample or MST sample at final stress
Photographs of test plugs before and after testing. These should include a scale marker (e.g. ruler). Post-test plug photographs of triaxial test samples should orient the core vertically and should clearly show the failure plane	To ensure shear failure has occurred

TABLE 12.5 Advantages and Drawbacks/Issues Associated with Triaxial Tests

Advantages	Drawbacks and Issues
• Data provides peak, and in brittle rocks residual, Mohr–Coulomb failure parameters under representative loading conditions for both analytical failure and numerical deformation models. • Elastic moduli provide estimates of formation and deformation under linear elastic loading. • Rock strength and deformation may be determined at a level of confinement particularly relevant to a given engineering situation.	• Mohr–Coulomb analysis assumes linear elastic behaviour over test stress range. The relationship between axial failure stress (σ_1) and confining stress (σ_3) can be non-linear, even at low confinements. An example is shown in Fig. 12.25. Extrapolating the linear Mohr–Coulomb function to predict σ_0 will tend to overestimate UCS. Thus, the extrapolated UCS values (which are derived from, and influenced by, the behaviour under stressed conditions) give larger values than when UCS is measured without confinement. This is exacerbated when the rock derives its strength principally from friction and not cohesion. The use of a polynomial fit to stress (σ_1) and confining stress (σ_3) data better describes the principal stress confining–failure stress relationships than Mohr–Coulomb and better predicts the unconfined strength. However, the Mohr–Coulomb parameters, by definition, must be determined using the linear function. Consequently, S_0 and θ are sensitive to the stress pairs used to derive the parameters. • The data obtained relate specifically to the confinement applied. Several tests on similar samples need to be conducted if peak strength parameters (i.e. cohesion and friction angle) are required. • If the outlet pressure control valve is blocked during elevated pore pressure testing, or there is a permeability barrier in the sample dissipation of excess pore water might occur much more slowly. Therefore, the short-term shear strength may be closer to undrained conditions. Under these load conditions, some of the load is carried by increased pore liquid pressure. • Low confining stresses may be required on very weak samples to prevent compactive failure during isostatic loading. • Tests are destructive and cannot be repeated.

Samples must sense the full amount of applied confining stress. The use of metal sleeves to jacket weak or poorly consolidated test samples is not recommended due to potential load coupling issues.

Where UCS and triaxial tests are performed on adjacent samples, the UCS derived from extrapolation to zero confining stress should be higher than direct UCS measurements under zero confinement conditions. Lower measured UCS at unconfined conditions can be a consequence of: the generally inelastic behaviour of the material; and sample surface irregularities and microcracks which do not impact on confined core strength measurements to the same degree.

A large difference between peak and residual strength is consistent with a strong to very strong rock (with high peak strength) that suffers brittle failure. This results in a significant loss of load-bearing capacity post-failure. In weaker rocks, the difference between peak and residual strength is less pronounced and, at higher stresses, can exhibit plastic behaviour.

The peak strength of a sample taken parallel to any laminations may differ to that if loading is perpendicular to the laminations. The strength of a dry sample will exceed that when saturated. Water-saturated samples are often weaker than oil-saturated samples.

Heterogeneity between adjacent SST samples may yield a wide scatter and inconsistencies in data. MST tests require fewer samples than a suite of SST triaxial tests and may overcome the problem of heterogeneity between adjacent samples. However, sample yield at lower (non-failure) MST confining stresses may affect final peak strength (at final stress station). The elastic properties, E and υ, cannot be meaningfully determined other than for the first loading stages, since the sample will have yielded.

As for UCS samples, and as discussed above, the measurement of Poisson's ratio is inherently more error prone and uncertain than the measurement of the Young's modulus.

Most post-yield plastic strain is normally well defined. However, some samples show an abrupt failure point where significant energy appears to be released at the moment of failure, and with significantly lower plastic strain between peak and yield. An example is shown in Fig. 12.26. This could be due to issues associated with the servo-control system or the feedback loop between the test gauges and the confining pump.

Triaxial tests on very weak sands can introduce problems. For example, Fig. 12.27 plots stress versus strain relationships for a weak sand that apparently indicates elastoplastic, ductile behaviour with no definitive peak strength and strain hardening post-yield. The radial stress was 1200 psi. This apparent elastoplastic behaviour may not necessarily represent the intact rock behaviour and may be a consequence of compactive failure prior to deviatoric stress loading in triaxial testing. In other words, the rock has failed in compaction while loading the sample isostatically to the desired radial stress. In some examples, the apparent inability to attain failure stress could be a result of

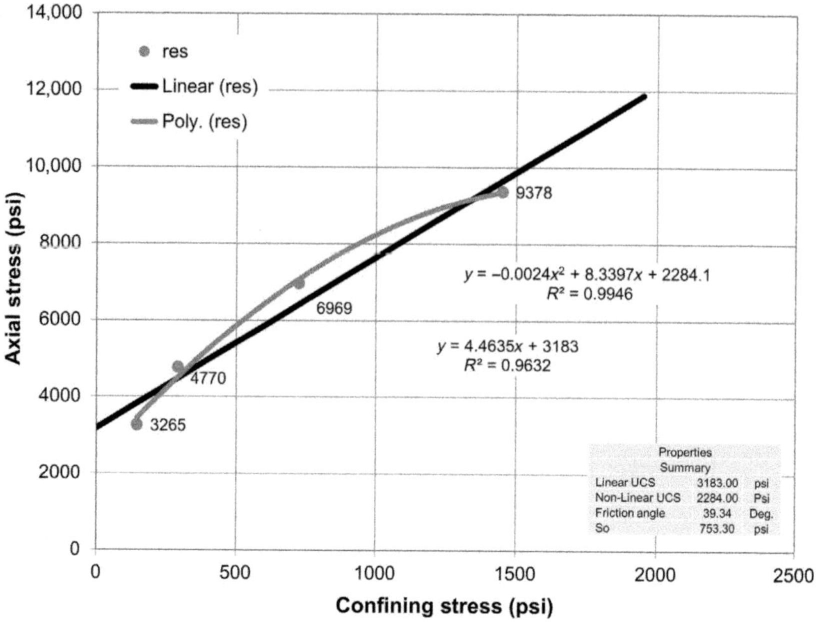

FIGURE 12.25 Example of non-linear principal stress plot.

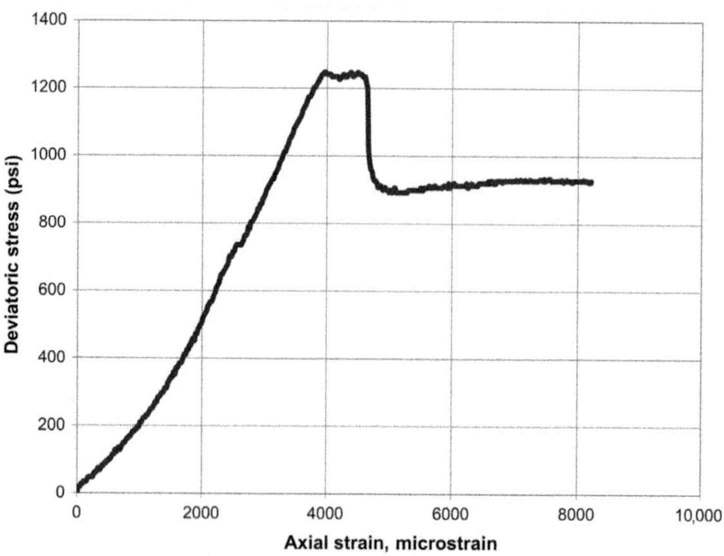

FIGURE 12.26 Example of apparently abrupt failure.

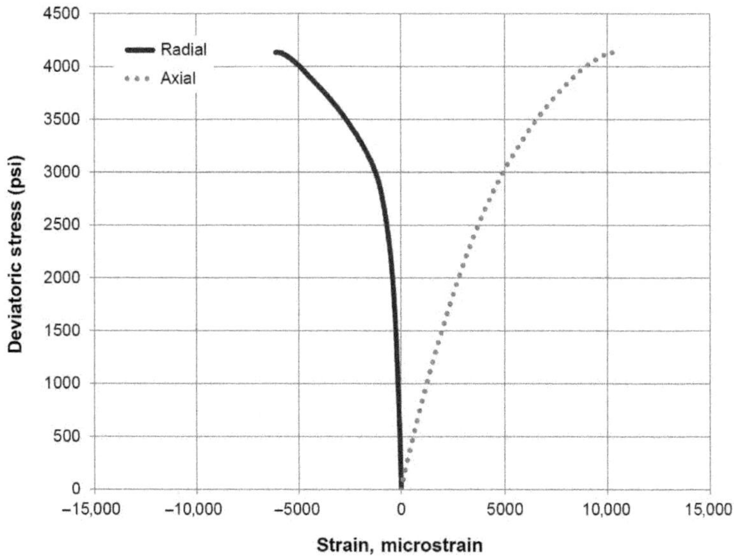

FIGURE 12.27 Example of elastoplastic ductile behaviour (radial stress = 1200 psi).

constraints in the strain measurement system which means that the test may have been terminated before peak strength was attained.

The Mohr–Coulomb plot for the set of four SST samples in this weak sand (Fig. 12.28) shows that the apparent peak stress does not increase significantly with radial stress on the Mohr circle plots and, as a result, the Mohr–Coulomb analysis predicts a very low friction angle (4.7°). This is untenable as even completely unconsolidated sands have friction angles typically in excess of 30°. This can be caused by making measurements on samples which have already suffered compactive failure on isostatic loading, or as a result of pore pressure failing to dissipate during loading, caused either by a blockage in the pore pressure outlet in the cell, or if the loading rate is too fast resulting in undrained conditions. Tests on permeable rocks are usually run on vertical (bedding plane perpendicular) samples at axial strain rates of $\sim 10^{-5}$/s. This is normally sufficient to allow the pore pressure to drain. However, if the sample is low permeability or has a permeability barrier across the sample (e.g. a vertical plug cut in a thinly laminated sand), or the pressure outlet is blocked by debris, then the pore pressure may not drain at higher loading rates. Commercial labs may not measure pore pressure during loading as the tests are carried out at ambient pressure. Under undrained conditions, some of the load is carried by increased pore liquid pressure. This means that the stress measurements represent effective stress but are mistaken for total stress, and the overall effect on Mohr–Coulomb analysis is to increase the apparent cohesive strength but significantly reduce the derived friction angle.

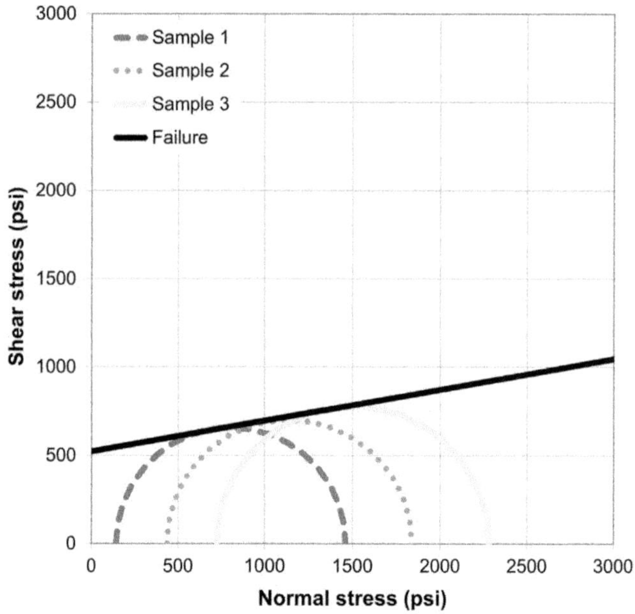

FIGURE 12.28 Mohr–Coulomb plot example for weak, low-permeability sand exhibiting a low friction angle.

12.5 TRIAXIAL TESTING OF SHALES

12.5.1 Purpose and Sample Requirements

The primary purpose of triaxial tests on shale samples is to determine failure parameters (e.g. Mohr–Coulomb cohesive strength and friction angle) and elastic moduli under representative reservoir loading/stress conditions for cap-rock integrity and wellbore stability calculations. Sample requirements are the same as for triaxial testing on permeable samples (sandstones and carbonates), but the low permeability and sensitivity to contacting fluids make shales difficult to handle correctly. As discussed in Chapter 2, special precautions must be taken to sample and preserve shale core, and special conditions must be applied during handling, preparation and testing.

12.5.2 Sample Preparation

Since shales are not normally the primary target, shale core samples are usually scarce and often ill preserved. In addition, shales have characteristic features which make them difficult to handle and test. Unloading from pressure and temperature can cause expansion, creation of microcracks, disking and reduced saturation.

Correct preservation and handling of shale cores are essential. These are discussed in detail in Chapter 2. Standard 'seal peel' wrapping techniques are adequate for at least a few months, but preserved samples should be taken immediately, and ideally at wellsite. Shale cores should never be frozen due to potential damage from micropore water expansion. CT scanning of preserved shale samples is essential for plug sample positioning and orientation. Once exposed, the core section should be characterised visually to examine crack patterns and how they might affect mechanical properties. UCS tests, in particular, can be affected by microcracks and fractures so quasi-UCS tests should be run with a small confining stress.

Plugs should be cut using mineral oil or air to avoid unwanted physiochemical reactions with water-based fluids. As shales can exhibit shear strength anisotropy with respect to bedding, plug orientation is important. A plug whose long axis is normal (perpendicular) to the bedding plane is defined as 90° orientation, whereas a plug with a long axis parallel to bedding is defined as 0° orientation (see Fig. 12.29).

Testing programmes may include shear strength tests for different sample orientations—from 0° to 90°, often at 30° increments—if the shale is mechanically anisotropic. An example of shear strength (in this case expressed as deviatoric strength, $\sigma_1 - \sigma_3$) as a function of plug orientation (inclination) with respect to bedding is shown in Fig. 12.30. For this example, both 0° and 90° plugs exhibit similar (maximum) shear strength, while 60° orientation samples show the minimum shear strength. Use of sonic velocity measurements made radially at equally spaced locations on the shale core section can help identify anisotropy and aid in plug selection and orientation. Usually, shale velocities are at a maximum propagating along the bedding planes, as indicated in Fig. 12.31.

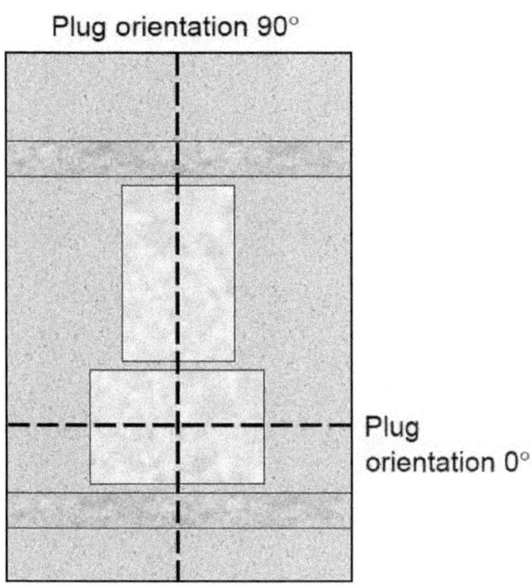

FIGURE 12.29 Plug orientation convention.

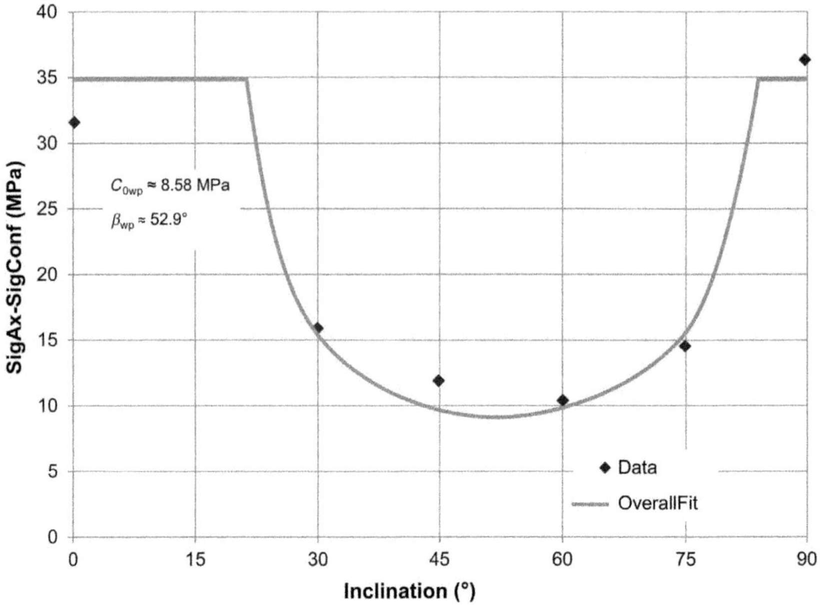

FIGURE 12.30 Deviatoric $(\sigma_1 - \sigma_3)$ shear strength as function of plug orientation (inclination) to bedding. 0° is parallel to bedding and 90° is normal to bedding.

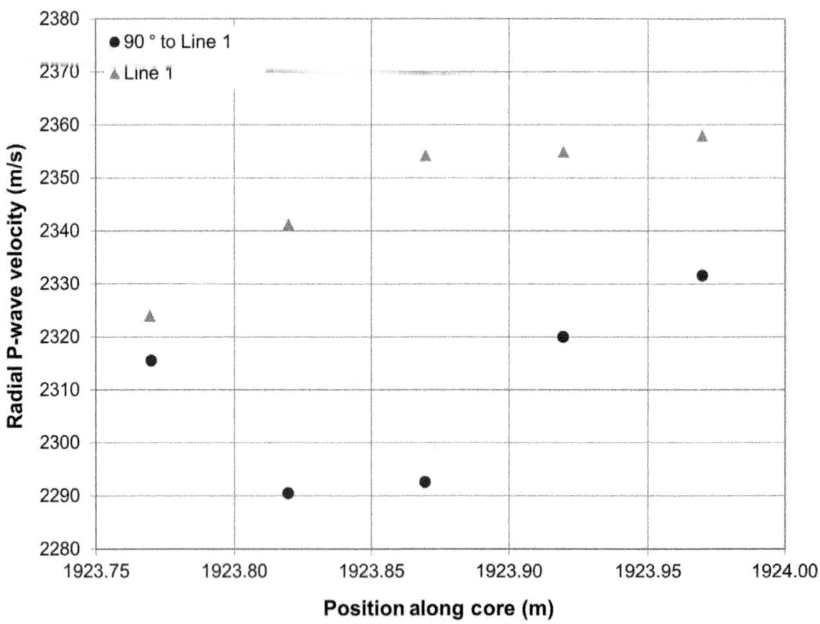

FIGURE 12.31 P-wave velocities measured radially along shale core section.

In addition, thin section, SEM and XRD tests are recommended to characterise the mineralogy, CEC tests to assess sensitivity and mercury injection tests to characterise the pore network.

Resaturating the shale samples to guarantee complete pore space saturation is essential to ensure consolidation prior to undrained testing. This can be achieved under controlled humidity conditions. However, this can take a considerable time to achieve and, as the samples may be initially dehydrated, this may induce changes in the shale structure which may also be exacerbated by incompatible saturants. Provided that the core has been well preserved, satisfactory saturation can be obtained from hydrostatic loading of the samples during the initial stage of the tests. The pore space saturant (typically a compatible saline brine) should be tested against shale offcuts to ensure compatibility.

12.5.3 Test Equipment

The test equipment used for shale strength testing is the same as that used for triaxial tests on permeable formations but does require to be modified for tests on low-permeability shales. To run the entire test under drained conditions may be extremely time consuming. Consequently, the triaxial loading stage in most shale tests is run in undrained mode. To reduce the test time, both axial and radial drainage of the sample should be included in the test set-up.

Figure 12.32 provides an example of the sample set-up and instrumentation modifications to a Hoek cell for shale testing. The sample is mounted on sintered plates and, in addition to measurement of external load, pressure and deformations, the pore pressure is measured on both ends of the sample and acoustic velocities are measured both radially and axially (optional).

12.5.4 Test Procedures

12.5.4.1 Quasi-UCS Tests

UCS tests can provide a useful indication of shale strength—particularly as a function of orientation with respect to bedding. However, due to potential surface artefacts, the tests should be run at nominal confining stress (say 0.2–0.5 MPa).

It is essential that these tests are run with axial deformation rate control at a low rate (10^{-6}/s for silty shales and 10^{-7}/s for muddy shales) to reduce potential rate effects on the results.

A stress hold period at about 1 MPa (145 psi) axial stress is generally included to check for possible consolidation effects, before unloading to 0.6 MPa and subsequently reloading to failure.

Quasi-UCS strength is the axial load at failure divided by the plug cross-sectional area.

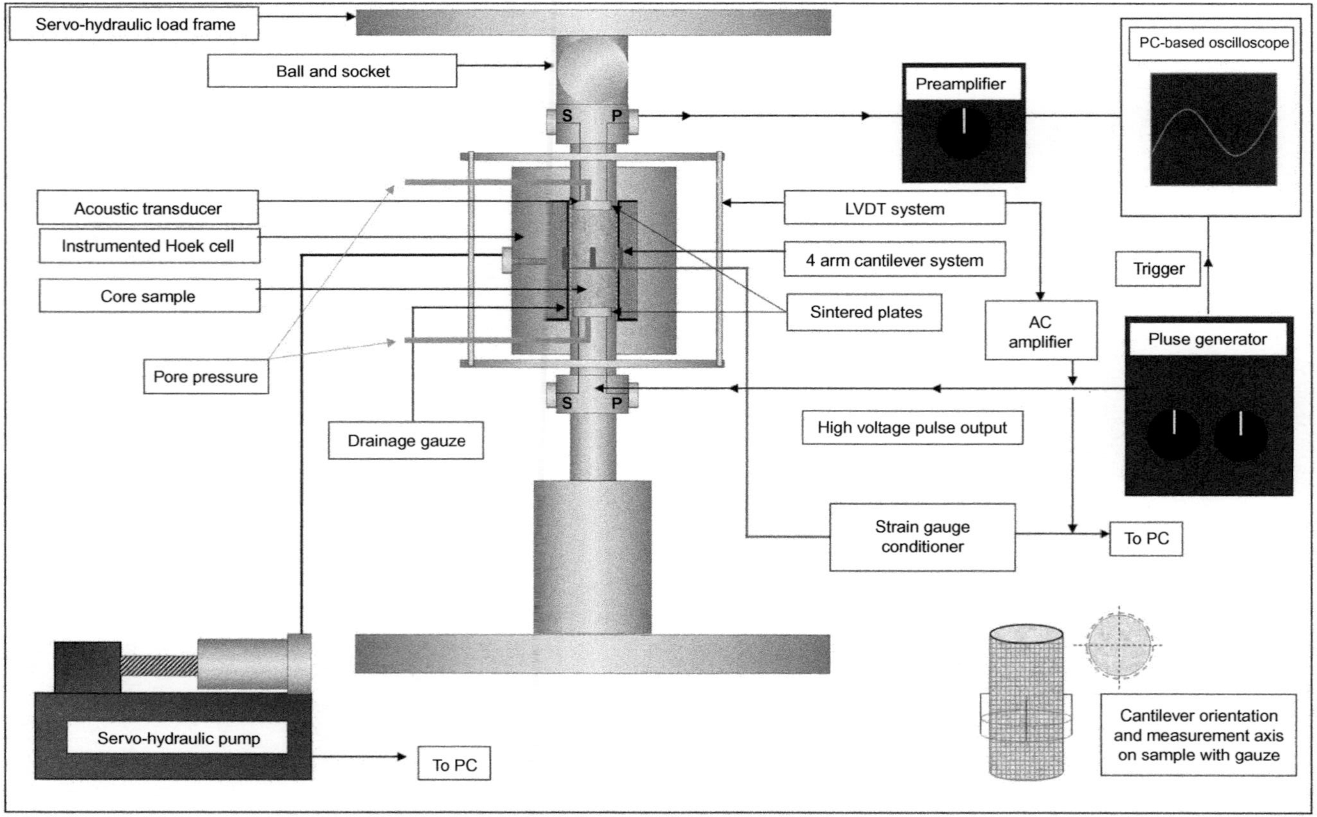

FIGURE 12.32 Example sample instrumentation for shale triaxial tests. *Courtesy of FracTech Laboratories.*

12.5.4.2 *Triaxial Tests*

Drained tests are not normally performed due to the time required to drain pore pressure fully on loading on such low-permeability material. Most tests are carried out using the consolidated-undrained method which normally consists of four stages:

1. Isostatic loading to low pressure (e.g. 145 psi) to introduce the contacting pore fluid (e.g. saline brine solution or mineral oil).
2. Isostatic loading with concurrent pore pressure ramp (maintaining constant net stress) to establish the main drained pore pressure level (e.g. 300 psi).
3. Consolidation stage: ramping of confining stress to target radial stress level, thereafter allowing for consolidation with constant (drained) pore pressure.
4. Undrained triaxial compression (deviatoric loading) at constant axial deformation rate: typically 10^{-6}–10^{-7}/s.

Figure 12.33 provides an illustration of the pressures and stresses at each stage of a triaxial experiment and the times typically required for stabilisation and loading. During Stage 1 (22 h), a small isostatic stress (1 MPa) was applied while introducing the pore fluid. Then the confining stress and pore pressure were increased in parallel to establish the initial pore pressure (2 MPa) before increasing the confining stress to the selected radial stress level (7 MPa) at about 33 h. After drained consolidation, the undrained triaxial phase started at about 44 h.

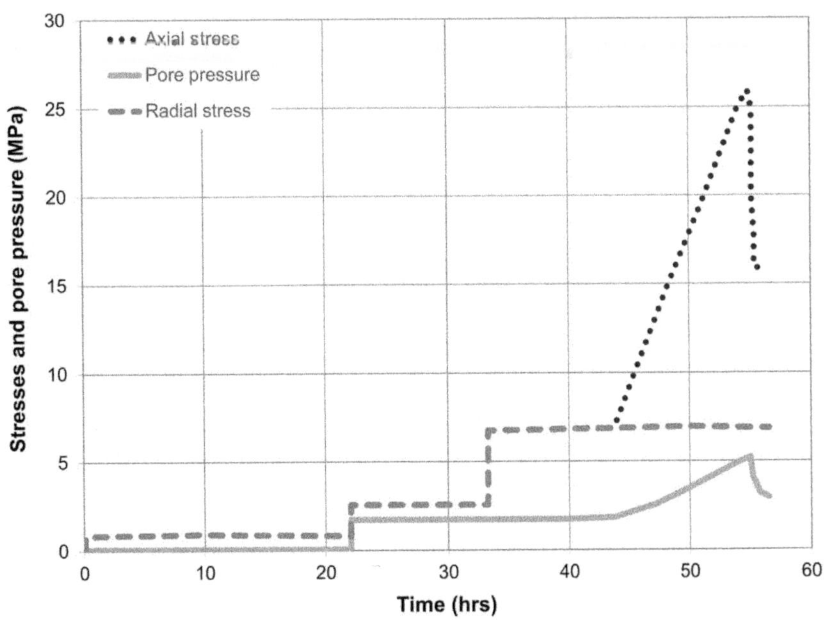

FIGURE 12.33 Illustration of stresses and pressure during consolidated-undrained triaxial test.

12.5.5 Data Utilisation

12.5.5.1 Consolidation Stage

To establish the degree of consolidation during the consolidation phase and the time required to reach 100% consolidation, the volumetric deformation is plotted as a function of the square root of time. An example is shown in Fig. 12.34. A straight line is fitted to the first part of the curve and has, by definition, a slope of 1.0. A second line is drawn with a slope lower than the first line by a factor of 1.15. The point where this second line crosses the consolidation curve is assumed to be 90% consolidation. From this, the volumetric deformation corresponding to 100% consolidation, and the time to achieve 100% consolidation, can be obtained.

12.5.5.2 Triaxial Parameters

The analytical techniques used to determine elastic moduli and peak Mohr–Coulomb failure parameters are similar to those for permeable samples. For example, Fig. 12.35 shows the principal stress plot (effective failure axial stress versus effective radial stress) for sets of two SST tests representing different sample orientations. Regression analysis is used to determine best-fit values for UCS and triaxial stress factor which can be used to define the Mohr–Coulomb parameters.

For shales, the p'–q analysis technique (based on soil mechanics principles) can be useful in assessing the degree of consolidation of the shales and potentially maximum stress exposure. In this analysis, the differential total stress, q, is:

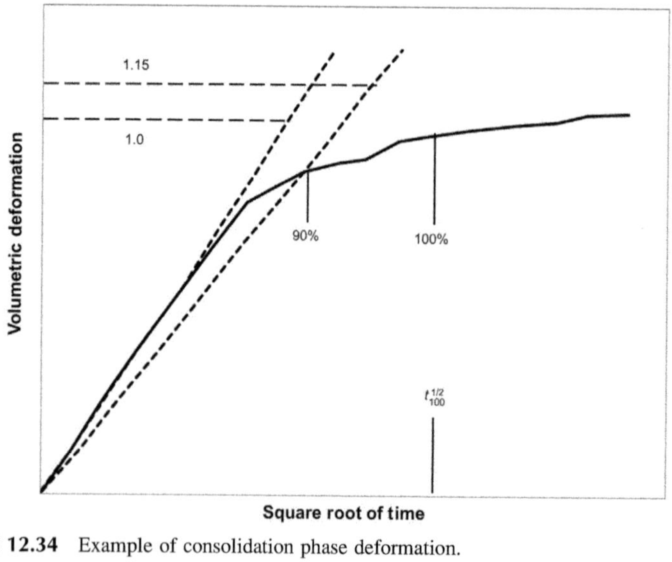

FIGURE 12.34 Example of consolidation phase deformation.

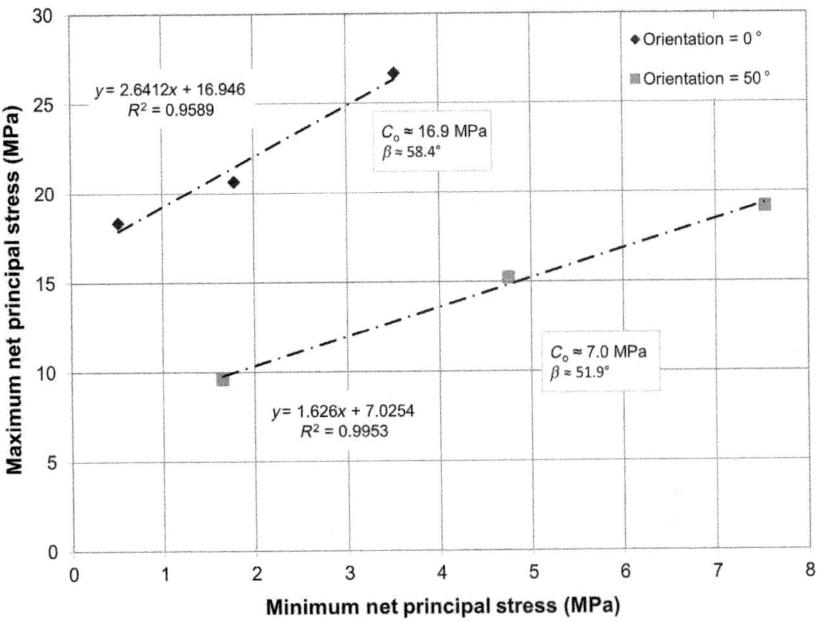

FIGURE 12.35 Principal stress plots for different sample orientations.

$$q = \sigma_1 - \sigma_3$$

and p' is the mean effective stress

$$p' = \frac{(\sigma_1 + 2\sigma_3)}{3} - p_p$$

Figure 12.36 provides the p'–q plot for the same (0° oriented) samples as Fig. 12.35. All three tests exhibit a behaviour which, in soil mechanics terms, is typical for over-consolidated material: the mean effective stress increases with loading. The peak strength values fall on a straight line (Hvorslev surface—shown dotted) in p'–q space. This can be translated into a Mohr–Coulomb failure criterion. However, if the loading curve becomes vertical, and starts to change direction this indicates that the normal consolidation level has been attained, that is: the maximum stress level that the shale has ever been exposed to.

However, shales differ from soils (clays) as they have some degree of cementation. This affects the response and hence can limit the application of these soil mechanics principles.

12.5.6 Data Reporting Requirements

The data that need to be reported by the test laboratory to ensure adequate quality control of the triaxial tests and test results are listed in Table 12.6.

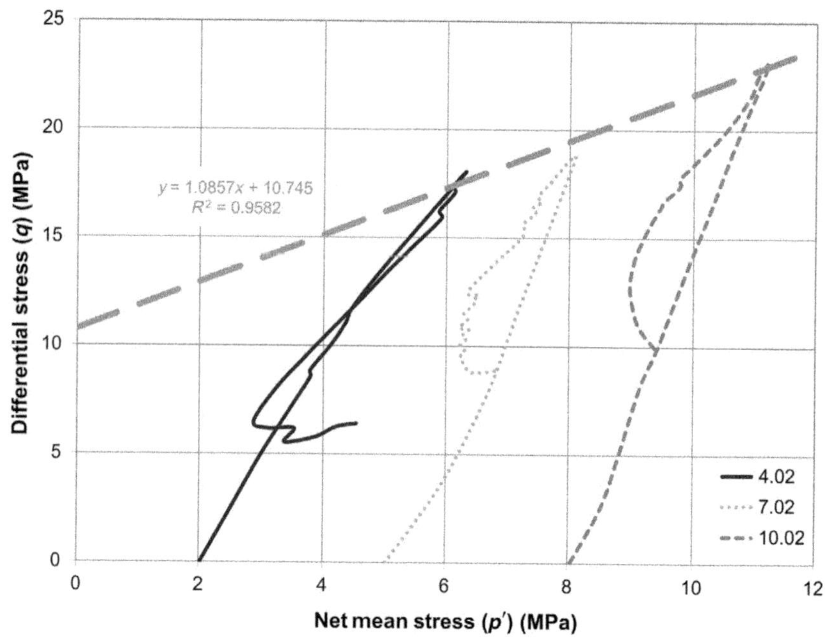

FIGURE 12.36 p'–q plot for 0° oriented samples.

12.5.7 Advantages and Drawbacks/Issues

Summaries of the advantages and drawbacks/issues associated with both triaxial tests on shales are provided in Table 12.7.

12.5.8 Shale Triaxial Test Quality Control Issues, Checks and Diagnostics

The pore pressure response during testing of shales may indicate whether the sample is fully saturated or not (essential for the maintenance of undrained conditions). Full details of the method are provided by Horsrud et al. (1998).

If shale porosity is required (e.g. for strength correlations), it is better determined by evaporating free water at 105 °C until constant sample weight, i.e.:

$$\phi_{\text{shale}} = \frac{\frac{(W_{\text{sat}} - W_{\text{dry}})}{\rho_{\text{w}}}}{\text{Vb}}$$

where W_{sat} and W_{dry} are saturated and dry weights, ρ_{w} is the water density and Vb is the sample bulk volume. Drying for conventional helium measurements can cause collapse of smectite.

Shale heterogeneity and potential anisotropy caused by bedding can result in significant differences in mechanical properties. Maximum strength is

TABLE 12.6 Data Reporting Requirements for Triaxial Tests on Shales

Data	Comments
Type of test	Drained or undrained/ consolidated
Brief description of procedures and experimental apparatus	Including how plugs were cut and prepared
Plug depth(s) and orientation	Angle to bedding planes. Parallel to bedding is defined as 0°
Plug length and diameter	Establish *L*:*D* ratio
Pore saturant	Oil or water/brine
Sample bulk density and pore saturant density	
Pore pressure, p_p	As a function of time during entire loading sequence
Total and effective confining stress, σ_3	As a function of time during entire loading sequence
Axial stress, σ_1	As a function of time during entire loading sequence
Data and plot of volumetric strain versus confining stress and versus time during isostatic loading and consolidation phase	Establish strain stability
Axial loading rate during deviatoric loading	For each sample
Peak (failure) stress	For each sample in SST
Any corrections that have been applied should the test sample be shorter than the ISRM standard	
Data and plots of axial stress versus axial strain, radial strain and volumetric strain during deviatoric loading. Data must clearly define post-failure behaviour	For each sample in SST
Calculated values of Poisson's ratio and Young's modulus over defined stress range (e.g. 40–60% peak stress)	Method used to calculate moduli
Mohr–Coulomb plot (normal stress vs. shear stress) with peak failure stress Mohr circle	Mohr circle for each sample in SST and each stress station in MST
Principal stress axis plot (peak axial stress vs. radial stress)	Analysis for each sample in SST and each stress station in MST
Mohr–Coulomb failure parameters for each plug set (including estimated UCS strength at zero confinement or at quasi-UCS confinement)	Cohesive strength, friction angle and estimated UCS

Continued

TABLE 12.6 Data Reporting Requirements for Triaxial Tests on Shales—Cont'd

Data	Comments
Deviatoric (differential) total stress versus net mean stress plot for each samples	p'–q plot
Description of failure mode (e.g. shear, planar shear, conjugate shear, etc.)	For each sample
Photographs of test plugs before and after testing. These should include a scale marker (e.g. ruler). Post-test plug photographs of triaxial test samples should orient the core vertically and should clearly show the failure plane	To ensure shear failure has occurred

TABLE 12.7 Advantages and Drawbacks/Issues Associated with Triaxial Tests on Shales

Advantages	Drawbacks and Issues
• Data provides Mohr–Coulomb failure parameters under representative loading conditions for both analytical failure and numerical deformation models. • Elastic moduli provide estimates of formation and deformation under loading. • Shale strength anisotropy with respect to bedding. • Consolidation analysis may be useful in establishing maximum stress that shales were exposed to.	• Sample availability is limited (few cores are taken in shales). • Samples must be preserved as soon as possible after coring (ideally at wellsite) to avoid any possibility of drying out and de lamination. • Incompatible drilling, plugging and saturating fluids may alter shale structure and cause mechanical damage. • Core must be carefully evaluated for presence of bedding anisotropy and microcracks. • Unconfined UCS tests are not recommended due to potential sample surface artefacts. If required, tests should be conducted at nominal confinement (quasi-UCS). • A higher level of test equipment specification and resources (compared to sands/carbonates) is required. • Mohr–Coulomb and elastic moduli analyses assume linear elastic behaviour over test stress range. S_0 and θ are sensitive to the stress pairs used to derive the parameters. • The data obtained relate specifically to the confinement applied. Several tests on similar samples need to be conducted if peak strength parameters (i.e. cohesion and friction angle) are required. • Low confining stresses may be required on very weak samples to prevent compactive failure during isostatic loading. • Tests are destructive and cannot be repeated.

usually developed at 0° to bedding and minimum strength at around 60° to bedding. Quasi-UCS tests and radial acoustic velocity tests may assist in assessing heterogeneity and characterise strength anisotropy. Application of strength anisotropy in wellbore stability calculations is limited by classical analytical failure models. Solutions that can accommodate anisotropy include the Willson et al. (2007) numerical model.

Some commercial rock mechanics labs may not access to the experience, equipment and resources to carry out triaxial tests on shales to the standards required.

12.6 THICK-WALL CYLINDER TESTS

12.6.1 Purpose and Sample Requirements

In the thick-wall (or hollow) cylinder tests (Antheunis et al., 1976; Veeken et al., 1991), a hollow cylinder of core is externally loaded by increasing the confining stress isostatically. External loading creates tangential stresses which act around the wall of the internal wall. When the tangential stress exceeds the effective strength of the rock, the internal hole starts to fail. An example of a collapsed thick-wall cylinder is shown in Fig. 12.37. Rocks which exhibit brittle failure under triaxial loading tend to fail with slit-like breakouts, whereas more ductile rocks will tend to exhibit wider and more stable breakouts. Rocks which tend to exhibit compactive failure tend to show a fairly uniform deformation so that the failure from the inner hole may not reach the outside of the sample.

During loading, the point at which the internal hole begins to spall material (yield stress) and then the stress at which the entire sample collapses (collapse stress) are determined. The geometry simulates the loading of a wellbore or perforation under downhole conditions and the stabilising effects of non-linear and/or plastic rock deformations are included in the measured collapse pressure. The collapse pressure is used to estimate the stress conditions in a well which will cause perforation or well collapse. Both static (standard) and dynamic (advanced) tests, with radial flow of fluid into the internal hole, are possible.

Standard TWC tests utilise 1″ or 1.5″ diameter plugs with a nominal length:outer diameter ratio of 2:1 and an outside diameter (OD) to inside hole diameter (ID) of 3:1, as indicated in Fig. 12.38. Ideally, the core section is plugged using a 0.33″ or 0.5″ diameter bit then over-cored using a 1.0″ or 1.5″ plugging bit to produce a hollow cylinder (Fig. 12.39). Although it is common to drill a central hole in the sample, this requires precise care. Larger diameter samples, often with higher OD:ID ratios, can also be tested for specific applications, such as advanced TWC tests.

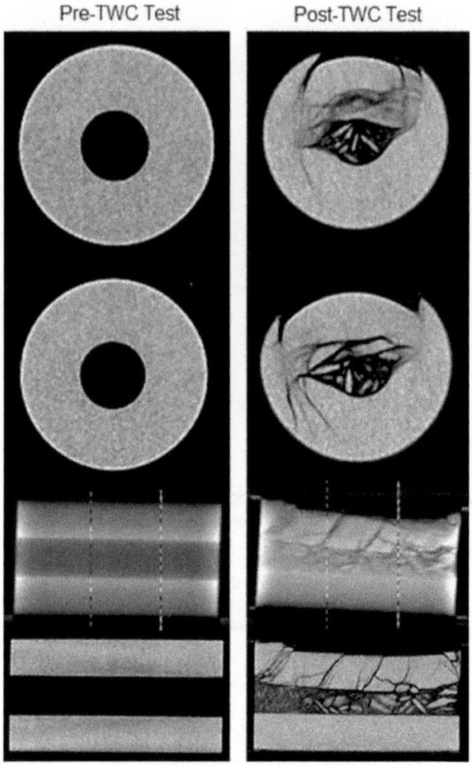

FIGURE 12.37 Example of pre- and post-test thick-wall cylinders. *Image courtesy of Schlumberger. All rights reserved. Copyright © 2015 Schlumberger.*

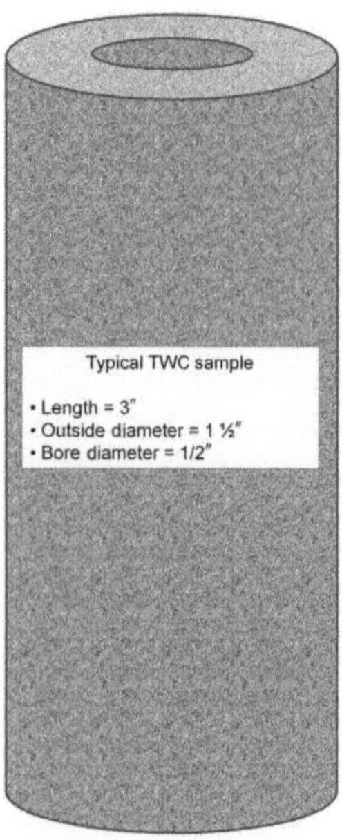

FIGURE 12.38 Standard TWC sample dimensions.

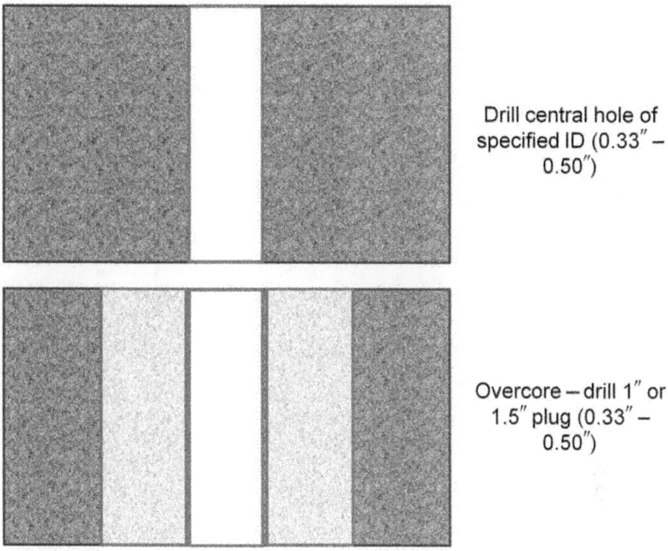

FIGURE 12.39 Overcoring to produce standard TWC sample.

12.6.2 Test Equipment

12.6.2.1 Standard TWC

Standard TWC collapse studies are conducted using a closed-loop servo-hydraulic, pressure intensifier in conjunction with a Hoek-type triaxial cell (Fig. 12.40). These are used to apply lateral stress to the TWC samples, with axial load being applied independently, via steel loading platens, using a closed-loop servo-controlled loading frame. The cell is fitted with radial strain measurement devices to determine external sample (volumetric) strain on loading.

The test equipment set-up should permit monitoring of internal hole dimensions during loading. These can include:

- endoscopes (visual deformation observation)
- internal calipers

but more commonly:

- monitoring expulsion volumes. This requires the internal hole to be full of liquid—normally mineral oil. At an isostatic confining stress, the stress concentration around the hole is sufficient to cause significant rock failure, and the sample then collapses radially inward. When the sample collapses, the volume in the pressure vessel suddenly increases. If the confining stress is increased at a constant rate, then the collapse pressure would be easily missed. However, by increasing at a constant fluid displacement rate, the collapse pressure is clearly evident as a sudden drop in the

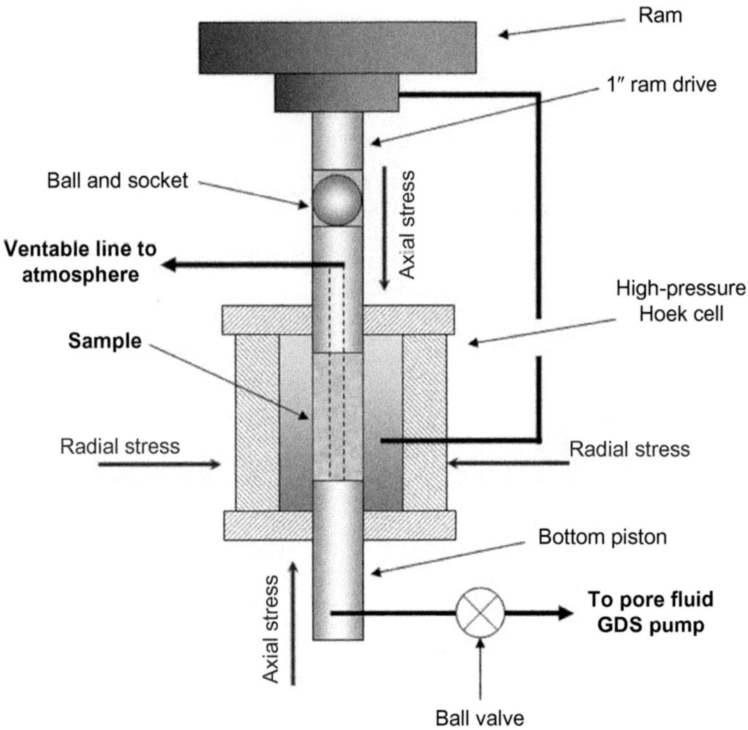

FIGURE 12.40 Hoek cell set-up for TWC tests. *Courtesy of FracTech Laboratories.*

confining stress. Fluid expelled measurements often provides an earlier indication of failure relative to external strain measurements. That is, breakouts can take place on the borehole wall without transmitting the deformation to the outside boundary.

- monitoring the piston stroke in the pressure intensifier applying the lateral stress, with hydraulic pressure in the triaxial cell being measured using a calibrated electronic pressure transducer. Axial loads applied to the specimens are measured using a calibrated electronic load cell. Output from both these measuring devices and the piston stroke in the pressure intensifier are monitored and recorded using a dedicated microprocessor system employing real-time displays, from which the moment of TWC collapse can be identified.

12.6.2.2 Advanced TWC

In advanced TWC testing, the cell design can allow for fluid (water, oil or gas) to be flowed radially through the sample into the internal hole (Papamichos et al., 2001; Ray et al., 2014). The test cell shown in Fig. 12.41 accommodates a 4″ OD core sample with a central hole. Oil, gas or water can be flowed

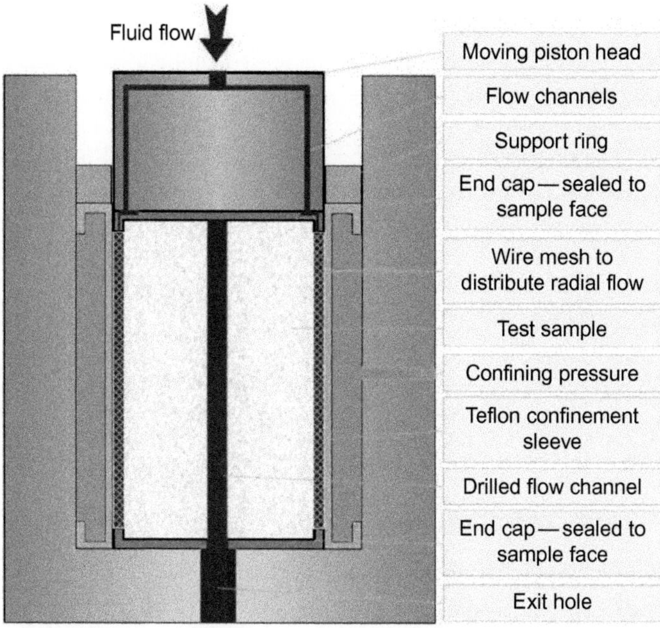

FIGURE 12.41 Advanced TWC test cell. *From Ray et al. (2014).*

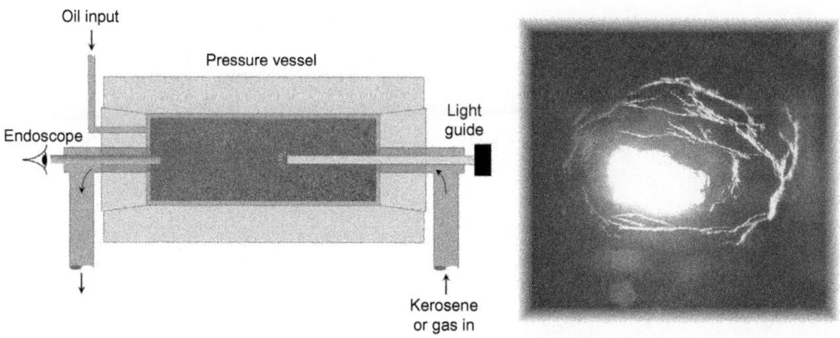

FIGURE 12.42 Advanced TWC set-up with endoscope (left). Endoscope highlights borehole deformation (right - *Courtesy of FracTech Laboratories*).

radially from the outside of the sample into the internal hole under controlled drawdown conditions. An endoscope and light source (Fig. 12.42) are often used to visualise the deformation of the internal borehole as a result of the tangential stresses acting on the wall of the hole. As the sample is loaded, solids spalling from the hole during loading can also be collected and weighed. In this way, solids production can be correlated with stresses acting on the internal hole, fluid type and drawdown, and these data can be used in sand failure prediction and sand production models for cased and perforated and open-hole wells.

12.6.3 Test Procedures (Standard TWC)

The radial stress and the axial stress are increased simultaneously and equally, so ensuring isostatic loading. Loading is continued at constant rate until complete collapse of the test specimen occurs, or the capacity of the test machine is reached. The ramp rate is no greater than 150 psi (1 MPa) per minute and, for volumetric measurements, the borehole is kept full of water or oil throughout the test. Pore pressure is kept constant at nominal pressures (e.g. 50 psi) by use of a servo-controlled pump to ensure drained behaviour in the samples. In advanced TWC tests oil, water or gas can be flowed radially through the sample into the internal hole under controlled drawdown during loading and any solids production is monitored.

Sample failure is determined by the measured internal volumetric strain (if calipers are used), external volumetric strain and from inner hole expulsion volumes.

12.6.4 Data Utilisation

The TWC strengths relate to:

- the yield stress that initiates internal wall failure (TWC Internal or yield stress), as defined by an increase in fluid volume expelled during constant loading; and
- the pressure that causes external wall failure (TWC External or collapse stress) in which the internal hole failure surface reaches the outside of the sample and the sample totally collapses.

An example of the volumetric displacement curves indicating both yield (initial) failure and external wall (catastrophic) failure stresses is shown in Fig. 12.43.

The internal wall (yield) failure stresses normally correspond to the onset of transient sand production from perforation tunnels. The external wall catastrophic failure (collapse strength) corresponds to the perforation failure condition that causes more continuous sand production. The difference between the stresses required to cause internal wall failure and total collapse depends on the material strength, deformation characteristics and the OD:ID ratio of the sample. Standard TWC tests are carried out on samples with a 3:1 OD:ID ratio. Figure 12.44 illustrates the effective increase in strength created by using higher OD:ID ratios. This plots the normalised TWC at collapse (i.e. TWC external for test sample divided by TWC external for a 3:1 sample) as a function of OD:ID ratio. For a 6:1 OD:ID ratio sample, the TWC collapse strength is divided by 1.17 to compare with 3:1 samples: that is the effective TWC strength is 1.17 times greater.

1.5″ diameter samples are preferred for TWC tests (Ewy et al., 2001) as the larger rock volume could make the sample more representative and as a

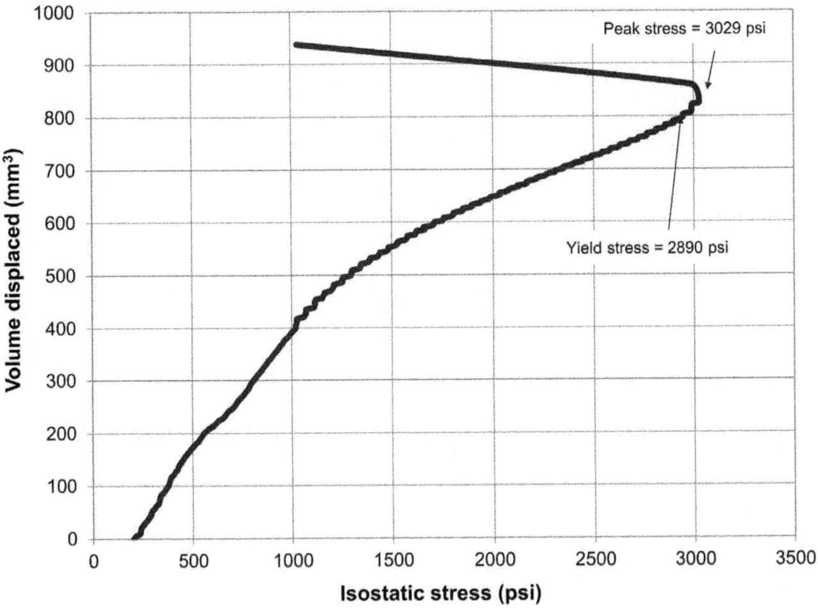

FIGURE 12.43 TWC internal and external wall failure example.

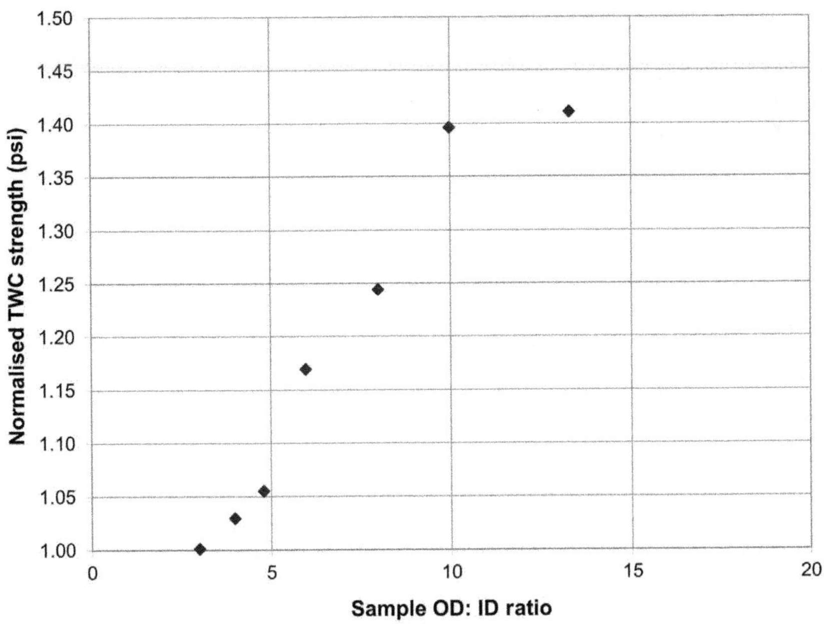

FIGURE 12.44 Effects of TWC sample OD:ID ratio.

slight offset in position of the central hole will not affect the result as much as for a smaller diameter plug. However, core condition and availability may dictate the use of 1″ samples. Even for a 'standard' 3:1 TWC sample, scale effects between 1″ and 1.5″ diameter samples must be considered when evaluating data. It is evident from work presented by van den Hoek et al. (2000) and Crook et al. (2003) that larger diameter holes are less stable than smaller perforations. For example, Fig. 12.45 plots external wall failure stress versus internal hole (perforation diameter) for TWC tests carried out on Berea and Castlegate outcrop sandstones. Clearly, a smaller hole (~8 mm for a 1″ diameter 3:1 OD:ID sample) is more stable than a larger hole (~13 mm for a 1.5″ diameter 3:1 OD:ID sample). At the wellbore scale, there is little difference in collapse strength between an 8½″ and a 12¼″ hole. The specific hole size–collapse strength relationship has to be evaluated empirically for a particular rock type.

12.6.5 Data Reporting Requirements

The data that need to be reported by the test laboratory to ensure adequate quality control of the TWC tests and test results are listed in Table 12.8.

12.6.6 Advantages and Drawbacks/Issues

Summaries of the advantages and drawbacks/issues associated with standard TWC tests are provided in Table 12.9.

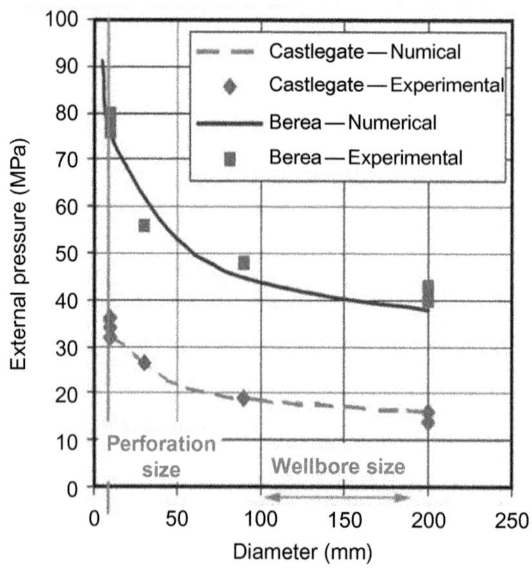

FIGURE 12.45 Hole size effect on TWC external failure stress. *After Crook et al. (2003).*

TABLE 12.8 Data Reporting Requirements for TWC Tests

Data	Comments
Type of test	Standard or advanced
Brief description of procedures and experimental apparatus	Including how plugs were cut and prepared (overcoring or drilling central hole)
Plug depth(s) and orientation	
Plug length and outside and inside hole diameter	Establish *L*:*D* and OD:ID ratios
Pore saturant	Oil or water/brine
Flowing fluid type, rate and drawdown	Radial flow in advanced TWC
Pore pressure, p_p	
Isostatic loading rate	MPa/min
Internal hole wall failure stress	Yield strength
External sample collapse stress	Collapse strength
Data and plot of internal volumetric strain versus isostatic loading	Advanced TWC
Data and plot of solids production versus isostatic loading and volumetric strain	Advanced TWC
Data and plots of volume displaced and external strain versus applied stress	Standard TWC
Any corrections that have been applied should the test sample be shorter than the ISRM standard	
Photographs of test plugs before and after testing. These should include a scale marker (e.g. ruler). Pre-test photographs must show both end faces and a longitudinal view	To ensure internal hole is centred
Three post-test plug photographs are required: two of each end face with the plug oriented horizontally so that any internal wall failure can be clearly seen (e.g. Fig. 12.46); and one with the plug oriented so that the external wall failure can be seen	To ensure shear failure has occurred

12.6.7 TWC Test Quality Control Issues, Checks and Diagnostics

Samples selected for TWC testing must be original and not subject to prior testing as pre-stressing and de-stressing can result in sample weakening.

TABLE 12.9 Advantages and Drawbacks/Issues with TWC Tests

Advantages	Drawbacks and Issues
• Relatively simple test to undertake though care has to be taken when preparing test samples to ensure internal hole is centred. • Direct simulation of perforation tunnel collapse. • Less influenced by plug artefacts and surface topology than UCS tests. • Method has been standardised to use small plugs. • Advanced TWC tests can provide important data on onset of solids production in simulated perforation tunnels, the effects of different fluids on collapse strengths, and can be used to calibrate analytical and numerical sand failure models.	• Sample preparation is not straightforward, and the test relies on a high standard of sample preparation (i.e. concentricity of bore in plug) to be meaningful. • The test is an over-simplification of reality, so is intended as an indicator of rock strength for use in first-look assessment of sanding potential. • Incorporation of TWC collapse strength into analysis and design is highly empirical, and normally needs a combination of other data to assess sand production risk. UCS and/or triaxial tests should also be carried out on adjacent samples. • Differences in scale (hole diameter and OD:ID ratio) must be considered in data evaluation and application. • Results of sanding analyses using TWC collapse strength need verification from numerical modelling. • Tests are destructive and cannot be repeated.

Samples must sense the full amount of applied confining stress. The use of metal sleeves to jacket weak or poorly consolidated test samples is not recommended due to load coupling issues.

For low-strength materials, external volumetric strain and internal hole fluid expulsion are in close agreement in terms of initiation of failure (solids production begins). Once failure begins, it propagates through the entire specimen immediately. This is attributed to plastic deformation around the internal hole. Due to its low strength, the material adjacent to the hole becomes plastic at an early stage of loading. As loading continues, failure (spalling of material into the inner hole or collapse of the inner hole) does not occur. Rather, the size of the plastic zone propagates outward to the external boundary. Once the entire sample becomes plastic, any deformation on the hole boundary is immediately transmitted to the outside of the sample, which means that both external strain and inner hole fluid expulsion measurements are in close agreement. Due to the simultaneous deformation occurring within the inner bore and the outside, the difference between the initiation of borehole failure and collapse strength (catastrophic failure) is not great. Conversely, for stronger and stiffer materials, the failure initiation pressure can be much lower than the final collapse pressure.

Water-saturated samples can be weaker than oil-saturated samples if they contain sensitive minerals.

Low-permeability materials (i.e. shales) may exhibit undrained or partially drained behaviour if tested at the recommended loading rates.

The internal hole must be concentric. If not, and it is eccentred as indicated in the plug photographs shown in Fig. 12.47, then the distance the plastic zone has to travel to the external surface of the sample after the onset of internal wall failure will be shorter at one end of the plug than the other, and the collapse strength will be lower than it should be for a concentric hole, especially if 1″ OD samples are tested.

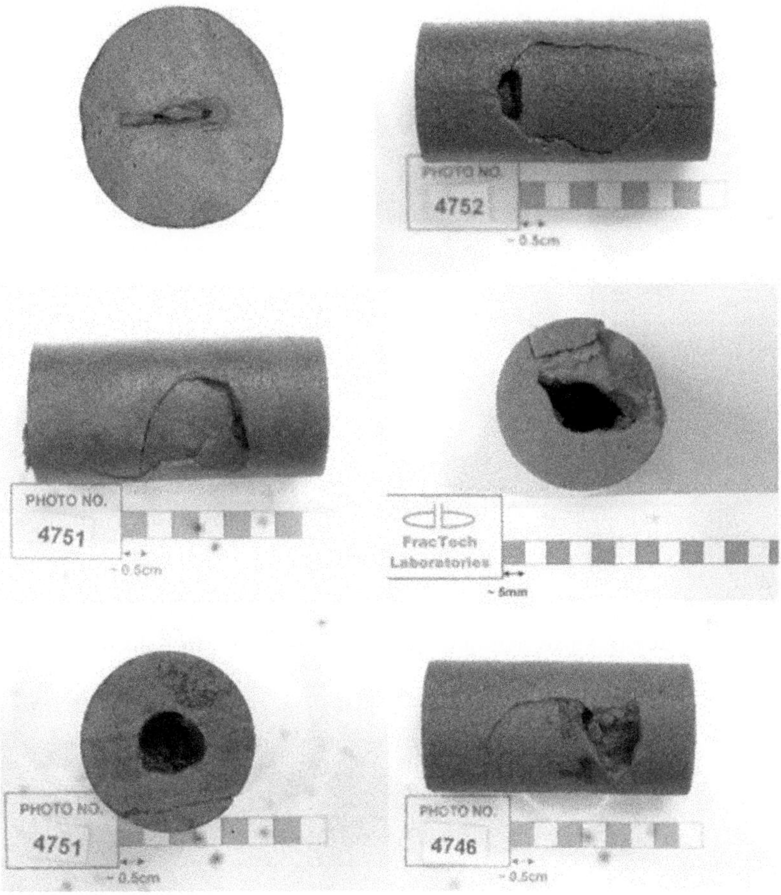

FIGURE 12.46 Examples of failed TWC samples. *Courtesy of FracTech Laboratories.*

FIGURE 12.47 Example of TWC samples with eccentred (offset) internal hole.

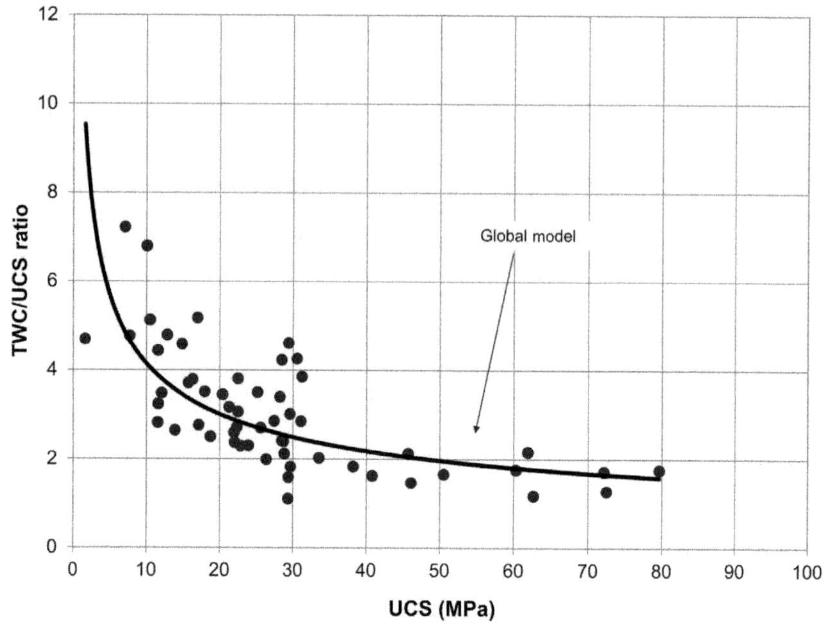

FIGURE 12.48 Generic TWC–UCS relationships.

Figure 12.48 shows data from a large database of TWC and UCS tests on discrete sample pairs. There is a clear relationship between the TWC/UCS ratio and UCS strength, although this can be a rock-type dependent. In stronger formations (>5000 psi UCS), the UCS/TWC ratio in this database appears constant at around 2. However, data acquired in very strong rock show TWC to UCS ratios approaching 1.5. This is related to the transmissibility of the applied radial stress through the sample which is much closer to 1 in a hard brittle rock. In weaker formations, the ratio increases with reducing strength and can often exceed 6 in intervals with UCS less than ~1000 psi. This is due to compaction effects which act to strengthen the sample in TWC tests but which are absent in UCS tests. Data from TWC/UCS pairs can be overlain on this plot and any significant deviations (uncertainties in UCS and/or TWC) are assessed and diagnosed.

12.7 TENSILE STRENGTH TESTS

12.7.1 Purpose and Sample Requirements

Tensile strength is a key parameter in borehole facture calculations. Two methods are normally used to measure it in the lab:

1. Direct test
2. Brazilian disk test

The Brazilian disk test provides an indirect estimate of tensile strength. The test can be conveniently performed on a core disk cut from a plug with a length (thickness) about ½ the diameter of the plug. The direct test uses a core plug about 1″ in diameter by around 2″ long. Normally, the test samples are not saturated but are tested dry.

12.7.2 Test Equipment and Procedures

12.7.2.1 Direct Test

Steel platens are bonded to each end of the sample using high-strength epoxy mortar. After the mortar has cured, the sample is loaded under tension (Fig. 12.49) using a hydraulic load frame with a calibrated load indicator. During the tests, load is applied to the steel platens, and hence the sample, via flexible couplings such that the sample experiences pure tension with no bending. The peak tensile load applied to the sample is recorded, from which the uniaxial tensile strength, σ_T, of the material is then calculated.

12.7.2.2 Brazilian Disk (Indirect) Test

A core disk is cut from a vertical plug sample to provide a target *L*:*D* ratio of 0.5. The disk is located between parallel steel platens incorporating soft (plywood) seating strips which act to spread the load along each edge and over a

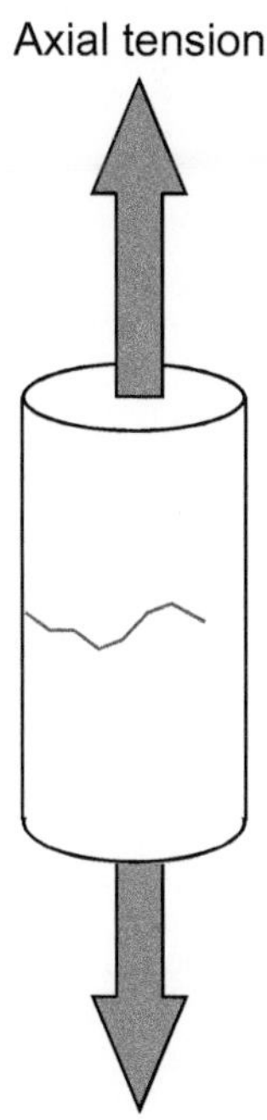

FIGURE 12.49 Direct tensile test.

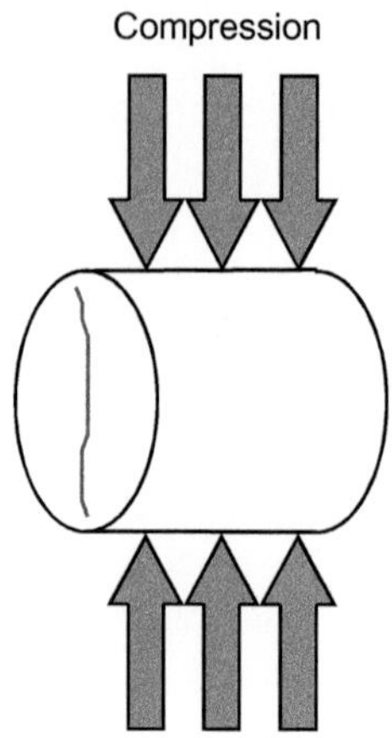

FIGURE 12.50 Brazilian disk test.

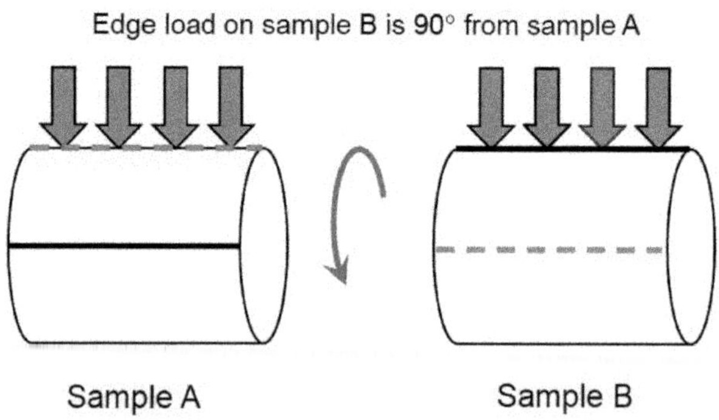

FIGURE 12.51 Brazilian disk test orientation.

small portion of the circumference (Fig. 12.50). Compressive loading is applied, using a closed-loop servo-controlled hydraulic loading frame incorporating a calibrated electronic load cell, at a nominal rate of 200 N/s.

The load at initial fracture is recorded using a microprocessor system.

In many cases, Brazilian disk tests are carried out on two paired samples (e.g. a 1″ by 1″ plug cut in half) to plug to assess tensile strength anisotropy. The B sample is loaded orthogonally to the A sample (Fig. 12.51).

12.7.3 Data Utilisation

The tensile strength, σ_T, is determined from the load at failure, L, divided by the diameter of the plug or disk, D, and the thickness of the plug or disk, t

$$\sigma_T = \frac{2L}{\pi Dt}$$

12.7.4 Data Reporting Requirements

The data that need to be reported by the test laboratory to ensure adequate quality control of the tensile strength tests and test results are listed in Table 12.10.

12.7.5 Advantages and Drawbacks/Issues

Summaries of the advantages and drawbacks/issues associated with tensile strength tests are provided in Table 12.11.

TABLE 12.10 Data Reporting Requirements for Tensile Strength Tests

Data	Comments
Type of test	Direct or Brazilian disk
Brief description of procedures and experimental apparatus	Including how plugs or disks were prepared
Plug depth(s) and orientation	
Plug/disk length and diameter	Establish *L*:*D* ratio
Pore saturant	If used
Loading rate	
Failure load	Tensile strength
Photographs of test plugs before and after testing. These should include a scale marker (e.g. ruler) Post-test plug photographs should clearly show the failure mechanism	To ensure tensile failure has occurred. An example is shown in Fig. 12.52

FIGURE 12.52 Example of tensile failure on plug end face.

TABLE 12.11 Advantages and Drawbacks/Issues with Tensile Strength Tests

Advantages	Drawbacks and Issues
• Direct measure of uniaxial tensile strength (direct test). • Requires little preparation and can be performed on full core (direct test). • Brazil test can be performed on offcuts from plug samples for other tests. • Suited to weak rocks (Brazil test). • Relatively low cost (Brazil test).	• Direct test is relatively involved and costly for the limited data obtained. • Brazil test is indirect measure of tensile strength but arguably more cost-effective and repeatable than Direct test. • Direct test is unsuitable for weak rocks, since the samples may easily break prematurely as they are being loaded into the test system. • Tests are destructive and cannot be repeated.

12.7.6 Tensile Strength Test Quality Control Issues, Checks and Diagnostics

Direct uniaxial tensile strength may differ to tensile strength data from the indirect Brazil test method. However, Brazil tensile strength is considered adequate for most rock engineering designs and analyses.

Normally, it is assumed that the tensile strength (TS) of a rock is around 1/10 to 1/12 of the UCS strength. That is the UCS/TS ratio is around 10–12. For many rocks, this supposition is not correct—the ratio is not constant but correlates with UCS strength. For example, Fig. 12.53 shows that the UCS/TS ratio increases with increasing UCS for relatively stiff and competent rocks. In general, the tensile strength for both sands and 'shales' ranges from 3% of UCS for high UCS (16,000 psi) to 8–10% for UCS ~6000 psi. This could also be a consequence of pervasive microfracturing in some samples but this might be expected to produce more variation in the data than is evident here.

12.8 ACOUSTIC VELOCITY (TRAVEL TIME) TESTS

12.8.1 Purpose and Sample Requirements

These tests are principally performed to determine compressional (P) and shear (S) wave sonic velocities (Vp and Vs) and compressional shear travel times (Δt_c and Δt_s) on formation core samples. These measurements can be used to calibrate Vp/Vs wave velocities from dipole sonic logs and to calculate the dynamic elastic Young's modulus (E_{dyn}) and Poisson's ratio (υ_{dyn}). The standard test method is described in ASTM (1976) and ISRM (1978).

Normally, the tests are performed on cylindrical plugs, with a nominal length:diameter ratio of 2:1. These may cleaned and dried samples which have been fully saturated in SFW and scheduled for porosity versus stress

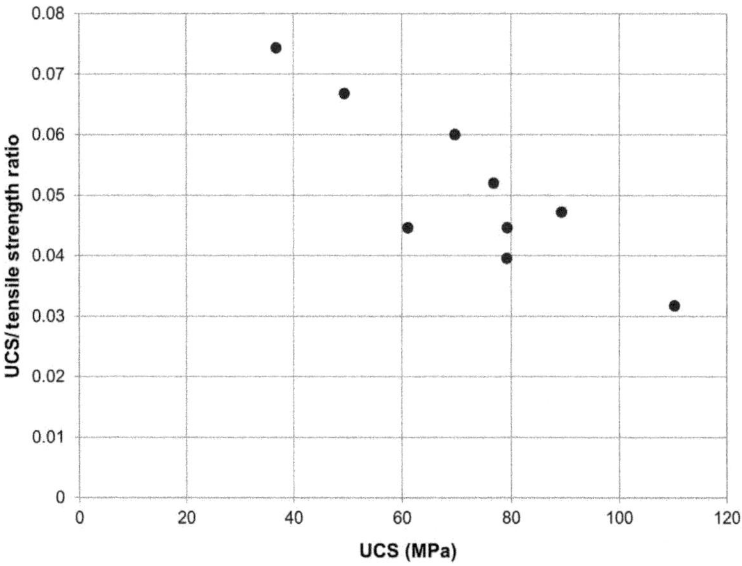

FIGURE 12.53 UCS and tensile strength relationships.

tests, or on triaxial test samples in which case both static (E_{sta} and υ_{sta}) elastic moduli, derived from stress–strain measurements, can be directly calibrated against the dynamic equivalents.

12.8.2 Sample Preparation

In general, the samples are prepared in a similar fashion to those for triaxial tests. However, if the dynamic elastic moduli are to be established in conjunction with porosity measurements, the samples are cleaned and dried, ensuring that the system selected causes minimal petrophysical damage to the core. The plugs are then saturated in synthetic formation water to ensure the pore volume is fully saturated. Since there can be a large change in the measured acoustic velocities at saturations between 95% and 100% of the pore volume, it is essential that the pore volume is 100% saturated.

12.8.3 Test Equipment

Typical ATT test equipment is illustrated by the schematic in Fig. 12.54.

The sample is located centrally in the sleeve of a Hoek cell. The samples are end-ground to improve the ultrasonic coupling between the transducer holders and the samples before ultrasound couplant is coated on to the faces of the ultrasonic loading platens. Ultrasonic coupling is further enhanced by use of thin lead foil between the samples and the loading platens. The platens are then inserted into the cell, and a syringe is used to apply a minimal confining stress to prevent the platens slipping. The cell and piston assembly are placed on top of a hydraulic ram within a reaction frame. Care needs to

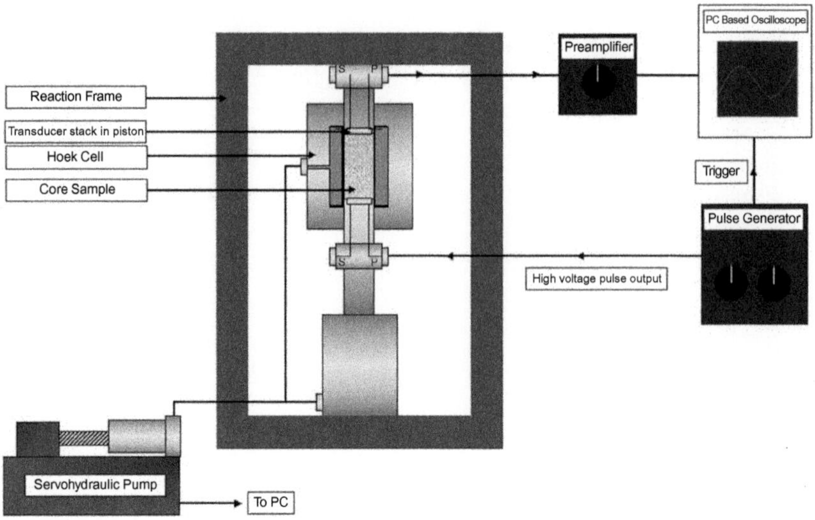

FIGURE 12.54 Example ATT test system schematic. *Courtesy of FracTech Laboratories.*

be taken to ensure the alignment marks indicating the direction of polarisation of the S-wave transducers lie within the same plane.

A servo-hydraulic pump is used to apply the required radial and axial confinement. Independent pressure controllers enable a deviatoric stress to be applied to the core.

A direct pulse transmission technique along the axes of the plugs is employed, using both P and S waves. A transducer energiser/pulser box is used to energise the transmitting 1 MHz P-wave transducer, typically with a 500 V peak/peak pulse of 1.5 μs length at a pulse repetition rate of 10 s^{-1}.

The signal from the receiving P-wave transducer is passed through a variable gain broadband preamplifier before being logged by a data acquisition system. The time base is triggered by a low-voltage pulse output by the pulser box which is synchronised to the energising pulse.

Signals from multiple pulses are electronically saved and combined later. The pulse generator is then interrupted, and the connections changed over to use the S-wave transducer. The same procedure is repeated for the 1-MHz S-wave transducer.

The system is initially calibrated by placing the platen face to face on top of the hydraulic ram within the reaction frame. The platens are then loaded to the required radial and axial stress, and P- and S-wave measurements are made. Where triaxial tests are used to calibrate log-derived dynamic elastic moduli with core-derived static elastic moduli, the confining stress should approximate the effective reservoir stress at virgin reservoir conditions. Loading rates are similar to those for triaxial testing.

For ATT tests where the core is saturated in formation water, the pore pressure is normally maintained at up to 500 psi pressure (to ensure 100% saturation), and the effective stress change is achieved by adjusting the total applied axial (vertical) and radial (horizontal) stresses to provide the equivalent effective reservoir isostatic stress.

12.8.4 Test Procedures

To avoid saturation anomalies, great care must be taken to minimise the possibility of air entering the sample on loading the test sample into the Hoek cell.

P- and S-wave pulses are initiated at one end of a cylindrical plug sample using an ultrasonic transmitter, and the times taken for them to reach an ultrasonic receiver at the opposite end of the sample are measured.

The sample is loaded to the specified test confining stress by simultaneously increasing both axial stress and radial stress while maintaining the specified vertical/horizontal stress ratio constant. The loading rate should not exceed 0.5 MPa/min. Pore volume expulsion is monitored until stable and cumulative expelled volume, $\mathrm{Vp_{ex}}$, is recorded. The stressed porosity can be obtained from

$$\phi_{\text{stress}} = \left[\frac{(\mathrm{Vp_i} - \mathrm{Vp_{ex}})}{(\mathrm{Vb_i} - \mathrm{Vp_{ex}})}\right]$$

given a knowledge of initial pore volume ($\mathrm{Vp_i}$) and bulk volume ($\mathrm{Vb_i}$). This calculation assumes grain compressibility is negligible. In some systems, radial and axial strain measurements are used to determine the volumetric strain and hence the change in bulk volume on loading.

The transducer energiser/pulser box is used to energise the transmitting 1 MHz P-wave transducer at a pulse repetition rate of 10 $\mathrm{s^{-1}}$ and the data from the data logger are recorded. The same procedure is repeated for the 1-MHz S-wave transducer. A minimum of three ultrasonic measurements (P and S waves) are made to ensure repeatability.

The P- and S-wave travel times determined for each sample are corrected for travel through the transducer holders, these corrections being determined by taking ultrasonic measurements with the transducer holders in face to face contact under an axial stress equal to that applied to the sample. The P- and S-wave velocities for the rock are computed using these times together with the length of the sample.

12.8.5 Data Utilisation

The dynamic elastic properties E_{dyn}, G_{dyn}, K_{dyn} and ν_{dyn} at stress conditions are computed from the corrected P- and S-wave velocities, expressed as travel times Δt_c and Δt_s, and from the bulk density, ρ_b, of the sample. The dynamic elastic moduli are determined from:

Poisson's ratio, ν_{dyn}	$\frac{\frac{1}{2}(\Delta t_s/\Delta t_c)^2-1}{(\Delta t_s/\Delta t_c)^2-1}$
Shear modulus, G_{dyn} (psi)	$1.34\times10^{10}\frac{\rho_b}{\Delta t_s^2}$
Young's modulus, E_{dyn} (psi)	$2G(1+\nu)$
Bulk modulus, K_b (psi)	$1.34\times10^{10}\rho_b\left(\frac{1}{\Delta t_c^2}-\frac{4}{3\Delta t_s^2}\right)$
Bulk compressibility, β (psi^{-1})	$\frac{1}{K_b}$

where Δt has units of μs/ft. and bulk density in g/cc.

12.8.6 Data Reporting Requirements

The data that need to be reported by the test laboratory to ensure adequate quality control of the tensile strength tests and test results are listed in Table 12.12.

12.8.7 Advantages and Drawbacks/Issues

Summaries of the advantages and drawbacks/issues associated with ATT tests are listed in Table 12.13.

12.8.8 ATT Test Quality Control Issues, Checks and Diagnostics

If porosity is to be determined in the ATT test then ambient helium porosity and ambient resaturation porosity must agree within ±0.2 p.u.

Compressional velocity, Vp, increases with increased saturation but shear wave velocity, Vs, is relatively unaffected by the degree of saturation in sample.

The computation of dynamic elastic parameters from Vp and Vs assumes isotropy. The dynamic moduli, E_{dyn} and ν_{dyn}, are not equivalent to the static moduli, E_{sta} and ν_{sta}, since acoustic pulse-derived and strain-derived elastic moduli are systematically different. This is due to strain magnitude. Acoustic measurements are reversible and therefore perfectly elastic. Laboratory measurements of elastic moduli impose large strains, most of which are irreversible. The moduli measured are therefore not purely elastic but introduce some additional irreversible deformation caused by friction. As a result, static strains are always larger than dynamic strains and the dynamic Young's modulus can be two to three times greater than the static modulus.

Cross-plot of the Vp/Vs ratio versus stressed porosity can highlight anomalous data, as indicated in Fig. 12.55. In addition, cross-plotting Δt_c (DTCO)

TABLE 12.12 Data Reporting Requirements for ATT Tests

Data	Comments
Brief description of procedures and experimental apparatus	
Test sample depth and orientation	
Sample length and diameter	Establish *L*:*D* ratio
Sample grain volume, bulk volume and calculated pore volume, porosity and grain density at ambient conditions	If porosity at stress is determined
Initial sample dry weight (prior to saturation), initial sample saturated weight (after saturation) and immersed weight in saturation fluid	If porosity at stress is determined
Sample grain loss procedures	How laboratory correct porosity for grain loss
Pore pressure, p_p	
P- and S-wave measurement data including arrival time plots	At each specified stress station
Calculated sample pore volume, porosity and bulk density at test pressure stations (calculated on basis of incompressible grain volume)	If porosity at stress is determined
Calculated Vp and Vs acoustic velocities (m/s) and Vp/Vs ratios; calculated Δt_c and Δt_s (μs/ft.) from inverse Vp and Vs data	At each specified stress station
Calculated dynamic moduli: E_{dyn}, G_{dyn}, K_{dyn} and ν_{dyn}	At each specified stress station
Calculated static moduli: E_{sta} and ν_{sta}	If ATT test in conjunction with triaxial loading
Photographs of test plugs before and after testing. These should include a scale marker (e.g. ruler)	

versus stressed porosity, as shown in the example in Fig. 12.56, and linear regression of the data allows calculation of matrix travel time, Δt_{ma} (DTMA), and fluid travel time, Δt_{fl} (DTFL). DTMA is determined from the intercept of the linear regression gradient extrapolated to 0% porosity, while DTFL is the intercept of the regression extrapolated to 100% porosity. These values can then be compared with published values for DTMA and DTFL. Potential reasons for differences between core ATT-derived and log analysis-derived DTMA and DTFL include the uncertainty in the porosity measurement at stress, inaccuracies or errors in acoustic transducer measurement/calibration, the difference

TABLE 12.13 Advantages and Drawbacks/Issues with ATT Tests

Advantages	Drawbacks and Issues
• Can be measured in the same loading operation as static elastic moduli (e.g. triaxial tests) depending on load cell configuration. • Non-destructive (provided applied stress does not exceed yield point in triaxial tests nor induce pore collapse under isostatic loading). • Enables correlation of log-derived acoustic elastic constants to core-derived acoustic parameters.	• The test is relatively involved and costly for the limited data obtained. • ATT dynamic elastic properties need to be scaled to static values for geomechanics applications. • The physics and environment of core ultrasonic measurements are not necessarily equivalent to wireline logs, and core measurements are adversely influenced by microcracks and differences in plug orientation and log/core test sonic frequencies. Scaling is better achieved by direct calibration of wireline dynamic elastic moduli against static core data.

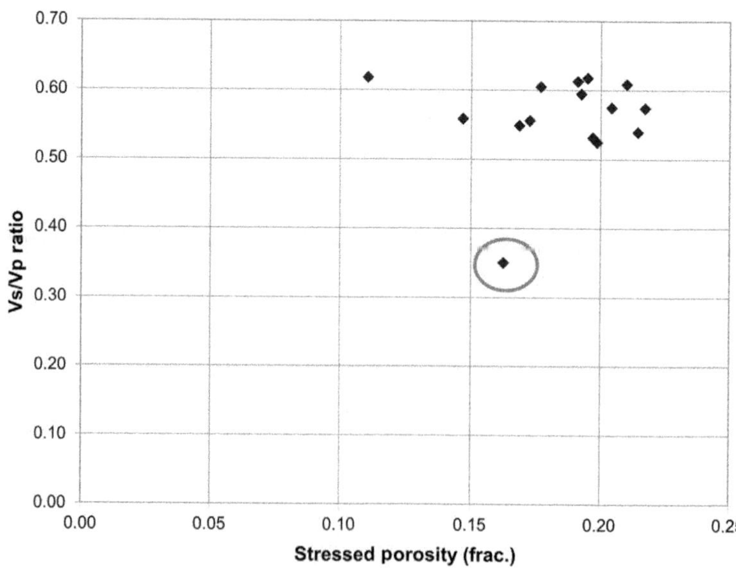

FIGURE 12.55 Vp/Vs ratio versus stressed porosity.

between the physics and environment of core ultrasonic measurements and wireline logs (especially signal frequency) and the probability that core measurements are adversely influenced by microcracks.

Most major SCAL laboratories provide acoustic velocity tests but procedures tend to be less rigorous than those adopted in rock mechanics laboratories.

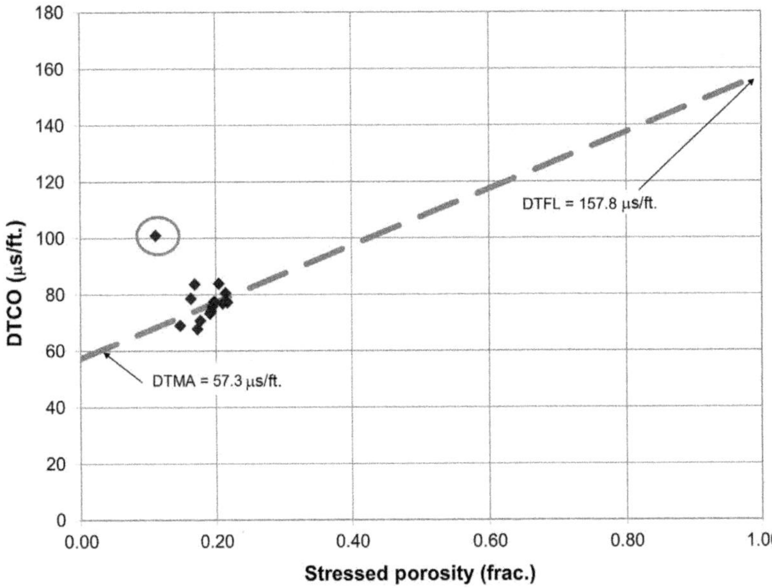

FIGURE 12.56 DTCO versus stressed porosity.

12.9 DSCA TESTS

12.9.1 Purpose and Sample Requirements

DSCA tests are carried out on core samples to determine the principal *in situ* stress ratios and, if vertical stress and pore pressure can be defined at the sample depth, the principal *in situ* stress magnitudes (Dyke, 1988; Ren and Roegiers, 1983). It is also possible to infer principal *in situ* stress orientations if oriented core is used, or the core can be oriented using image logs or palaeomagnetic reorientation (Hailwood and Ding, 1995). Data can be used directly but can also be used to calibrate or verify other *in situ* stress estimates from well tests such as extended leak-off tests and mini- or microfrac tests.

The DSCA tests are carried out on air-dried cube samples (approximately 25–40 mm) cut from full diameter core or core slabs. The test cube is referenced to a core orientation line which may either be an oriented core scribe line or an arbitrary orientation line which can be referenced to image log or palaeomagnetic reorientation lines. This is illustrated in Fig. 12.57.

12.9.2 Test Equipment

The cubes are instrumented normally with 8–12 electrical resistance strain gauges bonded at selected and known locations on the surface of each sample. Figure 12.58 shows a cube with four rosettes of strain gauges. The sample is sleeved in a flexible elastomer as shown in Fig. 12.59.

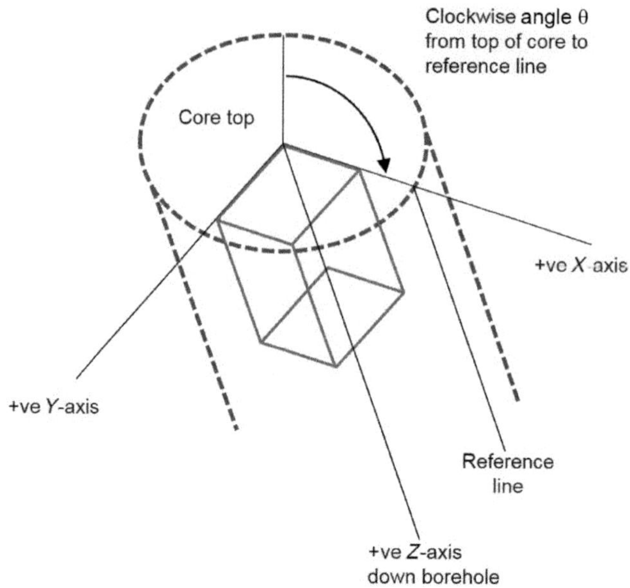

FIGURE 12.57 DSCA test sample orientation.

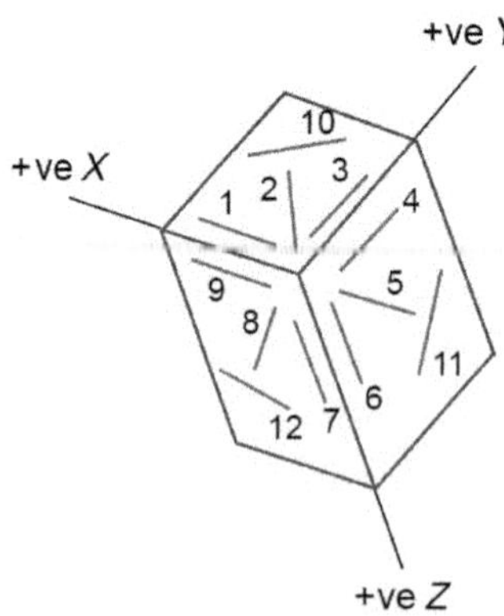

FIGURE 12.58 DSCA sample strain gauge rosettes (at 0°, 45°, 90° and 135°).

Isostatic closure stress is applied using an autoclave energised from a servo-hydraulic load frame operated in force control, and the stress–strain responses from each set of strain rosettes are recorded.

12.9.3 Test Procedures

The sample is subjected to a series of isostatic stress increments, up to an elevated boundary stress of around 70 MPa, and the corresponding stress–strain responses for each of the gauge locations are recorded. Measured strains are

FIGURE 12.59 DSCA sample in elastomer sleeve.

corrected for any pressure dependency in lead wires and gauges, using measurements of strain in two fused silica dummies, which are loaded simultaneously with the sample during each test. Relative strains are computed for each of the gauge locations.

As the sample is reloaded, microfractures created by stress relief on coring close first. Subsequently, continued loading causes deformation that is primarily a function of the intrinsic mechanical properties of the rock. By discriminating between intrinsic behaviour, at high hydrostatic stress, and induced behaviour (by the presence of open microfractures), the directional contribution due to microfractures can be determined. The direction of the maximum principal strain, ε_1, corresponds to a direction perpendicular to the strike of the orientation with the maximum microfracture density and, therefore, indicates the direction of the maximum *in situ* principal strain.

A typical stress–strain response from a DSCA test under isostatic loading (hydrostatic) is shown in Fig. 12.60. Initially, the sample is highly compliant because microcracks are open. As hydrostatic stress is applied, microcracks start to close and, at some point, a linear region is reached at which point all microfractures are considered to have closed. The point of deviation of a tangent fit to the closed fracture portion of this curve is taken to correspond to the effective closure stress in that direction in the field.

The cumulative strain in the different directions up to the point of microcrack closure is taken as being proportional to the effective stress acting in the field in that direction. A constant of proportionality is assumed and, in this way, attempts are made to correlate the magnitude of the cumulative strain to the effective stress state in the reservoir. A best-fit strain tensor is determined using a regression technique. A tensor of the *in situ* principal stresses (i.e. magnitudes and directions) is determined from the strain tensor.

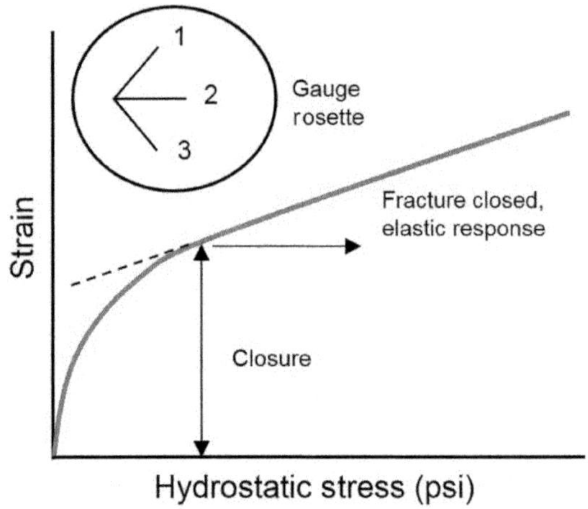

FIGURE 12.60 Idealised stress–strain profile in DSCA test.

12.9.4 Data Utilisation

From the stress closure profile of each test gauge, an effective closure stress in that direction can be determined. An azimuth and effective minimum and maximum principal stresses can then be estimated.

The derivation of principal stress magnitudes from laboratory-measured principal strains is difficult. In a simplified case, the principal stress directions and magnitudes can be inferred from elastic relations. A requirement, therefore, is to determine the elastic properties in the principal strain directions. A first-order approach to this is to infer elastic properties from directional bulk compressibilities, determined from individual strain gauges during the DSCA testing itself. Knowledge of the directionally dependent material properties allows the transformation from strain to stress. One basic assumption used in this calculation is that the stress and strain have a non-hysteretic relationship: that is, the amount of stress required to open microcracks is the same as that required to close them. The transformation yields a ratio (σ_1:σ_2:σ_3). This translates to maximum, minimum and intermediate stress. In normal stress regimes: $\sigma_1=\sigma_v$, $\sigma_2=\sigma_H$ and $\sigma_3=\sigma_h$.

As the tests are carried out at ambient pore pressure, the DSCA test analyses provide effective stress ratios, i.e.:

$$\frac{\sigma'_h}{\sigma'_v} \text{ and } \frac{\sigma'_H}{\sigma'_v} \text{ and } \frac{\sigma'_h}{\sigma'_H}$$

Thus, determination of total reservoir stresses also requires knowledge of the pore pressure gradient and one independent measure of *in situ* stress level. In this case, the vertical stress ($\sigma'_1=\sigma'_v$) at the core sample vertical depth is independently known from the density log. The total stress is then calculated from:

$$\sigma = \sigma' + \alpha p_{\mathrm{p}}$$

where the pore pressure (p_{p}) at the sample depth is calculated from the field gradient. In this example, the Biot poroelastic parameter is normally assumed to be equal to 1. The maximum total horizontal stress, σ_{H}, is determined from:

$$\sigma_{\mathrm{H}} = \left[\frac{\sigma'_{\mathrm{H}}}{\sigma'_{\mathrm{v}}}\right]\sigma'_{\mathrm{v}} + \alpha p_{\mathrm{p}}$$

and the total minimum horizontal stress from:

$$\sigma_{\mathrm{h}} = \left[\frac{\sigma'_{\mathrm{h}}}{\sigma'_{\mathrm{v}}}\right]\sigma'_{\mathrm{v}} + \alpha p_{\mathrm{p}}$$

The strain responses from the sets of gauges on the cube faces which are used to determine the strain at closure must be evaluated. The gauge reliability is generally classed as 'poor' or 'good'. The better data represents a situation when there is a full rosette suite in the same horizontal plane. An example of the strain gauge response with applied isostatic (hydrostatic) stress is shown in Fig. 12.61. Responses from Gauges 1–3 are classed as good quality, while Gauges 5 and 6 were classed as 'poor'. It was impossible to determine closure from Gauges 4 and 7.

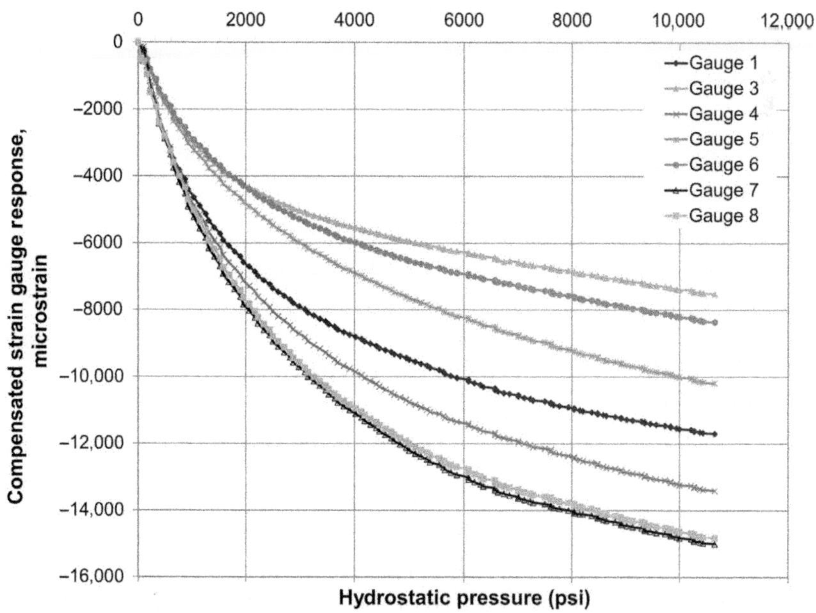

FIGURE 12.61 Example DSCA strain gauge response.

TABLE 12.14 Example DSCA Result Presentation

Vertical Depth (m TVDRKB)	Lithology	Vertical Stress Gradient (psi/ft.)	Pore Pressure Gradient (psi/ft.)	Strain Gauge Set	Maximum Horizontal Stress (psi/ft.)	Minimum Horizontal Stress (psi/ft.)	Effective Horizontal Stress Ratio	QC Flag
3434.90	Sand	1.022	0.514	1	0.867	0.823	1.05	
				2	0.864	0.665	1.30	
4100.80	Shale	1.022	0.514	1	0.905	0.738	1.23	
				2	0.968	0.772	1.25	
4107.20	Sand	1.022	0.514	1	1.068	0.811	1.32	
				2	1.114	0.605	1.84	
4190.30	Sand	1.022	0.514	1	0.930	0.827	1.12	
				2	0.935	0.833	1.12	

A typical presentation of specimen results for DSCA tests is summarised in Table 12.14 in terms of effective horizontal stress ratio. The sample vertical depth and both vertical stress and pore pressure gradients used in the total horizontal stress calculations are also provided. Biot's parameter, α, is assumed to be 1. The strain gauge sets which exhibit dubious or anomalous responses are shaded light grey and dark grey.

The direction of the maximum *in situ* principal strain is described by the angle it makes with arbitrary orientation line on the cube, θ. In this case, the core has not been oriented with respect to Grid North so it is not possible to infer a principal stress orientation.

12.9.5 Data Reporting Requirements

The data that need to be reported by the test laboratory to ensure adequate quality control of the DSCA tests and test results are listed in Table 12.15.

12.9.6 Advantages and Drawbacks/Issues

Summaries of the advantages and drawbacks/issues associated with DSCA tests are listed in Table 12.16.

12.9.7 DSCA Test Quality Control Issues, Checks and Diagnostics

The analysis assumes that:

- all of the microcracks are generated during unloading, due to the release of *in situ* compressive stresses, and thus are aligned with the directions of the principal stresses,
- the material is orthotropic—its mechanical properties are, in general, different along each axis and depend on the direction in which they are measured. For an orthotropic (and hence including transversely isotropic)

TABLE 12.15 Data Reporting Requirements for DSCA Tests

Data	Comments
Brief description of procedures and experimental apparatus	
Core orientation line (with respect to Grid or Magnetic North)	If core is or can be oriented
Test cube sample dimensions and sample location with respect to core orientations line	
Location of strain gauge rosettes with regard to vertical	
Data and plots of strain versus isostatic stress during loading from each strain gauge	With assessment of response quality
Principal *in situ* stress ratios	
Principal *in situ* stress orientations (if oriented core)	With respect to Grid or Magnetic North
Vertical stress, pore pressure and Biot factor used to determine principal *in situ* stress magnitudes	
Photographs of test samples before testing. These should include a scale marker (e.g. ruler)	

TABLE 12.16 Advantages and Drawbacks/Issues with DSCA Tests

Advantages	Drawbacks and Issues
• Semi-independent verification of other *in situ* stress estimate methods (e.g. hydraulic fracturing, XLOT, breakout analyses). • Can be run on old and slabbed core. • Requires minimal core material. • Significant cost advantage over rig-site measurements such as anelastic strain relaxation tests.	• The technique is only suited to reasonably competent crystalline materials, free of clay. Unsuited to weak and poorly cemented materials, soft carbonates, shales or highly anisotropic rocks. • Lengthy sample preparation process. • For the full *in situ* stress tensor (i.e. principal stress orientations and magnitudes), oriented core is required together with a good estimate of the vertical stress, pore pressure and Biot constant. • Results cannot be guaranteed, since not all rocks give an interpretable DSCA response.

material, the orientations of the orthotropy are assumed to coincide with the directions of the principal strains and
- the stress magnitudes required to close the microcracks are the same as those magnitudes that generated and opened the microcracks.

The core necessarily must have a certain range of properties for the DSCA tests to meet the assumptions and to provide feasible results. For example, if the core is very strong (stiff), the stress change may not be sufficient to generate microcracks. If the core is very weak, it may disaggregate on coring or testing losing the orientation of the fabric networks within the core. As a consequence of the dependency of interpretable stress–strain response on the core condition and strength properties, only around 50% of DSCA tests are deemed to be 'successful'.

The reliability of the strain gauge response is reflected in the QC Flag as shown in Table 12.14.

In view of this, DSCA data should not be used in isolation, but must be evaluated against other data from fracturing tests and geomechanical modelling.

12.10 PORE VOLUME COMPRESSIBILITY TESTS

12.10.1 Purpose and Compressibility Definitions

Modification of the reservoir effective stress due to production causes volumetric changes in the pore space in a reservoir. The engineering parameters quantifying these volumetric variations are compressibilities. Reliable compressibility values are essential for resource estimation, reservoir maintenance and drive assessments, as well as compaction and subsidence evaluations. Production forecasting is intimately related to a total system compressibility combining the compressibilities of liquid and gaseous phases in the pore space, the grain compressibility of the solid portions and the pore volume compressibility, often referred to as formation compressibility.

Three types of compressibilities are often cited in the characterisation of a porous medium:

- bulk compressibility, C_b, represents the relative changes in bulk volume of the medium;
- solid (grain) compressibility, C_g, represents the relative volumetric change of the solid portion of the medium; and
- pore volume compressibility, C_p, represents the relative change in pore volume.

C_{bc} is the variation in bulk volume (Vb) with change in confining stress (P_c) at constant pore pressure (p_p) and, therefore, represents the compressibility of the frame:

$$C_{bc} = \frac{1}{\mathrm{Vb}} \left\{ \frac{\partial \mathrm{Vb}}{\partial P_c} \right\}_{p_p}$$

C_{bp} is the variation in bulk volume with change in pore pressure at constant confining stress:

$$C_{bp} = \frac{1}{Vb}\left\{\frac{\partial Vb}{\partial p_p}\right\}_{P_c}$$

C_{pc} is the variation in pore volume (Vp) with change in confining stress at constant pore pressure:

$$C_{pc} = \frac{1}{Vp}\left\{\frac{\partial Vp}{\partial P_c}\right\}_{p_p}$$

C_{pp} describes the variation in pore volume with change in pore pressure at constant confining stress:

$$C_{pp} = \frac{1}{Vp}\left\{\frac{\partial Vp}{\partial p_p}\right\}_{P_c}$$

The interrelationship between these compressibilities can theoretically be derived if elasticity is assumed and porosity, ϕ, is known:

$$C_{bp} = C_{bc} - C_g$$

$$C_{pc} = \left[\frac{C_{bc} - C_g}{\phi}\right] \text{ and}$$

$$C_{pp} = \frac{C_{bc} - (1+\phi)C_g}{\phi}$$

As grain compressibility is small (around 0.16×10^{-6} psi^{-1} for sandstones), C_{bc} and C_{bp}; and C_{pc} and C_{pp} can be considered to be equal in high-compressibility formations.

Compressibility is determined in the laboratory under simulated reservoir stress conditions or under simulated depletion conditions. Data can also be interpreted to determine compaction factors for reservoir compaction and surface or mud line subsidence calculations.

12.10.2 Compressibility Test Loading Conditions

During depletion, the effective stress acting in the reservoir formation increases as the pore pressure depletes

$$\sigma' = \sigma - \alpha p_p$$

The most common stress condition for determination of pore volume compressibility in the SCAL laboratory is often referred to as the effective stress method where a fully saturated sample is confined under isostatic compression with a constant pore pressure, which is normally maintained at ambient conditions. Thus, the increase in effective stress caused by depletion is accomplished by increasing the total stress while maintaining the pore pressure constant. As the pore system compacts, liquid is expelled. As the test sample is not instrumented with strain gauges, pore volume compressibility is determined solely from measurements of expelled volume with increasing confining stress. The test methods are therefore the same as used to determine porosity compaction factor in combination with formation factor (see Chapter 8).

In the simulated depletion method, the sample is confined at a total stress equivalent to the reservoir total stress, and the pore pressure is ramped to the reservoir pressure. The pore pressure is then allowed to deplete, while the total stress is maintained constant, and the pore volume reduction is determined from the volumes withdrawn or sample strain deformation measurements under pore pressure control.

In theory, the pore volume compressibility from this effective stress test

$$C_{\text{pc}} = \frac{1}{\text{Vp}}\left\{\frac{\partial \text{Vp}}{\partial P_{\text{c}}}\right\}_{p_{\text{p}}}$$

should be similar to those from the simulated depletion test

$$C_{\text{pp}} = \frac{1}{\text{Vp}}\left\{\frac{\partial \text{Vp}}{\partial p_{\text{p}}}\right\}_{P_{\text{c}}}$$

if grain compressibility is small. That is:

$$C_{\text{pp}} = C_{\text{pc}} - C_{\text{g}}$$

In practice, the compressibilities determined using the effective stress method can be significantly higher than those determined from the simulated reservoir depletion method. This is primarily a consequence of excess pore volume associated with microcracks in the core sample. In the effective stress tests, the plug is loaded and the pore volume reduction is determined as a function of effective stress during initial loading. Any microcracks in the plug which are created as a result of stress release on core recovery will close during this loading cycle and water in the microcracks is expelled. This would appear as pore volume compaction. In the simulated depletion tests, microcrack closure will occur as the sample is loaded to the initial test conditions. The subsequent response of the sample on depletion therefore is only controlled by pore compaction. Figure 12.62 graphically illustrates the differences in compressibility determined by the effective stress and simulated depletion methods. The compressibilities from the effective stress method are artificially higher and exhibit a much less consistent response due to microcrack closure.

However, stress conditions during production are neither constant pore pressure nor isostatic compression. During production, reservoir rock is under mixed stress and strain boundary conditions. Generally, the vertical stress remains constant, while the pore pressure and the horizontal stresses change such that

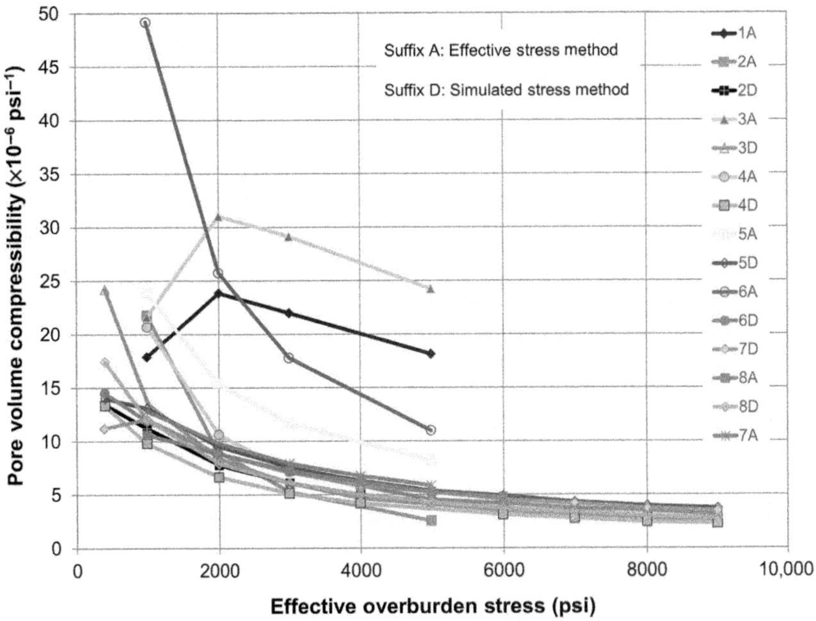

FIGURE 12.62 Pore volume compressibility from effective stress and simulated depletion methods.

the pre-existing lateral deformations are preserved: that is no additional lateral deformation takes place. Teeuw (1971) argued that if the reservoir height is small in relation to its width and has stiff lateral boundaries (faults) then, as the reservoir depletes, strain in the horizontal direction will be constrained by the stiff lateral boundaries, so the reservoir can only strain in the vertical direction. This reflects a *uniaxial strain* situation. Under isostatic loading, there is no constraint and the sample may strain axially as well as radially (laterally). Teeuw derived elastic relationships to correct from isostatically measured compressibilities to 'production condition' (uniaxial) compressibilities:

$$C_{\mathrm{pu}} = C_{\mathrm{pc}}\left[\frac{\alpha(1+v)}{3(1-v)}\right]$$

where α is Biot's factor and v is Poisson's ratio.

However, elastic relationships between compressibility and stress–strain behaviour for isostatic, triaxial compression and uniaxial strain conditions may only be true if the pore spaces are uniform and isotropic and retain self-similarity during loading. To measure the pore compressibility that truly reflects *in situ* conditions, the compressibilities should be measured under constant total axial stress or stress rate and zero radial strain conditions. The requirements to monitor and control loading rates and strains means that compressibility tests should only be carried out by rock mechanics laboratories with the necessary equipment, resources and capabilities.

Consequently, the principal compressibility tests are described below.

12.10.2.1 Depletion Compressibility Under Uniaxial Strain Conditions (K_0) Tests

The total axial stress is held constant during depletion, and the radial strain developed as the rock is loaded (pore pressure reduced) is constrained to zero or nominal values. This effectively means that the radial stress is required to maintain a zero strain condition, and the mean stress:

$$\sigma_{\text{mean}} = \frac{\sigma_{\text{axial}} + 2\sigma_{\text{radial}}}{3}$$

will not be constant during the experiment.

12.10.2.2 Rate Type Compaction Method Tests

When a reservoir is depleted, the rate of change of effective stress exerted on the reservoir formation is suddenly increased from that imposed on the burial process over a geologic time span. The change in loading rate will have a large influence on the *in situ* compaction behaviour. Obviously, in the laboratory, loading rates on depletion will be much higher than in the field. Compaction curves at different, but constant, loading rates form a fan of lines referred to as 'virgin' compaction curves. The lower the loading rate (slower the depletion), the more the sample will be compacted at a given stress level. When the loading rate is increased, the uniaxial compressibility initially becomes much lower. When the rate is decreased, the uniaxial compressibility initially becomes much higher.

Rate type compaction method (RTCM) tests are designed to evaluate rate effects on compressibility. Variable axial loading or pore pressure drawdown rates allow for correction of laboratory-measured compressibilities to values consistent with field production practices. Different drawdown rates are simulated by increasing the effective axial stress at different rates. Comparable testing can be done by reducing the pore pressure at different rates.

12.10.2.3 Creep Tests

Creep tests are long-term mechanical experiments for evaluating time-dependent effects of load and deformation. Creep tests are performed at a constant stress, or a combination of constant stress and deformation, while measuring the corresponding changes in deformation and/or stress. Creep segments are often incorporated as additional stages in compressibility tests. A common procedure is to hold confining pressure, axial stress and pore pressure constant (all servo-controlled) and monitor the deformation over time. Alternatively, a creep segment can maintain axial stress, pore pressure and radial strain constant. The latter procedure may best represent reservoir behaviour, although the former method provides information which can be interpreted more conveniently.

The time to maintain these boundary conditions is ascertained during the creep stage itself, from an assessment of the equilibration of the volumetric and/or axial strain (or stresses, if appropriate) in time.

In the following sections, general test procedures are described for the K_0 uniaxial loading tests.

12.10.3 Sample Preparation

The tests are carried out on cleaned and dried samples which have been fully saturated in SFW or mineral oil. Oil is preferred if the samples contain reactive or sensitive clays. 1.5″ diameter plug samples are preferred ideally with a *L:D* ratio of 2 or greater. Tests are normally performed on vertical samples if σ_v represents the maximum *in situ* principal stress.

The test plugs are end-ground, then flush cleaned or cleaned using a total immersion system. Unless otherwise specified, the plugs are surface dried at ambient conditions before being dried under humidified conditions at 60 °C and 40% relative humidity to constant weight (±0.01 g or better) in a humidity oven. Base porosity (unconfined helium porosity) and permeability are measured.

The samples are then saturated in synthetic formation water or light mineral oil. Measurements of saturated weight and immersed weight are used to determine saturated pore volume and bulk volume. The helium and resaturation porosities should agree within ±0.25 p.u.

12.10.4 Uniaxial K_0 Test Equipment

Compressibility tests are performed using a stiff compression machine with a closed-loop servo-controlled system incorporating an instrumented Hoek-type triaxial cell. Confining stress on the samples is generated and maintained using the triaxial cell in conjunction with a servo-controlled actuators and intensifiers.

Elevated temperature, to simulate actual *in situ* conditions, can also be applied in some laboratories. Pumps or accumulators are used to control sample pore pressure.

Axial and two radial strain gauges (90° apart) are used to determine axial, radial (ε_{rad}) and volumetric strains (ε_v) during initial loading and during the depletion phase. The volumetric strain is:

$$\varepsilon_v = \varepsilon_{rad(1)} + \varepsilon_{rad(2)} + \varepsilon_{axial}$$

12.10.5 Uniaxial K_0 Test Procedures

12.10.5.1 Initial Loading Conditions

The sample is instrumented and installed in the test cell. The data acquisition is initialised to obtain a complete record of the sample volumetric deformation.

The confining vessel is filled with laboratory oil and the sample is loaded isostatically to a nominal confining reference stress, typically 400 psi, at a rate of 1 psi/s. Strain is allowed to equilibrate for approximately 2 h. The pore volume and bulk volume used in subsequent calculations are often based on the values at this reference pressure.

The sample is flushed with the test saturant (SFW or oil) and the pore pressure line is connected to the pump or intensifier. The pore pressure is then ramped to 200 psi pore pressure at 1 psi/s.

12.10.5.2 *Initial* In Situ *Stress Loading Conditions*

The pore pressure and radial confining stress are ramped simultaneously to their target values at a rate of 1 psi/s under isostatic loading. For a normal stress regime, the target radial stress value represents the average of the total minimum and maximum horizontal stresses:

$$\sigma_{\mathrm{rad}} = \frac{\sigma_{\mathrm{H}} + \sigma_{\mathrm{h}}}{2}$$

The target pore pressure represents the reservoir value. During this loading segment, grain compressibility can be determined.

The confining (radial) stress and pore pressures are held constant, while the axial stress is increased to the *in situ* vertical stress condition so the sample is now under triaxial loading.

Axial and radial strain measurements during loading are used to determine the change in grain volume, bulk volume and, by difference, pore volume from ambient or nominal stress conditions to the initial stress conditions.

12.10.5.3 *Depletion Loading Conditions*

After allowing for strain equilibration with the sample held at the initial *in situ* stress and pore pressure levels, the pore pressure depletion phase is initiated. The confining stress control system is switched into uniaxial strain control so that supplementary radial strain is prevented. The uniaxial strain boundary conditions (no radial deformation) and constant total axial stress (constant overburden stress) are maintained while decreasing at a controlled rate of 5–6 psi/min to its target value: typically the reservoir abandonment pressure. Confining stress and mean stress will vary due to the imposed uniaxial strain boundary conditions.

Cumulative volumetric strain from the axial and radial strain gauge measurements are used to determine the change in grain volume, bulk volume and, by difference, pore volume from the initial *in situ* stress conditions during depletion.

Figure 12.63 plots cumulative volumetric strain during pore pressure depletion, and Fig. 12.64 provides an indication of the axial and radial (confining) stress, mean stress and pore pressure evolution during a typical K_0 test.

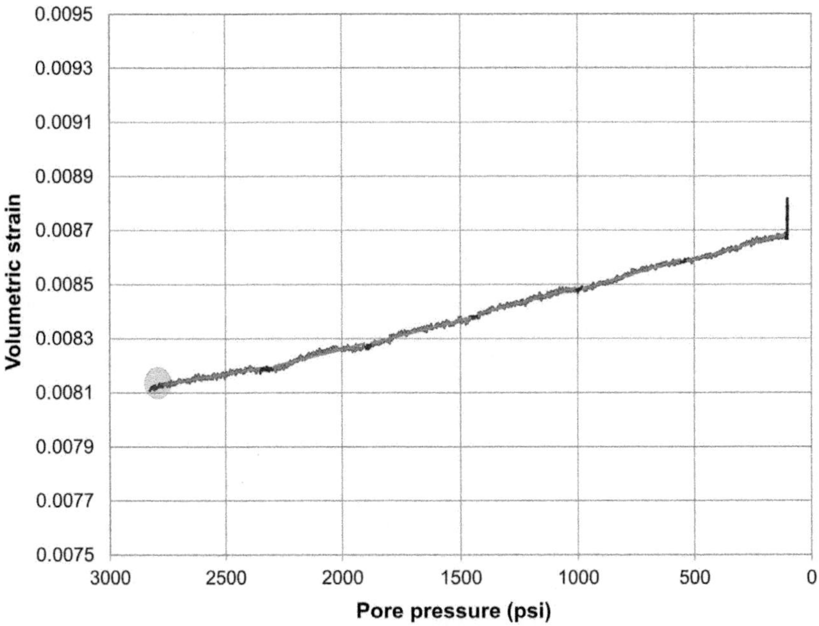

FIGURE 12.63 Cumulative volumetric strain during pore pressure depletion.

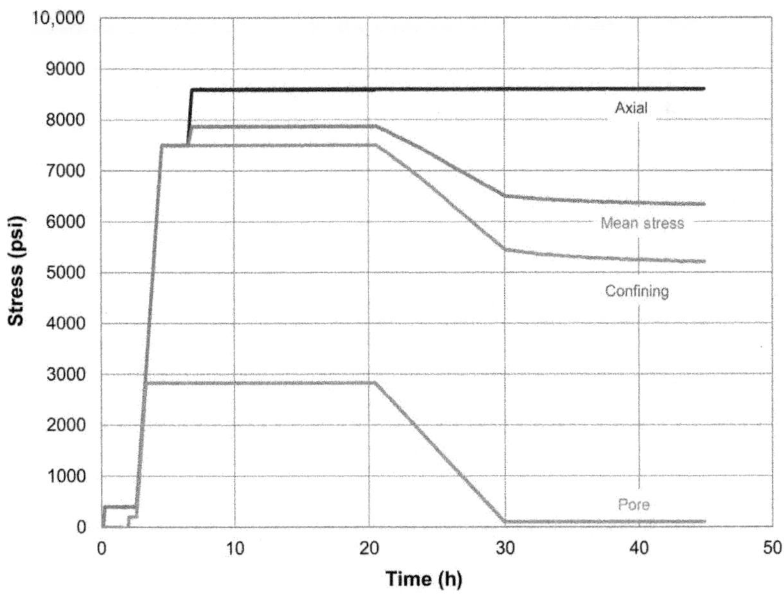

FIGURE 12.64 Stress and pore pressure evolution during typical uniaxial K_0 test.

12.10.6 Data Utilisation

12.10.6.1 Grain Compressibility and Biot's Factor

Applying isostatic confining pressure and pore pressure to the samples at identical, computer-controlled rates during isostatic loading to the initial target radial stress station provides a fundamental laboratory measurement of grain compressibility. As the total stress is increasing, and if the effective stress is held constant with a 200 psi differential between the isostatic confining pressure and the pore pressure, any sample deformation measured during this loading procedure is attributed to deformation of the grains alone. The grain compressibility, C_g, is determined as the relative change in sample bulk volume (Vb) with respect to the applied isostatic stress (confining pressure):

$$C_g = \frac{1}{\mathrm{Vg}}\left\{\frac{\partial \mathrm{Vb}}{\partial P_c}\right\}$$

where Vg is the grain volume at the nominal reference isostatic stress (e.g. 400 psi or ambient conditions). If another loading stage is also applied, where pore pressure is maintained constant and isostatic confining pressure is increased, it is possible to determine bulk compressibility during initial loading. The derivative of the volumetric deformation during this loading phase with respect to confining stress gives the isostatic bulk volume compressibility, C_b:

$$C_b = \frac{1}{\mathrm{Vb}}\left\{\frac{\partial \mathrm{Vb}}{\partial P_c}\right\}$$

where Vb is the bulk volume at the nominal reference isostatic stress (e.g. 400 psi). The grain and isostatic bulk compressibilities can then be used to determine Biot's parameter, α:

$$\alpha = 1 - \frac{C_g}{C_b}$$

12.10.6.2 Uniaxial Pore Volume Compressibility

As the sample is depleted, volumetric strain measurements (ε_v) are used to determine the change in sample bulk volume from the initial or reference value (Vb_0) measured at the reference ambient or nominal confining stress

$$\mathrm{Vb} = \mathrm{Vb}_0[1 - (\varepsilon_v - \varepsilon_{v@ref})]$$

where $\varepsilon_{v@ref}$ is the strain recorded at the initial reference confining stress.

The grain volume at any stage during loading is determined from:

$$\mathrm{Vg} = \mathrm{Vb}_0(1 - \phi_{ref})\left[1 - \left(\varepsilon_g - \varepsilon_{g@ref}\right)\right]$$

where ϕ_{ref} is the sample porosity at reference (ambient) conditions. The grain strain, ε_g, is determined from:

$$\varepsilon_g = \frac{\left[\sigma_{mean} - \left(p_p \phi_{ref}\right)\right] C_g}{1 - \phi_{ref}}$$

The pore volume at any time during loading to the initial *in situ* stress and during depletion is found from:

$$Vp = Vb - Vg$$

Figure 12.65 shows an example of calculated pore volume during the depletion phase.

The pore volume data are used to determine the pore volume compressibility during depletion loading under uniaxial conditions, C_{pu}, from:

$$C_{pu} = \frac{1}{Vp_i}\left\{\frac{\partial Vp}{\partial p_p}\right\}_{K_0}$$

The initial pore volume, Vp_i, is the pore volume calculated from volumetric and grain strain data at the initial *in situ* test conditions (total vertical and mean horizontal stress and reservoir pore pressure). The derivative $\delta Vp/\delta p_p$ is evaluated from cumulative data over specified pore pressure ranges. Figure 12.66 plots calculated uniaxial pore volume compressibility as a function of pore pressure during depletion.

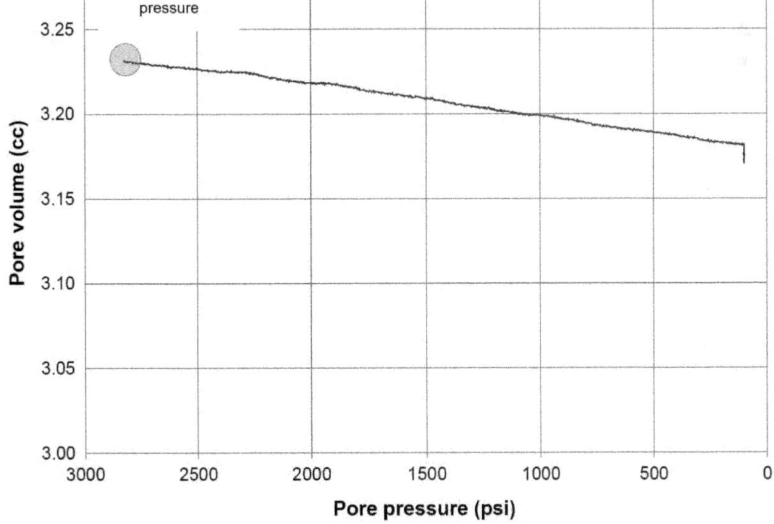

FIGURE 12.65 Pore volume reduction during depletion.

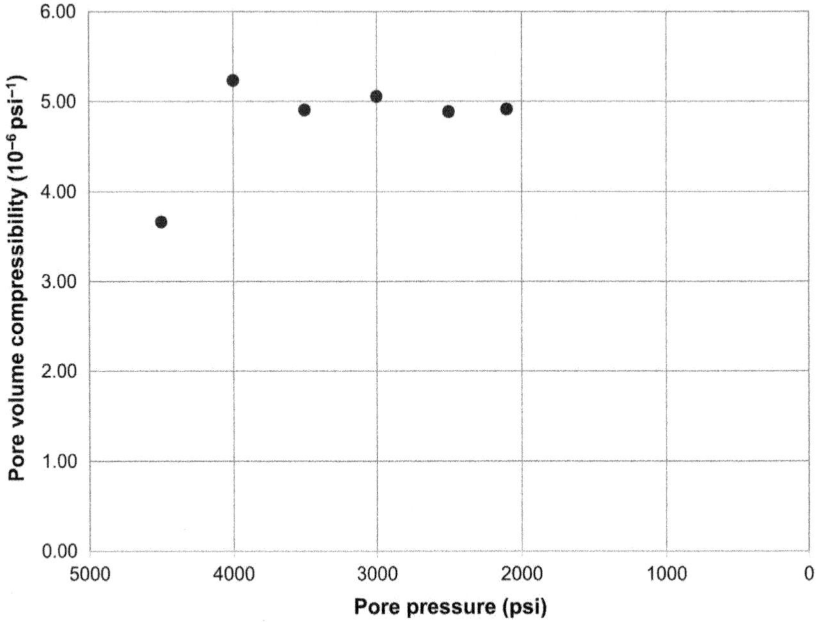

FIGURE 12.66 Calculated pore volume compressibility on depletion.

12.10.6.3 Compaction Coefficient

The effective stress change on depletion will cause a reduction in pore volume and bulk rock volume, and the reservoir will compact. Reservoir compaction, ΔH, during depletion can lead to strains acting on the production tubulars which, if excessive, can lead to buckling and failure. Compaction can be estimated using Geertsma's (1973) linear elastic model:

$$\Delta H = C_m H_{res} (P_i - P_{final})$$

where H_{res} is the net vertical reservoir height, P_i is the initial reservoir pressure and P_{final} is the final (depleted) reservoir pressure. The uniaxial compaction coefficient (C_m) is defined as the formation compaction per unit change in pore pressure reduction:

$$C_m = \frac{1}{3}\left[\frac{1+v}{1-v}\right](1-\beta)C_b$$

C_b is bulk compressibility and β is the ratio of rock matrix (grain) and rock bulk compressibility (C_g/C_b). Thus, grain compressibility (during initial loading) and bulk compressibility determined during the constant rate depletion phase of compressibility tests can be used to determine the compaction coefficient and reservoir compaction.

12.10.6.4 Elastic Properties

Young's modulus can be determined from axial strain and axial stress measurements during the depletion phase. Poisson's ratio cannot be determined during depletion as the radial strain is maintained to be zero on uniaxial loading. A pseudo Poisson's ratio can be obtained from:

$$\upsilon_{\text{pseudo}} = \frac{K_0}{(1+K_0)}$$

K_0 is the uniaxial stress factor required to maintain the zero lateral strain condition on depletion loading and is determined from:

$$K_0 = \frac{(\sigma_{\text{rad}} - p_{\text{p}})}{(\sigma_{\text{axial}} - p_{\text{p}})}$$

where σ_{axial} and σ_{rad} represent the axial and radial (confining) stresses.

12.10.7 Data Reporting Requirements

The data that need to be reported by the test laboratory to ensure adequate quality control of the compressibility tests and test results are listed in Table 12.17.

12.10.8 Advantages and Drawbacks/Issues

Summaries of the advantages and drawbacks/issues associated with pore volume compressibility tests are listed in Table 12.18.

12.10.9 Uniaxial K_0 Compressibility Test Quality Control Issues, Checks and Diagnostics

Samples must sense the full amount of applied confining pressure. The use of metal sleeves to jacket weak or poorly consolidated test samples is not recommended due to load coupling issues. Friable or unconsolidated samples are often subjected to extreme isostatic stress cycles before measurements are made.

Check that the sample helium porosity agrees with the test fluid-saturated porosity. The helium pore volumes at a nominal confining stress is used by some laboratories as the base porosity for subsequent stressed porosity calculations in the compressibility tests. For example, Fig. 12.67 plots normalised porosity (porosity at stress as a fraction of ambient condition porosity) versus stress for selected samples. The apparently large reduction in porosity between 0 and 400 psi signals errors in the lab helium pore volume.

Samples containing incipient fractures often exhibit anomalous pore volume compressibility behaviour. Closure of these fractures would influence both the porosity behaviour under initial stress (as evidenced by one sample in Fig. 12.67) and the subsequent pore volume compressibilities on depletion.

Unless material behaviour is elastic, corrections published in the literature (e.g. Teeuw, 1971) will not accurately 'convert' hydrostatic measurements to uniaxial strain equivalents.

TABLE 12.17 Data Reporting Requirements for Uniaxial K_0 Tests

Data	Comments
Brief description of procedures and experimental apparatus	
Sample permeability	At nominal confining stress
Sample helium grain volume and helium pore volume	At base or reference conditions
Sample saturated weight and immersed weight	After saturating in test fluid
Saturated pore volume and bulk volume (immersion)	At base or reference conditions
Test sample depth, dimensions and plug orientation	
Complete records of axial and two radial strain gauge data during entire loading and depletion sequence	Data versus axial and radial stresses
Axial stress, radial (confining) stress, mean stress and pore pressure evolution during testing	For example, Fig. 12.64
Grain compressibility and Biot factor	From initial isostatic loading
Calculated bulk volume, grain volume and pore volume during entire loading and depletion sequence	
Calculated cumulative pore volume compressibility versus pore pressure	For example, Fig. 12.66

TABLE 12.18 Advantages and Drawbacks/Issues with Uniaxial Compressibility Tests

Advantages	Drawbacks and Issues
• Uniaxial K_0 test provides a direct simulation of compressibility and compaction under simulated reservoir loading conditions (initial reservoir loading and depletion). • Better and more representative of reservoir effective stress-induced bulk volume and pore volume changes on depletion than effective stress SCAL tests. • Data can be used to estimate reservoir compaction.	• Complex test procedure requiring expertise, experience and resources. Viable results are only possible from fully strain gauge-instrumented samples. • Deformations on loading to initial conditions can be excessive in weak materials. • Elastic relationships between compressibility and stress–strain behaviour for hydrostatic, triaxial compression and uniaxial strain conditions may only be true if the pore spaces are uniform and isotropic and retain self-similarity during loading. • Sample with short $L{:}D$ ratios can exhibit considerably higher compactions than standard samples (2:1 or greater).

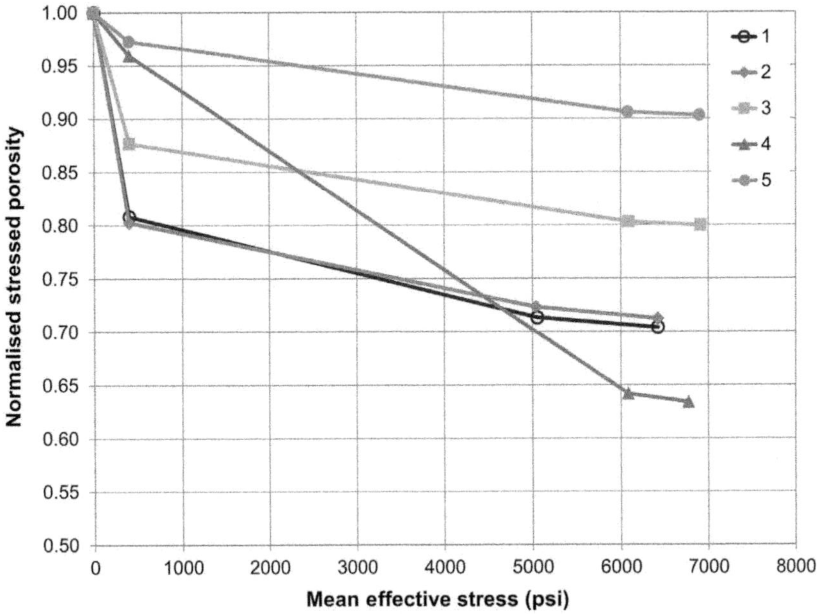

FIGURE 12.67 Normalised porosity versus mean stress from compressibility tests.

In weak sands, the pore pressure should be increased to a greater than the *in situ* level and then reduced to its virgin *in situ* level.

Most commercial SCAL laboratories offer pore volume compressibility testing under inappropriate loading conditions which results in unrepresentative and possibly erroneous or non-viable results. Compressibility tests must only be carried out on instrumented samples by rock mechanics laboratories using appropriate loading conditions. Pore volume compressibilities from constant pore pressure tests (effective stress increased by increasing confining stress) will be much higher than those under simulated depletion experiments (from initial reservoir stresses/pressures) and should not be used in compaction calculations.

Compressibility data should be evaluated against compaction coefficients obtainable from triaxial tests. This assumes linear elasticity. Bulk modulus, and its inverse, bulk compressibility (C_{bc}) can be calculated from Young's modulus (E) and Poisson's ratio (υ) obtained from the triaxial test data:

$$K = \frac{E}{3(1-2\nu)} \text{ and } C_{bc} = \frac{1}{K}$$

Pore volume compressibility, C_{pc}, can be estimated from:

$$C_{pc} = \frac{C_{bc} - C_g}{\phi}$$

Grain compressibilities and pore volume compressibilities should be checked against published values. C_g for sandstones are typically in the range from $0.16 \times 10^{-6}\,\text{psi}^{-1}$ to $0.2 \times 10^{-6}\,\text{psi}^{-1}$ (Morita et al., 1989). For carbonates, the mineral bulk moduli (inverse of grain compressibility) reported by Adam et al. (2006) for a range of carbonate samples ranges from 71 GPa (wackestone) to 85 GPa (packstone). Typical values for pure calcite and dolomite are 77 GPa (Mavko et al., 2009).

Pore volume compressibility from core tests on carbonates and sandstones is often correlated against porosity, for example, Hall (1953) and Newman (1973) models. The Hall correlation is:

$$C_{pp} = 1.87 \times 10^{-6} \phi^{-0.415}$$

for C_{pp} in microsips and ϕ as a fraction, and includes both sandstones and limestones. The Newman model for consolidated limestones is:

$$C_{pp} = \frac{a}{(1 + bc\phi)^{(1/b)}}$$

for Cpp in microsips and ϕ as a fraction, where $a = 0.8535$, $b = 1.075$ and $c = 2.202 \times 10^6$.

Akhoundzadeh et al. (2011) reported C_{pp} from 2.6 to 10.4×10^{-6} psi^{-1} on samples of pure limestone. They developed a robust correlation with porosity which is described by:

$$C_{pp} = \left(\frac{1}{0.367 + 0.099 \ln(\phi)} \right) \quad \text{for } C_{pp} \text{ in microsips and } \phi \text{ as a fraction}$$

When these three models are compared (Fig. 12.68), it is apparent that the-Hall model predicts lower C_{pp} for porosities less than 25% than the Newman

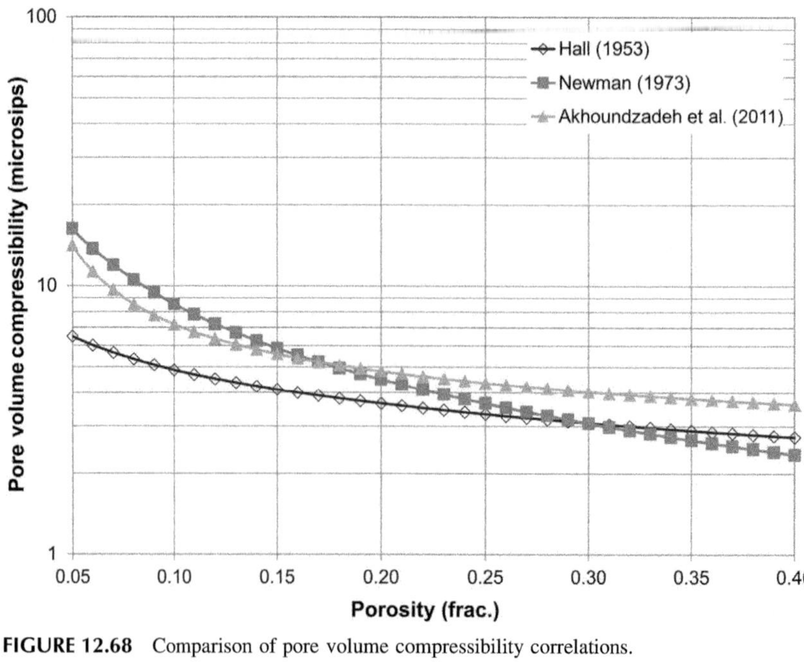

FIGURE 12.68 Comparison of pore volume compressibility correlations.

and Akhoundzadeh et al. correlations (for carbonates). Overall, the Akhoundzadeh et al. model predicts slightly higher compressibilities over a wider porosity range.

12.11 PARTICLE SIZE ANALYSIS TESTS

12.11.1 Purpose

Rock strength tests such as triaxial, UCS, and TWC tests are used to determine the risk of formation of shear failure under production conditions in a well. In a sandstone reservoir, if the rock is predicted to fail, sand may be produced to surface. This leads to a high risk of erosion of key well control components, such as chokes and flowlines, as well as lost production due to separator cleanouts. If sand production is expected, then some form of active sand control (filter screens or gravel packs) is required to prevent sand production to surface. In this case, particle size distribution analyses (PSA) are required on sand or sandstone samples to assist in the design, selection and specification of sand control systems. PSA on sands are also used in sedimentology and petrographic analyses.

Sand particles are three-dimensional objects. In order to provide a complete description of a particle, three parameters are required—length, breadth and height. Thus, it is impossible to describe a particle using a single number that equates to particle size. Therefore, most sizing techniques assume that the material being measured is spherical because a sphere is the only shape that can be described by a single number, its diameter, thus simplifying the way particle size distributions are represented.

Different measurement techniques can produce different results when measuring non-spherical particles. That said, any instrument or technique used for particle size analysis needs to generate data in a form that is relevant to the process. The method also needs to be reliable, simple to use and able to generate reproducible data.

There are several methods used to determine particle size.

12.11.1.1 Sedimentation

This is a traditional method widely used in the paint and ceramics industries. Equipment as simple as the Andreasen pipette or as complex as centrifuges and X-rays can be used in this method. The main advantage of this technique is that it determines particle size as a function of settling viscosity. However, as the density of the material is needed, this method is not good for emulsions where the material does not settle or for dense material that settles too quickly. It is also based on spherical particles, so can give large errors for particles with a large aspect ratio.

12.11.1.2 Image Analysis

This technology generates data by capturing direct images of each particle, providing users with the ultimate sensitivity and resolution. Image analysis systems are capable of high-resolution sizing ranging from 0.5 to 1000 μm. Subtle differences in particle size and shape can be accurately characterised using this method.

12.11.1.3 Acoustic Spectroscopy

Instead of using light, this technique employs ultrasound for collecting information on the particles that are dispersed in fluid. This can be done because dispersed particles absorb and scatter sound waves similar to light. Acoustic spectroscopy can be used to measure particle size distribution for any particle in a fluid system and can measure at very high particle concentrations.

However, the two most common methods used in core analysis for sand control completion design are:

Mechanical sieving (MPSA): This is historically the method used in most laboratories for sand control design purposes.

Laser particle size analysis (LPSA): This is the one of the most widely used particle sizing techniques and has become the standard method in many industries for characterisation and control. This type of particle size analyser relies on the fact that particles passing through a laser beam will scatter light at an angle that is directly related to their size. When particle size decreases, the observed scattering angle increases logarithmically. Scattering intensity is also subject to particle size, diminishing with particle volume. What this means is that large particles scatter light at narrow angles with high intensity, while small particles scatter at wider angles with low intensity. Advances in sophisticated data processing and automation have allowed this to become the dominant method used in industrial PSA determination although it is not yet fully accepted in the petroleum industry for sand control design and selection.

The MPSA and LPSA methods are described below.

12.11.2 Mechanical Particle Size Analysis

12.11.2.1 Sample Preparation

The test requires a minimum of 25 g of sand sample. However, larger samples provide more accurate results. No specific sample geometry is required as samples will be disaggregated. Samples are normally taken from plugs, plug trims, pieces of core or sidewall cores. Post-test UCS or TWC plugs are often used so that particle size can be correlated with rock strength. When taking samples from whole cores or plugs, the samples should be cleaved not cut to reduce fines production which will artificially increase the proportion of finer grains.

Ideally, the samples should be cleaned and dried before disaggregation to minimise potential loss of the lower size fraction (fines and clays). Samples are cleaned and dried using hot Soxhlet or total immersion Soxhlet and conventional oven drying.

Once cleaned and dried, the samples are disaggregated into grains using a mortar and pestle. Individual grains should not be crushed by using a rotary motion of the pestle. The grains are separated by using a back and forth motion. An optical microscope is used to ensure grains have been separated and not crushed. Digital images at relevant magnifications of the disaggregated samples should be provided. If the rock is well cemented, it might be necessary to use a rubber hammer to initiate disaggregation, but once the pieces are around 2 cm in size, it should be possible to continue with a mortar and pestle. The final sample dry weight is determined.

12.11.2.2 Test Equipment

Particle size analysis using test sieves entails the passing of particles through screens (sieves) of ever-decreasing size. To effectively conduct sieve testing operations beyond the simplest manual sieve test, several components are needed. The basic element is the test sieve (Fig. 12.69). Because of its normal construction, the tolerance of the sieve openings from one sieve to another of the same stated size becomes an important detail. ASTM and ISO provide standards dealing with test sieve construction and acceptable variations of sieve openings. These openings range in size from 5″ down to 20 μm.

The usual process of conducting a sieve test is to place a stack of sieves on a device called a Sieve Shaker (Fig. 12.69). A sample is loaded into the top sieve and the machine moves in both vertical and horizontal directions to 'shake' the sieve nest. The test is completed when no further material passes through the bottom sieve in the stack.

In the shaker, the vertical throwing motion is overlaid with a slight circular motion which results in distribution of the sample amount over the whole sieving surface. The particles are accelerated in the vertical direction (are

FIGURE 12.69 Test sieves (left - https://upload.wikimedia.org/wikipedia/commons/8/82/Laboratory_sieves_BMK.jpg) and sieve shaker (right - https://upload.wikimedia.org/wikipedia/commons/f/f5/Laborsiebmaschine_BMK.jpg).

thrown upwards). In the air, they carry out free rotations and interact with the openings in the mesh of the sieve when they fall back. If the particles are smaller than the openings, they pass through the sieve. If they are larger, they are thrown upwards again. The rotating motion while suspended increases the probability that the particles present a different orientation to the mesh when they fall back again, and thus might eventually pass through the mesh. Modern sieve shakers work with an electromagnetic drive which moves a spring–mass system and transfers the resulting oscillation to the sieve stack. Amplitude and sieving time are set digitally and are continuously observed by an integrated control unit. Therefore, sieving results are reproducible and precise (an important precondition for a significant analysis). Adjustment of parameters like amplitude and sieving time serves to optimise the sieving for different types of material.

Different labs use different standard sieve sets. Examples are presented below:

Lab A (μm)	Lab B (μm)
2000	2350
1000	1700
710	1400
500	1000
355	710
250	500
180	425
125	355
90	300
63	250
45	180
Pan	150
	125
	90
	75
	63
	53
	45
	38
	Pan

12.11.2.3 Test Procedures

A representative weighed sample (typically 25–50 g) is poured into the top sieve which has the largest screen openings. Each lower sieve in the column has smaller openings than the one above. At the base is a round pan, called the receiver.

The shaker shakes the column, usually for 15–30 min. After the shaking is complete, the material on each sieve is weighed. Each sieve is struck on the outside of the rim a few times and is cleaned using a bristle brush to free any grains trapped in the mesh openings. Each sieve is then weighed individually.

To check the sand has been fully disaggregated, the contents on the largest sieve should be viewed under an optical microscope. If some of the grains are aggregates then this fraction should be reground in the mortar and pestle. Re-sieve either by hand or on the sieve shaker and then repeat the grain inspection until all the grains are present as individual grains. The procedure is repeated for each sieve (i.e. check the nature of the sieve contents and regrind if necessary) to the 125-μm sieve. After this point, viewing the grains becomes impractical.

Each sieve and its contents are weighed individually and by difference, the weight percentage retained on each sieve is determined. The weight of the sample on each sieve is then divided by the total weight to give a percentage retained on each sieve.

The sample fraction that passes the smallest micron sieve is often retained for potential LPSA. The results of LPSA on this sample can be mathematically combined with mechanical sieve data to provide a full curve for fines content determination in high fines content material.

Most sieve analyses are carried out dry. But there are some applications which can only be carried out by wet sieving. This is the case when the sample tends to agglomerate (mostly <45 μm). In a dry sieving process, this tendency would lead to a clogging of the sieve meshes and this would make a further sieving process impossible. A wet sieving process is set up like a dry process: the sieve stack is clamped onto the sieve shaker and the sample is placed on the top sieve. Above the top sieve, a water-spray nozzle is placed which supports the sieving process additionally to the sieving motion. The rinsing is carried out until the liquid which is discharged through the receiver is clear. Sample residues on the sieves have to be dried and weighed. When it comes to wet sieving, it is very important that the sample does not react with water—that is no swelling, dissolving or reaction with the liquid.

12.11.2.4 Data Utilisation

The results are normally provided in tabular format and in graphical format. For example, Table 12.19 lists the cumulative weight percentage retained on each sieve. In graphical format, the data are conventionally plotted as decreasing grain size versus cumulative weight, as illustrated in Fig. 12.70. Marked on the figure are key sizes such as: D10, representing the size at which 10%

TABLE 12.19 Example of Tabular Particle Size Data Presentation

Screen Opening (μm)	Sample ID and Depth (ft.)										
	Well 2										
	2693.08	*2695.16*	*2696.97*	*2699.09*	*2704.15*	*2707.86*	*2710.84*	*2810.78*	*2814.75*	*2818.20*	*2824.86*
2350	0.147	10.946	0.000	1.332	0.000	0.000	0.000	4.580	0.226	0.000	0.000
1700	0.982	17.049	0.051	4.023	0.000	0.000	0.000	11.896	0.561	0.000	0.000
1400	2.994	20.616	0.898	7.313	2.485	0.020	0.505	21.632	1.295	0.805	0.000
1000	10.737	27.537	1.652	13.096	6.512	0.068	1.591	39.440	5.490	2.261	0.036
710	25.277	38.452	4.198	22.410	17.210	0.200	4.059	63.970	21.361	8.969	0.116
500	46.754	58.103	15.191	37.890	32.358	0.838	15.317	79.603	44.895	28.847	0.184
425	59.629	72.296	30.191	51.546	51.051	4.035	29.431	82.981	54.325	41.211	0.439
355	67.459	81.625	44.603	60.877	63.909	12.026	42.870	86.362	66.284	55.049	2.291
300	76.399	86.953	61.347	70.973	77.643	30.242	60.648	88.417	73.972	62.902	7.488
250	82.522	89.450	70.764	76.270	83.843	44.785	71.450	90.708	81.761	73.608	21.635
180	90.145	93.264	83.172	84.481	90.552	74.631	85.122	93.616	88.785	85.343	53.017
150	92.075	94.294	86.529	87.538	92.415	82.444	88.219	94.840	91.371	88.839	64.376
125	93.921	95.484	89.705	90.971	94.273	88.216	91.228	95.901	93.144	91.177	75.665
90	96.040	96.927	93.386	94.571	96.502	92.796	94.494	96.950	95.048	93.384	83.686
75	97.200	97.792	95.457	96.224	97.556	95.339	96.123	97.726	96.300	95.390	90.376
63	97.972	98.356	97.034	97.736	98.221	97.171	97.287	98.273	97.183	96.697	94.602
53	98.517	98.815	97.865	98.433	98.809	98.096	98.051	98.826	97.834	97.706	97.267
45	99.058	99.243	98.610	99.250	99.259	98.721	98.664	99.206	98.521	98.406	98.398
38	99.413	99.679	99.106	99.566	99.542	99.276	99.081	99.585	99.239	99.045	99.215

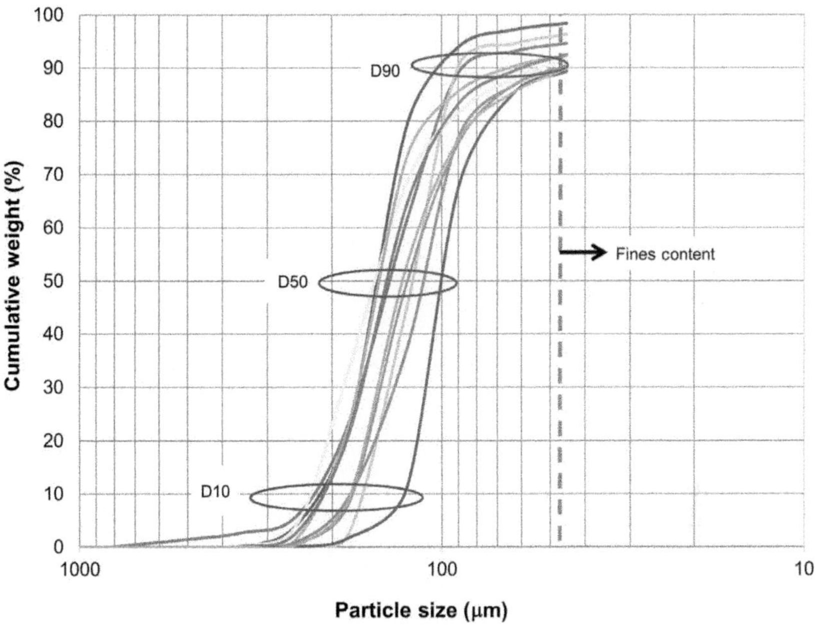

FIGURE 12.70 Example of graphical particle size data presentation.

of the particles are that size or larger; D50, representing the size where 50% of the grains are larger and 50% of the grains are smaller and D90 which is the size at which 90% of the particles are that size or larger.

The fines content is normally defined as the percentage of particles less than 44 or 45 μm. If the percentage of fines is high, then sieve analysis will not fully define the fines content and LPSA on the sub-45 μm fraction may be required.

The uniformity coefficient is defined as:

$$\mu = \frac{D40}{D90}$$

and the sorting coefficient as:

$$\phi = \frac{D10}{D95}$$

12.11.2.5 Data Reporting Requirements

The data that need to be reported by the test laboratory to ensure adequate quality control of the mechanical PSA tests and test results are listed in Table 12.20.

TABLE 12.20 Data Reporting Requirements for MPSA Tests

Data	Comments
Brief description of procedures and experimental apparatus	
Sample preparation methods	Sample source (e.g. plug or trim)
Digital images at relevant magnifications of the disaggregated samples	
Initial sample dry weight after disaggregation/crushing	
Weight and weight percentage retained on each sieve	
Semi-logarithmic particle size distribution chart, plotting cumulative fraction oversize as function of particle/sieve opening size, such that D0 is the maximum particle size and D100 is the minimum size	For example, Figs. 12.70 and 12.71
Tabulated cumulative particle size distribution data, detailing D10, D40, D50, D90, uniformity coefficient (D40/D90) and sorting coefficient (D10/D95)	

12.11.2.6 Advantages and Drawbacks/Issues

Summaries of the advantages and drawbacks/issues associated with mechanical PSA tests are listed in Table 12.21.

12.11.2.7 MPSA Test Quality Control Issues, Checks and Diagnostics

Compare the initial total sample dry weight with the sum of individual sieve sample weights to check for consistency.

When sampling, make sure only one rock type is being analysed on each test (especially in laminated sands), as bi-modal curves can be produced if different rock types are mixed in the same sample. An example is shown in Fig. 12.73. Sample 2 shows apparent poorer sorting and bi-modal particle size distribution due to laminations and not sorting.

Labs must carry out a thorough optical inspection of disaggregated sand retained in each sieve under a microscope (refer to Figs. 12.74–12.76) to check for crushed sand grains and grain accumulations that have not been completely disaggregated. The plot in Fig. 12.77 shows how sample preparation is critical and can lead to different results. Lab A failed to completely disaggregate the sample (example in Fig. 12.74), which resulted in a higher number of coarser particles. Figure 12.76 shows sand that has been properly disaggregated into its constituent grains.

TABLE 12.21 Advantages and Drawbacks/Issues with MPSA Tests

Advantages	Drawbacks and Issues
• Good and normally reliable method for medium and coarse sands where less than 5–10% of sample grain size will be smaller than 45 μm. • Sieve analysis will identify sand grains which are too large for accurate LPSA measurement (typically >750–800 μm). • Ideal for poorly consolidated sands as sample disaggregation issues are less significant. • In samples with low/no clay content, wet sieving (using water rather than dry sample) can be superior to dry sieving but test times are extended as sieve samples must be carefully dried prior to weight measurement.	• Correct sample preparation is critical to ensure samples have been correctly disaggregated and grain crushing has been prevented. • Relatively more expensive and time consuming than LPSA. • Not suitable for evaluation of poorly sorted fine sands, where more than 5–10% of the sample will be less than 45 μm. Calculating specific parameters (such as D90 and D95) will be impossible as fines 'tail' will be missing on PSD curve (Fig. 12.71). MPSA shows a greater cumulative percentage value for a given diameter than LPSA in poorly sorted fine sands (Fig. 12.72). • Sieves can easily blind off and usually the finer particles will need to be forced through, especially on the <90 μm sieves. • Screens can be potentially damaged during brushing and cleaning. • Fine particles can be poorly characterised by sieve analysis due to: finer grains becoming trapped on coarser particles from static and chemical forces; being lost as dust around the equipment; and plugging of sieve meshes.

If the sample contains significant amounts of common clay, then a PSA result, showing the sand to be well sorted, is probably inaccurate. If the sample contains negligible clay or cementation, then a PSA result showing the sand to be poorly sorted could be an indication of poor preparation (e.g. sample crushing).

Most core analysis laboratories offer this service but sample preparation is key to acquiring representative and reliable data, and their procedures and data must be carefully audited.

12.11.3 Laser Particle Size Analysis

12.11.3.1 Sample Preparation

The test requires a much smaller sand sample than MPSA, typically only 1–5 g. No specific sample geometry is required as samples will be disaggregated.

The samples are prepared for LPSA using the same careful procedures as for MPSA. However, obtaining a representative sample from the disaggregated material can be difficult. To try and overcome this problem, a riffler is

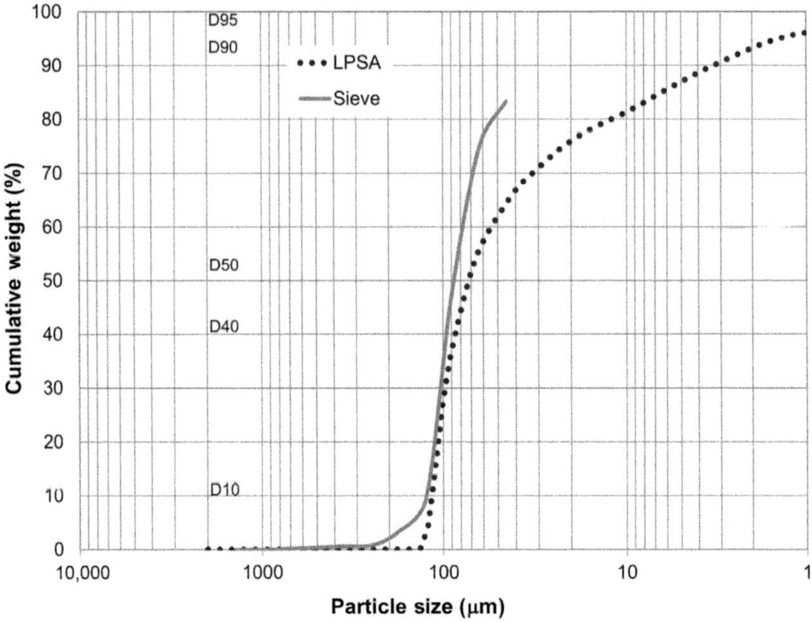

FIGURE 12.71 Example of particle size distribution curves.

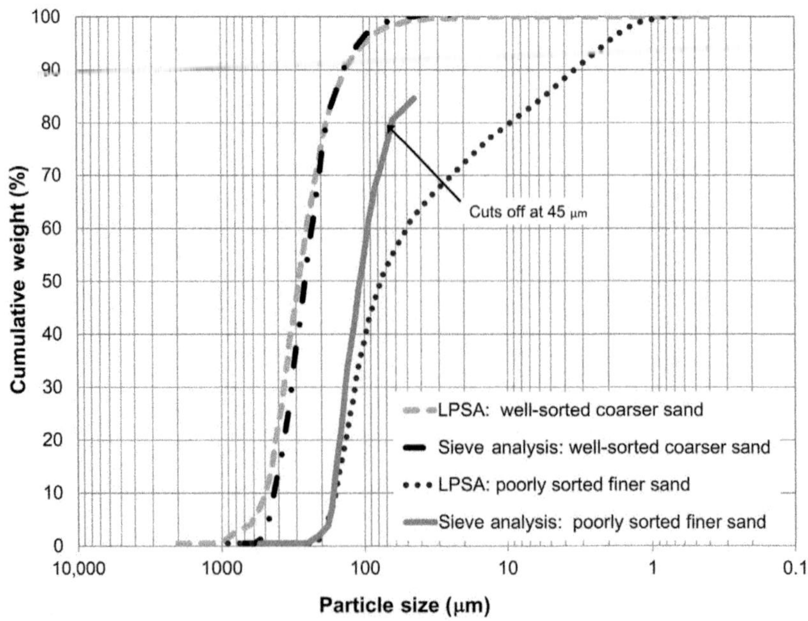

FIGURE 12.72 Sieve analysis versus LPSA results.

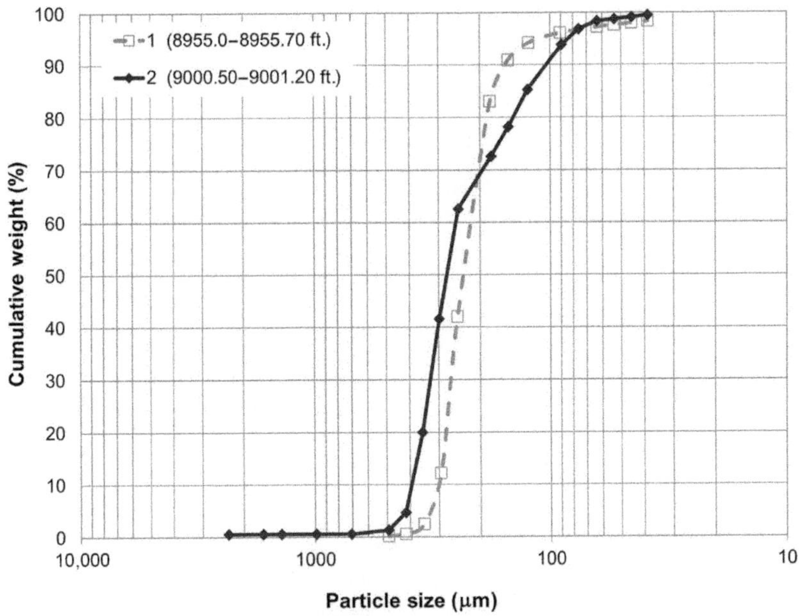

FIGURE 12.73 Example of bi-modal behaviour as a consequence of sampling in a section with different rock types.

FIGURE 12.74 Example of poorly disaggregated sand. *Courtesy of Corex UK.*

used to accurately divide the initial sample into two representative samples. A riffler is a device which has multiple ports directed into two different trays (Fig. 12.78). The sand sample is riffled down to the quantity required for an analysis. There should be two equal-sized samples in the riffler. Each sample should be measured twice for LPSA. In this way, four measurements are produced: duplicates on the same sample and the two different samples.

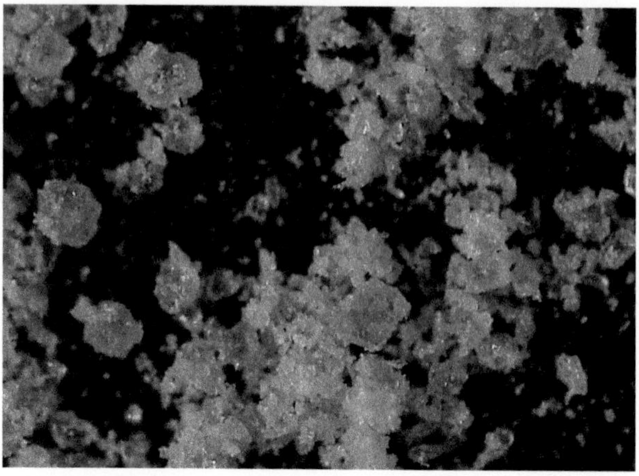

FIGURE 12.75 Example of crushed sand. *Courtesy of Corex UK.*

FIGURE 12.76 Example of correctly disaggregated sand. *Courtesy of Corex UK.*

Consequently, some indication of both the performance of the instrument and sample representativeness can be gained.

All samples are dispersed in 3% (weight by volume) SHMP (sodium hexametaphosphate) in deionised water and screened using a 710-μm sieve to remove any coarse material which the instrument is not able to analyse. If more than 5% by mass is collected in the 710-μm sieve, it is sieved separately through a set of mechanical sieves and the total is combined with the sub-710 μm LPSA fraction. SHMP can alter the chemical nature of clays through cation exchange and may artificially increase the fines content due to clay expansion and agglomeration. Where a sample contains swelling or sensitive clays, the LPSA is best performed using a non-aqueous fluid such as methanol.

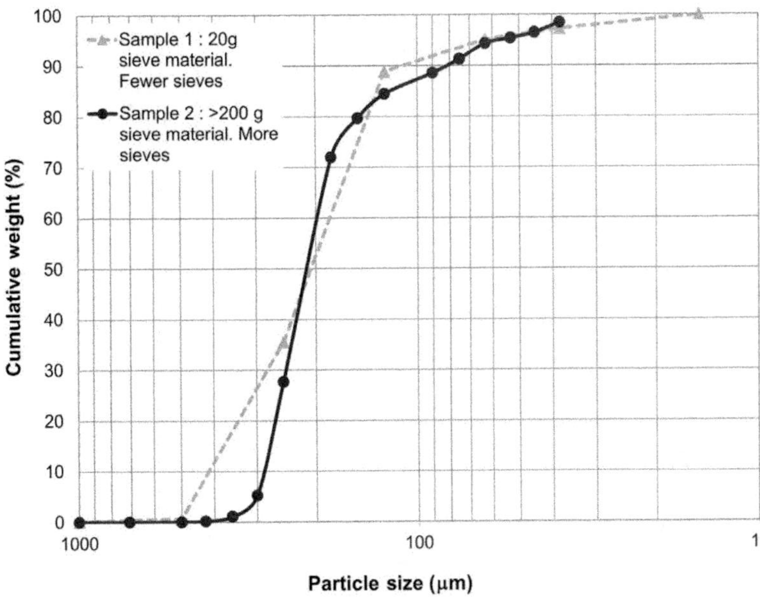

FIGURE 12.77 Different lab results on the same sample.

FIGURE 12.78 Example of a riffler. *Courtesy of Humboldt Manufacturing.*

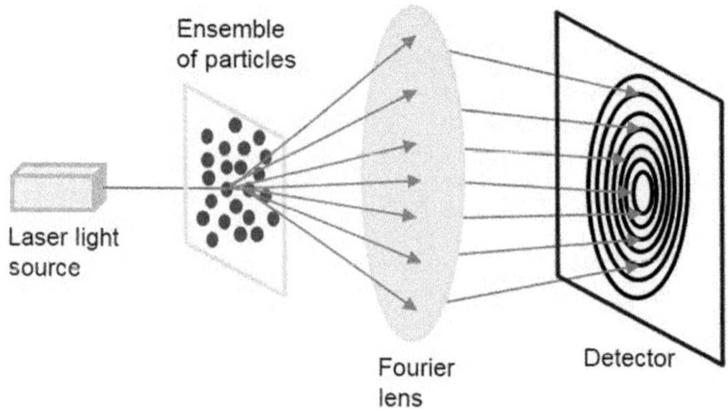

FIGURE 12.79 LPSA technique.

12.11.3.2 Test Equipment

In laser diffraction particle size analysis (Fig. 12.79), a representative cloud or 'ensemble' of particles passes through a broadened beam of laser light which scatters the incident light onto a Fourier lens. Detectors placed at fixed angles measure the intensity of light scattered at that position. The angle of diffraction increases as particle size decreases. Sizing particles using this technique depends upon accurate, reproducible, high-resolution light scatter measurements to ensure full characterisation of the sample.

The reproducibility of an analyser can be defined as the extent to which the same sample result is produced either on separate occasions on the same analyser or on different units of the same type of analyser. For laser diffraction particle size results, analyser-to-analyser reproducibility is primarily determined by how accurately the laser can be aligned, whereas measurement reproducibility on one instrument is primarily determined by how consistently the sample is prepared and how consistently the alignment can be maintained. Since the size of the material under test is determined by measuring the angles at which the light is scattered, an accurate and consistent 'angular zero' is vital. If the laser is not accurately aligned, the light scatter pattern will be observed at the incorrect angle, and hence systematic errors will creep into every measurement.

12.11.3.3 Test Procedures

The LPSA fraction is ultrasonically agitated at 25 watts for 1 min. After a further 1 min of circulation to allow air pockets to collapse, a series of measurements are made using the analyser. The resultant diffraction pattern is converted to a particle size distribution using a mathematical model (Mie or Fraunhofer Theory). The final result is reported on an Equivalent Spherical Diameter Volume basis. The statistical parameters obtained are calculated geometrically.

12.11.3.4 Data Utilisation

Data obtained from LPSA can be provided in the same format as for mechanical sieve analysis. This requires a grain density conversion as the LPSA data are based on volumetric analysis not weight analysis. Normally, a grain density of quartz (2.65 g/cc) is assumed but mixed mineralogy/heterogeneous sands can cause issues.

12.11.3.5 Data Reporting Requirements

In general, the same data reporting requirements for mechanical particle size distribution analysis apply (refer to Table 12.20).

12.11.3.6 Advantages and Drawbacks/Issues

Summaries of the advantages and drawbacks/issues associated with LPSA tests are listed in Table 12.22.

TABLE 12.22 Advantages and Drawbacks/Issues with LPSA Tests

Advantages	Drawbacks and Issues
• Good and normally reliable method for poorly sorted fine sands where a significant percentage of grain sizes is less than 45 μm. • Relatively cheaper and faster than MPSA.	• Correct sample preparation is critical to ensure samples have been correctly disaggregated and grain crushing has been prevented, and that the test samples are representative. • Representativeness of a small sample and quality of riffler can be an issue. This requires repeat tests on the same 'parent' sand sample. • May not be suitable for analysis of coarser grained sands unless done in conjunction with mechanical sieve tests (LPSA on sub-90 μm or sub-45 μm fraction). • Inappropriate for sand grains which are too large for accurate measurement (typically >750–800 μm). • If the fluid pump rate into the detection system is too high (turbulent flow), coarser particles can cause small currents (eddies), resulting in ripples being mismeasured as fines, i.e. fines overestimation. • Multi-scattering may occur if the particle concentration is too high, which can cause fines overestimation. • Signal-to-noise ratio may be insufficient if the particle concentration is too low. • Optical parts of the analyser can be damaged during cleaning. The diffraction unit in particular is subject to scratching. This will cause erroneous diffraction patterns and subsequent size distribution errors. • Failure to calibrate/service or incorrect calibration of the LPSA analyser can lead to significant errors in the results obtained. • Where a sample contains a swelling clay, the LPSA is best performed using a non-aqueous fluid such as methanol. • Assumes a single grain density for measured material in volumetric (not weight) percent calculations (e.g. in comparison with sieve results).

12.11.3.7 LPSA Test Quality Control Issues, Checks and Diagnostics

The LPSA tests are used for grain size analysis (sedimentology) and are crucial in the selection, sizing and operation of a sand control system and the prediction of long-term well production performance. They should not be considered to be standalone tests but instead need to be combined with other test results (especially mineralogy and petrographic tests) and mechanical PSA tests.

When sampling, make sure only one rock type is being analysed on each test (especially in laminated sands), as bi-modal curves can be produced if different rock types are mixed in the same sample.

Evaluate the representativeness of samples taking at least two halves and run two cycles for each half. Compare the four curves for overlap.

Test equipment needs to be calibrated/serviced regularly. Ask the lab for calibration/service certificates. Use of industry-standardised sand of known size can be used to calibrate the analyser and should be run on a regular basis. These sands (sand dust) are used in the filter industry to QC filters.

Evaluate fines overestimation caused by multi-scattering (check concentration) and small currents mis-measurement created by coarser particle transportation in turbulent flow (check pump rate).

Ensure the sample used is sufficient to give an acceptable signal (obscuration) but not too much to cause multiple scattering, resulting in an artificially high fines content. The correct level of obscuration must be checked but is usually ~20–30%.

Excessive stirring, to keep larger particles in suspension, will form air bubbles which will appear as coarse particles.

If budget allows, send similar samples to two different labs. Erroneous data are easily flagged.

Many commercial core analysis laboratories sub-contract LPSA services. The sub-contractor capabilities, resources and expertise should be verified with the core analysis company and approved by the client.

REFERENCES

Adam, L., Natzle, M., Brevik, I., 2006. Gassman fluid substitution and shear modulus variability in carbonates at laboratory seismic and ultrasonic frequencies. Geophysics 71 (6), F173–F183.

Akhoundzadeh, H., Moghadasi, A., Jamshid, J., Habibnia, B., 2011. Correlation of pore volume compressibility with porosity in one of the Iranian southern carbonate reservoirs. http://www.hsiran.com/ipec3-full-text/poster/3-132.pdf. 3rd Iranian Petroleum Engineering Congress, 1390.

Antheunis, D., Vriezen, P.B., Schipper, B.A., van der Vlis, A.C., 1976. Perforation collapse: failure of perforated friable sandstones. In: SPE European Spring Meeting, Amsterdam, Netherlands, 8–9 April, Paper SPE 5750.

ASTM, 1976. Standard Method for Laboratory Determination of Pulse Velocities and Ultrasonic Elastic Constants of Rock. American Society for Testing and Materials, West Conshohocken, PA, ASTM D2845-69.

Crook, T., Willson, S., Jian, G.Y., Owen, R., 2003. Computational modelling of the localized deformation associated with borehole breakout in quasi-brittle material. J. Pet. Sci. Eng. 38, 177–186.

Daniels, G., McPhee, C., McCurdy, P., Sorrentino, Y., 2012. Non-destructive strength index testing applications for sand failure evaluation. In: SPE Asia Pacific Oil and Gas Conference and Exhibition, Perth, Australia, 22–24 October, SPE Paper 158327.

Dyke, C.G., 1988. In Situ Rock Stress Indicators for Rock at Great Depth. Ph.D. thesis, University of London, Imperial College of Science and Technology, London, p. 361.

Ewy, R.T., Ray, P., Bovberg, C.A., Norman, P.D., Goodman, H.E., 2001. Openhole stability and sanding predictions by 3D extrapolation from hole collapse tests. SPE Drill. Complet. 16, 243–251.

Geertsma, J., 1973. Land subsidence above compacting oil and gas reservoirs. J. Pet. Technol. 25 (6), 734–744.

Hailwood, E.A., Ding, F., 1995. Palaeomagnetic reorientation of cores and the magnetic fabric of hydrocarbon reservoir sands. In: Turner, P., Turner, A. (Eds.), Paleomagnetic Applications in Hydrocarbon Exploration and Production. In: Geological Society Special Publication, vol. 98, pp. 245–258.

Hall, H.N., 1953. Compressibility of reservoir rocks. J. Pet. Technol. 5 (1), 17–19.

Horsrud, P., Sonstebo, E.F., Boe, P., 1998. Mechanical and physical properties of North Sea shales. Int. J. Rock Mech. Min. Sci. 35 (8), 1998.

ISRM, 1978. Suggested Methods for Determining Sound Velocity. International Society for Rock Mechanics, Lisboa, Portugal.

ISRM, 1979. Suggested Method for Determining the Uniaxial Compressive Strength and Deformability of Rock Materials. International Society for Rock Mechanics, Lisboa, Portugal.

Mavko, G., Mukerji, T., Dvorkin, J., 2009. The Rock Physics Handbook: Tools for Seismic Analysis of Porous Media, second ed. Cambridge University Press, Cambridge, ISBN: 9780521861366.

McCurdy, P., 2013. Core selection for successful sand management. In: Presentation at Society of Petroleum Engineers Advanced Technology Workshop on Practical Sand Management, Dubai, 22–24 April.

Morita, N., Whitfill, D.L., Nygaard, N., Bale, A., 1989. A quick method to determine subsidence, reservoir compaction and in-situ stress induced by reservoir depletion. J. Pet. Technol. 41 (1), 71–79.

Newman, G.H., 1973. Pore-volume compressibility of consolidated, friable, and unconsolidated reservoir rocks under hydrostatic loading. J. Pet. Technol. 25 (2), 129–134.

Papamichos, E., Vardoulakis, I., Tronvoll, J., Skjaerstein, A., 2001. Volumetric sand production model and experiment. Int. J. Numer. Anal. Methods Geomech. 25, 789–808.

Ray, P., Rijken, M., Cameron, J., Jones, C., El-Fayouni, A., 2014. Estimating sand production volume in oil and gas reservoir. In: SPE Annual Technical Conference and Exhibition, Amsterdam, Netherlands, 27–29 October, Paper SPE 170814.

Ren, N.K., Roegiers, J.C., 1983. Differential strain curve analysis—a new method for determining the pre-existing in situ stress state from rock core measurements. In: Proceedings of the 5th International Congress on Rock Mechanics, pp. F117–F127.

Smart, B.D.G., Somerville, J.M., Crawford, B.R., 1999. A rock test cell with true triaxial capability. Geotech. Geol. Eng. 17 (3–4), 157–176.

Suárez-Rivera, R., Stenebråten, J., Dagrain, F., 2002. Continuous scratch testing on core allows effective calibration of log-derived mechanical properties for use in sanding prediction evaluation. In: SPE/ISRM Rock Mechanics Conference, Irving, Texas, 20–23 October, SPE/ISRM Paper 78157.

Teeuw, D., 1971. Prediction of formation compaction from laboratory compressibility data. Soc. Pet. Eng. J. 11 (3), 263–271.

Tran, D.T., Pagoulatos, A., Sondergeld, C.H., Canh, N.N., Roegiers, J.C., 2010. Quantify uncertainty of rock failure parameters from laboratory triaxial testings using conventional and multistage approaches. In: Presented at 44th US Rock Mechanics Symposium, June, Paper ARMA 10-263.

van den Hock, P.J., Hertogh, G.M.M., Kooijman, A.P., de Bree, Ph., Kenter, C.J., Papamichos, E., 2000. A new concept of sand production prediction: theory and laboratory experiments. SPE Drill. Complet. 15 (4), 261–273.

Veeken, C.A.M., Davies, D.R., Kenter, C.J., Kooijman, A.P., 1991. Sand production prediction review: developing an integrated approach. In: SPE Annual Technical Conference and Exhibition, Dallas, Texas, 6–9 October, Paper SPE 22792.

Willson, S.M., Edwards, S.T., Crook, A., Bere, A., Moos, D., Peska, P., Last, N., 2007. Assuring stability in extended-reach wells—analyses, practices and mitigations. In: SPE/IADC Drilling Conference, Amsterdam, The Netherlands, 20–22 February, Paper SPE 105405.

RECOMMENDED READING

Aoki, H., Matsukura, Y., 2007. Estimating the unconfined compressive strength of intact rocks from Equotip hardness. Bull. Eng. Geol. Environ. 67, 23–29.

ASTM, 1986a. Standard Test Method for Triaxial Compressive Strength of Intact Rock Core Specimens Without Pore Pressure Measurement. American Society for Testing and Materials, West Conshohocken, PA, ASTM D 2664-86.

ASTM, 1986b. Standard Test Method for Unconfined Compressive Strength of Intact Rock Core Specimens. American Society for Testing and Materials, West Conshohocken, PA, ASTM D 2938-86.

ASTM, 1989. Standard Test Method for Direct Tensile Strength of Intact Rock Core Specimens. American Society for Testing and Materials, West Conshohocken, PA, ASTM D 2936-84.

ASTM, 1992. Standard Test Method for Splitting Tensile Strength of Intact Rock Core Specimens. American Society for Testing and Materials, West Conshohocken, PA, ASTM D 3967-92.

Ballard, T., Beare, S., 2003. Media sizing for premium sand screens: Dutch twill weaves. In: Society of Petroleum Engineers European Formation Damage Conference, The Hague, Netherlands, 13–14 May, Paper SPE 82244.

Ballard, T., Beare, S., et al., 2006. Sand retention testing—the more you do, the worse it gets. In: Society of Petroleum Engineers International Symposium and Exhibition on Formation Damage Control, Lafayette, Louisiana, USA, 15–17 February, Paper SPE 98308.

De Waal, J.A., Smits, R.M.M., 1985. Prediction of reservoir compaction and surface subsidence: field application of a new model. SPE Form. Eval. 3 (2), 347–356.

Fjaer, E., Holt, R.M., Horsrud, P., Raaen, A.M., Risnes, R., 1992. Petroleum-related rock mechanics. In: Developments in Petroleum Science, vol. 33. Elsevier, Amsterdam.

Hamilton, J.M., Shafer, J.L., 1991. Measurement of pore compressibility characteristics in rock exhibiting 'pore collapse' and volumetric creep. In: Society of Core Analysts Annual Conference, San Antonio, Texas, 21–22 August, SCA Paper 9124.

Hudson, J.A., (Ed.), 1993. Vol 3: Rock Testing and Site Characterization. In: Comprehensive Rock Engineering, vol. 3. Pergamon Press, Oxford.

ISRM, 1977. Suggested Method for Determining Tensile Strength of Rock Materials. International Society for Rock Mechanics, Lisboa, Portugal.

ISRM, 1979. Suggested Method for Determining the Uniaxial Compressive Strength and Deformability of Rock Materials. International Society for Rock Mechanics, Lisboa, Portugal.

ISRM, 1983. Suggested Method for Determining the Strength of Rock Materials in Triaxial Compression: Revised Version. International Society for Rock Mechanics, Lisboa, Portugal.

Jaeger, J.C., Cook, N.G.W., 1979. Fundamentals of Rock Mechanics, third ed. Chapman and Hall, London.

Lama, R.D., Saluja, S.S., 1974. Handbook on Mechanical Properties of Rocks: Testing Techniques and Results, vol. 1. Trans Tech Publications, University of Michigan.

Lama, R.D., Saluja, S.S., Vutukuri, V.S., 1978. Handbook on Mechanical Properties of Rocks: Testing Techniques and Results, vol. 4. Trans Tech Publications, University of Michigan.

Strickland, F.G., Ren, N.K., 1980. Use of differential strain curve analysis in predicting in situ stress state for deep wells. In: 21st U.S. Symposium on Rock Mechanics (USRMS), Rolla, Missouri, 27–30 May, Paper ARMA-80-0523.

Tiffin, D., King, G., Larese, R., Britt, L., 1998. New criteria for gravel and screen selection for sand control. In: Society of Petroleum Engineers Formation Damage Control Conference, Lafayette, Louisiana, 18–19 February, Paper SPE 39437.

Zimmerman, R.W., Somerton, W.S., King, M.S., 1986. Compressibility of porous rocks. J. Geophys. Res. 91 (12), 12765–12777.

Chapter 13

Example of a Core Analysis Programme

Chapter Outline

13.1 **Introduction** **781**
13.2 **Core Analysis Focal Point** **782**
13.3 **Design and Management** **783**
13.4 **Lithological Considerations for RCA and SCAL** **785**
13.4.1 Clean Consolidated Sandstones 785
13.4.2 Unconsolidated Sandstones 785
13.4.3 Shaly Sands 786
13.4.4 Carbonates 786
13.4.5 Vuggy Carbonates 786
13.4.6 Low-Permeability Reservoirs 787
13.4.7 Fractured Reservoirs 787
13.5 **Routine Core Analysis** **787**
13.5.1 Recommended RCA Tests 789
13.6 **SCAL Programme** **793**
13.6.1 Gas Field 793
13.6.2 Oil Field 801

13.1 INTRODUCTION

Core analysis, and/or processes that may affect core analysis results, begins during coring, continues into wellsite handling procedures and the rigours of shipment, and on into laboratory handling and procedures. At each stage, there is the potential for damage to, or alteration of, rock properties from possibly ill-conceived processes or use of improper fluids, for the particular reservoir lithology, mineralogy and properties being examined. It is therefore vital to plan the coring, core handling and core analysis procedures, specific to these reservoir properties (where known) and the required data acquisition.

Experience has evidenced that, although it is common practice for a number of the stakeholders to be involved during pre-coring meetings, not all stakeholders are always present. The acquisition of core can sometimes be undertaken without discussing the objective(s) of acquiring the core with some of the potential end-users (stakeholders). Thus, coring procedures are sometimes employed, which then preclude the measurement of necessary data and in the worst cases prevent the very analyses for which the core was intended.

Petrophysicists, geologists, reservoir engineers, drilling and completions engineers and team/asset managers all have a role to play in the planning team, which should also include the coring company, mud company, wellsite handling company and the laboratory that will carry out the work. Note that some of these tasks may overlap and be performed by a single company (for instance, either the coring company or the laboratory may often be involved with wellsite core handling). All the above are stakeholders in the coring and/or core project.

Planning provides the following benefits:

- quality improvement through planning and selection of the appropriate measurement suite;
- consensus-driven identification of core analysis needs obviates the need for later (and more expensive testing);
- defining time constraints so that the data can best impact on field development;
- roles and responsibilities of team members are clearly defined;
- improved sample selection and identification;
- greater confidence in, and better utilisation of, core data;
- improved project management;
- easier and more efficient presentation of core analysis plans and results to partners.

13.2 CORE ANALYSIS FOCAL POINT

It has been customary in some companies that the geologist or the petrophysicist is responsible for arranging the coring and core analysis programme. In some companies, reservoir engineers or completion/drilling engineers are responsible for controlling their own core testing (e.g. capillary pressure and relative permeability testing; rock mechanics tests). Recurring themes in many core laboratory audits are that the vendors feel that they are too often faced with conflicting and contradictory instructions from within the operator's different discipline functions. A single, knowledgeable, core analysis focal point, who understands the applications and limitations of core analysis tests, should be appointed. This should ensure that the data quality requirements are maintained and, more importantly, that the data are fit for purpose. The client focal point is the liaison between the client's different subsurface disciplines and the lab and is accountable for laboratory supervision and real-time quality control.

The laboratory should also appoint a project manager with the key responsibilities of client expectation management; organising and controlling the work; and documentation of the project requirements, test specifications, lab worksheets and analysed data. Peer review of both intermediate and final data is essential before delivery to the client.

13.3 DESIGN AND MANAGEMENT

The key steps in designing and then managing a core analysis programme are:

- Appoint the company focal point for the programme.
- Review the existing or legacy core analysis database. Are the data valid? Are there areas of concern or anomalies or suspicious core data in the database that need to be resolved? How well does the core, log and test data agree for the well in question and reservoir in general?
- Design and cost the programme with the assistance of the laboratory.
- Prepare a justification to management.
- Meet with drilling and wellsite engineers to review and specify core drilling, core recovery and wellsite handling, storage and transportation procedures.
- Design and specify the test and reporting procedures to be adopted in the scope of work. Detail any modifications that may be required to the test programme (for example, to resolve any anomalous or inconsistent results, as the data become available). Specify deliverables, milestones and project reporting requirements.
- Select laboratory contractor. Do not base the selection solely on price unless there is an overriding justification for this. The focal point should audit the tendering laboratories and assess their capabilities to perform the work to the required standard.
- Review project at regular intervals. Review contractor performance against initially set goals, objective and deliverables. Analyse and check the contractor's data as soon as possible after they are received.
- Prepare a final report on the routine core analysis (RCA) and/or special core analysis (SCAL) studies which will reconcile the core data with other well and reservoir data, and provide appropriately interpreted and reliable core analysis data that can be used for petrophysical and reservoir simulation models.

Regular monitoring of vendors' performance and the provision of, and checking, experimental data can ensure that any problems or unusual, anomalous or inconsistent results can be identified as soon as possible, so that they can be rectified before the test programme is completed. Thereafter, it is too late, and costs are often incurred in retesting which can lead to largely undeserved but lingering resentment over lab performance. Contractor supervision and quality control will ensure that complete records of the test methods and procedures together with laboratory experimental data are available, so that there is a complete audit trail.

Laboratory RCA and SCAL reports must include a detailed description of the work performed; the equipment and procedures used; and details of the methods used by the lab in analysing the data. Understanding the plug history is essential in QC of the SCAL data—particularly in formations sensitive to stress cycling and rock/fluid incompatibilities—but deciphering plug records from standard SCAL reports can be challenging. Figure 13.1 presents an example of a customisable single page plug sheet that charts the history of each sample through the plug preparation and testing sequence. Key results

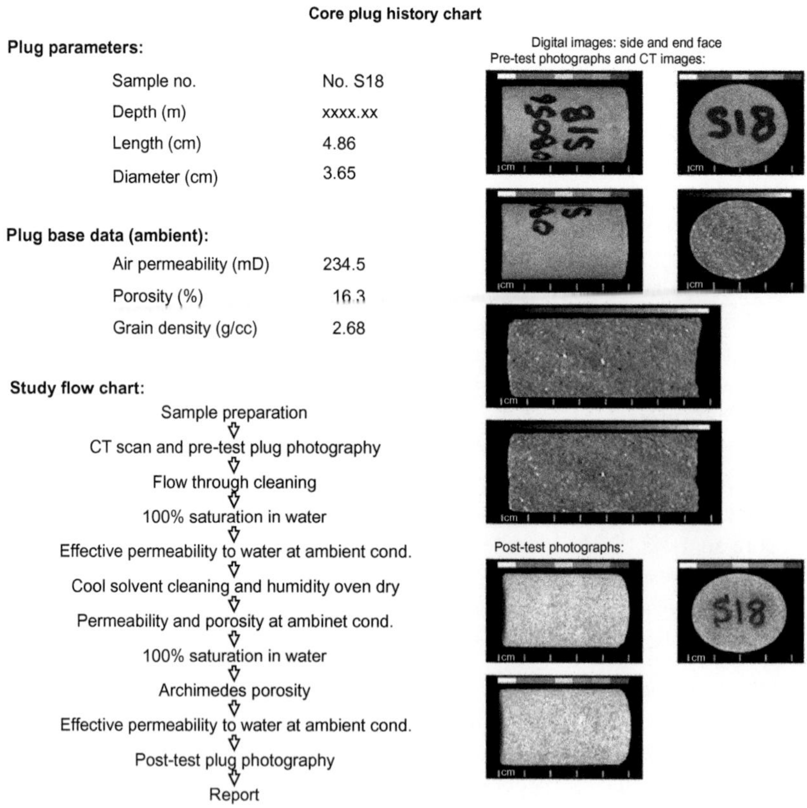

FIGURE 13.1 Example plug history sheet. *Courtesy of Woodside Energy Limited.*

can also be captured by this format. More effective knowledge sharing—e.g. highlighting of unusual or anomalous data in the report text, tables and figures—will allow the client to gain and benefit from the lab's undoubted expertise and experience in similar lithologies.

Relevant experimental data, details of appropriate instrument calibration data and the equations used to generate the analysed data from 'raw' measurements should also be provided. Labs often charge for compiling this information, and although the data might never be reviewed, they can prove invaluable in audit trailing and unitisation application.

In Senergy's experience, labs have enthusiastic, committed and highly experienced senior management teams, but they tend to be reactive and not proactive. This is not helped by the traditional master (client)–servant (lab) relationship. Engaging the labs through regular meetings and lab visits during ongoing projects means that they become more aligned with, and involved in, the client stakeholder objectives and better understand how important their data really are in field development planning and decisions.

This integrated approach will ensure that the end-user benefits from:

- appropriate and reliable core analysis data;
- improved integration of core test data with other well/reservoir data;
- cost savings through the elimination of wasteful or inappropriate testing.

13.4 LITHOLOGICAL CONSIDERATIONS FOR RCA AND SCAL

Core analysis considerations for different lithologies are indicated below.

13.4.1 Clean Consolidated Sandstones

- Coring is usually straightforward.
- All core analysis methods can be applied.
- Minimal Qv and cation-exchange capacity (CEC) analyses are required.
- Special core preparation precautions are normally not required beyond usual care and attention.
- Samples not usually sensitive to freshwater (e.g. coring and core plugging).

13.4.2 Unconsolidated Sandstones

- Requires specialist procedures for coring, core handling and transportation.
- Core stabilisation is required for transportation.
- Computer tomography (CT) scanning strongly recommended before plugging.

- Core sample preparation may be best done at stress by solvent flushing and lenient air drying.
- Most core analysis tests require specialist test methods to prevent plug damage.
- Most measurements should be done at stress.
- Core analysis planning is critical.

13.4.3 Shaly Sands

- Coring fluids should not cause clay swelling or migration.
- Core sample preparation for RCA and SCAL may differ depending upon data requirements (e.g. total porosity).
- Drill plugs with refined oil or synthetic formation brine.
- Qv and CEC measurements will be required.
- Air–brine and/or oil–brine capillary pressure measurements recommended to determine the effects of clay-bound water.
- Petrographic tests and clay typing recommended.
- Rock–fluid compatibility tests are recommended.
- Resistivity data may need correction for clay effects.
- Core analysis planning is critical.

13.4.4 Carbonates

- Coring is usually straightforward.
- Heterogeneity is expected between and within plugs.
- CT scanning strongly recommended before plugging.
- Whole-core analysis may be required.
- Qv and CEC measurements not normally required.
- Carbonates experience variable stress sensitivity.
- Compaction measurements important.
- Not usually freshwater sensitive (except chalks).

13.4.5 Vuggy Carbonates

- Coring can be difficult.
- CT scanning strongly recommended before plugging.
- Heterogeneity is expected between and within plugs.
- Whole-core analysis recommended to obtain representative properties.
- Bulk volume measurements suffer from mercury invasion.
- A high degree of mud invasion may occur.

13.4.6 Low-Permeability Reservoirs

- Sample properties are often difficult to measure.
- Planning is critical because all procedures take a significantly long time.

13.4.7 Fractured Reservoirs

- Coring is often difficult.
- CT scanning strongly recommended before plugging.
- Heterogeneity is expected between and within plugs.
- Whole-core analysis recommended to obtain representative properties.
- Sampling may be biased due to fragile nature of rock.
- A high degree of mud invasion may occur.

13.5 Routine Core Analysis

The following sections provide examples of RCA and SCAL programmes for both gas and oil reservoir core. These are only intended to provide a guide to the design decisions required and the potential programme workflows—they are NOT supposed to be applicable to every reservoir and every rock type. Each core analysis programme must be designed to meet the explicit data requirements, in the specific reservoir type, and specific lithology.

The design of a typical RCA programme is illustrated in the following examples from a competent rock on a gas or oil field.

The following requirements have been identified for an RCA programme on a typical well:

- core–log depth matching;
- water saturation (core drilled with OBM or doped WBM)
  - *NB*. OBM often has some water content that must also be doped;
- porosity and permeability (delicate clays present)
  - if delicate clays are known present, a subset may be required to undergo staged cleaning, to investigate effective versus total porosity effects;
- mercury injection capillary pressure curves (MICP);
- petrography;
- core photography;
- probe permeametry.

On completion of the RCA phase, SCAL, geomechanics and completion design programmes will be carried out.

The suggested general RCA programme workflow is provided in Fig. 13.2.

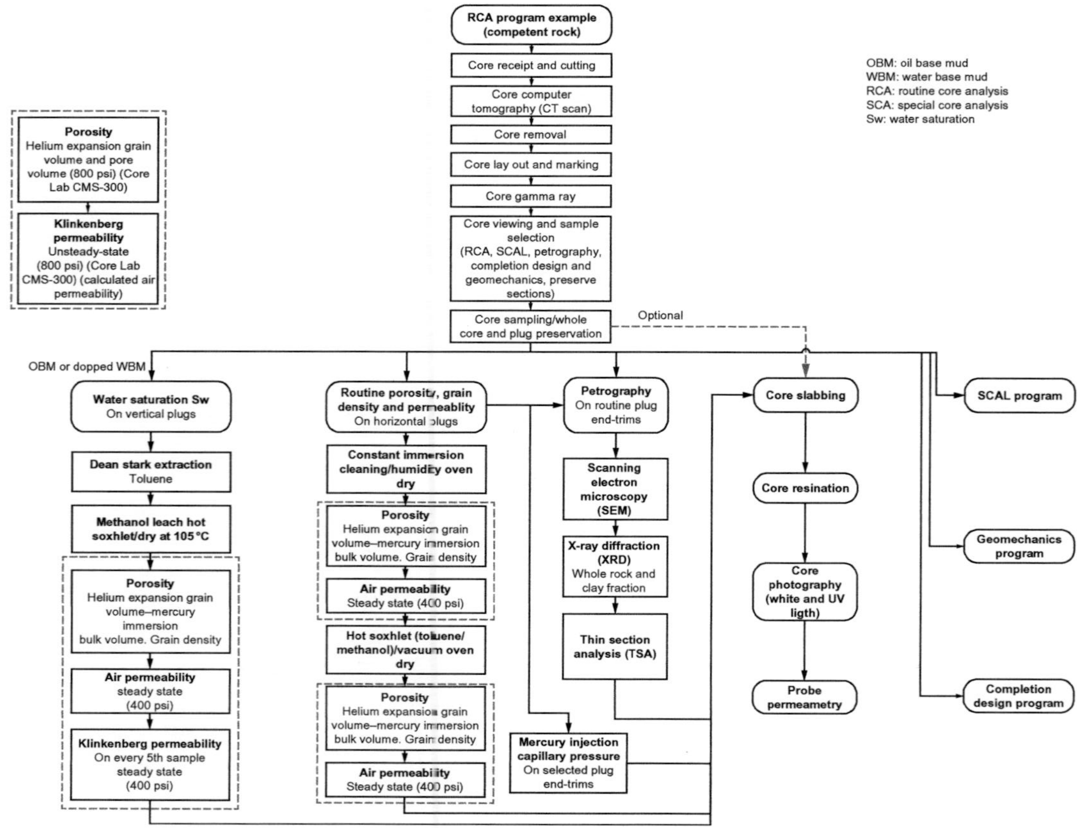

FIGURE 13.2 Example of an RCA programme workflow.

The anticipated tests and the data deliverables required for initial characterisation and to input for petrophysical and reservoir simulation models include (Table 13.1):

TABLE 13.1

Test	Purpose	Deliverable
Core CT scan	To document recovery, quality of recovered core, to locate any fracture zones, to evaluate any core damage, examine the degree of core heterogeneity, to observe bedding (if existing) for orientation and confirm suitability for further analyses. To aid in the selection of plugging sites (sample selection)	• Core photo CT scan images • Criteria for sample selection
Core gamma ray	To verify wellsite gamma ray, to correlate the depths of the well logs (core–log depth shifting) and to get an early idea of core recovery/quality	• Log of core depths versus core total and spectral gamma ray
Core viewing and sample selection	Inspect retrieved core to assess its quality and condition. Select sampling locations, sample dimensions, number of samples per test type, preserved sections, etc. using CT scans, well logs and core gamma ray as a guide The core analysis programme may need to be redesigned at this stage to take into account the actual core conditions This will need to be agreed among the different parties involved	• Core analysis programme is updated according to core characteristics • Tests are added or deleted accordingly • Lab personnel to recommend best sampling approach
Core sampling/ whole core and plug preservation	To take samples from the core following the core viewing/inspection/sample selection phase. Plugs to be preserved for SCAL and geomechanics tests	• Set of RCA plugs • Set of SCAL plugs (preserved) • Set of geomechanics plugs (preserved) • Whole-core preserved sections
Core slabbing (optional at this stage)	Allows for core preservation and preparation for core photography, probe permeametry and additional core logging. Recommended to slab the core once routine porosity and permeability data is available and it is decided that no further core plugging is required Slabbing may be required to reveal bedding for plug orientation, but slabbing before plugging limits the size of horizontal plugs	• Slab cuts depends on company/ regulatory authority requirements • One-third slab section (usually for regulatory authorities) • Two-third slab section (for future use)

Continued

TABLE 13.1—Cont'd

Test	Purpose	Deliverable
Water saturation (Sw) (Dean-Stark)	To obtain a direct measurement of extracted water volume (indirectly Sw), which can be used to develop quick-look saturation–height models (with MICP data) and to constrain saturation exponent estimates from resistivity index (RI) tests The technique is particularly suited to the case where the cores have been drilled with an oil-based mud and where the core water saturation has not been affected by mud filtrate flushing, or fluid flushing, on pressure expansion during core recovery. The core in the laboratory may be assumed to be at the same irreducible water saturation as in the reservoir. Plugs for this test are taken in the vertical direction, at the centre of the core were to minimise mud invasion Can be used with WBM provided mud doped with suitable tracers	• Water saturation (Sw) per plug (depth)
Routine porosity, grain density and permeability (horizontal plugs)	To measure base porosity and permeability of routine plugs Initially, the samples will be cleaned and dried following a mild procedure (constant immersion and humidity drying) as there are doubts about the influence of cleaning on delicate clays. Subsequently, a harsh cleaning and drying method will be followed; this is to allow for total porosity calibration with the logs. The methods and procedures for porosity and permeability measurement should be consistent with legacy data. For example, if lab propose unsteady-state Klinkenberg permeability measurements at 800 psi NCS and the legacy data were obtained using steady-state methods at 400 psi NCS, then the RCA programme should include both measurements on a representative subset of plugs to allow comparison between databases	• Porosity, permeability and grain density at ambient conditions for routine plugs cleaned following a mild preparation procedure • Porosity, permeability and grain density at ambient conditions for routine plugs cleaned following a harsh preparation procedure
MICP	At the RCA stage, MICP objective is to obtain relatively rapid drainage capillary pressure curves that can be used to develop a quick-look saturation–height model. Data can also be used during SCAL to design test pressure steps for drainage centrifuge capillary pressure. Data are also useful to design drilling muds for future drilling operations. Tests can be carried out on routine plug end-trims	• Pc versus Shg primary drainage capillary pressure curves • Entry pressures per sample

Petrography—scanning electron microscopy	SEM analysis provides visual information on clay and cement distribution, pore geometries and on the microstratigraphy of the mineralogical phases. Tests can be carried out on routine plug end-trims	• Full SEM description (including photomicrographs at varying magnification) per sample, documenting significant rock properties, pores, pore throats, clays, framework grains and cements • Pore controls on petrophysical parameters (including authigenic clays)
Petrography—X-ray diffraction	XRD analysis (whole rock and clay fraction (<2 μm)) provides semi-quantitative determination of sample mineralogy. Bulk and clay XRD analysis are performed to aid in reservoir description and evaluation of reservoir diagenesis. Identification of sensible clays and their proportion in the system can also provide useful information for sample preparation. Tests can be carried out on routine plug end-trims	• Whole rock and clay fraction composition of core samples (tabular format)
Petrography—thin-section analysis	TSA provides characteristics and distribution of detrital and authigenic minerals, including their grain shape, grain size, grain sorting and packing (textural features) and solution behaviour. The description and analysis permit the porosities/permeabilities to be traced on the basis of the properties of the detrital component and the plugging potential of authigenic minerals. Data may be semi-quantitatively substantiated by point counting. Tests can be carried out on routine plug end-trims	• Full TSA description (including photomicrographs) • Modal analysis (quantitative analysis of thin sections by point counting) • Mean grain size, sorting and point count determination of mineral abundances and porosity. Shale distribution, porosity varieties and abundances with a discussion of diagenesis, diagenetic sequence, porosity origins and development, controls on porosity and permeability

Continued

TABLE 13.1—Cont'd

Test	Purpose	Deliverable
Core slabbing	Allows for core preservation and preparation for core photography, probe permeametry and additional core logging	• Slab cuts depends on company/partners/regulatory authority requirements • One-third slab section (usually for authorities) • Two-third slab section (for future use)
Core resination	To preserve core integrity after slabbing. A 'biscuit' cut may be set in labelled trays. This can be used for core photography, probe permeametry, sedimentology analysis and visual inspection (easy to handle/store)	• Resinated slab (biscuit) in trays
Core photography (white and UV light (oil or gas condensate reservoir))	High-quality–high-resolution digital images of the core surface (resinated slab) are taken under white light to record sedimentary detail and ultraviolet light to reveal fluorescing hydrocarbons and minerals. Very useful for future core analysis QC and interpretation, as samples can be located without the need of physically viewing the core. Also useful for SCAL sample selection and for any future reference to the core	• White and UV light slabbed-core images (digital and printed format)

13.6 SCAL PROGRAMME

The following section provides examples of SCAL programmes for both gas and oil reservoir core. These have assumed that core was cut using WBM. They are intended only to provide a guide to the design decisions required and the potential programme workflows—they are NOT supposed to be applicable to every reservoir and every rock type. Each SCAL programme must be designed to meet the explicit data requirements, in the specific reservoir type, and specific lithology, for the specific coring process and necessary coring fluids used.

The design of possible SCAL programmes is illustrated by the following two examples: a gas field and an oil field.

13.6.1 Gas Field

The following requirements have been identified for the SCAL programme on this well:

Petrophysical parameters

- density log interpretation (porosity compaction transform);
- water saturation determination from resistivity logs (Archie parameters);
- saturation–height model (drainage capillary pressure);
- absolute ambient to effective overburden permeability transforms.

Dynamic parameters

- effective end-point permeability to gas at Swir (Kg@Swir);
- effective end-point permeability to brine at residual gas saturation, Sgr (Kw@Sgr);
- imbibition krg curve;
- initial versus residual gas saturation relationship.

The tests will be performed on preserved samples. The recommended general SCAL programme workflow is provided in Fig. 13.3.

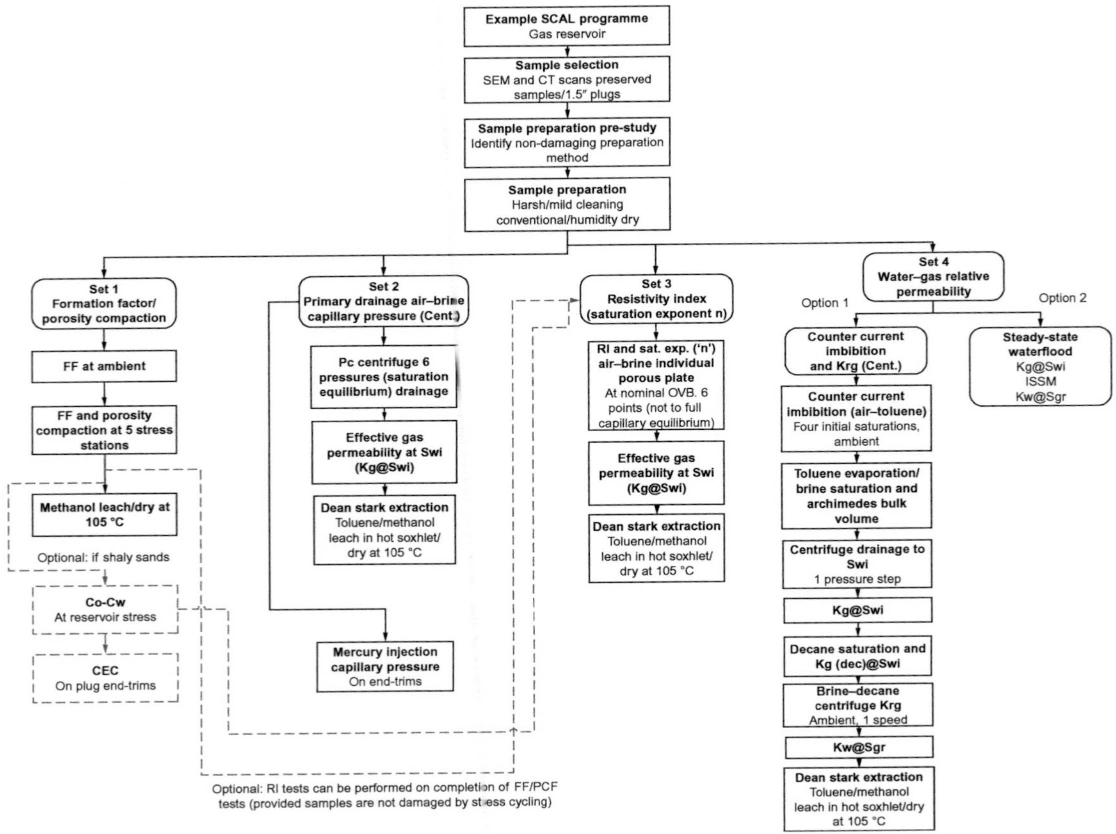

FIGURE 13.3 Example of a SCAL programme workflow—gas field.

The anticipated tests and the data deliverables required to minimise the current uncertainties in the petrophysical and reservoir simulation models include (Table 13.2):

TABLE 13.2

Test	Purpose	Deliverable
Core CT scan	To locate any fracture zones, evaluate any core/plug damage, examine the degree of core/plug heterogeneity, to observe bedding (if existing) and confirm suitability for further analyses. To aid in sample selection. CT scanning provides useful information on internal heterogeneity of the core (preserved section) and aids in assessing locations for cutting SCAL plug samples. If SCAL plugs have been preserved during the RCA programme, it is also recommended to carry out CT scan imaging on them	• Core photo CT scan images • Criteria for sample selection
Rock characterisation tests (SEM)	SEM analyses are essential in the design of the SCAL tests and in understanding and interpreting the results of the tests as well as assessing the requirement for a pre-study. This is normally carried out during the RCA programme	• Pore controls on petrophysical parameters (including authigenic clays)
Sample selection	The final SCAL test sample locations will be selected after review of the core CT scans and the log and RCA data from the well. The number of samples tested will depend on technical and budgetary objectives of the client	• Set of samples to be tested with designated measurements
Pre-screening study	Investigate optimum core preparation procedures. It is essential to preserve or prevent damage to any delicate detrital and authigenic components of the rock (e.g. clays); otherwise, the subsequent SCAL data will not represent the native formation properties. Illite, in particular, is a common pore occluding component in this formation A pre-cleaning study is scheduled, though this will depend on the results of the RCA programme. It is essential to identify the least damaging preparation method for the samples in order to maintain accuracy of the results. In the absence of delicate authigenic components, the optimum preparation method would be to Soxhlet clean the samples with methanol and conventional oven dry. If the difference is judged to be significant, all samples scheduled for main SCAL programme will be cleaned using a benign method (e.g. solvent flush cleaning and humidity drying)	• Optimum sample preparation procedure

Continued

TABLE 13.2—Cont'd

Test	Purpose	Deliverable
Porosity/formation factor tests at overburden (including optional brine permeability at overburden)	These tests will define Archie's cementation exponent 'm' at representative reservoir conditions. Data acquired at ambient conditions are not representative and, worse, suffer from laboratory artefacts. Porosity is also determined during the test sequence, and these measurements provide porosity compaction factors The equivalent hydrostatic stress at initial reservoir pressure must be calculated, and the test stress stations must be defined. Brine permeability can also be measured at each stress station (optional)	• Archie cementation exponent 'm' at reservoir stress (input for log calibration to calculate Sw) • Porosity compaction factor for density log interpretation • Absolute ambient to brine overburden permeability transform (optional)
CEC (destructive)	To determine the CEC of the rock (Qv) as an input to the Waxman–Smits shaly sands equation for log calibration. This is the most widely used technique because it is cheap and rapid. It can be easily done for large number of samples. The method can be used as a first pass to have an idea of the CEC of the sample. If the value is low, an Archie Sw model should be used. Test can be performed on end-trims or any available sample material—there are no specific shape requirements. However, sample preparation process tends to overestimate CEC	• CEC of the rock (Qv)
Multiple salinity method (Co–Cw) (non-destructive) (optional in shaly sands)	To determine the CEC of the rock (Qv) as an input to the Waxman–Smits shaly sands equation for log calibration. Qv can be determined at representative overburden stress in plugs on intact samples. CEC data used to select Co–Cw test samples	• CEC of the rock (Qv) on intact samples
Air–brine drainage capillary pressure (centrifuge) tests with final end-point effective gas permeability at Swi at overburden	Drainage Pc curve using appropriate fluid system (air–brine) Absolute to effective and ambient to overburden permeability transforms (option). Centrifuge tests are required to create a true capillary pressure curve, since porous plate tests would never attain residual saturation without resorting to extremely lengthy and costly desaturation periods. Moreover, since mercury injection can underestimate the inferred irreducible water saturation at high capillary pressures, the capillary pressure curve generated	• Saturation–height model (in conjunction with MICP data) • Kg at Swir • Absolute ambient to effective overburden permeability transforms

	by the centrifuge test will complement and will be merged with the air–mercury curves to yield fluid-appropriate drainage capillary pressure curves for saturation–height modelling Given the expected hydrocarbon column height, the tests must be performed using an ultra-centrifuge to achieve the equivalent capillary pressure On completion of the centrifuge test, the end-point effective gas permeability, Kg at Swir, is determined as a function of overburden pressure. These data will supplement similar data from water–gas relative permeability tests and can be used • to determine an absolute ambient to overburden effective permeability transform and • to convert all routine core air permeabilities to reservoir condition data This is essential when comparing core permeability with well-test permeability and in creating reservoir condition permeability maps Dean–Stark extraction is performed to corroborate the final plug water saturation	
MICP	Air–mercury injection tests are capable of defining the full capillary drainage pressure curves used to develop a saturation–height model, as well as providing useful pore size distribution (PSD) data. Although not strictly capillary pressure tests, as an irreducible wetting phase saturation is never attained, the tests provide an adequate description of saturation–height relationships in most gas reservoirs. It can be used in conjunction with centrifuge capillary pressure data and used as an input to design pressure steps	• Pc versus Shg primary drainage capillary pressure curves • Saturation–height model (in conjunction with centrifuge Pc data)
RI tests	These tests will provide data on Archie saturation exponent 'n' at nominal stress conditions. 'n' is not normally sensitive to stress, but tests are performed at a nominal overburden stress to minimise grain loss on handling and to minimise laboratory artefacts prevalent at ambient conditions	• Archie saturation exponent 'n' (input for log calibration to calculate Sw)

Continued

TABLE 13.2—Cont'd

Test	Purpose	Deliverable
	The maximum air–brine capillary pressure that is possible with the standard porous plate system is about 180 psi (which is equivalent to a gas column of around 270 ft.). This potential limitation coupled with the time required to achieve capillary saturation equilibrium in porous plate tests in low-permeability material means that the drainage process is unlikely to reach a truly equilibrated water saturation within a reasonable experimental time frame. However, the primary objective of these tests is to determine the RI–brine saturation relationship for each plug, not necessarily to determine an equilibrium air–brine capillary pressure relationship, since resistivity equilibrium is normally established much more rapidly than capillary pressure equilibrium Dean-Stark extraction is performed to corroborate the final plug water saturation As an option, RI tests can also be performed on completion of formation resistivity factor and porosity compaction tests, and provided samples have not been damaged by stress cycling If appropriate Dean–Stark data have been acquired in the RCA programme (core from intervals where water mobility is low; core drilled with OBM; core drilled with doped WBM and saturations corrected for invasion), then inverse modelling of core Sw versus log-derived Rt can be used to calculate 'n' under appropriate reservoir conditions. This may obviate the need for expensive and time consuming RI tests	
Residual gas CCI (countercurrent imbibition) tests	The tests provide a composite residual gas versus initial gas saturation relationship. The test mimics an imbibition process dominated by capillary forces and involves imbibition of liquid from various initial gas saturation values, covering the range in the productive reservoir zones. From the data generated in other reservoirs, CCI tests have proved to provide a simple yet reliable simulation of water imbibition processes but normally provide a maximum bound of Sgr due to gas trapping on plug immersion	• Sgr versus Sgi relationship

Water–gas imbibition relative permeability	To fully describe water movement in the reservoir as a function of gas production or reservoir depletion (e.g. aquifer influx, coning), imbibition water–gas curves are required. Imbibition water–gas tests are extremely difficult to do and are costly to perform. Dynamic tests using unsteady-state water-displacing gas methods can often yield a residual gas saturation that is too low due to the extremely favourable mobility ratio of the water flood and the imposition of viscous forces which overcome the capillary forces that trap gas at residual saturations. This is a particular problem in low-permeability samples where the viscous forces (even at low rates) are high. Steady-state tests, in contrast, often provide a Sgr that is too high All gas–liquid flood tests are prone to disequilibria, i.e., mass transfer between the gas and the liquid, leading to errors in end-point saturations. Differential pressure must be controlled as a function of the pore pressure (recommend $<2\%$) to reduce such effects The centrifuge technique proposed is a compromise with decane replacing the gas phase (both will be non-wetting in this formation). In centrifuge tests, the relative permeability obtained is that of the phase that is being displaced—not the displacing phase. Thus, water-displacing-decane is used to generate an imbibition cycle gas equivalent relative permeability. Decane is used as a replacement for gas due to compressibility and diffusion issues experienced when performing tests using gas. The IFT for water–decane systems at lab conditions is also close to that for gas–brine at reservoir conditions of ~ 50 dynes cm^{-1}. The use of a capillary dominated flood also better replicates the potential field conditions (aquifer influx) than a dynamic test Dean–Stark extraction is performed to corroborate the final plug water saturation	• Kg@Swir • krg (water imbibition) • Kw@Sgr

Continued

TABLE 13.2—Cont'd

Test	Purpose	Deliverable
Steady-state water–gas (imbibition) relative permeability	To determine intermediate krw and krg imbibition curves for comparison with centrifuge tests (krg only) Steady-state tests in which water and gas are injected simultaneously require non-invasive saturation monitorirg to determine accurate saturations All gas–liquid flood tests are prone to disequilibria, i.e., mass transfer between the gas and the liquid, leading to errors in end-point saturations. Differential pressure must be controlled as a function of the pore pressure (recommend <2%) to reduce such effects Requires careful design of test rates to avoid flow instabilities. Sgr may be too high but data can be scaled using end-point scaling Injection and production test data required for coreflood simulation Imbibition Pc data also required tc enable coreflood simulation	• Intermediate krw and krg imbibition curves

13.6.2 Oil Field

The following requirements have been identified for a possible SCAL programme on a well cored with WBM, observing little or no significant invasion.

NB. Note that use of OBM, and particularly SBM, may significantly alter test programme recommendations (for instance, fresh-state analyses are not normally recommended/possible, additional wettability analysis will usually be required: to verify clean state and to determine restored-state wettability, etc.)

Petrophysical parameters

- Density log interpretation (porosity compaction transform);
- Water saturation determination from resistivity logs (Archie parameters);
- Saturation–height model (drainage capillary pressure);
- Absolute ambient to effective overburden permeability transforms.

Dynamic parameters

- Wettability;
- Relative permeability;
- Imbibition capillary pressure.

Fluid properties, such as brine resistivity, brine and oil density and viscosity, and oil/water interfacial tension (IFT—of test fluids used and reservoir fluids), are required for laboratory material balance, permeability determination and capillary pressure data conversion to reservoir conditions.

Wettability data are essential in understanding the dynamic properties of the reservoir. Representative wettability is vital to ensure correct fluid dynamics during imbibition and secondary drainage processes. The use of WBM may allow plug samples to be tested in 'fresh state', which should provide a reasonable approximation of the native reservoir condition. Mud systems must be carefully considered whenever dynamic properties are required, since certain mud additives may irreversibly alter wetting conditions, leading to incorrect and meaningless data.

The tests will be performed on preserved samples.

Figure 13.4 provides a general diagrammatic workflow for the oil reservoir SCAL programme.

13.6.2.1 Recommended SCAL Tests – Oil Field

Since wettability is a key issue for capillary pressure and relative permeability experiments, these tests cannot proceed until the wettability tests have been completed and analysed. There is no constraint on the schedule for the non-wettability-dependent (petrophysics) tests.

The anticipated tests and the data deliverables required to minimise the possible uncertainties in the petrophysical and reservoir simulation models include (Table 13.3):

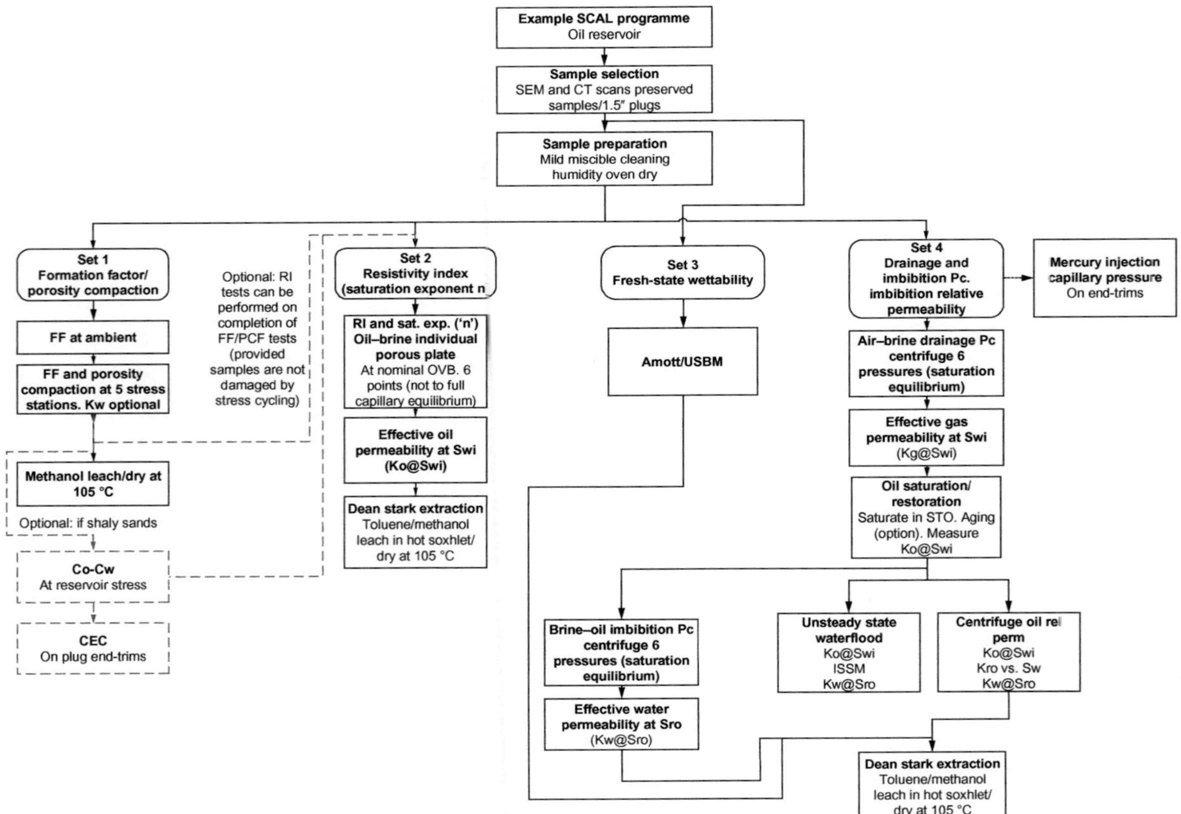

FIGURE 13.4 Example of a SCAL programme workflow—oil field

TABLE 13.3

Test	Purpose	Deliverable
Core CT scan	To locate any fracture zones, evaluate any core/plug damage, examine the degree of core/plug heterogeneity, observe bedding (if existing) and confirm suitability for further analyses. To aid in sample selection. CT scanning provides useful information on internal heterogeneity of the core (preserved section) and aids in assessing locations for cutting SCAL plug samples. If SCAL plugs have been preserved during the RCA programme, it is also recommended to carry out CT scan imaging on them to ensure suitability for testing	• Core photo CT scan images • Criteria for sample selection
Rock characterisation tests (SEM)	SEM analyses are essential in the design of the SCAL tests and in understanding and interpreting the results of the tests. This is normally carried out during the RCA programme	• Pore controls on petrophysical parameters (including authigenic clays)
Sample selection	The final SCAL test sample locations will be selected after review of the core CT scans and the log and RCA data from the well. The number of samples tested will depend on technical and budgetary objectives of the client	• Set of samples to be tested with designated measurements
Sample preparation	To remove saturating fluids and to retrieve water-wet samples effectively removing OBM and other contaminants. Although the cleaning procedures should render the core water wet, this cannot be guaranteed. This is attributable to difficulties in removing all adsorbed oil from a mixed wettability rock and is a particular problem when the native crude has a high resin and/or asphaltene content The success of the cleaning process to produce an initial water-wet state will depend on both the rock and original fluid types, and the chosen cleaning method and solvents Miscible flow-through (flush) cleaning represents the best approach because this method does the least damage to the core, and with the appropriate solvents will render most core material suitably water wet. Miscibility of the solvent mix with pore contaminants (oil/brine/salt/asphaltenes) is important to assure against potential damage from movement of interfacial tension forces over fine mineral structures, as is a controlled, low injection rate A toluene–chloroform–methanol solvent sequence acts simultaneously to remove oil and water from the core and dissolves crude oil components ranging from acidic to basic	• Samples prepared for testing

Continued

TABLE 13.3—Cont'd

Test	Purpose	Deliverable
	Hot Soxhlet extraction in toluene, because it vapourises water, permits the crude oil components to come in contact directly with mineral surfaces, which has been shown to alter restored wettability, whereas the flow-through cleaning process can prevent this. In addition, the flow-through process, followed by humidity drying, is less damaging to sensitive clays, and provided flow rates are maintained below critical rates	
Porosity/formation factor tests at overburden (including optional brine permeability at overburden)	These tests will define Archie's cementation exponent 'm' at representative reservoir conditions. Data acquired at ambient conditions are not representative and, worse, suffer from laboratory artefacts. Porosity is also determined during the test sequence, and these measurements provide porosity compaction factors Test pressure stations and the equivalent hydrostatic stress at initial reservoir pressure calculations need to be calculated. Brine permeability can also be measured at each pressure step (optional)	• Archie cementation exponent 'm' at reservoir stress (input for log calibration to calculate Sw) • Porosity compaction factor for density log interpretation • Absolute ambient to brine overburden permeability transform (optional)
CEC (destructive)	To determine the CEC of the rock (Qv) as an input to the Waxman–Smits shaly sands equation for log calibration. This is the most widely used technique because it is cheap and rapid. It can be easily done for large number of samples. The method can be used as a first pass to have an idea of the CEC of the sample. If the value is low, Archie model should be used. Test can be performed on end-trims or any available sample material—there are no specific shape requirements. However, sample preparation process tends to overestimate CEC	• CEC of the rock (Qv)
Multiple salinity method (Co–Cw) (non-destructive) (optional in shaly sands)	To determine the CEC of the rock (Qv) as an input to the Waxman–Smits shaly sands equation for log calibration. Qv can be determined at representative overburden stress in plugs on intact samples. CEC data used to select Co–Cw test samples	• CEC of the rock (Qv) on intact samples
RI tests	These tests will provide data on Archie saturation exponent 'n' at nominal stress conditions. 'n' is not normally sensitive to stress, but tests are performed at a nominal overburden stress to minimise grain loss on handling and to minimise laboratory artefacts prevalent at ambient conditions	• Archie saturation exponent 'n' (input for log calibration to calculate Sw)

	The test schedule calls for air–brine porous plate tests as it is assumed that the reservoir is water wet. If not, and as 'n' is wettability dependent, the tests should be carried out using a stock-tank oil (STO)—brine system at reservoir temperature (the samples effectively 'age' during primary drainage), but this is considerably more expensive and can take between 20 and 30 weeks to complete The maximum oil–brine capillary pressure that is possible with a standard porous plate system is about 100 psi. This potential limitation, coupled with the time required to achieve capillary saturation equilibrium in porous plate tests in low-permeability material, means that the drainage process is unlikely to reach a truly equilibrated water saturation within a reasonable experimental time frame. However, the primary objective of these tests is to determine the RI–brine saturation relationship for each plug, not necessarily to determine an equilibrium capillary pressure relationship, since resistivity equilibrium is normally established much more rapidly than capillary pressure equilibrium Dean–Stark extraction is performed to corroborate the final plug water saturation As an option, RI tests can also be performed on completion of Formation Resistivity Factor and Porosity Compaction tests, provided samples have not been damaged by stress cycling If appropriate Dean–Stark data have been acquired in the RCA programme (core from intervals where water mobility is low; core drilled with OBM; core drilled with doped WBM and saturations corrected for invasion) then inverse modelling of core Sw versus log-derived Rt can be used to calculate 'n' under appropriate reservoir conditions. This may obviate the need for expensive and time consuming RI tests	
Fresh-state wettability	To determine the wettability of the core retrieved from the reservoir as an aid for imbibition test core preparation. Wettability tests are to be carried out on fresh-state core. Cores drilled with WBM are not suspected to be contaminated with wettability-altering components from the mud filtrate. However, some wettability alteration from the native oil could occur as a result of loss of pressure and temperature on core recovery (which is unavoidable), or if the samples have not been preserved properly and have become exposed to air (which is avoidable). The potential for wettability contamination from the native oil exists, but since the core	• Fresh core wettability • Input to decide if ageing is required prior to imbibition tests

Continued

TABLE 13.3—Cont'd

Test	Purpose	Deliverable
	is likely to have been flushed towards Sro on filtrate invasion, excessive wettability alteration is not anticipated. Consequently, the results of the wettability tests on fresh-state core should provide a reasonable approximation of the native reservoir wettability. Tests on water-leg samples are recommended as the formation should be water wet so that wettability contamination from contact with mud, or due to core recovery, can be assessed The wettability tests are performed using the combined Amott/USBM method. This has two principal advantages over the standard USBM test: the resolution of the USBM method is improved by accounting for the saturation changes that occur at zero capillary pressure (spontaneous imbibition), and the Amott indices are also calculated On completion of the tests, the plugs are extracted in a Dean–Stark apparatus to provide an independent verification of the final water saturation (1 – Sro) calculated from material balance It is essential that the dynamic tests and imbibition capillary pressure tests are performed on samples whose wettability characteristics are representative of the reservoir zones. The results from the wettability study will therefore be used to decide the sample conditioning procedures for the both the dynamic and imbibition capillary pressure tests. If the fresh-state results indicate a water-wet tendency then the tests could be performed on cleaned-state core Although coring with WBM provides an opportunity to perform dynamic tests of fresh-state core (with potentially appropriate wettability), this is not recommended as: • quality control of the SCAL test saturation data is compromised as there is no opportunity to determine initial plug dry weights and pore volumes • WBM filtrate invasion is a primary drainage process, so subsequent fresh-state testing will represent secondary drainage and/or imbibition cycles Consequently, tests should be performed on cleaned core so that base parameters can be measured. The dynamic tests are then performed n cores saturated with mineral oil (if water wet) or on restored-state core	

Oil–brine drainage capillary pressure (centrifuge) tests	These data are used in saturation–height modelling. The proposed test protocols permit drainage capillary pressure to be determined during the plug conditioning procedures for imbibition Pc and relative permeability tests Centrifuge tests are used to determine capillary pressure curves in with correct fluid system, since porous plate tests may never attain residual saturation without resorting to extremely lengthy, and costly, desaturation periods	• Saturation–height model (in conjunction with MICP data)
MICP	Air–mercury injection tests are capable of defining the full capillary drainage pressure curves used to develop a saturation–height model, as well as providing useful PSD data. It is not strictly a capillary pressure test, as irreducible wetting phase saturation is never attained. It can be used in conjunction with centrifuge capillary pressure data and used as an input to design pressure steps	• Pc versus Shg primary drainage capillary pressure curves • Saturation–height model (in conjunction with centrifuge Pc data)
Oil saturation/ wettability restoration	To condition the samples to the 'appropriate' wettability prior dynamic testing. The samples, at Swi, are saturated in STO If the fresh-state wettability tests indicate a non-water-wet tendency, the samples are saturated in STO and base (end point) oil permeability (Ko at Swi) established prior to ageing. The samples are aged in STO at reservoir temperature and elevated pressure. To avoid excessive delays to the SCAL programme, a period of 500 h (20 days) ageing is advised Following ageing, the STO is replaced with refined mineral oil. Mineral oil can change the wettability of the core due to the non-equilibrium between the crude oil components adsorbed on the mineral surface and the composition of the mineral oil. The effect can cause either less oil-wetting tendency in the case of desorption or increased oil-wetting tendency in the case of asphaltene in the crude oil. However, this uncertainty must be balanced by practical considerations. STO is often difficult to handle (e.g. asphaltene precipitation, filtration and emulsion problems) and can produce problems during dynamic flood tests. If necessary, the crude oil may have to be displaced with decalin (or similar) to prevent precipitation from the crude oil before flushing with mineral oil. The mineral oil is selected or blended such that the ratio of oil/water viscosity at lab conditions matches the reservoir ratio If the wettability tests indicate a water-wet tendency, the samples may not require to be restored and are simply saturated in the mineral oil Base Ko at Swir is determined	• Samples with wettability restored for imbibition tests

Continued

TABLE 13.3—Cont'd

Test	Purpose	Deliverable
Brine–oil imbibition (centrifuge) capillary pressure tests (with final end-point effective brine permeability at Sro at overburden)	The primary imbibition cycle Pc data are initially acquired in an Amott cell (spontaneous imbibition) then in a centrifuge under brine (forced imbibition) There is a risk that the imbibition Pc curve in a water-wet sample may not be adequately defined by this test. Curve features, especially in high permeability samples, might be masked by rapid imbibition at low capillary pressures below the minimum rotor speed of the centrifuge. It is possible to perform the test using porous plate techniques, but these are slow and time consuming and, from experience, the integrity of the oil-wet membrane required for imbibition experiments cannot be guaranteed throughout the test period On completion of the centrifuge imbibition stage, with the core at Sro, the sample is removed, placed in a coreholder and flushed with brine (at a differential pressure less than the applied capillary pressure) and the end-point effective permeability to brine at ultimate oil saturation, Kw at Sro is determined. Dean-Stark extraction is used to verify the final Sw (1 − Sro)	• Imbibition capillary pressure curve
Unsteady-state imbibition waterflood	Oil and water relative permeability curves. Tests are performed at low rates (accepting that capillary end-effects could be present) using viscosity-matched fluids (where practicable) with *in situ* saturation monitoring (ISSM) to determine saturation distribution as a function of displacement force. End-point effective water permeability is measured before continuing flooding at an increased flow rate (bump flood)—usually a magnitude higher rate—to observe the effect of increased displacement force on the capillary end-effect, followed by another end-point effective permeability Core flood simulation must be performed to obtain relative permeability data corrected for laboratory end-effects (not observed deep reservoir) The principal goal is to determine reliable end-point permeability data, and intermediate relative permeability data that are hopefully corrected for end-effects	• Oil and water relative permeability curves corrected for end-effects

	On completion of the waterflood, the end-point permeability is determined. Dean-Stark extraction is used to verify the water saturation obtained from mass balance Test production data are required for coreflood simulation Imbibition Pc data are necessary for use in coreflood simulations to correct for the capillary end-effect ISSM data allow comparison of the simulated versus measured end-effect *Note*: USS method, although more representative of reservoir processes, is more susceptible to microscopic heterogeneity. SS method should be considered for less homogeneous core material. Many companies prefer the SS method over USS since data are obtained over a wider saturation range	
Centrifuge oil relative permeability	Centrifuge kro data should complement the unsteady-state tests (especially if simulated together with the USS and imbibition Pc data) and will help define the kro curve towards ultimate residual oil saturation The principal advantages of the centrifuge test are: • faster and more efficient drainage • no undesirable viscous force effects • the ability to measured very low Kr values (and so better define the 'tail' of the relative permeability curve towards residual saturations) and • usually lesser capillary end-effects The relative permeability obtained is the displaced phase only—not the displacing phase. Thus, brine-displacing-oil is used to generate the imbibition phase oil relative permeability, kro. Although krw data are not possible in the imbibition cycle, a final krw end-point value is obtained, if the sample is transferred to a coreholder and waterflooded on completion of the centrifuge test Uncertainties arise if the centrifuge technique is used to derive relative permeability data from systems exhibiting 'trapping' of residual oil. These uncertainties arise because of the difficulty associated with defining saturation distributions in	• Oil relative permeability to water imbibition curve and end points

Continued

TABLE 13.3—Cont'd

Test	Purpose	Deliverable
	centrifuged samples. As a 'rule of thumb', the Bond number (N_B: a dimensionless ratio of gravitational to capillary forces) should not exceed 10^{-4} Design of centrifuge speed is important, to ensure that the core is being driven towards residual oil but not at the cost of too little data • Too low spin speed (or N_B significantly lower than 10^{-4}) could produce incorrect relative permeability data, since end-effects are increased • Too high spin speed could produce fluids too rapidly, leading to a very limited saturation range (<5 saturation units) • Imbibition Pc data should be used to help determine the optimal speed, i.e., the lowest Pc approaching Sro (asymptoting data) On completion of the centrifuge test, the sample is removed, placed in a coreholder, flushed with SFW, and the end-point krw at ultimate Sro is determined Dean–Stark measurement is used to substantiate water saturation and hence Sro Test production data should be used in coreflood simulation to determine kro and krw together with USS data and imbibition Pc	

Index

Note: Page numbers followed by "*f*" indicate figures, and "*t*" indicate tables.

A

Absolute permeability, 221, 521
Acoustic spectroscopy, 762
Acoustic travel time (ATT) test
  advantages, 736, 738*t*
  checks and diagnostics, 736–738
  data reporting requirements, 736, 737*t*
  data utilisation, 735–736
  drawbacks/issues, 736, 738*t*
  equipment, 733–735
  procedures, 735
  purpose, 732–733
  quality control issues, 736–738
  sample preparation, 733
  sample requirements, 732–733
  system schematics, 734*f*
Air permeability, 244
ALS Liquid Trapper™ system, 31
Aluminium clam shell disposable core liners, 23–26, 24–25*f*, 26*t*
Aluminium solid disposable core liners, 23, 24*f*, 25–26, 26*t*
Amott-USBM (combination) wettability method, 338–339*t*, 339, 343*f*, 343*t*
  data reporting requirements, 341, 342*f*
  index calculation, 341
  sample preparation, 340
  test equipment, 340
  test procedures, 340–341, 340*f*
Amott (Amott-Harvey) wettability method, 322, 323*t*, 331–332
  data reporting requirements, 330–331
  index calculation, 329–330, 330*t*
  sample preparation, 323–325
  test conditions, 325–327
  test equipment, 327
  test procedures, 326*f*, 327–329
Annulus volume/conformance pressure correction, 361–363
  FRF test, 361–363
Archie's law, 349–352
ATT test. *See* Acoustic travel time (ATT) test
Authigenic clay minerals, 156–157, 158*f*
Azeotrope, 137

B

Barex™, 104–106
Barrier foil laminates, 104–105, 104*f*
Base permeability, 522
Biot's factor, 754
Bottom-hole assembly (BHA), conventional coring, 27
Boyle's law, 200, 205
Brazilian disk test, 729–730, 730*f*
Breakout analysis reservoir stress estimation, 308
Brine permeability, 521
  measurement, FRF test, 366
Brine preparation, 269–276
Bulk volume, 210–216
  advantages, 215*t*
  checks and diagnostics, 214–216
  data reporting requirements, 214*t*
  drawbacks/issues, 215*t*
  quality control issues, 214–216
  sample preparation, 211–212
  test equipment
    mercury immersion system, 213, 213*f*
    mercury pycnometer, 212–213, 212*f*
  test procedure
    mercury immersion system, 213–214
    mercury pycnometer, 213
Bulk volume irreducible fluid (BVI), 661

C

Capillary pressure (Pc)
  in carbonate reservoirs, 450
  centrifuge, 451, 453–470, 503*t*
  data corrections, 502–516
    clay-bound water corrections, 508–509
    closure corrections, 502–507
    fluid corrections, 510–513

Capillary pressure (Pc) (*Continued*)
J function, 513–516, 514–515*f*
stress corrections, 508
definition, 449, 523–524
distribution, 449–450
drainage, 640
drainage curve, 523–524, 524*f*
equilibration times, 501, 502*t*
imbibition test, 453
mercury-air method, 452, 484–501, 503*t*
negative, 525–526
porous plate method, 471–484, 503*t*
primary drainage, 450–453, 450*f*
saturation for, 450
semi-permeable membrane, 451, 471–484
water-oil imbibition, 526*f*
water-phase desaturation, 451
wettability, 453, 525
Carbonates
and chalks, 176
handling, 82
Carr–Purcell–Meiboom–Gill (CPMG), 657–658, 666–667
Cation-exchange capacity (CEC), 161–163, 163*t*, 164*f*
and Qv tests, 424–430
multiple salinity Co-Cw, 426–430
wet chemistry, 425–426
Waxman-Smits equation, 423
Cation mobility, 423
CEC. *See* Cation exchange capacity (CEC)
Cementation exponent
FRF test, data utilisation, 367–369, 369*f*
intrinsic, 430–431
plot presentation, 369*f*
Centrifuge
data interpretation methodology, 557–558
gas-oil relative permeability, 603–606
relative permeability
in experimental data, 641
measurement techniques, 547–548, 550
water-gas, 622–626, 624–625*t*
water-gas relative permeability, 622–623
equipment, 625
gas-equivalent imbibition, 623
imbibition *vs.* drainage, 623–624, 623*f*
key processes, 624
sample preparation, 623–624
test procedure, 625
water-oil relative permeability, 589–592, 589*t*, 591*t*
key processes, 590
sample preparation, 589–592
test procedures, 590–591
Centrifuge capillary pressure, 451, 454*t*
advantages, 463*t*, 503*t*
data reporting requirements, 462–470
drawbacks/issues, 503*t*
elevated temperature, 456
key processes, 455–457
Pc curve calculation, 460–462
QC checks, 463*t*
rotation head with CCTV camera, 457*f*
sample holder design, 457*f*
sample preparation, 454–455
test equipment, 456–457
test procedures, 458–462
test timing, 459
ultra-centrifuge, 456*f*
viscous-dominated desaturation, 459*f*
water saturation check, 464*f*
Centrifuge gas-oil displacement, 528
Chlorite group minerals, 161, 162*f*
Clam shell disposable core liners, 23–26, 24–25*f*, 26*t*
Clay-bound water (CBW)
corrections, capillary pressure, 508–509
NMR, 662
Clay damage mechanisms
clay swelling and migration, 167–168, 168*f*
core cleaning and drying, 165–166, 167*f*
Clay microporosity, 195–197
Clay minerals, 156
authigenic, 156–157, 158*f*
structure, 158–160, 159*f*
types, 160–161, 160–162*f*
Clay morphology, 163–165, 165*f*
Clay water models, 172
Cleaned state, 324
Closure corrections, capillary pressure, 502–507, 507*f*
Coal handling, 84
Combined porosity test. *See* Formation resistivity factor (FRF) tests
Compressibility test
loading conditions, 747–751
normalised porosity *vs.* mean stress, 759*f*
pore volume (*see* Pore volume compressibility tests)
Computational fluid dynamics (CFD) model, core trip rate, 55*f*
Conductive matrix, 405–406
Consolidated-undrained triaxial test, 711, 711*f*
Consolidation stage
deformation, 712*f*
triaxial test, 711–712
Contact angle goniometer, 316, 317–318*f*

Contact angle method, 315
data reporting requirements, 320
equilibrium, 319, 319*f*
measurement, 321*t*, 322*f*
results, 320
sample preparation, 316
test equipment, 317–318, 317–318*f*
test procedure, 319
Continuous injection resistivity index (CIRI) method
advantages, 385
design, 384
ISSM scans, 386*f*
Conventional core bits, 20–21, 20*f*
Conventional coring operations, 40–44
coring risks management, 42–44
coring team, 42, 43*t*
health, safety and environmental considerations, 41–42
Conventional full-diameter coring systems, 18–26
core barrels, 18, 19*t*, 21, 21*f*, 27
core bits, 18, 18*f*
conventional, 20–21, 20*f*
low-invasion, 20–21, 20*f*
core catcher, 22, 22*f*
disposable core liners, 22
aluminium clam shell, 23–26, 24–25*f*, 26*t*
aluminium solid, 23, 24*f*, 25–26, 26*t*
fibreglass, 22, 23*f*, 25, 26*t*
PDC bits, 18–19, 19*f*
Conventional (and vacuum) oven core drying, 146, 149*f*, 153*t*
Core analysis
advantages, 14–15
core plug history sheet, 14*f*
coring, core handling, 12
design and management, 783–785
focal point, 782–783
geomechanics tests, 13
planning, 782
programmes, 14
quality control procedures and diagnostics, 13–14, 14*f*
RCA (*see* Routine core analysis (RCA))
SCAL (*see* Special core analysis laboratory (SCAL) test)
Core analysis data
formation evaluation, 1–4
reservoir dynamic modelling, 5*t*
uncertainty
reasons and consequences, 4–7
reducing uncertainty, 8
volumetric calculations, 3*t*
Core analysis management framework
core analysis focal points, 9–11, 10*f*
planning and design, 8–9
programme design, 9
real-time quality control, 11
Core barrels
conventional full-diameter coring systems, 18, 19*t*, 21, 21*f*, 27
sponge, 30*f*
Core bits, 18, 18*f*
conventional, 20–21, 20*f*
low-invasion, 20–21, 20*f*
Core catcher, 22, 22*f*
conventional full-diameter coring systems, 22, 22*f*
Core cleaning
advantages, 146, 147–148*t*
azeotrope, 137
cool soxhlet extractor, 141
drawbacks/issues, 146, 147–148*t*
flush (flow) cleaning, 143–145, 144*f*, 148*t*
solvents, 136–138
boiling points and water solubility, 140*t*
soxhlet extraction, 138–141, 139*f*, 147*t*
total immersion soxhlet, 142–143, 142–143*f*, 148*t*
Core conditioning for porosity measurements
clay water models, 172
log and core porosity, 172–174, 173*f*, 175*f*
porosity evaluation, 170
total and effective porosity concepts, 170–172
V_{cl} models, 171–172
Core damage and core fluid/petrophysical property alteration, 48–66
fluid saturation alteration, 48–66, 49–51*f*
stress release, 52
pore pressure dissipation, 52–57, 53*f*, 53–54*t*, 55–56*f*
shear failure and compressive damage, 57, 58*f*
wettability alteration, 57–66
Core drying
advantages, 153–154*t*
checks and diagnostics, 153–156
conventional (and vacuum) oven drying, 146, 149*f*, 153*t*
critical point drying, 150–152, 151–152*f*, 155*t*, 157*f*
drawbacks/issues, 153–154*t*
flow-through drying, 152–153, 155*t*
humidity oven drying, 146–150, 149–150*f*, 154*t*
quality control issues, 153–156

Core imaging, 124, 125*f*
Core laboratory processing and screening
  conventional imaging, 121–124, 124*f*
  360° core imaging, 124, 125*f*
  core photography and imaging, 120–124
  corc plugging
    drill press and plugging fluids, 108–109, 109*f*
    plug allocation (*see* Plugging requirements, core analysis)
    plug orientation, 110–111, 110*f*, 111*t*
    plug samples, 106–108, 107–108*f*
    plug trimming, 116
  core receipt and cutting, 91
  core resination, 118–120, 121–123*f*
  core slabbing, 117–118, 119–120*f*
  core viewing and sample selection, 102–104
  CT scanning, 91–95, 92–94*f*
  flowchart, 90*f*
  gamma ray logging, 95–99, 96–99*f*
  removal from liners, 100–102, 100–103*f*
  sample preservation
    dry (*see* Dry preservation)
    wet, 106, 107*f*
Core plugging
  drill press and plugging fluids, 108–109, 109*f*
  plug allocation (*see* Plugging requirements, core analysis)
  plug orientation, 110–111, 110*f*, 111*t*
  plug samples, 106–108, 107–108*f*
  plug trimming, 116
  unconsolidated, 130–131, 130*f*
Core preparation
  carbonates and chalks, 176
  halite, 176–178, 177*f*
  wettability, 175–176
Core sample preservation
  dry (*see* Dry preservation)
  wet, 106, 107*f*
Core slabbing, 117–118, 119–120*f*, 129
Core stabilisation and preservation, 72–76
  foam and gypsum stabilisation, 74–75, 75*f*
  freezing, 73–74
  gypsum *vs*. foam, 76
  resin consolidation, 74
Core transportation, 79–80, 79–80*f*
Core trip rate CFD model, 55*f*
Corey exponents, 541
  gas-oil systems, 542
  water-gas systems, 542–543
  water-oil systems, 541, 541*t*
Coring fluids
  mud tracers
    oil-soluble, 48
    water-soluble, 46–48, 47*f*
  mud types, 45
Coring muds, 45–46
Coring report, 81
Coring systems, 17–18
  conventional full-diameter, 18–26, 18–21*f*, 23–24*f*
  gel, 28, 29*f*
  oriented, 35–37, 35–36*f*
  sidewall cores, 37–40, 37–40*f*
  wireline-retrievable cores, 27–28, 27*f*
Coring team, 42
  roles and responsibilities, 43*t*
CPD. *See* Critical point drying (CPD)
CPMG. *See* Carr–Purcell–Meiboom–Gill (CPMG)
Creep tests, 750–751
Critical gas saturation (Sgc)
  definition, 531
  gas-oil system, 530, 531*f*
  water-gas system, 530–531, 532*f*
Critical oil saturation
  definition, 535
  gas condensate systems
    ISSM scan, 534–535, 534*f*
    model fluid system, 535
    pressure depletion, 533–534
    relative permeability, 534
Critical point drying (CPD), 150–152, 151–152*f*, 155*t*, 157*f*
Critical water saturation (Swc)
  definition, 533
  water flow, 532
  water-gas fractional flow curve for, 533*f*
Crushed sand, 768, 772*f*
CT scanning
  core samples, 91–95, 92–94*f*
  weak/unconsolidated core processing, 125–129, 127–128*f*
CYDAR™, 563–564

D

Darcy's law, permeability measurement, 221–224, 222*f*, 223*t*
Dead crude oil, 278

Dean–Stark analysis, 187–189
advantages, 196*t*
checks and diagnostics, 195, 197*f*, 198*t*
data, 443
data reporting requirements, 193–194, 194*t*
drawbacks/issues, 196*t*
plugs, 78
quality control issues, 195, 197*f*, 198*t*
sample preparation, 189
saturation calculation, 192–193
test equipment, 189–190, 190–191*f*
test procedures, 190–192
Deuterium oxide (D_2O), 46
Differential strain curve analysis (DSCA) test, 678–679
advantages, 744, 745*t*
checks and diagnostics, 744–746
data reporting requirements, 744, 745*t*
data utilisation, 742–744
drawbacks/issues, 744, 745*t*
in elastomer sleeve, 741*f*
equipment, 739–740
procedures, 740–741
purpose, 739
quality control issues, 744–746
result presentation, 744*t*
sample orientation, 740*f*
sample requirements, 739
strain gauge response, 743*f*
strain gauge rosettes, 740*f*
stress–strain profile, 741, 742*f*
Direct tensile test, 729, 729*f*
Disaggregated sand, 771–772*f*
Discrete clays, 165
Disposable core liners, 22
aluminium clam shell, 23–26, 24–25*f*, 26*t*
aluminium solid, 23, 24*f*, 25–26, 26*t*
fibreglass, 22, 23*f*, 25, 26*t*
Double-layer models, 353
Drainage, 523–524
Drilling muds, 62–65, 63–64*f*, 65*t*
Dry preservation
barrier foil laminates, 104–105, 104*f*
hot wax/strippable plastics, 105–106, 105*f*
DSCA test. *See* Differential strain curve analysis (DSCA) test
Du Noüy Ring method
advantages, 287*t*
drawbacks/issues, 287*t*
IFT determination, 284–285, 285*f*
Dynamic reservoir models, 2
Dynamic test sample, selection procedure, 297–299, 298–299*f*

E

Effective permeability, 221, 521–522
Effective porosity, 661
Elastic moduli, 695–697
End-point effective permeability, 522
Entry pressure, 523–524
Equivalent effective isostatic stress method, 303–304
Equotip portable hardness tester, 674, 674–676*f*
Excess brine, 393–394

F

Fibreglass disposable core liners, 22, 23*f*, 25, 26*t*
FIT. *See* Formation integrity test (FIT)
Flow-through core drying, 152–153, 155*t*
Flow zone indicator (FZI), 297
Fluid corrections, capillary pressure, 510–513
Fluid preparation and characterisation
data reporting requirements, 280
gas, 279–280
oil, 276–279
mineral, 276–277
reservoir, 278–279
reservoir crude, 277–278
water samples and brine preparation, 269–276, 270*t*, 274*f*
resistivity, 273–275
special considerations, 275–276
viscosity and density, 275
Fluid saturation alteration, 48–66, 49–51*f*
Fluid saturation measurements
Dean–Stark analysis, 187–195, 190–191*f*, 194*t*, 196*t*, 197*f*, 198*t*
retort method, 183–186, 183–185*f*, 187*f*
Flush/flow core cleaning, 143–145, 144*f*, 148*t*
Foam *vs.* gypsum core stabilisation, 76
Forced imbibition, 525–526
brine, 328, 341
oil, 329, 341
Force tensiometry, IFT determination, 284
Forchheimer effect, non-Darcy flow, 228–229
Formation factor (F/FF). *See* Formation resistivity factor (FRF) tests
Formation integrity test (FIT), 309
Formation resistivity factor (FRF) tests
advantages, 369–370, 372*t*
ambient conditions, equipment, 356–357
checks and diagnostics, 371–378
data reporting requirements, 369, 370*t*
data utilisation
cementation exponent, 367–369, 369*f*

Formation resistivity factor (FRF) tests (*Continued*)
porosity compaction factor, 366–367, 373*f*, 375*f*
drawbacks/issues, 369–370, 372*t*
equipment, 356–359
procedures, 360–366
purpose of, 354–378
quality control issues, 371–378
sample selection and preparation, 354–355
stress (overburden) conditions
annulus volume/conformance pressure correction, 361–363
brine permeability measurements, 366
equipment, 358–359
pore collapse, 364–365
sample loading, 360
stressed porosity determination, 364
stress stations, 363–364
test procedures, 360–366
unloading, 366
Formation water (FW), 521
Fractional wet, 60, 314–315
Fresh state, core plugs, 323
Full-scale deflection (FSD), 249
FZI. *See* Flow zone indicator (FZI)

G

Gamma ray logging, 95–99, 96–99*f*
Gamma ray scanning, 129
Gas condensate systems
cost estimate of, 534–535
ISSM scan, 534–535, 534*f*
model fluid system, 535
pressure depletion, 533–534
relative permeability, 534
Gas-displacing-oil unsteady-state (USS) relative permeability, 527
Gas-driven solvent extraction, 136
Gas field, special core analysis
air–brine drainage capillary pressure tests, 795*t*
CEC, 795*t*
core CT scan, 795*t*
dynamic parameters, 793
MICP, 795*t*
multiple salinity method, 795*t*
petrophysical parameters, 793
porosity/formation factor tests at overburden, 795*t*
pre-screening study, 795*t*
residual gas CCI tests, 795*t*
RI tests, 795*t*
rock characterisation tests, 795*t*
sample selection, 795*t*
steady-state water–gas relative permeability, 795*t*
water–gas imbibition relative permeability, 795*t*
workflow, 794*f*
Gas-liquid drainage, 280
Gas-oil ratio (GOR) measurement, 278
Gas-oil relative permeability
centrifuge, 603–606
SS tests, 598–600, 598*t*, 602*t*
equipment, 600–601
key processes, 600
sample preparation, 600
test procedures, 601–602
USS tests, 592*t*, 596*t*
Corey exponents, 598
equipment, 594–595
example curve, 592, 593*f*
gas flow, in porous media, 592, 593*f*
key processes, 594
kro and kro refining and extending, 598
sample preparation, 594
test procedures, 595–597
typical setup, 595*f*
Gas-oil system, 280
Corey exponents, 542
critical gas saturation, 530, 531*f*
residual gas saturation
centrifuge, 528
immiscible gasflooding operations, 528
USS, 527
Gas saturation, determination, 184
Gel coring systems, 28, 29*f*
Geomechanics tests
acoustic travel time test, 732–738
description, 672–673
differential strain curve analysis test
advantages, 744, 745*t*
checks and diagnostics, 744–746
data reporting requirements, 744, 745*t*
data utilisation, 742–744
drawbacks/issues, 744, 745*t*
equipment, 739–740
procedures, 740–741
purpose, 739
quality control issues, 744–746
sample requirements, 739
particle size analysis test, 761–776
pore volume compressibility test
advantages, 757, 758*t*
checks and diagnostics, 757–761

data reporting requirements, 757, 758*t*
data utilisation, 754–757
drawbacks/issues, 757, 758*t*
loading conditions, 747–751
purpose, 746–747
quality control issues, 757–761
sample preparation, 751
uniaxial K_0 test, 751–753
sample selection and preparation, 673
dimensions of sample, 679–680
orientation of sample, 678–679
plugging fluids, 679
saturation of sample, 680–681
test sites, 674–678
tensile strength test, 728–732
thick-wall cylinder test
advanced, equipment, 719–720
advantages, 724, 726*t*
checks and diagnostics, 724
data reporting requirements, 724, 725*t*
data utilisation, 722–724
drawbacks/issues, 724, 726*t*
purpose, 717–718
quality control issues, 724
sample requirements, 717–718
standard, 719–720, 722
triaxial compression strength test, 687–705
advantages, 700, 702*t*
checks and diagnostics, 700–705
data reporting requirements, 700, 701*t*
drawbacks/issues, 700, 702*t*
elastic moduli, 695–697
equipment, 688
Mohr–Coulomb parameters, 697–700, 698*f*
multi-stage triaxial tests, 693–694
plug with irregular sides, 688*f*
purpose, 687
quality control issues, 700–705
sample requirements, 687
single-stage triaxial tests, 688–693
triaxial testing of shales
advantages, 714, 716*t*
checks and diagnostics, 714–717
consolidation stage, 712
data reporting requirements, 713, 715*t*
drawbacks/issues, 714, 716*t*
equipment, 709
preparation of sample, 706–709
purpose, 706
quality control issues, 714–717
quasi-UCS tests, 709–710
sample requirements, 706
triaxial parameters, 712–713
triaxial tests, 711
unconfined compressive strength tests, 681–687
Grain compressibility, 754
Grain loss, 391–393
Gross rock volume (GRV), 2
Gypsum *vs.* foam core stabilisation, 76

H

Halite, core preparation, 176–178, 177*f*
Health, safety and environmental (HSE) protection, during coring operations, 41–42
Heat-sealable plastic/aluminium laminate, 104, 104*f*
Helium grain volume and grain density
advantages, 203*t*
calculation, 202
checks and diagnostics, 203–204
data reporting requirements, 203*t*
drawbacks/issues, 203*t*
quality control issues, 203–204
sample preparation, 199
test equipment, 200–201, 200–201*f*
test procedures, 202
Helium pore volume
advantages, 207*t*
checks and diagnostics, 207–210, 208–210*f*
data reporting requirements, 206*t*
drawbacks/issues, 207*t*
pore volume and porosity calculation, 206
quality control issues, 207–210, 208–210*f*
sample preparation, 204
test equipment, 204–205, 204*f*
test procedures, 205
High pressure automated (HPMI) systems, 485
advantages, 496*t*
apparatus, 488*f*
drawbacks/issues, 496*t*
graphical output from, 493*f*
mercury injection-systems, 485
penetrometer, 489*f*
pore throat size frequency distribution, 492*f*
pore volume errors in, 498*f*
tabular output from, 494*f*
test equipment, 487–488
test procedures, 490
HOD. *See* Humidity oven drying (HOD)
Hoek cell, set-up for TWC, 720*f*

Hoek-type triaxial cell, 688–689, 689*f*
Hot soxhlet extraction, 141
Hot wax/strippable plastics, 105–106, 105*f*
Humidity oven drying (HOD), 146–150, 149–150*f*, 154*t*
Hydraulic fracturing test, 308
Hydrogen sulphide (H_2S) concentration, 42

I

IFT. *See* Interfacial tension (IFT)
Illite (mica) group minerals, 160–161, 161*f*
Image analysis technology, particle size analysis test, 761–762
Immersion soxhlet extraction, 142–143, 142–143*f*, 148*t*
Induction logs, 348
Inertial flow-corrected pressure plot, 234, 234*f*
In situ saturation monitoring (ISSM), 476, 477*f*
  scans, 534–535, 534*f*
  typical schematic, 551–552, 552*f*
In situ stress loading conditions, 752
Interfacial tension (IFT), 50, 281–285, 282*f*, 510–513, 512–513*f*
  determination methods, 282–285
    from compositional data, 282–283
    Du Noüy Ring, 284–285, 285*f*, 287*t*
    force tensiometry, 284
    pendant drop technique, 283–284, 283*f*, 287*t*
    Wilhelmy plate, 285, 286*f*, 287*t*
  forces, 141, 150–151
  SFW-mineral oil IFT *vs.* salt concentration, 281, 282*f*
Intermediate wet, 60
Intrinsic cementation exponent, 430–431
Intrinsic saturation exponent, 431–432
Inverse tangent method for saturation exponent calculation, 389–390
Irreducible water saturation, 523–524
  reservoir rock at, 60, 60*f*
Iso-propyl alcohol (IPA), 176
Isostatic stress, 300
  equivalent effective, 303–304

K

Kaolinite, 161, 162*f*
Klinkenberg effects, non-Darcy flow, 224–227, 225–228*f*
Klinkenberg permeability, 225–226, 227*f*
  calculations, 232
  data reporting, 244
  quality control issues, 235–239
Klinkenberg plot, 232

L

Laboratory measured end-point effective permeability (kr_{lab}), 522–523
Laser particle size analysis (LPSA) test. *See also* Mechanical particle size analysis (MPSA) test; Particle size analysis (PSA) tests
  advantages, 775, 775*t*
  checks and diagnostics, 776
  data reporting requirements, 775
  data utilisation, 774–775
  drawbacks/issues, 775, 775*t*
  equipment, 774
  procedures, 774
  quality control issues, 776
  sample preparation, 769–773
  sieve analysis *vs.*, 770*f*
Laterolog, 348–349
Leak-off tests (LOTs)
  interpretation, 309
  reservoir stress estimation, 308–309, 308*f*
Lenient air drying. *See* Flow-through core drying
Leverett J method, capillary pressure, 513–516, 514*f*
Linear variable differential transformers (LVDTs), 681
Liquefied gas extraction, 136
Liquid and gas retention coring systems, 28–30
  liquid capture systems, 31–33, 33*f*
  pressure-retaining systems, 33–34, 33*f*
  sponge coring system, 30–31, 30*f*, 32*f*
Liquid-gas imbibition, 280
Liquid nitrogen (LN2) drilling, 130–131, 130*f*
Liquid permeability, 225, 247–248
Liquid permeability measurements, steady-state, 247–248
  advantages, 252*t*
  checks and diagnostics, 250–252
  data reporting requirements, 251*t*
  drawbacks/issues, 252*t*
  quality control issues, 250–252
  sample preparation, 248
  saturation procedures, 248
  test procedures and permeability calculation, 248–250, 250*f*

Liquid saturation porosity
advantages, 219*t*
checks and diagnostics, 219
data reporting requirements, 218*t*
drawbacks/issues, 219*t*
porosity calculation, 217–218
quality control issues, 219
sample preparation, 216
test equipment, 216, 217*f*
test procedures, 217
Liquid trapper inner tube system, 31, 33*f*
LOTs. *See* Reservoir stress estimation
Low-invasion core bits, 20–21, 20*f*
Low pressure manual (LPM) systems, 485
advantages, 496*t*
apparatus, 486*f*
drawbacks/issues, 496*t*
pressure cell, 487*f*
test equipment, 486–487
test procedures, 489–490
LPSA test. *See* Laser particle size analysis (LPSA) test

M

Maximum residual gas saturation (Sgrm), 528
Mechanical particle size analysis (MPSA) test. *See also* Laser particle size analysis (LPSA) test; Particle size analysis (PSA) tests
advantages, 768, 769*t*
checks and diagnostics, 768–769
data reporting requirements, 767, 768*t*
data utilisation, 765–767
drawbacks/issues, 768, 769*t*
equipment, 763–764
procedures, 765
quality control issues, 768–769
sample preparation, 762–763
Mechanical sieving (MPSA), 762
Mercury-air capillary pressure, 452
advantages, 503*t*
clay-bound water corrections, 486
data reporting requirements, 493–501
drawbacks/issues, 503*t*
dual porosity systems, 498*f*
graphical output, 493
high pressure automated systems, 485
advantages, 496*t*
drawbacks/issues, 496*t*
test equipment, 487–488
test procedures, 490
loss of capillary contact, 483*f*
low pressure manual systems, 485
advantages, 496*t*
drawbacks/issues, 496*t*
test equipment, 486–487
test procedures, 489–490
in penetrometer, 488
permeability/rock quality, 490, 491*f*
pore throat size distribution, 490–492
QC checks, 496*t*
sample preparation, 486
volume injected *vs*. injection pressure curve, 499*f*
Mercury immersion system, 213–214, 213*f*
Mercury injection capillary pressure (MICP). *See* Mercury-air capillary pressure
Mercury pycnometer, 212–213, 212*f*
Mineral oil, 276–277
Mixed wet, 60, 314–315
Mohr circles, 697
definition, 698
for multi-stage triaxial tests, 693–694
plots, 698, 699*f*
Mohr–Coulomb theory
failure envelope, 698*f*
triaxial compression strength tests, 697–700
Montmorillonite (smectite) group minerals, 160, 160*f*
MPSA test. *See* Mechanical particle size analysis (MPSA) test
MST tests. *See* Multi-stage triaxial (MST) tests
Mud tracers
oil-soluble, 48
water-soluble, 46–48, 47*f*
Mud types, 45
Multiple salinity Co-Cw, 426–430
Multi-stage triaxial (MST) tests, 687
Mohr circles for, 693–694
schematic illustration, 694*f*
triaxial compression strength tests, 693–694

N

Native wettability alteration
cleaning methods, 540
core recovery, 538
drilling mud, 538
estimates, 537–538
oxidation of oil, 538, 539*f*
test conditions, 540
Net confining stress loading conditions, 301–302
Net overburden stress method, 302–303, 303*f*

Neutrally wet, 60, 314
fractional, 314–315
mixed, 314–315
NMR. *See* Nuclear magnetic resonance (NMR)
Non-Darcy flow
Forchheimer effect, 228–229
Klinkenberg effect, 224–227, 225–228*f*
Non-destructive membrane potential method, 424–430
Non-destructive method, 426–430
Non-polar solvents, 137–138
Non-uniform saturation, 394–398
Nuclear magnetic resonance (NMR)
clay-bound water, 662
core analysis
data reporting requirements, 667
key processes, 664
measurement parameters, 665–666
sample preparation, 663–664
synthetic formation water, 664
test equipment, 664–665
test procedures, 666–667
core measurement, 668
desk-top instrument, 665*f*
distribution of pore sizes, 656*f*
measurements, 655–657
permeability estimation of, 662–663
porosity from, 661
relaxation and pore size, 659–661, 660*f*
spin relaxation in rocks, 657–659
tools, 655
T_1/T_2 porosity, 659–661

O

OBM. *See* Oil-based mud (OBM)
Oil, 276–279
mineral, 276–277
reservoir, 278–279
reservoir crude, 277–278
Oil-based mud (OBM), 45, 47–48
filtrates, 20–21
tracer, 48
Oilfield core, cleaning process, 137
Oil field, special core analysis example
brine–oil imbibition capillary pressure tests, 803*t*
CEC, 803*t*
centrifuge oil relative permeability, 803*t*
core CT scan, 803*t*
dynamic parameters, 801
fresh-state wettability, 803*t*
MICP, 803*t*
multiple salinity method, 803*t*
oil–brine drainage capillary pressure, 803*t*
oil saturation/wettability restoration, 803*t*
petrophysical parameters, 801
porosity/formation factor tests, 803*t*
RI tests, 803*t*
rock characterisation tests, 803*t*
sample preparation, 803*t*
sample selection, 803*t*
unsteady-state imbibition waterflood, 803*t*
workflow, 802*f*
Oil initially in place (OIIP), in reservoir, 2
Oil reservoirs, 60
Oil saturation determination, sponge core, 32*f*
Oil-soluble mud tracers, 48
Oil-water system, 525–526, 526*f*
Oil wet, 60, 65–66, 314
Oven core drying, 146, 149*f*, 153*t*
Overburden conditions, FRF tests, 401. *See also* Stress (overburden) conditions, FRF tests

P

Palaeomagnetism, 36–37
Particle size analysis (PSA) tests
acoustic spectroscopy, 762
graphical, 767*f*
image analysis, 761–762
LPSA test
advantages, 775, 775*t*
checks and diagnostics, 776
data reporting requirements, 775
data utilisation, 774–775
drawbacks/issues, 775, 775*t*
equipment, 774
procedures, 774
quality control issues, 776
sample preparation, 769–773
sieve analysis *vs.*, 770*f*
MPSA test
advantages, 768, 769*t*
checks and diagnostics, 768–769
data reporting requirements, 767, 768*t*
data utilisation, 765–767
drawbacks/issues, 768, 769*t*
equipment, 763–764
procedures, 765
quality control issues, 768–769
sample preparation, 762–763
purpose, 761–762
sedimentation method, 761
PDC bit. *See* Polycrystalline diamond compact (PDC) bit

Pendant drop technique
advantages, 283–284, 287*t*
disadvantages, 287*t*
IFT determination, 283–284, 283*f*
Permeability measurements. *See also specific types of permeability measurements*
Darcy's law, 221–224, 222*f*, 223*t*
definitions, 221
estimation of NMR, 662–663
non-Darcy flow
Forchheimer effect, 228–229
Klinkenberg effects, 224–227, 225–228*f*
Plastic-aluminium laminate, 539*f*
Plugging requirements, core analysis, 112*f*
DS plugs, 114
formation damage test plugs, 113
rock mechanics test plugs, 114–115, 115–119*f*
routine core analysis plugs, 112
special core analysis plugs, 113, 113*f*
Poisson's ratio
elastic properties, 756–757
as function of confining stress, 696–697, 697*f*
measurement, 685–687
stress/strain relationships for, 683, 683*f*
Polar solvents, 137–138
Polycrystalline diamond compact (PDC) bit, 18, 19*f*
Pore collapse, 364–365
Pore-lining clays, 165
Pore pressure dissipation, 52–57, 53*f*, 53–54*t*, 55–56*f*
Pore size, nuclear magnetic resonance, 659–661
Pore throat size distribution (PSD), 490–492, 492*f*
Pore volume compressibility tests
advantages, 757, 758*t*
checks and diagnostics, 757–761
data reporting requirements, 757, 758*t*
data utilisation, 754–757
drawbacks/issues, 757, 758*t*
effective stress method, 747–748, 749*f*
loading conditions, 747–751
purpose, 746–747
quality control issues, 757–761
sample preparation, 751
simulated depletion method, 748
uniaxial K_0 test
depletion loading conditions, 752–753
equipment, 751
initial *in situ* stress loading conditions, 752
initial loading conditions, 751–752
Porosity
measurements, 195–221
accuracy and repeatability, 219–221, 220*f*
bulk volume, 210–216
grain density, 199–204
helium grain volume, 199–204
helium pore volume, 204–210
liquid saturation porosity, 216–219
from nuclear magnetic resonance, 661
at stress, 364, 374*f*
Porosity compaction factor (PCF), 361–362
vs. ambient porosity, 375*f*
FRF tests, data utilisation, 366–367, 373*f*, 375*f*
vs. stress, 373*f*
Porous plate method, 471–484, 471*t*
advantages and disadvantages, 480*t*, 503*t*
average and inlet face Sw-Pc curves, 466*f*
batch porous plate, 473, 473*f*
capillary pressure *vs.* water saturation
linear plot, 469*f*
log-log plot, 470*f*
semi-log plot, 469*f*
data reporting requirements, 479–484
desaturation, 471, 473
entry pressure, 473
individual cell porous plate, 474, 474*f*
input curves for audit dataset, 467*f*
Keraflux membranes, 478
key processes, 472–474
lab calculated Pc curves, 468*f*
loss of capillary contact, 482*f*
pseudo-equilibrium data, 478–479
QC checks, 480*t*
sample preparation, 472
short equilibration times in, 482*f*
test equipment, 472–473
test procedures
batch porous plate (ambient), 474–475, 480*t*
Dean-Stark test, 475, 478
individual porous plate (overburden), 476–478, 480*t*
ISSM monitoring, 476, 477*f*
saturation *vs.* time, 477*f*
Pressure-retaining coring systems, 33–34, 33*f*
Primary drainage process, 450–453, 450*f*, 458–462, 524, 532*f*
Probe (profile) permeability measurements, 252–253
advantages, 260*t*
checks and diagnostics, 257–262, 261*f*
data reporting requirements, 259*t*

Probe (profile) permeability measurements (*Continued*)
drawbacks/issues, 260*t*
quality control issues, 257–262, 261*f*
sample preparation, 253
steady-state permeability calculation, 255–256, 256*f*
test procedures
steady-state equipment, 253–255, 254–255*f*
unsteady-state equipment, 256–257, 258*f*
ProtecCore™, 104
PSA tests. *See* Particle size analysis (PSA) tests

Q
Quality control, core analysis
procedures and diagnostics, 13–14, 14*f*
real-time, 11
Quasi-UCS tests, 709–710, 714–717
Qv tests, CEC and, 424–430
multiple salinity Co-Cw, 426–430
wet chemistry, 425–426

R
Rate type compaction method (RTCM) tests, 750
RCA. *See* Routine core analysis (RCA)
Recovery factors, determination, 3–4
Relative permeability, 221
conversion, 523
Corey exponents, 541
gas-oil systems, 542
water-gas systems, 542–543
water-oil systems, 541, 541*t*
critical gas saturation, 634–635
definition, 531
equipment, 635–636
gas-oil system, 530, 531*f*
sample preparation, 635
test procedures, 636–639
water-gas system, 530–531, 532*f*
critical oil saturation
definition, 535
gas condensate systems, 533–535
critical water saturation
definition, 533
water flow, 532
water-gas fractional flow curve for, 533*f*
data interpretation methodology
capillary end effects, 558–561
centrifuge, 557–558
coreflood simulation, 562–565
flow rate, 558–561
SS tests, 557
definition, 522
experimental data, 639
centrifuge, 641
counter-current imbibition, 642
drainage capillary pressure, 640
samples and fluids, 639
USS and SS flood tests, 642
gas-oil, 568, 569–571*t*
centrifuge, 603–606
SS tests, 598–603
USS tests, 592–598
initial water saturation, 569–572, 573*t*
irreducible water saturation, 523–524
measurement techniques
centrifuge, 547–548, 550
curves, methodology function, 548, 548*f*
equipment, 548–550
in situ saturation monitoring, 551–554
SS technique, 545–548
USS technique, 543–545, 547
overburden stress, 568–569
sample selection, 565–568, 568*f*
water-gas
centrifuge, 622–626, 624–625*t*
SS test, 618–622, 618*t*, 620*t*
USS test, 613–618, 616*t*
water-oil, 568, 569–571*t*
centrifuge, 589–592
SS test, 584–589
USS test, 574–584
wettability, 525
native, 536–537
rock/gas/water system, 536
rock/oil-water system, 536, 536*t*
USBM tests, 540
water saturation, 525
Relaxation time, 655–656
Remaining oil saturation (ROS), 527
Reservoir crude oil, 277–278
Reservoir engineer, 1–2
Reservoir models
dynamic, 2
simulation, 1–2
static, 2
Reservoir net thickness, 2
Reservoir oil, 278–279
Reservoir simulation models, 1–2
Reservoir stress estimation
data sources
maximum horizontal stress, 310

pore pressure, 310
stress regime, 304–305, 305*f*
total minimum horizontal stress, 307–310
total vertical stress, 305–306, 306–307*f*
effective stress concepts, 300–301
equivalent effective isostatic stress, 303–304
net confining stress loading conditions, 301–302
net overburden stress method, 302–303, 303*f*
test cell design, 300, 300*f*
Residual gas saturation (Sgr)
disequilibria effects, 529–530
gas-oil system
centrifuge, 528
immiscible gasflooding operations, 528
USS, 527
imbibition, 525–526, 526*f*
with rock property, 529
water-gas relative permeability
CCI technique, 629–630
equipment, 631
key processes, 631
sample preparation, 631
test procedures, 631–633
water-gas system, 528–530
dynamic tests, 529–530
vs. Sgi relationships, 528–529, 529*f*
Sgrm, 528
water-oil system
forced imbibition, 525–526
HC components, 525
primary drainage, 525–526
ROS, 527
Srot, 526–527
Srow, 527
by water-oil imbibition capillary pressure, 525–526, 526*f*
water saturation, 525
wettability alterations, 525
Residual oil saturation, 525–527
Resistivity index tests
advantages, 409, 410–411*t*
ambient condition tests, equipment, 379–381
checks and diagnostics, 409–416
data assignment and utilisation
inverse tangent method, 389–390
saturation exponent, 387–388
water saturation calculation, 387
data reporting requirements, 406–408, 408*t*
drawbacks/issues, 409, 410–411*t*
equipment, 379–382
ambient condition tests, 379–381
overburden test, 381–382
laboratory artefacts
excess brine, 393–394
grain loss, 391–393
non-uniform saturation, 394–398
measurement, 378–416
procedures, 382–387
ambient condition tests (air-brine), 382–383
continuous injection resistivity index, 384–387
stress condition tests (air-brine and oil-brine), 384
quality control issues, 409–416
rock type and pore/fluid system controls
conductive matrix, 405–406
overburden stress, 401
rock texture/pore size modality, 398–401
saturation history, 401–403
wettability, 403–405
saturation exponent
laboratory artefacts, 390–398
rock type and pore/fluid system controls, 398–406
Resistivity logs, 348–349
induction logs, 348
laterolog, 348–349
Resistivity, synthetic formation water, 273–275
Restored state, core plugs, 324
Retort method
advantages and disadvantages, 188*t*
sample preparation, 183
saturation and porosity calculations, 186, 187*f*
test equipment, 183, 183–184*f*
test procedures, 184–186, 185*f*
RFT sample, water sample analysis from, 270*t*
Rock/fluid system, wettability of, 315
Rock quality index (RQI), SCAL samples, 297, 298*f*
Rocks
nuclear spin relaxation in, 657–659
resistivity, 360
Rocks handling
carbonates, 82
coal, 84
shales, 83–84
shaly sands, 83
unconsolidated core, 81–82
Rock texture/pore size modality, 398–401
Rock type and pore/fluid system controls
conductive matrix, 405–406
overburden stress, 401

Rock type and pore/fluid system controls (*Continued*)
rock texture/pore size modality, 398–401
saturation history, 401–403
wettability, 403–405
Routine core analysis (RCA), 7, 12, 181–182
carbonates, 786
clean consolidated sandstones, 785
fractured reservoirs, 787
low-permeability reservoirs, 787
MICP, 789*t*
petrography, 789*t*
requirements, 787
routine porosity, 789*t*
shaly sands, 786
tests, 789*t*
unconsolidated sandstones, 785–786
vuggy carbonates, 786
water saturation, 789*t*
workflow, 788*f*

S

SARA. *See* Saturates, aromatics, resins and asphaltenes (SARA)
Saran Wrap®, 105–106
Saturates, aromatics, resins and asphaltenes (SARA), 277
Saturation exponent, 387–388
alternative method, 439–444
Saturation history, 401–403
Saturation hysteresis, 525
Saturation Monitoring by Attenuation of X-rays (SMAX™)
calibration principles, 553*f*
core laboratories, 552, 553*f*
neutral-wet sample, 553, 554*f*
SBMs. *See* Synthetic oil-based muds (SBMs)
SCAL test. *See* Special core analysis laboratory (SCAL) test
SCORES, 564
Scratch test, 674–675, 677*f*
UCS strength profile from, 677*f*
SDR approach, 662
Sedimentation, PSA tests, 761
Semi-permeable membrane method, 451, 471–484
Sendra™, 562–563
Servo-hydraulic pump, 734
Sessile drop method, 316, 316*f*, 318–319
SFW. *See* Synthetic formation water (SFW)
Shales
handling, 83–84
triaxial testing, 711
advantages, 714, 716*t*
checks and diagnostics, 714–717
consolidation stage, 712
data reporting requirements, 713, 715*t*
drawbacks/issues, 714, 716*t*
equipment, 709
preparation of sample, 706–709
purpose, 706
quality control issues, 714–717
quasi-UCS tests, 709–710
sample requirements, 706
triaxial parameters, 712–713
Shaly sands, 352–353, 416–419
composite resistivity index, 407*f*
handling, 83
hydrocarbon-bearing, 421–422
water-bearing, 420–421
Sidewall coring systems, 37–40, 37–40*f*
Sieve analysis, 774–775
vs. LPSA, 770*f*
Signal-to-noise ratio (SNR), number of scans for, 666
Silver nitrate test, 366
Single-stage triaxial (SST) tests, 687
Mohr–Coulomb analysis for, 705
triaxial compression strength tests, 688–693
Sleeve conformance pressure (SCP), 361, 392
Slippage-corrected pressure plot, 232
Soxhlet extraction, core cleaning, 138–141, 139*f*
Special core analysis laboratory (SCAL) test, 12–13, 271, 364
carbonates, 786
clean consolidated sandstones, 785
CT scanning, 286–292, 290–291*f*, 292*t*, 293*f*
fractured reservoirs, 787
gas field
air–brine drainage capillary pressure tests, 795*t*
CEC, 795*t*
core CT scan, 795*t*
dynamic parameters, 793
MICP, 795*t*
multiple salinity method, 795*t*
petrophysical parameters, 793
porosity/formation factor tests at overburden, 795*t*
pre-screening study, 795*t*
residual gas CCI tests, 795*t*
RI tests, 795*t*

rock characterisation tests, 795*t*
sample selection, 795*t*
steady-state water–gas relative permeability, 795*t*
water–gas imbibition relative permeability, 795*t*
workflow, 794*f*
low-permeability reservoirs, 787
oil field
brine–oil imbibition capillary pressure tests, 803*t*
CEC, 803*t*
centrifuge oil relative permeability, 803*t*
core CT scan, 803*t*
dynamic parameters, 801
fresh-state wettability, 803*t*
MICP, 803*t*
multiple salinity method, 803*t*
oil–brine drainage capillary pressure, 803*t*
oil saturation/wettability restoration, 803*t*
petrophysical parameters, 801
porosity/formation factor tests, 803*t*
RI tests, 803*t*
rock characterisation tests, 803*t*
sample preparation, 803*t*
sample selection, 803*t*
unsteady-state imbibition waterflood, 803*t*
workflow, 802*f*
petrophysical properties, 292, 294
dynamic test sample selection procedures, 297–299, 298–299*f*
preserved sample selection bias, 294–295, 295–296*f*
rock quality index, 297, 298*f*
sample preservation and size, 295–297
photography, 286–292, 290–291*f*, 292*t*, 293*f*
results, 276*t*
sample provenance, 286
shaly sands, 786
tests, 7, 12–13, 276, 279
unconsolidated sandstones, 785–786
vuggy carbonates, 786
Spin relaxation in rocks, NMR, 657–659
Sponge core system, 30–31
advantage, 31
core barrel, 30*f*
oil saturation determination, 31, 32*f*
Spontaneous brine imbibition, 326*f*, 327, 340
Spontaneous imbibition (Spo), 525–526
Spontaneous oil imbibition, 329, 341
Static reservoir models, 2
Steady-state liquid (absolute) permeability measurements, 247–248
advantages, 252*t*
checks and diagnostics, 250–252
data reporting requirements, 251*t*
drawbacks/issues, 252*t*
permeability calculation, 248–250, 250*f*
quality control issues, 250–252
sample preparation, 248
saturation procedures, 248
test procedures, 248–250, 250*f*
Steady-state permeability measurements
advantages, 237*t*
checks and diagnostics, 235–239, 238*t*
data reporting requirements, 236*t*
drawbacks/issues, 237*t*
gas and Klinkenberg permeability calculations, 232
Klinkenberg effect evaluation, 232–235, 233*f*
non-Darcy flow evaluation, 232–235, 233*f*
quality control issues, 235–239, 238*t*
sample preparation, 229
test equipment, 229–231, 229–230*f*
test procedures, 231–232
Steady state (SS) tests
data interpretation methodology, 557
experimental data, 642
gas-oil relative permeability, 598–600, 598*t*, 602*t*
equipment, 600–601
key processes, 600
sample preparation, 600
test procedures, 601–602
measurement techniques, 545–548
water-gas relative permeability, 618–622, 618*t*, 620*t*
equipment, 619
key processes, 619
sample preparation, 619
test procedures, 619–620
water-oil relative permeability, 584–589, 584*t*, 588*t*
key processes, 586
sample preparation, 586
test procedures, 586–588
STO. *See* Dead crude oil
Stress (overburden) conditions, FRF tests
annulus volume/conformance pressure correction, 361–363
brine permeability measurements, 366
equipment, 358–359
pore collapse, 364–365

Stress (overburden) conditions, FRF tests (*Continued*)
  sample loading, 360
  stressed porosity determination, 364
  stress stations, 363–364
  test procedures, 360–366
  unloading, 366
Stress condition tests (air-brine and oil-brine), 384
Stress corrections, capillary pressure, 508
Surface tension, 283
Synthetic formation water (SFW), 168–170, 271, 664
  composition, 271*t*
  resistivity, 273–275
  result, 276*t*
Synthetic oil-based muds (SBMs), 45

T

Teeuw's method, 303–304
Tensile strength (TS) tests
  advantages, 730, 732*t*
  checks and diagnostics, 732
  data reporting requirements, 731, 731*t*
  data utilisation, 730
  drawbacks/issues, 730, 732*t*
  equipment, 729–730
  procedures, 729–730
  purpose, 728–729
  quality control issues, 732
  sample requirements, 728–729
  UCS and, 733*f*
Thick-wall cylinder (TWC) test
  advanced TWC
    equipment, 719–720
    set-up with endoscope, 721*f*
  advantages, 724, 726*t*
  checks and diagnostics, 724
  collapse, 718*f*
  data reporting requirements, 724, 725*t*
  data utilisation, 722–724
  drawbacks/issues, 724, 726*t*
  with eccentred internal hole, 727*f*
  Hoek cell set-up, 720*f*
  over-coring procedure for, 116*f*
  purpose, 717–718
  quality control issues, 724
  rock mechanics test plugs, 114, 115*f*
  sample requirements, 717–718
  standard TWC, 717
    equipment, 719–720
    overcoring to produce, 719*f*
    sample dimensions, 718*f*
    test procedures, 722
  and UCS, 728*f*
Threshold pressure, 523–524
Time inversion (TI), 666
Timur–Coates equation, 662–663
Tixier formula, 350
Total immersion soxhlet extraction, 142–143, 142–143*f*, 148*t*
Trapped gas saturation (Sgt). *See* Residual gas saturation (Sgr)
Triaxial compression strength (TCS) tests
  advantages, 700, 702*t*
  checks and diagnostics, 700–705
  data reporting requirements, 700, 701*t*
  data utilisation
    elastic moduli, 695–697
    Mohr–Coulomb parameters, 697–700, 698*f*
  drawbacks/issues, 700, 702*t*
  equipment, 688
  plug with irregular sides, 688*f*
  procedures
    multi-stage triaxial tests, 693–694
    single-stage triaxial tests, 688–693
  purpose, 687
  quality control issues, 700–705
  sample requirements, 687
Triaxial test, 688
  apparatus, 690*f*
  consolidated-undrained method, 711, 711*f*
  data reporting requirements for, 701*t*
  description, 711
  shales, 711
    advantages, 714, 716*t*
    checks and diagnostics, 714–717
    consolidation stage, 712
    data reporting requirements, 713, 715*t*
    drawbacks/issues, 714, 716*t*
    equipment, 709
    preparation of sample, 706–709
    purpose, 706
    quality control issues, 714–717
    quasi-UCS tests, 709–710
    sample requirements, 706
    triaxial parameters, 712–713
  on weak rock, 693*t*
  on weak sands, 703–705
Tritium, 46–47
True residual oil saturation (Srot), 526

U

Unconfined compressive strength (UCS) tests, 114–115
advantages, 683, 686*t*
checks and diagnostics, 684–687
data reporting requirements, 683, 685*t*
data utilisation, 682–683
drawbacks/issues, 683, 686*t*
equipment, 681–682
Equotip portable hardness tester, 674, 674–676*f*
procedures, 682
purpose, 681
quality control issues, 684–687
sample loading, 682*f*
sample requirements, 681
scratch tester, 674–675, 677*f*
shear failure surfaces, 686*f*
specific cutter energy and, 677*f*
strength profile from scratch test, 677*f*
tensile splitting of, 680*f*
TWC and, 728*f*
Unconsolidated core handling, 81–82
Unconsolidated core processing
CT scanning, 125–129, 127–128*f*
flowchart, 126*f*
gamma ray scanning, 129
photography, 133
plugging, 130–131, 130*f*
plug protection/mounting, 131–133, 132*f*
preservation, 133–134
receipt and cutting, 125
slabbing, 129
viewing and sample selection, 129–130
Uniaxial strain conditions (K_0) tests
data reporting requirements for, 758*t*
depletion compressibility under, 750
depletion loading conditions, 752–753
equipment, 751
initial loading conditions, 751–752
in situ stress loading conditions, 752
Uniformity coefficient, 767
Unloading stress (overburden) conditions, 366
Unsteady-state permeability measurements
advantages, 247*t*
checks and diagnostics, 246
data reporting requirements, 244–245, 245–246*t*
drawbacks/issues, 247*t*
quality control issues, 246
test equipment, 239–241, 240–241*f*
test procedures and permeability calculation, 241–244, 243*f*
Unsteady state (USS) tests
data interpretation methodology, 554–557
experimental data, 642
gas-oil relative permeability, 592*t*, 596*t*
Corey exponents, 598
equipment, 594–595
example curve, 592, 593*f*
gas flow, in porous media, 592, 593*f*
key processes, 594
refining and extending krg and kro, 598
sample preparation, 594
test procedures, 595–597
typical setup, 595*f*
measurement techniques, 543–545, 547
water-gas relative permeability, 613–618, 616*t*
co-current imbibition tests, 614
equipment, 615
imbibition tests, 613
intermediate data, 618
key processes, 614–615
sample preparation, 614
test procedures, 615–616
water-oil relative permeability, 574–584, 574*t*, 577*t*
curve shape, 579–580
key processes, 575
kro analysing, refining and extending, 579, 583–584
krw analysing, refining and extending, 579–583
sample preparation, 575
test procedures, 575–577
water saturation function, 584, 585*f*
USBM (U.S. Bureau of Mines) method, 323–324, 332–339, 333*t*
data reporting requirements, 337
index calculation, 336–337, 337*t*
key process, 334
sample preparation, 334
test equipment, 334
test procedure, 334–336, 335–336*f*

V

Viscous remnant magnetisation (VRM), 36–37
Vshale models, 352–353

W

Water analysis plugs, 78
Water-based muds (WBMs), 26, 47–48
filtrate invasion

Water-based muds (WBMs) (*Continued*)
in gas reservoir, 49*f*, 50
in oil reservoir, 49–50, 49–50*f*
mud types, 45
Water-bearing shaly sand, 420–421
Water-gas relative permeability
centrifuge, 622–623
equipment, 625
gas-equivalent imbibition, 623
imbibition *vs*. drainage, 623–624, 623*f*
key processes, 624
sample preparation, 623–624
test procedures, 625
in wetting phase, 623–624
imbibition centrifuge Pc
equipment, 627–628
key processes, 627
sample preparation, 627
residual gas saturation
CCI technique, 629–630
equipment, 631
key processes, 631
sample preparation, 631
test procedures, 631–633
SS tests, 618–619
equipment, 619
key processes, 619
sample preparation, 619
test procedures, 619–620
USS tests
co-current imbibition tests, 614
equipment, 615
imbibition tests, 613
intermediate data, 618
key processes, 614–615
sample preparation, 614
test procedures, 615–616
Water-gas system
Corey exponents, 542–543
critical gas saturation, 530–531, 532*f*
residual gas saturation, 528–530
dynamic tests, 529–530
vs. Sgi relationships, 528–529, 529*f*
Sgrm, 528
Water-oil relative permeability
centrifuge, 589–590, 589*t*, 591*t*
key processes, 590
sample preparation, 589–592
test procedures, 590–591
SS test, 584–586, 584*t*, 588*t*
key processes, 586
sample preparation, 586
test procedures, 586–588
USS test, 574*t*, 577*t*
curve shape, 579–580
key processes, 575
kro analysing, refining and extending, 579, 583–584
krw analysing, refining and extending, 579–583
sample preparation, 575
test procedures, 575–577
water saturation function, 584, 585*f*
Water-oil system
Corey exponents, 541, 541*t*
residual gas saturation
forced imbibition, 525–526
HC components, 525
primary drainage, 525–526
ROS, 527
Srot, 526–527
Srow, 527
by water-oil imbibition capillary pressure, 525–526, 526*f*
water saturation, 525
wettability alterations, 525
Water resistivity, 272
Water saturation
calculation, 387
interpretation
Archie's law, 349–352
core analysis parameters, 353–354
resistivity index, 351
saturation exponent, 351
shaly sands, 352–353
reservoir representative, 525
wettability, 525
Water-soluble mud tracers, 46–48, 47*f*
Water wet, 60, 314
definition, 60
plug samples, 63
rock, 65–66
Waxman-Smits equation
application workflow, 423–424
cation mobility, 423
CEC, 423
hydrocarbon-bearing shaly sand, 421–422
utilisation of, 422
water-bearing shaly sand, 420–421
Waxman-Smits model, 354
Waxman-Smits parameters
advantages, 433, 435–436*t*
CEC and Qv tests, 424–430
multiple salinity Co-Cw, 426–430
wet chemistry, 425–426
checks and diagnostics, 433–439

data reporting requirements, 433, 434*t*
data utilisation and assignment
intrinsic cementation exponent, 430–431
intrinsic saturation exponent, 431–432
drawbacks/issues, 433, 435–436*t*
quality control issues, 433–439
shaly sand models, 416–419
WBMs. *See* Water-based muds (WBMs)
Weak core processing. *See* Unconsolidated core processing
Welge equation, 555, 555*f*
Wellsite handling
core recovery on rig floor, 66–68, 66–67*f*
core stabilisation and preservation, 72–76
foam and gypsum stabilisation, 74–75, 75*f*
freezing, 73–74
gypsum *vs.* foam, 76
resin consolidation, 74
core transportation, 79–80, 79–80*f*
coring report, 81
gamma-ray scanning, 70, 70*f*
linear layout and marking, 68–69, 69*f*
liner cutting, 70–72, 71*f*
sampling
Dean-Stark and water analysis plugs, 78
methods, 77–78, 78*f*
reasons for, 77
shale preservation, 78–79
Wet chemistry, CEC, 425–426
Wet preservation, 106, 107*f*
Wettability, 403–405
capillary pressure, 525
core preparation, 175–176
native
alteration, 537–540
reservoir rock, 536, 537*t*
rock/gas/water system, 536
rock/oil-water system, 536, 536*t*
USBM tests, 540
water saturation, 525
Wettability alteration
definition, 59–60, 59*f*
description, 57, 59
drilling mud mechanism, 62–65, 63–64*f*, 65*t*
native reservoir wettability controls, 61–62
brine chemistry, 61
grain/mineral type, 61–62
oil type/composition, 61
pressure and temperature reduction, 65–66
Whole core analysis measurements
fluid saturations, 263
gas permeability, 263–266, 263*f*, 264*t*
porosity, 263
sample preparation, 262
Wilhelmy plate method
advantages and disadvantages, 287*t*
IFT determination, 285, 286*f*
Wireline-retrievable cores, 27–28, 27*f*

Y

Young–Laplace equation, 283
Young's modulus
elastic properties, 756–757
as function of confining stress, 696–697, 696*f*
measurement, 685–687
stress/strain relationships for, 683, 683*f*

CPI Antony Rowe
Chippenham, UK
2016-02-26 21:23